SI units

Quantity	Name	Symbol	Units
Base units			
Length	metre	m	
Mass	kilogram	kg	
Time	second	s	
Electric current	ampere	A	
Thermodynamic temperature	kelvin	K	
Amount of substance	mole	mol	
Derived units			
Force	newton	N	$kg\ m\ s^{-2}$
Pressure	pascal	Pa	$N\ m^{-2}$, $kg\ m^{-1}\ s^{-2}$
Energy	joule	J	$N\ m$, $kg\ m^2\ s^{-2}$
Power	watt	W	$J\ s^{-1}$, $m^2\ kg\ s^{-3}$
Electric charge	coulomb	C	$A\ s$
Electric potential difference	volt	V	$W\ A^{-1}$, $J\ C^{-1}$, $J\ A^{-1}\ s^{-1}$, $m^2\ kg\ s^{-3}\ A^{-1}$
Capacitance	farad	F	$C\ V^{-1}$, $s^4\ A^2\ m^{-2}\ kg^{-1}$
Electric resistance	ohm	Ω	$V\ A^{-1}$, $m^2\ kg\ s^{-3}\ A^{-2}$
Electric conductance	siemens	S	Ω^{-1}, AV^{-1}, $s^3\ A^2\ m^{-2}\ kg^{-1}$
Magnetic flux density	tesla	T	$Wb\ m^{-2}$, $V\ s\ m^{-2}$, $kg\ s^{-2}\ A^{-1}$
Magnetic flux	weber	Wb	$V\ s$, $m^2\ kg\ s^{-2}\ A^{-1}$
Magnetic inductance	henry	H	$Wb\ A^{-1}$, $V\ s\ A^{-1}$, $m^2\ kg\ s^{-2}\ A^{-2}$
Frequency	hertz	Hz	s^{-1}
Activity (radionuclide)	becquerel	Bq	s^{-1}

SI prefixes

Submultiple	Prefix	Symbol	Multiple	Prefix	Symbol
10^{-1}	deci	d	10	deca	da
10^{-2}	centi	c	10^2	hecto	h
10^{-3}	milli	m	10^3	kilo	k
10^{-6}	micro	μ	10^6	mega	M
10^{-9}	nano	n	10^9	giga	G
10^{-12}	pico	p	10^{12}	tera	T
10^{-15}	femto	f	10^{15}	peta	P
10^{-18}	atto	a	10^{18}	exa	E

Understanding Solids

Understanding Solids

The Science of Materials

Richard J. D. Tilley

Emeritus Professor, University of Cardiff

JOHN WILEY & SONS, LTD

Other Wiley Editorial Offices

John Wiley & Sons Inc., 111 River Street, Hoboken, NJ 07030, USA

Jossey-Bass, 989 Market Street, San Francisco, CA 94103-1741, USA

Wiley-VCH Verlag GmbH, Boschstr. 12, D-69469 Weinheim, Germany

John Wiley & Sons Australia Ltd, 33 Park Road, Milton, Queensland 4064, Australia

John Wiley & Sons (Asia) Pte Ltd, 2 Clementi Loop #02-01, Jin Xing Distripark, Singapore 129809

John Wiley & Sons Canada Ltd, 22 Worcester Road, Etobicoke, Ontario, Canada M9W 1L1

Wiley also publishes its books in a variety of electronic formats. Some content that appears in print may not
be available in electronic books.

Library of Congress Cataloging-in-Publication Data

Tilley, R. J. D.
        Understanding solids : the science of materials / Richard J. D. Tilley.
            p.   cm.
        Includes bibliographical references and index.
        ISBN 0-470-85275-5 (cloth) – ISBN 0-470-85276-3 (paper)
    1. Materials science.   2. Solids.   I. Title.
        TA403.T63 2004
        620.1′1–dc22   2004004221

British Library Cataloguing in Publication Data

A catalogue record for this book is available from the British Library

ISBN 0 470 85275 5 hardback
ISBN 0 470 85276 3 paperback

Typeset in 10/12 pt Times by Thomson Press, New Delhi, India
Printed and bound in Great Britain by Antony Rowe Ltd, Chippenham, Wilts
This book is printed on acid-free paper responsibly manufactured from sustainable forestry
in which at least two trees are planted for each one used for paper production.

For Anne

Contents

3   States of aggregation   61

9 Oxidation and reduction

Preface

This book originated in lectures to undergraduate students in materials science that were later extended to geology, physics and engineering students. The subject matter is concerned with the structures and properties of solids. The material is presented with a science bias and is aimed not only at students taking traditional materials science and engineering courses but also at those taking courses in the rapidly expanding fields of materials chemistry and physics. The coverage aims to be complementary to established books in materials science and engineering. The level is designed to be introductory in nature and, as far as is practical, the book is self-contained. The chapters are provided with problems and exercises designed to reinforce the concepts presented. These are in two parts. A multiple choice 'Quick Quiz' is designed to be tackled rapidly and aims to uncover weaknesses in a student's grasp of the fundamental concepts described. The 'Calculations and Questions' are more traditional, containing numerical examples to test the understanding of formulae and derivations that are not carried out in the main body of the text. Many chapters contain references to supplementary material (at the end of the book) that bear directly on the material but that would disrupt the flow of the subject matter if included within the chapter itself. This supplementary material is intended to provide more depth than is possible otherwise. Further reading sections allow students to take matters a little further. With only one exception, the references are to printed information. In general, it would be expected that a student would initially turn to the Internet for information. Sources here are rapidly located and this avenue of exploration has been left to the student.

The subject matter is divided into five sections. Part 1 covers the building blocks of solids. Chapters 1 and 2 centre on atoms and chemical bonding, and Chapter 3 outlines the patterns of structure that results. In this chapter, the important concepts of microstructure and macrostructure are developed, leading naturally to an understanding of why nanostructures possess unique properties. Defects that are of importance are also described here. The introductory material in Chapter 3 is further developed in Chapter 4, which covers phase relations, and Chapter 5 crystallography and crystal structures. Part 2, Chapter 6, is concerned with the traditional triumvirate of metals, ceramics and polymers, together with a brief introduction to composite materials. This chapter provides an overview of a comparative nature, focused on giving a broad appreciation of why the fundamental groups of materials appear to differ so much, and laying the foundations for why some, such as ceramic superconductors, seem to behave so differently from their congeners. Part 3 has a more chemical bias, and describes reactions and transformations. The principles of diffusion are outlined in Chapter 7, electrochemical ideas, which lead naturally to batteries, corrosion and electroplating, are described in Chapter 8. Solid-state transformations, which impinge on areas as diverse as shape-memory alloys, semiconductor doping and sintering are introduced in Chapter 9. Part 4 is a description of the physical properties of solids and complements the chemical aspects detailed in Part 3. The topics covered are those of importance to both science and technology: mechanical properties in Chapter 10; insulators in Chapter 11; magnetic properties in Chapter 12; electronic conductivity in Chapter 13; optical aspects in Chapter 14; and thermal, effects in Chapter 15. Part 5 is concerned with radioactivity. This topic is of enormous importance and, in particular, the disposal of nuclear waste in solid form is of pressing concern.

The material in all of the later sections is founded on the concepts presented in Part 1, that is, properties are explained as arising naturally from the atomic constituents, the chemical bonding, the microstructure and the defects present in the solid. This leads naturally to an understanding of why nanostructures have seemingly different properties from bulk solids. Because of this, nanostructures are not gathered together in one section but are considered throughout the book in the context of the better-known macroscopic properties of the material.

It is a pleasure to acknowledge the help of Dr A. Slade, Mrs Celia Carden and Mr Robert Hambrook of John Wiley, who have given continual encouragement and assistance to this venture. Ms Rachael Catt read the complete manuscript with meticulous care, exposed ambiguities and inconsistencies in both text and figures, and added materially to the final version. Mr Allan Coughlin read large parts of earlier drafts, clarified many obscurities and suggested many improvements. Mr Rolfe Jones has provided information and micrographs of solids whenever called upon. As always my family has been ever supportive during the writing of this book, and my wife Anne has endured the hours of being a computer widow without complaint. To all of these, my heartfelt thanks.

Richard J. D. Tilley

PART 1

Structures and microstructures

1

The electron structure of atoms

- What is a wavefunction and what information does it provide?

- Why does the periodic table summarise both the chemical and the physical properties of the elements?

- What is a term scheme?

1.1 Atoms

All matter is composed of aggregates of atoms. With the exception of radiochemistry and radioactivity (Chapter 16) atoms are neither created nor destroyed during physical or chemical changes. It has been determined that 90 chemically different atoms, the chemical elements, are naturally present on the Earth, and others have been prepared by radioactive transmutations. Chemical elements are frequently represented by symbols, which are abbreviations of the name of the element.

An atom of any element is made up of a small massive nucleus, in which almost all of the mass resides, surrounded by an electron cloud. The nucleus is positively charged and in a neutral atom this charge is exactly balanced by an equivalent number of electrons, each of which carries one unit of negative charge. For our purposes, all nuclei can be imagined to consist of tightly bound subatomic particles called neutrons and protons, which are together called nucleons. Neutrons carry no charge and protons carry a charge of one unit of positive charge. Each element is differentiated from all others by the number of protons in the nucleus, called the proton number or atomic number, Z. In a neutral atom, the number of protons in the nucleus is exactly balanced by the Z electrons in the outer electron cloud. The number of neutrons in an atomic nucleus can vary slightly. The total number nucleons (protons plus neutrons) defines the mass number, A, of an atom. Variants of atoms that have the same atomic number but different mass numbers are called isotopes of the element. For example, the element hydrogen has three isotopes, with mass numbers 1, called hydrogen; 2 (one proton and one neutron), called deuterium; and 3 (one proton and two neutrons), called tritium. An important isotope of carbon is radioactive carbon-14, that has 14 nucleons in its nucleus, 6 protons and 8 neutrons.

The atomic mass of importance in chemical reactions is not the mass number but the average mass of a normally existing sample of the element. This will consist of various proportions of the isotopes that occur in nature. The mass of atoms is of the order of 10^{-24} g. For the purposes of calculating the mass changes that take place in chemical reactions, it is most common to use the

Understanding solids: the science of materials. Richard J. D. Tilley
© 2004 John Wiley & Sons, Ltd   ISBNs: 0 470 85275 5 (Hbk) 0 470 85276 3 (Pbk)

mass, in grams, of one mole (6.022×10^{23}) of atoms or of the compound, called the molar mass. [A brief overview of chemical equations and the application of the mole are given in Section S1.1]. If it is necessary to work with the actual mass of an atom, as is necessary in radiochemical transformations (see Chapter 16), it is useful to work in atomic mass units, u. The atomic mass of an element in atomic mass units is numerically equal to the molar mass in grams. Frequently, when dealing with solids it is important to know the relative amounts of the atom types present as weights, the weight percent (wt%), or as atoms, the atom percent (at%). Details of these quantities and are given in Section S1.1.

The electrons associated with the chemical elements in a material (whether in the form of a gas, liquid or solid) control the important chemical and physical properties. These include chemical bonding, chemical reactivity, electrical properties, magnetic properties and optical properties. To understand this diversity, it is necessary to understand how the electrons are arranged and the energies that they have. The energies and regions of space occupied by electrons in an atom may be calculated by means of quantum theory.

Because the number of electrons is equal to the number of protons in the nucleus in a neutral atom, the chemical properties of an element are closely related to the atomic number of the element. An arrangement of the elements in the order of increasing atomic number, the periodic table, reflects these chemical and physical properties (Figure 1.1). The table is drawn so that the elements lie along a number of rows, called periods, and fall into a number of columns, called groups. The groups that contain the most elements (1, 2 and 13–18) are called main groups, and the elements in them are called main group elements. In older designs of the periodic table, these were given Roman numerals, I–VIII. The shorter groups (3–11) contain the

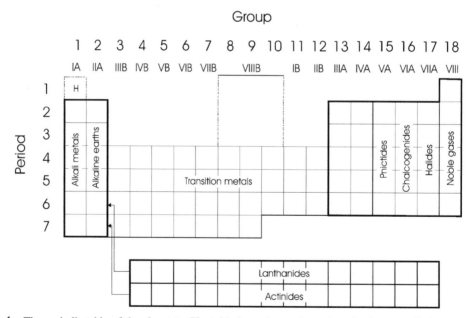

Figure 1.1   The periodic table of the elements. The table is made up of a series of columns, called groups, and rows, called periods. Each group and period is numbered. Elements in the same group have similar chemical and physical properties. The lanthanides and actinides fit into the table between groups 2 and 3, but are shown separately for compactness

transition metals. Group 12 is also conventionally associated with the transition metals. The blocks below the main table contain the inner transition metals. They are drawn in this way to save space. The upper row of this supplementary block contains elements called the lanthanides. They are inserted after barium, Ba, in Period 6 of the table. The lower block contains elements called the actinides. These are inserted after radium, Ra, in Period 7 of the table. The lightest atom, hydrogen, H, has unique properties and does not fit well in any group. It is most often included at the top of Group 1.

The chemical and physical properties of all elements in a single group are similar. However, the elements become more metallic in nature as the period number increases. The chemical and physical properties of the elements tend to vary smoothly across a period. Elements in Group 1 are most metallic in character, and elements in Group 18 are the least metallic. The properties of the elements lying within the transition metal blocks are similar. This family similarity is even more pronounced in the lanthanides and actinides.

The members of some groups have particular names that are often used. The elements in Group 1 are called the alkali metals; those in Group 2 are called the alkaline earth metals. The elements in Group 15 are called the pnictogens, and the compounds are called pnictides. The elements in Group 16 are called chalcogens and form compounds called chalcogenides. The elements in Group 17 are called the halogens, and the compounds that they form are called halides. The elements in Group 18 are very unreactive gases, called the noble gases.

Although the periodic table was originally an empirical construction, an understanding of the electron structure of atoms has made the periodic table fundamentally understandable.

1.2 The hydrogen atom

1.2.1 The quantum mechanical description of a hydrogen atom

A hydrogen atom is the simplest of atoms. It comprises a nucleus consisting of a single proton together with a single bound electron. The 'planetary' model, in which the electron orbits the nucleus like a planet, was initially described by Bohr in 1913. Although this model gave satisfactory answers for the energy of the electron, it was unable to account for other details and was cumbersome when applied to other atoms. In part, the problem rests upon the fact that the classical quantities used in planetary motion, position and momentum (or velocity), cannot be specified with limitless precision for an electron. This is encapsulated in the Heisenberg uncertainty principle, which can be expressed as follows:

$$\Delta x \, \Delta p \geq \frac{h}{4\pi}$$

where Δx is the uncertainty in the position of the electron, Δp the uncertainty in the momentum and h is the Planck constant. When this is applied to an electron attached to an atomic nucleus, it means that the exact position cannot be specified when the energy is known, and classical methods cannot be used to treat the system.

The solution to the problem was achieved by regarding the electron as a wave rather than as a particle. The idea that all particles have a wave-like character was proposed by de Broglie. The relationship between the wavelength, λ, of the wave, called the de Broglie wavelength, is given by:

$$\lambda = \frac{h}{p}$$

where h is the Planck constant and p is the momentum of the particle. In the case of an electron, the resulting wave equation, the Schrödinger equation, describes the behaviour of the electron well. The Schrödinger equation is an equation that gives information about the probability of finding the electron in a localised region around the nucleus, thus avoiding the constraints imposed by the Uncertainty Principle. There are a large number of permitted solutions to this equation, called wavefunctions, ψ, which describe the energy and probability of the location of the electron in any region around the proton nucleus. Each of the solutions contains three integer terms called quantum numbers. They are n, the principal quantum number, l, the orbital angular momentum quantum number and m_l, the magnetic quantum number. The names of the last two quantum numbers predate modern quantum chemistry. They are best regarded as labels rather than representing classical concepts such as the angular momentum of a solid body. Quantum numbers define the state of a system. The solutions to the wave equation can be written in a number of mathematically equivalent ways, one set of which is given in Table 1.1 for the three lowest-energy s orbitals.

Table 1.1   Some s wavefunctions

Orbital	Wavefunction
1s	$\psi = \frac{1}{\sqrt{\pi}} \left(\frac{Z}{a_0}\right)^{3/2} e^{-\sigma}$
2s	$\psi = \frac{1}{4\sqrt{2\pi}} \left(\frac{Z}{a_0}\right)^{3/2} (2 - \sigma)e^{-\sigma/2}$
3s	$\psi = \frac{1}{81\sqrt{3\pi}} \left(\frac{Z}{a_0}\right)^{3/2} (27 - 18\sigma + 2\sigma^2)e^{-\sigma/3}$

Note: Z is the atomic number of the nucleus; a_0 is the Bohr radius, 5.29×10^{-11} m; $\sigma = Zr/a_0$; and r is the radial position of the electron.

1.2.2   The energy of the electron

The principal quantum number, n, defines the energy of the electron. It can take integral values $1, 2, 3 \ldots$ to infinity. The energy of the electron is lowest for $n = 1$, and this represents the most stable or ground state of the hydrogen atom. The next lowest energy is given by $n = 2$, then by $n = 3$ and so on. The energy of each state is given by the simple formula:

$$E = \frac{-A}{n^2} \qquad (1.1)$$

where A is a constant equal to 2.179×10^{-18} J (13.6 eV) and E is the energy of the level with principal quantum number n. [The unit of energy 'electron volt' (eV) is frequently used for atomic

processes. $1\,\text{eV} = 1.602 \times 10^{-19}$ J.] The negative sign in the equation indicates that the energy of the electron is chosen as zero when n is infinite, that is to say, when the electron is no longer bound to the nucleus.

There is only one wave function for the lowest energy, $n = 1$, state. The states of higher energy each have n^2 different wavefunctions, all of which have the same energy, that is, there are four different wavefunctions corresponding to $n = 2$, nine different wave functions for $n = 3$ and so on. These wave functions are differentiated from each other by different values of the quantum numbers l and m_l, as explained below. Wavefunctions with the same energy are said to be degenerate.

It is often convenient to represent the energy associated with each value of the principal quantum number, n, as a series of steps or energy levels. This representation is shown in Figure 1.2. It is important to be aware of the fact that the electron can only take the exact energy values given by Equation (1.1).

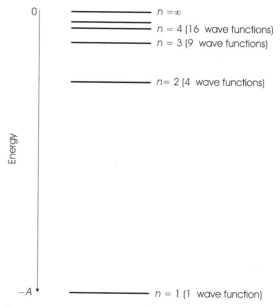

Figure 1.2   The energy levels available to an electron in a hydrogen atom. The energies are given by $-A/n^2$, and each level is n-fold degenerate. The lowest energy correspond to $n = 1$. The energy zero is taken at $n = \infty$, when the electron is removed from the atom

When an electron gains energy, it jumps from an energy level with a lower value of n to a level with a higher value of n. When an electron loses energy, it jumps from an energy level with a higher value of n to an energy level with a lower value. The discrete packets of energy given out or taken up in this way are photons of electromagnetic radiation (see Chapter 14). The energy of a photon needed to excite an electron from an energy E_1, corresponding to an energy level n_1, to an energy E_2, corresponding to an energy level n_2, is given by:

$$E = E_1 - E_2 = -2.179 \times 10^{-18} \left(\frac{1}{n_1^2} - \frac{1}{n_2^2} \right) \text{J}$$

$$= -13.6 \left(\frac{1}{n_1^2} - \frac{1}{n_2^2} \right) \text{eV} \qquad (1.2)$$

The energy of the photon emitted when the electron falls back from E_2 to E_1 is the same. The frequency ν, or the equivalent wavelength λ, of the photons that are either emitted or absorbed during these energy changes are given by the equation:

$$E = h\nu = \frac{hc}{\lambda} \qquad (1.3)$$

where h is the Planck constant and c is the speed of light. The energy needed to free the electron completely from the proton, which is called the ionisation energy of the hydrogen atom, is given by putting $n_1 = 1$ and $n_2 = \infty$ in Equation (1.2). The ionisation energy is 13.6 eV (2.179×10^{-18} J).

In the case of a single electron attracted to a nucleus of charge $+Ze$, the energy levels are given by:

$$E = -AZ^2/n^2 \qquad (1.4)$$

This shows that the energy levels are much lower in energy than in hydrogen, and that the ionisation energy of such atoms is considerably higher.

1.2.3   The location of the electron

The principal quantum number is not sufficient to determine the location of the electron in a hydrogen

Table 1.2 Quantum numbers and orbitals for the hydrogen atom

n	l	Orbital	m_l	Shell
1	0	1s	0	K
2	0	2s	0	L
	1	2p	$-1, 0, 1$	
3	0	3s	0	M
	1	3p	$-1, 0, 1$	
	2	3d	$-2, -1, 0, 1, 2$	
4	0	4s	0	N
	1	4p	$-1, 0, 1$	
	2	4d	$-2, -1, 0, 1, 2$	
	3	4f	$-3, -2, -1, 0, 1, 2, 3$	

atom. In addition, the two other interdependent quantum numbers, l and m_l are needed:

- l takes values of $0, 1, 2, \ldots, n - 1$;

- m_l takes values of $0, \pm 1, \pm 2, \ldots, \pm l$.

Each set of quantum numbers defines the state of the system and is associated with a wavefunction. For a value of $n = 0$ there is only one wavefunction, corresponding to $n = 0$, $l = 0$ and $m_l = 0$. For $n = 2$, l can take values of 0 and 1, and m then can take values of 0, associated with $l = 0$, and -1, 0 and $+1$, associated with $l = 1$. The combinations possible are set out in Table 1.2.

The probability of encountering the electron in a certain small volume of space surrounding a point with coordinates x, y and z is proportional to the square of the wavefunction, ψ^2. With this information, it is possible to map out regions around the nucleus where the electron is most likely to be encountered. These regions are referred to as orbitals and, for historical reasons, they are given letter symbols. Orbitals with $l = 0$ are called s orbitals, those with $l = 1$ are called p orbitals, those with $l = 2$ are called d orbitals and those with $l = 3$ are called f orbitals. This terminology is summarised in Table 1.2.

The orbitals with the same value of the principal quantum number form a shell. The lowest-energy shell is called the K shell, and corresponds to $n = 1$.

The other shells are labelled alphabetically, as set out in Table 1.2. For example, the L shell corresponds to the four orbitals associated with $n = 2$.

There is only one s orbital in any shell, 1s, 2s and so on. There are three different p orbitals in all shells from $n = 2$ upwards, corresponding to the m_l values of 1, 0 and −1. Collectively they are called 3p, 4p and so on. There are five d orbitals in the shells from $n = 3$ upwards, corresponding to the m_l values 2, 1, 0 −1, −2. Collectively they are called 3d, 4d, 5d and so on. There are seven different f orbitals in the shells from $n = 4$ upwards, corresponding to the m_l values 3, 2, 1, 0, −1, −2, −3. Collectively they are called 4f, 5f and so on.

1.2.4   Orbital shapes

The probability of encountering an electron in an s orbital does not depend upon direction. A surface of constant probability of encountering the electron is spherical. Generally, s orbitals are drawn as spherical boundary surfaces that enclose an arbitrary volume in which there is a high probability, say 95 %, that the electron will be found, as in Figures 1.3d and 1.3e. However, the probability of encountering an s electron does vary with distance from the nucleus, as shown in Figures 1.3a–1.3c for the 1s, 2s and 3s orbitals. The positions of the peaks in Figure 1.3 represent regions in which the probability of encountering the electron is greatest. These peaks can be equated with the shells described in Table 1.2. As can be seen from Figure 1.3, the maximum probability of finding an electron is further from the nucleus for an electron in a 3s orbital than it is for an electron in a 2s orbital. Thus, electrons with higher energies are most likely to be found further from the nucleus.

Because the other wavefunctions depend upon three quantum numbers it is more difficult to draw them in two-dimensions. These wavefunctions can be divided into two parts, a radial part, with similar probability shapes to those shown in Figure 1.3, multiplied by an angular part. The maximum probability of finding the electron depends on both the radial and angular part of the wavefunction. The resulting boundary surfaces have complex shapes.

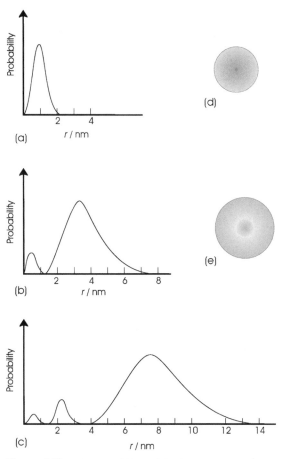

Figure 1.3   The probability of finding (a) a 1s, (b) a 2s and (c) a 3s electron at a distance r from the nucleus; the boundary surfaces of (d) the 1s and (e) the 2s orbitals

For many purposes, however, it is sufficient to describe the boundary surfaces of the angular part of the wavefunctions.

The boundary surfaces of the angular parts of each of the three p orbitals are approximately dumbbell shaped and lie along three mutually perpendicular directions, which it is natural to equate to x, y and z axes, as sketched in Figures 1.4a–1.4c. The corresponding orbitals are labelled $n\,p_x$, $n\,p_y$ and $n\,p_z$, for example $2\,p_x$, $2\,p_y$ and $2\,p_z$. Note that if a p orbital contains only one electron, it occupies both lobes. Similarly, when two electrons are accommodated in a p orbital they also occupy both lobes. The probability of encountering a p

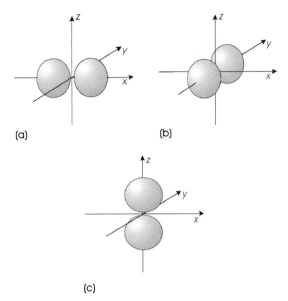

(a)

(b)

(c)

Figure 1.4 The boundary surfaces of the p orbitals: (a) p_x, (b) p_y and (c) p_z. The sign of the wave function is opposite in each lobe of the orbital

electron on the perpendicular plane that separates the two halves of the dumbbell is zero, and this plane is called a nodal plane. The sign of the wavefunction is of importance when orbitals overlap to form bonds. The two lobes of each p orbital are labelled as + and −, and the sign changes as a nodal plane is crossed. The radial probability of encountering an electron in a p orbital is zero at the nucleus, and increases with distance from the nucleus. The maximum probability is further from the nucleus for an electron in a 3p orbital than for an electron in a 2p orbital, and so on, so that 3p orbitals have a greater extension in space than do 2p orbitals.

The electron distribution of an electron in either the d or f orbitals is more complicated than those of the p orbitals. There are five d orbitals, and seven f orbitals. The shapes of the angular part of the 3d set of wavefunctions is drawn in Figure 1.5. Three of these have lobes lying between pairs of axes: d_{xy}, between the x and y axes (Figure 1.5a); d_{xz}, between

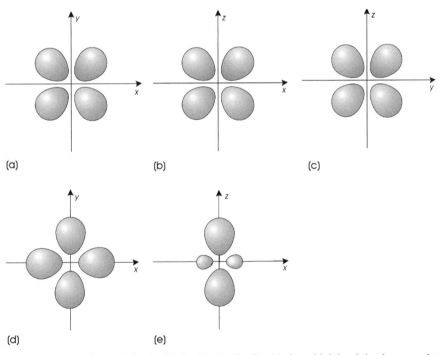

(a)

(b)

(c)

(d)

(e)

Figure 1.5 The boundary surfaces of the d orbitals: (a) d_{xy}, (b) d_{xz}, (c) d_{yz}, with lobes lying between the axes, and (d) $d_{x^2-y^2}$ and (e) d_{z^2}, with lobes lying along the axes

the x and z axes (Figure 1.5b); and d_{yz}, between the y and z axes (Figure 1.5c). The other two orbitals have lobes along the axes: $d_{x^2-y^2}$ pointing along the x and y axes (Figure 1.5d) and d_{z^2} pointing along the z axis (Figure 1.5e). Except for the d_{z^2} orbital, two perpendicular planar nodes separate the lobes and intersect at the nucleus. In the d_{z^2} orbital, the nodes are conical surfaces.

1.3 Many-electron atoms

1.3.1 The orbital approximation

If we want to know the energy levels of an atom with a nuclear charge of $+Z$ surrounded by Z electrons, it is necessary to write out a more extended form of the Schrödinger equation that takes into account not only the attraction of the nucleus for each electron but also the repulsive interactions between the electrons themselves. The resulting equation has proved impossible to solve analytically but increasingly accurate numerical solutions have been available for many years.

The simplest level of approximation, called the orbital approximation, supposes that each electron moves in a potential due to the nucleus and the average field of all the other electrons present in the atom. That is, as the atomic number increases by one unit, from Z to $(Z + 1)$, an electron is added to the atom and feels the potential of the nucleus diluted by the electron cloud of the original Z electrons. This means that the electron experiences an effective nuclear charge, Z_{eff}, which is considered to be located as a point charge at the nucleus of the atom. Compared with hydrogen, the energy levels of all of the orbitals drop sharply as Z_{eff} increases (Figure 1.6). Even when one reaches lithium ($Z = 3$) the 1s orbital energy has decreased so much that it forms a chemically unreactive shell. This is translated into the concept of an atom as consisting of unreactive core electrons surrounded by a small number of outermost valence electrons, which are of chemical significance. Moreover, the change of energy as Z increases justifies the approx-

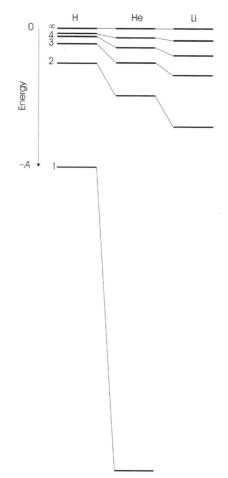

Figure 1.6 The schematic decrease in energy of the orbitals of the first three elements in the periodic table – hydrogen, helium and lithium – as the charge on the nucleus increases

imation that the valence electrons of all atoms are at similar energies.

In fact, the effective nuclear charge is different for electrons in different orbitals. This has the effect of separating the energy of the ns, np, nd and nf orbitals (where n represents the principal quantum number, say 4), which are identical in hydrogen. It is found that for any value of n, the s orbitals have the lowest energy and the three p orbitals have equal and slightly higher energy, the five d orbitals have equal and slightly higher energy again and the seven f orbitals have equal and slightly higher energy again (Figure 1.7). However, the energy differences between the higher energy orbitals are very small, and this simple ordering is not followed exactly for heavier atoms (see Section S1.2.2).

In the orbital approximation, the electrons move in the potential of a central point nucleus. This potential does not change the overall form of the angular part of the wavefunction and hence the shapes of the orbitals are not changed from the shapes found for hydrogen. However, the radial part of the wavefunction is altered, and the extension of the orbitals increases as the effective nuclear charge increases. This corresponds to the idea that heavy atoms are larger than light atoms.

1.3.2   Electron spin and electron configuration

The results presented so far, derived from solutions to the simplest form of the Schrödinger equation, do not explain the observed properties of atoms exactly. In order to account for the discrepancy the electron is allocated a fourth quantum number called the spin quantum number, s. The spin quantum number has a value of $\frac{1}{2}$. Like the orbital angular momentum quantum number, the spin of an electron in an atom can adopt one of two different directions, represented by a quantum num-ber, m_s, which take values of $+\frac{1}{2}$ or $-\frac{1}{2}$. These two spin directions have considerable significance in chemistry and physics and are frequently repre-sented by arrows: ↑ for spin up, or α, and ↓ for spin down, or β. Although the spin quantum number was originally postulated to account for certain experimental observations, it arises naturally in more sophisticated formulations of the Schrödinger equation that take into account the effects of rela-tivity.

The electron configuration of an atom is the description of the number of electrons in each orbital, based upon the orbital model. This is usually given for the lowest energy possible, called the ground state. To obtain the electron configura-tion of an atom, the electrons are fed into the orbitals, starting with the lowest-energy orbital, 1s, and then proceeding to the higher-energy orbitals so as fill them up systematically from the 'bottom' up (Figure 1.7). This is called the Aufbau (or building up) principle. Before the configurations can be constructed, it is vital to know that each orbital can hold a maximum of two electrons, which must have opposite values of m_s, either $+\frac{1}{2}$ or $-\frac{1}{2}$. This fundamental feature of quantum mechanics is a result of the Pauli exclusion principle. No more than two electrons can occupy a single orbital and, if they do, the spins must be different, that is, spin up and spin down. Two electrons in a single orbital are said to be spin paired.

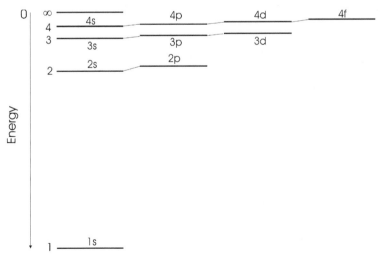

Figure 1.7 Schematic of the energy levels for a light many-electron atom

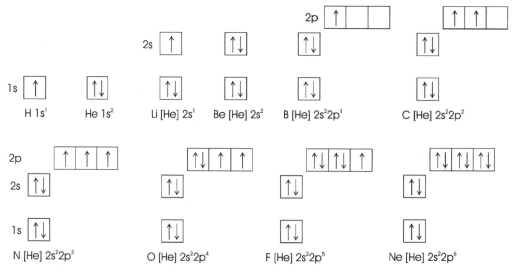

Figure 1.8   The building up of the electron configurations of the first 10 atoms in the periodic table

The electron configurations of the first few elements, derived in this way, are built up schematically in Figure 1.8.

- H: hydrogen has only one electron, and it will go into the orbital of lowest energy, the 1s orbital. The four quantum numbers specifying this state are $n = 1$, $l = 0$, $m_l = 0$ and $m_s = +\frac{1}{2}$. The electron configuration is written as $1s^1$.

- He: helium has two electrons. The first will be allocated to the 1s orbital, as in hydrogen. The lowest energy will prevail if the second is be allocated to the same orbital. This can be done provided m_s has a value of $-\frac{1}{2}$ and the electrons are spin paired. The electron configuration is written $1s^2$. There is only one orbital associated with the $n = 1$ quantum number, hence the shell corresponding to $n = 1$, the K shell is now filled, and holds just 2 electrons.

- Li: lithium has three electrons. Two of these can be placed in the 1s orbital, which is then filled. The next lowest energy corresponds to the 2s orbital, and the third electron is allocated to this. It will have quantum numbers $n = 2$, $l = 0$, $m_l = 0$ and $m_s = +\frac{1}{2}$, and the electron configura-

tion is written $1s^2\ 2s^1$. This configuration is often written in a more compact form as [He] $2s^1$. The part of the configuration written [He] refers to the core electrons, which generally do not take part in chemical reactions. The electron outside the core is the chemically reactive valence electron.

- Be: beryllium has four electrons. The first three are allocated as for lithium. The fourth can be allocated to the 2s orbital, with quantum numbers $n = 2$, $l = 0$, $m_l = 0$ and $m_s = -\frac{1}{2}$, giving an electron configuration $1s^2\ 2s^2$, or [He] $2s^2$. Note that the L shell is not filled, because there are three p orbitals still available to the $n = 2$ quantum number.

- B: boron has five electrons. The first four of them will be distributed as in beryllium. The fifth must enter a 2p orbital, with $n = 2$, $l = 1$, $m_l = 0$ or ± 1. The electron can be assigned the four quantum numbers $n = 2$, $l = 1$, $m_l = +1$ and $m_s = \frac{1}{2}$, and the configuration $1s^2\ 2s^2\ 2p^1$, or [He] $2s^2$ $2p^1$.

- C: carbon has six electrons. The first five are allocated as for boron. The sixth electron also enters a p orbital. There is a choice here. The electron can go into the already half-occupied orbital or into one of the empty orbitals. The lowest-energy situation is that in which the electron goes into an unoccupied orbital. This situation is expressed in Hund's First Rule. When electrons have a choice of several orbitals of equal energy, the lowest-energy, or ground-state, configuration corresponds to the occupation of separate orbitals with parallel spins rather than one orbital with paired spins. Thus, as the first p electron has a spin of $+\frac{1}{2}$, the second electron also has a spin of $+\frac{1}{2}$. This gives an electron distribution $1s^2 2s^2 2p^2$, or [He] $2s^2 2p^2$.

- N: nitrogen has one more electron, and following the rules laid down it is allocated to the remaining empty p orbital, with quantum numbers $n = 2$, $l = 1$, $m_l = -1$ and $s = +\frac{1}{2}$, giving a configuration $1s^2 2s^2 2p^3$, or [He] $2s^2 2p^3$.

- O: oxygen has eight electrons. The first seven are placed as in nitrogen. The eighth electron must be added to one of the half-filled p orbitals. The quantum numbers for the new electron will be $n = 2$, $l = 1$, $m_l = 1$ and $s = -\frac{1}{2}$, giving a distribution of $1s^2 2s^2 2p^4$, or [He] $2s^2 2p^4$.

- F: fluorine possesses one more electron than oxygen, and so we expect its state to be $n = 2$, $l = 1$, $m_l = 0$ and $s = -\frac{1}{2}$, that is, $1s^2 2s^2 2p^5$, or [He] $2s^2 2p^5$.

- Ne: neon has six electrons, filling the 2p orbitals. The state occupied by the last electron is represented by $n = 2$, $l = 1$, $m_l = -1$ and $s = -\frac{1}{2}$, and the distribution by $1s^2 2s^2 2p^6$, or [He] $2s^2 2p^6$. This is often written as [Ne]. The L shell is now filled. It has been shown that all filled shells have only radial symmetry and are especially stable.

To summarise, the building up procedure we have used is called the Aufbau principle. Each electron occupies one electron state, represented by four quantum numbers, one of which represents the spin of the electron. Each orbital can contain two electrons with opposite spins. In a set of orbitals of equal energy, electrons tend to keep apart and so make the total electron spin a maximum.

The electron configurations of the rest of the elements are derived in the same way. The M shell ($n = 3$), with a maximum capacity of 18 electrons, consists of one 3s orbital, three 3p orbitals and five 3d orbitals. The N shell ($n = 4$) can hold 32 electrons in one 4s, three 4p, five 4d and seven 4f orbitals. The maximum number of electrons in each shell is $2n$, where n is the principal quantum number.

The electron configurations of the elements in the first two periods are listed in Section S1.2.1. The configuration of the ground state depends upon the energy of the orbitals and the interaction of the electrons. These effects are very delicately balanced in the heavier atoms, so that a strict Aufbau arrangement does not always hold good. For example, chromium (Cr) has an electron configuration of [Ar] $3d^5 4s^1$, in contrast to that of its neighbours – vanadium (V) with [Ar] $3d^3 4s^2$, and manganese (Mn) with [Ar] $3d^5 4s^2$ – both of which have the 4s shell filled. This indicates that the half-filled d orbital has a special stability that can influence the configuration.

1.3.3 The periodic table

The periodic table, described in Section 1.1, was an empirical construction. However, it is fundamentally understandable in terms of the electron configurations just discussed. The chemical and many physical properties of the elements are simply controlled by the valence electrons. The valence electron configuration varies in a systematic and repetitive way as the various shells are filled. This leads naturally to both the periodicity and the repetitive features displayed in the periodic table (Figure 1.9).

Figure 1.9(a) shows the relationship between the outer orbitals that are partly filled and the position of the element in the periodic table. The filled shells are very stable configurations and take part in chemical reactions only under extreme conditions. The atoms with this configuration, the noble gases,

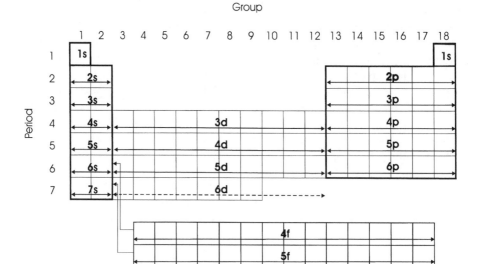

(a)

Period 2

Group 1/I	Group 2/II	Group 13/III	Group 14/IV	Group 15/V	Group 16/VI	Group 17/VII	Group 18/VIII
lithium	beryllium	boron	carbon	nitrogen	oxygen	fluorine	neon
Li 3	Be 4	B 5	C 6	N 7	O 8	F 9	Ne 10
2p —	2p —	2p ↑	2p ↑↑	2p ↑↑↑	2p ↑↓↑↑	2p ↑↓↑↓↑	2p ↑↓↑↓↑↓
2s ↑	2s ↑↓	2s ↑↓	2s ↑↓	2s ↑↓	2s ↑↓	2s ↑↓	2s ↑↓
core 2	core 2	core 2	core 2	core 2	core 2	core 2	core 2

Period 3

Group 1/I	Group 2/II	Group 13/III	Group 14/IV	Group 15/V	Group 16/VI	Group 17/VII	Group 18/VIII
sodium	magnesium	aluminium	silicon	phosphorus	sulphur	chlorine	argon
Na 11	Mg 12	Al 13	Si 14	P 15	S 16	Cl 17	Ar 18
3p —	3p —	3p ↑	3p ↑↑	3p ↑↑↑	3p ↑↓↑↑	3p ↑↓↑↓↑	3p ↑↓↑↓↑↓
3s ↑	3s ↑↓	3s ↑↓	3s ↑↓	3s ↑↓	3s ↑↓	3s ↑↓	3s ↑↓
core 10	core 10	core 10	core 10	core 10	core 10	core 10	core 10

Period 4

Group 1/I	Group 2/II	Group 13/III	Group 14/IV	Group 15/V	Group 16/VI	Group 17/VII	Group 18/VIII
potassium	calcium	gallium	germanium	arsenic	selenium	bromine	krypton
K 19	Ca 20	Ga 31	Ge 32	As 33	Se 34	Br 35	Kr 36
4p —	4p —	4p ↑	4p ↑↑	4p ↑↑↑	4p ↑↓↑↑	4p ↑↓↑↓↑	4p ↑↓↑↓↑↓
4s ↑	4s ↑↓	4s ↑↓	4s ↑↓	4s ↑↓	4s ↑↓	4s ↑↓	4s ↑↓
core 18	core 18	core 28	core 28	core 28	core 28	core 28	core 28

Period 5

Group 1/I	Group 2/II	Group 13/III	Group 14/IV	Group 15/V	Group 16/VI	Group 17/VII	Group 18/VIII
rubidium	strontium	indium	tin	antimony	tellurium	iodine	xenon
Rb 37	Sr 38	In 49	Sn 50	Sb 51	Te 52	I 53	Xe 54
5p —	5p —	5p ↑	5p ↑↑	5p ↑↑↑	5p ↑↓↑↑	5p ↑↓↑↓↑	5p ↑↓↑↓↑↓
5s ↑	5s ↑↓	5s ↑↓	5s ↑↓	5s ↑↓	5s ↑↓	5s ↑↓	5s ↑↓
core 36	core 36	core 46	core 46	core 46	core 46	core 46	core 46

Period 6

Group 1/I	Group 2/II	Group 13/III	Group 14/IV	Group 15/V	Group 16/VI	Group 17/VII	Group 18/VIII
caesium	barium	thallium	lead	bismuth	polonium	astatine	radon
Cs 55	Ba 56	Tl 81	Pb 82	Bi 83	Po 84	At 85	Rn 86
6p —	6p —	6p ↑	6p ↑↑	6p ↑↑↑	6p ↑↓↑↑	6p ↑↓↑↓↑	6p ↑↓↑↓↑↓
6s ↑	6s ↑↓	6s ↑↓	6s ↑↓	6s ↑↓	6s ↑↓	6s ↑↓	6s ↑↓
core 54	core 54	core 78	core 78	core 78	core 78	core 78	core 78

(b)

Figure 1.9   (a) The relationship between the electron configuration of atoms and the periodic table arrangement, and (b) part of the periodic table, giving the valence (outer-electron) structure of the main group elements

are placed in Group 18 of the periodic table. A 'new' noble gas appears each time a shell is filled. Following any noble gas is an element with one electron in the outermost s orbital. These are the alkali metals, and once again, a 'new' alkali metal is found after each filled shell. The alkali metals are in Group 1 of the periodic table. Similarly, the alkaline earth elements, listed in Group 2 of the periodic table, all have two valence electrons, both in the outermost s orbital. Thus, the periodic table simply expresses the Aufbau principle in a chart.

The outermost electrons take part in chemical bonding. The main group elements are those with electrons in outer s and p orbitals giving rise to strong chemical bonds (Figure 1.9b). The valence electron configuration of all the elements in any group is identical, indicating that the chemical and physical properties of these elements will be very similar. The d and f orbitals are shielded by s and p orbitals from strong interactions with surrounding atoms and do not take part in strong chemical bonding. The electrons in these orbitals are responsible for many of the interesting electronic, magnetic and optical properties of solids. Because of their importance, the electron configurations of the 3d transition metals and the lanthanides are set out in Sections S1.2.2 and S1.2.3.

1.4 Atomic energy levels

1.4.1 Electron energy levels

Spectra are a record of transitions between electron energy levels. Each spectral line can be related to the transition from one energy level to another. It is found that an ion in a magnetic field has a more complex spectrum, with more lines present, than has the same ion in the absence of the magnetic field. The presence of additional lines in the spectrum of an atom or ion when in a magnetic field is called the Zeeman effect. A similar, but different, complexity, called the Stark effect, arises in the presence of a strong electric field. The electron configurations described are not able to account for all of the observed transitions, and to derive the possible

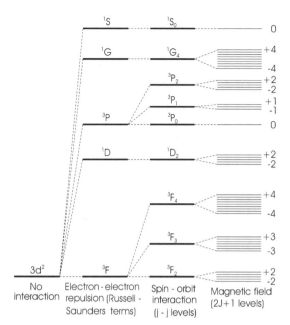

Figure 1.10 The evolution of the energy levels of an atom with a $3d^2$ electron configuration, taking into account increasing electron–electron and other interactions, the energy scales are schematic

energy levels appropriate to any electron configuration a more complex model of the atom is required. These steps are outlined below and are summarised in Figure 1.10.

1.4.2 The vector model

In this model, classical ideas are grafted onto the quantum mechanics of the atom. The quantum number l is associated with the angular momentum of the electron around the nucleus. It is represented by an angular momentum vector, *l*. Similarly, the spin quantum number of the electron, s, is associated with a spin angular momentum vector, *s*. (Vectors in the following text are specified in bold type and quantum numbers in normal type.) In the vector model of the atom, the two angular momentum vectors are added together to get a total angular momentum for the atom as a whole. This is then related to the electron energy levels of the atom.

There are two ways of tackling this task. The first of these makes the approximation that the electrostatic repulsion between electrons is the most important energy term. In this approximation, called Russell–Saunders coupling, all of the individual *s* vectors of the electrons are summed vectorially to yield a total spin angular momentum vector *S*. Similarly, all of the individual *l* vectors for the electrons present are summed vectorially to give a total orbital angular momentum vector, *L*. The vectors *S* and *L* can also be summed vectorially to give a total angular momentum vector *J*. Note that the convention is to use lower-case letters for a single electron and upper-case for many electrons.

The alternative approach to Russell–Saunders coupling is to assume that the interaction between the orbital angular momentum and the spin angular momentum is the most important. This interaction is called spin–orbit coupling. In this case, the *s* and *l* vectors for an individual electron are added vectorially to give a total angular momentum vector *j* for a single electron. These values of *j* are then added vectorially to give the total angular momentum vector *J*, for the whole atom. The technique of adding *j* values to obtain energy levels is called *j–j* coupling.

Broadly speaking, Russell–Saunders coupling works well for lighter atoms, and *j–j* coupling for heavier atoms. In reality, the energy levels derived from each scheme represent approximations to those found by experiment, which may be regarded as intermediate between the two.

1.4.3   Terms and term schemes

For almost all purposes, the Russell–Saunders coupling scheme is adequate for the specification of the energy levels of an isolated many-electron atom. In general, it is not necessary to work directly with the vectors *S*, *L* and *J*. Instead, many electron quantum numbers (not vectors), *S*, *L* and *J*, are used to label the energy levels in a simple way. The method of derivation is set out in Section S1.3.1. The value of *S* is not used directly but is replaced by the spin multiplicity, $2S + 1$. Similarly, the total angular momentum quantum number, *L*, is replaced

Table 1.3 The correspondence of *L* values and letter symbols

L	Symbol
0	S
1	P
2	D
3	F
4	G
5	H

by a letter symbol similar to that used for the single electron quantum number *l*. The correspondence is set out in Table 1.3. After $L = 3$ (F) the sequence of letter is alphabetic, omitting J. Be aware that the symbol S has two interpretations, as the symbol when $L = 0$ (upright S) and as the value of total spin (italic *S*).

The combinations are written in the following form:

$$^{2S+1}L$$

This is called a term symbol. It represents a set of energy levels, called a term in spectroscopic parlance. States with a multiplicity of 1 are called singlet states, states with a multiplicity of 2 are called doublet states, states with a multiplicity of three are called triplets, states with a multiplicity of 4 are called quartets and so on. Hence, ^{1}S is called singlet S, and ^{3}P is called triplet P.

The energies of the terms are difficult to obtain simply, and they must be calculated by using quantum mechanical procedures. However, the lowest-energy term, the ground-state term, is easily found by using the method described in Section S1.3.2.

The term symbol does not account for the true complexity found in most atoms. This arises from the interaction between the spin and the orbital momentum (spin–orbit coupling) that is ignored in Russell–Saunders coupling. A new quantum number, *J*, is needed. It is given by:

$$J = (L + S), (L + S - 1) \ldots, |L - S|$$

where $|L - S|$ is the modulus (absolute value, taken to be positive whether the expression within the vertical lines is negative or positive) of L and S. Thus the term ^3P has J values given by:

$$J = (1+1), (1+1-1), \ldots, |1-1| = 2, 1, 0.$$

The new quantum number is incorporated, as a subscript to the term, now written $^{2S+1}L_J$, and this is no longer called a term symbol but a level. Each value of J represents a different energy level. It is found that a singlet term always gives one energy level, a doublet two, a triplet three and so on. Thus, the ground-state term ^3P is composed of three levels, $^3P_0$, $^3P_1$ and $^3P_2$. The separation of these energy levels is controlled by the magnitude of the interaction between L and S. Hund's third rule (see Section S1.3.2 for the first and second rules) allows the values of J to be sorted in order of energy. The level with the lowest energy is that with the lowest J value if the valence shell is up to half full and that with the highest J value if the valence shell is more than half full.

In a magnetic field, each of the $^{2S+1}L_J$ levels splits into $(2J+1)$ separated energy levels. The spacing between the levels is given by $g_J \mu_B B$, where g_J is the Landé g-value, given by

$$g_J = \frac{1 + [J(J+1) - L(L+1) + S(S+1)]}{2J(J+1)}$$

μ_B is a fundamental physical constant, the Bohr magneton, and B is the magnetic induction (more information on this is found in Chapter 12, on magnetic properties)

The way in which these levels of complexity modify the energy levels of a $3d^2$ transition metal atom or ion is drawn schematically in Figure 1.10. At the far left-hand side of the figure, the electron configuration is shown. This is useful chemically, but is unable to account for the spectra of the atom. Russell–Saunders coupling is a good approximation to use for the 3d transition metals, and the terms that arise from this are given to the right-hand side of the configuration in Figure 1.10. In Russell–Saunders coupling the electron–electron repulsion is considered to dominate the interactions. The terms are spilt further if spin–orbit coupling (j–j coupling) is introduced. The number of levels

that arise is the same as the multiplicity of the term, $2S+1$. Finally, the levels are split further in a magnetic field. In this case $2J+1$ levels arise. The magnitude of the splitting is proportional to the magnetic field, and the separation of each of the new energy levels is the same.

Note that in a heavy atom it might be preferable to go from the electron configuration to levels derived by j–j coupling and then add on a smaller effect for electron–electron repulsion (Russell–Saunders coupling) before finally including the magnetic field splitting. In real atoms, the energy levels determined experimentally are often best described by an intermediate model between the two extremes of Russell–Saunders and j–j coupling.

Answers to introductory questions

What is a wavefunction and what information does it provide?

A wavefunction, ψ, is a solution to the Schrödinger equation. For atoms, wavefunctions describe the energy and probability of location of the electrons in any region around the proton nucleus. The simplest wavefunctions are found for the hydrogen atom. Each of the solutions contains three integer terms called quantum numbers. They are n, the principal quantum number, l, the orbital angular momentum quantum number and m_l, the magnetic quantum number. These simplest wavefunctions do not include the electron spin quantum number, m_s, which is introduced in more complete descriptions of atoms. Quantum numbers define the state of a system. More complex wavefunctions arise when many-electron atoms or molecules are considered.

Why does the periodic table summarise both the chemical and the physical properties of the elements?

The periodic table was originally formulated empirically. However, it is fundamentally understandable in terms of electron configurations. The chemical and many physical properties of the elements are dominated by the valence electrons. The valence electron configuration varies in a systematic and repetitive way as the various shells are filled. This filling corresponds to the groups of the periodic table.

The filled shells are very stable configurations and take part in chemical reactions only under extreme conditions. The atoms with this configuration, the noble gases, are placed in Group 18 of the periodic table. A 'new' noble gas appears each time a shell is filled. Following any noble gas is an element with one electron in the outermost s orbital. These are the alkali metals and, once again, a 'new' alkali metal is occurs after each filled shell. The alkali metals are found in Group 1 of the periodic table. Similarly, the alkaline earth elements, listed in Group 2 of the periodic table, all have two valence electrons, both in the outermost s orbital. The transition metals have partly filled d orbitals, and the lanthanides partly filled f orbitals. Thus, the periodic table simply expresses the Aufbau principle in a chart, which itself accounts for the periodic variation of properties.

What is a term scheme?

A term scheme is a representation of an energy level in an isolated many-electron atom, derived via the Russell–Saunders coupling scheme. In general, a term scheme is written as a collection of many-electron quantum numbers S and L. The value of S is not used directly but is replaced by the spin multiplicity, $2S + 1$. Similarly, the total angular momentum quantum number, L, is replaced by a letter symbol similar to that used for the single-electron quantum number l. The term scheme is written ^{2S+1}L. States with a multiplicity of 1 are called singlet states, states with a multiplicity of 2 are called doublet states, those with a multiplicity of three are called triplets, those with a multiplicity 4 are called quartets and so on. Hence, ^1S is called singlet S, and ^3P is called triplet P.

Further reading

Elementary chemical concepts and an introduction to the periodic table are clearly explained in the early chapters of:

P. Atkins, L. Jones, 1997, *Chemistry*, 3rd edn, W.H. Freeman, New York.

D.A. McQuarrie, P.A. Rock, 1991, *General Chemistry*, 3rd edn, W.H. Freeman, New York.

The outer electron structure of atoms is described in the same books, and in greater detail in:

D.F. Shriver, P.W. Atkins, C. H. Langford, 1994, *Inorganic Chemistry*, 2nd edn, Oxford University Press, Oxford, Ch. 1.

The quantum mechanics of atoms is described lucidly in:

D.A. McQuarrie, 1983, *Quantum Chemistry*, University Science Books, Mill Valley, CA.

An invaluable dictionary of quantum mechanical language and expressions is:

P.W. Atkins, 1991, *Quanta*, 2nd edn, Oxford University Press, Oxford.

Problems and exercises

Quick quiz

1 What is the name of the element whose symbol is Pb?
   (a) Tin
   (b) Phosphorus
   (c) Lead

2 What is the name of the element whose symbol is Hg?
   (a) Mercury
   (b) Silver
   (c) Holmium

3 What is the name of the element whose symbol is Ag?
   (a) Argon

(b) Silver

(c) Mercury

4 What is the chemical symbol for gold?
   (a) Au

   (b) G

   (c) Sb

5 What is the chemical symbol for potassium?
   (a) P

   (b) Po

   (c) K

6 An isotope is:
   (a) The nucleus of the atom

   (b) A subatomic particle

   (c) An atom with a specified number of protons and neutrons

7 The periodic table contains how many periods?
   (a) 18

   (b) 7

   (c) 14

8 Iodine is an example of:
   (a) A halogen

   (b) A chalcogen

   (c) An alkaline earth metal

9 Magnesium is an example of:
   (a) A transition metal

   (b) An alkaline earth metal

   (c) An alkali metal

10 Nickel is an example of:
   (a) A transition metal

   (b) A lanthanide

   (c) An actinide

11 A wavefunction is:
   (a) A description of an electron

   (b) An atomic energy level

   (c) A solution to the Schrödinger equation

12 An orbital is:
   (a) A bond between an electron and a nucleus

(b) A region where the probability of finding an electron is high

(c) An electron orbit around an atomic nucleus

13 The Pauli principle leads to the conclusion that:
   (a) The position and momentum of an electron cannot be specified with limitless precision

   (b) Only two electrons of opposite spin can occupy a single orbital

   (c) No two electrons can occupy the same orbital

14 The configuration of an atom is:
   (a) The number of electrons around the nucleus

   (b) The electron orbitals around the nucleus

   (c) The arrangement of electrons in the various orbitals

15 The outer electron configuration of the noble gases is:
   (a) $ns^2 np^6$

   (b) $ns^2 np^6 (n+1)s^1$

   (c) $ns^2 np^5$

16 The valence electron configuration of the alkali metals is:
   (a) ns^2

   (b) np^1

   (c) ns^1

17 The valence electron configuration of carbon is:
   (a) $1s^2 2p^2$

   (b) $2s^2 2p^2$

   (c) $2s^2 2p^4$

18 The valence electron configuration of calcium, strontium and barium is:
   (a) $ns^2 np^2$

   (b) ns^2

   (c) $(n-1)d^1 ns^2$

19 What atom has filled K, L, M and N shells?
   (a) Argon

   (b) Krypton

   (c) Xenon

20 How many electrons can occupy orbitals with $n = 3$, $l = 2$?
  (a) 6 electrons
  (b) 10 electrons
  (c) 14 electrons

21 How many permitted l values are there for $n = 4$?
  (a) One
  (b) Two
  (c) Three

22 How many electrons can occupy the 4f orbitals?
  (a) 14
  (b) 10
  (c) 7

23 Russell–Saunders coupling is:
  (a) A procedure to obtain the energy of many-electron atoms
  (b) A description of atomic energy levels
  (c) A procedure to obtain many-electron quantum numbers

24 A term symbol is:
  (a) A label for an atomic energy level
  (b) A label for an orbital
  (c) A description of a configuration

25 The many-electron quantum number symbol D represents:
  (a) $L = 1$
  (b) $L = 2$
  (c) $L = 3$

26 An atom has a term ^1S. What is the value of the spin quantum number, S?
  (a) ½
  (b) 0
  (c) 1

27 An atom has a term ^1S. What is the value of the orbital quantum number, L?
  (a) 2
  (b) 1
  (c) 0

Calculations and questions

1.1 The velocity of an electron crossing a detector is determined to an accuracy of $\pm 10\,\mathrm{m\,s^{-1}}$. What is the uncertainty in its position?

1.2 A particle in an atomic nucleus is confined to a volume with a diameter of approximately $1.2 \times 10^{-15}\,A^{1/3}\,\mathrm{m}$, where A is the mass number of the atomic species. What is the uncertainty in the velocity of a proton trapped within a sodium nucleus with a mass number of 23?

1.3 The wavelength of an electron in an electron microscope is 0.0122 nm. What is the electron velocity?

1.4 The velocity of a proton scattered by an energetic cosmic ray is $2 \times 10^3\,\mathrm{m\,s^{-1}}$. What is the wavelength of the proton?

1.5 What velocity must a proton attain in a proton microscope to have the same wavelength as green light in an optical microscope (350 nm)?

1.6 The velocity of an argon (Ar) atom at the surface of the Earth is $397\ \mathrm{m\ s^{-1}}$. What is its wavelength?

1.7 The velocity of a krypton (Kr) atom at the Earth's surface is $274\,\mathrm{m\,s^{-1}}$. What is its wavelength?

1.8 What energy is required to liberate an electron in the $n = 3$ orbital of a hydrogen atom?

1.9 What is the energy change when an electron moves from the $n = 2$ orbit to the $n = 6$ orbit in a hydrogen atom?

1.10 Calculate the energy of the lowest orbital (the ground state) of the single-electron hydrogen-like atoms with $Z = 2$ (He^+) and $Z = 3$ (Li^{2+}).

1.11 What are the frequencies (ν) and wavelengths (λ) of the photons emitted from a hydrogen atom when an electron makes a transition from $n = 4$ to the lower levels $n = 1, 2$ and 3?

1.12 What are the frequencies (ν) and wavelengths (λ) of the photons emitted from a hydrogen atom when an electron makes a transition from $n = 5$ to the lower levels $n = 1, 2$ and 3?

1.13 What are the frequencies (ν) and wavelengths (λ) of photons emitted when an electron on a Li^{2+} ion makes a transition from $n = 3$ to the lower levels $n = 1$ and 2?

1.14 What are the frequencies (ν) and wavelengths (λ) of photons emitted when an electron on a He^+ ion makes a transition from $n = 4$ to the lower levels $n = 1, 2$ and 3?

1.15 Sodium lights emit light of a yellow colour, with photons of wavelength 589 nm. What is the energy of these photons?

1.16 Mercury lights emit photons with a wavelength of 435.8 nm. What is the energy of the photons?

1.17 What are the possible quantum numbers for an electron in a 2p orbital?

1.18 Starting from the configuration of the nearest lower noble gas, what are the electron configurations of C, P, Fe, Sr and W?

1.19 Titanium has the term symbol ^3F. What are the possible values of J? What is the ground-state level?

1.20 Phosphorus has the term symbol ^4S. What are the possible values of J? What is the ground-state level?

1.21 Scandium has a term symbol ^2D. What are the possible values of J? What is the ground-state level?

1.22 Boron has a term symbol ^2P. What are the possible values of J? What is the ground-state level?

1.23 What is the splitting g_J, for sulphur, with a ground state $^3P_2$?

1.24 What is the splitting g_J, for iron, with a ground state $^5D_4$?

1.25 Sketch the 1s, 2s and 3s orbitals, roughly to scale, from the wavefunctions given in Table 1.1. [Note: answer is not provided at the end of this book.]

1.26 Older forms of the periodic table have Group 1 divided into IA, containing the alkali metals, Li, Na, K, Rb and Cs, and IB, containing the metals Cu, Ag and Au; Group 2 was divided into IIA, containing Be, Mg, Ca, Sr and Ba, and IIB, containing Zn, Cd and Hg; and Group 3 was divided in to IIIA, containing Sc, Y and the lanthanides, and IIIB, containing B, Al, Ga and In. Why should this be? [Note: answer not provided at the end of this book.]

1.27 Draw a diagram equivalent to Figure 1.10 for a chlorine atom, with a ground state $^2P_{3/2}$. [Note: answer is not provided at the end of this book.]

2

Chemical bonding

- What are the principle geometrical consequences of ionic, covalent and metallic bonding?

- What orbitals are involved in multiple bond formation between atoms?

- What are allowed energy bands?

Theories of chemical bonds have three important roles. First, they must explain the cohesion between atoms. In addition, they must account for the concept of chemical valence. Valence is the notion of the 'combining power' of atoms. Chemists have long known that atoms show a characteristic valence, depicted as little hooks in textbooks of 100 years ago. Hydrogen and chlorine had a valence of one (i.e. one hook each); oxygen had a valence of two, nitrogen three and carbon four. Although this concept gives correct chemical formulae – water (H_2O), ammonia (NH_3), methane (CH_4), and so on – the fundamental understanding of valence had to wait for the advent of quantum theory. In addition to explaining cohesion and valence, one of the important aspects of any theory of bonding is to explain the geometry of molecules and solids. For example, why is a water molecule angular, and why

does salt (NaCl) exist as crystals and not as small molecules?

It is important to remember that chemical bonds describe the electron density between the atomic nuclei. They are not best considered as rigid sticks or hooks. It is not surprising, therefore, that the most rigorous way to obtain information about the chemical bonds in a solid is to calculate the interaction energies of the electrons on the atoms that make up the material. Fortunately, for many purposes, trends in the chemical and physical properties of solids can usually be understood with the aid of simple models. Three ideas normally suffice to describe strong chemical bonds, called ionic, covalent and metallic bonding.[1] In this chapter, the origins of cohesion, valence and geometry are discussed for each of these three bonding models.

2.1 Ionic bonding

2.1.1 Ions

Ions are charged species that form when the number of electrons surrounding a nucleus varies slightly

[1]Remember that chemical bonds are never pure expressions of any one of these concepts, and the chemical and physical properties of solids can be explained only by applying selected aspects of all of these models to the material in question. The fact that a solid might be discussed in terms of ionic bonding sometimes and in terms of metallic bonding at other times simply underlines the inadequate nature of the models.

Understanding solids: the science of materials.   Richard J. D. Tilley
© 2004 John Wiley & Sons, Ltd   ISBNs: 0 470 85275 5 (Hbk) 0 470 85276 3 (Pbk)

from that required for an electrically neutral atom. The result can be a positively charged particle, a cation, if there are too few electrons, or a negatively charged particle, an anion, if there are too many. Metals tend to lose electrons and form cations – for example, Na^+, Mg^{2+} and Al^{3+}. The charge on the ions, written as a superscript, is equal to the number of electrons lost. Nonmetals tend to form anions – for example, F^-, O^{2-}, N^{3-}. The charge on the ions, written as a superscript, is equal to the number of electrons gained. Groups of atoms can also form ions. These are normally found as anions – for example, carbonate (CO_3^{2-}) and nitrate (NO_3^-) ions. Ions are called monovalent if they carry a charge of ±1, divalent if they carry a charge of ±2, trivalent if they carry a charge of ±3 and so on. This does not depend upon the number of atoms in an ion. Thus, both Zn^{2+} and CO_3^{2-} are regarded as divalent ions. The size and shapes of ions is deferred until later in this chapter.

2.1.2 Ionic bonding

Central to the idea of ionic bonding is that positive and negative ions attract each other. The resulting ion pair will be held together by electrostatic attraction. Such a bond is called an ionic bond. Key features of ionic bonding are that electrostatic interactions are long-range and nondirectional. The electrostatic attraction will tend to decrease the distance between oppositely charged ions continuously. At some interionic distance, the electron clouds of the ions begin to interact and lead to repulsion between the ions. Ultimately, the two opposing energies will balance and the ions will adopt an equilibrium separation. At this point, the overall bonding energy is the difference between the attractive and repulsive terms:

ionic bonding energy = electrostatic attraction
— repulsive energy;

that is,

$$E_{bond} = E_{electro} - E_{rep}$$

An advantage of the ionic bonding model is that these energies can be calculated. This, in turn, allows one to estimate other properties of ionic solids, including mechanical properties.

2.1.3 Madelung energy

The electrostatic potential energy between a pair of ions can be calculated if the ions are replaced by appropriate point charges. Thus the electrostatic energy of a pair of monovalent ions such as Na^+ and Cl^-, which we can define as E_e, is given by:

$$E_e = \frac{(+e)(-e)}{4\pi\varepsilon_0 r} = \frac{-e^2}{4\pi\varepsilon_0 r} \tag{2.1}$$

where the point charges on the interacting species are $\pm e$, the distance separating the charges is r, and ε_0 is the vacuum permittivity. The value of ε_0 is $8.854 \times 10^{-12}\,F\,m^{-1}$, e is measured coulombs and r is in metres. The negative charge arises because one ion has a positive charge and one a negative charge. The energy is zero when the ions are infinitely far from each other, and a negative overall energy means a stable pairing (Figure 2.1).

Although it is obvious that a pair of oppositely charged ions will be attracted, it is by no means clear that a collection of ions will hold together, because ions with the same charge repel each other

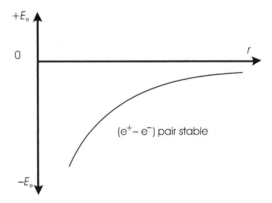

Figure 2.1 The attractive potential energy between a pair of monovalent ions, E_e, as a function of interionic separation, r. The energy is set at zero for ions that are at infinite separation

just as those with opposite charges attract each other. The resultant overall attraction or repulsion will depend on the number of ions and their location relative to one another. The computation of the energy of a cluster of point charges replacing real ions requires several steps.

- Step 1: calculate the total interaction energy, with use of an equation similar to Equation (2.1), between 'ion 1' and all the other ions in the cluster; the interaction is given a plus or minus sign depending on whether the ions have the same or opposite charges.

- Step 2: repeat this summation for all the other ions in the cluster.

- Step 3: divide the total energy calculated by two, as each ion will be counted twice.

The energy so derived is called the Madelung energy of the cluster.

It is found that the electrostatic energy of an ionic crystal has a form identical to that of Equation (2.1) multiplied by a constant that arises from the geometry of the crystal, the arrangement of the ions in space, and a term representing the charges on the ions:

$$E_e = \left(\frac{-e^2}{4\pi\varepsilon_0 r}\right) \times (\text{geometry}) \times (\text{ionic charges})$$
$$(2.2)$$

The term reflecting the geometry of the structure is called the Madelung constant. Equation (2.2) is the electrostatic energy per pair of ions. The energy is most conveniently expressed per mole of compound. Thus, the electrostatic energy per mole of a crystal of the *halite* (NaCl) structure, containing equal numbers of ions of charge $+Ze$ and $-Ze$, is:

$$E_e = N_A \left(\frac{-e^2}{4\pi\varepsilon_0 r}\right) \alpha Z^2 \qquad (2.3)$$

where N_A is Avogadro's constant, α is the Madelung constant of the *halite* structure (equal to 1.748), and

r is the nearest equilibrium distance between neighbouring ions in the crystal. As in Equation (2.1), the negative sign arises because the charge on the cations is $+Ze$ and the charge on the anions is $-Ze$. An overall negative value of the electrostatic energy means that ionic *halite* structure crystals are stable. This equation is applicable to all crystals with the *halite* structure, irrespective of the ions that make up the crystal, and can be used with solids as diverse as NaCl itself ($Z_1 = Z_2 = 1$), magnesium oxide (MgO; $Z_1 = Z_2 = 2$) or lanthanum phosphide (LaP; $Z_1 = Z_2 = 3$).

In a structure in which the ions have different charges, such as the *fluorite* structure of CaF$_2$, the charge contribution is more complicated. In the case of a compound $M_m X_n$ the electrostatic energy is given by:

$$E_e = -N_A \frac{e^2}{4\pi\varepsilon_0 r} \alpha(Z_M Z_X)\frac{m+n}{2} \qquad (2.4)$$

where, to maintain charge neutrality, $m Z_M = n Z_X$. The Madelung constant defined by this equation is called the reduced Madelung constant. Alternative definitions are used in some sources. The differences are explained in Section S1.4. Madelung constants for all the common crystal structures have been calculated. Some are listed in Table 2.1. Surprisingly, the reduced Madelung constant is very similar for a wide range of structures, and is equal to 1.68 ± 0.08, or about 5 %, as is apparent from Table 2.1. This means that the approximate electrostatic energy of any crystal structure can be estimated as long as the chemical formula is available, by using Equation [2.4] and a value of 1.68 for α.

Table 2.1 Reduced Madelung constants, α

Structure	Formula	Example	α
Halite	M^+X^-	NaCl	1.748
Caesium chloride	M^+X^-	CsCl	1.763
Sphalerite	$M^{2+}X^{2-}$	ZnS	1.638
Wurtzite	$M^{2+}X^{2-}$	ZnO	1.641
Fluorite	$M^{2+}X_2^-$	CaF$_2$	1.68
Rutile	$M^{4+}X_2^{2-}$	TiO$_2$	1.60

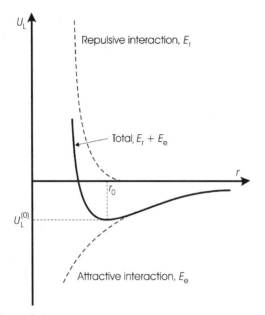

Figure 2.2 The total potential energy, U_L, between monovalent ions as a function of the ionic separation, r. The total energy is the sum of the attractive and repulsive potential energy terms. The lattice energy, $U_L^{(0)}$, corresponds to the minimum in the total energy curve, reached at an interionic separation of r_0

2.1.4 Repulsive energy

Ions are not simply point charges and as they are brought together their closed electron shells begin to overlap and, for quantum mechanical reasons, repulsion sets in. This increases sharply as the interionic distance, r, decreases until, neglecting other forces, a balance is obtained with the electrostatic attractive forces (Figure 2.2). The repulsive potential energy, E_r, can be formulated in a number of ways. One of the first to be used was an empirical expression of the type

$$E_r = \frac{B}{r^n} \qquad (2.5)$$

where B and n are constants. The value of n can be derived from compressibility measurements. Larger ions are more compressible and have larger values

Table 2.2 Values of the constant n [Equation (2.5)]

Ion configuration	Example	n
He	Li^+	5
Ne	Na^+, F^-	7
Ar	K^+, Cl^-	9
Kr	Rb^+, Br^-	10
Xe	Cs^+, I^-	12

of n. Some values are given in Table 2.2. An average value is used for ionic combinations that have different electron configurations. For example, a value of 6 can be used for the compound LiF. Other ways of describing the repulsive energy are given in the Section 2.1.5.

2.1.5 Lattice energy

The total potential energy of an ionic crystal, which is often referred to as the lattice energy, U_L, per mole, may be represented as the sum of the electrostatic and repulsive energy terms. For a *halite* structure crystal, MX, by summing Equations (2.3) and (2.5), we obtain the lattice energy, U_L, per mole:

$$U_L = E_e + E_r = \frac{-N_A \alpha Z^2 e^2}{4\pi\varepsilon_0 r} + \frac{N_A B}{r^n} \qquad (2.6)$$

The energy is a function of the distance between the ions, r, and at equilibrium this energy must pass through a minimum (Figure 2.2). Thus, we can write:

$$\frac{dU_L}{dr} = \frac{N_A \alpha Z^2 e^2}{4\pi\varepsilon_0 r^2} - \frac{n N_A B}{r^{n+1}} = 0$$

This allows the constant B to be eliminated, to give:

$$U_L^{(0)} = \left(\frac{N_A \alpha Z^2 e^2}{4\pi\varepsilon_0 r_0^2}\right)\left(1 - \frac{1}{n}\right) \qquad (2.7)$$

where $U_L^{(0)}$ is the equilibrium value of the lattice energy and r_0 is the equilibrium value of the interionic separation. Values of the lattice energy can be calculated by using experimental values for the equilibrium separation of the ions, r_0. The results are in good agreement with experimental determinations of lattice energy.

The advent of high-speed computers has made the calculation of lattice energies and other aspects of an ionic bonding model straightforward. The approach is similar to that outlined above. The lattice energy is derived by summing electrostatic interactions and including a repulsive potential, just as outlined. The advantage of computer routines is that it is possible to include effects such as crystal vibration and terms such as ionic polarisation as well as more sophisticated repulsive potentials. These repulsive potentials are called pair potentials.

Two forms are commonly employed. One is an empirical expression of the type:

$$E_r = +N_A B \exp\left(\frac{-r}{r^*}\right) \qquad (2.8)$$

where B and r^* are constants that are structure-sensitive. Values of r^* can be derived from compressibility measurements. Linking Equation (2.8) with the electrostatic energy term, and eliminating the constant B, as above, gives an equation for the lattice potential energy called the Born–Mayer equation:

$$U_L^{(0)} = \left(\frac{N_A \alpha Z^2 e^2}{4\pi\varepsilon_0 r_0}\right)\left(\frac{1 - r^*}{r_0}\right) \qquad (2.9)$$

where the symbols have the same meaning as before. Another equation combines aspects of Equations (2.5) and (2.8). It is a form of a more general *Buckingham potential*, and is written as:

$$E_r = +B \exp\left(\frac{-r}{\rho}\right) - \frac{C}{r^6} \qquad (2.10)$$

where B, C and ρ are parameters that vary from one ion pair to another, and are determined empirically.

2.1.6   The formulae and structures of ionic compounds

In order to understand the valence of ions it is necessary to consider the electronic configuration in more detail. The gain or loss of electrons is most often such as to produce a stable closed-shell configuration, found in the noble gas atoms of Group 18 of the periodic table. Hence, atoms to the left-hand side of the periodic table tend to lose electrons. For example, sodium (Na), with a configuration [Ne] $3s^1$, forms a sodium ion (Na^+), with configuration [Ne]. Atoms on the right-hand side of the periodic table tend to gain electrons to form a noble gas configuration. For example, chlorine (Cl), with a configuration [Ne] $3s^2 3p^5$, readily gains an electron to form an anion (Cl^-), with a configuration [Ar].

Ions that occur in the middle of the periodic table have configurations that are different from that of the noble gases. Elements following the d-block transition metals tend to have an outer electron configuration d^{10}. For example, the electron configuration of silver (Ag) is [Kr] $5s^1 4d^{10}$. To gain a noble gas configuration, the silver atom would have to lose 11 electrons or gain 7 electrons. Each of these alternatives is energetically unreasonable. However, if the silver atom loses the single 5s electron it will still have a closed-shell format, with a filled d^{10} shell outermost. This configuration is relatively stable, and the univalent ion Ag^+, with a configuration [Kr] $4d^{10}$, is stable. The other elements in the group – copper (Cu) and gold (Au) – are similar. They also have the configuration [noble gas] d^{10}. The elements zinc (Zn), cadmium (Cd) and mercury (Hg), with a [noble gas]$d^{10}s^2$ outer electron configuration, tend to lose the s electrons to form Zn^{2+}, Cd^{2+} and Hg^{2+} ions with a configuration [noble gas] d^{10}.

Atoms in at the lower part of Groups 13, 14 and 15 are able to take two ionic states. For example, tin (Sn) has an outer electron configuration [Kr] $4d^{10} 5s^2 5p^2$. Loss of the two p electrons will not leave the ion either with a noble gas configuration or with a d^{10} configuration but it will still possess a series of closed shells that is moderately stable. This is the Sn^{2+} state, with a configuration of [Kr] $5s^2 6d^{10}$. However, loss of the two s electrons will produce the stable configuration [Kr] $6d^{10}$ of Sn^{4+}. The atoms that behave in this way are characterised by two valence states, separated by a charge difference of +2. The examples are indium [In (1+, 3+)], thallium [Tl (1+, 3+)], tin [Sn (2+, 4+)], lead [Pb (2+, 4+)], antimony [Sb (3+, 5+)]

and bismuth [Bi (3+, 5+)]. When present, the pair of s electrons has important physical and chemical effects, and ions with this configuration are called lone-pair ions.

The transition metal ions generally have a number of d electrons in their outer shell, and because the energy difference between the various configurations is small, the arrangement adopted will depend upon a variety of external factors, such as the geometry of the crystal structure (see also Chapter 12 and Section S4.5). The lanthanides have an incomplete 4f shell of electrons, and the actinides an incomplete 5f shell. In these elements, the f orbitals are shielded from the effects of the surrounding crystal structure. The d and f electrons control many of the important optical and magnetic properties of solids.

The formula of an ionic compound follows directly from the idea that cations have integer positive charges, anions have integer negative charges and ionic compounds are neutral. Consider a crystal of sodium chloride, NaCl. Each Na^+ cation has a charge of $+1e$. Each Cl^- anion has a charge of $-1e$. As crystals of sodium chloride are neutral, the number of Na^+ ions and Cl^- ions must be equal. The chemical formula is $Na_n Cl_n$, that is, NaCl. Similarly, a magnesium Mg^{2+} ion united with an oxygen O^{2-} ion will form a compound of formula MgO, magnesium oxide. It is necessary for two monovalent (M^+) cations to combine with a divalent (X^{2-}) anion to form a neutral unit M_2X – for example, sodium oxide (Na_2O). Similarly, a divalent (M^{2+}) cation will need to combine with two monovalent (X^-) anions to give neutral MX_2 – for example, magnesium chloride ($MgCl_2$). Trivalent (M^{3+}) cations need three monovalent anions – for example, aluminium chloride ($AlCl_3$). Two trivalent cations need to combine with three divalent anions to give a neutral unit – for example, aluminium oxide (Al_2O_3).

2.1.7 Ionic size and shape

The concept of allocating a fixed size to each ion is an attractive one and has been extensively utilised. Ionic radii are generally derived from X-ray crystal-lographic structure determinations (Chapter 5). This technique only gives a precise knowledge of the distances between the atoms in an ionic crystal. To derive ionic radii, it is assumed that the individual ions are spherical and in contact. The radius of one commonly occurring ion, such as the oxygen ion, O^{2-}, is taken as a standard. Other consistent radii can then be derived by subtracting the standard radius from measured interionic distances.

The ionic radius quoted for any species depends upon the standard ion by which the radii were determined. This has led to a number of different tables of ionic radii. Although these are all internally self-consistent, they have to be used with thought. Additionally, cation radius is found to be sensitive to the surrounding coordination geometry. The radius of a cation surrounded by six oxygen ions in octahedral coordination is different from that of the same cation surrounded by four oxygen ions in tetrahedral coordination. Similarly, the radius of a cation surrounded by six oxygen ions in octahedral coordination is different from that of the same cation surrounded by six sulphur ions in octahedral coordination. Ideally, tables of cationic radii should apply to a specific anion and coordination geometry. Representative ionic radii are given in Figures 2.3(a) and 2.3(b).

Several trends in ionic radius are apparent:

- Cations are usually smaller than anions, the main exceptions being the largest alkali metal and alkaline earth metal cations, all larger than the fluorine ion F^-. The reason for this is that removal of electrons to form cations leads to a contraction of the electron orbital clouds as a result of the relative increase in nuclear charge. Similarly, addition of electrons to form anions leads to an expansion of the charge clouds as a result of a relative decrease in the nuclear charge.

- The radius of an ion increases with atomic number.

- The radius decreases rapidly with increase of positive charge for a series of isoelectronic ions such as Na^+, Mg^{2+}, Al^{3+}, all of which have the electronic configuration [Ne]. Note that the real

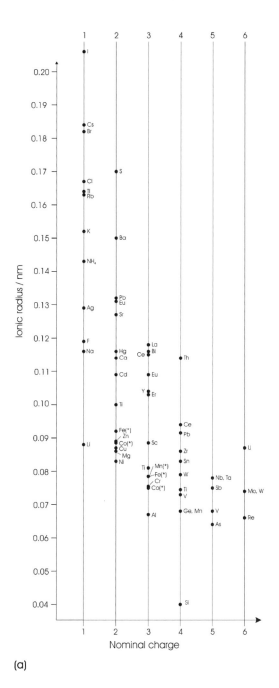

(a)

charges on cations in solids are generally smaller than the formal ionic charges expressed in isolated ions, and the effect will be smaller in solids than the tables of ionic radii suggest.

- Successive valence increases decrease the radius. For example, Fe^{2+} is larger than Fe^{3+}.

- An increase in negative charge has a smaller effect than an increase in positive charge. For example, F^- is similar in size to O^{2-}, and Cl^- is similar in size to S^{2-}.

Although the majority of the ions of elements can be considered to be spherical, the lone-pair ions are definitely not so. These ions – In^+, Tl^+, Sn^{2+}, Pb^{2+}, Sb^{3+} and Bi^{3+} – tend to be surrounded by an irregular coordination polyhedron of anions. This is often a distorted trigonal bipyramid, and it is hard to assign a unique radius to such ions.

Complex ions, such as CO_3^{2-} and NO_3^-, are not spherical, although at high temperatures rotation often makes them appear spherical.

2.1.8 Ionic structures

Ionic bonding is nondirectional. The main structural implication of this is that ions simply pack together to minimise the total lattice energy. There have been many attempts to use this simple idea to predict the structure of an ionic crystal in more detail. This approach was of great importance in the early days of X-ray crystallography, where the investigator had more or less to guess at a model structure to start with by using chemical and physical intuition, and any help that could be obtained from the ionic model was to be welcomed. At present, X-ray techniques allow structures to be solved without such input.

The early structure-building rules, based on ionic bonding guidelines, are still of value, however, in understanding some of the patterns underlying the multiplicity of crystal structures that are known. A simple assumption is that crystals are built of hard spherical ions linked by nondirectional ionic bonding. In terms of this idea, a structure is made up of

Figure 2.3 Ionic radii for ions commonly found in solids: (a) graphical representation; (b) periodic table. Note: a superscript *, indicates a high-spin configuration (Section S4.5); cation radii are those for ions octahedrally coordinated to oxygen, except where marked with a t, which are for ions in tetrahedral coordination

Charge state: +1	+2	+3	+4 (+3)	+5 (+4) [3+]	+6 (+4) [3+]	+6 (+4) [3+] {2+}	+4 (+3) [2+]	+4 (+3) [2+]	+4 (+2)	+2 (+1)	+2	+3	+4 (2+)	+5 (3+)	-2 (6+)	-1
Li 0.088	Be 0.041(t)											B 0.026(t)	C	N	O 0.126	F 0.119
Na 0.116	Mg 0.086											Al 0.067	Si 0.040(t)	P	S 0.170	Cl 0.167
K 0.1521	Ca 0.114	Sc 0.0885	Ti 0.0745 (0.081)	V 0.068 (0.073) [0.078]	Cr [0.0755]	Mn (0.068) [0.079*] {0.097*}	Fe (0.079*) [0.079*]	Co (0.075*) [0.092*]	Ni (0.083)	Cu 0.087 (0.108)	Zn 0.089	Ga 0.076	Ge 0.068	As 0.064	Se (0.043 t)	Br 0.182
Rb 0.163	Sr 0.1217	Y 0.104	Zr 0.086	Nb 0.078	Mo 0.074	Tc (0.078)	Ru 0.076	Rh 0.0755	Pd (0.100)	Ag (0.129)	Cd 0.109	In 0.094	Sn 0.083 (0.105)	Sb 0.075	Te (0.068)	I 0.206
Cs 0.184	Ba 0.150	La 0.1185	Hf 0.085	Ta 0.078	W 0.074 (0.079)	Re 0.066	Os 0.077	Ir 0.077	Pt 0.077 (0.092)	Au	Hg 0.116	Tl 0.1025	Pb 0.0915 (0.132)	Bi 0.086 (0.116)	Po	At

(b)

Figure 2.3   (*Continued*)

large spherical anions packed in such a way as to fill the space available optimally. Cations fit into positions between the large anions. Large cations tend to be surrounded by a cubic arrangement of anions, medium-sized cations by an octahedral arrangement of anions, and small cations by a tetrahedron of anions. The smallest cations are surrounded by a triangle of anions. Local charge neutrality should occur, as far as possible. These and other ways of looking at ionic structures are described more fully in the sources listed in the Further Reading section at the end of this chapter.

2.2   Covalent bonding

2.2.1   Molecular orbitals

Covalent bonds form when an unpaired electron in an atomic orbital on one atom interacts with an unpaired electron in an atomic orbital on another atom. The electrons, which are initially completely localised on the parent atoms, are now shared between the two, in a molecular orbital. This constitutes a covalent bond. The electrons have become delocalised. As two electrons are involved, covalent bonds are also called electron-pair bonds. Covalent bonds are strongest when there is maximum overlap between the contributing atomic orbitals. Covalent bonds are, therefore, strongly directional, and covalent bonding successfully explains the geometry of molecules.

An example of the way in which electron sharing comes about can be given by considering the hydrogen molecule, H_2. An isolated hydrogen atom has a single electron in a spherical 1s orbital. As distance between the atoms is reduced, two different kinds of interaction are possible, depending on whether the spins of the electrons in the s orbitals of the two atoms are parallel or opposed. If the spins of the electrons on the two atoms are opposed, as the interatomic distance is reduced both electrons begin to experience attraction from both nuclei. There is also electrostatic repulsion between the two electrons, but the attraction preponderates, bonding is said to occur and the nuclei are pulled together. A (covalent) bond forms. It is found that the electron density, which was originally spherically distributed around each atom (Figure 2.4a) is now concentrated between the nuclei (Figure 2.4b). If

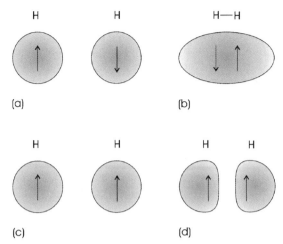

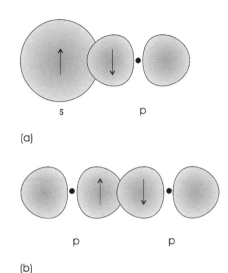

Figure 2.4 Isolated hydrogen atoms have spherically symmetrical 1s orbitals, each containing one s electron, represented as an arrow. Two atoms can have electrons in (a) an antiparallel or (c) parallel arrangement; (b), if the electrons have antiparallel spins the electron density accumulates between the nuclei to form a covalent bond; (d) if the electrons have parallel spins the electron density is low between the nuclei and no bond forms

Figure 2.5 A covalent σ bond formed by the overlap of (a) an s orbital and an end-on p orbital when the two electrons have antiparallel spins and (b) two end-on p orbitals when the electrons have antiparallel spins

the spins of the two electrons are parallel, the Pauli exclusion principle stipulates that it is energetically unfavourable for the electron clouds to overlap. The electron density avoids the internuclear region (Figure 2.4d), and bonding does not occur. The consequences of this 'antibonding' alternative are considered further below.

Two p orbitals end-on to each other and each containing a single electron can interact in very much the same way (see Figure 2.5b). The same is true for a combination of half-filled s and end-on p orbitals (Figure 2.5a).

A molecular orbital formed by s orbitals, end-on p orbitals or by s and p orbitals has rotational symmetry about the bond axis, which is the line joining the two nuclei contributing the electrons. As a result, a cross-section through the orbital looks like an s orbital and, in recognition of this symmetry relationship, such molecular orbitals are termed σ orbitals. The bonds formed by σ molecular orbitals are often called σ bonds.

A different type of molecular orbital can be formed between two p orbitals, each with a single electron and with opposed spins, approaching each other sideways on (Figure 2.6). In this case, the 'pile-up' of the electron density occurs either side of the nodal plane in which the two nuclei are situated. In this configuration bonding can also occur, but the molecular orbital looks like a p orbital in cross-section, and such molecular orbitals are termed π orbitals. The bonds formed by π molecular orbitals are called π bonds.

It is important to note that the designation of a bond as σ or π does not depend on the type of orbital forming the bond, only the geometry of overlap of the orbitals.

2.2.2 The energies of molecular orbitals in diatomic molecules

In order to be sure that a bond actually forms between two atoms linked by a molecular orbital it is necessary to calculate the energies of the molecular orbitals and then allocate electrons to them. In essence, the approximate Schrödinger equation

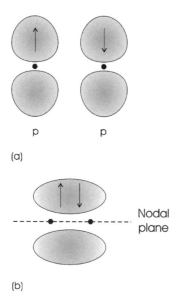

(a)

Nodal plane

(b)

Figure 2.6 (a) Two sideways-on p orbitals containing electrons with antiparallel spins; (b) a π bond formed by the sideways-on overlap of p orbitals. The electron density is concentrated above and below the plane containing the nuclei, and is zero in this plane, called a nodal plane

for the molecule must be solved. This process is similar to the method used for solving the electron configuration of many-electron atoms. An approach called molecular orbital theory is usually chosen for this task. In this, the molecular orbital is obtained by adding together contributions from all of the atomic orbitals involved. This is called the linear combination of atomic orbitals, or LCAO, method. Thus for two identical atoms, each contributing one orbital – say two hydrogen atoms each contributing an s orbital – the molecular orbitals are given by:

$$\psi(\text{molecule}) = c_1 \chi_1 + c_2 \chi_2$$

where c_1 and c_2 are parameters that have to be determined, and χ_1 and χ_2 are the wavefunctions on atom 1 and atom 2. The values of the parameters and the energy of the molecular orbitals are calculated by using standard methods (see the further reading section).

The calculations show that when two atomic orbitals interact, two molecular orbitals form, one with a higher energy than the original pair and one with a lower energy than the original pair. The molecular orbital of lower energy than the parent atomic orbitals is the one with the greatest concentration of electron density between the nuclei (Figure 2.4b). These orbitals are called bonding orbitals. The molecular orbital of higher energy than the parent atomic orbitals is the one in which the electron density is concentrated in the region outside of the line joining the nuclei (Figure 2.4d). Such orbitals are antibonding orbitals.

The energies of the two molecular orbitals are given as follows:

$$E_{\text{bond}} = \alpha + \beta$$
$$E_{\text{abond}} = \alpha - \beta$$

The term α, called the Coulomb integral, is related to the Coulomb energy of the electrons in the field of the atoms and in general is a function of the nuclear charge and the type of orbitals involved in the bond. By definition, the Coulomb energy is regarded as negative. [Note that the Coulomb integral and the Madelung constant, confusingly, both use the same symbol, α; take care not to equate the two terms.] The term β is called the resonance integral, or interaction integral, and in general is a function of the atomic number of the atoms, the orbital types and the degree of overlap of the orbitals. In the case where electron density 'piles up' between the nuclei, β is negative. Thus, the lower energy bonding orbital corresponds to E_{bond} and the higher energy antibonding orbital corresponds to E_{abond}.

Consider again the situation when two hydrogen atoms interact. The two 1s orbitals give two molecular orbitals, one bonding and one antibonding (Figure 2.7). To stress the links with the atomic orbitals, these are called $\sigma 1s$, which is the bonding orbital, and $\sigma^* 1s$, which is the antibonding orbital. When two hydrogen atoms meet, both electrons will occupy the bonding, $\sigma 1s$, orbital provided that they have opposed spins. This will be the lowest-energy configuration, or ground state, of the pair, and a

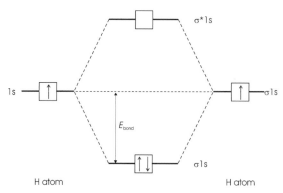

Figure 2.7 The close approach of two hydrogen atoms, each with an electron in a 1s orbital, leads to the formation of two molecular orbitals, a bonding $\sigma 1s$ molecular orbital and an antibonding $\sigma^* 1s$ orbital. In the H_2 molecule, both electrons occupy the bonding orbital, and a strong bond with energy $2 E_{bond}$ results

covalently bonded hydrogen molecule, H_2, will form. The bond energy will be $2 E_{bond}$.

To explain the electron configuration and bonding in other diatomic molecules, the method used to obtain the electron configuration of atoms is copied. Electrons are fed into the available molecular orbitals by using the Aufbau (building-up) principle to obtain the lowest-energy ground state. As before, start at the orbital of lowest energy and work up, feeding two electrons with opposing spin into each orbital and following the Pauli principle and Hund's rules (see Sections 1.3.2 and S1.3.2). This can be illustrated by considering the series of diatomic molecules made up from identical atoms, called homonuclear molecules.

Following hydrogen, the next molecule of this type to consider – singly ionised di-helium, He_2^+ – contains three electrons. The bonding $\sigma 1s$ orbital is full, as a molecular orbital can only contain two electrons of opposed spins, and so the third electron will go into the antibonding orbital $\sigma^* 1s$. Because the energy of this orbital is higher than that of the two isolated atoms, the extra electron will have the effect of partly cancelling the bonding induced by the filled $\sigma 1s$ bonding orbital. We thus expect a weaker and longer bond compared with that of H_2, but the molecule can be expected to form.

When two helium atoms interact there are four electrons to place in the orbitals and so both the $\sigma 1s$ and the $\sigma^* 1s$ orbitals will be filled. The effect of the filled antibonding orbital completely negates the effect of the filled bonding orbital. No energy is gained by the system and so He_2 does not form.

To derive the electron configurations of the other homonuclear X_2 molecules, formed from the elements of the second period of the periodic table, Li_2 to Ne_2, exactly the same procedure is followed. That is, electrons from the separate atomic orbitals are allocated to the molecular orbitals from the lowest energy upwards, remembering that the $\sigma 1s$ and $\sigma^* 1s$ orbitals are filled and constitute an unreactive core. The interaction of the 2s outer orbitals will form $\sigma 2s$ and $\sigma^* 2s$ orbitals. In addition, the 2p orbitals can overlap to form molecular orbitals. End-on overlap, as drawn in Figure 2.5(b), produces $\sigma 2 p_x$ and $\sigma^* 2 p_x$ molecular orbitals. The sideways on overlap of a pair of p orbitals, as in Figure 2.6, forms one $\pi 2 p_y$ bonding orbital, one $\pi 2 p_z$ bonding orbital, one $\pi^* 2 p_y$ antibonding orbital and one $\pi^* 2 p_z$ antibonding orbital. The energy of the orbitals is sketched in Figure 2.8(a) for molecules as far as dinitrogen, N_2. The difference in energy between the $\sigma 2 p_x$ and $\pi 2p$ orbitals is small and gradually changes along the series, so that the $\sigma 2 p_x$ orbital drops below the $\pi 2p$ orbitals for the last three molecules – O_2, F_2 and the hypothetical Ne_2 – drawn in Figure 2.8(b).

The molecular configurations of the homonuclear diatomic molecules can now be obtained by using the Aufbau principle. The first to consider is dilithium Li_2. The electron configuration of lithium, (Li), is $[He]2s^1$. Both 2s electrons will occupy the lowest available bonding orbital, and a stable molecule will form. The next element, beryllium, (Be), has an electron configuration $[He] 2s^2$. An attempt to form the molecule Be_2 will necessitate placing two electrons in the bonding orbital and two in the lowest available antibonding orbital. No stable molecule will form. The next atom, boron, (B), has an electron configuration $[He] 2s^2 2p^1$, and electrons now enter the bonding $\pi 2p$ orbitals. The repetitive filling continues with the other elements, with the result given in Table 2.3.

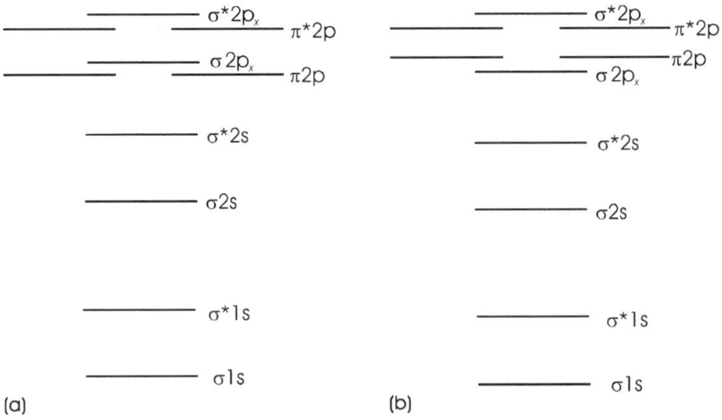

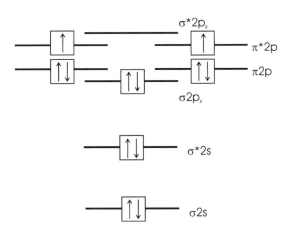

Figure 2.8 (a) Schematic molecular orbital energy level diagram for homonuclear diatomic molecules H_2 to N_2; (b) schematic energy level diagram for the homonuclear diatomic molecules O_2 to Ne_2

(a)   (b)

Table 2.3 The electron configurations of some homonuclear diatomic molecules

Molecule	Ground-state configuration	Bond length/ nm	Bond energy/ kJ mol^{-1}
Li_2	$[He_2]\,(\sigma 2s)^2$	0.267	101
Be_2	$[He_2](\sigma 2s)^2(\sigma^* 2s)^2$	–	–
B_2	$[Be_2](\pi 2p)^2$	0.159	289
C_2	$[Be_2](\pi 2p)^4$	0.124	599
N_2	$[Be_2](\pi 2p)^4(\sigma 2p_x)^2$	0.110	941
O_2	$[Be_2](\pi 2p)^4(\sigma 2p_x)^2\,(\pi^* 2p)^2$	0.121	494
F_2	$[Be_2](\pi 2p)^4(\sigma 2p_x)^2\,(\pi^* 2p)^4$	0.142	154
Ne_2	$[Be_2](\pi 2p)^4(\sigma 2p_x)^2(\pi^* 2p)^4(\sigma^* 2p_x)^2$	–	–

– Molecule is not formed.
Note: $[He_2] = (\sigma 1s)^2(\sigma^* 1s)^2$; $[Be_2] = (1\sigma)^2(1\sigma^*)^2(2\sigma)^2(2\sigma^*)^2$.

An important verification of the molecular orbital theory was provided by the oxygen molecule, O_2. This molecule had long been known to be paramagnetic; a puzzling property. However, the electron configuration given in Table 2.3 shows that the two electrons with highest energy have to be placed in separate orbitals (Figure 2.9). These unpaired electrons make the molecule paramagnetic (see Chapter 12).

2.2.3 Bonding between unlike atoms

When a molecular orbital, whether of σ or π type, is formed between atoms of two different elements, A

Figure 2.9 The ground-state electron configuration of O_2. Each oxygen atom contributes eight electrons, and each orbital up to the $\pi 2p$ set contains paired electrons. The last two electrons occupy separate $\pi^* 2p$ orbitals, with parallel spins

and X, then the energy levels of the initial atomic orbitals will differ, as will their extensions in space. One can construct an appropriate molecular orbital energy diagram for this situation, as in Figure 2.10. This corresponds to the case where element A is more metallic (or less electronegative, see p. 35) in character than element X. The bonding energy E_b is now with respect to the average energy of the uninteracting A and X atoms: $\frac{1}{2}(E_A + E_X)$. It is found that the X atom contributes most to the bonding molecular orbital, and the atom A more to the antibonding molecular orbital. The bonding

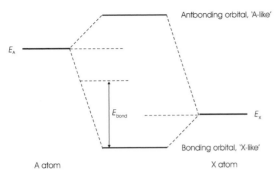

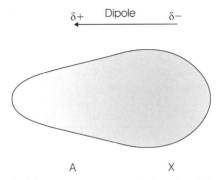

Figure 2.10 Molecular orbitals formed by a more metallic atom *A* and a less metallic atom *X*. The nonmetallic element contributes more to the bonding orbital, which is said to be *X*-like. The more metallic atom contributes more to the antibonding orbital, which is said to be *A*-like

Figure 2.11 The electron density in a bonding molecular orbital between two dissimilar atoms is distorted so that the end nearer to the nonmetallic atom attracts more of the charge cloud. The bond is then an electric dipole, with charges $\delta+$ and $\delta-$, represented by an arrow pointing from the negative to the positive charge

molecular orbitals are often said to be '*X*-like' in character, and the antibonding orbitals '*A*-like' in character.

A bonding molecular orbital concentrates electronic charge density in the region between the bonded nuclei (subject, in the case of π bonding, to the limitation set by the nodal plane). If the two nuclei are different, they will have different effective nuclear charges. This will cause the concentration of charge to shift to increase the screening of the higher effective charge and decrease that of lower effective charge, until both have become equalised. Therefore, the symmetrical build up of electron density shown in Figures 2.4 and 2.6 will become modified to that in Figure 2.11.

Obviously with a very large difference in effective nuclear charge, one would have something approaching ions being formed, both electrons of the molecular orbital becoming almost completely associated with the *X* atom, giving it nearly unit negative charge, whereas the *A* atom would have almost unit positive charge.

A covalent bond in which the electron pair is distributed unevenly is sometimes called a polar covalent bond. The bond will have one end that carries a small positive charge, written $\delta+$, and the other end a small negative charge, $\delta-$. The charge separation gives rise to an internal electric dipole (Figure 2.11) and such molecules are called polar

molecules. An electric dipole is a vector quantity and is drawn as an arrow pointing from the negative charge to the positive.

A polyatomic molecule may contain a number of polar covalent bonds. For example, water (H_2O) is a polar molecule as the two O—H bonds form dipoles pointing towards the hydrogen atoms. However, not all molecules containing several dipoles are polar, as the dipoles within the molecule, the internal dipoles, may sum to zero.

2.2.4   Electronegativity

The idea of atoms possessing a tendency to attract electrons is rather useful, and the electronegativity, χ, of an element represents a measure of its power to attract electrons during chemical bonding. Atoms with a low electronegativity are called electropositive elements. These are the metals, and when bonded they do not have a strong tendency to attract electrons and so tend to form cations. Atoms with a high electronegativity, called electronegative elements, tend to attract electrons in a chemical bond and tend to form anions. The magnitude of the partial charges, $\delta+$, $\delta-$, in a polar molecule is dependent on the electronegativity difference between the two atoms involved.

Table 2.4   Electronegativity values

H	2.2														
Li	1.0	Be	1.5	B	2.0	C	2.5	N	3.0	O	3.5	F	4.0
Na	1.0	Mg	1.2	Al	1.5	Si	1.8	P	2.1	S	2.5	Cl	3.0
K	0.8	Ca	1.0	Ga	1.5	Ge	1.8	As	2.0	Se	2.4	Br	2.8
Rb	0.8	Sr	1.0	In	1.5	Sn	1.8	Sb	1.8	Te	2.1	I	2.5
Cs	0.8	Ba	0.9	Tl	1.5	Pb	1.7	Bi	1.8				

Source: adapted from selected values of Gordy and Thomas, taken from W. B. Pearson, 1972, *The Crystal Chemistry and Physics of Metals and Alloys*, Wiley-Interscience.

Electronegativity values have been derived in a number of ways. The first of these was by Pauling and made use of thermochemical data to obtain a scale of relative values for elements. Most electronegativity tables since then have also contained relative values, which do not have units.

In general, the most electronegative atoms are those on the right-hand side of the periodic table, typified by the halogens fluorine and chlorine (Table 2.4). The least electronegative atoms, which are the most electropositive, are those on the lower left-hand side of the periodic table, such as rubidium, Rb, and caesium, Cs (Table 2.4). Covalent bonds between strongly electronegative and strongly electropositive atoms would be expected to be polar.

2.2.5   Bond strength and direction

So far, the energetic aspects of covalent bonds have been considered by using molecular orbital theory. Molecular orbital theory is equally well able to give exact information about the geometry of molecules. However, a more intuitive understanding of the geometry of covalent bonds can be obtained via an approach called valence bond theory. (Note that both molecular orbital theory and valence bond theory are formally similar from a quantum mechanical point of view, and either leads to the same result.)

Valence bond theory starts with the idea that a covalent bond consists of a pair of electrons shared between the bound atoms. Two resulting ideas make it easy to picture covalent bonds. The first of these is the concept that that the direction of a bond will be such as to make the orbitals of the bonding electrons overlap as much as possible. The second is that the strongest bonds are formed when the overlapping of the orbitals is at a maximum. On this basis, we expect differences in bond-forming power for s, p, d and f orbitals since these orbitals have different radial distributions. The relative scales of extension for 2s and 2p orbitals are 1 and $\sqrt{3}$ respectively (Figure 2.12). The shapes of the p orbitals leads to the expectation that p orbitals should be able to overlap other orbitals better than s orbitals and hence that bonds involving p orbitals should generally be stronger than bonds involving s orbitals. If there is a choice between s or p orbitals, use of p orbitals should lead to more stable compounds.

The geometry of many molecules can be qualitatively explained by these simple ideas. Consider the bonding in a molecule such as hydrogen chloride, HCl. The hydrogen atom will bond via its s orbital end on to the one half-filled p orbital on the chlorine atom. The hydrogen nucleus will lie along the axis of the 2p orbital since this gives the maximum overlap for a given internuclear spacing (Figure 2.13). Consider the situation in a water molecule,

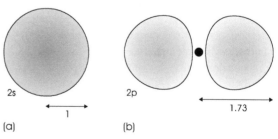

Figure 2.12   The relative extension of (a) a 2s orbital (1.0) and (b) a 2p orbital ($\sqrt{3}$, 1.73)

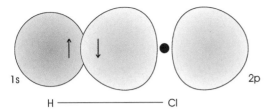

Figure 2.13 The covalent bond in a linear molecule such as hydrogen chloride (HCl) is formed by the overlap of a 1s orbital on the hydrogen atom with the $2p_x$ orbital on the chlorine atom. The electron spins in each orbital must be antiparallel for a bond to form

H_2O. The two hydrogen atoms form bonds with two different half-filled p orbitals on the oxygen atom. As these lie at 90° to each other, the molecule should be angular, with an H—O—H angle of 90°. Similarly, the molecule of ammonia, NH_3, involves bonding of the hydrogen 1s orbitals to the three 2p orbitals that lie along the three Cartesian axes. The shape of the molecule should mimic this, with the three hydrogen atoms arranged along the three Cartesian axes, to form a molecule that resembles a flattened tetrahedron. To a rough approximation, these molecular shapes are correct, but they are not precise enough. For example, the actual H—O—H angle is 104.5°, considerably larger than 90°. To explain the discrepancy it is necessary to turn to a more sophisticated concept.

2.2.6 Orbital hybridisation

Although water and ammonia provide examples of the disagreement between the simple ideas of orbital overlap and molecular geometry, the most glaring example is provided by carbon. From what has been said so far, one would expect carbon, with an electron configuration of $1s^2 2s^2 2p^2$, to form compounds with two p bonds at 90° to one another. That is to say, in reaction with hydrogen, following the same procedure as above, a molecule of formula CH_2 should form and have the same 90° geometry as water. Now the common valence of carbon is four and, as early as the latter half of the 19th century, organic chemists established beyond doubt that in

the small molecules formed by carbon the four bonds are directed away from the carbon atom towards the corners of a tetrahedron. The orbital picture so-far presented clearly breaks down when applied to carbon. This discrepancy between theory and experiment has been resolved by introducing the concept of orbital hybridisation.

Hybridisation involves combining orbitals in such a way that they can make stronger bonds (with greater overlap) than the atomic orbitals depicted earlier. To illustrate this, suppose that we have one s and one p orbital available on an atom (Figure 2.14a). These could form two bonds, but neither orbital can utilise all of its overlapping ability when another atom approaches. However, an s and a p orbital can combine, or *hybridise*, to produce two new orbitals pointing in opposite directions, (Figure 2.14b). Each resulting hybrid orbital is composed of one large lobe and one very small lobe, which can be thought of as the positive s orbital adding to the positive lobe of the p orbital to produce a large lobe, and the positive s orbital adding to the negative lobe

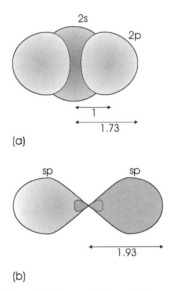

Figure 2.14 (a) The 2s and $2p_x$ orbitals on an atom and (b) the two sp hybrid orbitals formed by combining the 2s and $2p_x$ orbitals. Each hybrid orbital has a large lobe and a small lobe. The extension of the large lobe is 1.93, compared with 1.0 for a 2s orbital and 1.73 for a $2p_x$ orbital. The orbitals point directly away from each other

of the p orbital to give a small lobe. The overlapping power of the new combination is found to be significantly larger than that of s or p orbitals, because the extension of the hybrid orbitals is 1.93, compared with 1.0 for an s orbital and 1.73 for a p orbital. Although it requires energy to form the hybrid configuration, this is more than recouped by the stronger bonding that results, as discussed in more detail below with respect to the tetrahedral bonding of carbon. Since the hybrid orbitals are a combination of one s and one p orbital, they are called sp hybrid orbitals. The large lobe on each of the hybrid orbitals can be used for bond formation, and bond angles of 180° are expected.

The idea can be illustrated with the atom mercury, Hg. The outer electron configuration of mercury is $6s^2$. The filled electron shell would not be able to form a bond at all. However, the outer orbital energies are very close in these heavy atoms, and little energy is required to promote an electron from the 6s orbital to one of the 6p orbitals. In this configuration, the orbitals can combine to form two sp hybrid orbitals. Mercury makes use of sp hybrid bonds in the molecule $(CH_3)_2Hg$. In this molecule, the Hg forms two covalent bonds with carbon atoms. The C–Hg–C angle is 180°. The strong bonds that can then form more than repay the energy expenditure involved in hybridisation. The linear geometry of sp hybrid bonds is further illustrated with respect to bonding in a molecule of ethyne (acetylene, C_2H_2), described below.

It is a general rule of hybrid bond formation that the same number of hybrid orbitals form and can be used for bonding as the number of atomic orbitals used in the initial combination. Thus, one s and one p orbital yield two sp hybrid orbitals. One s orbital and two p orbitals yield three new sp^2 hybrid orbitals for bond formation. For maximum overlap we expect these orbitals to point as far away from each other as possible, so forming bonds at angles of 120° (Figure 2.15). The sp^2 hybrid orbitals have an overlapping power of about twice that of s orbitals. This type of bonding is found in a number of trivalent compounds of boron, for example BCl_3, which has bond angles of 120°. It is also commonly encountered in borosilicate glasses, in which the boron atoms are linked to three oxygen atoms at the

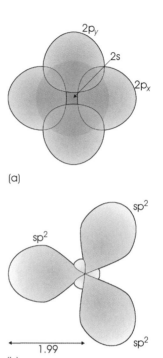

(a)

(b)

Figure 2.15 (a) The 2s, $2p_x$ and $2p_y$ orbitals on an atom and (b) the three sp^2 hybrid orbitals formed by combining the three original orbitals. Each hybrid orbital has a large lobe and a small lobe. The extension of the large lobe is 1.99, compared with 1.0 for a 2s orbital and 1.73 for a $2p_x$ orbital. The orbitals are arranged at an angle of 120° to each other and point towards the vertices of an equilateral triangle

corners of an equilateral triangle by sp^2 hybrid bonding orbitals.

It is now possible to return to the case of carbon. As mentioned above, it is certain that carbon forms four bonds in many of its compounds. The outer electron configuration of carbon is $2s^2\ 2p^2$. If one electron is promoted from the filled $2s^2$ orbital into the empty p orbital, sp^3 hybrid orbitals are possible. Calculation shows that the resulting four bonds will point towards the corners of a tetrahedron, at angles of 109° to each other (Figure 2.16). These angles are the tetrahedral angles found for methane, CH_4, carbon tetrachloride, CCl_4 and many other carbon compounds.

The hybrid orbitals have an overlapping power of twice the overlapping power of s orbitals. Therefore,

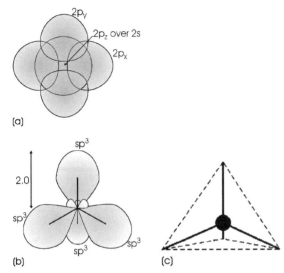

(a)

(b)

(c)

Figure 2.16 (a) The 2s, $2p_x$, $2p_y$ and $2p_z$ orbitals on an atom; (b) the four sp^3 hybrid orbitals formed by combining the four original orbitals. Each hybrid orbital has a large lobe and a small lobe. The extension of the large lobe is 2.0, compared with 1.0 for a 2s orbital and 1.73 for a $2p_x$ orbital. (c) The orbitals are at an angle of 109.5° to each other and point towards the vertices of a tetrahedron

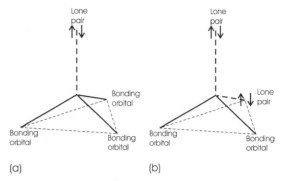

(a)   (b)

Figure 2.17 The sp^3 hybrid bonds in (a) nitrogen and (b) oxygen. In part (a), three bonds form (full lines), as in NH_3, and in part (b), two bonds, form (full lines), as in H_2O. The remaining orbitals are filled with electron pairs, called lone pairs

the bonds formed by sp^3 hybrid orbitals are extremely strong. The C–C bond energy in diamond, the hardest of all solids, is 245 kJ mol^{-1}. For these orbitals to form, one electron must be promoted from the filled 2s orbital to the empty 2p orbital. The energy of the latter process is approximately 400 kJ for the change $1s^2 2s^2 2p^2$ to $1s^2 2s\, 2p^3$. This is an energetic process, but the energy loss is more than made up by the greater overlap achieved, the stronger bonds that result and, importantly, the number of bonds that form with the rearranged orbitals. For example, carbon, with an electron configuration $2s^2 2p_x\, 2p_y$, might form two p bonds of perhaps 335 kJ each to hydrogen atoms, which would liberate perhaps 670 kJ, whereas four sp^3 bonds of 430 kJ each would liberate 1725 kJ. The energy of the latter process is clearly sufficient to accommodate the electron promotion energy.

Hybridization explains the geometry of ammonia (NH_3) and water (H_2O) and similar compounds.

Nitrogen has an outer electron configuration of $2s^2 2p^3$, and oxygen has an outer electron configuration of $2s^2 2p^4$. Although bonding to three or two atoms, respectively, is possible, using the available p orbitals, as described above, stronger bonds result if hybridisation occurs. In both atoms, the s and p orbitals form sp^3 hybrids. In the case of nitrogen (Figure 2.17a) there are five electrons to be allocated. Three of these go into separate sp^3 hybrid orbitals and form three partly filled orbitals. These can be used for bonding, as in NH_3. The other two electrons fill the remaining orbital. This cannot be used for bonding as it is filled and is said to contain a lone pair of electrons. These lone-pair electrons add significant physical and chemical properties to the ammonia molecule.

A similar situation holds for water. There are now six electrons on the oxygen atom to allocate to the four sp^3 orbitals. In this case, two orbitals are filled, and accommodate lone pairs of electrons, and two remain available for bonding (Figure 2.17b). The two lone pairs occupy two corners of the tetrahedron, and the two bonding orbitals point to the other corners of the tetrahedron. The H–O–H bonding angle should now be the tetrahedral angle, 109°. As in the case of ammonia, the lone pairs contribute significant physical and chemical properties to the molecules. The geometry of the molecules is not quite tetrahedral. The H–O–H angle is 104.5° and

Table 2.5 The geometry of some hybrid orbitals

Coordination number	Orbital configuration	Geometry	Example
2	sp	Linear	$HgCl_2$
3	sp^2	Trigonal	BCl_3
4	sp^3	Tetrahedral	CH_4
	dsp^2	Square planar	$PdCl_2$
6	d^2sp^3	Octahedral	SF_6

not 109°. Qualitatively, it is possible to say that the presence of the lone pairs distorts the perfect tetrahedral geometry of the hybrid orbitals. Quantitatively, it indicates that the hybridisation model needs further modification.

Hybridisation is not a special effect in which precise participation by, for example, one s and three p orbitals produces four sp^3 hybrid orbitals. Continuous variability is possible. The extent to which hybridisation occurs depends on the energy separation of the initial s and p orbitals. The closer they are energetically, the more complete will be the hybridisation. Hybridisation can also occur with d and f orbitals. Hybridisation is no more than a convenient way of viewing the manner in which the electron orbitals interact during chemical bonding. The shape of various hybrid orbitals is given in Table 2.5.

2.2.7 Multiple bonds

In the previous discussion, it was taken for granted that only one bond forms between the two atoms involved. However, one of the most characteristic features of covalent compounds is the presence of multiple bonds between atoms. Multiple bonds result when atoms link via σ and π bonds at the same time.

Multiple bonding occurs in the nitrogen molecule, N_2. Traditionally, nitrogen is described as trivalent, and the molecule is depicted as N≡N, with three bonds linking the two atoms to each other. This is explained in the following way. The outer electron configuration of nitrogen is $2s^2 2p^3$. Instead of

forming hybrid orbitals, the three p orbitals on each nitrogen atom can interact to create three bonds. As two nitrogen atoms approach each other, one pair of these p orbitals, say the p_x orbitals, combine in an end-on fashion, to form a σ bond. The other two p orbitals, p_y and p_z, can overlap in a sideways manner to form two π bonds. Each individual π bond has two lobes, one lobe to one side of the internuclear axis and one lobe to the other. The two π bonds comprise four lobes altogether, surrounding the internuclear axis (Figure 2.18). Note that the traditional representation of three bonds drawn as lines N≡N, does not make it clear that two different bond types exist in N_2.

In the case of the oxygen molecule, O_2, conventionally written O=O, a similar state of affairs is found. Oxygen atoms are regarded as divalent, and the molecule consists of two oxygen atoms linked by a double bond. The outer electron configuration of oxygen is $2s^2 2p^4$. One p orbital will be filled with an electron pair and takes no part in bonding. Only the two p orbitals, p_x and p_y, are available for bonding. Close approach of two oxygen atoms will allow the p_x orbitals to overlap end on to form a σ bond and the p_y orbitals to overlap in a sideways fashion to form a π bond (Figure 2.19). As with nitrogen, the conventional representation of the double bond, O=O, does not reveal that two different bond types are present.

Multiple bonding is of considerable importance in carbon compounds and figure prominently in the chemical and physical properties of polymers. Two compounds need to be examined, ethyne (acetylene, C_2H_2) and ethene (ethylene, C_2H_4).

The organic compound ethyne, C_2H_2, combines hybridisation with multiple bond formation. The formula is conventionally drawn as HC≡CH, in which three bonds from each quadrivalent carbon atom link to another carbon atom and the other bond links to hydrogen. Because there are two p electrons available on carbon it would be possible to write down a bonding scheme involving only σ bonds, one between the two carbons and one between a carbon and a hydrogen atom. However, this is not in accord with the properties of the molecule. First, the carbon atom so described would be divalent not quadrivalent. Second, experiments show that the

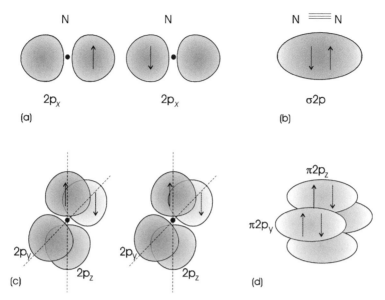

Figure 2.18 Bonding in N_2; each nitrogen atom has an unpaired electron in each of the three 2p orbitals. Overlap of the $2\,p_x$ orbitals, (part a) results in a σ bond (part b). Overlap of the $2\,p_y$ and $2\,p_z$ orbitals (part c) results in the formation of two π bonds (part d). The conventional representation of a triple bond, N≡N, does not convey the information that there are two different bond types

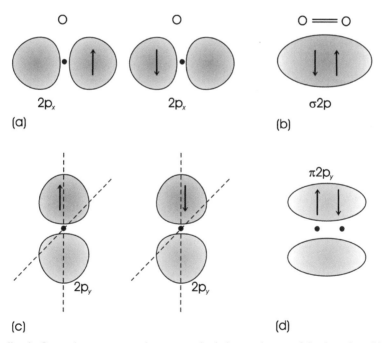

Figure 2.19 Bonding in O_2; each oxygen atom has an unpaired electron in two of the three 2p orbitals. Overlap of the $2\,p_x$ orbitals (part a) results in a σ bond (part b). Overlap of the $2\,p_y$ orbitals (part c) results in the formation of a π bonds (part d). The conventional representation of a double bond, O=O, does not convey the information that there are two different bond types

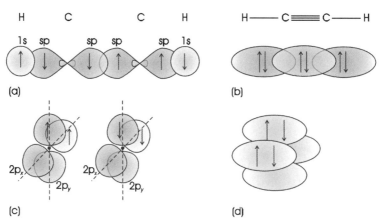

Figure 2.20   Bonding in ethyne (acetylene), C_2H_2. Overlap of the 1s orbitals of H with the sp hybrid orbitals on C (part a) results in a σ bonded molecule (part b). Overlap of the $2p_y$ and $2p_z$ orbitals on C (part c) results in the formation of two π bonds (part d). The conventional representation of the triple bond as C≡C, does not convey the information that there are two different bond types

carbon – carbon bond in ethyne is much stronger than a normal single carbon – carbon σ bond. Third, the H – C – C – H angles are not 90°, as they would be in the simple picture, but are 180°, and the molecule is linear (Figure 2.20a). The bonding in this molecule is best treated in terms of hybridisation. As described above, the $2s^2 2p^2$ configuration of carbon is changed to $2s2p^3$ by promotion of one of the s electrons to a p orbital. The next stage is the formation of a pair of sp hybrid orbitals, using up the s orbital and the p_x orbital. These form the linear molecular skeleton, joined by σ bonds (Figure 2.20b). The p_y and p_z orbitals on the carbon atoms each hold one electron. These can overlap sideways on, exactly as in the N_2 molecule, to form two π bonds, and complete the triple bond (Figure 2.20c,d).

The physical and chemical properties of ethene, C_2H_4, drawn in a conventional fashion in Figure 2.21, are not well explained by models involving only σ bonds. Instead, the $2s^2 2p^2$ configuration of carbon is changed to $2s2p^3$ by promotion of one of the s electrons to a p orbital. In the next step, sp^2 hybrid orbitals form on each carbon atom, leaving one unpaired electron in the unaltered p_z orbital. The three sp^2 hybrid orbitals on each carbon atom lie at angles of 120° to each other. These are used to

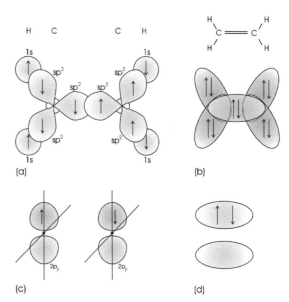

Figure 2.21   Bonding in ethene (ethylene), C_2H_4. Overlap of the 1s orbitals of H with the sp^2 hybrid orbitals on C (part a) results in a σ bonded molecule (part b). Overlap of the $2p_y$ orbitals on C (part c), results in the formation of a π bonds (part d). The conventional representation of the double bond as C=C, does not convey the information that there are two different bond types

bond to two hydrogen atoms and one carbon atom in a triangular arrangement, to form the σ bonded skeleton of the molecule (Figure 2.21a,b). The observed value for the H–C–H angles of ethylene is $117°$, which is close to the $120°$ value for sp^2 bonds. The remaining p_z orbital on the two carbon atoms overlaps sideways on to form a π bond (Figure 2.21c,d).

The exposed electrons in the π bonds contribute significantly to the properties of the molecules. In particular, they endow them with a high refractive index and high chemical reactivity.

2.2.8 Resonance

The bonding in molecules containing multiple bonds can often be drawn in a number of alternative ways. The classical example of this is the molecule benzene, C_6H_6. The two conventional ways of representing the scheme of bonding in this molecule are drawn in Figure 2.22(a). The molecule is planar, the carbon atoms are arranged in a perfect hexagon and the hydrogen atoms, one attached to each carbon, are omitted. The properties of benzene are best explained if the bonding is considered to be a blend of the two schemes (as well as of other more energetic structures not shown). The resultant is called a resonance hybrid, often drawn as in Figure 2.22(b), for the reason given below.

The bonding that can give rise to this is closely related to that in ethene. Each carbon atom forms sp^2 hybrids, and six carbon atoms link together and to six hydrogen atoms to produce the σ bond skeleton of the hexagonal molecule (Figure 2.22c). The remaining p_z orbitals, one on each carbon atom, overlap sideways on to form π bonds with lobes that extend above and below each of the carbon atoms in the plane of the hexagon (Figures 2.22d and 2.22e).

In this concept, the skeleton of σ bonds is invariant. These bonds, (like each of the others that have been discussed above) are said to be localised, which means that they are limited to the region between two atoms. Unlike the σ bonds, the π bonds spread out between all of the contributing atoms, to give delocalised orbitals, and are not restricted to lie between pairs of atoms. To indicate that the bonding

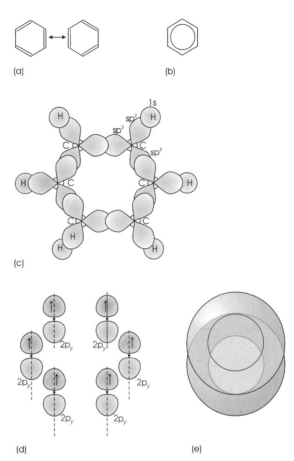

Figure 2.22 Bonding in benzene, C_6H_6. (a) Conventional bonding diagrams for the two main alternative resonance hybrid structures of the benzene molecule. The lines indicate bonds between carbon atoms, which lie at the vertices of a hexagon. Each carbon is linked to one hydrogen atom (omitted) and two carbon atoms. (b) The benzene molecule is often drawn as a hexagon enclosing a circle, to indicate the resonance nature of the bonding. The bonding is more complicated than the conventional diagrams indicate. Overlap of the 1s orbitals of H with the sp^2 hybrid orbitals on C (part c) results in a σ bonded hexagonal molecule. Overlap of the $2p_y$ orbitals on C (part d), results in the formation of π bonds (part e). The lobes of the π bonds lie above and below the plane of the C–H planar structure, and delocalisation means that the orbitals are spread equally over the ring. The representation of the molecule in part (b) attempts to convey some of this information

in a molecule such as benzene is a superposition of localised and delocalised bonding schemes, it is often represented as in Figure 2.22(b), where the hexagon represents the σ bond skeleton of the molecule and the circle drawn within the hexagon represents the delocalised π bonds. Often it is convenient to use localised bond representations of the structures contributing to resonance hybrids. The fact that the real situation is one of resonance is indicated by the use of double-headed arrows, $\leftrightarrow$, between the resonance hybrid structures (Figure 2.22a). (Note that these are not the only resonance hybrid structures that can be drawn for benzene, only the two that contribute most to the stability of the molecule.)

There is a distinction between resonance hybrids and hybrid orbitals. Resonance hybrids are the result of blending the bonding in molecules (or fragments of molecules) to give a bonding picture that better mirrors the chemical and physical properties of the molecules. Hybrid orbitals are a blend of orbitals on a single atom. Resonance is found in inorganic and organic molecules and is often revealed by intense colour or unusual electronic properties.

2.3   Metallic bonding

2.3.1   Bonding in metals

Ionic and covalent bonds are theories of chemical valence. A theory of metallic bonding has not only to explain this aspect of the linking of atoms but also typically 'metallic' properties. Metallic bonds must possess the following characteristics. The bonds act between identical and different metallic atoms, as is revealed by the formation of numerous alloy structures as well as the structures of the elements. The bonds act between many atoms, as metal atoms in crystals often have either 8 or 12 neighbours. The bonds are maintained in the liquid state, as liquid metals retain the distinguishing properties of crystalline metals. The bonds must permit easy electron transfer throughout the structure. In addition, the theory should account for the

fact that the majority of the elements in the periodic table are metals.

The present-day understanding of metals and metallic bonding has arisen from a combination of two different approaches. In the first of these, electrons travel more or less freely through the structure. Formally, this is called free-electron theory. A second approach, more chemical in nature, considers metallic bonds as a spreading out or delocalisation of covalent bonds. This delocalisation was mentioned in Section 2.2.8 with respect to the π orbitals of benzene, which extend over the whole molecule. In metals, the delocalisation is thought to extend throughout the solid. Formally, this is called the tight-binding theory. To understand metallic materials in the broadest sense, a combination of both approaches must be used. Moreover, accurate calculations use more sophisticated models than either the free-electron or the tight-binding approach. The application of these ideas, called band theory, successfully explains the detailed physical properties of metals.

The chemical approach is considered initially. Free-electron theory is discussed in Sections 2.3.5 and 2.3.6, and an account of the important electronic properties of solids is postponed to Chapter 13.

2.3.2   Chemical bonding

From the point of view of the molecular orbital model of chemical bonding, when two atoms approach each other the outer electrons interact and atomic orbitals become extended over both nuclei to form molecular orbitals. When two similar atomic orbitals interact, one on each atom, two molecular orbitals form, one of lower energy than either of the atomic orbitals, the bonding orbital, and one of higher energy, the antibonding orbital (Figure 2.23). The same principle has been found to apply no matter how many atoms are involved. If, instead of two atoms, there are three, three orbitals form. One will be a lower-energy bonding orbital, one a higher-energy antibonding orbital and one will have the same energy as the original atomic orbitals and is called a nonbonding orbital. With

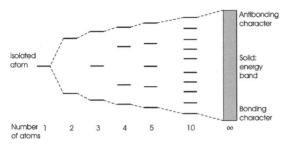

Figure 2.23 The development of an energy band from delocalised molecular orbitals. Each atom in the molecule contributes one molecular orbital. An isolated atom has sharp energy levels (left-hand side). As the number of atoms increases, the number of discrete orbitals merges into a band of closely spaced energy levels (right-hand side)

four atoms, each contributing one atomic orbital, four molecular orbitals, two of higher-energy antibonding orbitals and two of lower-energy bonding orbitals, are produced. If there are N atoms in the solid, each contributing one atomic orbital, N molecular orbitals will form, half at lower energies and half at higher energies.

There is an important point that emerges from energy calculations. The separation between the highest-energy and the lowest-energy rapidly becomes constant as the number of atoms in the solid increases. However, as the number of energy levels that have to fit into this energy space is equal to the number of atoms in the solid, the energy levels get closer together. When a large number of atoms are included, it is simpler to say that an energy band has been created, as illustrated on the far right-hand side of Figure 2.23. Within the energy band, the energy levels are so closely spaced that it is not too bad an approximation to think that a continuum of energies is found. The upper part of the band will correspond to antibonding orbitals and is said to have an antibonding character. The lower part of the band will correspond to bonding orbitals and have a bonding character.

The origin of the cohesive energy of a metal can now be obtained. Suppose that the energy band spreads out symmetrically from the original atomic orbital energy and that the energy levels in the band are all equally spaced. (This is *not* true, but it is a

good starting point.) The electrons will feed into the energy levels in a band, following the Aufbau principle, with two electrons of opposing spins in each level. For bonding to occur, an *atomic* orbital must contain a single unpaired electron. Therefore, each atom in the progression illustrated in Figure 2.23 will contribute one electron to the energy band. When all atoms have contributed an electron the energy band will be half filled. The cohesive energy for a metal is due to this part filling of the energy band, because the electrons preferentially occupy the energy levels with a bonding character.

The bonding energy of a diatomic molecule is obtained by feeding two electrons into the bonding molecular orbital (Figures 2.7 and 2.24), to give a bonding energy of $2E_{bond}(mol)$. The average energy of an electron in a half-filled energy band, E_{av}, can be equated (very roughly) to $E_{bond}(mol)$. The bonding energy of a metal will be obtained by feeding two electrons into the average energy level in the band, to give a value for the bond energy in a metal, $2E_{bond}(metal)$, approximately equal to $2E_{bond}(mol)$ (Figure 2.24). The bond energy of a metal then should be of the same order of magnitude as the bond energy of a diatomic molecule of the metal. For example, the bond energy of the molecule Li_2 is calculated to be about $140\,kJ\,mol^{-1}$, and the energy

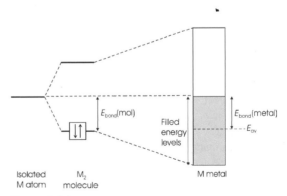

Figure 2.24 The bonding energy in a molecule, M_2, compared with an isolated M atom, is due to filling the bonding molecular orbital while the antibonding orbital remains empty. In a solid metal, the bonding energy is due to filling the lower bonding character energy levels while leaving the upper antibonding character energy levels empty

of evaporation of Li metal, which should be similar to the cohesive energy, is about $100\,\text{kJ}\,\text{mol}^{-1}$.

An accurate value for the cohesive energy of a metal will require knowledge of the number and distribution of the energy levels in the band as well as of other factors that have not been described, and will vary from metal to metal. Band theory, described in Section 2.3.8, is needed to take this further and to account for the differences between one metal and another.

2.3.3   Atomic orbitals and energy bands

The gradual broadening of atomic orbitals into energy bands will happen to each of the atomic orbitals on each of the atoms in the crystal. The sharp energy levels applicable to an atom thus become translated into a set of bands. The separation of the bonding and antibonding molecular orbitals, and thus the strength of a covalent bond, is related to the degree to which the atomic orbitals are able to overlap. Similarly, the energy spread in an energy band is related to the possibility of orbital overlap and so to the extension of each orbital. The filled core electron orbitals are compact and are shielded from any interactions with orbitals on other atoms by the valence electron orbitals. No significant overlap of these orbitals with orbitals on other atoms is possible. Thus, core electron orbitals hardly broaden at all and form very narrow energy bands little different from those in a free atom. In contrast, the outer orbitals of adjacent atoms will interact strongly and the energy bands will be broad. In this respect, outer p orbitals are more condensed than are outer s orbitals and produce narrower bands. As an example, Figure 2.25 shows the schematic development of energy bands for a typical alkali metal such as sodium (Na, $3s^1$). Energy bands that can be related directly to atomic orbitals are often given a similar terminology. Thus, the uppermost band in Figure 2.25 is called an s band or the 3s band.

Any d and f orbitals present are always shielded beneath outer s and p orbitals. They interact only weakly with orbitals on other atoms and form narrow d or f bands. The d and f electrons are

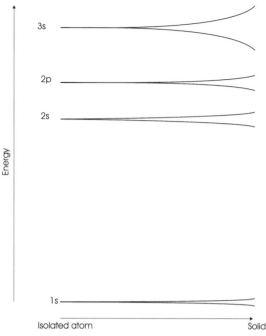

Figure 2.25   The development of energy bands from atomic orbitals for sodium metal. Isolated atoms (left-hand side) have sharp energy levels. In a solid, these are broadened into energy bands. Outermost orbitals broaden more than inner orbitals

sometimes best described as existing in narrow bands delocalised over the solid and sometimes as being localised in sharp atomic orbitals on individual atoms. Changes in temperature or pressure can make the behaviour of the electrons switch from one state to another. The d and f orbitals and bands do not play an important part in bonding but dominate the magnetic and optical properties of many solids.

2.3.4   Divalent and other metals

The cohesive energy in a metal such as sodium arises from the half-filled 3s band (Figure 2.25). An immediate difficulty arises when the neighbouring element to sodium, magnesium (Mg), with an electron configuration [Ne] $3s^2$, is considered. If each magnesium atom contributes both s electrons to the s band then the band will be filled. The bonding and antibonding contributions will be equal, and no

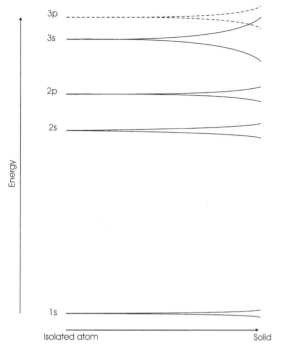

Figure 2.26 The development of energy bands from atomic orbitals for magnesium metal; the outermost 3s and 3p bands overlap in the metal

cohesion will result. A solid will not condense from the gas. This does not agree with commonplace observations; magnesium is clearly a metallic solid under normal temperatures and pressures. The answer to the puzzle is that the 3p orbitals, although empty on an isolated atom, broaden into a band that overlaps the 3s band (Figure 2.26). As magnesium atoms coalesce, electrons may initially enter the 3s band, but ultimately the highest-energy electrons in the 3s band spill over into lower-energy levels in the 3p band. The combined s and p band can hold eight electrons per atom. As each magnesium atom contributes two, these will mainly occupy bonding levels in partially filled bands. A solid metal will result.

The spreading of the energy bands associated with the outer orbitals increases for the heavier elements, simply because of their larger size and consequently greater interaction. Because of this, the outer orbitals on most of the heavier elements in the periodic table are transformed into rather wide, overlapping, bands in the solid state. In addition, a glance at the electronic configurations of these heavier atoms reveals that, for most, the bands will be partly filled. Thus, the molecular orbital model explains why so many of the elements can condense together to form metallic solids.

Because the energy bands are derived from delocalised molecular orbitals the electrons in the energy band are delocalised over all of the atoms in the solid. The strong bonding direction imparted by the overlap of atomic orbitals to form covalent bonds is now lost. Metallic bonds are therefore expected to be nondirectional. Moreover, as the outer electron orbitals are now transformed into bands, the remaining core electron orbitals, whether broadened or not, are of a spherical shape. Thus, metals are characterised by the packing of spherical atoms. Spheres can pack most efficiently when each is surrounded by a large number (8 to 12) nearest neighbours (see Chapter 5). Thus, the model explains the high apparent valence of metals in crystals. Moreover, as the outer electrons are lost to individual atoms, one metallic element is much like another, so that alloys will be expected to form readily. Similarly, the development of energy bands is dependent upon the close approach of the atoms and not on the relative configuration of the atoms. This means that a liquid metal should have similar properties to a solid metal.

The most important characteristic of a metal is its ability to conduct electricity. The molecular orbital theory contains the idea that the electrons are delocalised throughout the crystal, which suggests that delocalised electrons are responsible for electrical conductivity. To develop this idea quantitatively it is necessary to use band theory.

2.3.5 The classical free-electron gas

Band theory started with a gross approximation: the idea that a metal contained a 'gas' of completely free electrons, uninfluenced by the other electrons or anything else. The model, an adaptation of the kinetic theory of gases, was very successful. The electrons were supposed to be 'flying about' in the metal, in every direction, at random. The imposition of a voltage caused the random motion to be replaced by a drift in average motion that was equated to the electric current that flowed. This simple model predicted that a metal should obey Ohm's law and that the resistivity of the metal should increase slightly with temperature, as it does (see also Chapter 13).

However, it could not explain some important properties. In particular, the theory was unable to account for the fact that metals had a similar specific heat to insulators. If an electron gas occurs in metals, but not in insulators, it could be proved that metals should have a molar specific heat of about $4.5R$ whereas insulators would have a molar specific heat of only $3R$, where R is the gas constant.

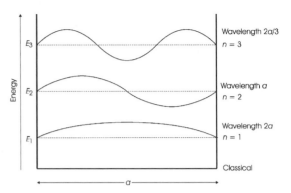

Figure 2.27 The energy levels of an electron confined to a line of length a. Quantum mechanics restricts the electron wave to fit into the length of the line, thus limiting the energy of the electron to discrete values corresponding to $n = 1, 2, 3$ and so on

2.3.6 The quantum free-electron gas

The advent of quantum theory suggested that the electrons in the metal, although still free, should be treated like waves. If this was so, the Schrödinger equation could be used to determine the energy of an electron in a metal. The only variables to be specified were the dimensions of the solid metal containing the electron and the potential energy of the electron in the metal. Mathematically, this is much simpler than the corresponding equations for molecular orbitals and is analogous to the hydrogen atom as described in Chapter 1.

As a first step, suppose that the potential energy is constant throughout the metal. To illustrate the results, take a one-dimensional case in which a single electron is confined to a line of length a. (This is often referred to as a 'particle in a box'.) The potential energy of the electron is set at zero on the line (or in the box) and at infinity at the extremities, so that the electron wave is completely trapped. The solutions in such a case show that the only allowed electron waves are standing waves, with nodes at the fixed ends of the line (Figure 2.27). The allowed wavelengths, λ, are given by:

$$\lambda = \frac{2a}{n} \qquad (2.11)$$

where n is a quantum number that can take integer values $1, 2, 3, \ldots$ and so on. The wave equation describing this situation is:

$$\psi = A \, \sin\left(\frac{n\pi x}{a}\right) = A \, \sin\left(\frac{2\pi x}{\lambda}\right) = A \, \sin kx \qquad (2.12)$$

where A is a constant, $\sqrt{(2/a)}$, and k is called the wavenumber. It has quantised values given by:

$$k = \frac{2\pi}{\lambda} = \frac{n\pi}{a} \qquad (2.13)$$

(In three dimensions $\mathbf{k}$ is a vector quantity, and is called the wave vector.)

The electron is confined not only to certain wavelengths but also to quantised energies, E_n:

$$E_n = \frac{k^2 h^2}{8\pi^2 m} = \frac{n^2 h^2}{8 m a^2} \qquad (2.14)$$

where E_n is the energy of the wave associated with quantum number n, and m is the mass of the electron. The form of the energy versus k graph is parabolic, as:

$$E = B k^2$$

where B is a constant, $h^2/8\pi^2 m$ (Figure 2.28).

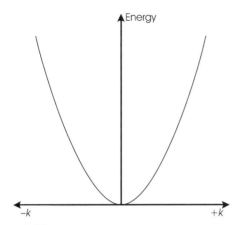

Figure 2.28   The relationship between the energy of an electron, E, confined to a line, and the wavenumber, k. The curve is a parabola as E is proportional to k^2

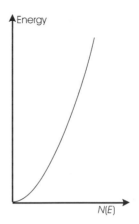

Figure 2.29   The form of the density of states, $N(E)$, for a free electron in a metal

The energy of a single electron in a rectangular block of metal with sides of length a, b and c, lying along the three axes x, y and z, is:

$$E(n_x, n_y, n_z) = \frac{h^2}{8m}\left(\frac{n_x^2}{a^2} + \frac{n_y^2}{b^2} + \frac{n_z^2}{c^2}\right) \qquad (2.15)$$

where n_x, n_y and n_z are the quantum numbers along the x, y, and z axes, equivalent to n in the one-dimensional case. For a cubic block of sides length a, this reduces to:

$$E(n_x, n_y, n_z) = \frac{h^2}{8ma^2}\left(n_x^2 + n_y^2 + n_z^2\right) \qquad (2.16)$$

As described in Chapter 1, wavefunctions with the same energy are said to be degenerate. Thus, for a cubic container the solutions $(n_x = 1,\ n_y = 1,\ n_z = 2)$, $(n_x = 1,\ n_y = 2,\ n_z = 1)$ and $(n_x = 2,\ n_y = 1,\ n_z = 1)$ are all of the same energy and so are degenerate. (This is analogous to the situation between the s, p and d orbital energies for a hydrogen atom.)

The degeneracy means that the number of energy levels within a particular energy range is not a constant but increases with increasing values of $\mathbf{k}$. In order to solve the problem of the electron contribution to the specific heat of a metal it is necessary to discover how many energy levels (i.e.

wavefunctions, or orbitals) occur in any particular range of energy dE between the energies E and $E + \delta E$. The result for a cube of volume V is:

$$N(E) = 2\pi V\left(\frac{8m}{h^2}\right)^{3/2} E^{1/2}. \qquad (2.17)$$

This is called the density of states function, $N(E)$. This curve, like the curve of E against k, is also parabolic in form (Figure 2.29). The number of energy levels in a small range of energy increases sharply as the energy increases. As each energy level can accommodate two electrons, the number of electrons in an energy interval between E and $E + \delta E$ is double that in Equation (2.17).

2.3.7   The Fermi energy and Fermi surface

The discussion in the previous section considered the energy of a single electron in the metal. Extension to many electrons employs a similar strategy to the orbital approximation in Chapter 1. The 'one-electron' energy levels calculated above are populated with the additional electrons, following the Pauli exclusion principle and the Aufbau principle. Thus, each level can hold just two electrons, with differing spins. To determine the overall distribution, electrons are allocated two at a time to the

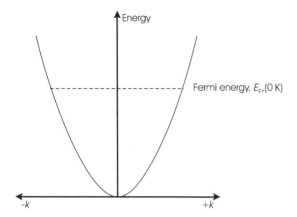

Figure 2.30 The Fermi energy, E_F, is the uppermost energy level filled at 0 K. At higher temperatures, the electrons are distributed over nearby energy levels and the boundary becomes less sharp

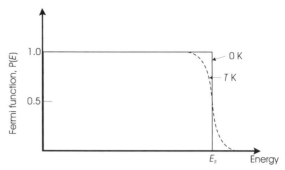

Figure 2.31 The Fermi function, P(E), giving the probability of the occupation of an energy level as a function of the energy, E. At 0 K, the probability is 1.0 up to the Fermi energy, E_F, and thereafter is 0. At higher temperatures, the curve changes smoothly from 1.0 to 0.

energy levels, starting with the lowest energy up to a maximum energy governed by the number of electrons available. At absolute zero, the uppermost occupied level, labelled E_F in Figure 2.30, is at the Fermi energy. In three dimensions this highest filled energy level takes the form of a surface, the Fermi surface. The value of the Fermi energy can be found by integrating Equation (2.17) between the limits of $E = 0$ and $E = E_F$. This leads to the result:

$$E_F = \left(\frac{h^2}{8\,m}\right)\left(\frac{3\,N}{\pi\,V}\right)^{2/3}$$

where h is the Planck constant, m represents the electron mass and N is the number of free electrons in a volume V of metal.

At temperatures above absolute zero, the electrons are liable to gain energy. However, it is not possible for electrons in lower levels to move into adjacent levels, as these are already filled. Only those electrons with the highest energies, near to the Fermi surface, can move to higher energy levels. The vast body of electrons remains untouched by the rise of temperature because no empty energy levels are available to them. The statistics that govern the distribution of electrons between the energy levels are called Fermi – Dirac statistics. Particles obeying these statistics, such as electrons,

are called fermions. The distribution of electrons at temperatures other than 0 K is given by the Fermi function:

$$P(E) = \left[\exp\left(\frac{E - E_F}{k\,T}\right) + 1\right]^{-1} \quad (2.18)$$

where P(E) is the probability that an energy level E is occupied by an electron at temperature T (Figure 2.31; see also Chapter 15 and Section S4.12). At $T = 0\,\text{K}$, P(E) = 1 for $E < E_F$, and P(E) = 0 for $E > E_F$. We also note from Equation (2.18) that at $E = E_F$, P(E) = $\frac{1}{2}$. Thus the Fermi level in metals above 0 K can be defined as the energy for which P(E) = $\frac{1}{2}$.

The variation of the population of electrons in the metal with energy E, $N_e(E)$, is simply given by the density of states function, N(E), multiplied by 2 (as each energy level can hold two electrons) multiplied by the probability that the energy level will be occupied (the Fermi function), P(E). That is:

$$N_e(E) = 4\,\pi\,V\left(\frac{8\,m}{h^2}\right)^{3/2} E^{1/2}\left[\exp\left(\frac{E - E_F}{k\,T}\right) + 1\right]^{-1}$$

This curve, which is a superposition of the density of states curve (Figure 2.29) and the Fermi function (Figure 2.31) is drawn in Figure 2.32.

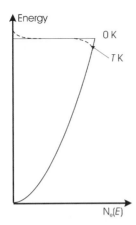

Figure 2.32 The population density, $N_e(E)$, for free electrons in an energy band at 0 K and a higher temperature, T K

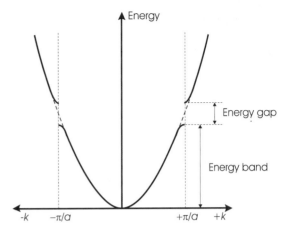

Figure 2.33 The one-dimensional free-electron curve of energy, E, versus wavenumber k; it is 'broken up' into energy bands separated by energy gaps because of the periodic potential of atomic nuclei of separation a

Because only a few higher energy levels will be populated by electrons as the temperature rises, the electrons do not contribute appreciably to the specific heat. Thus, the electronic heat capacity is almost negligible, and a major drawback of the classical theory has been corrected.

2.3.8 Energy bands

In the free-electron approach, the electrons were not supposed to interact with one another or the atomic nuclei present. (In fact, the calculations are made for a single electron, as in the hydrogen atom.) This is obviously a gross approximation. Later theories suppose that the electrons move in a potential that is more representative of the crystal structure. The potential is described as a periodic potential. This means that it mirrors the positions of the atoms in the solid. It is designed so that the probability of encountering the electron is low at atomic nuclei and high between atoms.

The first solution to the way in which a periodic potential modifies the free electron model was given by Bloch in 1928. The solution of the Schrödinger equation was found to consist of the free-electron wavefunctions multiplied by a function with the same periodicity as the crystal structure. These wave equations are called Bloch functions. A number of sophisticated potential models are now used routinely to calculate the properties of metals (see the Further Reading section).

Regardless of the exact potential used, certain generalisations are found always to hold. First, if the potential is weak, the wavefunctions must be similar to the free-electron wavefunctions. Second, regardless of the potential, at small values of the wave vector, k, wavefunctions are almost the same as the free-electron wavefunctions. Small values of k correspond to electron waves with a long wavelength and low energy. At this extreme, the energy levels are closely spaced and vary with k in a parabolic fashion. The electron is 'not aware' of the atoms in the solid.

As the wavenumber gets larger, the energy versus k curve is distorted away from a parabolic shape. Ultimately, no solutions to the wave equation are found to describe the situation, and the energy versus k curve is broken (Figure 2.33).

The origin of this break is well understood. Electron waves in a solid behave in a similar way to X-rays and other waves in any medium. When a wave encounters an object much smaller than its wavelength, it barely changes the wave propagation. However, when a wave encounters an object of the same dimensions as the wavelength, considerable interaction occurs. The wave is diffracted (covered in more detail in Chapter 14). In the case of electron waves in a solid, most waves will pass straight through the solid. However, waves with a wavelength similar to the repeat distance between the atoms in the solid will be diffracted by the structure.

Although diffraction is a complicated process, in the present situation the effects of diffraction can be equated to reflection. The circumstances for diffraction to occur are given by Bragg's law:

$$n\lambda = 2d\sin\theta \qquad (2.19)$$

where n is an integer taking values 1, 2, 3, and so on; λ is the wavelength of the wave; d is the spacing of the planes of atoms; and θ is the diffraction angle (Figure 2.34; see also Chapter 5).

In terms of the wave vector, k, the Bragg equation can be written as:

$$\frac{\pi n}{k} = d\sin\theta \qquad (2.20)$$

An electron wave normally incident upon planes of atoms of spacing a has a diffraction angle of 90° (Figure 2.35). From Equation (2.20), with n taking

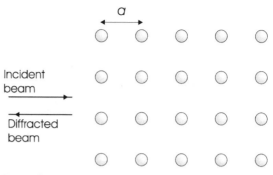

Figure 2.35   An electron wave normal to planes of atoms of interplanar spacing a will be diffracted back on itself when the wavelength λ is equal to $2na$ and the wave vector k is equal to $2n\pi/a$

integer values, we find that diffraction of the wave occurs at values of the wave vector given by:

$$k = \pm\frac{\pi}{a}, \quad \pm\frac{2\pi}{a}, \quad \pm\frac{3\pi}{a}, \dots \qquad (2.21)$$

Waves with these wave vectors will not be able to pass through the crystal.

The more accurate one-dimensional E versus k curve for this situation is drawn in Figure 2.36. When the wavelength of the electron wave corresponds to those given by Equation (2.21) the wave

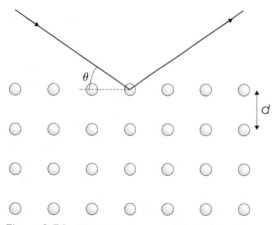

Figure 2.34   Electron waves are diffracted by planes of atoms separated by the interplanar spacing d when the Bragg equation ($n\lambda = 2d\sin\theta$) is obeyed

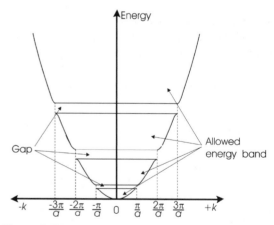

Figure 2.36   Energy bands for electrons in a one-dimensional crystal with atom spacing a. Energy gaps occur when the electron wave is diffracted and cannot pass through the crystal

cannot propagate through the crystal. Propagation will occur again when the wave vector increases slightly. Substitution of these values into the Schrödinger equation reveals that the change in wave vector is accompanied by an energy jump from a lower value to a higher value. It is not possible for an electron to have an energy value lying between the two extremes. This is described by saying that an electron can exist within a band of allowed energies. These bands are separated from each other by bands of forbidden energies.

2.3.9   Brillouin zones

The discontinuities in the energy versus k curve resulting from electron diffraction mark the boundaries of Brillouin zones. The first Brillouin zone in the one-dimensional case (Figure 2.36) extends from $k = \pi/a$ to $k = -\pi/a$. The second Brillouin zone extends from $k = \pi/a$ to $k = 2\pi/a$ on the positive side of the graph and from $k = -\pi/a$ to $k = -2\pi/a$ on the negative side (Figure 2.37). The third, fourth and subsequent Brillouin zones can be similarly located. The concept of a Brillouin zone is an abstract concept, as the zones exist in a space

defined by the wave vector and the energy of the electron. The wave vector is proportional to the velocity and the momentum of the electron, and these zones are sometimes described as existing in velocity or momentum space.

In three dimensions the geometry becomes more complicated. Planes of atoms, with spacing b, at an angle to the planes in Figure 2.35, will generate a set of Brillouin zones that extend from $k = \pi/b$ to $k = -\pi/b$, and so on. The positions of the Brillouin zone boundaries thus mirror the structure of the solid. A real crystal will therefore have a set of nested Brillouin zones that have polyhedral shapes. These shapes are closely related to the symmetry of the crystal structure. The Fermi energy will now be represented by a surface in energy versus k space. Far from the Brillouin zone boundaries, this surface is spherical. However, the shape is distorted near to the Brillouin zone boundaries. The Fermi surface can exist in more than one Brillouin zone if it lies below the energy gap in one direction and above the energy gap in another. Brillouin zone theory is central to the explanation of the dynamics of electrons in metals.

2.3.10   Alloys and noncrystalline metals

Alloys resemble pure metallic elements. In a solid in which metallic bonding is of importance the outer electrons on the atoms held together are shared communally. Provided all of the atoms present in the alloy can lose outer electrons in this way, the material will resemble a metal. Moreover, the loss of the outer electrons removes the main chemical distinguishing feature of an element. Thus, in alloys the individual chemical variations between the components will be suppressed. A significant difference between elemental metals and alloys, however, will be the band structure and the extent to which the bands are filled with electrons. For example, sodium has a half-filled 3s band whereas magnesium has partly filled overlapping 3s and 3p bands. The band structure of sodium – magnesium alloys is therefore expected to depend on alloy composition. In general, this is true for all alloys, and the chemical and physical properties of alloys

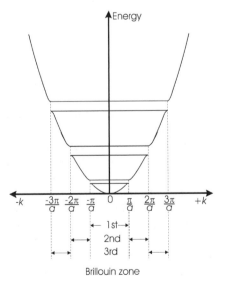

Figure 2.37   The first three Brillouin zones for a one-dimensional crystal with atom spacing a

are closely related to the composition of the alloy and to how the electrons donated by the metal atoms interact with the Brillouin zone boundaries.

An important aspect of metallic bonding is that it operates in the liquid state. Mercury is no less a metal when liquid than when it is solid. In this context, the properties of a liquid metal that are characteristically metallic are due to the partly filled upper band of energy levels.

The molecular orbital approach to metallic bonds explains this. The spreading of atomic orbitals into energy bands is not limited to an ordered crystalline array of atoms but hinges on close approach of atoms. The main features of the energy bands in a material are derived from the geometry of the nearest-neighbour atoms only. For many metals this geometry does not change greatly in a non-crystalline material compared with that of a crystal. For example, silicon atoms bond to four neighbours arranged tetrahedrally in crystals and in noncrystalline silicon. In the crystalline state the tetrahedra are linked together in a completely ordered way throughout the solid, whereas in the noncrystalline state the tetrahedra do not show this long-range order. Similarly, the number of nearest neighbours to any sodium atom in a crystal of sodium metal (six) is similar to the number in the liquid state. Mercury, although having a more complex crystal structure, has six near neighbours in the solid and the liquid. Thus, the energy band arrangement in a crystal or a liquid will be similar.

The filling of the uppermost energy band is dependent on the atoms involved. Again, this will not depend on the order of the atoms, only on the overlap of the orbitals to allow collective pooling of electrons.

Thus, the simple models presented above still apply to liquid metals as well as to metallic glasses and other noncrystalline materials. However, Brillouin zones will not occur in liquids or glasses, as they are features of crystalline arrays.

2.3.11 Bands in ionic and covalent solids

In the discussion of ionic and covalent bonding, the spreading of atomic orbitals into bands has been

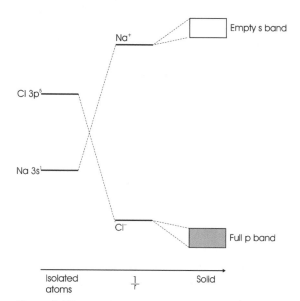

Figure 2.38 Energy bands of isolated atoms in sodium chloride (NaCl). As the spacing between the atoms, r, decreases, first ionisation occurs and then energy bands develop. These are narrow and widely separated, as expected from the ionic model of bonding

ignored. It is reasonable to question this and re-examine ionic and covalent solids.

Consider the archetypal ionic compound, sodium chloride (NaCl). A sketch of the development of bonding in this compound is shown in Figure 2.38. To the left, the constituents are isolated from each other. The lowest energy corresponds to Na and Cl atoms in their respective ground states. (For clarity, the core energy levels of both atoms have been ignored and only relative changes in energy are included in Figure 2.38.) As the interatomic separation decreases, a critical separation is reached at which energy is gained by the transfer of an electron from sodium to chlorine to form Na^+ and Cl^- ions. As the ions condense into a crystal the energy of the Cl^- ions falls below that of the Na^+ ions. This occurs at an interatomic separation of approximately 1.0 nm. However, calculations show that even at this close spacing the energy levels are still similar to those on isolated ions. As the interatomic spacing decreases further, the electron orbitals overlap. The electrons with opposed spins will tend to lose energy whereas those with parallel spins will tend to gain energy. These are the familiar bonding and antibonding interactions and cause the narrow energy levels to broaden into bands. However, the ionic electron distribution is unchanged. Thus the band that develops on the Cl^- ions, the 3p band, is full, and the band that develops on the Na^+ ions, the 3s band, remains empty. Thus the ionic model is still a good model for the bonding in NaCl.

In the case of a covalently bonded crystal such as germanium (Ge), with an outer electron configuration of [Ar] $4s^2 4p^2$, a modification to the molecular orbital model is again needed. As an example, consider germanium crystals, which are composed of a tetrahedral array of germanium atoms linked by sp^3 hybrid bonding. In the case of isolated atoms, use of atomic orbitals and energy levels is appropriate (Figure 2.39). As the atoms are brought together, these orbitals interact to form four degenerate sp^3 hybrid orbitals. Each sp^3 orbital on an atom contains one electron, and overlap with adjacent orbitals causes strong bonds with a tetrahedral geometry to form. As the atoms approach further, each single sp^3 energy level widens into a narrow band, the lower half of which is bonding and the upper part of which is antibonding. The band derives from the overlap of four sp^3 energy levels

and is able to contain a maximum of eight electrons. As each atom donates one electron per sp^3 energy level, the band is half full. Continued approach causes the antibonding part of the band to increase in energy because of the repulsion between the parallel electrons and causes the bonding part to decrease in energy because of the favourable interaction between the spin-paired electrons. At a critical separation, the single band is split into two bands separated by an energy gap. When the chemistry of the material is discussed; the lower of these bands is usually called a bonding band and the upper an antibonding band. In semiconductor physics they are referred to as the valence band and conduction band, respectively. The electrons, one in each sp^3 orbital, that combine in a covalent bond end up in the lowest bonding band, which will be completely full. The upper antibonding band will be completely empty. The cohesive energy in these crystals is, in fact, due to the appearance of the energy gap between the upper (empty) and lower (filled) bands.

The elements carbon (diamond), silicon, germanium, and one form of tin, α-tin, all crystallise with the same structure. The size variation of this series of atoms modifies the band formation in a predictable way. The smallest atom, carbon, has the smallest degree of orbital interaction and hence the narrowest bands. The energy gap is therefore large, and diamond has a very high cohesive strength. As the atoms increase in size, the orbital interactions increase, bands widen, the gap narrows and the cohesive energy falls. Unlike diamond, α-tin is a weak solid and is easily crushed.

Although this description is quite different from that of covalent bonding, the separation of the energy bands in the solid is similar to the separation of the bonding and antibonding molecular orbitals that would be found on a diatomic C_2, Si_2, Ge_2 or Sn_2 molecule with a similar interatomic separation to that of the atoms in the crystal. The covalent bond picture is, therefore, quite similar to the band picture. In the covalent model, the bond energy is related to the separation of the molecular orbitals. In the solid, the cohesive energy is related to the energy of the bands that have developed from the molecular orbitals. With this limitation in mind,

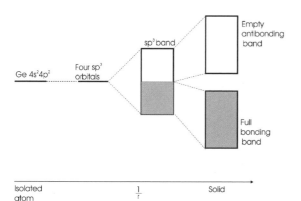

Figure 2.39 Energy bands from isolated atoms in germanium (Ge). As the spacing between the atoms, r, decreases, first orbital hybridisation occurs and then energy bands develop. These are broad and separated by a narrow energy gap, as expected from the covalent model of bonding

the covalent bonding model is adequate for many solids.

Answers to introductory questions

What are the principle geometrical consequences of ionic, covalent and metallic bonding?

Ionic bonding is long-range in operation, and the electrostatic force that controls the attraction or repulsion is the same in all directions. Thus, the most important geometrical feature of ionic bonding is that it is nondirectional. Atoms linked by ionic forces pack together to minimise the various attractive and repulsive forces so that cations are surrounded mainly by anions, and anions by cations.

Atoms that are linked by electron-pair bonds are positioned so that orbital overlap is maximised. The orbitals used are also sensitive to bond overlap and hybridisation, so that atomic orbitals frequently mix to give hybrid orbitals with greater overlapping power. The shapes of atomic orbitals and hybrid orbitals are quite definite and point in fixed directions. This leads to the fact that covalent bonding is directional. From a geometrical point of view, the array of covalent bonds in a solid resembles a net.

The formation of metallic bonds requires that the atoms lose outer electrons to a common electron band that runs throughout the solid. This means that atoms lose a large part of their chemical identity and resemble round spheres. The geometrical consequence of this is that atoms pack together like spheres and the bonding is nondirectional. Atoms simply pack together to minimise the space occupied.

What orbitals are involved in multiple bond formation between atoms?

Atoms linked by σ and π bonds at the same time are linked by multiple bonds. For example, multiple bonding in the nitrogen molecule, N_2 ($N\equiv N$) comprises the p_x orbitals combined in an end-on fashion to form a σ bond, and the p_y and p_z orbitals combined in a sideways manner to form two π bonds. Note that the traditional representation of three bonds drawn as three horizontal lines does not make it clear that two different bond types exist in N_2. In the case of the oxygen molecule, O_2 (O=O) the p_x orbitals overlap end on to form a σ bond, and the p_y orbitals overlap in a sideways fashion to form a π bond. As with nitrogen, the conventional representation of the double bond does not reveal that two different bond types are present.

Multiple bonding is of considerable importance in carbon compounds and figures prominently in the chemical and physical properties of polymers.

What are allowed energy bands?

The electron energy levels in a solid are so closely spaced that they can be treated as a continuum. This forms an energy band. The simple electron gas model of a solid has only one band of energies. When the potential due to the atomic nuclei is taken into account the single free-electron band is broken up into a number of 'allowed' energy bands separated by regions that have no energy levels available to the electrons. Allowed energy bands are thus closely spaced sets of energy electron levels that can be occupied collectively by the electrons surrounding the atoms in a solid.

Further reading

The following references greatly expand the material in this chapter. The first two textbooks below merit reading for a broad-based introduction to the ideas contained in this chapter.

L. Pauling, 1960, *The Nature of the Chemical Bond*, 3rd edn, Cornell University Press, Ithaca, NY.

W.B. Pearson, 1972, *The Crystal Chemistry and Physics of Metals and Alloys*, Wiley-Interscience, New York.

P.W. Atkins, 1981, *Quanta*, 2nd edn, Oxford University Press, Oxford.

W.A. Harrison, 1980, *Electronic Structure and the Properties of Solids*, W.H. Freeman, New York.

A. Cottrell, 1988, *Introduction to the Modern Theory of Metals*, The Institute of Metals, London.

A critical account of the ionic model, particularly with respect to calculations and structures, is given by:

M. O'Keeffe, 1981, 'Some Aspects of the Ionic Model of Crystals', in *Structure and Bonding in Crystals, Volume 1*, eds M.O'Keeffe and A. Navrotsky, Academic Press, London, pp. 299 – 322.
M. O'Keeffe, B.G. Hyde, 1985, 'An Alternative Approach to Non-molecular Crystal Structures', *Structure and Bonding* **61**, 77.

Textbook errors in the discussion of lattice energies are discussed in:

D. Quane, 1970, 'Textbook Errors, 98: Crystal Lattice Energy and the Madelung Constant', *Journal of Chemical Education* **47**, 396.

Ionic radii are discussed and tabulated in:

R.D. Shannon, C.T. Prewitt, 1969, 'Effective Ionic Radii in Oxides and Fluorides', *Acta Crystallogr.*, **B25** 925, 1969; 1970, 'Revised Values of Effective Ionic Radii', *Acta Crystallogr.*, **B26** 1046.
R.D. Shannon, 1976, 'Revised Effective Ionic Radii and Systematic Studies of Interatomic Distances in Halides and Chalcogenides', *Acta Crystallogr.* **A32** 751.

The calculation of crystal energetics is described in:

C.R.A. Catlow, 1987, 'Computational Techniques and the Simulation of Crystal Structures', in *Solid State Chemistry Techniques*, eds A.K. Cheetham and P. Day, Oxford University Press, Oxford, Ch. 7.

Problems and exercises

Quick quiz

1 The concept of the 'valence' of an atom refers to:
   (a) The charge on an ion
   (b) The strength of the chemical bonds formed
   (c) The combining ability of an atom

2 A cation is:
   (a) An atom that has gained a small number of electrons
   (b) An atom that has lost a small number of electrons
   (c) A charged atom stable in solutions

3 An anion is:
   (a) An atom that has lost a small number of electrons
   (b) An atom that has gained a small number of electrons
   (c) A charged atom stable in solutions

4 Which of the following is a trivalent cation?
   (a) V^{3+}
   (b) V^{4+}
   (c) V^{5+}

5 Ionic bonds are formed by which of the processes?
   (a) Electron sharing
   (b) Electron transfer
   (c) Electron delocalisation

6 A key feature of an ionic bond is that it is:
   (a) Strongly directional
   (b) Completely nondirectional
   (c) Acts between identical atoms

7 The Madelung energy is:
   (a) The bond energy of the crystal
   (b) The lattice energy of an ionic crystal
   (c) The electrostatic energy of an ionic crystal

8 The reduced Madelung constant of crystals:
   (a) Varies widely from one structure to another
   (b) Is similar for all structures
   (c) Depends on the charges on the ions in the structure

9 The lattice energy of an ionic solid is:
   (a) The sum of electrostatic and repulsive energies
   (b) The minimum of the electrostatic and repulsive energies
   (c) The difference between the electrostatic and repulsive energies

10 Lone pair ions have:
    (a) A pair of electrons in outer orbitals
    (b) A pair of s electrons outside a closed shell
    (c) A pair of p electrons outside of a closed shell

11 Cations are generally:
    (a) Smaller than anions
    (b) Bigger than anions
    (c) The same size as anions

12 A cation gets:
    (a) Larger as the charge on it increases
    (b) Smaller as the charge on it increases
    (c) Remains the same whatever the charge

13 Covalent bonds are formed by which of the processes?
    (a) Electron sharing
    (b) Electron transfer
    (c) Electron delocalisation

14 Covalent bonds are:
    (a) Strongly directional
    (b) Completely nondirectional
    (c) Variable in direction

15 Covalent bonds are also called:
    (a) Molecular orbitals
    (b) Hybrid orbitals
    (c) Electron-pair bonds

16 A σ bond is characterised by:
    (a) Being formed between identical atoms
    (b) Being formed by two s orbitals
    (c) Radial symmetry about the bond axis

17 A π bond is characterised by:
    (a) Being formed by two p orbitals
    (b) Reflection symmetry about the bond axis
    (c) Being formed between different atoms

18 Bonding molecular orbitals have:
    (a) The same energy as antibonding orbitals
    (b) A lower energy than antibonding orbitals
    (c) A higher energy than antibonding orbitals

19 An asterisk marks:
    (a) Antibonding orbitals
    (b) Bonding orbitals
    (c) Nonbonding orbitals

20 A polar covalent bond:
    (a) Forms between different sized atoms
    (b) Involves ions linked with molecular orbitals
    (c) Has the electron pair unevenly distributed in the molecular orbital

21 The electronegativity of an atom is a measure of its tendency to:
    (a) Attract electrons
    (b) Repel electrons
    (c) Form a covalent bond

22 Strongest covalent bonds form:
    (a) When the orbitals are of the same type
    (b) When the orbitals overlap maximally
    (c) When the orbitals are symmetrically arranged

23 A hybrid orbital is:
    (a) An overlapping pair of orbitals
    (b) A combination of orbitals on adjacent atoms
    (c) A combination of orbitals on a single atom

24 sp^2 hybrid orbitals:
    (a) Point towards the vertices of a tetrahedron
    (b) Point towards the vertices of a triangle
    (c) Point towards the vertices of a cube

25 Multiple bonds between two similar atoms:
    (a) Always consist of the same types of molecular orbital
    (b) Always consist of the same types of hybrid orbital
    (c) Always consist of different types of molecular orbital

26 Metallic bonds are formed by which of the following processes?
    (a) Electron sharing
    (b) Electron transfer
    (c) Electron delocalisation

27 Metallic bonds:
(a) Only act between the same metallic elements
(b) Only act between different metallic elements
(c) Act between any metal atoms

28 An energy band is due to the spreading out of energy levels as a result of:
(a) The close approach of atoms
(b) Chemical bonding
(c) Molecular orbitals

29 The cohesive energy of a metal is due to:
(a) A partly filled energy band
(b) An empty energy band
(c) An overlap of energy bands

30 A metallic bond is predominantly:
(a) Strongly directional
(b) Completely nondirectional
(c) Partly directional

31 Wavefunctions are said to be degenerate if:
(a) They have different energies
(b) They overlap
(c) They have the same energy

32 The Fermi energy is:
(a) The uppermost energy level filled
(b) The highest energy in a band
(c) The energy of an electron in a band

33 Energy gaps in crystals can be thought of being as being a result of:
(a) The diffraction of electron waves
(b) The bonding of electrons
(c) The localisation of electron waves

34 Energy bands can form:
(a) Only in crystals
(b) Only in solids
(c) In solids and liquids

35 Energy bands form:
(a) Only in metals

(b) In all solids
(c) Only in electrical conductors

Calculations and questions

2.1 Write out the electron configuration of the ions Cl^-, Na^+, Mg^{2+}, S^{2-}, N^{3-}, Fe^{3+}.

2.2 Write out the electron configuration of the ions F^-, Li^+, O^{2-}, P^{3-}, Co^{2+}.

2.3 Write the symbols of the ions formed by oxygen, hydrogen, sodium, calcium, zirconium and tungsten.

2.4 Write the symbols of the ions formed by iron, chlorine, aluminium, sulphur, lanthanum, and tantalum.

2.5 Use Equation (2.7) to calculate the lattice energy of the ionic oxide CaO, which has the *halite* (NaCl) structure; $\alpha = 1.75$, $n = 9$, $r_0 = 0.240$ nm.

2.6 Use Equation (2.7) to calculate the lattice energy of the ionic halides NaCl and KCl, which have the *halite* (NaCl) structure; $\alpha = 1.75$, $n = 9$, $r_0(NaCl) = 0.281$ nm, $r_0(KCl) = 0.314$ nm.

2.7 Use Equation (2.7) to calculate the lattice energy of the ionic halides NaBr and KBr, which have the *halite* (NaCl) structure; $\alpha = 1.75$, $n = 9$, $r_0(NaBr) = 0.298$ nm, $r_0(KBr) = 0.329$ nm.

2.8 Determine the number of free electrons in gold, assuming that each atom contributes one electron to the 'electron gas'. The molar mass of gold is 0.19697 kg mol^{-1}, and the density is 19281 kg m^{-3}.

2.9 Determine the number of free electrons in nickel, assuming that each atom contributes two electrons to the 'electron gas'. The molar mass of nickel is 0.05869 kg mol^{-1}, and the density is 8907 kg m^{-3}.

2.10 Determine the number of free electrons in copper, assuming that each atom contributes one electron to the 'electron gas'. The molar mass of copper is $0.06355 \text{ kg mol}^{-1}$, and the density is 8933 kg m^{-3}.

2.11 Determine the number of free electrons in magnesium, assuming that each atom contributes two electrons to the 'electron gas'. The molar mass of magnesium is $0.02431 \text{ kg mol}^{-1}$, and the density is 1738 kg m^{-3}.

2.12 Determine the number of free electrons in iron, assuming that each atom contributes two electrons to the 'electron gas'. The molar mass of iron is $0.05585 \text{ kg mol}^{-1}$, and the density is 7873 kg m^{-3}.

2.13 Using Equation (2.16) calculate the lowest energy level of a free electron in a cube of metal with edge length 1 cm and compare this with the thermal energy at room temperature, given by kT, where k is the Boltzmann constant and T is the absolute temperature.

2.14 Estimate the Fermi energy of silver; the molar mass of silver is $0.1079 \text{ kg mol}^{-1}$, the density is 10500 kg m^{-3} and each silver atom contributes one electron to the 'electron gas'.

2.15 Estimate the Fermi energy of sodium; the molar mass of sodium is $0.02299 \text{ kg mol}^{-1}$, the density is 966 kg m^{-3} and each sodium atom contributes one electron to the 'electron gas'.

2.16 Estimate the Fermi energy of calcium; the molar mass of calcium is $0.04408 \text{ kg mol}^{-1}$, the density is 1530 kg m^{-3} and each calcium atom contributes two electrons to the 'electron gas'.

2.17 Estimate the Fermi energy of aluminium; the molar mass of aluminium is $0.02698 \text{ kg mol}^{-1}$, the density is 2698 kg m^{-3}, and each aluminium atom contributes three electrons to the 'electron gas'.

2.18 Covalent bonding describes well the bonding in small molecules. Explain how covalent bonding leads naturally to the concept of definite molecular geometry. [Note: answer is not provided at the end of this book.]

2.19 The electronic configuration of atomic sodium is $1s^2\,2s^2\,2p^6\,3s^1$, and that of magnesium is $1s^2\,2s^2\,2p^6\,3s^2$. Explain why crystals of both of these elements behave as metals. [Note: answer is not provided at the end of this book.]

2.20 Both the covalent theory and the ionic bond theory can be used to guess the likely structure of a crystal. What coordination would be expected for the metal atoms or ions in the compounds ZnO and ZnS if they were covalent or ionic? In fact, the coordination polyhedra in both crystals are tetrahedral. What conclusions can you draw about the bonding in these solids? Regardless of coordination number, the ionic radii can be taken as: Zn^{2+}, 0.089 nm, O^{2-}, 0.124 nm, S^{-2}, 0.170 nm. [Note: answer is not provided at the end of this book.]

2.21 Draw the wavefunctions and the probability of locating an electron at a position x for an electron trapped in a one-dimensional potential well with zero internal potential and infinite external potential. [Note: answer is not provided at the end of this book.]

2.22 Make accurate plots of the Fermi function that expresses the probability of finding an electron at an energy E for temperatures of 0 K, 300 K, 100 K and 5000 K. [Note: answer is not provided at the end of this book.]

3

States of aggregation

- What type of bonding causes the noble gases to condense to liquids?

- What is the scale implied by the term 'nano-structure'?

- What line defects occur in crystals?

Aggregation is not solely due to the strong chemical bonds described in Chapter 2. Even noble gas atoms experience weak interatomic forces that lead to liquefaction and, except for helium, solidification at low temperatures. Although these interactions are weak in terms of bond energy, they are of vital importance, especially in living organisms. They also lead to the formation of magnetic domains (Chapter 12) and should not be despised.

3.1 Formulae and names

3.1.1 Weak chemical bonds

The strengths of chemical bonds vary widely. Table 3.1 lists the forces between atoms in a solid. The strong chemical bonds, covalent, ionic and metallic, have been described in Chapter 2. The strongest of

the weak bonds involves dipoles. Permanent dipoles are usually found on molecules containing two atoms with very different electronegativities, as described in Chapter 2. For example, a molecule of HCl has a region of positive charge, $\delta+$, associated with the hydrogen atom, and a region of negative charge, $\delta-$, associated with the chlorine atom (Figure 3.1a). The dipole moment of the molecule is 3.60×10^{-30} C m (see Chapter 11 for more information on units.) Water, an angular molecule, also has a permanent dipole moment, of 6.17×10^{-30} C m. The dipole is directed away from the oxygen atom and is augmented by the two lone pairs of electrons (Figure 3.1b). The charges making up the permanent dipole can interact with ions, and ion–dipole interaction energies, of the order of $15 \, \text{kJ mol}^{-1}$ are found. The hydration of cations, in water solution and solid hydrates, is mainly a result of ion–dipole effects.

Permanent dipoles can also interact with the charges on other dipoles in dipole–dipole interactions. These are of the order of $2 \, \text{kJ mol}^{-1}$ for fixed molecules, but the interaction is reduced to about one tenth of this value if the molecules bearing the dipoles are free to rotate.

The only elements that exist as atoms under normal conditions are the noble gases, in group 18 of the periodic table. These all have the outer electron structure ns^2np^6 and, at normal temperatures, they exist as monatomic gases. On cooling, helium (He), the lightest, turns to a liquid (with very curious properties) at 4.2 K, the lowest known

Understanding solids: the science of materials.   Richard J. D. Tilley
© 2004 John Wiley & Sons, Ltd   ISBNs: 0 470 85275 5 (Hbk) 0 470 85276 3 (Pbk)

Table 3.1 Forces between atoms, ions and molecules

Type of bond	Approximate energy/kJ mol^{-1}	Species involved
Covalent	350	Atoms with partly filled orbitals
Ionic	250	Ions only
Metal	200	Metal atoms
Ion–dipole	15	Ions and polar molecules
Dipole–dipole	2	Stationary polar molecules
Dipole–dipole	0.3	Rotating polar molecules
Dispersion	2	All atoms and molecules
Hydrogen bond	20	N, O or F plus H

boiling point of an element. Helium can only be turned into a solid by applying pressure. The other members of the family can be liquefied and solidified by cooling.

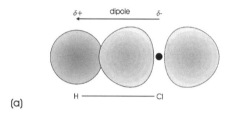

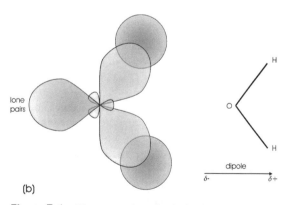

Figure 3.1 The permanent dipole in the molecule (a) HCl and (b) H_2O arises from unequally shared electrons in a covalent bond. In H_2O the lone pair of electrons also contributes to the dipole. The small charges that constitute the dipole are represented by $\delta+$ and $\delta-$. An electric dipole is represented by an arrow pointing from negative to positive

This condensation is due to weak interactions between the outer electrons on the atoms. Fleeting instantaneous fluctuations in the electron cloud surrounding these atoms create momentary dipoles, which are regions with a slight positive charge relative to regions of slight negative charge. These instantaneous charges lead to a weak attraction, which occurs between all atoms and molecules, including otherwise neutral atoms. The resultant force of attraction is called the London or dispersion force, and the interaction is called van der Waals bonding. The bond energy is approximately 2 kJ mol^{-1}. This force is responsible for the liquid states of most molecular species, such as H_2, benzene and the noble gases. The strength of the interaction increases as the size (mass and radius) of the atoms or molecules increases. Because of this, large molecules tend to exist as solids, smaller ones as liquids and light molecules as gases. This trend is exemplified by the smooth increase in boiling point of the saturated hydrocarbons, which have a series formula C_nH_{2n+2} (Figure 3.2).

These weak interactions can be represented by potential energy curves similar to those described in Chapter 2 as acting between atoms. A commonly used form of the interaction energy between a pair of atoms or molecules is the Lennard-Jones potential, $V(r)$:

$$V(r) = Ar^{-12} - Br^{-6}$$

where A and B are constants and r is the distance between the atoms or molecules. The first term on

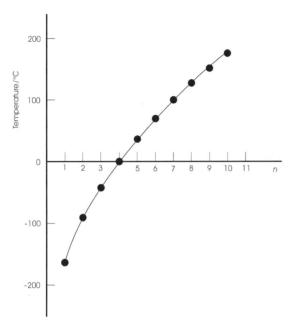

Figure 3.2 The boiling points of the saturated hydrocarbons, of formula C_nH_{2n+2}, plotted against the value of n, which is proportional to the size of the molecule

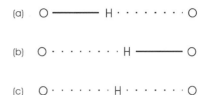

Figure 3.3 The hydrogen bond between two oxygen atoms can be thought of as a strong bond to one oxygen atom and a weak bond to the other. Two alternatives are possible (parts a and b). At high temperatures in solids the hydrogen atom is often found midway (on average) between the oxygen atoms (part c)

the right-hand side of this equation is a repulsive energy term and the second is an attractive energy. The potential energy, $V(r)$ passes through a minimum, V_{min}, at a distance r_{min}. Under normal circumstances, this would represent the bonding energy of a pair of atoms or molecules, at an equilibrium separation of r_{min}. The Lennard-Jones potential can be written in terms of V_{min} as:

$$V(r) = 4\,V_{min}\left[\left(\frac{r_0}{r}\right)^6 - \left(\frac{r_0}{r}\right)^{12}\right]$$

where r_0 is the value of r when $V(r)$ is zero.

Thermal energy is taken to be of the order of kT, where k is the Boltzmann constant and T is the absolute temperature. In cases where the energy of the bond, V_{min}, is greater than kT, one can expect pairs of atoms or molecules to be stable and a liquid phase to condense. When V_{min} is less than kT the bond would be expected to be too weak to hold the pair together and a gas is likely.

A hydrogen bond is a weak bond formed when a hydrogen atom lies between two highly electronegative atoms – fluorine, oxygen, chlorine or nitrogen. The bond results from the interaction of the small positive charge, $\delta+$, found in dipolar molecules containing hydrogen, with the partial charge of $\delta-$ located on the electronegative partner. It is naturally linked to the exposed lone-pair electrons on atoms such as oxygen and nitrogen. The hydrogen bond is usually drawn as a dotted bond between the electronegative atoms (Figure 3.3). This representation emphasises the fact that the hydrogen atom has an ambiguous position in the bond. At low temperatures it adopts a position nearer to one or other of the electronegative atoms, and at high temperatures it is found midway between them. The bonding to the nearer atom is then described as a normal covalent σ bond, and the bonding to the further atom is the hydrogen bond. In general the two links, O—H and H$\cdots$O, for example, are not in the same straight line. The angle between them commonly deviates from 180° by 10° to 20°, and sometimes by much more.

Because these bonds are comparatively weak it follows that they not only are easily ruptured but also are formed with equal ease. Thus, hydrogen bonds form in all appropriate materials at normal temperatures. Hydrogen bonding is important in many hydrogen-containing compounds such as water (H_2O), hydrogen fluoride (HF), ammonia (NH_3) and potassium hydrogen fluoride (KHF_2). The existence of hydrogen bonds dramatically changes many of the properties of the material. For example, HF, H_2O and NH_3 are characterised by melting points, boiling points and molar heats of vaporisation that are abnormally high in comparison with those of similar elements. The fact that water is liquid on Earth at normal temperatures is largely because of hydrogen bonding. In living organisms, hydrogen bonding is of great importance in controlling the folded (tertiary) structure of proteins. This tertiary structure largely determines the biological activity of the molecule, and mistakes in the folding can lead to serious illness. Hydrogen bonding endows solids with significant physical properties, such as ferroelectric behaviour, described in Chapter 11.

The range over which these forces are significant varies widely. Covalent bonds act over a few nanometres only. Interactions that are essentially electrostatic in nature, as in ionic bonds, operate over larger distances, and are proportional to $1/r$, where r is the interionic distance. Ion–dipole interactions decrease more rapidly, being proportional to $1/r^2$. Dipole–dipole interactions vary as $1/r^3$ for static dipoles and as $1/r^6$ for rotating dipoles. Dispersion forces also decrease as $1/r^6$.

3.1.2 Chemical names and formulae

Compounds are broadly classified as alloys, inorganic compounds or organic compounds. Alloys are metallic materials composed of varying proportions of metallic elements. Organic compounds are compounds of carbon, and make up the living world. Inorganic solids comprise everything else, such as rocks and minerals.

Molecules are groups of atoms that are linked together by chemical bonds to form recognisable units in stable structures. The formulae of molecules are written as a set of atomic symbols, with the number of atoms given a subscript to the atomic symbol. Examples are: water, H_2O; methane, CH_4; ammonia, NH_3. The molecules that are important for life, such as proteins and DNA (deoxyribonucleic acid) are extremely large. Polymers are very large molecules formed from smaller molecules, called monomers.

Although the bonds between the atoms within molecules are strong, those between molecules are usually much weaker and are of the types described above. Small molecules tend to exist as gases at room temperature, whereas larger molecules exist as liquids or form solids. The formula of a molecular solid is the same as the formula of the molecules that make up the solid.

Not all solids and liquids are molecular. Many inorganic solids and liquids are built of ions or uncharged atoms. For such solids, the formula often simply expresses the ratio of the atomic species present. For example: crystalline rock salt, NaCl, also called halite or sodium chloride, contains equal numbers of sodium and chlorine atoms, although the total number of each will depend on the size of the sample. Similarly, fool's gold, FeS_2, also called iron pyrites or iron sulphide, always contains twice as many sulphur atoms as iron atoms, although no molecules of FeS_2 exist in the crystals.

In some types of solid, the number of atoms present is not given by simple whole numbers. This is especially so for metallic alloys, where the composition range is closely dependent on temperature. For example, common brass can have a composition anywhere between Cu_6Zn_4 and $Cu_{4.5}Zn_{5.5}$ at 800 °C. Inorganic solids are less prone to having a variable composition than are alloys, but many important examples are known. For example, iron monoxide never attains the composition FeO at atmospheric pressure but has a composition closer to $Fe_{0.945}O$. These types of solid are called nonstoichiometric compounds and their composition range is usually temperature dependent.

Many inorganic compounds are called by a mineral name; for example, the oxide magnesium aluminate, $MgAl_2O_4$, is found as the mineral spinel, and synthetic magnesium aluminate is also referred to as spinel. The mineral name is often applied to a family of compounds all with the same structure. That is, both copper aluminate, $CuAl_2O_4$, and nickel gallate, $NiGa_2O_4$, are called *spinels* because they have the same crystal structure as $MgAl_2O_4$. Because of the relationship with crystal structures, use of mineral names often simplifies matters when solids that may have an imprecise chemical formula are being discussed.

A number of compounds, especially refractory oxides (which are stable at high temperatures), are called by older chemical names, such as calcia for calcium oxide, CaO; magnesia for magnesium oxide, MgO; titania for titanium dioxide, TiO_2; zirconia for zirconium dioxide, ZrO_2; silica for silicon dioxide, SiO_2; and alumina for aluminium oxide, Al_2O_3. Physicists often refer to crystalline colourless aluminium oxide as sapphire, although sapphire is blue. The correct mineral name for the colourless form of aluminium oxide is corundum.

Organic compounds have an elaborate naming system, necessary because of the enormous complexity exhibited by these molecules. The rudiments are explained in Section S2.1.

3.1.3 Polymorphism and other transformations

As the temperature rises, the vibrational energy of the solid becomes similar in magnitude to the bond energy holding the atoms together, and a number of transformations take place. The temperature at which these occur varies with the pressure applied to the system, and not all changes may be possible at normal atmospheric pressure. The changes can be depicted schematically on a diagram that shows the phases present as a function of the temperature and pressure (Figure 3.4). Diagrams of this type are called phase diagrams or existence diagrams. They are discussed in more detail in Chapter 4.

In the solid, changes of crystal structure, known as polymorphism, frequently occur. For example, the asymmetrical environment of hydrogen atoms in hydrogen bonds in solids does not persist at higher temperatures, and lattice vibrations tend to cause the hydrogen atoms to occupy an 'average' position midway between the neighbouring nitrogen, oxygen or fluorine atoms, leading to a new crystal structure. Such transitions are important in a number of ferro-electric materials, (see Section 11.3).

The rise in temperature corresponds to greater and greater atomic vibration about a mean low-

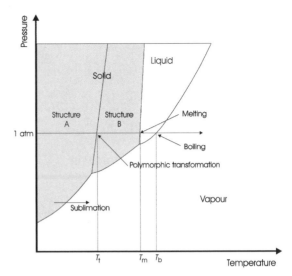

Figure 3.4 Phase diagram of a pure substance. As the temperature increases the solid can change from one structure to another (polymorphism), and transform directly to the vapour (sublimation). Normally a solid passes initially to a liquid (melting) and then to a vapour (boiling)

temperature equilibrium position of the atoms. At some point, these vibrations become so great that the atoms or molecules, although still linked together, are able to move about fairly freely. This corresponds to the liquid state.

Increased temperature will allow atoms or small groups of atoms to break away from the surface to form gaseous species. Ultimately, the whole of the liquid may be vaporised. The actual structure of the species making up the vapour will depend, largely, on the bonding in the solid. Metals often give monatomic vapours. Solids that are predominantly ionic or covalent often vaporise as small charged or neutral fragments containing small numbers of ions or atoms. Tungsten trioxide, WO_3, for example, vaporises to yield the molecular fragments $(WO_3)_n$, where n takes values of 1, 2 and 3.

In some solids, the energy to form the liquid state is similar to the energy to form gaseous species. In this case, the solid may transform directly to the vapour without the intervention of a liquid state. This process is called sublimation. Solid iodine, which consists of I_2 molecules linked by van der Waals bonds, transforms directly to a vapour of I_2 molecules when heated.

3.2 Macrostructures, microstructures and nanostructures

3.2.1 Structures and microstructures

The shape of an object reflects its function. The shape of a container is different from the shape of a blade, and the purposes of the two objects are readily discriminated by eye (Figure 3.5a). The properties of an object that fit it to its functional use are based on a scale that can be called the macrostructure (Figure 3.5b). For example, a container may be glazed or porous.

Many of the measured properties of bulk materials are dominated by structures at a scale somewhere between millimetres and micrometers, called the microstructure of the material (Figure 3.5c). For example, good-quality ceramics have a microstructure that is a mixture of crystallites in glass. Much of materials processing is centred on the production of the correct microstructure in the finished product. The architecture of older silicon chips were somewhere between macrostructure and microstructure. Modern chips have architecture at a smaller

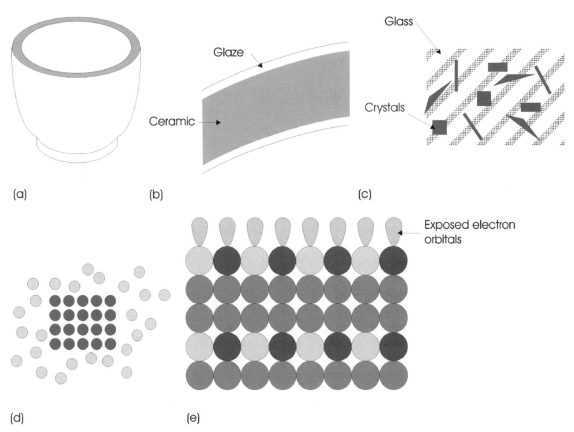

Figure 3.5 Structure and scale: (a) gross shape, a porcelain bowl; (b) the macrostructure of the bowl consists of surface glaze and ceramic body; (c) the microstructure of the ceramic consists of crystals in a glass matrix; (d) the nanostructure of the ceramic consists of atom arrays, which are ordered in the crystals and disordered in the glass; (e) the surface structure consists of exposed atoms of several types and unpaired electron orbitals

scale, somewhere between microstructure and nanostructure.

The atomic structure of the object, called the nanostructure, is at a more fundamental level again. Atomic structure lists the atoms present, their positions and whether they are ordered, as crystals, or disordered as glasses (Figure 3.5d).

The environment of an atom in the surface of a solid differs from that of the bulk (Figure 3.5e). If several atom types make up the crystal, then some surfaces will preferentially contain atoms of one species and other surfaces other species. If nothing else, the bonding of surface atoms is incomplete, meaning that electron orbitals are exposed to outside influences, leading to enhanced reactivity. Surfaces are at the heart of many chemical processes, such as heterogeneous catalysis or corrosion. They also play an important part in the operation of many electronic devices. Recently, there has been great effort put into the production of devices that are close to the atomic scale, less than about 10 nm. This area is known as nanotechnology.

Table 3.2 summarises these relationships.

3.2.2 Crystalline solids

Crystals are solids in which all of the atoms occupy well-defined locations, ordered across the whole of the material (Figure 3.6a). Considering that chemical bonds tend to operate over only a few interatomic distances, it is rather surprising that so much of the solid state is crystalline. Nevertheless, this is

so, and it is only with difficulty that many ordinary solids can be prepared in a noncrystalline form. Crystallography and a description of crystalline solids are to be found in Chapter 5. Single crystals are used for fundamental investigations of solid properties. Single crystals are mandatory for semiconductor devices. Single-crystal turbine blades are fabricated for superior performance. Crystals often show cleavage on certain planes, indicating that some planes of atoms are linked by weaker bonds.

Polycrystalline solids are composed of many interlocking small crystals (Figure 3.6b). Most metals and ceramics in their normal states are polycrystalline. The small crystals are often called grains, especially in metallurgy. The properties of polycrystalline materials are often dominated by the boundaries between the crystallites, called grain boundaries.

3.2.3 Noncrystalline solids

Noncrystalline solids do not have long-range order of the atoms in the structure (Figure 3.6c). There is usually some short-range order, extending over a few atom radii, but no correlation of atom positions at longer distances. There are three types of noncrystalline solid of most importance: glasses, polymers and amorphous solids.

A glass is normally defined as an inorganic substance, mostly transparent, that has passed from a high-temperature liquid state to a solid without the formation of crystals. The best-known glasses are

Table 3.2 Scales of structure

Description	Dimension (m)	Example	Analytical tool
Gross shape	≥ 1	Ceramic vessel	Visual examination
Macrostructure	10^{-2}	Surface glaze, underlying material	Magnifying glass, optical microscope
Microstructure	10^{-4}	Crystallites and noncrystalline material	Microscopy (optical, electron)
Crystal structure	10^{-9}	Crystals, glass	Diffraction (X-ray, neutron, electron)
Atomic structure	10^{-10}	Elemental compositions, atomic topography	Spectroscopy, atomic force microscopy

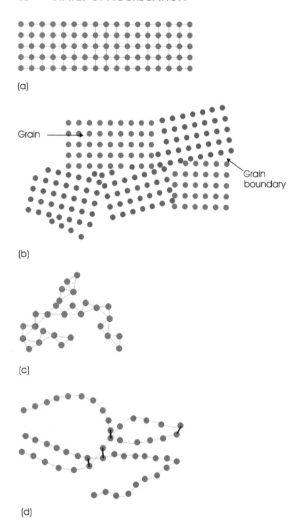

Grain

Grain
boundary

(a)

(b)

(c)

(d)

Figure 3.6 Nanostructures: (a) a single crystal, (b) a polycrystalline array, (c) a glass and (d) polymer chains linked by hydrogen bonds (shown as bars)

manufactured from silicon dioxide, SiO_2, mixed with other oxides. They are called silicate glasses. There are a number of naturally occurring silicate glasses, including obsidian (a volcanic rock which is black because of iron oxide impurities), pumice (a glassy froth), flint and opal. These all show the typical glass properties of hardness and brittleness. However, metals and organic compounds can also solidify as a glass. For example, by boiling and cooling

crystalline sugar, an organic molecular compound, one can form glasses called 'boiled sweets' or toffee, depending on the other ingredients included. [If additives are used to make the melt partly crystallise during cooling, the product is fudge.] In the case of metals, a glass state is much harder to achieve. The molten material must be cooled extremely rapidly, within a time-span of the order of 10^{-6} s. This is achieved by squirting a fine jet of molten metal against a rapidly rotating copper disc that has been cooled by water or liquid nitrogen.

There is no one structure of glass any more than there is one crystal structure. Almost any solid can be produced in a glassy state if the melt is cooled sufficiently quickly. To some extent, glass can be thought of a product of kinetics, and the structure of a glass can depend on the rate at which the liquid is cooled. Theories of glass structure and formation must consider this (see also Section 6.3).

Glasses are described as supercooled liquids. This status of glass is revealed by the behaviour on warming. Glasses do not have a melting point. Instead, they continually soften from a state that can be confidently defined as solid to a state that can be defined as a viscous liquid.

Glasses containing several components are often found to be inhomogeneous at a scale of about 10^{-6} m. Composition variations occur that can be detected by electron microscopy. These composition variations can arise in the melt, when the various components of the system do not mix completely, rather like, but not as extreme as, oil and water. They can also arise on cooling, when some components separate by a process called *spinodal decomposition*. The degree of inhomogeneity found in the glass will depend on both of these factors.

Polymers are a class of substances that consist of very large molecules, macromolecules, built up from many multiples of small molecules, monomers. They can be synthetic (polythene, nylon) or natural (protein, rubber), and occur widely in nature as vital components of living organisms. Most polymers, both natural and synthetic, have a framework of linked carbon atoms. These are strong because the carbon atoms are linked by covalent bonds. The long molecules themselves are linked by some of the weak bonds listed in Table 3.1 and are usually present in a disordered state (Figure 3.6d). A sheet of a solid transparent polymer such as methyl methacrylate (Plexiglas® or Perspex®) is very difficult to tell from a sheet of window glass by sight alone because the structure of these polymeric solids is noncrystalline.

Solids evaporated and then condensed onto cool surfaces usually do not crystallise and are said to be in an amorphous state. Amorphous coatings of this type are widely used in the electronics and optics industries. Such compounds will generally transform into a crystalline state if sufficient energy is supplied to allow crystallisation to occur.

Aerogels are ultra-low-density solids that have a microstructure of highly porous foam. The interconnected pores have a size of less than 100 nm and the structure can be described as of a fractal nature, with the smallest characteristic dimension being of the order of 10 nm. Aerogels have been made from many materials, but silica aerogels are the best known. These have extremely low densities, with porosities of 99.9 % available. Because of this, the physical properties of aerogels vary considerably from that of the parent material. For example, the thermal conductivity of a silica areogel is 10^{-2}–10^{-3} that of ordinary silica glass, the refractive index varies (depending on the porosity) from 1.002 to 1.3, compared with 1.5 for silica glass, and the speed of sound drops to 100–300 m s^{-1} compared with 5000 m s^{-1} in silica glass. These materials find uses ranging from thermal insulation to high-energy nuclear particle detectors.

3.2.4 Partly crystalline solids

Although most solids turn out to be crystalline, there are important groups that are partly crystalline and partly disordered. For example, glasses are not stable thermodynamically. Given enough time, a glass will crystallise. The process of glass crystallisation is called devitrification. Opal glass is a silica (SiO_2) glass prepared so that it has partly recrystallised to give a glassy matrix containing small crystallites dispersed through the bulk. These crystallites reflect light from their surfaces to create the opacity of the solid. Glass ceramics are deliberately

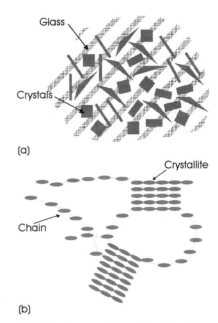

Figure 3.7 Partly crystalline solids: (a) a glass ceramic or porcelain and (b) a partly crystalline polymer

(virtually completely) recrystallised during processing to give a material with the formability of glass and the enhanced mechanical properties of a polycrystalline ceramic. Porcelain is a material consisting of a glassy matrix in which small crystals of other oxides are embedded (Figure 3.7a).

Polymers also show a natural tendency to crystallise, which is thwarted to a greater or lesser extent by the structure of the polymer molecules. Most polymers have a chain-like form. It is possible to change the average chain length created during polymerisation, and longer chains are more difficult to crystallise. The presence or absence of sidegroups attached to the chain also has a considerable effect on the ease with which a polymer chain can crystallise. The partly crystalline structure of many linear polymers (Figure 3.7b) is typified by one of the simplest of polymers, polyethylene (polythene). Polyethylene molecules are 10^4 monomer units or more long and resemble thin strings. If the liquid is cooled reasonably quickly, the chains remain in an extended form. The material has a low density, a low refractive index and is very flexible. It resembles a glass. However, if the polyethylene is cooled slowly from the melt some chains can fold up into crystalline regions 10–20 nm thick. These crystalline regions are of higher density and refractive index compared with the noncrystalline parts. Most polyethylene is a mixture of crystalline and amorphous regions, which is why it appears milky.

Many factors control the degree of crystallinity of a polymer, and these are carefully controlled in production to obtain the correct mix of crystalline and noncrystalline material.

3.2.5   Nanostructures

The nanostructure of a material is its structure at an atomic scale. Nanoparticles and nanostructures generally refer to structures that are small enough that chemical and physical properties are observably different from the normal or 'classical' properties of bulk solids. For example, the energy levels of isolated atoms are sharp, whereas atoms in a solid contribute to an energy band, as described in Section 2.3.2 (see Figure 2.23, page 45). At some stage, as the solid is imagined to fragment into smaller and smaller units, the energy levels must change from typically bulk-like bands to more atom-like sharp levels.

The dimension at which this transformation becomes apparent depends on the phenomenon investigated. In the case of thermal effects, the boundary occurs at approximately the value of thermal energy, kT, which is about 4×10^{-21} J. In the case of optical effects, nonclassical behaviour is noted when the scale of the object illuminated is of the same size as a light wave, say about 5×10^{-7} m (see Section 14.11). For particles such as electrons the scale is determined by the Heisenberg uncertainty principle, at about 3×10^{-8} m (see Section 13.3).

The areas where this transition has been most apparent are microelectronics and optoelectronics. As the dimensions of microelectronic circuit elements have decreased, nanostructures are increasingly in evidence. A thin layer of a solid will have bulk properties modified towards atom-like properties in a direction normal to the layers. A thin layer of a semiconductor sandwiched between layers of a different semiconductor is called a quantum well (Figure 3.8a). In this structure, the electrons are essentially confined to the two-dimensional plane of the layers by the difference in the band structures of the two materials. They are regarded as two-dimensional from the point of view of microelectronics. Similar devices built up from several alternating layers of semiconductors are called multiple quantum well structures, or superlattices (Figure 3.8b). Structures that are small on an atomic scale in two directions are called quantum wires (Figure 3.8c). In these structures, the electrons are confined in two dimensions by the band structure of the surrounding materials and, from a microelectonics perspective, they are one-dimensional conductors. A cluster of atoms has properties rapidly approaching that of the isolated atoms. Electrons are confined to a localised region of space, and the structure is called a quantum dot (Figure 3.8d). These are

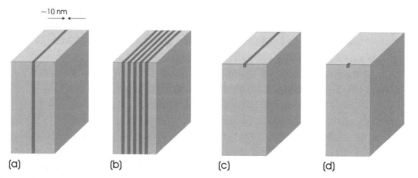

~10 nm

(a)   (b)   (c)   (d)

Figure 3.8   Electronic nanostructures: (a) a quantum well, (b) a series of quantum wells, to form a multiple quantum well structure or quantum superlattice, (c) a quantum wire and (d) a quantum dot

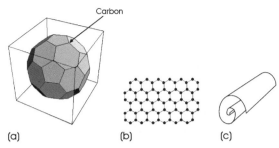

Carbon

(a)   (b)   (c)

Figure 3.9 The truncated icosahedral structure of a C_{60} 'buckyball'; a carbon atom is situated at each vertex. (b) The hexagonal structure of a single sheet of carbon atoms arranged as in graphite. (c) Carbon nanotubes, which consist of similar sheets rolled up into a variety of configurations

regarded as zero-dimensional electronic structures. These structures are fabricated by using the standard techniques of the semiconductor industry. The electronic and optical consequences of these restricted nanostructures are described in more detail in Chapters 13 and 14.

Among the most commonly investigated nanoparticles are the forms of carbon called fullerenes and carbon nanotubes. Fullerenes are roughly spherical assemblies of carbon atoms linked by strong covalent bonds. The first example to be characterised, C_{60}, was called Buckminsterfullerene, as the structure (Figure 3.9a) resembled the geodesic dome structure developed by R. Buckminster Fuller. The structure of C_{60} is a truncated icosahedron, and is built of faces made up of pentagons and hexagons. A carbon atom is found at each vertex of the structure. Fullerenes have the electronic properties of quantum dots. Carbon nanotubes can be thought of as a layer of carbon atoms of the sort found in graphite (Figure 3.9b) coiled into a tube (Figure 3.9c; see also Section 5.3.7 for the structure of graphite). Carbon nanotubes behave as quantum wires. The electronic and optical properties of fullerenes and nanotubes can be modified by encapsulating other atoms, especially metal atoms, into the structure.

In the case of optoelectronics, the aim is to utilise light in an analogous role to electrons. The optical equivalents to transistors are photonic crystals. These structures interact with light in controlled and predetermined ways. Many are based on the structure of a natural 'photonic crystal', the gemstone opal. Opals contain regular arrays of spherical silica particles with a diameter similar to the wavelength of light. This gives rise to the flashing colours in natural stones. Photonic crystals are discussed in more detail in Chapter 14.

Nanostructures exist at many levels, and as the example of opal suggests, many of these have been discovered in nature. For example, typical spider thread (there are many species of spider and many types of thread) is a polymeric material that has remarkable properties, that have been likened to the ability of a fishing net to capture a ballistic missile. These properties result from the nanostructure of the thread, which consists of interleaved crystalline and noncrystalline regions.

3.3   The development of microstructures

The development of the correct microstructure in a manufacturing process is of prime importance. It is the increasing mastery of this ability that has marked out the progression of ancient and modern civilisations. In early times, this control was achieved by trial and error. The resulting recipes were then closely guarded by tradespeople or trade guilds. The control of microstructures in modern times has come to depend on a precise knowledge of the science behind the chemical and physical changes that are taking place. This is typified by the rise of metallurgy concurrent with the development of the modern steel industry, some 100 or so years ago. Currently, the production of nanostructures and nanodevices requires considerable scientific and engineering skills.

3.3.1   Solidification

Many solids, especially metals, are produced from liquid precursors, and control of solidification is important in the development of the appropriate microstructure. Rapid solidification can lead to amorphous or poorly crystalline products. Slow cooling can lead to the formation of large crystals or single crystals. As these observations indicate, the microstructure formed often depends on the rate of solidification.

There are two important steps involved. Nucleation is the initial formation of tiny crystallites. As a liquid cools, small volumes tend to take on a structure similar to that of crystals, which will ultimately form. This occurs especially at mould edges, on dust particles and so on, which act as sites for nucleation. Nucleating agents can be added deliberately to cause this to happen. The formation of nuclei is suppressed during glass formation. If only one nucleus forms, a single crystal is produced. If many nuclei form, a polycrystalline solid results.

Crystal growth follows nucleation and contributes greatly to the development of microstructure. The resulting solid will usually contain crystals of different compounds, as in the rock granite, which is composed mainly of mica, quartz and feldspars (Figure 3.10a). Pure metals and alloys are also normally polycrystalline (Figure 3.10b). Many crystals grow from the melt with a branching shape or morphology that resembles a tree in form. These are called dendrites and the growth is called dendritic growth (Figure 3.10c). The shape of the dendritic crystal reflects the internal symmetry of the crystal structure (see Chapter 5 for more information on crystal symmetry). Cubic metals usually have 'side arms' perpendicular to the long growth axis, whereas in hexagonal crystals the side arms are at angles of 60°. This gives snowflakes and frost, which are dendritic ice crystals, their definitive form.

The ultimate microstructure of a solid will depend on how quickly different crystal faces develop. This controls the overall shape of the crystallites, which may be needle-like, 'blocky' or one of many other shapes. The shapes will also be subject to the constraint of other nearby crystals. The product will be a solid consisting of a set of interlocking grains. The size distribution of the crystallites will reflect the rate of cooling of the solid. Liquid in contact with the cold outer wall of a mould may cool quickly and give rise to many small crystals. Liquid within the centre of the sample may crystallise slowly and produce large crystals. Finally, it is important to mention that the microstructure will depend sensitively on impurities present. This aspect is discussed in Chapters 4 and 8.

(a)

(b)

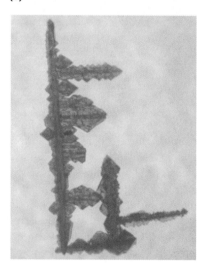

(c)

Figure 3.10 Optical micrographs of polycrystalline solids. (a) Granite, composed of interlocking crystals of mica (black), quartz and feldspars (colourless); the mica crystals are approximately 2 mm in width. (b) Aluminium, consisting of interlocking grains up to 1 cm in width. (c) Rutile, TiO_2, with a dendritic form; each 'branch' is a separate crystal, and the whole group forms a polycrystalline solid. The branches are perpendicular to the main crystal, revealing the underlying symmetry of the structure. The crystal is approximately 5 mm long

3.3.2 Processing

Processing refers to the treatment of a solid to alter the microstructure and external form. It is a large subject and of considerable importance in industry. Just a few processing routes are mentioned here. More information is given in the publications in the Further Reading section.

Working and heat treatment are techniques applied mainly to metals. When a metal is hammered, rolled or deformed it is referred to as working. Cold metals get harder on working as the process introduces large numbers of defects (see below) and strain energy into the sample. If the metals are heated to about half their melting point (called annealing) they can partly recrystallise and release the strain energy. This causes the metal to become softer and more ductile.

Thermoplastic polymers can easily be melted and moulded into flexible shapes. The rigidity and strength of the product can be improved by cross-linking between the polymer chains. One of the first deliberate cross-linking processes was the vulcanisation of rubber, which is used in car tyre manufacture. The process transforms sticky, soft rubber into a hard, flexible material.

The devitrification of glass to produce glass ceramics, mentioned in Section 3.2.4, is typical of processing in the glass and ceramic industries. Here, the processing aim is to overcome the brittleness typical of glasses while retaining good chemical inertness.

Sintering is widely used to make polycrystalline ceramic bodies. A powder is compressed and heated at a temperature below the melting point to produce a strong polycrystalline solid (see Section 8.4). This comes about by atomic mobility (diffusion). The presence of traces of liquid helps the process greatly. Many electrical and electronic components are produced by sintering. Some metal parts are also made via this method, and the subject area is called powder metallurgy. The main aim of processing in this general area is to produce a high-density solid with little porosity. An associated aim is to reduce dimensional changes, especially shrinkage, which preclude the use of powders to form solids with precise engineering tolerances.

Dehydration, or, more exactly, fluid phase removal, is used to form the ultraporous microstructure of aerogels. Normally, when a gel (e.g. ordinary gelatine) is dehydrated, the material shrinks and collapses. As fluid in the pores evaporates, a meniscus forms which generates large surface-tension forces. These cause the pore structure to disintegrate. The formation of aerogels is a typical processing problem: how is it possible to remove fluid while maintaining the porous microstructure? In original work, high pressures and temperatures were used to take the fluid in the material above its critical temperature (see Section 4.1.1). In this state, the fluid does not exert surface tension, and it is possible to remove it without collapse of the solid framework. A variety of related processing methods are now used for aerogel production.

3.4 Defects

Defects in crystalline solids are important because they modify important properties. For example, just a trace of chromium impurity changes colourless aluminium oxide into ruby. Metals are ductile when linear defects called dislocations are free to move. Crystals dissolve and react at increased rates at points where dislocations intersect the surface of the crystal. Thus, it is necessary to have an idea of the types of defect that form and the role that they play in the control of properties in order to understand the behaviour of solids.

3.4.1 Point defects in crystals of elements

Crystals of the solid elements, such as silicon, contain only one atom type. The simplest localised defect that we can imagine in a crystal is a 'mistake' at a single atom site. These defects are called point defects.

Two types of point defect can occur in a pure crystal: atoms can be absent from a normally occupied position, to give what are called vacancies, or an atom may be incorporated at a position not normally occupied, called an interstitial atom, and sometimes a self-interstitial atom (Figure 3.11).

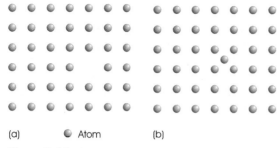

(a)   ● Atom   (b)

Figure 3.11 Point defects in pure crystals, such as silicon: (a) a vacancy and (b) an interstitial (shown here as a self-interstitial)

Such vacancies and interstitials, which occur in even the purest of materials, are called intrinsic defects.

For these defects to be stable, the Gibbs energy of a crystal containing defects must be less than the Gibbs energy of a crystal without defects (see also Section S3.2). The Gibbs energy varies with the number of point defects present (Figure 3.12). Initially, a population of defects lowers the Gibbs energy, but, ultimately, a large number of point defects results in an increase in Gibbs energy. The minimum in the energy curve represents the equilibrium situation that will exist at a given temperature. Thermodynamics allows the position of the

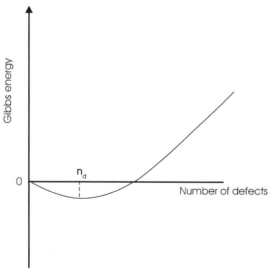

Figure 3.12 The Gibbs energy of a crystal as a function of the number of point defects present. At equilibrium, n_d defects are present in the crystal

minimum and the approximate associated number of point defects present in a crystal to be calculated. The number of defects is expressed by the formula:

$$n_d \approx N \exp\left(\frac{-\Delta H}{kT}\right) \qquad (3.1)$$

where n_d is the number of defects per unit volume, N is the number of sites affected by defects per unit volume, ΔH is the enthalpy (loosely, the heat energy) needed to form a single defect, k is the Boltzmann constant and T is the temperature (in kelvin).

The fraction of atom sites which contain a defect, n_d/N, at any temperature, can be calculated if the enthalpy of defect formation, ΔH, is known.

$$\frac{n_d}{N} \approx \exp\left(\frac{-\Delta H}{RT}\right) \qquad (3.2)$$

where ΔH is measured in J mol^{-1}.

To obtain the absolute number of defects in the solid, it is necessary to know the number of atoms, N, in a unit volume of the crystal. This value is often obtained from the crystal structure of the compound. The crystal structure is described in terms of the unit cell, which is a small representative volume of the crystal (see Chapter 5). We may write:

$$N = \frac{\text{number of atoms in the unit cell}}{\text{unit cell volume}}$$

For example, the unit cell of silicon is cubic, with a side of 0.5431 nm, and contains 8 atoms of silicon, which allows N to be found.

Alternatively, it is possible to obtain the same information from the density of the material, ρ. The relative molar mass of an element contains N_A atoms, where N_A is Avogadro's constant. The value of N is then given by:

$$N = \frac{\rho \times N_A}{\text{molar mass}}$$

No material is completely pure, and foreign atoms will be present. If these are undesirable or acciden-

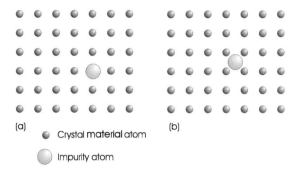

(a)   ⬤ Crystal material atom   (b)

⬤ Impurity atom

Figure 3.13 Impurity or dopant point defects in a crystal: (a) substitutional and (b) interstitial

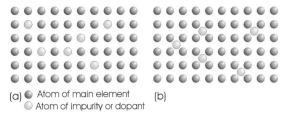

(a) ⬤ Atom of main element   (b)
○ Atom of impurity or dopant

Figure 3.14 (a) A random substitutional solid solution and (b) a random interstitial solid solution

tal, they are known as impurities, but if they have been added deliberately, to change the properties of the material on purpose, they are called dopants. Foreign atoms can rest on sites normally occupied by the parent atom type to form substitutional defects. Foreign atoms may also occupy positions not normally occupied in the crystal, to create interstitial impurities or interstitial dopants (Figure 3.13). There is no simple thermodynamic formula for the number of impurities present in a crystal.

3.4.2 Solid solutions

Some compounds, especially alloys, have quite extensive composition ranges. Close to the parent composition the structure can be thought of in terms of impurity defects in a crystal. When quite large numbers of impurity atoms enter a crystal, without changing the crystal structure, the resultant phase is referred to as a solid solution.

A substitutional solid solution is a mixture of two similar elements in which one atom substitutes on the sites of the other atoms in the structure. In the copper–nickel system, both parent phases adopt the same crystal structure. When both atom types are present, they occupy random positions in the crystal to form a substitutional solid solution (Figure 3.14a). Near to pure copper it is possible to say that the nickel atoms form substitutional impurity defects, and near to pure nickel it is possible to say that copper forms substitutional impurity defects. Substitutional alloys generally have lower thermal

and electrical conductivity than do the pure elements, and are harder and stronger.

Interstitial solid solutions form when small atoms enter spaces between the atoms in a crystal (Figure 3.14b). The interstitial impurities must be small, with a radius less than about 60 % of the parent-structure atoms if an interstitial solid solution is to form. They are typically elements from the first row of the periodic table, such as carbon and nitrogen. Steel is the most important interstitial alloy and consists of interstitial carbon atoms in crystals of iron. Interstitial alloys are usually very hard materials, often used as hard coatings on surfaces liable to excessive wear, such as drill bits.

Both of these solid solutions can become ordered, and this frequently occurs in alloys if they are heated for lengths of time at moderate temperatures. In this case, the crystal adopts a new structure that is no longer regarded as containing impurity defects (see Chapter 5).

3.4.3 Schottky defects

The situation in ionic compounds is slightly more complex than in metals because the ionic charges must remain balanced when point defects are introduced into the crystal. Take the compound sodium chloride, which contains equal numbers of sodium (Na^+) and chlorine (Cl^-) ions and has the chemical formula NaCl. To separate out the effects of the anions from that of the cations it is convenient to refer to the anion sublattice for the Cl^- array and to the cation sublattice for the Na^+ array.

Vacancies on the cation sublattice will change the composition of the compound. As the constituents

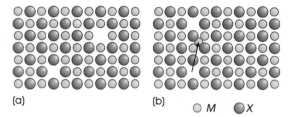

(a)   (b)   ○ M   ● X

Figure 3.15 Point defects in an ionic crystal of formula MX: (a) Schottky defects and (b) Frenkel defects

are charged, this will also alter the charge balance in the crystal. If x vacancies occur, the formula of the crystal will now be $Na_{1-x}Cl$ and the overall material will have an excess negative charge of x^-, because the number of chloride ions is greater than the number of sodium ions by this amount. The formula could be written $[Na_{1-x}Cl]^{x-}$. If x vacancies are found on the anion sublattice, the material will take on an overall positive charge, because the number of sodium ions now outnumbers the chlorine ions, and the formula becomes $[NaCl_{1-x}]^{x+}$. Now, crystals of sodium chloride do not normally show an overall negative or positive charge or have a formula different from NaCl. This means that equal numbers of vacancies must occur on both sublattices.

The defects arising from balanced populations of cation and anion vacancies in any crystal (not just NaCl) are known as Schottky defects (Figure 3.15a). Any ionic crystal of formula MX must contain equal numbers of cation vacancies and anion vacancies. In such a crystal, one Schottky defect consists of one cation vacancy plus one anion vacancy. (These vacancies need not be near to each other in the crystal.) The number of Schottky defects in a crystal of formula MX is equal to one half of the total number of vacancies. In crystals of more complex formulae, charge balance is also preserved. The ratio of two anion vacancies to one cation vacancy will hold in all compounds of formula MX_2, such as titanium dioxide, TiO_2. Schottky defects in this material will introduce twice as many anion vacancies as cation vacancies into the structure. In crystals with a formula M_2X_3, a Schottky defect will consist of two vacancies on the cation sublattice and three vacancies on the anion sublattice. In Al_2O_3,

for example, two Al^{3+} vacancies will be balanced by three O^{2-} vacancies.

Under equilibrium conditions, the Gibbs energy of a crystal, G, is lower if it contains a small population of Schottky defects, similar to the situation shown in Figure 3.12. This means that Schottky defects will always be present in crystals at temperatures above 0 K, and hence Schottky defects are intrinsic defects. The approximate number of Schottky defects, n_S, in crystal with a formula MX, at equilibrium, is given by:

$$n_S \approx N \exp\left(\frac{-\Delta H_S}{2kT}\right) \quad (3.3)$$

where N is the number of M atoms per unit volume (equal to the number of nonmetal atoms per unit volume) affected by Schottky defects, ΔH_S is the enthalpy (loosely, the heat energy) required to form one defect, k is the Boltzmann constant and T is the temperature (in kelvin). Sometimes, Equation (3.3) is written in the form:

$$n_S \approx N \exp\left(\frac{-\Delta H_S}{2RT}\right) \quad (3.4)$$

In this case, ΔH_S is in J mol^{-1} and represents the enthalpy required to form 1 mole of Schottky defects, and R is the gas constant (J K^{-1} mol^{-1}). The fraction of vacant sites in a crystal as a result of Schottky disorder is given by:

$$\frac{n_S}{N} \approx \exp\left(\frac{-\Delta H_S}{2RT}\right) \quad (3.5)$$

where ΔH_S is measured in J mol^{-1}. (Remember that this formula applies only to materials with a composition MX.) The formation enthalpies of Schottky defects in some alkali halides are given in Table 3.3.

The fraction of vacant sites at any temperature can be calculated if the enthalpy of defect formation, ΔH_S, is known. To obtain the absolute number of defects in the solid, it is necessary to know the number of atoms, N, of the appropriate type in a unit volume of the crystal. As with elements, this

Table 3.3 The formation enthalpy of Schottky defects, ΔH_S, in some alkali halide compounds of formula MX

Compound	ΔH_S/J
LiF	3.74×10^{-19}
LiCl	3.39×10^{-19}
LiBr	2.88×10^{-19}
LiI	1.70×10^{-19}
NaF	3.87×10^{-19}
NaCl	3.75×10^{-19}
NaBr	2.75×10^{-19}
NaI	2.34×10^{-19}
KF	4.35×10^{-19}
KCl	4.06×10^{-19}
KBr	3.73×10^{-19}
KI	2.54×10^{-19}

Note: all compounds listed have the *halite* structure

value is often obtained from the unit cell (see Chapter 5):

$$N = \frac{\text{number of appropriate atoms in unit cell}}{\text{unit cell volume}}$$

Alternatively, it is possible to obtain the same information from the density of the material, ρ. The relative molar mass of a compound contains N_A atoms of each type, where N_A is the Avogadro constant. The atomic density of each atom type present in the crystal is given by:

$$N = \frac{\rho \times N_A}{\text{molar mass of compound}}$$

3.4.4 Frenkel defects

It is possible to imagine a defect in ionic crystals similar to the interstitial defects described above. Such defects are known as Frenkel defects. In this case, an ion from one sublattice moves to a normally empty place in the crystal, leaving a vacancy behind. One Frenkel defect consists of an interstitial ion plus a vacancy (Figure 3.15b). Because the total number of ions present does not change, there is no

need for charge balance to be considered. For example, a Frenkel defect on the anion sublattice in calcium fluoride, fluorite, CaF_2, consists of one F^- ion displaced to an interstitial site. It is not necessary to displace two ions to form the Frenkel defect.

Frenkel defects occur in silver bromide, AgBr. In this compound some of the silver ions (Ag^+) move from the normal positions to sit at usually empty places to generate interstitial silver ions and leave behind vacancies on some of the usually occupied silver sites. The bromide ions (Br^-) are not involved in the defects. (Frenkel defects in AgBr make possible black and white and colour photography on photographic film.)

The presence of a small number of Frenkel defects reduces the Gibbs energy of a crystal and so Frenkel defects are intrinsic defects. The formula for the equilibrium concentration of Frenkel defects in a crystal is similar to that for Schottky defects. There is one small difference compared with the Schottky defect equations: the number of interstitial positions that are available to a displaced ion, N^*, need not be the same as the number of normally occupied positions, N, from which the ion moves. The number of Frenkel defects, n_F, present in a crystal of formula MX at equilibrium is given by:

$$n_F \approx (NN^*)^{1/2} \exp\left(\frac{-\Delta H_F}{2kT}\right) \qquad (3.6)$$

where ΔH_F is the enthalpy of formation of a single Frenkel defect, k is the Boltzmann constant and T is the absolute temperature (in kelvin). This is also often expressed in molar quantities:

$$n_F \approx (NN^*)^{1/2} \exp\left(\frac{-\Delta H_F}{2RT}\right) \qquad (3.7)$$

where the values for ΔH_F are in units of $J\,mol^{-1}$, and R is the gas constant. The formation enthalpies of Frenkel defects in some compounds of formulae MX and MX_2 are given in Table 3.4.

The fraction of interstitials, $n_F/(NN^*)^{1/2}$, at any temperature, can be calculated if the enthalpy of defect formation, ΔH_F is known. To obtain the

Table 3.4 The formation enthalpy of Frenkel defects, ΔH_F, in some compounds of formula MX and MX_2

Compound[a]	ΔH_F/J	Compound[b]	ΔH_F/J
AgCl	2.32×10^{-19}	CaF_2	4.34×10^{-19}
AgBr	1.81×10^{-19}	SrF_2	2.78×10^{-19}
β-AgI	0.96×10^{-19}	BaF_2	3.06×10^{-19}

[a]Frenkel defects on the cation sublattice of *halite* structure compounds.
[b]Frenkel defects on the anion sublattice of *fluorite* structure compounds.

absolute number of defects in the solid, it is necessary to know the number of ions affected by Frenkel disorder, N, and the number of sites that can accept an interstitial, N^*, in a unit volume of the crystal. These values can be assessed by a consideration of the crystal structure of the compound. The two numbers N and N^* are not usually identical, but in cases where they are, the formulae in Equations (3.6) and (3.7) become identical to those for Schottky defects, Equations (3.3) and (3.4).

3.4.5 Nonstoichiometric compounds

Although molecules have a fixed formula and composition, many nonmolecular solids are found to exist over a range of compositions. This variation is considered normal in alloys but unusual in nonmetallic compounds such as oxides. However, not all such solids have a definite and fixed formula, especially at high temperatures. Nonmetallic materials with a composition range are called nonstoichiometric compounds. Two ways in which this composition variation can occur are described below.

Zirconia, ZrO_2, is an important oxide as it remains inert and stable at temperatures of up to 2500 °C and finds uses in many high-temperature applications. Unfortunately, pure zirconia fractures when cycled repeatedly from high to low temperatures because the crystal structure changes at approximately 1100 °C. This shortcoming is overcome by reacting zirconia with calcia, CaO. The material that is formed exists over a wide composition range. This nonstoichiometric phase, which has

a cubic unit cell, is called calcia-stabilised zirconia. Although it is not stable to room temperature, transformations to other structures are very slow at lower temperatures. For practical purposes calcia stabilised zirconia can be cycled from high to low temperatures without problem. This significant modification in the properties of zirconia is brought about by the introduction of defects into the crystal.

As with Schottky defects, it is important that charge balance is maintained during the reaction of ZrO_2 and CaO. Cubic calcia-stabilised zirconia crystallises with the *fluorite* (CaF_2) structure. In the present case, the parent material is ZrO_2. The stabilised phase has Ca^{2+} cations in some of the positions that are normally filled by Zr^{4+} cations, that is, cation substitution has occurred (Figure 3.16). As the Ca^{2+} ions have a lower charge than the Zr^{4+} ions, the crystal will show an overall negative charge if we write the formula as $Ca_x^{2+}Zr_{1-x}^{4+}O_2$. The crystal compensates for the extra negative charge by leaving some of the anion sites unoccupied. For exact neutrality, the number of vacancies on the anion sublattice needs to be identical to the number of calcium ions in the structure. Thus, each Ca^{2+} added to the ZrO_2 produces an oxygen vacancy at the same time. The formula of the crystal is $Ca_x^{2+}Zr_{1-x}^{4+}O_{2-x}$.

Calcia-stabilised zirconia provides a good example of the consequences of incorporating an ion with

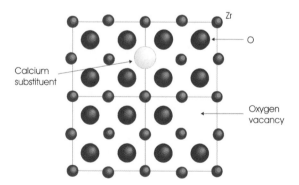

Figure 3.16 The structure of calcia stabilised zirconia. The crystal contains a number of Ca^{2+} ions substituted for Zr^{4+} ions. Each Ca^{2+} ion is accompanied by an O^{2-} vacancy to maintain charge neutrality. The unit cell is outlined

a lower valence into a crystal structure. When a material is 'doped' with substitutional impurity cations of lower charge, anion vacancies are a common method of achieving charge balance. This has a significant effect on the properties of the solid, as the diffusion coefficient of oxygen ions is greatly increased, to the extent that the material is widely used as a solid electrolyte in electrochemical cells and sensors (see Section 9.3.7).

In cases where a cation of higher valence is substituted for the native cation in a crystal, charge balance will also be disturbed. This can occur, for example, if Mg^{2+} impurities are present in a crystal of NaCl. The Mg^{2+} cations will occupy Na^+ sites forming substitutional impurity defects. In order to maintain charge balance, each Mg^{2+} impurity must be balanced by a vacancy on the cation sublattice. As a rule, in a crystal that contains substitutional impurity cations of a higher charge, cation vacancies tend to form in the sublattice of the lower charged ions.

It is also possible to vary the composition of a solid by introducing extra atoms into spaces within the crystal. This process is called interpolation. The likelihood of finding that a nonstoichiometric composition range is due to the presence of interpolated atoms in a crystal will depend on the openness of the structure and the size of the impurity. One of the most important groups of materials in this category, the interstitial alloys, has been described in Section 3.4.2 and is considered further in Section 6.1.5.

Another group of materials that use interpolation are layered structures, in which atoms are taken in between weakly held layers. The resulting compounds, often called insertion compounds, or intercalation compounds, are finding increasing use in batteries (see Section 9.3.6). These are typified by disulphides such as TiS_2 and NbS_2. The structure of TiS_2 is of the CdI_2 (cadmium iodide) type (Figure 3.17). It is made up of weakly bound layers of composition TiS_2. Small atoms such as lithium (Li) can enter the structure between these layers to form nonstoichiometric phases with a general formula Li_xTiS_2. Because the bonding between the layers is weak this process is easily reversible, so the compound can act as a convenient reservoir of Li atoms in lithium batteries. For reasons of efficiency,

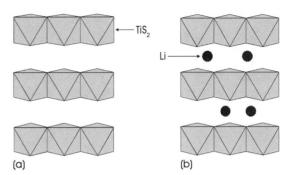

Figure 3.17 (a) The structure of TiS_2, composed of layers of TiS_6 octahedra, linked by weak bonds. (b) Insertion of Li (or other) atoms between the layers

TiS_2 has now been superseded by other compounds in modern batteries.

Another group of insertion compounds of importance are derived from graphite. The layers of carbon atoms in graphite are only weakly linked by van der Waals bonding. Atoms and molecules are able to enter the structure between the layers, into the so-called van der Waals gap. As with Li_xTiS_2, lithium inserted into graphite, Li_xC, is used as a battery material (see Section 9.3.6). The material that forms when fluorine is incorporated, CF_x, is used as a solid lubricant (see Section 10.3.1). The strongly electronegative fluorine atoms bond covalently to the carbon atoms, which forces the layers further apart, with the result that CF_x is a better solid lubricant than is pure graphite.

3.4.6 Edge dislocations

It has long been known that the theoretical strength of a metal crystal is far greater than the strength normally observed. Moreover, metals can be deformed easily and retain the new shape, a process called plastic deformation, whereas ceramic solids fracture under the same conditions. The typical mechanical properties of metals are due to the presence of linear defects called dislocations.

Although there are many different types of dislocations, they can all be thought of as combinations of two fundamental types. Edge dislocations consist

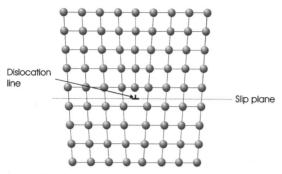

Figure 3.18   An edge dislocation, consisting of an extra half plane of atoms inserted in a crystal. The dislocation line is perpendicular to the plane of the figure, and is marked by ⊥. The slip plane is the crystal plane of preferred movement of the dislocation

of an extra half-plane of atoms inserted into the crystal (Figure 3.18). These dislocations are instrumental in allowing metals to undergo plastic deformation. The dislocation line is marked by ⊥. It runs normal to the plane of the drawing. The slip plane, lying perpendicular to the dislocation line, is the plane along which the dislocation moves. If a shear force is applied to the crystal the dislocation glides (moves), to reduce the shear, resulting in a permanent deformation, recognised as the step in the crystal profile (Figure 3.19; see also Section 10.1.9). In essence, only one line of bonds is broken each time the dislocation is displaced by one atomic spacing, and the stress required is relatively small. If the same deformation were to be produced in a

perfect crystal, large numbers of bonds would have to be broken simultaneously, requiring much greater stress. The presence of edge dislocations successfully explains the deformation properties of metals. Similarly, if dislocation movement is impeded or impossible, the material becomes hard and brittle. This is so in ceramics, in which dislocation movement is impeded, at least in part, by the charges on the ions.

The disruption to the crystal introduced by a dislocation is characterised by the Burgers vector. The Burgers vector of a dislocation is determined by drawing a circuit in the crystal, from atom to atom, in a region of crystal away from the defect (Figure 3.20). This is called a Burgers circuit. The Burgers circuit starts and ends on the same atom in a perfect crystal (Figure 3.20a), but will not do so if the circuit contains a dislocation (Figure 3.20b). The vector describing this failure, running from the start atom to the end atom, is the Burgers vector, b, of the dislocation. The Burgers vector of an edge dislocation is perpendicular to the dislocation line.

A perfect dislocation has a Burgers vector equal to an atom-to-atom vector in the crystal. As the energy of a dislocation is proportional to b^2, the most common dislocations in metals have small Burgers vectors.

3.4.7   Screw dislocations

Screw dislocations, the second important category of dislocation, looks rather like a spiral staircase.

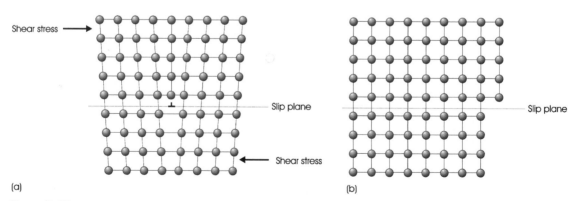

(a)   (b)

Figure 3.19   Application of a shear stress to a crystal containing an edge dislocation (part a) can cause the dislocation to move out of the crystal to leave a surface step (part b)

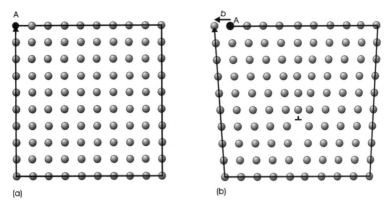

Figure 3.20 (a) A Burgers circuit in a perfect crystal (heavy line) will be closed; (b) the same circuit in a crystal containing a dislocation will remain open. The vector required to close the circuit, *b*, is the Burgers vector

The dislocation can be formed (conceptually) by cutting halfway through a crystal and sliding the regions on each side of the cut parallel to the cut, to create spiralling atom planes (Figure 3.21). The dislocation line is the central axis of the 'staircase'. The Burgers vector of a screw dislocation is parallel to the dislocation line. Screw dislocations play an important part in crystal growth.

3.4.8 Partial and mixed dislocations

Dislocations can be imaged and their Burgers vectors determined by using transmission electron microscopy. This technique has shown that many dislocations have a Burgers vector that is less than the repeat distance of the structure. These are called partial dislocations.

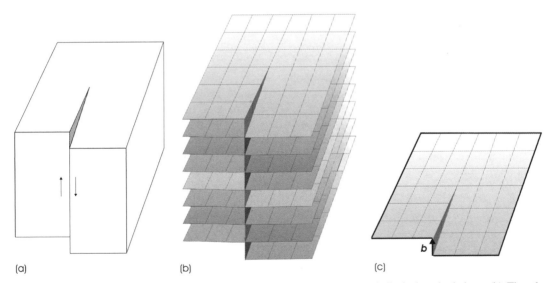

Figure 3.21 (a) A screw dislocation can be formed by cutting a crystal and displacing the halves. (b) The planes around a screw dislocation spiral around the dislocation line like a spiral staircase. (c) In a screw dislocation the Burgers vector, *b*, is parallel to the dislocation line

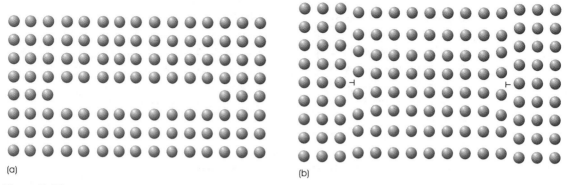

Figure 3.22 (a) Aggregation of vacancies on a plane in a crystal. (b) Formation of an edge dislocation loop by collapse of the crystal in the vacancy region

A dislocation line separates a region of crystal that has moved relative to an adjoining part. This disruption means that the dislocation must either end on the surface of a crystal or else form a closed loop. Dislocation loops have been found to occur frequently in crystals. One way in which a dislocation loop can form is by the aggregation of vacancies on a plane in a metal crystal (Figure 3.22). Vacancy populations are relatively large at high temperatures and, if a metal is held at a temperature near to its melting point, these defects can migrate from site to site (see Section 7.4). If sufficient vacancies aggregate, a dislocation loop can form. The dislocation that delineates the loop is an edge dislocation. As dislocation loops grow, they can change character, so that at one part of the loop the Burgers vector is parallel to the dislocation line and at another part it is parallel to it. That is, the character of the dislocation changes from pure edge to pure screw. Elsewhere on the loop, the dislocation has an intermediate character and is called a mixed dislocation.

3.4.9 Multiplication of dislocations

Dislocations are introduced into solids when they crystallise. In addition, dislocations form in materials subjected to stress. There are a number of mechanisms by which new dislocations can form, but most require an existing dislocation to become

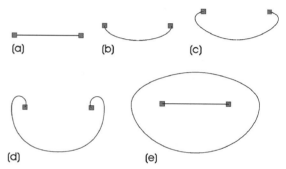

Figure 3.23 A dislocation pinned at each end (part a) can respond to stress by bowing out (parts b–d) to form a dislocation loop and reform the pinned dislocation (part e)

trapped, or pinned, in the crystal, so that glide can no longer occur. One of these involves a length of dislocation pinned at each end (Figure 3.23a). This is called a Frank–Read source. When stress is applied to this defect, the pinned dislocation cannot glide to relieve the stress, but it can bulge out from the pinning centres to achieve the same result (Figure 3.23b). Further stress increases the degree of bulging (Figures 3.23c and 3.23d), until both sides of the bulge can unite to form a dislocation loop and a new dislocation between the pinning centres (Figure 3.23e). A Frank–Read source can continually emit dislocation loops during stress, thus significantly multiplying the dislocation density in a crystal.

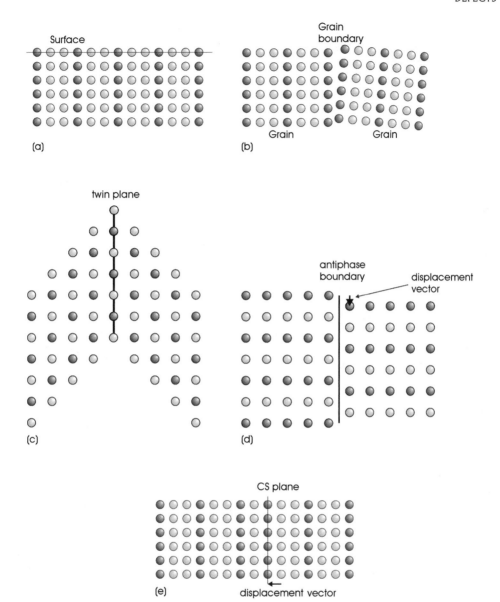

Figure 3.24   Surfaces and boundaries in a crystal: (a) the external surface; (b) grain boundaries; (c) a twin plane; (d) an antiphase boundary and (e) a crystallographic shear plane

3.4.10   Planar defects

Planar defects are two-dimensional surface defects. In many cases, the most important planar defects in a solid are the external surfaces (Figure 3.24a). These may dominate chemical reactivity, and solids designed as catalysts, or, for example, for use in gas filters, must have large surface areas in order to function. Rates of reaction during corrosion are frequently determined by the amount of surface exposed to the corrosive agent.

Grain boundaries are boundaries between crystallites in a polycrystalline array (Figure 3.24b). The energy of these boundaries, much of which is surface energy, depends on the planes of the boundaries. Annealing (i.e. heating for extended periods at temperatures high enough to allow for extensive atom diffusion) will cause rearrangement to occur, leading to lower-energy configurations. In these cases there is often a relationship between the crystallography of the material and the boundary planes. Grain boundaries are invariably weaker than the crystal matrix, and the mechanical strength of many solids is a result of the presence of grain boundaries and does not reflect the intrinsic strength of the crystallites making up the solid. In metals, grain boundaries prevent dislocation motion and reduce the ductility. Grain boundaries also increase the electrical resistance of a polycrystalline solid compared with that of a single crystal and introduce scattering and opacity into otherwise transparent solids.

Twin boundaries are boundaries in a crystal in which the crystal matrix on one side of the boundary mirrors the crystal matrix on the other (Figure 3.24c). The mirror plane, or twin plane, may not be identical to the plane along which the two mirror-related parts of the crystal join, which is called the composition plane. Twin boundaries affect mechanical, optical and electronic properties of materials in a similar way to grain boundaries.

Antiphase boundaries (APBs) are boundaries within a crystal across which one part of the crystal has been displaced with respect to the other side (Figure 3.24d). The vector describing the displacement of the two parts of the crystal is parallel to the boundary plane. When the displacement vector of the boundary is at an angle to the interface so that there is a collapse of the crystal, the boundary is called a crystallographic shear (CS) plane. In these latter boundaries, one or more planes of atoms are removed with respect to the parent structure, as shown in Figure 3.24(e), and the composition of the crystal changes. Regular arrays of crystallographic shear planes in a crystal lead to a series of new compounds, called a homologous series. For example, removal of oxygen from titanium dioxide (rutile), TiO_2, causes arrays of CS planes to form. When these are ordered, members of the homologous series Ti_nO_{2n-1}, Ti_9O_{17} and $Ti_{10}O_{19}$, for

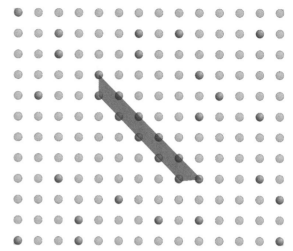

Figure 3.25 A precipitate formed by clustering of atoms in a crystal

instance, are produced. The value of n in a homologous series formula is a function of the spacing of the crystallographic shear planes. Large values of n correspond to wide spacing.

3.4.11   Volume defects: precipitates

Volume defects are regions of an impurity phase in the matrix of a material. Precipitates in a solid represent most volume defects (Figure 3.25). Precipitates form in a variety of circumstances. Solid solutions are often not stable at low temperatures, and decreasing the temperature of a solid solution slowly will frequently lead to the formation of precipitates of a new crystal structure within the matrix of the solid solution. Glasses are inherently unstable, and a glass may slowly recrystallise. In this case, precipitates of crystalline material will appear. Precipitates have important effects on the mechanical, electronic and optical properties of solids. Precipitation hardening is an important process used to strengthen metal alloys. In this technique, precipitates are induced to form in the alloy matrix by carefully controlled heat treatment. These precipitates interfere with dislocation movement and have the effect of significantly hardening the alloy.

Answers to introductory questions

What type of bonding causes the noble gases to condense to liquids?

The noble gases, in group 18 of the periodic table, have the outer electron structure ns^2np^6. This stable electron configuration does not lend itself to strong bonding and, at normal temperatures, the noble gases exist as monatomic gases. The condensation of these gases into liquids is a result of weak interactions between the outer electrons on the atoms. Fleeting instantaneous fluctuations in the electron cloud surrounding these atoms create momentary dipoles. These instantaneous charges lead to a weak attraction, which occurs between all atoms and molecules, including the noble gases. The resultant force of attraction is called the London or dispersion force and the interaction is called van der Waals bonding. The bond energy is approximately 2 kJ mol^{-1}. The strength of the interaction increases as the size of the atoms increases, so that the boiling points of the noble gases form a series: He, $-268.9\,°C$; Ne, $-246.1\,°C$; Ar, $-185.9\,°C$; Kr, $-153.2\,°C$; Xe, $-108.1\,°C$; Rn, $-61.7\,°C$.

What is the scale implied by the term 'nanostructure'?

The nanostructure of a material is its structure at an atomic scale. Nanoparticles and nanostructures generally refer to structures that are small enough that chemical and physical properties are observably different from the normal or 'classical' properties of bulk solids. The dimension at which this transformation becomes apparent depends on the phenomenon investigated. In the case of thermal effects, the boundary occurs at approximately the value of thermal energy, kT, which is about 4×10^{-21} J. In the case of optical effects, nonclassical behaviour is noted when the scale of the object illuminated is of the same size as a light wave, say about 5×10^{-7} m. For particles such as electrons, the scale is determined by the Heisenberg uncertainty principle, at about 3×10^{-8} m.

Nanostructures exist at many levels and many of these have been discovered in nature.

What line defects occur in crystals?

The line defects that occur in crystals are called dislocations. There are many different types of dislocations but they all can be thought of as combinations of two fundamental types: edge and screw dislocations. Edge dislocations consist of an extra half-plane of atoms inserted into the crystal. The dislocation line is the termination of this half-plane. The Burgers vector of an edge dislocation lies perpendicular to the dislocation line. These dislocations are instrumental in allowing metals to undergo plastic deformation. Screw dislocations are analogous to a spiral staircase. The dislocation can be formed in principle by cutting halfway through a crystal and sliding the regions on each side of the cut parallel to the cut, to create spiralling atom planes. The dislocation line is the central axis of the 'staircase'. The Burgers vector of a screw dislocation is parallel to the dislocation line. Screw dislocations play an important part in crystal growth.

Dislocations that have a Burgers vector that is less than the repeat distance of the structure are called partial dislocations. Dislocations that have a Burgers vector that is neither parallel nor perpendicular to the dislocation line are called mixed dislocations. Dislocation loops are closed, and occur within the bulk of a crystal. One way in which a dislocation loop can form is by the aggregation of vacancies on a plane in a metal crystal. As dislocation loops grow, they can change character, so that at one part of the loop the Burgers vector is parallel to the dislocation line and at another part it is parallel to it. That is, the character of the dislocation changes from pure edge to pure screw. Elsewhere on the loop, the dislocation has an intermediate character.

Further reading

W.D. Callister, 2000, *Materials Science and Engineering, an Introduction*, 5th edn, John Wiley & Sons, New York.

J. Eckert, G.D. Stucky, A.K. Cheetham, 1999, 'Partially disordered Inorganic Materials', *Materials Research Society Bulletin* **24** (May) 31.

W.D. Kingery, 1987, 'A Role for Ceramic Materials Science in Art, History and Archeology', *Journal of Materials Education* **9** 679–718.

W.D. Kingery, H.K. Bowen, D.R. Uhlmann, 1960, *Introduction to Ceramics*, Wiley-Interscience, New York.

A.M. Kraynik, 2003, 'Foam Structure: From Soap Froth to Solid Foams', *Materials Research Society Bulletin* **28** 275.

H. Megaw, 1973, *Crystal Structures*, Saunders, Philadelphia, PA.

W.F. Smith, 1993, *Foundations of Materials Science and Engineering*, 2nd edn, McGraw-Hill, New York.

R.J.D. Tilley, 1998, *Principles and Applications of Chemical Defects*, Stanley Thornes, Cheltenham, Glos.

Problems and exercises

Quick quiz

1 Hydrogen bonding is found in:
(a) Solid and liquid hydrogen

(b) Hydrocarbons

(c) Compounds containing oxygen and hydrogen

2 Van der Waals bonds are a result of:
(a) Dispersion forces

(b) Ion–dipole forces

(c) Dipole–dipole forces

3 Nonstoichiometric compounds have:
(a) A mineral composition

(b) A variable composition

(c) A metallic composition

4 A refractory oxide is:
(a) An oxide that is difficult to prepare

(b) An oxide that is rare

(c) An oxide that is resistant to high temperatures

5 Alumina is another name for:
(a) Silicon oxide

(b) Aluminium oxide

(c) Zirconium oxide

6 Sublimation is characteristic of compounds containing:
(a) Ionic bonds

(b) Metallic bonds

(c) Van der Waals bonds

7 The microstructure of a solid is at a scale of:
(a) $1–10^{-3}$ m

(b) $10^{-3}–10^{-6}$ m

(c) $10^{-7}–10^{-9}$ m

8 A polycrystalline sample is:
(a) A polymer

(b) A powder

(c) Composed of small crystals

9 A glass is:
(a) A noncrystalline inorganic solid

(b) A crystalline inorganic solid

(c) A solid containing silica

10 Sintering is a process that involves:
(a) Cross-linking polymer chains

(b) Heating powdered ceramics or metals until the particles unite

(c) Cold working a solid to make it harder

11 An important step in the fabrication of opal glass and glass ceramics is:
(a) Devitrification

(b) Sintering

(c) Quenching

12 In a quantum well, electrons are confined in:
(a) Three dimensions

(b) Two dimensions

(c) One dimension

13 Fullerenes are:
(a) Large carbon molecules with a roughly spherical form

(b) Large carbon molecules with a roughly tubular form

(c) Large carbon molecules with a chain-like form

14 Intrinsic point defects are:
   (a) Always present in a solid
   (b) A result of impurities in a solid
   (c) Grains in a solid

15 The number of different intrinsic point defects possible in a single crystal of a pure element is:
   (a) One
   (b) Two
   (c) Three

16 The number of intrinsic point defects expected to predominate in a crystal is:
   (a) One
   (b) Two
   (c) Three

17 A Schottky defect in a crystal of potassium bromide, KBr, consists of:
   (a) A potassium vacancy and a bromide interstitial
   (b) A potassium vacancy and a bromide vacancy
   (c) A potassium interstitial and a potassium vacancy

18 A Frenkel defect in a crystal of silver bromide, AgBr, consists of:
   (a) A silver vacancy and a bromide interstitial
   (b) A silver vacancy and a bromide vacancy
   (c) A silver interstitial and a silver vacancy

19 Calcia-stabilised zirconia contains:
   (a) Interstitial defects
   (b) Substitutional defects
   (c) Interpolation defects

20 The Burgers vector and dislocation line are normal to each other in:
   (a) A screw dislocation
   (b) A partial dislocation
   (c) An edge dislocation

21 The Burgers vector and dislocation line are parallel to each other in:
   (a) A screw dislocation
   (b) A partial dislocation
   (c) An edge dislocation

22 A dislocation loop has:
   (a) No Burgers vector
   (b) A Burgers vector always normal to the loop
   (c) A Burgers vector that changes along the periphery of the loop

23 Dislocations are:
   (a) Planar defects
   (b) Line defects
   (c) Point defects

24 Dislocation movement occurs by:
   (a) Motion in the slip plane
   (b) Motion perpendicular to the slip plane
   (c) Motion of the slip plane

25 A twin boundary in a solid is an example of:
   (a) A line defect
   (b) A planar defect
   (c) A volume defect

26 A precipitate is a:
   (a) Point defect
   (b) Planar defect
   (c) Volume defect

Calculations and questions

3.1 The potential energy between a pair of atoms or molecules is given approximately by a Lennard-Jones potential, which has the form

$$V(r) = Ar^{-12} - Br^{-6}$$

where $V(r)$ is the potential energy, A and B are constants, and r is the separation of the atoms or molecules. The term Ar^{-12}

represents the weak attractive London force, and the term $B r^{-6}$ represents a repulsive term arising from the interaction of the core electrons. For a pair of argon atoms, $A = -1.78 \times 10^{-134}\,J\,m^{12}$ and $B = -1.08 \times 10^{-77}\,J\,m^6$.

Plot the attractive and repulsive potential energies. [Note: not shown in answers at the end of this book.] Estimate the minimum potential energy of the pair, which can be considered the bonding energy of a molecule of argon, and the interatomic separation of this molecule.

3.2   By equating the thermal energy (kT, where k is the Boltzmann constant and T is the temperature, in kelvin), with the bonding energy, estimate the temperature at which argon atoms are likely to start to form pairs, and so form a liquid, by using the data in Question 3.1. Compare this with the boiling point of argon (see Question 3.22).

3.3   If the atoms in liquid argon are surrounded by 12 nearest neighbours on average, estimate the energy of evaporation of the liquid.

3.4   The Lennard-Jones constants for neon (Ne), are $A = -4.39 \times 10^{-136}\,J\,m^{12}$, and $B = -9.30 \times 10^{-79}\,J\,m^6$. Calculate the bonding energy and equilibrium separation of a pair of neon atoms.

3.5   The Lennard-Jones constants for helium (He), are $A = -4.91 \times 10^{-137}\,J\,m^{12}$, and $B = -4.16 \times 10^{-80}\,J\,m^6$, and for xenon (Xe), are $A = -2.54 \times 10^{-133}\,J\,m^{12}$, and $B = -5.66 \times 10^{-77}\,J\,m^6$. Using these values and those in Questions 3.1 and 3.4, estimate the Lennard-Jones constants for krypton (Kr).

3.6   Derive the relationship

$$V(r) = 4V_{min}\left[\left(\frac{r_0}{r}\right)^6 - \left(\frac{r_0}{r}\right)^{12}\right]$$

from

$$V(r) = Ar^{-12} + Br^{-6}.$$

[Note: derivation is not provided in answers at the end by this book.]

3.7   The enthalpy of formation of vacancies in pure nickel is $\Delta H = 97.3\,kJ\,mol^{-1}$. What is the fraction of sites vacant at $1100\,°C$?

3.8   The enthalpy of formation of vacancies in pure copper is $\Delta H = 86.9\,kJ\,mol^{-1}$. What is the fraction of sites vacant at $1084\,°C$?

3.9   The enthalpy of formation of vacancies in pure gold is $\Delta H = 123.5\,kJ\,mol^{-1}$. The density of gold is $19281\,kg\,m^{-3}$. What number of atom positions is vacant at $1000\,°C$?

3.10   The enthalpy of formation of vacancies in pure aluminium is $\Delta H = 72.4\,kJ\,mol^{-1}$. The density of aluminium is $2698\,kg\,m^{-3}$. What number of atom positions is vacant at $600\,°C$?

3.11   Calculate how the fraction of Schottky defects in a crystal of KCl varies with temperature if the value of ΔH_S is $244\,kJ\,mol^{-1}$. [Note: the answers at the end of this book gives values for $T = 500\,K$; $T = 1000\,K$.]

3.12   Calculate the number of Schottky defects in a crystal of KCl at $800\,K$. The unit cell of this material is cubic with a cell edge of $0.629\,nm$. Each unit cell contains 4 K^+ and 4 Cl^- ions.

3.13   The enthalpy of formation of a Frenkel defect in silver bromide, AgBr, is $1.81 \times 10^{-19}\,J$. Estimate the fraction of interstitial silver atoms owing to Frenkel defect formation in a crystal of AgBr at $300\,K$.

3.14   Silver bromide, AgBr, has a cubic unit cell with an edge of $0.576\,nm$. There are four silver atoms in the unit cell; assume that there are four interstitial positions available for silver atoms. Calculate the absolute number of interstitial defects present per cubic metre at $300\,K$.

3.15   Calculate the enthalpy of formation of Frenkel defects in sodium bromide, NaBr, using the data on the number of defects, n_F, present given in Table 3.5.

Table 3.5 Number of Frenkel defects, n_F, as a function of temperature, T

T/K	n_F/m^{-3}
200	1.428×10^2
300	7.257×10^{10}
400	1.636×10^{15}
500	6.693×10^{17}
600	3.687×10^{19}
700	6.468×10^{20}
800	5.538×10^{21}
900	2.943×10^{22}

3.16 The energy of formation of Schottky defects in a crystal of calcium oxide, CaO, is given as 9.77×10^{-19} J. Calculate the number of Schottky defects, n_S, present in CaO at 1000 °C and 2000 °C. How many vacancies are present at these temperatures? CaO has a density of 3300 kg m^{-3}.

3.17 Calculate the number of Schottky defects in a crystal of magnesium oxide, MgO, at 1500 °C. The enthalpy of formation of Schottky defects in MgO is 96.5 kJ mol^{-1}, and the density of MgO is 3580 kg m^{-3}.

3.18 Calculate the number of Frenkel defects present in a crystal of silver chloride, AgCl, at 300 K, given that the material has a cubic unit cell of edge 0.555 nm that contains four silver atoms. Assume that the interstitial atoms occupy any of eight tetrahedral sites in the unit cell. The enthalpy of formation of a Frenkel defect in AgCl is 2.69×10^{-19} J.

3.19 Calculate the number of vacancies in a crystal of nickel oxide, NiO, at 1000 °C, given that the enthalpy of formation of Schottky defects is 160 kJ mol^{-1}, and the density is 6670 kg m^{-3}.

3.20 The fraction of Schottky defects in nickel oxide, NiO, at 1000 °C is 1.25×10^{-4}. The cubic unit cell contains four nickel atoms and has a cell edge of 0.417 nm. Calculate the number of nickel vacancies present.

3.21 The number of Schottky defects in lithium fluoride, LiF, which has a cubic unit cell containing four lithium and four fluorine atoms, with a cell edge of 0.4026 nm, is 1.12×10^{22} m^{-3} at 600 °C. Calculate the activation energy for the formation of these defects.

3.22 The melting points and boiling points of the noble gases are given in Table 3.6. Explain

Table 3.6 Melting points (Mpts) and boiling points (Bpts) of the noble gases

Element	Mpt/°C	Bpt/°C
Helium	N.A.	−268.9
Neon	−248.6	−246.1
Argon	−189.4	−185.9
Krypton	−117.4	−153.2
Xenon	−111.8	−108.0
Radon	−71	−61.7

N.A. Not applicable.

these trends and why the melting and boiling points are so close. [Note: answer is not provided at the end of this book.]

3.23 A total of 9 mol% of Y_2O_3 is mixed with 91 mol% ZrO_2 and heated until a uniform product with high oxygen ion conductivity is obtained. The resulting crystal is a stabilised zirconia with the formula $Y_xZr_yO_z$. Determine x, y and z, explaining your answer.

3.24 CaO forms a solid solution with Bi_2O_3 to give a material with a high anionic conductivity. If 10 mol% CaO is reacted with 90 mol% Bi_2O_3, what is the formula of the final solid and what are the numbers and types of vacancies created?

3.25 What defects will form in the crystals made by adding small amounts of compound A to compound B:
(a) A = LiBr, B = CaBr$_2$?
(b) A = CaBr$_2$, B = LiBr?
(c) A = MgO, B = Fe$_2$O$_3$?
(d) A = MgO, B = NiO?

3.26 What defects will form in the crystals made by adding small amounts of compound A to compound B:

(a) A = $CdCl_2$, B = NaCl?

(b) A = NaCl, B = $CdCl_2$?

(c) A = Sc_2O_3, B = ZrO_2?

(d) A = ZrO_2, B = HfO_2?

3.27 Show that the number of metal atom sites in a crystal of composition MX is given by

$$N = \frac{\rho \times N_A}{\text{relative molar mass of } MX},$$

where ρ is the density of MX, and N_A is the Avogadro constant. Sodium chloride, NaCl, has a density of 2165 kg m^{-3}. The unit cell, which is cubic, with a cell of edge 0.563 nm, contains four Na and four Cl atoms. Calculate the number of atoms of sodium per cubic metre by using (a) the unit cell dimensions and (b) by using the density and molar mass. [Note: derivation is not provided in answers at the end of this book.]

4

Phase diagrams

- What is a binary phase diagram?

- What is a peritectic transformation?

- What is the difference between carbon steel and cast iron?

4.1 Phases and phase diagrams

A phase is a part of a system that is chemically uniform and has a boundary around it. Phases can be solids, liquids and gases, and, on passing from one phase to another, it is necessary to cross a phase boundary. Liquid water, water vapour and ice are the three phases found in the water system. In a mixture of water and ice it is necessary to pass a boundary on going from one phase, say ice, to the other, water.

Phase diagrams are diagrammatic representations of the phases present in a system under specified conditions, most often composition, temperature and pressure. Phase diagrams relate mostly to equilibrium conditions. If a diagram represents non-equilibrium conditions it is called an existence diagram. Phase diagrams can also give guidance on the microstructures that form on moving from one region on a phase diagram to another. This aspect is described in Chapter 8. Phase diagrams essentially display thermodynamic information, and phase diagrams can be constructed by using thermodynamic data. The conditions limiting the existence and coexistence of phases is given by the thermodynamic expression called the phase rule, originally formulated by Gibbs. Some aspects of the phase rule are described in Section S1.5.

The phases that are found on a phase diagram are made up of various combinations of components. Components are simply the chemical substances sufficient for this purpose. A component can be an element, such as carbon, or a compound, such as sodium chloride. The exact components chosen to display phase relations are the simplest that allow all phases to be described.

4.1.1 One-component (unary) systems

In a one-component, or unary, system, only one chemical component is required to describe the phase relationships, for example, iron (Fe), water (H_2O) or methane (CH_4). There are many one-component systems, including all of the pure elements and compounds. The phases that can exist in a one-component system are limited to vapour, liquid and solid. Phase diagrams for one-component systems are specified in terms of two variables, temperature, normally specified in degrees centigrade,

Understanding solids: the science of materials.   Richard J. D. Tilley
© 2004 John Wiley & Sons, Ltd   ISBNs: 0 470 85275 5 (Hbk) 0 470 85276 3 (Pbk)

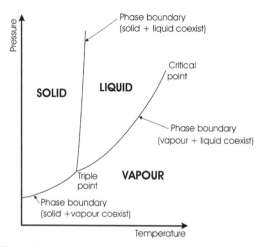

Figure 4.1 The generalised form of a one-component phase diagram

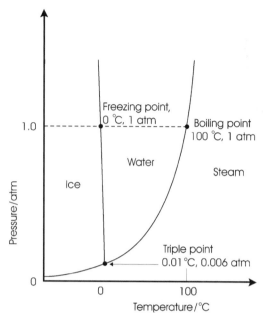

Figure 4.2 The approximate phase diagram for water; not to scale

and pressure, specified in atmospheres (1 atmosphere $= 1.01325 \times 10^5$ Pa).

A generalised one-component phase diagram is drawn in Figure 4.1. The ordinate (y-axis) specifies pressure and the abscissa (x-axis) the temperature. The areas on the diagram within which a single phase exists are labelled with the name of the phase present. The phase or phases occurring at a given temperature and pressure are read from the diagram. The areas over which single phases occur are bounded by lines called phase boundaries. On a phase boundary, two phases coexist. If the phase boundary between liquid and vapour in a one-component system is followed to higher temperature and pressures, ultimately it ends. At this point, called the critical point, at the critical temperature and the critical pressure, liquid and vapour cannot be distinguished. A gas can be converted to a liquid by applying pressure only if it is below the critical temperature. At one point, three phases coexist at equilibrium. This point is called the triple point. If there is any change at all in either the temperature or the pressure, three phases will no longer be present. The triple point is an example of an invariant point.

Perhaps the most important one-component system for life on Earth is that of water. A simplified phase diagram for water is drawn in Figure 4.2. The three phases found are ice (solid), water (liquid) and steam (vapour). The ranges of temperature and pressure over which these phases are found are read from the diagram. For example, at 1 atm pressure and 50 °C, water is the phase present. In a single-phase region, both the pressure and the temperature can be changed independently of one another without changing the phase present. For example, liquid water exists over a range of temperatures and pressures, and either property can be varied (within the limits given on the phase diagram) without changing the situation.

On the phase boundaries, two phases coexist indefinitely, ice and water, water and steam, or ice and steam. If a variable is changed, the two-phase equilibrium is generally lost. In order to preserve a two-phase equilibrium, one variable, either pressure or temperature, can be changed at will, but the other must also change, by exactly the amount specified in the phase diagram, to maintain two phases in coexistence and so to return to the phase boundary.

The critical point of water, at 374 °C and 218 atm, is the point at which water and steam become identical. The triple point is found at 0.01 °C and 0.006 atm (611 Pa). At this point and only at this

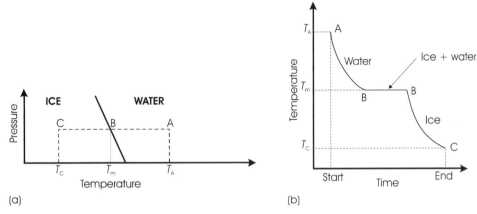

(a)   (b)

Figure 4.3   (a) A small part of the water phase diagram and (b) the cooling curve generated as a uniform sample of water cools from temperature A (liquid) to temperature C (solid; ice)

point the three phases water, ice and steam occur together. Any change in either the temperature or the pressure destroys the three-phase equilibrium.

The slopes of the phase boundaries give some information about the change of boiling and freezing points as the pressure varies. For example, the phase boundary between water and steam slopes upwards to the right. This indicates that an increase in pressure will favour liquid compared with vapour, and that the boiling point of water increases with increasing pressure. The ice–water phase boundary slopes upwards towards the left. This indicates that an increase in pressure will favour the liquid over the solid. An increase in pressure will cause the water to freeze at a lower temperature, or ice to melt. This is one reason for supposing that liquid water might be found at depths under the surface of some of the cold outer moons in the solar system.

A phase diagram can be used to explain the pattern of temperature changes observed as a substance cools (Figure 4.3). For example, a sample of water at A will cool steadily until point B, on the water–ice phase boundary, is reached. The slope of the temperature versus time plot, called a cooling curve, will change smoothly. At point B, if there is any further cooling, ice will begin to form and two phases will be present. The temperature will now remain constant, and more and more ice will form until all of the water has become ice. This follows

directly from the phase rule (see Section S1.5). Thereafter, the ice will then cool steadily again to point C, and a smooth cooling curve will be found.

This form of cooling curve will be found in any one-component system as a sample is cooled slowly through a phase boundary, so that the system is always at equilibrium. Normal rates of cooling are faster, and experimental curves often have the form shown in Figure 4.4. The property at the dip in the curve is called supercooling or undercooling. Supercooling reflects the fact that energy is needed to cause a microscopic crystal nucleus to form. In a very clean system, in which dust and other nucleating agents are absent, supercooling can be appreciable.

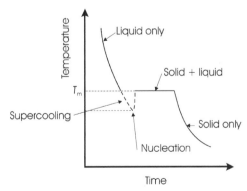

Figure 4.4   A cooling curve showing supercooling

Glasses, which form in systems in which nucleation is difficult or prevented, are called supercooled liquids, because they reach the solid state before crystallisation.

A change in slope of a cooling curve is an indication that the system is passing across a phase boundary, irrespective of the complexity of the system. Cooling curves are therefore useful in mapping out the presence of phase boundaries and in the construction of phase diagrams.

4.2   Binary phase diagrams

4.2.1   Two-component (binary) systems

Binary systems contain two components, for example, $Fe + C$, $NaNbO_3 + LiNbO_3$, $Pb + Sn$. The added component means that three variables are needed to display a phase diagram. The variables are usually chosen as temperature, pressure and composition. A binary phase diagram thus needs to be plotted as a three-axis figure (Figure 4.5a). A single phase will be represented by a volume in the diagram. Phase boundaries form two-dimensional surfaces in the representation, and three phases will coexist along a line in the phase diagram.

However, as most experiments are carried out at atmospheric pressure, a planar diagram, using temperature and composition as variables, is usually

sufficient (Figure 4.5b). These sections at a fixed pressure are called isobaric phase diagrams, although often this is not stated explicitly. In metallurgical phase diagrams, compositions are usually expressed as weight percentages (wt%). That is, the total weight is expressed as 100 g (or in kilograms and so on), and the amount of each component is given as x g and $(100 - x)$ g. In chemical work, atom percentages (at%) or mole percentages (mol%) are used. In these cases, the amounts of each component are given by x atoms (or moles) and $(100 - x)$ atoms (or moles). In these constant-pressure diagrams, the temperature is specified in degrees centigrade. A single phase occurs over an area in the figure, and phase boundaries are drawn as lines. A point in such a binary phase diagram defines the temperature and composition of the system.

In all of the binary phase diagrams discussed here, it is assumed that pressure is fixed at 1 atm. The sources of the experimental phase diagrams that have been adapted for this chapter are given in the Further Reading section.

4.2.2   Simple binary diagrams: nickel–copper

The simplest form of two-component phase diagram is exhibited by components that are very similar in chemical and physical properties. The nickel–copper

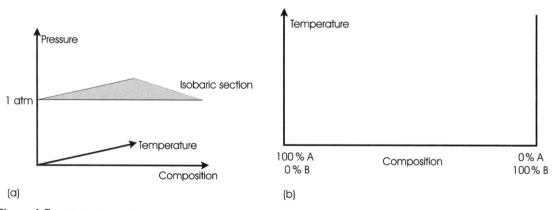

Figure 4.5   (a) A three-axis pressure–temperature–composition frame, required to display the phase relations in a binary system, and (b) isobaric sections, in which the pressure is fixed and only temperature and composition are used

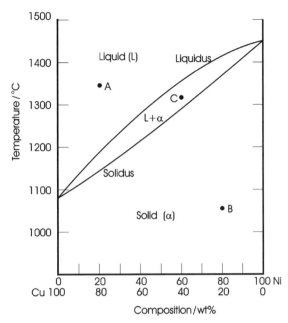

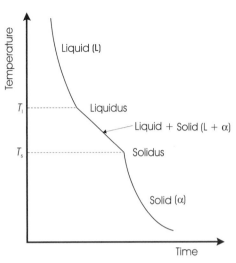

Figure 4.7 A cooling curve for a sample passing through a two-phase liquid + solid region

Figure 4.6 The nickel–copper (Ni–Cu) phase diagram at atmospheric pressure

(Ni–Cu) system provides a good example (Figure 4.6). At the top of the diagram, corresponding to the highest temperatures, one homogeneous phase, a liquid phase, occurs. In this liquid, the copper and nickel atoms are mixed together at random. In the copper-rich part of the diagram (left-hand side), the liquid can be considered as a solution of nickel in molten copper, and in the nickel-rich region (right-hand side), the liquid can be considered as a solution of copper in liquid nickel.

At the bottom of the diagram, corresponding to the lowest temperatures, another homogeneous phase, a solid, called the α phase, is found. Just as in the liquid phase, the copper and nickel atoms are distributed at random and, by analogy, such a material is called a solid solution. Because the solid solution exists from pure copper to pure nickel it is called a *complete* solid solution. (The physical and chemical factors underlying solid solution formation are described in Section 6.1.3.)

Between the liquid and solid phases, phase boundaries delineate a lens-shaped region. Within this area solid (α) and liquid (L) coexist. The lower phase boundary, between the solid and the liquid + solid region is called the solidus. The upper phase boundary, between the liquid + solid region and the liquid only region is called the liquidus.

The cooling curve of the liquid through the two-phase region shows an arrest, just as in a one-component system. However, in this case the change of slope of the cooling curve is not so pronounced. Moreover, breaks in the smooth curve occur as the sample passes both the liquidus and the solidus (Figure 4.7). Carefully interpreted cooling curves for samples spanning the whole composition range can be used to map out the positions of the solidus and liquidus.

The most obvious information found in the diagram is the phase or phases present at any temperature. Thus, suppose that a mixture of 50 g copper and 50 g nickel is heated. At 1400 °C, one phase will be present, a homogeneous liquid. At 1100 °C, one phase will also be present, a homogeneous solid, the α phase. At 1250 °C two phases are present, liquid (L) and solid (α).

The composition of any point in the diagram is simply read from the composition axis. Thus, point A in Figure 4.6 has a composition of 80 wt% copper (and thus 20 wt% nickel). Point B has a composition of 20 wt% copper (and thus 80 wt% nickel). Point C

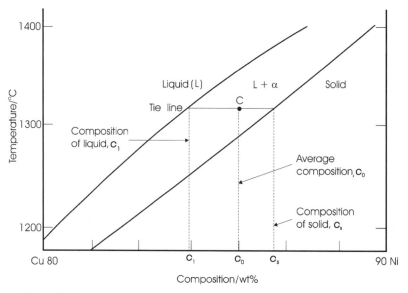

Figure 4.8   Part of the nickel–copper (Ni–Cu) phase diagram; not to scale

has an *average* composition of 40 wt% copper (and thus 60 wt% nickel). The average is quoted for point C because there are two phases present, solid and liquid. To determine the composition of each of these phases it is simply necessary to draw a line parallel to the composition axis, called a tie line. The composition of the solid phase is read from the diagram as the composition where the tie line intersects the solidus. The composition of the liquid is read from the diagram as the composition where the tie line intersects the liquidus (Figure 4.8). The composition of the liquid phase, c_l, is approximately 51 wt% copper, and that of the solid, c_s, is approximately 33 wt% copper.

The amounts of each of the phases in a two-phase region can be calculated using the lever rule (Figure 4.8). The fraction of solid phase x_s, is given by:

$$x_s = \frac{c_0 - c_l}{c_s - c_l} \quad (4.1)$$

The fraction of liquid phase, x_l, is given by:

$$x_l = \frac{c_s - c_0}{c_s - c_l} \quad (4.2)$$

In these equations, c_0 is the average composition of the sample, c_s the composition of the solid phase present in the two-phase mixture, and c_l the com-

position of the liquid phase present in the two-phase mixture. These compositions are read from the composition axis as described above. Note that if the composition scale is uniform, these amounts can simply be measured as a distance.

4.2.3   Binary systems containing a eutectic point: lead–tin

The vast majority of binary phase diagrams are more complex than the example described above. Typical of many is the diagram of the lead–tin (Pb–Sn) system (Figure 4.9).

At high temperatures, the liquid phase is a homogeneous mixture of the two atom types, lead and tin. However, the mismatch in the sizes of the lead and tin atoms prevents the formation of a complete homogeneous solid solution in the crystalline state. Instead, partial solid solutions occur at each end of the phase range, close in composition to the parent phases. The solid solutions, also referred to a terminal solid solutions, are normally called α, which is found on the lead-rich side of the diagram, and β, found on the tin-rich side. These solid solutions adopt the crystal structure of the parent phases. Thus, the α phase has the same crystal

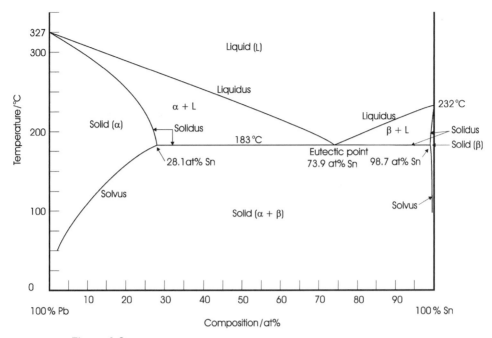

Figure 4.9   The lead–tin (Pb–Sn) phase diagram at atmospheric pressure

structure as lead, and the tin atoms are distributed at random within the crystal as defects. The β phase has the same crystal structure as that of pure tin, and the lead atoms are distributed at random within the crystal as defects. The extent of solid solution in the α phase is much greater than that in the β phase, as the smaller tin atoms are more readily accommodated in the structure of the large lead atoms than are lead atoms in the tin structure. The extent of the solid solution increases with temperature for both phases. This is because increasing temperature leads to greater atomic vibration, which allows more flexibility in the accommodation of the foreign atoms.

The overall composition of a crystal in the solid solution region is simply read from the composition axis, as in the nickel–copper system. The amount of the phase present is always 100 %. Thus point A in Figure 4.10 corresponds to a homogeneous α-phase solid of composition 15 at% tin, 85 at% lead, $Pb_{0.85}Sn_{0.15}$, at a temperature of 200 °C.

Between the partial solid solutions, in the solid, a two-phase region exists. This is a mixture of the two solid solutions, α and β, in proportions depending

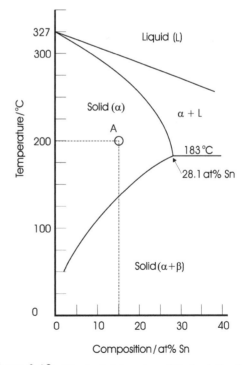

Figure 4.10   The lead-rich region of the lead–tin phase diagram

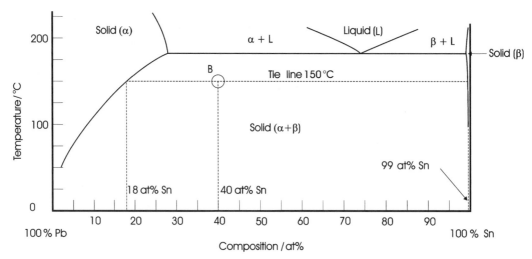

Figure 4.11   The central region of the lead–tin phase diagram

on the overall composition of the system. The phase boundaries between the solid solutions and the two-phase region are called the solvus lines. The overall composition of any sample is read from the composition axis. The compositions of the two phases present are given by the compositions at which the tie line intersects the appropriate solvus, drawn at the appropriate temperature. Thus, in Figure 4.11, the overall composition of point B is 40 at% tin, 60 at% lead. The composition of the α phase is 18 at% tin, 72 at% lead, and the composition of the β phase is 99 at% tin and 1 at% lead, at 150 °C. The amounts of the two phases are found by application of the lever rule, using the compositions just quoted. Thus:

$$\text{Amount of } \alpha \text{ phase} = \frac{99 - 40}{99 - 18} = 72.8 \,\%;$$

$$\text{Amount of } \beta \text{ phase} = \frac{40 - 18}{99 - 18} = 27.2 \,\%.$$

The liquidus has a characteristic shape, meeting the solidus at the eutectic point. The eutectic composition, which is the overall composition at which the eutectic point is found, solidifies at the lowest temperature in the system, the eutectic temperature. At this point (and only at this point, as explained below) a liquid transforms directly into a solid,

consisting of a mixture of α and β phases. The eutectic point in the lead–tin system is at 73.9 at% tin and a temperature of 183 °C.

A eutectic point, in any system, is characterised by the coexistence of three phases, one liquid and two solids. At a eutectic transformation, a liquid transforms directly into two solids on cooling:

$$\text{L(l)} \rightarrow \alpha(\text{s}) + \beta(\text{s}).$$

The eutectic point is therefore analogous to a triple point in a one-component system and, like a triple point, it is also an invariant point. The three phases can be in equilibrium only at one temperature and composition, at a fixed pressure (see Section S1.5). The reaction that occurs on cooling or heating through a eutectic point is called an invariant reaction. A cooling curve shows a horizontal break on passing through a eutectic.

Solidification over the rest of the phase diagram involves the passage through a two-phase solid + liquid region. For example, a composition on the lead-rich side of the eutectic, on passing through the liquidus, will consist of solid α phase plus liquid. A composition on the tin-rich side of the eutectic, on passing through the liquidus, will consist of solid β phase together with liquid. The composition of the two phases is obtained by

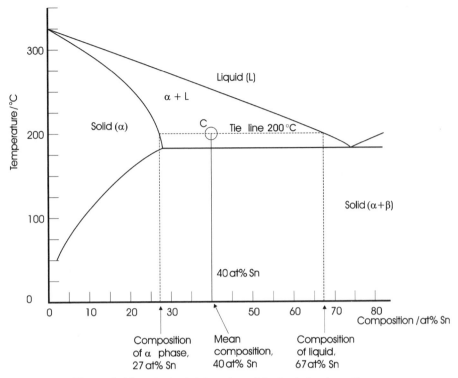

Figure 4.12   The lead-rich region of the lead–tin phase diagram

drawing a tie line at the appropriate temperature and reading from the composition axis. The amounts of the solid and liquid phases are obtained by noting the average composition and using the lever rule. For example, point C in Figure 4.12 corresponds to an overall of composition 40 at% tin. On slow cooling to 200 °C the sample will consist of liquid of composition 67 at% tin and solid α phase with a composition of 27 at% tin. The amounts of these two phases can be obtained via the lever rule, as above.

On slowly cooling a sample from a homogeneous liquid through such a two-phase region, it is seen that, as the temperature falls, the composition of the solid follows the left-hand solidus and the composition of the liquid follows the liquidus. When the eutectic temperature is reached, the remaining liquid will transform to solid with a composition equal to the eutectic composition. At this stage, the solid will contain only solid α phase and solid β

phase. Further slow cooling will not change this, but the compositions of the solid α phase and solid β phase will evolve, as the compositions at a given temperature always correspond to the compositions at the ends of the tie lines.

The microstructure of the solid will reflect this history, as discussed in Chapter 8 on reactions and transformations.

4.2.4   Solid solution formation

Not all systems have parent structures that show solid solution formation. Solid solution formation is generally absent if the crystal structures and compositions of the parent phases are quite different from each other. In general, the phase diagrams of metallic systems, drawn schematically in Figure 4.13(a), are similar in form to the lead–tin diagram.

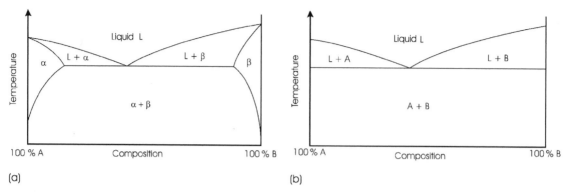

Figure 4.13   (a) A typical binary metallurgical phase diagram; (b) a typical ceramic (nonmetallic) phase diagram

The likelihood of forming a substitutional solid solution between two metals will depend on a variety of chemical and physical properties, which are discussed in Chapter 6 (see the Hume-Rothery solubility rules in Section 6.1.4). Broadly speaking, substitutional solid solution in metallic systems is more likely when:

- the crystal structures of the parent phases are the same;

- the atomic sizes of the atoms present are similar;

- the electronegativities of the metals are similar.

When oxide phase diagrams are considered, the valence is also important, as charge neutrality must be maintained in the solid solution. Thus, the similar oxides Al_2O_3, Cr_2O_3 and Fe_2O_3, all with similar sized cations and the same crystal structures and formulae (i.e. cation valence) would be expected to form extensive solid solutions, similar to that found in the nickel–copper system. Compounds containing cations with widely differing sizes, that adopt quite different crystal structures, such as B_2O_3 and Y_2O_3, would be expected to have almost no mutual solubility, even though the valence of the cations is the same. In such cases, the phase diagrams have a form similar to that in Figure 4.13(b). Compounds with different formulae often form intermediate phases, as discussed in the next section.

4.2.5   Binary systems containing intermediate compounds

Many binary mixtures react to produce a variety of compounds. In the context of phase diagrams, the reactants are known as the parent phases and the reaction products are called intermediate phases. For example, Figure 4.14 shows the phase diagram for the binary system comprising the parent ceramic phases calcium silicate ($CaSiO_3$), also called wollastonite, and calcium aluminate ($CaAl_2O_4$). Calcium aluminate and calcium silicate react to form a compound called gehlenite, $Ca_2Al_2SiO_7$, at high temperatures:

$$CaSiO_3 + CaAl_2O_4 \rightarrow Ca_2SiAl_2O_7$$

Gehlenite is the single intermediate phase in this system. None of the phases has any composition range, unlike the alloys described above, and such compounds are often called line phases.

The diagram is exactly like two of the phase diagrams in Figure 4.13(b) joined side by side. Thus, exactly the same methods as described above can be used to obtain quantitative information. In each two-phase region, the composition of the two phases present is obtained by drawing tie lines, and the relative amounts of the two phases are determined by use of the lever rule.

The phase diagram shows that gehlenite melts without any changes occurring. This feature is

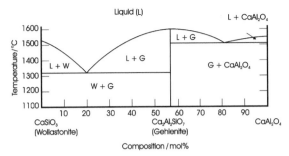

Figure 4.14 The wollastonite–calcium aluminate (Ca-SiO₃–CaAl₂O₄) phase diagram showing the intermediate phase gehlenite, $Ca_2Al_2SiO_7$

called congruent melting. It also reveals that every composition, except that of the parent phases and the intermediate phase, corresponds to a two-phase mixture. There are no extensive single-phase regions.

Not all intermediate compounds show congruent melting. Many intermediate compounds transform into a liquid at a peritectic point. On heating through a peritectic point, a solid transforms to a liquid plus another solid of a different composition:

$$\alpha(s) \rightarrow \beta(s) + L(l).$$

The solid is said to melt incongruently. As an example, Figure 4.15(a) shows a hypothetical ceramic system with an intermediate phase of composition AB_2, which melts incongruently at a peritectic point into liquid + solid B. At a peritectic point, three phases coexist. The point is thus an invariant point, and the reaction is an invariant reaction. The diagram also shows a eutectic point between pure A and the compound AB_2.

As described above, metallic systems invariably contain alloys with significant composition ranges (Figure 4.15b). Here the parent phases form terminal solid solutions α near to parent A, and β near to parent B. The intermediate alloy, labelled γ, has a composition close to AB_2. (The first intermediate phase is usually labelled γ in metallurgical phase diagrams.) The phase range of this material can be thought of as made up of 'terminal solid solutions' of A in AB_2, and B in AB_2. The γ phase melts incongruently at the peritectic point, and a eutectic point is found between the α and γ phases.

4.2.6 The iron–carbon phase diagram

The systematic understanding of the iron–carbon (Fe–C) phase diagram at the end of the 19th century and the early years of the 20th century, was at the heart of the technological advances that characterise these years. This is because steel is an alloy of carbon and iron, and knowledge of the iron–carbon phase diagram allowed metallurgists to fabricate on demand steels of known mechanical properties. Apart from this historical importance, the phase

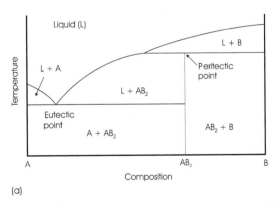

(a)

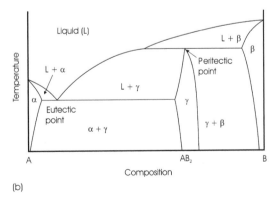

(b)

Figure 4.15 (a) A hypothetical ceramic (nonmetallic) phase diagram containing a peritectic point and (b) a hypothetical metallurgical phase diagram containing a peritectic point

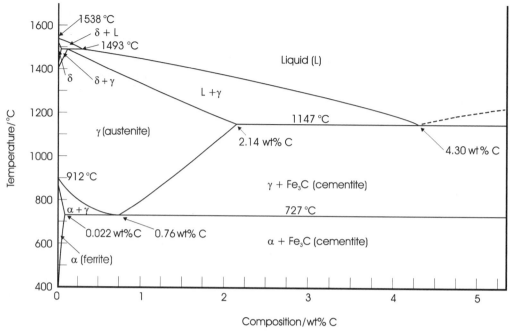

Figure 4.16   The iron-rich region of the iron–carbon existence diagram. The phase cementite (Fe$_3$C; not shown) occurs at 6.70 wt% carbon. This phase is a nonequilibrium phase and does not occur on the equilibrium phase diagram, which is between iron and graphite. The diagram is not to scale, and the α-ferrite and δ-ferrite phase fields have been expanded for clarity

diagram shows a number of interesting features in its own right.

The low-carbon region of the phase diagram is the region relevant to steel production. The version most used is that in which the composition axis is specified in wt% carbon (Figure 4.16). In fact, this is not the equilibrium phase diagram of the system. The intermediate compound cementite, Fe$_3$C, is metastable and slowly decomposes. The true equilibrium is between iron and graphite. However, cementite is an important constituent of steel, and the rate of decomposition is slow under normal circumstances, so that the figure drawn is of most use for practical steelmaking. Cementite occurs at 6.70 wt% carbon and has no appreciable composition range.

On the left-hand side of the diagram, the forms of pure iron are indicated. Pure iron has a melting point of 1538 °C. Below the melting point, pure iron adopts one of three different crystal structures

(called allotropes) at atmospheric pressure. Below a temperature of 912 °C, α-iron, which has an A2 (body-centred cubic) structure (see Section 5.3.4), is stable. This material can be made magnetic below a temperature of 768 °C. The old name for the non-magnetic form of iron, which exists between temperatures of 768 °C and 912 °C, was β-iron, but this terminology is no longer in use. Between the temperatures of 912 °C and 1394 °C the allotrope γ-iron is stable. This phase adopts the A1 (face-centred cubic) structure (see Section 5.3.3). At the highest temperatures, between 1394 °C and the melting point, 1538 °C, the stable phase is called δ-iron. The structure of δ-iron is the same as that of α-iron. It is rare that low-temperature and high-temperature polymorphs share the same crystal structure.

Between the pure iron allotropes and the intermediate phase cementite, a number of solid solution regions occur. The extent of these depends on the

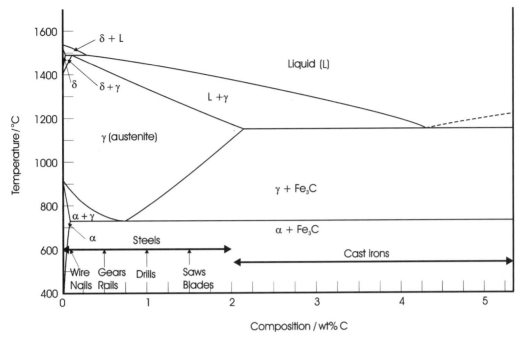

Figure 4.17 The iron–carbon existence diagram showing the approximate composition ranges for various plain carbon steels and cast irons

crystal structure of the iron. Only a small amount of carbon can enter the body-centred cubic structure of α-iron as interstitial defects between the matrix of iron atoms. At 727 °C, this amounts to 0.022 wt% carbon. This solid solution is called ferrite, or sometimes α-ferrite if it needs to be differentiated from the high-temperature solid solution. Much more interstitial carbon can be taken into solid solution in face-centred cubic γ-iron, to a maximum of 2.14 wt% carbon at 1147 °C. This material is called austenite. The amount of interstitial carbon that can enter the body-centred cubic structure of δ-iron is larger than that in α-iron, amounting to 0.09 wt% carbon at 1493 °C. This material is also called ferrite, but is generally called δ-ferrite to distinguish it from α-ferrite.

4.2.7 Steels and cast irons

The phase diagram allows us to understand the difference between plain carbon steels, that is,

alloys of carbon and iron only, and cast irons. Plain carbon steels contain less than about 2 wt% carbon (Figure 4.17), although commercial steels rarely contain much more than 1.4 wt% carbon. They can be heated to give a homogeneous austenite solid solution. In this condition, they can be worked or formed as a homogeneous material. Low-carbon steel, with less than 0.15 wt% carbon, is ductile, not very hard and is used for wires. Mild carbon steel, containing 0.15–0.25 wt% carbon, is harder and less ductile. It is used for cables, chains, nails and similar objects. Medium-carbon steel, containing 0.20–0.60 wt% carbon, is used for nails, girders, rails and structural steels. High-carbon steels, containing 0.61–1.5 wt% carbon, still well inside the austenite phase region, are used in applications requiring greater hardness, such as knives, razors, cutting tools and drill bits. Recently, ultrahigh-carbon steels, containing between 1 wt% and 2 wt% carbon, have been studied and found to accept extreme deformation before fracture. This is called superelasticity (see Section 10.1.3).

When the carbon content is greater than about 2 wt% and less than about 5 wt% carbon, the material cannot be heated to give a homogeneous solid solution. At all temperatures below the eutectic temperature of 1148 °C the solid is a mixture of austenite and cementite or ferrite and cementite (Fe$_3$C). The effect of this is that the materials are hard, brittle and resist deformation. The material can be cast into the desired shape, and is referred to as cast iron. Commercial cast irons rarely contain much more than about 4.5 wt% carbon.

Some elements, especially silicon, which occurs naturally as an impurity in iron ores, promote the transformation of cementite into graphite. White cast iron has less than 1.3 wt% silicon, and all of the carbon is present as Fe$_3$C. It is very brittle and too hard to be machined. It has to be ground to shape. Gray cast iron has between 2 wt% and 5 wt% silicon. The carbon is mainly in the form of graphite flakes embedded in the metal. It is easy to cast and machine but cracks tend to form at the graphite flakes. Black cast iron is made from white cast iron by prolonged heating at about 900 °C. During the heating, the cementite is transformed to rosette-shaped graphite nodules rather than flakes. This material offers a good compromise between machinability and strength.

4.2.8 Invariant points

There are three invariant points in the iron–carbon diagram (Figures 4.16 and 4.17). A eutectic point is found at 4.30 wt% carbon and 1148 °C. At a eutectic point, a liquid transforms to two solids on cooling:

$$L(l) \rightarrow \alpha(s) + \beta(s)$$

A similar feature, but involving only solid phases, occurs at the lowest temperature of the austenite phase field, 727 °C, and at 0.76 wt% carbon. This is also an invariant point, called a eutectoid point. At a eutectoid point, a solid transforms to two solids on cooling:

$$\gamma(s) \rightarrow \alpha(s) + \beta(s)$$

The eutectoid transformation, which takes place as austenite is cooled below the eutectoid temperature, is of great importance in steelmaking. (See Section 8.2.4.)

The phase diagram also contains a peritectic point, at the highest temperature of the austenite phase region, 1493 °C. At this peritectic point, austenite transforms into liquid and δ-ferrite on heating:

$$\alpha(s) \rightarrow \gamma(s) + L(l)$$

Similar transformations, in which only solid phases occur, are called peritectoid transformations. On heating through a peritectoid point, we find:

$$\gamma(\text{solid}) \rightarrow \alpha(\text{solid}) + \beta(\text{solid})$$

4.3 Ternary systems

4.3.1 Ternary phase diagrams

Ternary systems have three components. These require five-axis coordinate systems to display the phase relations – three for the compositions, one for pressure and one for temperature. In practice, the three components are arranged at the vertices of an equilateral triangle, and the composition of each component is indicated along the sides of the triangle. The temperature axis is drawn normal to the composition plane, to form a triangular prism (Figure 4.18a). It is necessary to draw a different prism for each pressure. Working phase diagrams are normally sections through the prism at a chosen value of temperature and a pressure of 1 atm (Figure 4.18b). The diagrams are called isothermal sections. The composition of the components, in wt%, at% or mol%, are measured along the three sides of the equilateral triangle. Each of the three faces of the prism is the binary phase diagram A–B, A–C or B–C. In Figure 4.18(a), these are simple eutectic diagrams rather like Figure 4.13(b).

The compositions in isothermal sections are most easily plotted with use of triangular graph paper (Figure 4.19). The composition of a point on one of the edges is read directly from the diagram. For

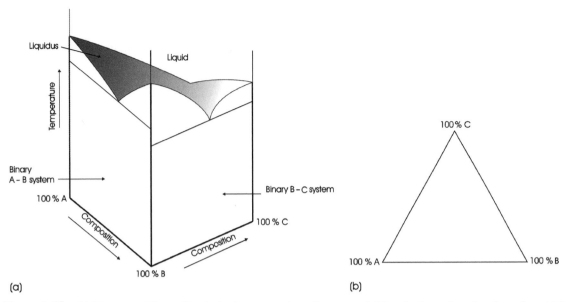

(a)   (b)

Figure 4.18   (a) The general form of an isobaric ternary phase diagram and (b) an isothermal section through part (a)

example, point D in Figure 4.19 represents a composition of 60 % A and 40 % C. The material consists of solid A + solid C. The amounts of the two phases can be determined via the lever rule, as explained below.

The composition of an internal point such as E is also found from the composition axes. The point

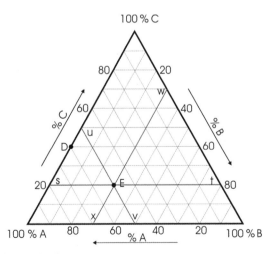

Figure 4.19   Representations of compositions on an isothermal three-component phase diagram

lies on the line s–t. This line is the locus of all points with a composition of 20 % C, and so E corresponds to 20 % C. Similarly it lies on line u–v, corresponding to the locus of all points containing 50 % A, and on line w–x, corresponding to the locus of all points containing 30 % B. The composition at E is therefore 50 % A, 30 % B and 20 % C.

Phases, with or without composition ranges, are plotted in an analogous way to those on binary phase diagrams. A simple ternary phase diagram is drawn in Figure 4.20. It represents the system WO_3–WO_2–ZrO_2 at approximately 1400 °C, which is part of the ternary W–Zr–O system. There are no phases lying within the body of the phase diagram, and all of the phases represented are to be found in the appropriate binary phase diagrams. It is seen that the area of the diagram is divided up into triangles. This is because a point in such a diagram must represent three solid phases if it lies within a triangle, two if it lies on a triangle edge, or one if it lies at a triangle vertex. Thus, point G represents a composition containing WO_2, WO_3 and ZrO_2 at 1400 °C. A composition represented by point H would consist of the phases ZrW_2O_8 and $W_{18}O_{49}$.

The amount of a phase present at a vertex of a triangle is 100 %. The amounts of the phases

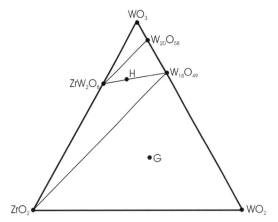

Figure 4.20 The simplified WO_3–WO_2–ZrO_2 phase diagram

present for points lying on the side of a triangle can be determined by the lever rule. For example, the amounts of the two phases present at point I in Figure 4.21 are given by:

$$\text{Amount of } W_{18}O_{49} = \frac{a}{a+b}$$
$$\text{Amount of } ZrO_2 = \frac{b}{a+b}$$

The amounts of the three phases present at point J can be determined by an extension of the lever rule. The phase triangle made up of $W_{18}O_{49}$–WO_2–WO_3

is called a tie triangle, by analogy with the tie line of binary systems. Lines are drawn connecting the point J to the vertices of the tie triangle, as shown in Figure 4.21. The amounts of the phases are then given by the lengths of these lines. For example, at point J:

$$\text{Amount of } W_{18}O_{49} = \frac{c}{c+d}$$
$$\text{Amount of } ZrO_2 = \frac{e}{e+f}$$
$$\text{Amount of } WO_2 \text{ present} = \frac{g}{g+h}$$

This method is called the triangle rule. Note that, just as in the lever rule, we assume that the composition scales are linear. If they are not, actual compositions must be used, not distances. However, this is rarely the case in ternary diagrams, which always use a linear scale for the composition axes.

The approximate phase diagram of a more complex system, MgO–Al_2O_3–SiO_2, is drawn in Figure 4.22. Ceramic bodies are rarely pure phases, and cordierite ceramics, for example, have compositions over a region around the composition of cordierite itself, as shown by the right-hand shaded area. In addition, the composition of some of these phases is

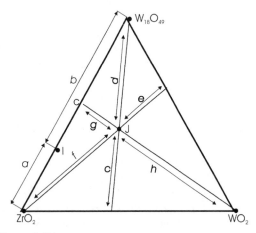

Figure 4.21 The method of determination of compositions on an isothermal phase diagram

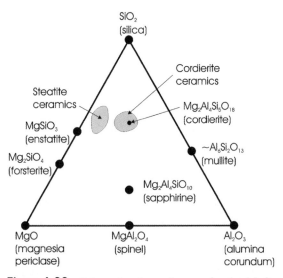

Figure 4.22 Schematic phase diagram for the MgO–Al_2O_3–SiO_2 system, which contains several important ceramic materials

open to confusion. Sapphirine, for example, is sometimes written as $Mg_2Al_2SiO_{10}$, and sometimes as $Mg_4Al_{10}Si_2O_{23}$, a fact that indicates the variable composition of many minerals. Additionally, mullite is a nonstoichiometric compound with a variable composition, and the point drawn for this compound on Figure 4.22 is simply representative of this phase. Steatite ceramics, also noted, have a composition range but are associated with the mineral talc, which does not occur in the phase diagram, as it contains hydroxyl, which is not one of the components.

Answers to introductory questions

What is a binary phase diagram?

Binary systems contain two components, for example, $Fe + C$. A binary phase diagram displays the phase relations found in a binary system. The diagram is plotted with three variables, temperature (°C), pressure (atm) and composition. In metallurgical phase diagrams, compositions are usually expressed as weight percentages (wt%). In chemical work, atom percentages (at%) or mole percentages (mol%) are used. A single phase in a binary phase diagram will be represented by a volume in the diagram. Phase boundaries form two-dimensional surfaces in the representation, and three phases will coexist along a line in the phase diagram.

As most experiments are carried out at atmospheric pressure, a planar diagram, using temperature and composition as variables, is sufficient for many purposes. These sections at a fixed pressure are called isobaric phase diagrams. In these constant-pressure diagrams a single phase occurs over an area in the figure, and phase boundaries are drawn as lines. A point in such a binary phase diagram defines the temperature and composition of the system.

What is a peritectic transformation?

On heating through a peritectic point, a solid transforms to a liquid plus another solid of a different composition:

$$\text{solid A} \rightarrow \text{solid B} + \text{liquid}.$$

The solid is said to melt incongruently. Many intermediate compounds transform into a liquid at a peritectic point. At a peritectic point, three phases coexist. The point is thus an invariant point, and the reaction is an invariant reaction.

What is the difference between carbon steel and cast iron?

The difference between plain carbon steels (i.e alloys of carbon and iron only) and cast irons can be understood from the iron–carbon phase diagram (Figure 4.17). Plain carbon steels contain less than about 2 wt% carbon, although commercial steels rarely contain much more than 1.4 wt% carbon. The significance of this is that it corresponds to the austenite phase range. This means that steels can be heated to give a homogeneous austenite solid solution. In this condition, they can be worked or formed as a homogeneous material.

When the carbon content is greater than about 2 wt% and less than about 5 wt% carbon, the material cannot be heated to give a homogeneous solid solution. At all temperatures below the eutectic temperature of 1148 °C the solid is a mixture of austenite and cementite, or ferrite and cementite. The effect of this is that the materials are hard, brittle and resist deformation. The material can be cast into the desired shape, and is referred to as cast iron. Commercial cast irons rarely contain much more than about 4.5 wt% carbon.

Further reading

American Ceramic Society, *Phase Diagrams for Ceramicists*, Volume 1 (1964) to Volume 10 (1994); this is a continuing series, with changing editors, published by the American Ceramic Society, Westerville, OH.

P.W. Brown, 1998, 'Phase Change in a One-component System', *Journal of Materials Education* **20** 43.

P.W. Brown, 1999, Interpreting Ternary Phase Diagrams', *Journal of Materials Education* **21** 203.

E.H. Ehlers, 1972, *The Interpretation of Geological Phase Diagrams*, W.H. Freeman, San Francisco, CA.

T.B. Massalski, editor in chief, 1990, *Binary Alloy Phase Diagrams*, 2nd edn, Volumes 1–3, ASM International, Materials Park, OH.

R. Powell, 1978, *Equilibrium Thermodynamics in Petrology*, Harper and Row, London.

P. Villars, A. Prince, H. Okamoto, 1995, *Ternary Phase Diagrams*, Volumes 1–10, ASM International, Materials Park, OH.

Problems and exercises

Quick quiz

1 A phase is:
   - (a) A compound in a system
   - (b) An element in a system
   - (c) A homogeneous part of a system enclosed in a boundary

2 A component is:
   - (a) A compound in a phase diagram
   - (b) An element in a phase diagram
   - (c) An essential substance used to construct a phase diagram

3 The phase diagram of a one-component system is described in terms of:
   - (a) One variable
   - (b) Two variables
   - (c) Three variables

4 How many phases coexist at a triple point in the iron phase diagram:
   - (a) One?
   - (b) Two?
   - (c) Three?

5 On a phase boundary in the sulphur system:
   - (a) One phase exists
   - (b) Two phases exist
   - (c) Three phases exist

6 At a critical point in a unary system:
   - (a) Vapour and liquid cannot be distinguished
   - (b) Solid and liquid cannot be separated
   - (c) Freezing cannot occur

7 A cooling curve changes slope when:
   - (a) An invariant point is passed
   - (b) A phase boundary is crossed
   - (c) Crystals form

8 A binary system is one in which:
   - (a) There are two components present
   - (b) There are two variables needed
   - (c) A solid and a liquid are present

9 The number of variables needed to specify phase relations in a binary system are:
   - (a) Two
   - (b) Three
   - (c) Four

10 The liquidus is a boundary that separates:
   - (a) Two different liquids
   - (b) The liquid from the solid phase
   - (c) The liquid from the solid + liquid region

11 The solidus is a boundary that separates:
   - (a) A solid from a liquid
   - (b) A solid from a solid + liquid region
   - (c) Two different solid phases

12 A tie line is drawn:
   - (a) Parallel to the pressure axis
   - (b) Parallel to the temperature axis
   - (c) Parallel to the composition axis

13 A binary phase diagram in which the pressure is always constant is:
   - (a) An isobaric section
   - (b) An isostatic section
   - (c) An isothermal section

14 The solvus line on a phase diagram separates:
   - (a) A solid from a solid + liquid region
   - (b) A solid solution from a two-solid region
   - (c) A solid from a solid solution region

15 A eutectic point on a binary phase diagram is:
   - (a) An invariant point

(b) A triple point

(c) A critical point

16 On cooling a homogeneous liquid sample through a eutectic point:

(a) A liquid plus a solid forms

(b) Two solid phases form

(c) A homogeneous solid forms

17 A line phase on a phase diagram is a phase that:

(a) Exists along a phase boundary

(b) Has a composition range along a tie line

(c) Has no apparent composition range

18 Solid phases that melt congruently melt to form:

(a) A liquid of the same composition and a solid of a different composition

(b) A solid of the same composition and a liquid of a different composition

(c) A liquid of the same composition

19 Heating a solid through a peritectic point produces:

(a) A solid with a different composition, and a liquid

(b) A solid with the same composition, and a liquid

(c) A liquid with a different composition

20 Pure iron has:

(a) Two allotropes

(b) Three allotropes

(c) Four allotropes

21 Steel is an alloy of iron and carbon in which the carbon occupies:

(a) Substitutional sites

(b) Interstitial sites

(c) Vacancies

22 Ferrite is:

(a) An allotrope of iron

(b) An intermediate phase

(c) An iron–carbon alloy

23 Austenite has:

(a) No appreciable composition range

(b) A narrow composition range compared with ferrite

(c) A wide composition range compared with ferrite

24 Austenite has the same crystal structure as:

(a) α-iron

(b) β-iron

(c) γ-iron

25 Steel is an alloy of iron that has a composition less than:

(a) The maximum austenite composition

(b) The maximum ferrite composition

(c) The maximum cementite composition

26 Cast irons generally have compositions of iron and:

(a) Exactly 2 wt% carbon

(b) More than 2 wt% carbon

(c) Less that 2 wt% carbon

27 On cooling a homogeneous solid phase through a eutectoid point it forms:

(a) Two solid phases

(b) A solid and a liquid phase

(c) A homogeneous single solid phase

28 On heating a homogeneous solid through a peritectoid point it forms:

(a) A solid and a liquid phase

(b) A homogeneous single solid phase

(c) Two solid phases

29 A ternary system is one in which there are:

(a) Three components

(b) Three variables

(c) Three phases

30 To represent all possible phase relations, a ternary system needs:

(a) Five axes

(b) Four axes

(c) Three axes

31 A triangular representation of the phase relations in a ternary system is:
    (a) An isothermal section
    (b) An isobaric section
    (c) An isobaric and isothermal section

32 Within a tie triangle on a ternary phase diagram, a point will correspond to:
    (a) Two solid phases
    (b) Three solid phases
    (c) Four solid phases

Calculations and questions

4.1 A copper–nickel alloy is made up with 28 g copper in 100 g alloy. What is the atom percentage of nickel in the alloy?

4.2 A solid solution of aluminium oxide (Al_2O_3) and chromium oxide (Cr_2O_3) has a composition $Al_{0.70}Cr_{1.30}O_3$. What mass of Cr_2O_3 needs to be weighed out to prepare 100 g of sample?

4.3 A copper–zinc alloy is made up with 35 g copper in 100 g alloy (35 wt%). What is the atom percentage of zinc in the alloy?

4.4 A solder contains 50 wt% tin (Sn) and 50 wt% lead (Pb). What are the atom percentages of tin and lead in the solder?

4.5 An intermetallic compound in the titanium–aluminium (Ti–Al) system is found at 78 wt% Ti and 22 wt% Al. What is the approximate formula of the compound?

4.6 An equilibrium sample of a copper–nickel (Cu–Ni) alloy with a composition of 65 wt% nickel is prepared. With reference to the Cu–Ni phase diagram shown in Figures 4.6 and 4.8:
    (a) What is the atom percentage of copper present in the alloy?
    (b) On heating the alloy from room temperature, at what temperature does liquid first appear?
    (c) What is the composition of the liquid?

    (d) At what temperature does the solid disappear?
    (e) What is the composition of the final solid?

4.7 For the sample in the previous question, the alloy is held at a temperature of 1340 °C:
    (a) What phase(s) are present?
    (b) How much solid is present, if any?
    (c) How much liquid is present, if any?

4.8 An equilibrium sample of a copper–nickel (Cu–Ni) alloy with a composition of 37 wt% nickel is held at a temperature of 1250°C:
    (a) What are the amounts of solid and liquid present?
    (b) If the density of the solid is $8.96 \times 10^{-3}\ kg\,m^{-3}$, and the density of the liquid is 90 % of that of the solid, calculate the volume percentage of the solid and liquid present

4.9 The phase diagram of the aluminium oxide (Al_2O_3) and chromium oxide (Cr_2O_3) system is given in Figure 4.23. An equilibrium

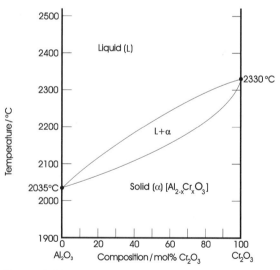

Figure 4.23 Phase diagram of the aluminium oxide (Al_2O_3) and chromium oxide (Cr_2O_3) system, for Questions 4.9–4.12

sample of composition 50 mol% Cr_2O_3 is prepared.

(a) What is the weight percentage of Al_2O_3 present?

(b) The sample is held at 2200 °C. What is the composition of the solid phase present?

(c) How much liquid phase is present?

(d) At what temperature will the last of the solid disappear?

(e) What will the composition of this solid be?

4.10 With reference to the phase diagram of the aluminium oxide (Al_2O_3) and chromium oxide (Cr_2O_3) system given in Figure 4.23, a sample of composition 30 mol% Cr_2O_3 is held at 2300 °C and then slowly cooled.

(a) At what temperature does solid first appear?

(b) What is the composition of the solid

(c) At what temperature does the liquid finally disappear?

(d) What is the composition of the last drop of liquid?

4.11 With reference to the phase diagram of the aluminium oxide (Al_2O_3) and chromium oxide (Cr_2O_3) system given in Figure 4.23, a sample of composition 30 mol% Cr_2O_3 is held at 2180 °C.

(a) What is the composition of any solid phase present?

(b) What is the composition of any liquid phase present?

(c) How much of each phase is present?

4.12 With reference to the phase diagram of the aluminium oxide (Al_2O_3) and chromium oxide (Cr_2O_3) system given in Figure 4.23, a sample of composition 30 mol% Cr_2O_3 is held at 2100 °C.

(a) What is the composition of any solid phase present?

(b) What is the composition of any liquid phase present?

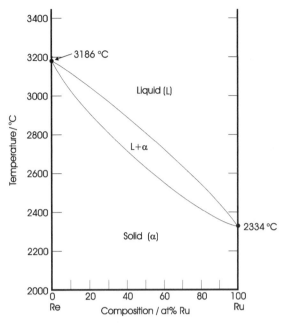

Figure 4.24 Phase diagram of the ruthenium–rhenium (Ru–Re) system, for Questions 4.13 and 4.14

(c) How much of each phase is present?

4.13 The phase diagram of the ruthenium–rhenium (Ru–Re) system is given in Figure 4.24.

(a) What are the melting points of pure Re and pure Ru?

(b) A sample of composition 70 at% Ru is made up. What weights have to be added to prepare 100 g of sample?

(c) This alloy is held at a temperature of 2200 °C. What phases are present and what are their compositions?

(d) The alloy is held at 3000 °C. What phases are present and what are their compositions?

4.14 The phase diagram of the ruthenium–rhenium (Ru–Re) system is given in Figure 4.24. A sample of composition 60 at% Ru is made up and held at 2700 °C.

(a) What is the composition of any solid phase present?

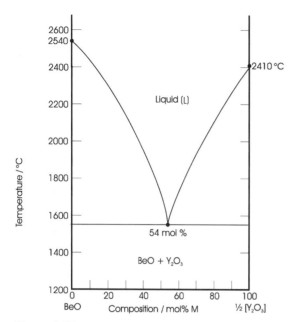

Figure 4.25 Phase diagram of the BeO–Y$_2$O$_3$ system, for Questions 4.15–4.17

(b) What is the composition of any liquid phase present?

(c) How much of each phase is present?

4.15 The phase diagram of the BeO–Y$_2$O$_3$ system is shown in Figure 4.25. Explain why this figure differs from that of the lead–tin system, Figure 4.9, and comment on the composition axis chosen. [Note: answer is not provided at the end of this book.]

4.16 With respect to the phase diagram for BeO–Y$_2$O$_3$, shown in Figure 4.25, a composition is made up with 80 mol% BeO and held at 2000 °C.
(a) What weights of the components are needed to make 100 g of sample?
(b) How much solid is present?
(c) What is the composition of the solid?
(d) How much liquid is present?
(e) What is the composition of the liquid?

4.17 With respect to the sample in Question 4.16, the material is cooled to 1400 °C.

(a) What phases are present?
(b) What are the compositions of the phases?
(c) What are the proportions of each phase present?

4.18 With respect to the lead–tin (Pb–Sn) phase diagram shown in Figures 4.9–4.12, a sample is made up with 40 at% tin.
(a) What weights of lead and tin are needed to make 100 g solid?
(b) What phase(s) is (are) present when the sample is held at 300 °C?
(c) What is (are) the composition(s) of the phase(s)?
(d) How much of each phase is present at 300 °C?

4.19 With respect to the lead–tin (Pb–Sn) phase diagram shown in Figures 4.9–4.12, the sample made up with 40 at% tin is cooled slowly to 250 °C.
(a) What phases are present at 250 °C
(b) What is the composition of each phase?
(c) How much of each phase is present?

4.20 With respect to the lead–tin (Pb–Sn) phase diagram shown in Figures 4.9–4.12, the sample made up with 40 at% tin is cooled further to 100 °C.
(a) What phases are present at 100 °C?
(b) What is the composition of each phase?
(c) How much of each phase is present?

4.21 With respect to the iron–carbon (Fe–C) diagram shown in Figures 4.16 and 4.17, an alloy with a composition of 1.5 wt% carbon is homogenised by heating for a long period at 1000 °C.
(a) What phase(s) is (are) present?
(b) How much of each phase is present?
(c) What is (are) the composition(s) of the phase(s)?

4.22 With respect to the iron–carbon (Fe–C) diagram shown in Figures 4.16 and 4.17, the alloy with a composition of 1.5 wt% carbon is homogenised by heating for a long period at

1000 °C and is subsequently cooled slowly to 800 °C.

(a) What phase(s) is (are) present?

(b) How much of each phase is present?

(c) What is (are) the composition(s) of the phase(s)?

4.23 With respect to the iron–carbon (Fe–C) diagram shown in Figures 4.16 and 4.17, an alloy with a composition of 5 at% carbon is homogenised by heating for a long period at 1350 °C.

(a) What is the composition of the liquid phase present?

(b) How much of the liquid phase is present?

(c) What is the composition of the solid phase present?

(d) How much solid phase is present?

4.24 With respect to the phase diagram of the WO_3–WO_2–ZrO_2 system shown in Figure 4.20, a sample is made up of an equimolar mixture of WO_3, WO_2 and ZrO_2, (1:1:1) and heated at 1100 °C to equilibrium.

(a) What phases are present?

(b) How much of each phase is present?

4.25 With respect to the phase diagram of the WO_3–WO_2–ZrO_2 system shown in Figure 4.20, a sample is made up of a mixture of 80 mol% WO_3, 10 mol% WO_2 and 10 mol% ZrO_2 (8:1:1) and heated at 1100 °C to equilibrium.

(a) What phases are present?

(b) How much of each phase is present?

4.26 According to Figure 4.22, what phases are present in steatite ceramics and cordierite ceramics? [Note: answer is not provided at the end of this book.]

5

Crystallography and crystal structures

- How does a lattice differ from a structure?

- What is a unit cell?

- What is meant by a (1 0 0) plane?

Crystallography describes the ways in which atoms and molecules are arranged in crystals. Many chemical and physical properties depend on crystal structure, and an understanding of crystallography is essential if the properties of materials are to be understood.

In earlier centuries, crystallography developed via two independent routes. The first of these was observational. It was long supposed that the regular and beautiful shapes of mineral crystals were an expression of internal order, and this order was described by the classification of external shapes, the *habit* of crystals. All crystals could be classified into one of 32 crystal classes, belonging to one of seven crystal systems. The regularity of crystals, together with the observation that many crystals could be cleaved into smaller and smaller units, gave rise to the idea that all crystals were built up from elementary volumes, that came to be called unit cells, with a shape defined by the crystal system. A second route, the mathematical descrip-

tion of the arrangement of arbitrary objects in space, was developed in the latter years of the 19th century. Both of these play a part in helping us to understand crystals and their properties. The two approaches were unified with the exploitation of X-ray and other diffraction methods, which are now used to determine crystal structures on a routine basis.

5.1 Crystallography

5.1.1 Crystal lattices

Crystal structures and crystal lattices are different, although these terms are frequently (and incorrectly) used as synonyms. A crystal structure is built of atoms. A crystal lattice is an infinite pattern of points, each of which must have the same surroundings in the same orientation. A lattice is a mathematical concept. If any lattice point is chosen as the origin, the position of any other lattice point is defined by

$$P(u\,v\,w) = u\boldsymbol{a} + v\boldsymbol{b} + w\boldsymbol{c}$$

where $\boldsymbol{a}$, $\boldsymbol{b}$ and $\boldsymbol{c}$ are vectors, called basis vectors, and u, v and w are positive or negative integers. Clearly, there are any number of ways of choosing $\boldsymbol{a}$, $\boldsymbol{b}$ and $\boldsymbol{c}$, and crystallographic convention is to choose vectors that are small and reveal the underlying symmetry of the lattice. The parallelepiped

Understanding solids: the science of materials. Richard J. D. Tilley
© 2004 John Wiley & Sons, Ltd ISBNs: 0 470 85275 5 (Hbk) 0 470 85276 3 (Pbk)

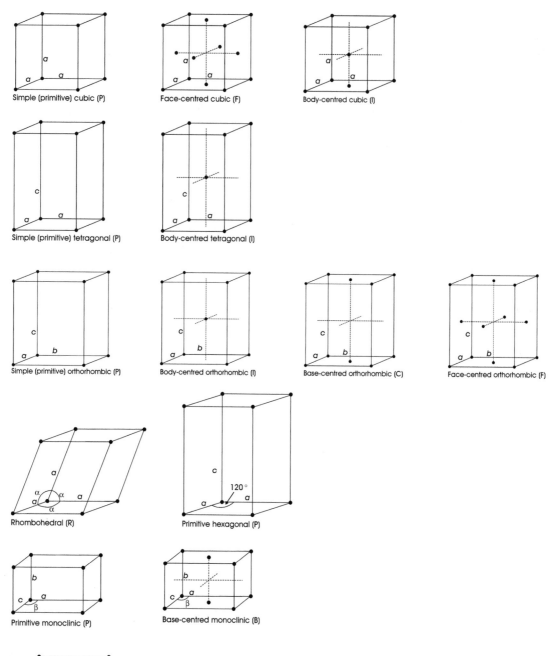

Figure 5.1   The 14 Bravais lattices. Note that the lattice points are *not* atoms. The axes associated with the lattices are shown, and are described in Table 5.1. The monoclinic lattices have been drawn with the *b* axis vertical, to emphasise that it is normal to the plane containing the *a* and *c* axes

formed by the three basis vectors a, b and c defines the unit cell of the lattice, with edges of length a_0, b_0, and c_0. The numerical values of the unit cell edges and the angles between them are collectively called the lattice parameters or unit cell parameters. The unit cell is not unique and is chosen for convenience and to reveal the underlying symmetry of the crystal.

There are only 14 possible three-dimensional lattices, called Bravais lattices (Figure 5.1). Bravais lattices are sometimes called direct lattices. The smallest unit cell possible for any of the lattices, the one that contains just one lattice point, is called the primitive unit cell. A primitive unit cell, usually drawn with a lattice point at each corner, is labelled P. All other lattice unit cells contain more than one lattice point. A unit cell with a lattice point at each corner and one at the centre of the unit cell (thus containing two lattice points in total) is called a body-centred unit cell, and labelled I. A unit cell with a lattice point in the middle of each face, thus containing four lattice points, is called a face-centred unit cell, and labelled F. A unit cell that has just one of the faces of the unit cell centred, thus containing two lattice points, is labelled A-face-centred if the faces cut the a axis, B-face-centred if the faces cut the b axis and C-face-centred if the faces cut the c axis.

The external form of crystals, the internal crystal structures and the three-dimensional Bravais lattices need to be defined unambiguously. For this purpose, a set of axes is used, defined by the vectors a, b and c, with lengths a_0, b_0, and c_0. These axes are chosen to form a right-handed set and, conventionally, the axes are drawn so that the a axis points out from the page, the b axis points to the right and the c axis is vertical (Figure 5.1). The angles between the axes are chosen to be equal to or greater than $90°$ whenever possible. These are labelled α, β and γ, where α lies between b and c, β lies between a and c, and γ lies between a and b. Just seven different arrangements of axes are needed in order to specify all three-dimensional structures and lattices (Table 5.1), these being identical to the crystal systems derived by studies of the morphology of crystals.

The unique axis in the monoclinic unit cell is the b axis. It would be better to choose the c axis, as

Table 5.1 The crystal systems

System	Unit cell parameters
Cubic (isometric)	$a = b = c$; $\alpha = 90°$, $\beta = 90°$, $\gamma = 90°$
Tetragonal	$a = b \neq c$; $\alpha = 90°$, $\beta = 90°$, $\gamma = 90°$
Orthorhombic	$a \neq b \neq c$; $\alpha = 90°$, $\beta = 90°$, $\gamma = 90°$
Monoclinic	$a \neq b \neq c$; $\alpha = 90°$, $\beta \neq 90°$, $\gamma = 90°$
Triclinic	$a \neq b \neq c$; $\alpha \neq 90°$, $\beta \neq 90°$, $\gamma \neq 90°$
Hexagonal	$a = b \neq c$; $\alpha = 90°$, $\beta = 90°$, $\gamma = 120°$
Rhombohedral	$a = b = c$; $\alpha = \beta = \gamma \neq 90°$ $a' = b' \neq c'$; $\alpha' = 90°$, $\beta' = 90°$, $\gamma' = 120°$

then the unique axis in tetragonal, hexagonal and orthorhombic crystals with a polar axis (see below) would all have the same designation. However, convention is now fixed and monoclinic unit cells are usually described with the b axis as unique. Rhombohedral unit cells are often specified in terms of a different (bigger) hexagonal unit cell.

5.1.2 Crystal structures and crystal systems

All crystal structures can be built up from the Bravais lattices by placing an atom or a group of atoms at each lattice point. The crystal structure of a simple metal and that of a complex protein may both be described in terms of the same lattice, but whereas the number of atoms allocated to each lattice point is often just one for a simple metallic crystal it may easily be thousands for a protein crystal. The number of atoms associated with each lattice point is called the *motif*, the *lattice complex* or the *basis*. The motif is a fragment of structure that is just sufficient, when repeated at each of the lattice points, to construct the whole of the crystal. A crystal structure is built up from a lattice plus a motif.

The axes used to describe the structure are the same as those used for the direct lattices, corresponding to the basis vectors lying along the unit cell edges. The position of an atom within the unit cell is given as x, y, z, where the units are a_0 in a direction along the a axis, b_0 along the b axis, and c_0 along the c axis. An atom with the coordinates

$(\frac{1}{2}, \frac{1}{2}, \frac{1}{2})$ is at the body centre of the unit cell, that is $\frac{1}{2} a_0$ along the a axis, $\frac{1}{2} b_0$ along the b axis and $\frac{1}{2} c_0$ along the c axis.

Different compounds which crystallise with the same crystal structure, for example the two alums, $NaAl(SO_4)_2.12H_2O$ and $NaFe(SO_4)_2.12H_2O$, are said to be isomorphous[1] or isostructural. As noted in Section 3.1.3, sometimes the crystal structure of a compound will change with temperature and with applied pressure. This is called polymorphism. Polymorphs of elements are known as allotropes. Graphite and diamond are two allotropes of carbon, formed at different temperatures and pressures.

Because repetition of the unit cell must reproduce the crystal, the atomic contents of the unit cell must also be representative of the overall composition of the material. It is possible to determine the density of a compound by dividing the total mass of the atoms in the unit cell by the unit cell volume, described in more in Section 5.3.2.

5.1.3   Symmetry and crystal classes

The shape and symmetry of crystals attracted the attention of early crystallographers and, until the internal structure of crystals could be determined, was an important method of classification of minerals. The external shape, or habit, of a crystal is described as isometric (like a cube), prismatic (like a prism, often with six sides), tabular (like a rectangular tablet or thick plate), lathy (lath-like) or acicular (needle-like). An examination of the disposition of crystal faces, which reflected the symmetry of the crystal, led to an appreciation that all crystals could not only be allocated to one of the seven crystal systems but also to one of 32 crystal classes.

The crystal class mirrors the internal symmetry of the crystal. The internal symmetry of any isolated object, including a crystal, can be described by a combination of axes of rotation and mirror planes,

all of which will be found to intersect in a point within the object. There are just 32 combinations of these symmetry elements, each of which is a *crystallographic point group*. The point group is equivalent to the crystal class of a crystal, and the terms are often used interchangeably.

Point groups are used extensively in crystal physics to relate external and internal symmetry to the physical properties that can be observed. For example, the piezoelectric effect (see Section 11.2.2) is found only in crystals that lack a centre of symmetry. A unit cell with a centre of symmetry at a position $(0, 0, 0)$ is such that any atom at a position (x, y, z) is accompanied by a similar atom at $(-x, -y, -z)$. Crystallographic notation writes negative signs above the symbol to which they apply, thus: $(\bar{x}, \bar{y}, \bar{z})$. Crystals that do not possess a centre of symmetry have one or more polar directions and polar axes. A polar axis is one that is not related by symmetry to any other direction in the crystal. That is, if an atom occurs at $+z$ on a polar c axis, there is no similar atom at $-z$. This can be illustrated with reference to an SiO_4 tetrahedron, a group that lacks a centre of symmetry (Figure 5.2). The oxygen atom at $+z$ on the c axis is not paired with a similar oxygen atom at $-z$.

The symmetry of the internal structure of a crystal is obtained by combining the point group symmetry with the symmetry of the lattice. It is found that 230 different patterns arise. These are called *space groups*. Every crystal structure can be assigned to

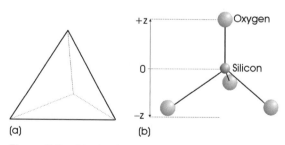

Figure 5.2   (a) An ideal tetrahedron. All faces are composed of equilateral triangles. (b) An ideal tetrahedral (SiO_4) unit. A silicon atom lies at the tetrahedron centre, and four equispaced oxygen atoms are arranged at the tetrahedral vertices. An oxygen atom at $+z$ does not have a counterpart at $-z$, and the unit is not centrosymmetric

[1]This description originally applied to the same external form of the crystals rather than the internal arrangement of the atoms.

a space group. The space group, because it is concerned with the symmetry of the crystal structure, places severe restrictions on the placing of atoms within the unit cell. The determination of a crystal structure generally starts with the determination of the correct space group for the sample. Further information on the importance of symmetry in crystal structure analysis will be found in the Further Reading section at the end of this chapter.

5.1.4 Crystal planes and Miller indices

The facets of a well-formed crystal or internal planes through a crystal structure are specified in terms of Miller Indices. These indices, h, k and l, written in round brackets $(h\,k\,l)$, represent not just one plane but the set of all parallel planes $(h\,k\,l)$. The values of h, k and l are the fractions of a unit cell edge, a_0, b_0 and c_0, respectively, intersected by this set of planes. A plane that lies parallel to a cell edge, and so never cuts it, is given the index 0 (zero). Some examples of the Miller indices of important crystallographic planes follow.

A plane that passes across the end of the unit cell cutting the a axis and parallel to the b and c-axes of the unit cell has Miller indices $(1\,0\,0)$ (Figure 5.3a). The indices indicate that the plane cuts the cell edge running along the a axis at a position $1\,a_0$ and does not cut the cell edges parallel to the b or c axes at all. A plane parallel to this that cuts the a cell edge in half, at $a_0/2$, has indices $(2\,0\,0)$ (Figure 5.3b). Similarly, parallel planes cutting the a cell edge at $a_0/3$ would have Miller indices of $(3\,0\,0)$ (Figure 5.3c). Remember that $(1\,0\,0)$ represents all of the set of other identical planes as well. There is no need to specify a plane $(100,00)$, it is simply $(1\,0\,0)$. Any general plane parallel to $(1\,0\,0)$ is written $(h\,0\,0)$.

A general plane parallel to the a and c axes, perpendicular to the b axis, and so only cutting the b cell edge, has indices $(0\,k\,0)$ (Figure 5.3d), and a general plane parallel to the a and b axes and perpendicular to the c axis, and so cutting the c cell edge, has indices $(0\,0\,l)$ (Figure 5.3e).

Planes that cut two edges and parallel to a third are described by indices $(h\,k\,0)$, $(0\,k\,l)$ or $(h\,0\,l)$. Figures 5.4(a)–(c) show, respectively: (1 1 0), intersecting

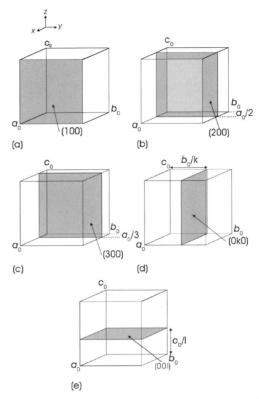

Figure 5.3 Miller indices of crystal planes. (a). $(1\,0\,0)$; (b). $(2\,0\,0)$; (c). $(3\,0\,0)$; (d). $(0\,k\,0)$; (e). $(0\,l\,0)$

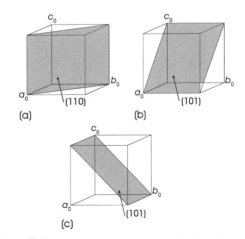

Figure 5.4 Miller indices of crystal planes in cubic crystals: (a) (1 1 0), (b) (1 0 1) and (c) (0 1 1)

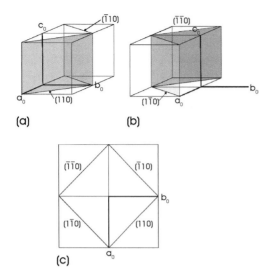

Figure 5.5 Miller indices of crystal planes in a cubic crystal: (a) (1 1 0) and ($\bar{1}$ 1 0), (b) (1 $\bar{1}$ 0) and ($\bar{1}$ $\bar{1}$ 0) and (c) projection down the c axis, showing all four equivalent {1 1 0} planes

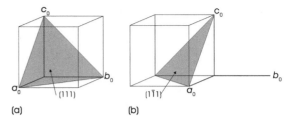

Figure 5.6 Miller indices in cubic crystals: (a) (1 1 1) and (b) (1 $\bar{1}$ 1)

the cell edges in $1\,a_0$ and $1\,b_0$ and parallel to c; (1 0 1), intersecting the cell edges in $1\,a_0$ and $1\,c_0$ and parallel to b; and (0 1 1), intersecting the cell edges in $1\,b_0$ and $1\,c_0$ and parallel to a.

Negative intersections are written with a negative sign over the index and are pronounced 'bar h', 'bar k' and 'bar l'. For example, there are four planes related to the (1 1 0) plane. As well as the (1 1 0) plane, a similar plane also cuts the b axis in $1\,b_0$, but the a axis is cut in a negative direction, at $-a_0$ (Figure 5.5a). Two other related planes, one of which cuts the b axis at $-b_0$, and so has Miller indices (1 $\bar{1}$ 0), pronounced '(one, bar one, zero)' and the other, with Miller indices ($\bar{1}$ $\bar{1}$ 0), are drawn in Figure 5.5(b). Because the Miller indices ($h\,k\,l$) refer to a set of planes, ($\bar{1}$ $\bar{1}$ 0) is equivalent to (1 1 0), as the position of the axes is arbitrary. Similarly, the plane with Miller indices (1 $\bar{1}$ 0) is equivalent to ($\bar{1}$ 1 0) (Figure 5.5c).

This notation is readily extended to cases where a plane cuts all three unit cell edges (Figure 5.6). An easy way to determine Miller indices is given in Section S1.6.

In crystals of high symmetry there are often several sets of ($h\,k\,l$) planes that are identical, from the point of view of both symmetry and of the atoms lying in the plane. For example, in a cubic crystal, the (1 0 0), (0 1 0) and (0 0 1) planes are identical in every way. Similarly, in a tetragonal crystal, (1 1 0) and ($\bar{1}$ 1 0) planes are identical. Curly brackets, {$h\,k\,l$}, designate these related planes. Thus, in the cubic system, the symbol {1 0 0} represents the three sets of planes (1 0 0), (0 1 0) and (0 0 1). Similarly, in the cubic system, {1 1 0} represents the six sets of planes (1 1 0), (1 0 1), (0 1 1), ($\bar{1}$ 1 0), ($\bar{1}$ 0 1), and (0 $\bar{1}$ 1), and the symbol {1 1 1} represents the four sets (1 1 1), (1 1 $\bar{1}$), (1 $\bar{1}$ 1) and ($\bar{1}$ 1 1).

5.1.5 Hexagonal crystals and Miller–Bravais indices

The Miller indices of planes parallel to the c axis in crystals with a hexagonal unit cell, such as magnesium, can be ambiguous (Figure 5.7). In this representation, the c cell edge is normal to the plane of the page. Three sets of planes, imagined to be perpendicular to the plane of the figure, are shown. From the procedure just outlined, the sets have the following Miller indices: A, (1 1 0); B, (1 $\bar{2}$ 0); and C, ($\bar{2}$ 1 0). Although these seem to refer to different types of plane, clearly they are identical from the point of view of atomic constitution. In order to eliminate this confusion, four indices, ($h\,k\,i\,l$), are often used to specify planes in a hexagonal crystal. These are called Miller–Bravais indices and are

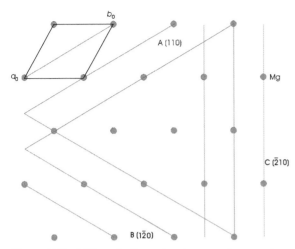

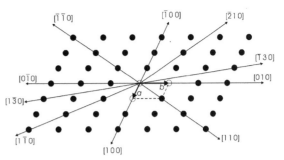

Figure 5.8 Directions in a lattice. The directions do not take into account the length of the vectors, and the indices are given by the smallest integers that lie along the vector direction

Figure 5.7 Miller indices in hexagonal crystals. Although the indices appear to represent different types of plane, in fact they all are identical

used only in the hexagonal system. The index i is given by:

$$h + k + i = 0,$$

or

$$i = -(h + k)$$

In reality this third index is not needed. However, it does help to bring out the relationship between the planes. Using four indices, the planes are: A, $(11\bar{2}0)$; B, $(1\bar{2}10)$; and C, $(\bar{2}110)$. Because it is a redundant index, the value of i is sometimes replaced by a dot, to give indices $(hk.l)$. This nomenclature emphasises that the hexagonal system is under discussion without actually including a value for i.

5.1.6 Directions

The response of a crystal to an external stimulus, such as a tensile stress, electric field and so on, is usually dependent on the direction of the applied stimulus. It is therefore important to be able to specify directions in crystals in an unambiguous fashion. Directions are written generally as $[uvw]$ and are enclosed in square brackets. Note that the symbol $[uvw]$ means all parallel directions or vectors.

The three indices u, v and w define the coordinates of a point with respect to the crystallographic a b and c axes. The index u gives the coordinates in terms of a_0 along the a axis, the index v gives the coordinates in terms of b_0 along the b axis and the index w gives the coordinates in terms of c_0 along the c axis. The direction $[uvw]$ is simply the vector pointing from the origin to the point with coordinates u, v, w (Figure 5.8). For example, the direction $[100]$ is parallel to the a unit cell edge, the direction $[010]$ is parallel to the b cell edge, and $[001]$ is parallel to the c cell edge. Because directions are vectors, $[uvw]$ is not identical to $[\bar{u}\bar{v}\bar{w}]$, in the same way that the direction 'north' is not the same as the direction 'south'. Remember, though, that, *any* parallel direction shares the symbol $[uvw]$, because the origin of the coordinate system is not fixed and can always be moved to the starting point of the vector (Figure 5.9). A north wind is always a north wind, regardless of where you stand.

As with Miller indices, it is sometimes convenient to group together all directions that are identical by virtue of the symmetry of the structure. These are represented by the notation $\langle uvw \rangle$. In a cubic crystal, $\langle 100 \rangle$ represents the six directions $[100]$, $[\bar{1}00]$, $[010]$, $[0\bar{1}0]$, $[001]$, $[00\bar{1}]$.

A zone is a set of planes, all of which are parallel to a single direction, called the zone axis. The

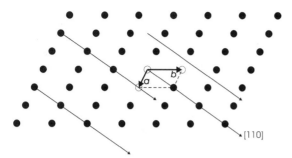

Figure 5.9   Parallel directions. These all have the same indices, [110]

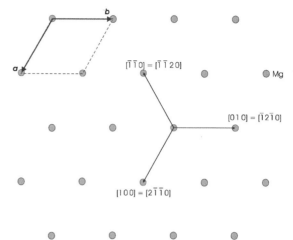

Figure 5.10   Directions in the basal (0 0 1) plane of a hexagonal crystal structure, given in terms of three indices, $[u\,v\,w]$, and four indices, $[u'\,v'\,t\,w']$

zone axis $[u\,v\,w]$ is perpendicular to the plane $(u\,v\,w)$ in cubic crystals but *not* in crystals of other symmetry.

It is sometimes important to specify a vector with a definite length perhaps to indicate the displacement of one part of a crystal with respect to another part, as in an antiphase boundary or crystallographic shear plane. In such a case, the direction of the vector is written as above, and a prefix is added to give the length. The prefix is usually expressed in terms of the unit cell dimensions. For example, in a cubic crystal, a displacement of two unit cell lengths parallel to the b axis would be written $2\,a_0[0\,1\,0]$.

As with Miller indices, to specify directions in hexagonal crystals a four-index system, $[u'\,v'\,t\,w']$ is sometimes used. The conversion of a three-index set to a four-index set is given by the following rules:

$$[u\,v\,w] \rightarrow [u'\,v'\,t\,w']$$

$$u' = \frac{n}{3}(2\,u - v)$$

$$v' = \frac{n}{3}(2\,v - u)$$

$$t = -(u' + v')$$

$$w' = n\,w$$

In these equations, n is a factor sometimes needed to make the new indices into smallest integers. Thus directions $[0\,0\,1]$ always transform to $[0\,0\,0\,1]$. The three equivalent directions in the basal $(0\,0\,0\,1)$ plane of a hexagonal crystal structure such as

magnesium, Figure 5.10, are obtained by using the above transformations. The correspondence is:

$$[1\,0\,0] = [2\,\bar{1}\,\bar{1}\,0]$$
$$[0\,1\,0] = [\bar{1}\,2\,\bar{1}\,0]$$
$$[\bar{1}\,\bar{1}\,0] = [\bar{1}\,\bar{1}\,2\,0]$$

The relationship between directions and planes depends on the symmetry of the crystal. In cubic crystals (and *only* cubic crystals) the direction $[h\,k\,l]$ is normal to the plane $(h\,k\,l)$.

5.1.7   The reciprocal lattice

Many of the physical properties of crystals, as well as the geometry of the three-dimensional patterns of radiation diffracted by crystals, are most easily described by using the reciprocal lattice. Each reciprocal lattice point is associated with a set of crystal planes with Miller indices $(h\,k\,l)$ and has coordinates $h\,k\,l$. The position of the $h\,k\,l$ spot in the reciprocal lattice is closely related to the orientation of the $(h\,k\,l)$ planes and to the spacing between these planes, d_{hkl}, called the interplanar spacing. Crystal structures and Bravais lattices, sometimes called the direct lattice, are said to occupy real space, and the reciprocal lattice occupies reciprocal space. The reciprocal lattice is defined in terms of three basis vectors labelled $a*$, $b*$ and $c*$. The lengths of the basis vectors of the reciprocal lattice are:

$$a* = \frac{1}{d_{100}} \quad b* = \frac{1}{d_{010}} \quad c* = \frac{1}{d_{001}}$$

For cubic, tetragonal and orthorhombic crystals, these are:

$$a* = \frac{1}{a_0} \quad b* = \frac{1}{b_0} \quad c* = \frac{1}{c_0}$$

For crystals of other symmetries, the relationship between the direct and reciprocal lattice distances is more complex (see Section S1.7)

The reciprocal lattice of a crystal is easily derived from the unit cell. For cubic cells, the reciprocal lattice axes are parallel to the direct lattice axes, which themselves are parallel to the unit cell edges, and the spacing of the lattice points hkl, along the three reciprocal axes, is equal to the reciprocal of the unit cell dimensions, $1/a_0 = 1/b_0 = 1/c_0$ (Figure 5.11). For some purposes it is convenient to multiply the length of the reciprocal axes by a constant.

Thus, physics texts usually multiply the axes given in Figure 5.11 by 2π, and crystallographers by λ, the wavelength of the radiation used to obtain a diffraction pattern. The derivation of the reciprocal lattice for symmetries other than cubic is given in Section S1.8.

5.2 The determination of crystal structures

Crystal structures are determined by using diffraction (see Section 14.7.3). The extent of diffraction is significant only when the wavelength of the radiation is very similar to the dimensions of the object that is irradiated. In the case of crystals, radiation with a wavelength similar to that of the spacing of the atoms in the crystal will be diffracted. X-ray diffraction is the most widespread technique used for structure determination, but diffraction of electrons and neutrons is also of great importance, as these reveal features that are not readily observed with X-rays.

The physics of diffraction by crystals has been worked out in detail. It is found that the incident radiation is diffracted in a characteristic way, called a diffraction pattern. If the positions of the

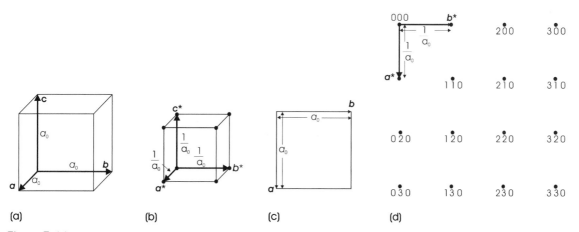

Figure 5.11 The direct lattice and reciprocal lattice of a cubic crystal: (a), (c) the direct lattice, specified by vectors a, b and c, with unit cell edges a_0 ($a_0 = b_0 = c_0$); (b), (d) the reciprocal lattice, specified by vectors $a*$, $b*$ and $c*$, with unit cell edges $1/a_0$, ($1/a_0 = 1/b_0 = 1/c_0$). The vector $a*$ is parallel to a, $b*$ parallel to b and $c*$ parallel to c. The vector from 000 to hkl in the reciprocal lattice is perpendicular to the (hkl) plane in a cubic crystal

diffracted beams are recorded, they map out the reciprocal lattice of the crystal. The intensities of the beams are a function of the arrangements of the atoms in space and of some other atomic properties, especially the atomic number of the atoms. Thus, if the positions and the intensities of the diffracted beams are recorded, it is possible to deduce the arrangement of the atoms in the crystal and their chemical nature.

5.2.1 Single-crystal X-ray diffraction

In this technique, which is the most important structure determination tool, a small single crystal of the material, of the order of a fraction of a millimetre in size, is mounted in a beam of X-rays. The diffraction pattern used to be recorded photographically, but now the task is carried out electronically. The technique has been used to solve enormously complex structures, such as that of huge proteins, or DNA.

Problems still remain, though, in this area of endeavour. Any destruction of the perfection in the crystal structure degrades the sharpness of the diffracted beams. This in itself can be used for crystallite size determination. Poorly crystalline material gives poor information, and truly amorphous samples give virtually no crystallographic information this way.

5.2.2 Powder X-ray diffraction and crystal identification

A common problem for many scientists is to determine which compounds are present in a polycrystalline sample. The diffraction pattern from a powder placed in the path of an X-ray beam gives rise to a series of cones rather than spots, because each plane in the crystallite can have any orientation (Figure 5.12a). The positions and intensities of the diffracted beams are recorded along a narrow strip (Figure 5.12b), and the diffracted beams are often

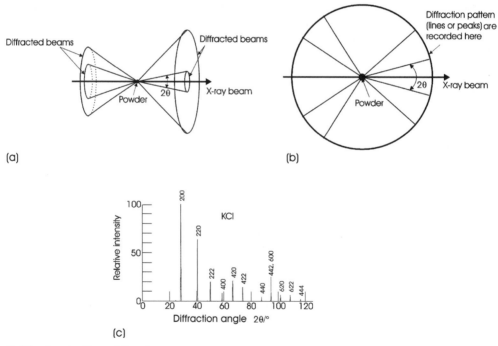

Figure 5.12 Powder X-ray diffraction: (a) a beam of X-rays incident on a powder is diffracted into a series of cones; (b) the intensities and positions of the diffracted beams are recorded along a circle, to give a diffraction pattern. (c) The diffraction pattern from powdered potassium chloride, KCl, a cubic crystal. The numbers above the 'lines' are the Miller indices of the diffracting planes

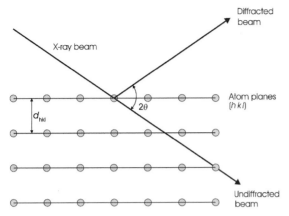

Figure 5.13   The geometry of Bragg reflection from a set of crystal planes, $(h\,k\,l)$, with interplanar spacing $d_{h\,k\,l}$

called lines (Figure 5.12c). The position of a difracted beam (not the intensity) is found to depend only on the interplanar spacing, $d_{h\,k\,l}$, and the wavelength of the X-rays used. Bragg's Law, Equation (5.1), gives the connection between these quantities:

$$\lambda = 2\,d_{h\,k\,l}\,\sin\theta \qquad (5.1)$$

where λ is the wavelength of the X-radiation, $d_{h\,k\,l}$ is the interplanar spacing of the $(h\,k\,l)$ planes and θ is the diffraction angle (Figure 5.13). (Although the geometry of Figure 5.13 is identical to that of reflection, the physical process occurring is diffraction.) The relationship is simplest for cubic crystals. In this case, the interplanar spacing is given by:

$$\frac{1}{d_{h\,k\,l}^2} = \frac{h^2 + k^2 + l^2}{a_0^2}$$

hence, $d_{h\,k\,l} = a_0, a_0/\sqrt{2}, a_0/\sqrt{3}, a_0/\sqrt{4}$, etc., where a_0 is the cubic unit cell lattice parameter.

The positions of the lines on the diffraction pattern of a single phase can be used to derive the unit cell dimensions of the material. The unit cell of a solid with a fixed composition is a constant. If the solid has a composition range, as in a solid solution or an alloy, the cell parameters will vary. Vegard's law, first propounded in 1921, states that the lattice parameter of a solid solution of two phases with similar structures will be a linear function of the lattice parameters of the two end members of the composition range (Figure 5.14a):

$$x = \frac{a_{ss} - a_1}{a_2 - a_1}$$

where a_1 and a_2 are the lattice parameters of the parent phases, a_{ss} is the lattice parameter of the solid solution, and x is the mole fraction of the parent phase with lattice parameter a_2. This 'law' is simply an expression of the idea that the cell parameters are a direct consequence of the sizes of the component atoms in the solid solution. Vegard's law, in its ideal form (Figure 5.14a), is almost never obeyed exactly. A plot of cell parameters that lies below the ideal line (Figure 5.14b) is said to show a negative deviation from Vegard's law, and a plot that lies

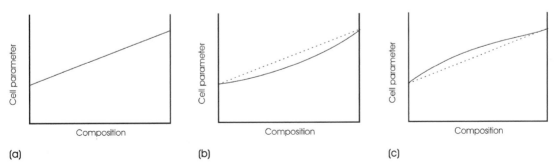

Figure 5.14   Vegard's law relating unit cell parameters to composition: (a) ideal Vegard's law behaviour; (b) a negative deviation from Vegard's law; and (c) a positive deviation from Vegard's law

above the ideal line (Figure 5.14c) is said to show a positive deviation form Vegard's law. In these cases, atomic interactions, which modify the size effects, are responsible for the deviations. In all cases, a plot of composition versus cell parameters can be used to determine the composition of intermediate compositions in a solid solution.

When the intensity and the positions of the diffraction pattern are taken into account, the pattern is unique for a single substance. The X-ray diffraction pattern of a substance can be likened to a fingerprint, and mixtures of different crystals can be analysed if a reference set of patterns is consulted. This technique is routine in metallurgical and mineralogical laboratories. The same technique is widely used in the determination of phase diagrams.

The experimental procedure can be illustrated with reference to the sodium fluoride–zinc fluoride (NaF–ZnF_2) system. Suppose that pure NaF is mixed with it a few percent of pure ZnF_2 and the mixture heated at $600\,^\circ C$ until reaction is complete. The X-ray powder diffraction pattern will show the presence of two phases: NaF, which will be the major component, and a small amount of a new compound (point A, Figure 5.15). A repetition of the experiment, with gradually increasing amounts of ZnF_2, will yield a similar result, but the amount of the new phase will increase relative to the amount of NaF until a mixture of $1NaF$ plus $1ZnF_2$ is

heated. At this composition, only one phase will be indicated on the X-ray powder diagram. It has the composition $NaZnF_3$.

A slight increase in the amount of ZnF_2 in the reaction mixture again yields an X-ray pattern that shows two phases to be present. Now, however, the compounds are $NaZnF_3$ and ZnF_2 (point B, Figure 5.15). This state of affairs continues as more ZnF_2 is added to the initial mixture, with the amount of $NaZnF_3$ decreasing and the amount of ZnF_2 increasing until pure NaF_2 is reached. Careful preparations reveal the fact that NaF or ZnF_2 appear alone on the X-ray films only when they are pure, and $NaZnF_3$ appears alone only at the exact composition of one mole NaF plus one mole ZnF_2. In addition, over all the composition range studied, the unit cell dimensions of each of these three phases will be unaltered.

An extension of the experiments to higher temperatures will allow the whole of the solid part of the phase diagram to be mapped.

5.2.3   Neutron diffraction

Neutron diffraction is very similar to X-ray diffraction in principle but is quite different in practice, because neutrons need to be generated in a nuclear reactor. One advantage of using neutron diffraction is that it is often able to distinguish between atoms that are difficult to distinguish with X-rays. This is because the scattering of X-rays depends on the atomic number of the elements, but this is not true for neutrons and, in some instances, neighbouring atoms have quite different neutron-scattering capabilities, making them easily distinguished. Another advantage is that neutrons have a spin and so interact with unpaired electrons in the structure. Thus neutron diffraction gives rise to information about the magnetic properties of the material. The antiferromagnetic arrangement of the Ni^{2+} ions in nickel oxide, for example, was determined by neutron diffraction (see Section 12.3.3).

5.2.4   Electron diffraction

Electrons are charged particles and interact very strongly with matter. This has two consequences for

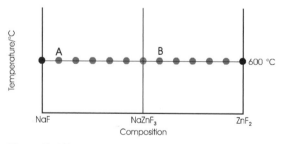

Figure 5.15 The determination of phase relations using X-ray diffraction. The X-ray powder patterns will show a single material to be present only at the exact compositions NaF, $NaZnF_3$ and ZnF_2. At points such as A, the solid will consist of NaF and $NaZnF_3$. At points such as B, the solid will consist of $NaZnF_3$ and ZnF_2. The proportions of components in the mixtures will vary across the composition range

structure determination. First, electrons will pass only through a gas or very thin solids. Second, each electron will be diffracted many times in traversing the sample, making the theory of electron diffraction more complex than the theory of X-ray diffraction. The relationship between the position and intensity of a diffracted beam is not easily related to the atomic positions in the unit cell. Moreover, delicate molecules are easily damaged by the intense electron beams needed for a successful diffraction experiment. Electron diffraction, therefore, is not used in the same routine way as X-ray diffraction for structure determination.

Electrons, however, do have one advantage. Because they are charged they can be focused by magnetic lenses to form an image. The mechanism of diffraction as an electron beam passes through a thin flake of solid allows defects such as dislocations to be imaged with a resolution close to atomic dimensions. Similarly, diffraction (reflection) of electrons from surfaces of thick solids allows surface details to be recorded, also with a resolution close to atomic scales. Thus although electron diffraction is not widely used in structure determination it is used as an important tool in the exploration of the microstructures and nanostructures of solids.

5.3 Crystal structures

5.3.1 Unit cells, atomic coordinates and nomenclature

Irrespective of the complexity of a crystal structure, it can be constructed by the packing together of unit cells. This means that the positions of all of the atoms in the crystal do not need to be given, only those in a unit cell. The minimum amount of information needed to specify a crystal structure is thus the unit cell type, the cell parameters and the positions of the atoms in the unit cell. For example, the unit cell of the rutile form of titanium dioxide has a tetragonal unit cell, with cell parameters, $a_0 = b_0 = 0.459\,\text{nm}$, $c_0 = 0.296\,\text{nm}$.[2]

---

[2] The unit cell dimensions are often specified in terms of the Ångström unit, Å, where $10\,\text{Å} = 1\,\text{nm}$.

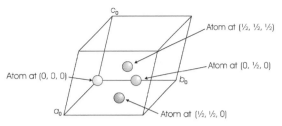

Figure 5.16 The positions of atoms in a unit cell. Atom positions are specified as fractions of the cell edges, not with respect to Cartesian axes

The (x, y, z) coordinates of the atoms in each unit cell are expressed as fractions of a_0, b_0 and c_0, the cell sides. Thus, an atom at the centre of a unit cell would have a position specified as $(\frac{1}{2}, \frac{1}{2}, \frac{1}{2})$, irrespective of the type of unit cell. Similarly, an atom at each corner of a unit cell is specified by $(0, 0, 0)$. The normal procedure of stacking the unit cells together means that this atom will be duplicated at every other corner. For an atom to occupy the centre of a face of the unit cell, the coordinates will be $(\frac{1}{2}, \frac{1}{2}, 0)$, $(0, \frac{1}{2}, \frac{1}{2})$ and $(\frac{1}{2}, 0, \frac{1}{2})$, for C-face-centred, A-face-centred and B-face-centred cells, respectively. Atoms on cell edges are specified at positions $(\frac{1}{2}, 0, 0)$, $(0, \frac{1}{2}, 0)$ or $(0, 0, \frac{1}{2})$, for atoms on the a b and c axes, respectively. Stacking of the unit cells to build a structure will ensure that atoms appear on all of the cell edges and faces. These positions are illustrated in Figure 5.16.

A unit cell reflects the symmetry of the crystal structure. Thus, an atom at a position (x, y, z) in a unit cell may require the presence of atoms at other positions in order to satisfy the symmetry of the structure. For example, a unit cell with a centre of symmetry will, of necessity, require that an atom at (x, y, z) be paired with an atom at $(-x, -y, -z)$. To avoid long repetitive lists of atom positions in complex structures, crystallographic descriptions usually list only the minimum number of atomic positions which, when combined with the symmetry of the structure, given as the space group, generate all the atom positions in the unit cell. Additionally, the Bravais lattice type and the motif are often specified as well as the number of formula units in the unit cell, written as Z. Thus, in the unit cell of rutile, given above, $Z = 2$. This means that there are two TiO_2 units in the unit cell; that is, two titanium atoms and four oxygen atoms. In the following sections, these features of nomenclature will be developed in the descriptions of some widely encountered crystal structures.

A vast number of structures have been determined, and it is very convenient to group those with topologically identical structures together. On going from one member of the group to another the atoms in the unit cell differ, reflecting a change in chemical compound, and the atomic coordinates and unit cell dimensions change slightly, reflecting the difference in atomic size. Frequently, the group name is taken from the name of a mineral, as mineral crystals were the first solids used for structure determination. Thus all solids with the *halite* structure have a unit cell similar to that of sodium chloride, NaCl. This group includes the oxides NiO, MgO and CaO (see Section 5.3.9). Metallurgical texts often refer to the structures of metals using a symbol for the structure. These symbols were employed by the journal *Zeitschrift für Kristallographie*, in the catalogue of crystal structures Strukturberichte Volume 1, published in 1920, and are called Strukturberichte symbols. For example, all solids with the same crystal structure as copper are grouped into the A1 structure type. These labels remain a useful shorthand for simple structures but become cumbersome when applied to complex materials, when the mineral name is often more convenient (e.g. see the *spinel* structure, Section 5.3.10).

5.3.2   The density of a crystal

The atomic contents of the unit cell give the composition of the material. The theoretical density of a crystal can be found by calculating the mass of all the atoms in the unit cell. (The mass of an atom is its molar mass divided by the Avogadro constant; see Section S1.1). The mass is divided by the unit cell volume. To count the number of atoms in a unit cell, we use the following information:

- an atom within the cell counts as 1;
- an atom in a face counts as 1/2;

- an atom on an edge counts as 1/4;
- an atom on a corner counts as 1/8.

A quick method to count the number of atoms in a unit cell is to displace the unit cell outline to remove all atoms from corners, edges and faces. The atoms remaining, which represent the unit cell contents, are all within the boundary of the unit cell and count as 1.

The measured density of a material gives the average amount of matter in a large volume. For a solid that has a variable composition, such as an alloy or a nonstoichiometric phase, the density will vary across the phase range. Similarly, an X-ray powder photograph yields a measurement of the average unit cell dimensions of a material and, for a solid that has a variable composition, the unit cell dimensions are found to change in a regular way across the phase range. These two techniques can be used in conjunction with each other to determine the most likely point defect model to apply to a material. As both techniques are averaging techniques they say nothing about the real organisation of the defects, but they do suggest first approximations.

The general procedure is to determine the unit cell dimensions, the crystal structure type and the real composition of the material. The ideal composition of the unit cell will be known from the structure type. The ideal composition is adjusted by the addition of extra atoms (interstitials or substituted atoms) or removal of atoms (vacancies) to agree with the real composition. A calculation of the density of the sample assuming either that interstitials or vacancies are present is then made. This is compared with the measured density to discriminate between the two alternatives.

5.3.2.1   Example: iron monoxide

The method can be illustrated by reference to iron monoxide. Iron monoxide, often known by its mineral name of wüstite, has the *halite* (NaCl) structure. In the normal *halite* structure, there are four metal and four nonmetal atoms in the unit cell, and compounds with this structure have an ideal composition $MX_{1.0}$ (see Section 5.3.9 for further information on the *halite* structure). Wüstite has a composition that is always oxygen-rich compared with the ideal formula of $FeO_{1.0}$. Data for an actual sample found an oxygen:iron ratio of 1.058, a density of 5728 kg m^{-3} and a cubic lattice parameter, a_0, of 0.4301 nm[3]. The real composition can be obtained by assuming either that there are extra oxygen atoms in the unit cell, as interstitials (Model A), or that there are iron vacancies present (Model B).

Model A   Assume that the iron atoms in the crystal are in a perfect array, identical to the metal atoms in *halite*, and that an excess of oxygen is due to interstitial oxygen atoms present in addition to those on the normal anion positions. The *ideal* unit cell of the structure contains four iron atoms and four oxygen atoms and so, in this model, the unit cell must contain four atoms of iron and $4(1 + x)$ atoms of oxygen. The unit cell contents are Fe_4O_{4+4x} and the composition is $FeO_{1.058}$.

- The mass of 1 unit cell in model A, m_A, is

$$m_A = \frac{1}{N_A}\{(4 \times 55.85) + [4 \times 16 \times (1 + x)]\}$$
$$= \frac{1}{N_A}[(4 \times 55.85) + (4 \times 16 \times 1.058)]\,g$$
$$= 4.834 \times 10^{-25}\,kg.$$

- The volume, v, of the cubic unit cell is given by a_0^3, thus:

$$v = (0.4301 \times 10^{-9})^3\,m^3 = 7.9562 \times 10^{-29}\,m^3.$$

- The density, ρ, is given by the mass, m_A, divided by the volume, v:

$$\rho = \frac{4.834 \times 10^{-25}\,kg}{7.9562 \times 10^{-29}\,m^3}$$
$$= 6076\,kg\,m^{-3}.$$

Model B   Assume that the oxygen array is perfect and identical to the nonmetal atom array in the

[3] The data are from the classical paper by E.R. Jette and F. Foote; *Journal of Chemical Physics*, volume 1, page 29, 1933.

halite structure. As there are more oxygen atoms than iron atoms, the unit cell must contain some vacancies on the iron positions. In this case, one unit cell will contain four atoms of oxygen and $(4 - 4x)$ atoms of iron. The unit cell contents are $Fe_{4-4x}O_4$, and the composition is $Fe_{1/1.058}O_{1.0}$ or $Fe_{0.945}O$.

- The mass of one unit cell in model B, m_B, is

$$m_B = \frac{1}{N_A}\{[4 \times (1 - x) \times 55.85] + (4 \times 16)\}$$
$$= \frac{1}{N_A}[(4 \times 0.945 \times 55.85) + (4 \times 16)]\,g$$
$$= 4.568 \times 10^{-25}\,kg.$$

- The density, ρ, is given by m_B divided by the volume, v, to yield

$$\rho = \frac{4.568 \times 10^{-25}\,kg}{7.9562 \times 10^{-29}\,m^3}$$
$$= 5741\,kg\,m^{-3}$$

Conclusion   The difference in the two values is surprisingly large. The experimental value of the density, 5728 kg m^{-3}, is in good accord with that for model B, in which vacancies on the iron positions are assumed. This indicates that the formula should be written $Fe_{0.945}O$.

5.3.3   The cubic close-packed (A1) structure

- General formula: M; example: Cu.

- Lattice: cubic face-centred, $a_0 = 0.360$ nm.

- $Z = 4$ Cu.

- Atom positions: $(0, 0, 0)$; $(\frac{1}{2}, \frac{1}{2}, 0)$; $(0, \frac{1}{2}, \frac{1}{2})$; $(\frac{1}{2}, 0, \frac{1}{2})$.

There are four lattice points in the face-centred unit cell, and the motif is one atom at $(0, 0, 0)$. The structure is typified by copper (Figure 5.17). The cubic close-packed structure is adopted by many metals (see Figure 6.1, page 152) and by the noble

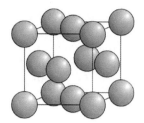

Figure 5.17 The A1 structure of copper

gases, Ne(s), Ar(s), Kr(s), Xe(s). This structure is often called the face-centred cubic (fcc) structure but, from a crystallographic point of view, this name is not ideal[4] and it is convenient to use the Strukturbericht symbol, A1. Each atom has 12 nearest neighbours and, if the atoms are supposed to be hard touching spheres, the fraction of the volume occupied is 0.7405. More information on this structure is given in Section 6.1.1

5.3.4 The body-centred cubic (A2) structure

- General formula: M; example: W.

- Lattice: cubic body-centred, $a_0 = 0.316$ nm.

- $Z = 2$ W.

- Atom positions: $(0,0,0)$; $(\frac{1}{2},\frac{1}{2},\frac{1}{2})$.

There are two lattice points in the body-centred unit cell, and the motif is one atom at $(0,0,0)$. The structure is adopted by tungsten, W (Figure 5.18)

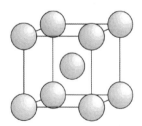

Figure 5.18 The A2 structure of tungsten

[4]See H.D. Megaw, *Crystal Structures*, Saunders, London, 1973.

and by many other metallic elements (see Figure 6.1, page 152). This structure is often called the body-centred cubic (bcc) structure. As with the A1 structure, this is not a good name (see Footnote 4) and it is better to refer to the Strukturbericht symbol, A2. In this structure, each atom has eight nearest neighbours and six next-nearest neighbours at only 15 % greater distance. If the atoms are supposed to be hard touching spheres, the fraction of the volume occupied is 0.6802. This is less than that for either the A1 structure (Section 5.3.3) or the A3 structure (Section 5.3.5), both of which have a volume fraction of occupied space of 0.7405. The bcc structure is often the high-temperature structure of a metal that has a close-packed structure at lower temperature. More information on this structure is given in Section 6.1.1.

5.3.5 The hexagonal (A3) structure

- General formula: M; example: Mg.

- Lattice: primitive hexagonal, $a_0 = 0.321$ nm, $c_0 = 0.521$ nm.

- $Z = 2$ Mg.

- Atom positions: $(0,0,0)$; $(\frac{1}{3},\frac{2}{3},\frac{1}{2})$.

The lattice is primitive, and so there is only one lattice point in each unit cell. The motif is two atoms, one atom at $(0,0,0)$ and one atom at $(\frac{1}{3},\frac{2}{3},\frac{1}{2})$. The structure is represented by magnesium, Mg (Figure 5.19). If the atoms are supposed to be hard touching spheres, the fraction of the volume occupied is 0.7405 and the ratio c/a is

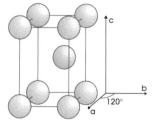

Figure 5.19 The A3 structure of magnesium

equal to $\sqrt{8}/\sqrt{3}, = 1.633$. Many metals adopt the A3 structure, some over a limited temperature range (see Figure 6.1, page 152). More information on this structure is given in Section 6.1.1.

5.3.6 The diamond (A4) structure

- General formula: M; example: C.

- Lattice: cubic face-centred, $a_0 = 0.356$ nm.

- $Z = 8$ C.

- Atom positions: $(0, 0, 0)$; $\left(\frac{1}{4}, \frac{1}{4}, \frac{1}{4}\right)$; (repeat in the face-centred pattern).

There are four lattice points in the face-centred unit cell, and the motif is two atoms, one atom at $(0, 0, 0)$ and one at $\left(\frac{1}{4}, \frac{1}{4}, \frac{1}{4}\right)$. The structure is adopted by diamond and, in it, each carbon atom is bonded to four other carbon atoms that are arranged tetra-hedrally around it (Figure 5.20). The bonds, of length 0.154 nm, are extremely strong sp^3 hybrids. The crystal can be regarded as a giant molecule.

The elements silicon ($a_0 = 0.542$ nm) and germanium ($a_0 = 0.564$ nm) also have the same structure as diamond. Grey tin has the same structure ($a_0 = 0.649$ nm) below a temperature of 13.2 °C.

5.3.7 The hexagonal (graphite), A9 structure

- General formula: C; graphite.

- Lattice: primitive hexagonal, $a_0 = 0.246$ nm, $c_0 = 0.671$ nm.

- $Z = 4$ C.

- Atom positions: $(0, 0, 0)$; $\left(0, 0, \frac{1}{2}\right)$; $\left(\frac{1}{3}, \frac{2}{3}, 0\right)$; $\left(\frac{2}{3}, \frac{1}{3}, \frac{1}{2}\right)$.

The lattice is primitive, and so there is only one lattice point in each unit cell. The motif is four atoms, at the positions specified above.

Graphite is a form of elemental carbon. The bonding in this material is closely related to that in benzene. The structure is made up of planar layers of carbon atoms bonded via sp^2 hybrid orbitals to give a strong framework with a hexagonal geometry (Figure 5.21). Because the bonding electrons are held in place, this skeleton is insulating. Above and below the layers, delocalised π bonds form a cloud of delocalised electrons which, because the layers are stacked directly on top of each other, repel each other strongly. This results in a large interlayer distance (0.335 nm) compared with the C–C distance in the plane

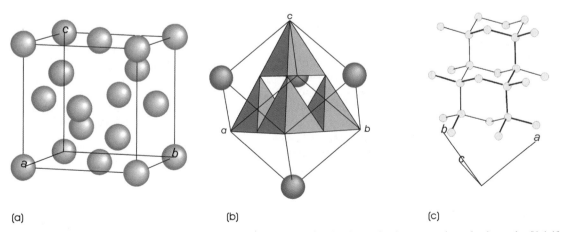

(a)   (b)   (c)

Figure 5.20 The A4 structure of diamond: (a) atoms in the unit cell, (b) projection approximately down the [1 1 1] direction to show the structure as carbon-centred tetrahedra; and (c) a similar projection to (b), to reveal the tetrahedral bond geometry

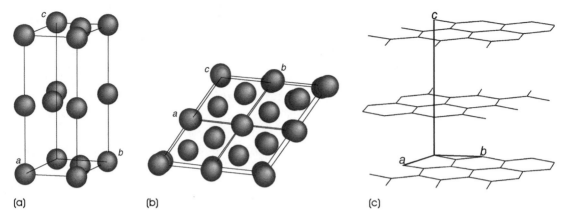

Figure 5.21 The A9 structure of graphite: (a) perspective view of the structure; (b) Projection approximately down the $[0\,0\,1]$ direction; (c) the structure drawn as hexagonal sheets, in which a carbon atom lies at each hexagonal vertex

(0.141 nm). The bonding between layers is very weak, made up of a van der Waals interaction between the delocalised electrons. Hence, although each layer of graphite is strong, the layers slide over one another easily. Graphite is easily cleaved in this direction and is a good dry lubricant. The delocalised electrons between the layers are similar to electrons in a metal, and these make graphite an electronic conductor parallel to the layers.

5.3.8 The structure of boron nitride

Boron nitride, BN, has a very similar structure to graphite but the hexagonal sheets are composed of boron and nitrogen atoms. It is white, chemically inert and a good insulator. The electronic structure of boron is $1s^2 2s^2 2p^1$ and that of nitrogen is $1s^2 2s^2 2p^3$. In order to generate the BN structure, the boron and the nitrogen atoms must form sp^2 hybrid bonds in the B–N sheets. In the case of boron, sp^2 hybridisation uses all of the outer electrons. In the case of nitrogen, sp^2 hybrid orbitals are formed by promoting one of the $2s^2$ electrons to give the configuration $1s^2 2s^1 2p_x^1 2p_y^1 2p_z^2$. After formation of the sp^2 orbitals, the remaining two p electrons are located in the (filled) p_z orbital.

The σ bonding in the BN sheets that results is strong and similar to the bonding in the graphite sheets. However, π bonding between the full $2p_z$ orbitals of nitrogen and the empty $2p_z$ orbitals of boron is not possible. This is because the orbital energies of boron and nitrogen are too dissimilar for a large energy gain, and hence bonding, to take place. Thus, no delocalised electrons are present in the structure, and BN is an insulator. Because of this, the boron and nitrogen atoms in alternate layers 'avoid' each other. This allows for the more efficient packing of the filled p_z orbitals, and the layers are closer than they are in graphite. Nevertheless, the lack of bonding between the layers still means that BN retains the easy cleavage of graphite and is still a good dry lubricant.

5.3.9 The halite (rock salt, sodium chloride, B1) structure

- General formula: MX; example: NaCl.

- Lattice: cubic face-centred, $a_0 = 0.563$ nm.

- $Z = 4$ NaCl.

- Atom positions: Na at $(0,0,0)$; Cl at $\left(\frac{1}{2},\frac{1}{2},\frac{1}{2}\right)$; (repeat in the face-centred pattern).

There are four lattice points in the face-centred unit cell, and the motif is one sodium atom at $(0,0,0)$ and one chlorine atom at $\left(\frac{1}{2},\frac{1}{2},\frac{1}{2}\right)$. In this structure, called the *halite, rock salt* or *sodium chloride*

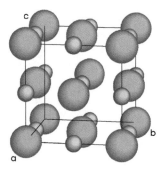

Figure 5.22   The B1 *halite* structure of NaCl

structure, each ion is surrounded by six ions of the opposite type at the corners of a regular octahedron (Figure 5.22).

This structure is extremely common and is adopted by many oxides, sulphides, halides and nitrides, with a formula *MX*.

5.3.10   The spinel (H1₁) structure

- General formula: AB_2X_4; example: $MgAl_2O_4$.

- Lattice: face-centred cubic, $a_0 = 0.809$ nm.

- $Z = 8$ $MgAl_2O_4$.

- Atom positions: there are 56 atoms in the unit cell, the positions of which will not be listed here.

There are four lattice points in the face-centred unit cell, and the motif is two $MgAl_2O_4$ complexes. This structure is named after the mineral spinel, $MgAl_2O_4$. The oxygen atoms in the crystal structure are in the same relative positions as the chlorine atoms in eight unit cells of halite, stacked together to form a $2 \times 2 \times 2$ cube (Figure 5.23). Thus in the cubic unit cell of spinel there are 32 oxygen atoms, to give a unit cell content of $Mg_8Al_{16}O_{32}$. The magnesium and aluminium atoms are inserted into this array in an ordered fashion. To a good approximation, all of the magnesium atoms are surrounded by four oxygen atoms in the form of a tetrahedron and are said to occupy tetrahedral positions, or sites, in the structure. Similarly, to a good approximation, the aluminium atoms are surrounded by six oxygen atoms and are said to occupy octahedral positions or sites (see below). When the structure is viewed down the [1 1 1] direction the oxygen atoms can be seen to form cubic close-packed layers, emphasising the relationship with the *halite* (NaCl) structure (see below).

The mineral spinel, $MgAl_2O_4$, has given its name to an important group of compounds with the same

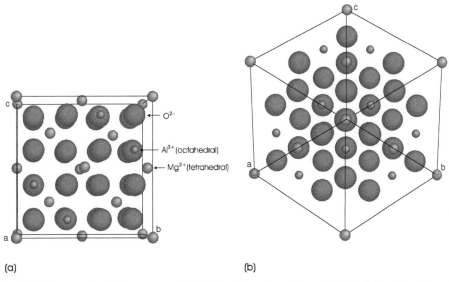

(a)   (b)

Figure 5.23   The H1₁ normal spinel structure of $MgAl_2O_4$: (a) projected down the [1 0 0] direction; (b) projected down the [1 1 1] direction (projected)

structure, collectively known as *spinels*, which includes halides, sulphides and nitrides as well as oxides. Only oxide spinels are considered here. These are often regarded as ionic compounds. The formula of the oxide spinels, AB_2O_4, is satisfied by a number of combinations of cations, the commonest of which is A^{2+} and B^{3+}, typified by Mg^{2+} and Al^{3+} in spinel itself.

In each unit cell there are the same number of octahedral sites as there are oxygen ions, that is, 32, and twice as many tetrahedral sites as oxygen ions, that is 64. However, not all of these can be occupied. The A^{2+} and B^{3+} cations are inserted into this array in an ordered fashion, filling half of the available octahedral positions and an eighth of the available tetrahedral positions. This means that there are 8 occupied tetrahedral sites and 16 occupied octahedral sites in a unit cell.

There are two principle arrangements of cations found. If the 8 A^{2+} ions per unit cell are confined to the available tetrahedral sites, these are filled completely. The 16 B^{3+} ions are then confined to the octahedral sites. This cation distribution is often depicted as $(A)[B_2]O_4$, with the tetrahedral cations in round brackets and the octahedral cations in square brackets. This is called the *normal spinel* structure, and *spinels* with this arrangement of cations are said to be normal spinels. If the 8 A^{2+} ions are placed in half of the available 16 octahedral sites, half of the B^{3+} ions must be placed in the remaining octahedral sites and the other half in the tetrahedral sites. This can be written as $(B)[AB]O_4$. This arrangement is called the *inverse spinel* structure and compounds with this cation arrangement are said to be inverse spinels.

In reality, very few *spinels* have exactly the normal or inverse structure, and these are sometimes called mixed *spinels*. The cation distribution between the two sites is a function of a number of parameters, including temperature. This variability is described by an occupation factor, λ, which gives the fraction of B^{3+} cations in tetrahedral positions. A normal *spinel* is characterised by a value of λ of 0, and an inverse *spinel* by a value of λ of 0.5. The spinel $MgAl_2O_4$ has a value of λ of 0.05, and so is quite a good approximation to a normal *spinel*.

Cubic $A^{2+}Fe^{3+}O_4$ ferrites form an important group of magnetic oxides that have an inverse *spinel* structure, $(Fe^{3+})[A^{2+}Fe^{3+}]O_4$. Lodestone, or magnetite, Fe_3O_4, is an inverse *spinel* with a more correct formula of $(Fe^{3+})[Fe^{2+}Fe^{3+}]O_4$. The tetrahedral and octahedral sites in cubic ferrites provide two magnetic substructures, which give these oxides very flexible magnetic properties that can be tailored by varying the cations and the distribution between the octahedral and tetrahedral sites (see Section 12.3.5). In reality, the distribution of the cations in cubic ferrites is rarely perfectly normal or inverse, and the distribution tends to vary with temperature. Thus processing conditions are important if the desired magnetic properties are to be obtained.

5.4   Structural relationships

A list of atomic positions is often not very helpful when a variety of structures have to be compared. In this section two ways of looking at structures that facilitate comparisons are described. In the first of these, structures are described in terms of being built up by packing together spheres, and, in the other, in terms of polyhedra linked by corners and edges.

5.4.1   Sphere packing

The structure of many crystals can conveniently be described in terms of an ordered packing of spheres, representing spherical atoms or ions. Although there are an infinite number of ways of doing this, only two main arrangements, called closest (or close) packing, are sufficient to describe many crystal structures. These structures are made up of close-packed layers of spheres. Each close-packed layer consists of a hexagonal arrangement of spheres just touching each other to fill the space as much as possible (Figure 5.24).

These layers of spheres can be stacked in two principal ways to generate the structures. In the first of these, a second layer fits into the dimples in the first layer, and the third layer is stacked in dimples

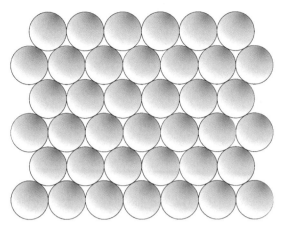

Figure 5.24 A single close-packed layer of spheres

on top of the second layer to lie over the first layer (Figure 5.25). This sequence is repeated indefinitely. If the position of the spheres in the first layer is labelled A, and the positions of the spheres in the second, B, the complete stacking is described by the sequence ABABAB. ... The structure has a hexagonal symmetry and unit cell. The a and b axes lay in the close-packed A sheet, and the hexagonal c axis is perpendicular to the stacking and runs from one A sheet to the next. There are two spheres (two atoms)

in a unit cell, at positions $(0,0,0)$ and $\left(\frac{1}{3}, \frac{2}{3}, \frac{1}{2}\right)$. If the spheres just touch, the relationship between the sphere radius, r, and the lattice parameter a_0, is:

$$2r = a_0$$

The ratio of c_0/a_0 in this ideal sphere packing is 1.633. The structure is identical to the A3 structure, described in Section 5.3.5. In most real structures the c_0/a_0 ratio departs from this ideal value of 1.633.

The second structure of importance is also formed by two layers of spheres, A and B, as before. The difference lies in the position of the third layer, which is not above either A or B, and is given the position label C (Figure 5.26). This three-layer stacking is repeated indefinitely, thus: ABCABC. ...

Although this structure can be described in terms of a hexagonal unit cell, the structure turns out to be cubic, and this description is always chosen. In terms of the cubic unit cell, there are atoms at the corners of the cell and in the centre of each of the faces. The close-packed layers lie along the $[1\,1\,1]$ direction (Figure 5.27). The spacing of the close-packed planes for an ideal packing, d_{cp}, is a third of the body diagonal of the cubic unit cell (i.e. $a_0/\sqrt{3}$). If the spheres just touch, the relationship between

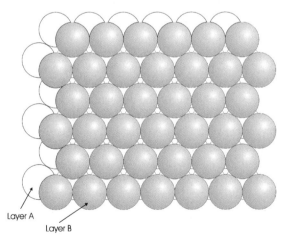

Layer A
Layer B

Figure 5.25 Hexagonal closest packing of spheres. All layers are identical to those in Figure 5.24. The first layer is labelled A, the second, B. Subsequent layers follow the sequence ... ABAB ...

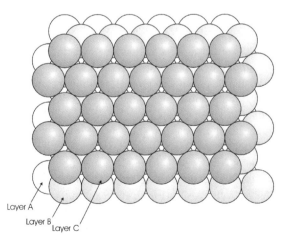

Layer A
Layer B
Layer C

Figure 5.26 Cubic closest packing of spheres. All layers are identical to those in Figure 5.24. The first layer is labelled A, the second B and the third C. Subsequent layers follow the sequence ... ABCABC ...

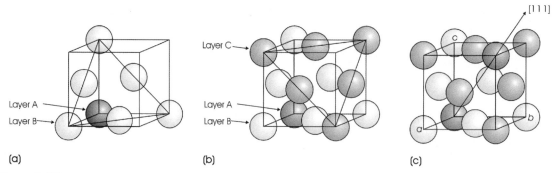

Figure 5.27 The cubic A1 structure in terms of cubic closest packing. The layers in Figure 5.26 lie perpendicular to the [1 1 1] direction and form (1 1 1) planes. (a) The first two layers, A and B; (b) the first three layers, A, B and, C; and (c) the fourth layer, completed by an atom at the top corner; this is identical to the position $(0, 0, 0)$ in the unit cell and so is part of an A layer

the sphere radius, r, and the lattice parameter a_0, is:

$$r = \frac{a_0}{\sqrt{8}}$$

The relationship between the spacing of the close-packed planes of spheres, d_{cp}, the cell parameter a_0, and r, is therefore:

$$d_{cp} = \frac{a_0}{\sqrt{3}} = \frac{r\sqrt{8}}{\sqrt{3}} = 1.63299\, r$$

The cubic close-packed structure is identical to the A1 structure described in Section 5.3.3.

Both the hexagonal closest packing of spheres and the cubic closest packing of spheres result in the (equally) densest packing of spheres. The fraction of the total volume occupied by the spheres, when they touch, is 0.7405.

The sphere arrangements described are only two of an infinite number of ways for spheres to be stacked. It is surprising, for example, that the sequence ABAC is not commonly encountered in metal structures, although this is the structure of the metal lanthanum (La). In addition, a number of more complex arrangements have been found, especially in the compounds silicon carbide (SiC) and zinc sulphide (ZnS).

5.4.2 Ionic structures in terms of anion packing

The ionic model suggests that because many ions have a closed shell of outer electrons they can be regarded as spherical. Moreover, because anions have gained electrons whereas cations have lost electrons it seems reasonable to regard anions as larger than cations. Packing the large anions together and inserting small cations into the gaps in the anion array so formed can reveal relationships between structures that are otherwise difficult to recognise.

The problem of packing spherical anions, (neglecting charges), is the same as the geometric problem as packing spheres. In such structures, the spheres do not fill all the available volume. There are small holes between the spheres that occur in layers between the sheets of spheres. These holes, which are called interstices, interstitial sites or interstitial positions, are of two types (Figure 5.28). In one type of position, three spheres in the lower layer are surmounted by one sphere in the layer above, or vice versa. The geometry of this site is that of a tetrahedron. The other position is made up of a lower layer of three spheres and an upper layer of three spheres. The shape of the enclosed space is not so easy to see but is found to have an octahedral geometry.

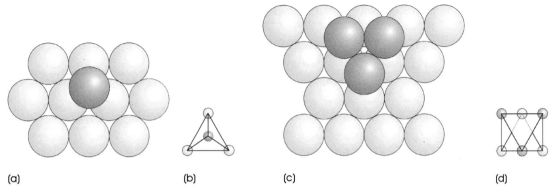

Figure 5.28 Tetrahedral and octahedral sites in cubic closest packed arrays of spheres: (a) a tetrahedral site between two layers and (b) the same site drawn as a polyhedron; (c) an octahedral site between two layers and (d) the same site drawn as a polyhedron

In the two closest packed sequences, ... ABCABC ... and ... ABAB ..., there are $2N$ tetrahedral interstices and N octahedral interstices for every N anions. Structures are derived by placing cations into the interstices, making sure that the total positive charge on the cations is equal to the total negative charge on the anions. The formula of the structure can be found by counting up the numbers of ions of each sort present.

Consider the structures that arise from a cubic close-packed array of X anions. If every octahedral position contains an M cation there are equal numbers of cations and anions in the structure. The formula of the compounds with this structure is MX, and the structure corresponds to the *halite* (NaCl, B1) structure. In the case of halide anions, X^-, to maintain charge balance, each cation must have a charge of $+1$, and the alkali halide, MX, structure results. Should oxygen anions, O^{2-}, form the anion array, the cations must necessarily have a charge of $2+$, to ensure that the charges balance, and the oxides will have a formula MO, but still retain the *halite* structure. This structure is possessed by a number of oxides, including MgO, CaO, SrO, BaO, MnO, FeO, CoO, NiO and CoO.

Should the anions adopt hexagonal close packing and all of the octahedral sites contain a cation, the hexagonal analogue of the *halite* structure is produced. In this case, the formula of the crystal is again MX. The structure is the *nicolite* (NiAs) structure and is adopted by a number of alloys and metallic sulphides, including NiAs, CoS, VS, FeS and TiS.

If only a fraction of the octahedral positions in the hexagonal packed array of anions is filled, a variety of structures results. The *corundum* structure is adopted by the oxides $\alpha\text{-}Al_2O_3$, V_2O_3, Ti_2O_3 and Fe_2O_3. In this structure two-thirds of the octahedral sites are filled in an ordered way. Of the structures that form when only half of the octahedral sites are occupied, those of rutile (TiO_2) and $\alpha\text{-}PbO_2$ are best known. The difference between the two structures lies in the way in which the cations are ordered. In the rutile form of TiO_2 the cations occupy straight rows of sites whereas in $\alpha\text{-}PbO_2$ the rows are staggered.

A large number of structures can be generated by the various patterns of filling the octahedral or tetrahedral interstices. The number can be extended if both types of position are occupied. One important structure of this type is the *spinel* structure, discussed in Section 5.3.10. The oxide lattice can be equated to a cubic close-packed array of oxygen ions, and the cubic unit cell contains 32 oxygen atoms. There are, therefore, 32 octahedral sites and 64 tetrahedral sites for cations. The unit contains only 16 octahedral (B cations) and 8 tetrahedral (A cations), which are distributed in an ordered way over these positions to give the formula AB_2O_4, where $A = M^{2+}$, and $B = M^{3+}$. The formula of the

Table 5.2 Some structures in terms of anion packing

Fraction of sites occupied		Sequence of anion layers	
tetrahedral	octahedral	. . . ABAB . . .	. . . ABCABC . . .
0	1	NiAs (nicolite)	NaCl (halite)
$\frac{1}{2}$	0	ZnO, ZnS (wurtzite)	ZnS (sphalerite or zinc blende)
0	$\frac{2}{3}$	Al_2O_3 (corundum)	–
0	$\frac{1}{2}$	TiO_2 (rutile), α-PbO_2	TiO_2 (anatase)
$\frac{1}{8}$	$\frac{1}{2}$	Mg_2SiO_4 (olivine)	$MgAl_2O_4$ (spinel)

cubic spinels is therefore $A_8B_{16}O_{32}$ or, as usually written, AB_2O_4. If all the A cations are tetrahedral sites we have a normal *spinel*, $(A)[B_2]O_4$. If they are in octahedral sites we have an inverse *spinel*, $(B)[AB]O_4$.

Structures containing cations in tetrahedral sites can be described in exactly the same way. In this case, there are twice as many tetrahedral sites as anions and so, if all sites are filled, the formula of the solid will be M_2X. When half are filled this becomes MX, and so on.

A survey of some of the structures that can be linked in this way is given in Table 5.2.

5.4.3 Polyhedral representations

It is often necessary to focus on the surroundings of a particular atom or ion in a solid and, for this purpose, structures drawn in terms of polyhedra are helpful. The polyhedra selected are generally metal–nonmetal coordination polyhedra. These are composed of a central metal atom surrounded by nonmetal atoms. By reducing the nonmetal atoms to points and then joining the points by lines one is able to construct the polyhedral shape. These polyhedra are then linked together to build up the complete structure. This representation has already been used as a way of describing the structure of diamond (Figure 5.20).

The advantage of using polyhedral representations of solids is that family relationships can be illustrated clearly. The disadvantage is that important structural details are often ignored, especially when polyhedra are idealised.

The complex families of silicates are best compared if the structures are described in terms of linked tetrahedra. The tetrahedral shape used is the idealised coordination polyhedron of the $[SiO_4]$ unit (Figure 5.29). Each silicon atom is linked to four oxygen atoms by tetrahedrally directed sp^3 hybrid bonds. For example, Figure 5.30 shows the way in which the $[SiO_4]$ tetrahedra are linked in the commonest form of silica (SiO_2), quartz. The $[SiO_4]$ units are very strong and persist during physical and chemical reactions, so that structural transformations are also often easily visualised in terms of the rearrangement of the $[SiO_4]$ tetrahedra.

Octahedral coordination is frequently adopted by the important 3d transition metal ions. In this coordination polyhedron, each cation is surrounded by six anions, to form an octahedral $[MO_6]$ group (Figure 5.31). The structure of rhenium trioxide, ReO_3, in terms of linked $[ReO_6]$ octahedra, has the appearance of a three-dimensional chessboard (Figure 5.32a). This structure is similar to that of tungsten trioxide, WO_3, but in the latter compound the octahedra are distorted slightly, so that the symmetry is reduced from cubic in ReO_3 to monoclinic in WO_3. The idealised ABO_3 *perovskite* structure is similar, but has the large A cation in the centre of the cage of octahedra (Figure 5.32b). Most real *perovskites*, such as barium titanate, $BaTiO_3$, are built of slightly distorted $[MO_6]$

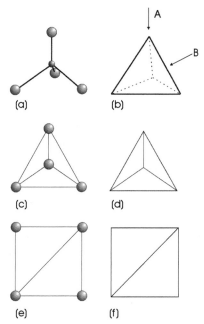

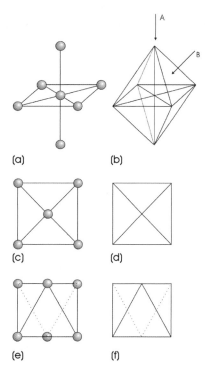

Figure 5.29 Representations of tetrahedra found in crystal structure diagrams: (a) a 'ball-and-stick' diagram of a tetrahedron, with a central silicon atom surrounded by four oxygen atoms and (b) its representation as a polyhedron. The views in parts (c), and (d) are the equivalent to those in parts (a) and (b), along the direction A in part (b), in which one tetrahedral vertex is uppermost. The views in parts (e) and (f) are the equivalent to those in parts (a) and (b), along the direction B in part (b), in which one tetrahedral edge is towards the observer

Figure 5.31 Representations of octahedra found in crystal structure diagrams: (a) a 'ball-and-stick' diagram of an octahedron, with a central metal atom surrounded by six oxygen atoms and (b) its representation as a polyhedron. The views in parts (c) and (d) are equivalent to those in parts (a) and (b), along the direction A in part (b), in which one octahedral vertex is uppermost. The views in parts (e) and (f) are the equivalent to those in parts (a) and (b), along the direction B in part (b), in which one octahedral face is towards the observer

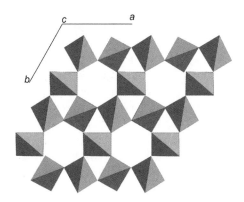

Figure 5.30 The structure of the high-temperature form of SiO_2, β-quartz, drawn as corner-shared tetrahedra projected down the hexagonal c axis (normal to the plane of the page). This projection obscures the fact that the tetrahedra form three-dimensional spirals, not rings

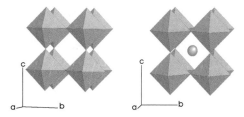

Figure 5.32 (a) The cubic ReO_3 structure represented as corner-shared ReO_6 octahedra; (b) the idealised cubic *perovskite* ABO_3 structure. The framework is identical to that in part (a) and consists of corner-shared BO_6 octahedra, containing an A cation in the central cage site (note $B \neq$ boron)

octahedra, which reduces the symmetry from cubic to orthorhombic or monoclinic.

Answers to introductory questions

How does a lattice differ from a structure?

Crystal structures and crystal lattices are different, although these terms are frequently (and incorrectly) used as synonyms. A crystal structure is built of atoms. A crystal lattice is an infinite pattern of points, each of which must have the same surroundings in the same orientation. A lattice is a mathematical concept. If any lattice point is chosen as the origin, the position of any other lattice point is defined by

$$P(u\,v\,w) = u\mathbf{a} + v\mathbf{b} + w\mathbf{c}$$

where $\mathbf{a}$, $\mathbf{b}$ and $\mathbf{c}$ are vectors, called basis vectors, and u, v and w are positive or negative integers. Clearly there are any number of ways of choosing $\mathbf{a}$, $\mathbf{b}$ and $\mathbf{c}$, and crystallographic convention is to choose vectors that are small and reveal the underlying symmetry of the lattice. There are only 14 possible three-dimensional lattices, called Bravais lattices.

All crystal structures can be built up from the Bravais lattices by placing an atom or a group of atoms at each lattice point. The crystal structure of a simple metal and that of a complex protein may be described in terms of the same lattice, but whereas the number of atoms allocated to each lattice point is often just one for a simple metallic crystal it may easily be thousands for a protein crystal. The number of atoms associated with each lattice point is called the motif, the lattice complex or the basis. The motif is a fragment of structure that is just sufficient, when repeated at each of the lattice points, to construct the whole of the structure. A crystal structure is built up from a lattice plus a motif.

What is a unit cell?

In a lattice, if any lattice point is chosen as the origin, the position of any other lattice point is

defined by

$$P(u\,v\,w) = u\mathbf{a} + v\mathbf{b} + w\mathbf{c}$$

where $\mathbf{a}$, $\mathbf{b}$ and $\mathbf{c}$ are vectors, called basis vectors, and u, v and w are positive or negative integers. Clearly, there are any number of ways of choosing $\mathbf{a}$, $\mathbf{b}$ and $\mathbf{c}$, and crystallographic convention is to choose vectors that are small and reveal the underlying symmetry of the lattice. The parallelepiped formed by the three basis vectors $\mathbf{a}$, $\mathbf{b}$ and $\mathbf{c}$ defines the unit cell of the lattice, with edges of length a_0, b_0 and c_0. The numerical values of the unit cell edges and the angles between them are collectively called the lattice parameters or unit cell parameters. The unit cell is not unique and is chosen for convenience and to reveal the underlying symmetry of the crystal.

As a crystal structure can be derived from the addition of atoms to each lattice point, a crystal structure can also be described in terms of a unit cell. The complete crystal is built up by a regular stacking of unit cells to fill space. The axes used to describe the crystal structure are the same as those used for the lattices, corresponding to the basis vectors lying along the unit cell edges. The position of an atom within the unit cell is given as (x, y, z), where the units are a_0 in a direction along the a axis, b_0 along the b axis, and c_0 along the c axis.

What is meant by a (1 0 0) plane?

The facets of a well-formed crystal or internal planes through a crystal structure are specified in terms of Miller Indices. These indices, h, k and l, written in round brackets, $(h\,k\,l)$, represent not just one plane but the set of all parallel planes, $(h\,k\,l)$. The values of h, k and l are the fractions of a unit cell edge, a_0, b_0 and c_0, respectively, intersected by this family of planes. A plane that lies parallel to a cell edge, and so never cuts it, is given the index 0 (zero). A plane that passes across the end of the unit cell cutting the a axis and parallel to the b and c axes of the unit cell has Miller indices $(1\,0\,0)$. The indices indicate that the plane cuts the cell edge running along the a axis at a position $1\,a_0$, and does not cut the cell edges parallel to the b or c axes at all. Because the unit cell may be chosen anywhere in the structure, $(1\,0\,0)$ means all planes that intersect all the unit cells in the structure in the way specified. There is no need to specify a plane $(1\,0\,0,0\,0)$, it is simply $(1\,0\,0)$.

Further reading

D.M. Adams, 1974, *Inorganic Solids*, John Wiley & Sons, Chichester.

F.D. Bloss, 1971, *Crystallography and Crystal Chemistry*, Holt Rinehart and Winston. New York.

H.D. Megaw, 1973, *Crystal Structures*, Saunders, Philadelphia, PA.

M.O'Keeffe, 1977, 'On the Arrangements of Ions in Crystals', *Acta Crystallogr.* **A33**, 924.

M.O'Keefe, B.G. Hyde, 1985, 'An Alternative Approach to Non-molecular Crystal Structures with Emphasis on the Arrangement of Solids', *Structure and Bonding*, **61**, 77.

A.F. Wells, 1984, *Structural Inorganic Chemistry*, 5th edn, Oxford University Press, Oxford.

Problems and exercises

Quick quiz

1 A crystal structure is:
   (a) A three-dimensional ordered array of points
   (b) A three-dimensional ordered array of atoms
   (c) A three-dimensional unit cell

2 A lattice is:
   (a) A three-dimensional ordered array of points
   (b) A three-dimensional ordered array of atoms
   (c) A three-dimensional unit cell

3 The basis vectors in a lattice define:
   (a) The unit cell
   (b) The crystal structure
   (c) The atom positions

4 The number of Bravais lattices is:
   (a) 12

   (b) 13
   (c) 14

5 A face-centred (F) unit cell contains:
   (a) One lattice point
   (b) Two lattice points
   (c) Four lattice points

6 A single face-centred unit cell with a lattice point in the plane cutting the b axis is:
   (a) A face-centred
   (b) B face-centred
   (c) C face-centred

7 A crystal system is:
   (a) A set of axes
   (b) A lattice
   (c) A unit cell

8 A tetragonal unit cell is defined by:
   (a) $a = b = c$
   (b) $a = b \neq c$
   (c) $a \neq b \neq c$

9 A crystal class summarises:
   (a) The internal symmetry of an object
   (b) The unit cell of the crystal
   (c) The type of crystal

10 A point group is identical to:
    (a) A crystal structure
    (b) A crystal lattice
    (c) A crystal class

11 The Miller indices $(h\,k\,l)$ represent:
    (a) A single plane in a crystal structure
    (b) A set of parallel planes in a crystal structure
    (c) A family of planes related by symmetry in a crystal structure

12 An $(h\,0\,0)$ plane in a crystal structure is:
    (a) Parallel to the a and b axes
    (b) Parallel to the b and c axes
    (c) Parallel to the a and c axes

13 A $(1\,1\,0)$ plane in a crystal cuts:
    (a) The a and b axes

(b) The b and c axes

(c) The a and c axes

14 $\{h\,k\,l\}$ represents:
    (a) A set of parallel planes

    (b) A group of nonequivalent planes

    (c) A family of symmetry-related planes

15 Miller–Bravais indices $(h\,k\,i\,l)$ are used with:
    (a) All noncubic crystals

    (b) Hexagonal crystals

    (c) Primitive crystals

16 A direction in a crystal structure is represented by:
    (a) $\{u\,v\,w\}$

    (b) $[u\,v\,w]$

    (c) $\langle u\,v\,w \rangle$

17 A family of directions related by symmetry is represented by:
    (a) $\{u\,v\,w\}$

    (b) $[u\,v\,w]$

    (c) $\langle u\,v\,w \rangle$

18 Bragg's Law for X-ray diffraction is:
    (a) $\lambda = 2\,d_{hkl}\,\sin\theta$

    (b) $\lambda = d_{hkl}\,\sin\theta$

    (c) $\lambda = 2\,d_{hkl}\,\cos\theta$

19 The atom coordinates $\left(\frac{1}{2},\frac{1}{2},\frac{1}{2}\right)$ represent an atom at:
    (a) The centre of a unit cell

    (b) The face centres of a unit cell

    (c) A position $\frac{1}{2}$ nm from the unit cell origin

20 In a cubic close-packed (A1) unit cell there:
    (a) Is one atom

    (b) Are two atoms

    (c) Are four atoms

21 In a body-centred cubic (A2) unit cell there:
    (a) Is one atom

    (b) Are two atoms

    (c) Are four atoms

22 In a hexagonal (A3) unit cell there:
    (a) Is one atom

    (b) Are two atoms

    (c) Are four atoms

23 The symbol Z gives the number of:
    (a) Formula units in a unit cell

    (b) Atoms in a unit cell

    (c) Positions in a unit cell

24 The lattice parameters give:
    (a) The crystal symmetry

    (c) The lattice symmetry

    (c) The dimensions of the unit cell edges

25 The *halite* (sodium chloride, B1) structure has:
    (a) Four atoms in a unit cell

    (b) Four MX units in a unit cell

    (c) Four NaCl units in a unit cell

26 The sphere packing giving rise to a hexagonal structure is:
    (a) ... ABABAB ...

    (b) ... ABCABC ...

    (c) ... ABACABAC ...

27 In a cubic close-packed array of N spheres there are:
    (a) N tetrahedral interstices

    (b) $2\,N$ tetrahedral interstices

    (c) $4\,N$ tetrahedral interstices

28 In a hexagonal close-packed array of N spheres there are:
    (a) N octahedral interstices

    (b) $2\,N$ octahedral interstices

    (c) $4\,N$ octahedral interstices

29 A tetrahedron is a polyhedron with:
    (a) Four triangular faces

    (b) Six triangular faces

    (c) Eight triangular faces

30 An octahedron is a polyhedron with:
    (a) Four triangular faces

(b) Six triangular faces

(c) Eight triangular faces

31 The structure of lithium oxide can be thought of as having anions in a cubic close-packed array with lithium ions in all of the tetrahedral positions; the formula of the oxide is:

(a) Li_2O

(b) LiO

(c) LiO_2

32 The alloy nickel arsenide has a structure in which all of the arsenic atoms are in a hexagonal close-packed array and the nickel atoms occupy all of the octahedral positions; the formula of nickel arsenide is:

(a) Ni_3As

(b) Ni_2As

(c) NiAs

33 The wurtizite structure of zinc sulphide has the sulphur atoms in a hexagonal close-packed array and the zinc atoms occupying half the tetrahedral positions; the formula of the sulphide is:

(a) Zn_2S

(b) ZnS

(c) ZnS_2

Calculations and questions

5.1 Sketch and define cubic, tetragonal and orthorhombic unit cells. [Note: answer is not provided at the end of this book.]

5.2 The lines on Figure 5.33(A) represent planes in a cubic crystal parallel to the c axis, which is normal to the plane of the page. The circles mark the corners of the cubic unit cell. Index planes (a)–(d).

5.3 The lines on Figure 5.33(B) represent planes in a cubic crystal parallel to the c axis, which is normal to the plane of the page. The circles mark the corners of the cubic unit cell. Index planes (a)–(d).

5.4 The lines on Figure 5.33(C) represent planes in a cubic crystal parallel to the c axis, which is normal to the plane of the page. The circles mark the corners of the cubic unit cell. Index planes (a)–(d).

5.5 The lines on Figure 5.33(D) represent planes in a cubic crystal parallel to the c axis, which is normal to the plane of the page. The circles mark the corners of the cubic unit cell. Index planes (a)–(d).

5.6 Sketch the $(1\,1\,1)$ and $(1\,\bar{1}\,1)$ planes in a cubic crystal. [Note: answer is not provided at the end of this book.]

5.7 How many planes belong to $\{1\,1\,0\}$ in a cubic crystal? List them.

5.8 How many planes belong to $\{1\,1\,1\}$ in a cubic crystal? List them.

5.9 How many planes belong to $\{h\,h\,0\}$ in a cubic crystal? List them.

5.10 How many planes belong to $\{h\,k\,0\}$ in a cubic crystal? List them.

5.11 The lines on Figure 5.34(A) represent directions in a cubic crystal in the plane of the page. The circles mark the corners of the cubic unit cell. Index directions (a)–(e).

5.12 The lines on Figure 5.34(B) represent directions in a cubic crystal in the plane of the page. The circles mark the corners of the cubic unit cell. Index directions (a)–(e).

5.13 The lines on Figure 5.34(C) represent directions in a cubic crystal in the plane of the page. The circles mark the corners of the cubic unit cell. Index directions (a)–(e).

5.14 The lines on Figure 5.34(D) represent directions in a cubic crystal in the plane of the page. The circles mark the corners of the cubic unit cell. Index directions (a)–(e).

5.15 How many directions does $\langle 1\,0\,0 \rangle$ represent? List them.

5.16 How many directions does $\langle 1\,1\,0 \rangle$ represent? List them.

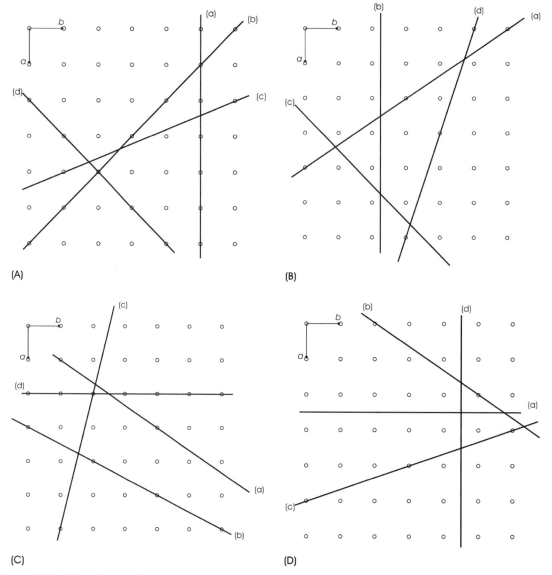

Figure 5.33 Planes in a cubic crystal, parallel to the *c* axis (normal to the plane of the page); circles mark the corners of the cubic unit cell: (A) for Question 5.2, (B) for Question 5.3, (C) for Question 5.4 and (D) for Question 5.5

5.17 What is the angle between (1 1 0) and [1 1 0] in a cubic crystal?

5.18 What is the angle between (1 3 2) and [1 3 2] in a cubic crystal?

5.19 Sketch the reciprocal lattice of a cubic crystal with $a_0 = 5$ nm. [Note: answer is not provided at the end of this book.]

5.20 Nickel has an A1 (face-centred cubic) structure, with $a_0 = 0.352$ nm. A powder sample

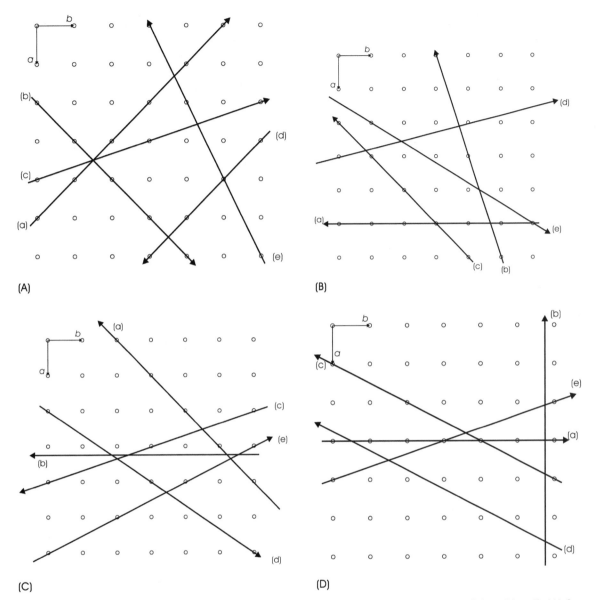

Figure 5.34   Directions in a cubic crystal, in the plane of the page; circles mark the corners of the cubic cell: (A) for Question 5.11, (B) for Question 5.12, (C) for Question 5.13 and (D) for Question 5.14

is irradiated with X-rays with a wavelength of 0.1542 nm. What angles would the diffracted beams from the $(1\,1\,1)$, $(2\,2\,0)$ and $(4\,0\,0)$ planes make with the incident beam direction?

5.21 Tantalum has an A2 (body-centred cubic) structure, with $a_0 = 0.3303$ nm. A powder sample is irradiated with X-rays with a wavelength of 0.1542 nm. What angles would the diffracted beams from the $(1\,1\,0)$, $(2\,1\,1)$ and

(3 1 0) planes make with the incident beam direction?

5.22 A sample of the cubic alloy β-brass (an alloy of copper and zinc) gives an X-ray powder pattern in which, when the X-ray wavelength is 0.229 nm, the first three reflections are: (1 0 0), $\theta = 22.9°$; (1 1 0), $\theta = 33.35°$; (1 1 1), $\theta = 42.35°$. Calculate the lattice parameter of the brass.

5.23. A sample of the cubic *spinel* $CuAl_2O_4$ gives an X-ray powder pattern in which, when the X-ray wavelength is 0.1541 nm, the first three reflections are: (1 1 1), $\theta = 9.51°$; (2 0 0), $\theta = 11.00°$; (2 2 0), $\theta = 15.65°$. Calculate the lattice parameter of the spinel.

5.24 A mineral sample contains a crystalline oxide with a formula $NiAl_2O_x$, where x is uncertain. The crystals gave an X-ray powder pattern with reflections characteristic of a cubic material. The first reflection, (1 1 1), was at $\theta = 9.55°$ when the X-ray wavelength was 0.1541 nm. Confirm that the sample is the *spinel* nickel aluminate, $NiAl_2O_4$, which has $a_0 = 0.8048$ nm.

5.25 The unit cell size of CaO is 0.48105 nm, and that of SrO is 0.51602 nm. Both adopt the *halite* structure type. Estimate the composition of a crystal of formula $Ca_xSr_{1-x}O$, which was found to have a unit cell of 0.5003 nm.

5.26 A mixed cubic *spinel* $ZnAl_{2-x}Ga_xO_4$ is made up by heating together $ZnAl_2O_4$ ($a_0 =$ 0.8086 nm) and $ZnGa_2O_4$ ($a_0 = 0.8328$ nm). The X-ray powder pattern, taken using radiation of wavelength 0.1541 nm, gave the first reflection, (1 1 1), at a position $\theta = 9.435°$. Estimate the value of x and give the composition.

5.27 The cubic unit cell of iridium is drawn in Figure 5.35(a). What are the atomic coordinates? What is the unit cell type?

5.28 The cubic unit cell of CsCl is drawn in Figure 5.35(b). What are the atomic coordinates of each ion?

5.29 The cubic unit cell of perovskite, $CaTiO_3$, is drawn in Figure 5.35(c). What are the atomic coordinates of the atoms?

5.30 The coordinates of cubic nickel oxide are

$$Ni : (0,0,0), \left(\tfrac{1}{2},\tfrac{1}{2},0\right), \left(0,\tfrac{1}{2},\tfrac{1}{2}\right), \left(\tfrac{1}{2},0,\tfrac{1}{2}\right);$$
$$O : \left(\tfrac{1}{2},\tfrac{1}{2},\tfrac{1}{2}\right), \left(0,0,\tfrac{1}{2}\right), \left(\tfrac{1}{2},0,0\right), \left(0,\tfrac{1}{2},0\right).$$

Sketch the unit cell [Note: not shown in answers at the end of this book.]. What is the formula of the oxide? What is the structure type?

5.31 A copper–gold alloy has a cubic structure. The atom positions are:

$$Au : (0,0,0);$$
$$Cu : \left(0,\tfrac{1}{2},\tfrac{1}{2}\right), \left(\tfrac{1}{2},0,\tfrac{1}{2}\right), \left(\tfrac{1}{2},\tfrac{1}{2},0\right).$$

Sketch the unit cell [Note: not shown in answers at the end of this book] and determine the formula of the alloy.

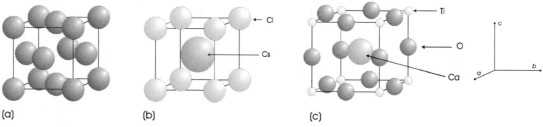

Figure 5.35   The cubic unit cells of (a) iridium, (b) CsCl and (c) perovskite ($CaTiO_3$)

5.32 Aluminium has a cubic A1 structure, with a lattice parameter of 0.361 nm. Estimate the density of the metal.

5.33 Tungsten has a cubic A2 structure, with a lattice parameter of 0.31651 nm. Estimate the density of the metal.

5.34 Magnesium has an A3 structure, with hexagonal lattice parameters of $a_0 = 0.320$ nm, $c_0 = 0.520$ nm. Estimate the density of the metal.

5.35 Copper has a cubic A1 structure and a density of 8.96×10^3 kg m^{-3}. What is the length of the unit cell edge?

5.36 A sample of calcia-stabilised zirconia was prepared by heating 85 mol% ZrO_2 with 15 mol% CaO at 1600 °C. The material had a cubic unit cell with a lattice parameter, a_0, of 0.5144 nm and a measured density of 5485 kg m^{-3}. Calculate the theoretical density of the sample assuming that it contains either anion vacancies or cation interstitials and hence determine whether interstitials or vacancies are more likely to be present in the

structure. [The parent structure is cubic *fluorite*, in which each unit cell contains four metal positions and eight nonmetal positions, to give an overall composition of MX_2.]

5.37 The unit cell of a zirconium sulphide, with a measured composition of 77 Zr:100 S, is of the *halite* (B1) type, with $a_0 = 0.514$ nm. The measured density is 4.80×10^3 kg m^{-3}. Calculate the theoretical density of ideal ZrS with the B1 structure assuming that it contains either zirconium vacancies or sulphur interstitials, and give an opinion on the defect structure of the real material.

5.38 Will the density of a crystal go up or down if it contains: (a) Schottky defects; (b) Frenkel defects; (c) vacancies; (d) interstitials?

5.39 An iron titanium oxide has the anions in a hexagonal close-packed array. The iron and titanium atoms each occupy a third of the octahedral sites available in an ordered array. What is the formula of the oxide? What is the likely parent structure of the oxide?

PART 2

Classes of materials

6

Metals, ceramics, polymers and composites

- Are hydrides alloys or ceramics?

- Are glasses liquids?

- Are polymers glasses?

- Why are plastic bags difficult to degrade?

Traditionally, materials have been divided into three major groups – metals, ceramics and polymers. Metallic materials are made up of pure metals, for example, titanium, iron or copper, and a vast number of alloys, including the historically important alloys bronze, brass and steel. Ceramics bring to mind porcelain, silicon carbide, glass and synthetic gemstones such as ruby and zirconia. Polymers are mainly compounds of carbon and include the familiar materials poly(vinylchloride), polyethylene and nylon as well as important biological molecules such as DNA.

In addition to these major divisions, two others should be mentioned – composites and biomaterials. Composites are important materials that are combinations of compounds from more than one of the groups listed above. They are of importance because they have superior engineering properties to the separate compounds. For example, glass fibre (ceramic) reinforced epoxy resin (polymer) has mechanical properties superior to either of the separate components. One of the most important of composites is concrete, which is a composite of cement and stony material called aggregate.

Biomaterials are naturally occurring materials with important properties, such as wood, silk and bone. They are invariably composites, made of more than one material type. Because of the superior properties of many biomaterials, much effort is placed into trying to recreate these materials synthetically, as biomimetics.

At first sight, metals, ceramics and polymers have little in common. This is because of two main factors – the chemical bonding holding the atoms together and the microstructure of the solids themselves – that are quite different in representative examples of each material. However, the difference is illusory. Many ceramics can be considered as metals, for example the ceramic superconductors. Many polymers show electronic conductivity greater than metals and have use in lightweight batteries and electronic devices. The material in this and later chapters will allow these apparent anomalies to be understood.

6.1 Metals

Roughly speaking, about three quarters of the elements can be regarded as metallic, and metals form

---

Understanding solids: the science of materials. Richard J. D. Tilley
© 2004 John Wiley & Sons, Ltd ISBNs: 0 470 85275 5 (Hbk) 0 470 85276 3 (Pbk)

the largest group of materials in the periodic table. Because of the variation in the outer electron configuration that this implies, one might expect that a large variety of metallic structures would form and that the structures would vary in a predictable way across the periodic table. It is rather surprising, therefore, to find that the majority of metallic elements possess one of only three structures. This fact arises because the outer electrons of metals are distributed throughout the crystal structure and the core that remains is, to a very good approximation, spherical. The crystal structures of many metals can then be approximated to those described by sphere packing.

Alloys, materials made up of two or more metallic elements, show a much greater variety of structures. In this chapter two basic but important types of alloy structure are described. Substitutional alloys frequently have a structure similar to that of a simple metal, but with several metal atoms distributed over the atom positions. One metal atom type is said to substitute for the other. In interstitial alloys, one atom type fits into the structure between the 'parent' metal atoms. Both of these alloy types are of great importance.

6.1.1 The crystal structures of pure metals

Most pure metals adopt one of three crystal structures: A1, copper structure (cubic close-packed; Section 5.3.3); A2, tungsten structure (body-centred cubic; Section 5.3.4); or A3 magnesium structure (hexagonal close-packed; Section 5.3.5). The structures found at room temperature (25 °C) and atmospheric pressure are listed in Figure 6.1. The difference in energy between these structures is small, and changes in structure are commonly induced by changes in temperature and pressure. The different forms are called allotropes. Some changes induced by increased temperature are given in Table 6.1.

It is surprising, in view of the many structures that are derived from either the hexagonal (ABAB) or cubic (ABCABC) closest packing (see Section 5.4.1), that so few complex arrangements occur. Cobalt is one metal that shows this behaviour. Below about 435 °C the structure is a disordered random stacking of A, B and C planes of metal atoms. It can be transformed into the A3 structure by careful annealing at lower temperatures, and this transforms to the A1 structure above 435 °C.

Li	Be											B	C
A2	A3												
0.3509	a 0.2286												
	c 0.3585												
Na	Mg											Al	Si
A2	A3											A1	
0.4291	a 0.3209											0.4050	
	c 0.5211												
K	Ca	Sc	Ti	V	Cr	Mn	Fe	Co	Ni	Cu	Zn	Ga	Ge
A2	A1	A3	A3	A2	A2		A2		A1	A1	A3		
0.5321	0.5588	a 0.3309	a 0.2951	0.3024	0.3885		0.2867		0.3524	0.3615	a 0.2665		
		c 0.5268	c 0.5686								c 0.4947		
Rb	Sr	Y	Zr	Nb	Mo	Tc	Ru	Rh	Pd	Ag	Cd	In	Sn
A2	A1	A3	A3	A2	A2	A3	A3	A1	A1	A1	A3	0.1663	
0.5705	0.6084	a 0.3648	a 0.3232	0.3300	0.3147	a 0.2738	a 0.2706	0.3803	0.3890	0.4086	a 0.2979		
		c 0.5732	c 0.5148			c 0.4393	c 0.4282				c 0.5620		
Cs	Ba	La	Hf	Ta	W	Re	Os	Ir	Pt	Au	Hg	Tl	Pb
A2	A2		A3	A2	A2	A3	A3	A1	A1	A1		A3	A1
0.6141	0.5023		a 0.3195	0.3303	0.3165	a 0.2761	a 0.2734	0.3839	0.3924	0.4078		a 0.3457	0.4950
			c 0.5051			c 0.4458	c 0.4392					c 0.5525	

Figure 6.1 The crystal structures of the metallic elements. Note: unit cell parameters are in nanometres. The figure given for cubic structures is a_0; A1 = copper (cubic close-packed) structure; A2 = tungsten (body-centred cubic) structure; A3 = magnesium (hexagonal close-packed) structure

Table 6.1 Allotropic structures of metals

Element	Room-temperature structure	High-temperature structure	Transition temperature/ °C
Ca	A1	A2	445
Sr	A1	A2	527
Sc	A3	A2	1337
Ti	A3	A2	883
Zr	A3	A2	868
Hf	A3	A2	1742
Y	A3	A2	1481
Fe	A2	A1	912
Co	(A3)	A1	435

The metals lanthanum (La), praseodymium (Pr) and neodymium (Nd) adopt mixed closest packing that has an ABAC repeat. Samarium (Sm), has a packing repeat BABCAC.

To the right-hand side of Figure 6.1, the simple structures are no longer found. These elements were once called the semimetals. In them, the outer electrons are not completely lost to the structure, and the shapes of the electron orbitals begin to influence bonding. This first becomes noticeable in the anomalous metal, mercury (Hg). The structure can be thought of as the A1 structure compressed along one body diagonal, so that the crystal structure becomes rhombohedral. Similarly, indium has a slightly distorted A1 structure.

Stronger bonding effects are found within the carbon group. At normal temperatures and pressures, the bonding in carbon (graphite) is a mixture of sp^2 and weaker van der Waals bonding. At high pressures, graphite transforms to the diamond structure, in which the atoms are linked by sp^3 hybrid bonds arranged tetrahedrally. The diamond structure is adopted by silicon and germanium at normal temperatures and pressures. Tin is a borderline solid from the point of view of bonding effects. At temperatures below 13.2 °C, the allotrope α-tin (grey tin) is stable. This has the diamond structure built with sp^3 hybrid bonding. At temperatures above 13.2 °C the stable structure is β-tin (white tin). This is the normal metallic form of tin. Metallic tin is not stable at most normal temperatures. How-

ever, the transition from white to grey tin is slow, and the metallic form is stabilised by metallic impurities, so that tin is normally found in the metallic form. Although white tin is metallic, the structure is complex and not simply related to the A1, A2 or A3 structures, revealing the importance of bonding effects. With lead, the increased atomic size leads to extensive outer-electron delocalisation. The solid is metallic and the structure is the A1 type.

In the semimetals antimony, arsenic and bismuth, bonding effects are more pronounced, and the structures are not related to the simple structures of most metals. Bismuth, the heaviest, is the most 'metallic', and phosphorus, lying above antimony in the periodic table, is not even considered to be a semimetal.

6.1.2 Metallic radii

If we assume that the structures of metals are made up of touching spherical atoms, it is quite easy, knowing the structure type and the size of the unit cell, to work out metallic radii. The relationship between the cell edge, a_0, for cubic crystals – a_0 and c_0, for hexagonal crystals – and the radius of the component atoms, r, for the three common metallic structures is given below.

For the A1, copper structure [face-centred cubic (fcc)] the atoms are in contact along a cube face diagonal, so that

$$r = \frac{a_0}{2\sqrt{2}}$$

The separation d of the close-packed atom planes (along a cube body diagonal) is

$$d = \frac{a_0}{\sqrt{3}} = d_{111}$$

Each atom has 12 nearest neighbours.

For the A2, tungsten structure [body-centred cubic (bcc)] the atoms are in contact along a cube body diagonal, so that

$$r = \frac{\sqrt{3}\,a_0}{4}$$

Each atom has 8 nearest neighbours.

For the A3, magnesium structure [hexagonal close packed (hcp)] the atoms are in contact along the a axis, hence,

$$r = \frac{a_0}{2}$$

The separation of the close packed atom planes is $c_0/2$. The ratio of c_0/a_0 in an ideal close-packed structure is $\sqrt{8}/\sqrt{3} = (1.633)$. Each atom has 12 nearest neighbours.

As in the case of ionic radii, the radius determined experimentally is found to depend on the number of nearest neighbours – the coordination number – of the atom in question. Both the A1 (fcc) and A3 (hcp) structures have 12 nearest neighbours [i.e. a coordination number of 12 (CN12)], and the radius determined will be appropriate to that coordination. The A2 (bcc) structure has 8 nearest neighbours [i.e. a coordination number of 8 (CN8)], and it is necessary to convert the radii measured, r (CN8), directly into those appropriate to 12 coordination, r (CN12), in order to obtain a self-consistent set of values. The conversion can be made by using the empirical formula

$$r(\text{CN12}) = 1.032\, r\,(\text{CN8}) - 0.0006$$

where the radii are measured in nanometres. Figure 6.2 shows metallic radii for coordination number 12. The majority of the elements shown have the A1, A2 or A3 structure or close approximations to them. A few important elements, notably manganese (Mn), gallium (Ga) and tin (Sn), which have complex structures, are also included. The radii for these latter elements are derived from a comparison of the interatomic distances in many alloys with appropriate structures.

There are a number of trends to note. In the 'well-behaved' alkali metals and alkaline earth metals the radius of an atom increases smoothly as the atomic number increases. The d transition metals all have rather similar radii as one passes along the period, and these generally increase with atomic number going down a group. The same is true for the lanthanides and actinides.

6.1.3 Alloy solid solutions

Alloys are important because they often show superior properties to pure elements, especially mechanical properties. There are large numbers of alloys, many of which have unusual and complex structures. Here we will mention only two sorts of

Li 0.1562	Be 0.1128											B	C
Na 0.1911	Mg 0.1602											Al 0.1432	Si
K 0.2376	Ca 0.1974	Sc 0.1641	Ti 0.1462	V 0.1346	Cr 0.1282	Mn 0.1264	Fe 0.1274	Co 0.1252	Ni 0.1246	Cu 0.1278	Zn 0.1349	Ga 0.1411	Ge
Rb 0.2546	Sr 0.2151	Y 0.1801	Zr 0.1602	Nb 0.1468	Mo 0.1400	Tc 0.1360	Ru 0.1339	Rh 0.1345	Pd 0.1376	Ag 0.1445	Cd 0.1568	In 0.1663	Sn 0.1545
Cs 0.2731	Ba 0.2243	La 0.1877	Hf 0.1580	Ta 0.1467	W 0.1408	Re 0.1375	Os 0.1353	Ir 0.1357	Pt 0.1387	Au 0.1442	Hg	Tl 0.1716	Pb 0.1750

Figure 6.2 Metallic radii (nm) for metallic elements in 12 coordination

alloy – substitution and interstitial solid solutions – with structures closely related to the pure metals just discussed. Substitutional solid solutions have a structure identical to one of the metals involved – the parent structure – with the other alloy-forming atoms, or foreign atoms, simply occupying positions in the structure normally occupied by the parent atoms. Interstitial solid solutions are formed when very small atoms percolate the parent structure and sit in normally unoccupied interstitial positions. The foreign atoms in each of these examples can be regarded as being in solution in the matrix of the parent metal.

6.1.3.1   Substitutional solid solutions

The likelihood of a substitutional solid solution forming between two metals will depend on a variety of chemical and physical properties. A large number of alloy systems were investigated by Hume-Rothery, in the first part of the 20th century, with the aim of understanding the principles that controlled alloy formation. His findings with respect to substitutional solid solution formation are summarised in the empirical Hume-Rothery solubility rules. The likelihood of obtaining a solid solution between two metals is highest when:

1. the crystal structure of each element of the pair is identical;

2. the atomic sizes of the atoms do not differ by more than 15 %;

3. the elements do not differ greatly in electronegativity (otherwise they will form compounds with each other), implying that they should be near to each other in the periodic table;

4. the elements have the same valence, implying that they should lie in the same group of the periodic table.

Although formulated a century ago, these 'rules' remain of value for predicting substitutional alloy formation.

The rules predict, for example, that nickel–copper (Ni–Cu, see Section 4.2.2) and copper–gold (Cu–Au) should form extensive substitutional solid solutions. What they do not predict is the likelihood that the atoms in the solid solution will order. In such cases a new phase, an ordered solid solution, will form. This happens in many systems, especially when random substitutional solid solutions are annealed at lower temperatures. For example, a copper–gold alloy heated at temperatures between the melting point, (about 890 °C), and 410 °C, and then rapidly cooled, will have a random distribution of copper and gold atoms over the sites of the A1 (fcc) structure (Figure 6.3a). However, if the alloy is heated at a temperature of about 400 °C for some time, the atoms of copper and gold will order into a new arrangement. The ordering that occurs depends on the composition of the alloy, and two main patterns have been characterised: Cu_3Au and CuAu.

The structure of the copper-rich alloy phase, Cu_3Au, is shown in Figure 6.3(b). The gold atoms are located at the corners of the cubic unit cell and the copper atoms at the face centres. The ordered structure of the CuAu alloy, which contains equal numbers of atoms (Figure 6.3c) has alternating $\{1\,0\,0\}$ planes composed of either copper or gold

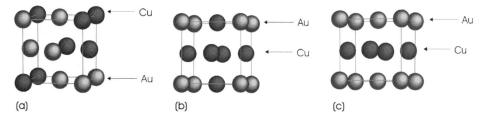

(a)   (b)   (c)

Figure 6.3   The cubic crystal structures of (a) disordered CuAu, (b) ordered Cu_3Au and (c) ordered CuAu I

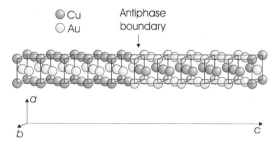

Figure 6.4   The ordered structure of CuAu II. The c axis is approximately $10 \times$ the cubic unit cell c axis of CuAu I

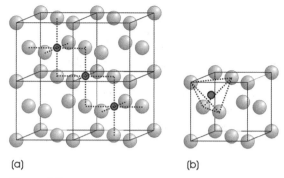

(a)   (b)

Figure 6.5   (a) Octahedral and (b) tetrahedral sites in the A1 structure. Four unit cells are drawn in part (a), and not all equivalent sites are marked

atoms. This structure is called the CuAu I structure. Arrangements that are more complex also occur. For example, in the CuAu II structure (Figure 6.4) the structure 'slips' every five CuAu unit cells, to give an overall repeat along the c axis of approximately 10 times the cubic cell parameter. The plane of the displacement is an antiphase boundary. [Note that 'antiphase' means 'out of register', and the word 'phase' is used in a different sense from that in phase diagrams.]

Large numbers of ordered structures can be formed by appropriate annealing conditions, and these structures play a large part in determining the physical, especially mechanical, properties of the alloys. Annealing is therefore a very important process in the production of materials with specific properties.

6.1.3.2   Interstitial solid solutions

Just as atoms should ideally be of similar sizes to form extensive substitutional solid solutions, to form an interstitial solid solution the radius of the foreign atom should be less than about half of the atomic radius of the atoms of the parent structure. Traditionally, the interstitial alloys most studied are those of the transition metals with carbon and nitrogen, as the addition of these atoms to the crystal structure increases the hardness of the metal considerably. Steel remains the most important traditional interstitial alloy from a world

perspective, consisting of carbon atoms distributed at random in interstitial sites within the A1 (fcc) structure of iron to form the phase austenite. More recently, hydrogen storage has become important and, today, interstitial alloys formed by incorporation of hydrogen into metals are of considerable interest.

The sites that are available for foreign atoms in interstitial alloys are of tetrahedral or octahedral geometry. The representation of the A1 (fcc) or A3 (hcp) structures as sphere packings (Section 5.4.1) shows that there are twice as many tetrahedral sites and the same number of octahedral sites present as metal atoms.

In a unit cell of the A1 structure, containing four metal atoms, the octahedral sites lie at the midpoints of each of the cell edges, with a further site at the cell centre (Figure 6.5a). Each site on a cell edge is shared by 4 cells, and there are 12 cell edges, hence each cell contains $\frac{1}{4} \times 12$ sites at the edges plus one at the cell centre, making four in total. The tetrahedral sites are found in each quarter of the unit cell, making 8 in all (Figure 6.5b). Two metal atoms are found per unit cell of the A3 structure and hence four tetrahedral sites and two octahedral sites per unit cell occur. In both of these structures, all of the tetrahedral sites and all of the octahedral sites have identical geometries (Figures 6.6a and 6.6b).

The A2 (bcc) structure also has 12 tetrahedral and 6 octahedral sites available. The octahedral sites lie

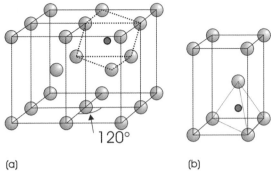

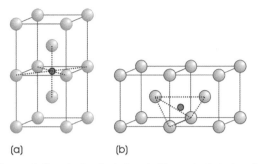

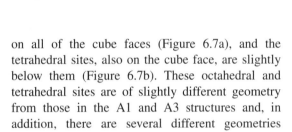

Figure 6.6 (a) Octahedral and (b) tetrahedral sites in the A3 structure. Four unit cells are drawn in part (a). Not all equivalent sites are marked

Figure 6.7 (a) Octahedral and (b) tetrahedral sites in the A2 structure. Two unit cells are drawn. Not all equivalent sites are marked

on all of the cube faces (Figure 6.7a), and the tetrahedral sites, also on the cube face, are slightly below them (Figure 6.7b). These octahedral and tetrahedral sites are of slightly different geometry from those in the A1 and A3 structures and, in addition, there are several different geometries found for the tetrahedral positions.

In all three structures the octahedral sites are larger and can accommodate carbon and nitrogen atoms. The tetrahedral sites are smaller, and only hydrogen commonly uses these positions.

The process by which these interstitial alloys form is similar in all systems. A reactive gas, typically hydrogen (H_2) for hydrides, methane (CH_4) for carbides, or ammonia (NH_3) for nitrides, decomposes on the metal surface. The atoms formed can then enter the structure, to occupy sites at random. The phases formed are often called α phases. Continued reaction leads to the formation of new structures, either by the ordering of the impurity atoms, as described for substitutional alloys (Section 6.1.3.1), or by more extensive structural rearrangements, as in cementite (Fe_3C).

Although the size of the foreign atoms is of importance, the sites that are occupied and the degree of occupancy of the available sites also depends critically on chemical interactions between the species. For example, in the α phase NbH_x only those tetrahedral sites with a particular geometry are occupied. The limiting composition of the α phase, approximately $NbH_{0.1}$, is achieved when only a

fraction of these particular sites is filled. Chemical interactions then lead to the nucleation of the hydride NbH.

6.1.4 Metallic glasses

As mentioned in Section 3.2.3, metals can be made into a noncrystalline form by cooling at a rate of approximately 10^5–$10^6 \, K \, s^{-1}$. The first noncrystalline metallic material to be made in this way had a composition of $Au_{75}Si_{25}$. It was later confirmed that these rapidly cooled solids were glasses, with definite glass transition temperatures (see Section 6.3.1).

In order to make a metallic glass by rapid cooling it is necessary to disrupt the natural tendency of the liquid to crystallise. This can be achieved in a number of ways. First, a system having a deep eutectic point (see Section 4.2.3) is helpful, because then the liquid can cool significantly before solidification starts. Second, a system containing several metals, each of which crystallises in a different structure, will deter crystallisation. Last, mixtures of atoms with widely differing sizes crystallise less readily. This has led to a general formula for metallic glasses of $T_{70-90}(SM)_{0-15}(NM)_{10-30}$, where T is one or more transition metals, (SM) is one or more semimetals such as silicon (Si) or germanium (Ge), and (NM) is one or more nonmetals such as phosphorus (P) or carbon (C). These solids, which are usually produced in the form of ribbons, are used in transformer cores. A widely used magnetic material for this purpose is METGLAS®,[1] with a composition of $Fe_{40}Ni_{40}P_{14}B_6$. Other examples of transformer core materials include $Fe_{86}B_8C_6$ and $FeB_{11}Si_9$.

Studies of more complex systems have made it possible to fabricate glassy metals using much slower cooling rates, down to 10 K s^{-1}. This has opened the door to the production of glassy metals in bulk form compared to the ribbons normally produced by very rapid cooling. The same principles as mentioned above are used, but systems have tended to become more complex. For example, bulk glasses of interest in a variety of magnetic applications have formulae of which $Fe_{72}Al_5Ga_2P_{11}C_6B_4$ is typical. Another recently developed metallic glass, Vitreloy, which has superior mechanical properties, has a composition $Zr_{1.65}Ti_{0.55}Cu_{1.125}Ni_{0.9}B_{0.225}$. It is used, among other things, for golf-club heads.

6.1.5   The principal properties of metals

As with all materials, the observed properties arise via the interaction of bonding and microstructure. The typically metallic properties of good electrical and thermal conductivity are consequences of the metallic bond. The free electrons can move throughout the metal under the imposition of only a very low driving force, voltage or thermal energy, and the magnitude of the electrical conductivity in a metal is closely related to that of the thermal conductivity. The precise relationship is called the Wiedemann–Franz law, which states that the ratio of thermal conductivity to electrical conductivity is given as follows:

$$\frac{\text{thermal conductivity}}{\text{electrical conductivity}} = \frac{3Tk^2}{e^2}$$

where T is the temperature (in K), k is the Boltzmann constant and e is the charge on the electron.

The high reflectivity of metals is also due to the free electrons. When light photons strike the metal surface, those electrons near to the Fermi surface can absorb the photons, as plenty of empty energy states lie nearby. However, the electrons can just as easily fall back to the lower levels originally occupied, and the photons are re-emitted. A detailed explanation of reflectivity of a metal requires knowledge of the exact shape of the Fermi surface and the number of energy levels (density of states) at the Fermi surface.

When two dissimilar metals are joined, electrons will flow from the higher Fermi energy to the lower. This gives rise to thermoelectric effects and to the operation of thermocouples. Less directly, the Fermi energy is related to the extent to which metals corrode.

Alloying with other metals or nonmetals in small amounts will not change these physical attributes drastically but the foreign atoms generally impede electron transport. Alloys therefore tend to have poorer electrical conductivity than do pure metals. If a new phase forms, the Fermi surface and the density of states at the Fermi surface will change and the electrical conductivity will alter abruptly. For example, the incorporation of hydrogen within magnesium will initially produce an interstitial alloy with metallic properties but an inferior electronic conductivity. Additional hydrogen leads to the formation of the hydride MgH_2, which is transparent and nonmetallic.

An important mechanical property of a metal is that of ductility, meaning that a metal can be deformed easily without breaking and can retain the deformed shape indefinitely. Pure metals in particular are soft and easily drawn into wires or hammered into foils. This property can be attributed to the crystal structures of the metals, which consist of packing of more or less spherical atoms. These can readily roll over each other to produce the deformation. There are no strong localised bonds to be broken, and so the metallic bonding does not hinder the deformation. However, metallic bonding does occur and, generally, calculations show that a pure metal deforms more easily than the metallic bonds should permit. The conflict is resolved by recognition of the role that defects play in the

[1]METGLAS is a registered tradename of AlliedSignal Inc.

crystals. Dislocations allow deformation to occur without the necessity of breaking significant numbers of metallic bonds at any moment.

Associated with the easy deformability of a metal is the fact that metals can easily be hardened. The explanation again lies with the dislocations present in the crystals. The trapping of these dislocations, called pinning, prevents easy deformation, and hence the metal becomes harder. At its simplest level, this can come about by the introduction of impurities and the formation of substitutional or interstitial alloys, a fact used empirically since the Bronze Age. However, more effective hardening comes about if precipitates form in the crystal. These, brought about by alloying and annealing, produce very hard metals. Steels, already far harder than pure iron, are hardened further by the incorporation of carbon or nitrogen followed by heat treatment that causes these elements to combine and form precipitates in the parent crystals. Alloy compositions that result in large amounts of second phase formation are hardest, as dislocation movement becomes completely impossible, with the consequence that the metal becomes brittle. Thus cast iron, which contains numerous precipitates of cementite and graphite, is very brittle, as are metals that contain large quantities of hydrogen.

In this short section, only the principal properties of metals and a simple appreciation of the origins of these properties are outlined. Far more detail will be found in Part 4, Chapters 10–15.

6.2   Ceramics

Ceramics are inorganic materials fabricated by a high-temperature chemical reaction. Most ceramics are oxides, but the term is also used for silicides, nitrides and oxynitrides, hydrides and other inorganic materials. Ceramics are regarded as chemically inert materials that are hard, brittle thermal and electronic insulators.

It is convenient to consider ceramics that are essentially silicates, called traditional ceramics, separately from all of the others. This latter group comprises engineering ceramics, with important mechanical properties, electroceramics, when electronic properties are emphasised, or glasses (noncrystalline ceramics). Traditional ceramics are used in utility applications such as brickwork and drainage pipes as well as for porcelain and other fine decorative ceramic objects. Engineering ceramics are used to extend the operating range available to metallic components. They are valued for high-temperature stability and for extreme hardness. Typical uses include: hard surface coatings on metallic components [titanium nitride (TiN), tungsten carbide (WC) and diamond]; inert high-temperature components (valves, cylinder liners, ceramic shielding and blankets, and furnace linings); high-speed cutting-tool inserts; and as abrasives [alumina (Al_2O_3), silicon carbide (SiC) and diamond]. Electroceramics are very-high-purity materials that possess unique electronic properties, varying from insulating to superconducting. Electroceramics form the active element in many gas sensors, temperature sensors, batteries and fuel cells. Ceramic magnets are widely used in motors. Ceramics are also important in fluorescent lighting and as components of computer displays.

6.2.1   Bonding and structure of silicate ceramics

The Earth's crust is composed very largely of silicates, as are the majority of semiprecious gemstones. Natural silicates are minerals that were formed from a complex molten magma and it is therefore not surprising that they are of variable composition. To each mineral is ascribed an 'ideal' composition – the composition that it would have if it were homogeneous. Isomorphous replacement, in which some cations are replaced by others of similar size, although not necessarily of the same charge, is common in minerals. Thus, the cations Na^+, Mg^{2+}, Ca^{2+}, Mn^{2+} and Fe^{3+} are readily interchangeable, as are the anions O^{2-}, F^- and OH^-. Aluminium, which is to the left of silicon in the periodic table, occupies a special role in silicate chemistry. Aluminium can replace silicon in silicates and does so in a random manner and to an indefinite extent. Isomorphous replacement produces substitutional defects in the crystal structure.

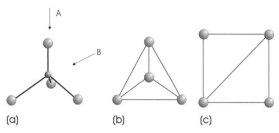

Figure 6.8 (a) Perspective view of a silicon – oxygen SiO_4 tetrahedron, (b) projection down A and (c) projection down B

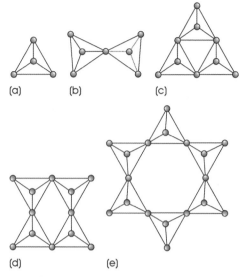

Figure 6.9 Corner-linked $[SiO_4]$ units found in ionic silicates: (a) isolated $[SiO_4]^{4-}$, (b) $[Si_2O_7]^{6-}$, (c) $[Si_3O_9]^{6-}$, (d) $[Si_4O_{12}]^{8-}$ and (e) $[Si_6O_{18}]^{12-}$

The mineral hornblende provides an illustrative example of isomorphous replacement. The ideal composition of this silicate is Ca_2Mg_2-$(Si_4O_{11})_2$ $(OH)_2$. A typical analysis of a naturally occurring sample might well show that up to a quarter of the silicon is replaced by aluminium; most of the Mg^{2+} replaced by Fe^{2+}, together with smaller amounts of Fe^{3+}, Mn^{2+} and Ti^{4+}, and about a third of the Ca^{2+} replaced by a mixture of Na^+ and K^+.

Silicon is a small atom with an electronic structure [Ne] $3s^2$ $3p^2$. Silicon lies below carbon in the periodic table and, like carbon, makes use of sp^3 hybrid bonds, which are arranged tetrahedrally. In silicates, each silicon atom is usually linked to oxygen to form an $[SiO_4]$ tetrahedron (Figure 6.8 and Figure 5.29, page 139). The bonds are very strong, and silicon–oxygen tetrahedra are stable and vary very little in size. The Si–O distances are always close to 0.162 nm, and the O–O distances to 0.27 nm. In terms of an ionic model, the $[SiO_4]$ tetrahedral group has an overall charge of −4, and is written $[SiO_4]^{4-}$.

These tetrahedra form the basic structural unit in silicate chemistry, and the $[SiO_4]$ unit dominates silicate chemistry and physics. They are found as isolated units, or condensed to form $[SiO_4]$ chains, sheets or three-dimensional networks. In forming these $[SiO_4]$ structures, only the vertices of the tetrahedra are shared, never edges or faces.

As a first approximation, silicates can be divided into three groups.

6.2.1.1   Three-dimensional silicates containing isolated silicate groups (ionic silicates)

These show little anisotropy of properties; that is, properties are the same in all crystallographic directions. There are a number of important divisions within this group, based on the geometry of the silicate groups. These are: isolated $[SiO_4]^{4-}$ units; pairs, $[Si_2O_7]^{6-}$; three-membered rings, $[SiO_9]^{6-}$; four-membered rings, $[Si_4O_{12}]^{8-}$; and six-membered rings, $[Si_6O_{18}]^{12-}$ (Figure 6.9).[2] Some examples of ionic silicates are listed in Table 6.2.

6.2.1.2   Silicates containing chains or sheets of silicate tetrahedra (extended anion silicates)

There are thousands of these compounds, many of which are valuable minerals. Even the simplest structure, a single chain of tetrahedra linked by corners,

[2] The figures in this section are descriptive. In real structures the silicon–oxygen tetrahedra are often slightly distorted, and crystal structure studies should be consulted when precise detail is required.

Table 6.2   A summary of silicate structures

Structure	Formula	Mohs Hardness	Examples
Isolated silicate groups:			
Monomer	$[SiO_4]^{4-}$	8–5	Mg_2SiO_4, forserite, (*olivines*)
			$Ca_3Cr_2(SiO_4)_3$, uvarovite, (*garnets*)
Dimer	$[Si_2O_7]^{6-}$	5	$Sc_2Si_2O_7$, thortveitite
Three-ring	$[Si_3O_9]^{6-}$	7–4	$BaTi(Si_3O_9)$, benitoite
Four-ring	$[Si_4O_{12}]^{8-}$	7–4	$Ca_3Al_2(BO_3)(Si_4O_{12})(OH)$, axinite
Six-ring	$[Si_6O_{18}]^{12-}$	6–4	$Be_3Al_2(Si_6O_{18})$, beryl
			$NaMg_3Al_6(BO_3)_3(Si_6O_{18})(OH)_4$, tourmaline
Chains:			
Single	$[SiO_3]^{2-}$	7–4	$MgSiO_3$, enstatite, (*pyroxenes*)
Double	$[Si_4O_{11}]^{6-}$	5	$Ca_2Mg_5Si_8O_{22}(OH)_2$, tremolite, (*amphiboles*)
Sheets:			
Single silicate layer	$[Si_2O_5]^{2-}$	3–1	$Na_2Si_2O_5$
Double silicate layer	$[SiO_2]$	3–1	$CaAl_2Si_2O_8$ (half Si replaced by Al)
Single silicate plus single	$[Si_2O_5]$	3–1	$Al_2(OH)_4Si_2O_5$, kaolinite, (*clays*)
hydroxide layer	plus hydroxide		$Mg_3(OH)_4SiO_5$, chrysotile, (*clays*)
Single silicate plus	$[Si_4O_{10}]$	3–1	$Al_2(OH)_2Si_4O_{10}$, pyrophyllite, (*clays*)
double hydroxide layer	plus hydroxide		$Mg_3(OH)_2Si_4O_{10}$, talc, (*clays*)
Single silicate	$[Si,AlO_{10}]$	3–1	$KAl_2(OH)_2Si_3AlO_{10}$, muscovite, (*micas*)
plus double hydroxide			$KMg_3(OH)_2Si_3AlO_{10}$, phlogopite, (*micas*)
Networks:			
Silicate	$[SiO_2]$	8	SiO_2, quartz
Aluminosilicate	$[(Si,Al)_4O_8]$	7–5	$KAlSi_3O_8$, *feldspars*

can adopt several different configurations, one of which is drawn in Figure 6.10(a). The formula of the single chains is $[SiO_3]^{2-}$, leading to compounds of formula $MSiO_3$, the *pyroxenes*. Two single pyroxene chains can join by linking half of the free vertices to form a double chain, of formula $[Si_4O_{11}]^{6-}$ (Figure 6.10b). These double chains are found in the *amphiboles*, a group including several forms of asbestos. Linkage of the free vertices in the plane of Figure 6.10(b) leads to single silicate layers (Figures 6.11a and 6.11b), and linkage of the free vertices that lie above the plane of Figure 6.11(a) will form double silicate layers (Figure 6.11c).

These latter structures are rather rare in mineralogical terms, but silicate layers in which the free vertices link to a layer of magnesium or aluminium hydroxide octahedra to produce composite layers

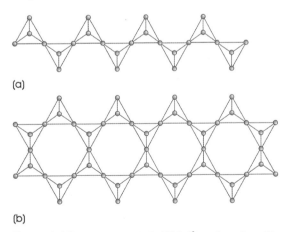

(a)

(b)

Figure 6.10   (a) Single-chain $[SiO_3]^{2-}$ strings, found in *pyroxenes* and (b) double-chain $[Si_4O_{11}]^{6-}$ strings found in *amphiboles*

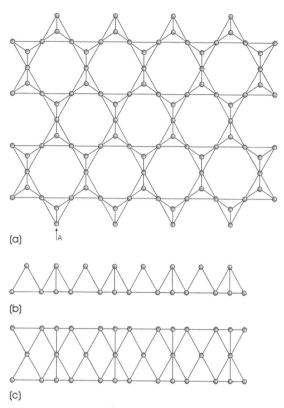

(a)

(b)

(c)

Figure 6.11 (a) A single sheet of corner-linked [SiO$_4$] tetrahedra, (b) the same sheet viewed along A and (c) double sheets formed by joining two layers of the type shown in part (a) by the free tetrahedral vertices, one over the other, viewed along A

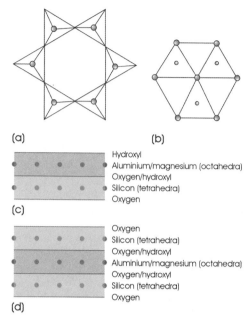

(a)   (b)

Hydroxyl
Aluminium/magnesium (octahedra)
Oxygen/hydroxyl
Silicon (tetrahedra)
Oxygen

(c)

Oxygen
Silicon (tetrahedra)
Oxygen/hydroxyl
Aluminium/magnesium (octahedra)
Oxygen/hydroxyl
Silicon (tetrahedra)
Oxygen

(d)

Figure 6.12 The structure of clays and related minerals: (a) a single layer of corner-linked [SiO$_4$] tetrahedra (the upper layer of apical oxygen atoms are drawn as spheres); (b) a basal close-packed layer of oxygen atoms in Mg(OH)$_2$ or Al(OH)$_3$ (large spheres) with cations in the octahedral sites above the layer (small spheres); (c) a composite layer formed by uniting a single silicate layer as in part (a) and a single hydroxide layer as in part (b); (d) a composite sandwich structure formed by uniting two silicate layers as in part (a), with a single hydroxide layer, as in part (b)

form an enormous group that includes many clays. These composite layers are possible because the positions of the oxygen atoms at the vertices of a silicate layer matches well the geometry of the close-packed array of oxygen atoms that occurs in the hydroxides Mg(OH)$_2$ and Al(OH)$_3$ (Figures 6.12a and 6.12b). In both of these structures, the cations are in octahedral sites (Figure 6.12b). The match between the oxygen arrays allows a single silicate layer to link to a single hydroxide layer (Figure 6.12c) or two silicate layers to sandwich a hydroxide layer (Figure 6.12d). The single silicate plus hydroxyl layer is found in clays such as kaolinite and in minerals such as chrysotile asbestos (Figure 6.13a). Double silicate sandwich layers

enclosing a hydroxyl layer are found in minerals such as talc (Figure 6.13b). If aluminium replaces some of the silicon, the silicate layers take on a negative charge. This is counterbalanced by cations placed between the layers, to form the micas (Figure 6.13c).

6.2.1.3 Silicates with [SiO$_4$] tetrahedra linked by all vertices to form a three-dimensional covalent network

This type of network can adopt a number of different conformations, typified by the polymorphs of silica, SiO$_2$. The structure of the commonest of

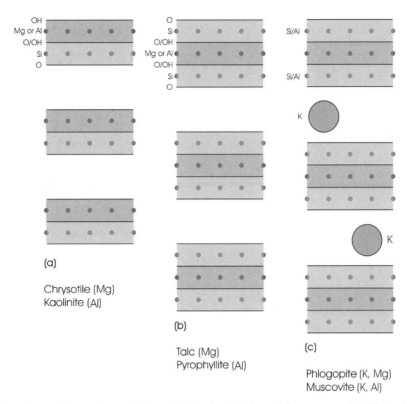

Figure 6.13   Structures of some clay and mica related minerals, formed from composite silicate–hydroxide layers

these polymorphs, quartz, is shown schematically in Figure 5.30 (page 139). When aluminium replaces some of the silicon, the framework takes on a negative charge, compensated for by the insertion of cations into the network. This produces the large group of minerals, the aluminosilcates, which includes the important *feldspars*, which make up many strong rocks, and the *zeolites* and *ultramarines*. All of these are of considerable industrial importance.

6.2.2 Bonding and structure of nonsilicate ceramics

Almost every inorganic oxide that does not contain silicon, as well as many carbides and nitrides, can be thought of, to some extent, as a nonsilicate ceramic. The bonding across this great variety of compounds varies widely, and no general bonding type predominates.

A large number of important ceramics adopt the *halite* (NaCl, B1) structure (Section 5.3.9). These include the oxides magnesium oxide (MgO) and nickel oxide (NiO) and many carbides and nitrides with a formula MX, such as titanium carbide (TiC) and titanium nitride (TiN). The oxides are often considered as ionic solids. The carbides and nitrides have metallic properties.

Alumina, Al_2O_3, is an oxide considered ionic in character. It is able to withstand high temperature and is used in laboratory furnaceware. A number of other important oxides, including Fe_2O_3 and Cr_2O_3, adopt the same (*corundum*) structure. The structure is most easily described as a hexagonal close-packed array of oxygen ions, with Al^{3+} ions distributed in an ordered fashion over two-thirds of the available octahedral sites (Figure 6.14). A number

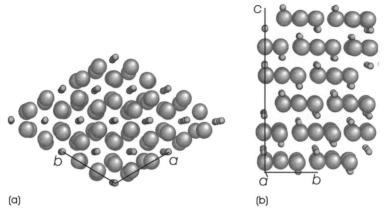

Figure 6.14   The structure of corundum, Al_2O_3: (a) projection down the [0 0 1] direction and (b) projection close to the [1 0 0] direction

of well-known gemstones consist of corundum doped with small amounts of transition metal impurities. Ruby is corundum containing about a half percent of Cr^{3+} substituted for Al^{3+}, and sapphire contains a small amount of Ti^{4+} and Fe^{2+} substituted for the Al^{3+}. Although the colour in these precious stones is due to the substitutional defects present, the exact hue exhibited is related to the unit cell parameter of the matrix.

Zirconia, ZrO_2, is an important ceramic because it is able to withstand high temperatures. At room temperature, the structure contains irregular polyhedra formed by seven oxygen ions surrounding each Zr^{4+} cation. The structure is monoclinic. At a temperature of about 1100 °C it becomes tetragonal,

a change caused by the coordination polyhedra becoming more regular. At temperatures of about 2300 °C the structure becomes cubic and adopts the fluorite, CaF_2, structure (Figure 6.15). In this structure, each Zr^{4+} ion is surrounded by eight oxide ions at the corners of a cube. The structure can be visualised as an array of ZrO_8 cubes, each linked by all edges, in a three-dimensional checkerboard array. The monoclinic to tetragonal crystallographic transformation at approximately 1100 °C involves the crystals in a significant volume change. This causes cracking and weakness in ceramic components and makes it impossible to use zirconia in its pure state for high-temperature components. However, the cubic form can be stabilised by the

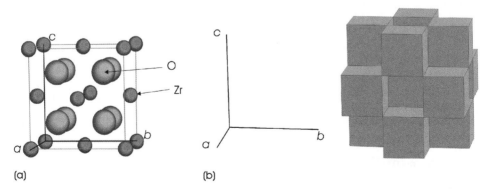

Figure 6.15   The cubic ZrO_2 (*fluorite*, CaF_2-type) structure. (a) a single unit cell and (b) the structure drawn as a stacking of (ZrO_8) cubes linked by edge sharing

addition of impurities such as calcia, CaO, in which Ca^{2+} ions substitute for Zr^{4+} ions in the structure (see Section 3.4.5). The resulting calcia-stabilised zirconia remains cubic from room temperature to the melting point, above 2300 °C, allowing it to be used for high-temperature applications.

The structures of a number of other important nonsilicate ceramics are discussed elsewhere, (Chapters 5, 8–14 and 16).

6.2.3 The preparation and processing of ceramics

Many of the characteristic properties of ceramics arise during manufacturing. Ceramic bodies are made by high-temperature firing (heating) routes that induce chemical reactions to take place, during which the final material is produced.

Traditional ceramics are mainly made from mixtures of clays, silica (often extracted as flint) and feldspars (especially $K_2Al_2Si_6O_{16}$ and $Na_2Al_2Si_6O_{16}$). Low-quality structural products such as bricks and pipes are made directly from the appropriate clay. Higher-quality ceramics such as porcelain are made from carefully controlled amounts of specific clay, flint and feldspar. The use of these three major components has led to the name triaxial whitewares for these materials.

Clay-based traditional ceramics are first formed into the shape desired by using traditional potters' techniques or automated processes. The resulting shapes are then heated to 1000 °C or higher. At these temperatures, the clays initially react to lose water and subsequently hydroxyl ions. The residues react to give a mixture of new phases, including glasses. Considerable shrinkage usually occurs during these reactions, and the production of an object with a precise size and shape remains a problem.

Engineering ceramics and electroceramics are usually pure single phases. Many are made by (a) milling the components to produce a fine powder; (b) pressing this into the desired shape; and (c) firing (heating to a high temperature). This third step is important for the production of the final shape in a useful form, and heating cycles are often complex. Initial heating is usually at a low tempera-

ture, 100–400 °C, to remove water and to burn off binder (an organic material used to cement the dry grains together). At a later stage of processing, the material is heated to a final temperature, in excess of 1000 °C, to sinter the particles and allow any chemical reactions to occur (see Section 8.4). In this final stage, some components may melt and in some cases (i.e. porcelain) one of the products is a glass. This is called vitrification.

Hard ceramic surface coatings on metallic components can be made by heating the metal in an appropriate gaseous atmosphere. Reaction takes place at the metal surface and atoms from the gaseous component diffuse into the surface layer. Thus, if titanium is heated in nitrogen gas a layer of titanium nitride (TiN) will form on the surface as a hard layer.

6.2.4 The principal properties of ceramics

The principal properties of ceramics arise from a combination of chemical bonding and the atomic defects and microstructure resulting from the fabrication techniques described above.

The bonding, whether described as ionic or covalent, is strong, which ensures that the solids are chemically inert and often stable to high temperatures (refractory). Refractory ceramics are widely used in furnaces and other high-temperature equipment. In addition, in aerospace applications refractory components are used both externally, to protect the outside of re-entry modules, and internally, in rocket motors.

The lack of free electrons endows basic ceramics with poor thermal and electronic conductivity. The chemical flexibility of ceramics, however, allows them to be selectively doped with other ions. In particular, doping with transition metal or lanthanide ions generates a wide variety of colours and can radically alter electronic and magnetic properties. Thus, insulators can be transformed into superconductors. How this comes about is described in Part 4, Chapters 10–15.

Ceramics do not deform very easily at ordinary temperatures, as strong chemical bonds must be broken and, unlike metals, dislocation movement is severely hampered. Generally, stress results in brittle fracture, especially under impact (see Section 10.1.7). However, in large part, the brittle nature of ceramics arises not from the bonding or dislocation density but in the microstructure. Surface flaws cause many ceramics to fail catastrophically when flexed, and pores in the structure, resulting from the chemical reactions that take place during high-temperature fabrication, are another source of weakness. Although ceramics have a rather low tensile strength and readily fracture when stretched they are much stronger when compressed. This property, coupled with the strong bonding, makes ceramics hard, and they are used as abrasives, cutting tools and hard coatings. Mechanical failure under compression also occurs because of structural defects, including voids and pores, large grains and foreign inclusions, rather than by bond breaking.

6.3   Glass

Most glasses are oxides, but specialised glasses such as metallic glasses, are becoming increasingly important. Glass is usually understood to mean a hard, transparent, fairly strong, corrosion-resistant material, in which the main component is silica, SiO_2. These silicate glasses are the only glasses discussed in this Chapter. There are a number of naturally occurring silicate glasses, including obsidian (a volcanic rock which is black as a result of iron oxide impurities), pumice (a glassy froth), flint and opal. These all show the typical glass properties of hardness and brittleness.

Silicate glass production marks an early stage in civilisation. Faience was made by the Egyptians thousands of years ago. This material was made from moulded sand coated with 'natron', a residue of minerals left after flooding of the River Nile, consisting mainly of calcium carbonate ($CaCO_3$), sodium carbonate (Na_2CO_3), common salt ($NaCl$) and copper oxide (CuO). The object was heated to about $1000\,^\circ C$, at which point the alkali coating reacted to form a glassy exterior with a blue colour imparted by the copper oxide.

Improvements in glass technology were carefully guarded secrets of medieval guilds, and it is only relatively recently that high-quality transparent glass has been readily available. Great advances in the manufacture of silica glass of high transparency were brought about with the development of optical fibre based communication systems towards the end of the 20th century.

Some widely used silicate glasses are listed in Table 6.3.

6.3.1   Bonding and structure of silicate glasses

There is no one structure of glass any more than there is one structure of a crystal, and almost any solid can be produced in a glassy state if the melt is cooled sufficiently quickly. To some extent, glass can be thought of a product of kinetics, and the

Table 6.3   Some silicate glasses

Name	Typical composition	Important property	Principal uses
Soda glass	15 % Na_2O: 85 % SiO_2	Cheap	Window glazing
Soda-lime glass	72 % SiO_2: 14 % Na_2O: 14 % CaO	Cheap	Window glazing
Borosilicate (Pyrex®)	80 % SiO_2: 13 % B_2O_3: 7 % Na_2O	Low coefficient of expansion	Cooking ware, laboratory ware
Crown glass	9 % Na_2O: 11 % K_2O: 5 % CaO: 75 % SiO_2	Low refractive index	Optical components
Flint glass	45 % PbO: 55 % SiO_2	High refractive index	Optical components, 'crystal' glass
Lead glass	Up to 80 % PbO: SiO_2	Absorbs radiation	Radiation shielding
Silica	100 % SiO_2	Very low coefficient of thermal expansion	Optical components, laboratory ware, optical fibre

structure of a glass may depend on the rate at which the liquid is cooled.

The main structural unit in silicate glasses is the $[SiO_4]$ group (Figure 6.8). The covalent bonds between the central silicon atom and the oxygen atoms are very strong, and the liquid and the solid states of silica and silicates contain large numbers of $[SiO_4]$ tetrahedra. These tetrahedra link to one another by sharing corner oxygen atoms to form discrete or interpenetrating chains and rings. In the solid these are locked in place whereas in the liquid they are continually changing orientation.

To form a crystalline solid, a liquid must first form crystal nuclei. In the case of metals, this is extremely easy, as the spherical atoms can quickly pack into arrays. In the case of silicates, the entangled chains are difficult to rearrange into an ordered crystalline array. This process often needs bonds to be broken as well as a rearrangement of tetrahedra. Because of this, nucleation is very slow, and cooling the melt at even slow rates can result in the formation of a glass. The structure of a silicate glass was thus envisaged by Zachariasen, in 1932, as an irregular intertwining of chains of corner-linked $[SiO_4]$ tetrahedra to form a loose random network (Figure 6.16).

A silicate glass can also be thought of as a liquid that has been cooled to below the crystallisation point without crystallisation occurring. This state is referred to as supercooling, and glass is often described as a supercooled liquid. However, it is important to keep in mind that glass is *not* a liquid but a solid with a structure that does not show any long-range order. This status of glass is revealed by the behaviour on heating, because glasses do not have a sharp melting point. Instead, they continually soften from a state which can be confidently defined as solid to a state which can be defined as a viscous liquid. In place of a melting point, glasses can be characterised by a glass transition temperature, T_g. The glass transition temperature is determined by plotting the specific volume (the volume per unit mass) of the glass as a function of temperature (Figure 6.17). Both the high-temperature and low-temperature regions of such a plot are usually linear. The value of T_g is given by the intersection of the extrapolated high-temperature and low-temperature

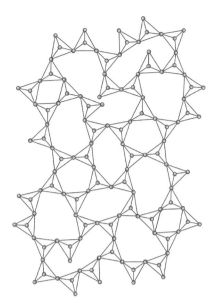

Figure 6.16   The random network structure of corner-linked $[SiO_4]$ tetrahedra in a silicate glass

lines. Above the glass transition temperature, the material can be considered a liquid whereas below the glass transition temperature it is considered a solid. The glass transition temperature is not a fixed material property but varies with the cooling rate. Nevertheless, it is a useful material parameter and gives guidance concerning the softening and working temperature of a particular glass.

The random network model works well for silicate glasses. It is based on the idea that the cations forming the network are linked to three or four oxygen atoms to form a strong polyhedron, and each oxygen atom is linked to one cation or, at most, two cations. This is equivalent to noting that polyhedra share some corners with neighbouring polyhedra. Small ions, which adopt triangular or tetrahedral metal–oxygen coordination polyhedra, should therefore take part in glass formation. These ions, typically B^{3+}, Ge^{4+}, Al^{3+}, Be^{2+} and P^{5+}, are known as *network formers*. Large cations, which tend to disrupt the ability of the $[SiO_4]$ tetrahedra to crystallise in regular arrays, also enhance glass formation. These ions, known as *network modifiers*, are typified by K^+, Na^+, Mg^{2+} or Ca^{2+}. Other cations, those with higher valence

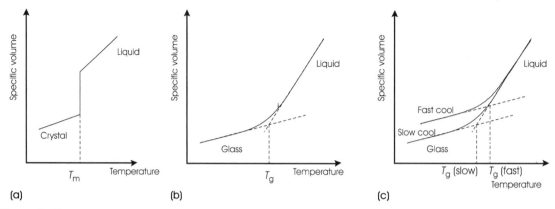

Figure 6.17 Specific volume versus temperature curves: (a) a crystalline solid with a melting point T_m; (b) a glass, with a glass transition temperature T_g; and (c) the effects of cooling rate on glass transition temperature

and a coordination typically of 6, are called *intermediates*. Typical examples are Ti^{4+}, Cu^{2+} and Zn^{2+}. Although intermediate ions alter the properties of a glass (e.g. by adding colour) they do not have a direct role to play in glass formation. Note that these three divisions are not mutually exclusive and some ions fall into more than one category. For example, Al^{3+} is regarded both as a network former, when in triangular or tetrahedral positions, and as an intermediate, when in octahedral coordination.

6.3.2 Glass deformation

One of the most curious properties of glass, and one that differentiates glass from metals and crystalline

ceramics, is the fact that it can be deformed readily in a semimolten state by traditional techniques such as glass blowing. This is because glasses behave as very viscous liquids at moderate temperatures. This state can be manipulated into the desired shape, which is retained in solid form on cooling.

The viscosity of a glass is an important physical parameter, as it defines the temperature regimes over which the glass can be worked. For example, the working range of a glass is the range of viscosities over which normal processing takes place, usually between viscosity values of 10^3 Pa s and $10^{6.6}$ Pa s. The important viscosity ranges for glass manipulation are given in Table 6.4, together with the approximate temperature at which these values are reached for an ordinary soda-lime glass.

Table 6.4 Glass viscosity

Glass condition	η/Pa s	Comment	Temperature/°C for soda-lime glass
Melting temperature	10	Glass becomes fluid and a homogeneous melt is achievable	1450
Working point	10^3	Glass is easily deformed but retains its shape	1000
Softening point	$10^{6.6}$	Glass deforms under its own weight	700
Annealing point	10^{12}	Residual stress in a thin plate can be removed in 15–20 minutes	550
Strain point	$10^{13.5}$	Fracture–plastic deformation boundary (see Section 10.1.10)	500

Note: η = viscosity. All figures given in this table are approximate.

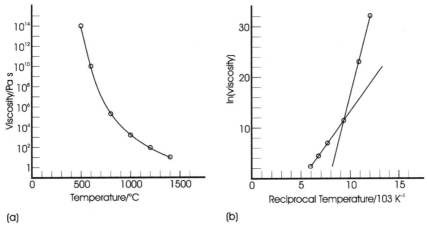

Figure 6.18 (a) Viscosity versus temperature for a typical soda-lime glass and (b) plot of ln (viscosity) versus reciprocal temperature for the same glass

[The temperature at which any glass reaches the requisite viscosity depends on its composition.] The variation of viscosity with temperature for a soda-lime glass is plotted in Figure 6.18(a).

Because of the practical importance of viscosity, there have been several attempts to define the relationship between viscosity and temperature by a mathematical equation. The simplest of these is an Arrhenius equation:

$$\eta = \eta_0 e^{E/RT} \qquad (6.1)$$

where η is the viscosity (Pa s), η_0 is a constant, E is the energy for viscous flow, often called the activation energy, R is the gas constant and T the temperature (in kelvin). This equation well describes activated processes, such as diffusion (see Section 7.4). In such processes, the reaction or transformation is possible only when the reactants or participants overcome an energy barrier, the activation energy. Arrhenius behaviour is confirmed by a plot of ln η versus $1/T$:

$$\ln \eta = \ln \eta_0 + \frac{E}{RT} \qquad (6.2)$$

The plot should be linear with a slope of E/R. Materials that conform to Equation (6.2) are said to

exhibit Arrhenian behaviour. This is not generally found to occur for glasses. The data in Figure 6.18(a) is plotted in this form in Figure 6.18(b). It is seen that the low-temperature and high-temperature parts of the graph are good fits to Equation (6.2), but the two slopes are quite different. At high temperatures, the slope is lower than at low temperatures. Such behaviour is typical of glasses and is referred to as non-Arrhenian behaviour.

Several other equations have been suggested to overcome the shortcomings of the Arrhenius equation. The most widely used is the Vogel–Fulcher–Tamman equation (also called the Vogel–Fulrath equation):

$$\eta = A \, \exp\left(\frac{B}{T - T_0}\right)$$

where A, B and T_0 are empirical constants, and T is the temperature (in kelvin). This leads to the relation;

$$\ln \eta = \ln A + \frac{B}{T - T_0}$$

It has been found that liquids that readily form glasses exhibit non-Arrhenian behaviour whereas those that do not readily form glasses show Arrhenian behaviour. Melts that are Arrhenian are classified as strong whereas those that are non-Arrhenian are classified as fragile. The fragility of a liquid, its fragility index, can be related to the structure of the liquid and gives guidance on the ease of glass formation in a system (for more information on this development, see the Further Reading section at the end of this chapter).

6.3.3 Strengthened glass

Glass is strong in compression but notoriously weak in tension. Freshly drawn glass fibres, for example, are stronger than steel, but, in particular, attack of the surface by water vapour in the atmosphere causes the strength of the fibre to fall dramatically. This weakness is usually attributed to small flaws in the glass surface, called Griffiths flaws (see Section 10.1.7). Under tension the flaws generate cracks, which open, allowing the material to fracture. The various mechanisms for strengthening glass are ways to prevent the cracks from propagating through the solid.

Tempered glass is about four times stronger than ordinary glass, and, when it fractures, it breaks into small, blunt pieces rather than jagged shards. The glass is strengthened by rapidly cooling the hot surface with air jets. Initially, the glass is cut to shape, and surface flaws and rough edges removed by grinding and polishing. The glass is then heated to 620 °C. High-pressure jets of air, in a predetermined array, rapidly cool the surface in several seconds. During this quenching process, the outside of the glass sheet becomes rigid, as the outside temperature drops below the glass transition temperature. The inside, though, is still above this temperature as it cools more slowly than the outside. As the inside cools, it tends to shrink away from the solid outer surfaces. This results in the centre of the sheet being in tension, whereas the outside of the sheets are being pulled in and so are in compression. These opposing tensile and compressive stresses are the strengthening mechanism. As glass usually fails because of the generation of surface cracks, a surface under a compressive force is much harder to break. Carefully controlled patterns of stresses are generated, dependent on the final use of the material.

Chemical strengthening aims to mimic the tension and compression distribution just described by using chemical means. The methods used are successful but tend to be more expensive than air cooling and are used only in applications in which cost is a secondary consideration. The principal of the method is selectively to replace some of the metal ions in the glass to achieve tension or compression. For example, if a soda-lime glass is placed in a bath of molten potassium nitrate (KNO_3) the Na^+ in the glass surface is replaced by the larger K^+. This causes a surface compressive stress and an interior tensile stress, and the glass is strengthened. This process is used to produce aircraft glazing and lenses. Similarly, if the Na^+ is replaced by Li^+, which is smaller, the surface is under tension and the centre under compression. This process is used in the fabrication of glass for use as laser material.

6.3.4 Glass ceramics

A glass ceramic is a solid that is largely crystalline, made by the crystallisation or devitrification of a glass object of the desired shape. Glass ceramics are therefore composite materials that consist of crystals and some glass. They combine the ease of production of glass with much enhanced thermal and mechanical properties. In this section, the microstructures of some glass ceramics are described, and the way in which the superior properties are achieved is outlined.

The transformation of the initial glass object into a largely crystalline object of the same shape and size is carried out by a thermal treatment. This induces the precipitation of crystal nuclei, the growth of crystals on the nuclei and the development of an almost fully crystalline final product (Figure 6.19). It is important that each of these steps is carefully controlled if the desired glass ceramic is to result. In addition, the crystallisation must occur in a glass that is viscous enough not to sag or distort during the transformation, so the viscosity must also be closely controlled. As nucleation, crystal growth

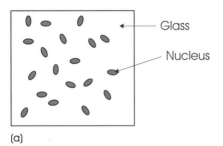

(a)

- Glass
- Nucleus

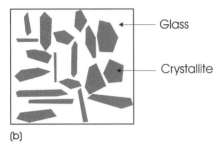

(b)

- Glass
- Crystallite

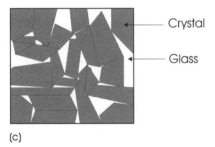

(c)

- Crystal
- Glass

Figure 6.19 Nucleation (part a) and growth (parts b and c) in a glass ceramic

and viscosity are all temperature-sensitive the correct conditions to give a satisfactory product must be achieved by a precise juggling of the chemical components used.

Good mechanical properties are achieved because the solid consists of a mass of interlocking crystals. Any surface flaws cannot easily propagate through the solid, as the passage of a crack is blocked by crystals in its path. Good thermal properties are achieved by ensuring that the crystals that form have very low coefficients of expansion, thus making the material resistant to thermal shock (see Sections 15.1.6–15.1.8). The optical properties of

the solid can also be manipulated. If the crystals are kept to a dimension below that of the wavelength of light, the solid will be transparent to visible light. If the crystal dimensions are larger, the solid can be, for instance, opaque to visible light but transparent to microwaves or radar waves. Crystallites of a modest size that are well dispersed in a glass residue give a translucent solid, typified by porcelain.

The two most important factors in glass ceramic production are the composition of the melt and the microstructure of the final product. These are interrelated, of course. The composition controls the ability of the substance to form a glass with the correct viscosity and workability, as the starting solid is completely glassy in nature. Composition also controls what nuclei can form in the glass and the types of crystal that can grow. Most crystals have a definite crystal habit, and this factor greatly influences the microstructure of the final solid.

The simplest glass ceramics, from a microstructural viewpoint, are the transparent ultrafine-grained materials. These consist essentially of the high-temperature form of quartz (Figure 5.30, page 139). The preparation of these materials starts with a silica melt that also contains some zirconia (ZrO_2), titania (TiO_2), alumina (Al_2O_3) and magnesia (MgO). The melt is formed into the desired shape and then heat treated, with the viscosity remaining at a value great enough to prevent sag or deformation. At this temperature, $ZrTiO_4$ crystals are the first to nucleate and, on these, crystals of the high-temperature form of quartz, stable between 573 °C and 870 °C, grow. The presence of the aluminium in the melt means that the quartz is not pure SiO_2, but some aluminium substitutes for silicon in the crystals. In order to maintain charge neutrality, some Mg^{2+} is also incorporated into the quartz, and this stabilises the structure to room temperature. The final solid is a mass of crystals less than 60 nm in size, together with some residual glass, which cements them together. The small crystal size means that the solid is transparent, and the very low thermal expansion of quartz means that the material is resistant to thermal shock. The many small grains in the structure prevent crack growth, so that the ceramic is also strong.

Other compositions and heat treatments result in glass ceramics with different microstructures. The amount of glass present and the type of crystalline phase present will have a profound effect on the resultant properties. For example, in materials used as rocket nose cones, the structure contains mainly cordierite ($Mg_2Al_2Si_5O_{18}$), an aluminosilicate with a very low thermal expansion coefficient. The residual glass is in the form of isolated volumes, so that flow is not easy when the material is heated. Machinable glass ceramics can also be prepared by suitable control of chemistry and crystallisation. These materials contain a crystalline phase consisting of interlocking plates of mica, with a composition approximating to fluorphlogopite ($KMg_3AlSi_3O_{10}F_2$). The presence of the interlocking plates prevents the solid from splintering when it is cut. These are used for precision ceramic components that can be manufactured only via a machining stage.

6.4   Polymers

Polymers are long chain-like giant molecules (macromolecules) made by the linkage of large numbers of small repeating molecules called monomers. Short chain lengths formed in the course of synthesis or degradation of polymers are called oligomers. The majority of polymers, and the only ones considered here, are compounds of carbon. Polymers are very widespread and can be synthetic (e.g. nylon) or natural (e.g. rubber). They form vital components of living organisms, and the most important molecule, DNA, is a polymer of amino acids. Colloquially, polymers are often called plastics. More precisely, plastics are sometimes defined as polymers that can be easily formed at low temperatures, and sometimes as a pure polymer together with a nonpolymeric additive, which may be solid, liquid or gas.

There are two main divisions of polymeric materials: thermoplastic and thermosetting. Thermoplastic materials can be formed repeatedly; that is, they can be melted and reformed a number of times. Thermosetting materials can be formed only once; they cannot be remelted. They are usually strong,

and are typified by resins. A further group of polymers merits mention: elastomers. Elastomers can be deformed a considerable amount and return to their original size rapidly when the force is removed.

The properties of polymers depend both on the details of the carbon chain of the polymer molecule and on the way in which these chains fit together. The chain form can be linear, branched or cross-linked, and a great variety of chemical groups can be linked to the chain backbone. The chains can be carefully packed to form crystals, or they can be tangled in amorphous regions. Amorphous polymers tend not to have a sharp melting point, but soften gradually. These materials are characterised by a glass transition temperature, T_g, (Section 6.3.1), and in a pure state are often transparent.

Although polymers are associated with electrically insulating behaviour, the increasing ability to control both the fabrication and the constitution of polymers has led to the development of polymers that show metallic conductivity superior to that of copper (see Section 13.2.8) and to polymers that can conduct ions well enough to serve as polymer electrolytes in batteries and fuel cells (see Sections 9.2.5 and 9.3.7).

In this section, polymers will be discussed largely from a structural and microstructural point of view. Several typical and differing polymers are used as examples: polyethylene (polythene), nylon and epoxy resins. In addition, elastomers are described, as they differ in a fundamental way from other materials.

6.4.1   The chemical structure of some polymers

Polymers were once grouped mainly in terms of the overall chemistry of the polymerisation reaction. Molecules that simply added together were called addition polymers, and those that joined and at the same time eliminated one or more small molecules were called condensation polymers. These designations, which derived from organic chemistry, have now largely been replaced by groupings that reflect the mechanism of the polymerisation rather than the overall chemical reaction. [Section S2.1 gives a synopsis of relevant organic chemistry terminology.]

(a) $H_2C = CH_2$

(b) $H_2\overset{|}{C} - \overset{|}{C}H_2$

(c) $H_2\overset{|}{C} - \overset{|}{C}H_2$
$\quad\quad H_2\overset{|}{C} - \overset{|}{C}H_2$
$\quad\quad\quad\quad H_2\overset{|}{C} - \overset{|}{C}H_2$
$\quad\quad\quad\quad\quad\quad H_2\overset{|}{C} - \overset{|}{C}H_2$

Scheme 6.1 The polymerisation of ethene (ethylene): (a) a single monomer molecule, (b) double bond opening and (c) monomer linkage to form the polymer chain

In this section the older common names, rather than the systematic names, are frequently used for organic compounds, as they have largely been retained in polymer names.

In order to link a large number of molecules, or monomers, together it is necessary for each end of the molecules to be made chemically reactive. Generally, this involves breaking chemical bonds to yield two reactive half-bonds. The simplest starting point for a discussion of this process is the monomer molecule ethene (ethylene):

$$CH_2 = CH_2$$

Schematically, the monomer ethene can be linked to other ethene monomers if the double bond is opened and the resulting broken bonds are linked together, in an addition reaction (Scheme 6.1). The chemical formula of the resulting polymer, called polyethylene or polythene, is:

$$[CH_2{-}CH_2]_n$$

where n takes a value of several thousand. [Note that the industrial preparation of polyethylene (and of all the polymers described here) is quite different from the scheme illustrated. Very skilled chemistry has been employed to achieve the precise properties that polymers now display.] The polymer chain is constructed from (CH_2) units. In these, the carbon atoms are bonded to two hydrogen atoms and two carbon atoms using strong sp^3 hybrid bonds

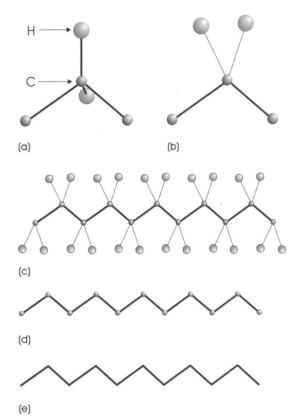

Figure 6.20 (a, b) The structure of the CH_2 unit in a polymer chain, in two orientations (the four bonds arising at the carbon atom are arranged tetrahedrally), (c) a chain of linked CH_2 units in a polymer chain and (d, e) representations of the chains with H atoms omitted

(Figure 6.20). The carbon–carbon bonds are free to rotate, which allows the polymer chain to coil into ordered or disordered regions. Note that polyethylene, like all polymers, does not have a definite chemical formula. The number of (CH_2) units in the chain is influenced by preparation conditions. Polymers with low average values of n have different physical properties from those in which n is larger (see below).

Other related polymers are formed by replacing one or more of the hydrogen atoms in the monomer ethene with a chemical group of atoms, represented by X. Hence the formula of the monomer becomes $CH_2{=}CHX$ when one hydrogen is replaced. The

Table 6.5   Addition polymers

Monomer	Polymer	Names	Uses
$CH_2=CH_2$ (ethene, ethylene)	$(CH_2-CH_2)_n$	Polyethylene, polythene, PE	Squeeze bottles, food bags, dishes, insulation, coatings
$CH_2=CHCl$ (vinyl chloride)	$(CH_2-CHCl)_n$	Poly(vinylchloride), PVC	Pipes, floor covering, insulation, adhesives, films, credit cards
$CH_2=CCl_2$ (vinylidene chloride)	$(CH_2-CCl_2)_n$	Poly(vinylidene chloride)	Food wraps, fibres, cling film
$CH_2=CHCH_3$ (propylene)	$[CH_2-CH\ (CH_3)]_n$	Polypropylene, PP	Pipes, valves, carpets
$CH_2=CHC_6H_5$ (styrene)	$[CH_2-CH\ (C_6H_5)]_n$	Polystyrene, PS	Jugs, cups, packaging, styrofoam, appliance parts
$CH_2=CHCN$ (acrylonitrile)	$[CH_2-CH\ (CN)]_n$	Polyacrylonitrile, PAN, Orlon®, Acrilan®	Fabrics, carpets, high-impact plastics
$CH_2=CHCOOCH_3$ (vinyl acetate)	$[CH_2-CH\ (CH_3\ COO)]_n$	Poly(vinyl acetate), PVA	Wood adhesives, paper coatings, latex paints
$CF_2=CF_2$ (tetrafluoroethene)	$(CF_2-CF_2)_n$	Polytetrafluoroethylene, PTFE, Teflon	Nonstick coating, electrical insulation, bearings
$CH_2=C(CH_3)COOCH_3$ (methyl methacrylate)	$[CH_2-C(CH_3)\ COOCH_3]_n$	Poly(methyl methacrylate) Perspex®, Lucite®, Plexiglas®	Substitute glass, acrylic paints, pipes
$CH_2=CH-CH=CH_2$ (1,3-butadiene)	$(CH_2-CH\ -CH-CH_2)_n$	Polybutadiene, buna rubber	Tyres, hoses, pond liners

principle of polymerisation remains precisely the same, however, although the details are modified by the size and location of the side-group *X*. Some examples of these polymers are given in Table 6.5.

A typical condensation reaction can take place between an acid group, —COOH, and a hydroxyl group, —OH, to form a larger molecule and to 'split out' water:

$$R-COOH + R'-OH \rightarrow R-COO-R' + H_2O$$
[6.1]

where R and R′ represent different carbon chains. In molecules with only one acid or hydroxyl group, as written in Reaction [6.1], the reaction stops after the first step. In order to create a polymer, the monomers of condensation polymers need *two* reactive functional groups on each monomer.

The polyesters, which generally start from terephthalic acid and the alcohol ethylene glycol, are an important group of polymers made in this way. The reaction is between acid groups and hydroxyl groups (Scheme 6.2). The product, a polyester

called poly(ethylene terephthalate), or PET, is widely used to make shatter-proof bottles.

A second group of polymers that form in a similar way are the thermoplastic polyamides, better known as nylons. The principal reaction is between an acid group, —COOH, and an amide group, —NH₂:

$$R-COOH + R'-NH_2 \rightarrow R-CONH-R' + H_2O$$

where R and R′ represent different carbon chains. Once again, in order to form a polymer, each monomer must have a reactive group at each end of the molecule. The schematic structure and the reaction scheme for the formation the commonest type of nylon, called nylon 6,6, is drawn in Scheme 6.3(a). In this scheme, the carbon skeleton of the polymer chain is drawn as a zigzag (see Figure 6.20) and the carbon atoms are numbered. The name, nylon 6,6, indicates that there are six carbon atoms in each section of the repeat unit of the polymer.

Amino acids are molecules with an acid group at one end and an amide group at the other, separated

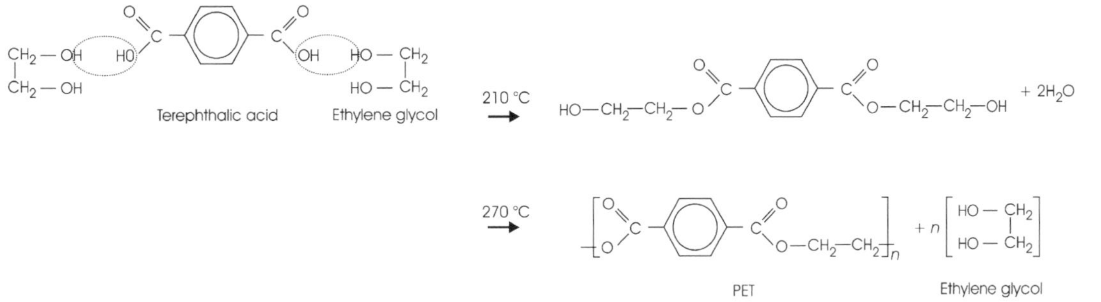

Scheme 6.2   The production of the polyester poly(ethylene terephthalate), PET. The linking of the terephthalic acid and ethylene glycol molecules occurs at 210 °C, and polymerisation and regeneration of ethylene glycol at 270 °C

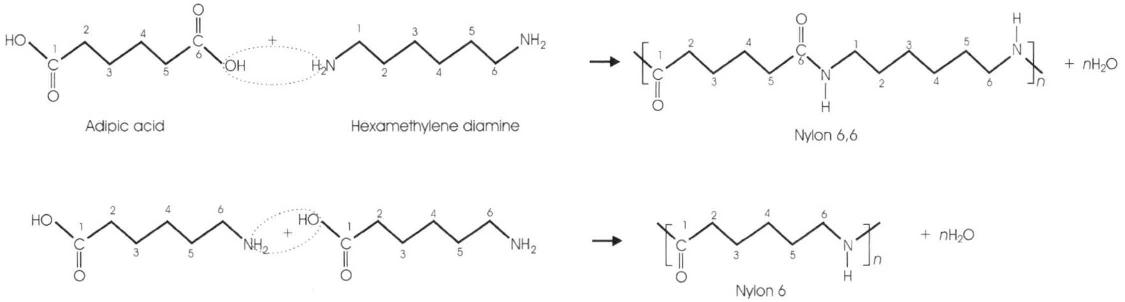

Scheme 6.3   (a) The formation of nylon 6,6; (b) the formation of nylon 6. The figures give the number of carbon atoms in the polymer chain between nitrogen atoms

by a carbon chain, $HOOC-R-NH_2$. These molecules can also polymerise by linking head to tail, to generate a nylon. The amino acid analogue of nylon 6,6 is nylon 6 (Scheme 6.3b). [Note that nylon 6 is not actually made from the amino acid shown, but from a molecule called ε-caprolactam.]

Nylon 6,6 is an example of an even–even nylon, sometimes just abbreviated to 'even'. It is easy to envision other even–even polymers, such as nylon 4,4. Similarly, it is possible to think of odd–odd, or just 'odd', nylons, such as nylon 5,5. Although these nylons have very similar chemical properties they differ in important electrical aspects, and the nature of the chain determines whether these plastics can be used to make piezoelectric components (see Section 11.2.3).

A rather similar chemical reaction produces widely used thermosetting polymers typified by epoxy resins. In this group of materials, instead of a reactive double bond a reactive group, the epoxide group (Scheme 6.4), is opened and used to link monomers. Epoxy resin adhesives normally come as two-part mixes. The resin component contains small and medium-sized molecules with an epoxy group at each end, called diepoxy molecules (Schemes 6.4b and 6.4c). The resin is set by adding a cross-linking agent, or 'catalyst'. This material is a diamine (Scheme 6.4d). These join the epoxy-containing molecules together to form a strong cross-linked network. Once the network has been formed it is very difficult to disrupt, and epoxy resins are typical thermoset polymers. Note that there is no reaction product on polymerisation, which means that only a small dimensional change occurs as the precursors harden. This is of importance in applications where shrinkage or expansion would create difficulties, as in the original application of these materials, for dental fillings.

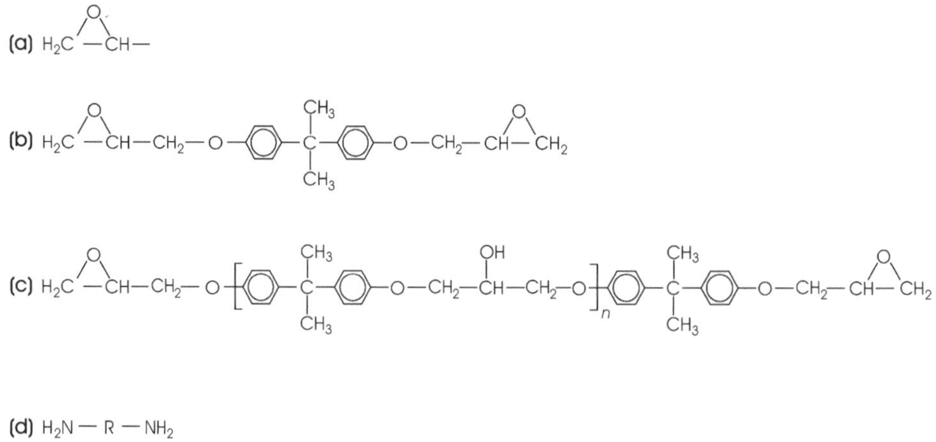

(a)

(b)

(c)

(d) $H_2N - R - NH_2$

Scheme 6.4 The structure of molecules in epoxy resins: (a) the epoxy group; (b) a small diepoxy molecule; (c) a small polymer molecule (with *n* up to about 25), found in the resin part of a two-part epoxy adhesive mix; and (d) a diamine linking group ('catalyst') in which R represents a short chain of CH_2 groups

All cross-linked polymers tend to be difficult to disrupt, and they can be regarded as one giant molecule. This makes them of use when durability is needed. Many of the polymers mentioned above can be modified so as to form extensive cross-linking by increasing the number of reactive groups on the monomers. For example, the polyester formed by phthalic acid and glycerol (Scheme 6.5) can cross-link to form a polymer that is used in bake-on car paints.

6.4.2   Microstructures of polymers

Although the chemical makeup of polymers influences properties, by far the most important aspect,

from this point of view, is that of the microstructure. The microstructure of polymers can be considered to derive from that of a long chain of strongly linked carbon atoms. The most important aspects of polymer microstructure that need to be considered are chain length, chain branching, chain side-groups (which contribute to chain stiffness) and the strength of cross-links between chains. The degree of crystallinity of the polymer, which depends on the factors just listed, is also of considerable importance. In fact, the strength of most thermoplastic polymers depends on the degree of crystallinity of the material. In this section, these basic microstructural features are described. The way in which these are achieved during manufacture is covered in Section 6.4.3, and properties are discussed in terms of microstructure in Section 6.4.5.

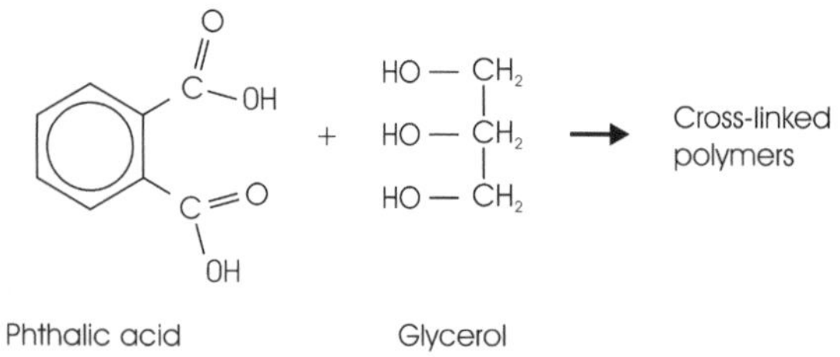

6.4.2.1   Molar mass

Polymers consist of long chains of varying length that cannot be characterised by a constant molar mass. Indeed, cross-linked polymers can be thought of as a single molecule, so that the molar mass is the total mass of the object. However, it is helpful to have a measure of the degree of polymerisation or the distribution of chain lengths in a polymer, and

Phthalic acid   Glycerol

Scheme 6.5 Reaction of molecules with several active groups can give rise to cross-linked polymers: the reaction of phthalic acid and glycerol

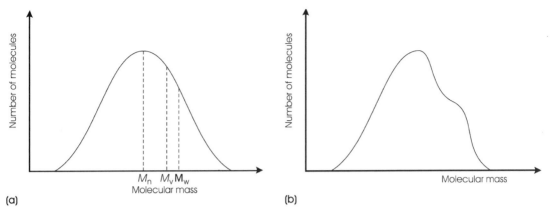

Figure 6.21   (a) Idealised distribution of molecular mass of a polymer (M_n = number average molecular mass; M_v = viscosity molecular mass; M_w = weight average molecular mass); (b) example of a real mass distribution, more complex than the simple distribution in part (a)

this is given in terms of an average molar mass. Ideally, the distribution of chain lengths in a sample of a polymer will take on the shape of a bell-shaped curve. As the chain length is reflected in the mass of the molecule, this information is often given as a graph of the number of molecules against the molar mass (Figure 6.21a). Generally, the distribution departs from a simple bell shape (Figure 6.21b) and the form of the curve is often characteristic of a particular reaction mechanism.

In order to quantify the chain length, the molar mass (or molecular weight) must be defined statistically. The number average molar mass, M_n, is given by:

$$M_n = \sum_i x_i M_i$$

$$x_i = n_i \left[\sum_i n_i \right]^{-1}$$

where x_i is the fraction of the total number of chains within the chosen molar mass range, and n_i is the number of molecules with a molar mass M_i. The number average molar mass corresponds to the peak in a bell-shaped distribution curve. An alternative measure, the weight average molar mass, M_w, takes into account the fact that most of the mass of the

sample resides in bigger molecules:

$$M_w = \sum_i w_i M_i$$

$$w_i = n_i M_i \left[\sum_i n_i M_i \right]^{-1}$$

where w_i is the mass fraction of each type of molecule present within the chosen molar mass range. The molar mass can also be determined experimentally. A method frequently used is to measure the viscosity of a solution containing the polymer. A greater viscosity indicates longer chains and a higher molar mass. The molar mass determined in this way, M_v, lies between M_n and M_w (Figure 6.21a).

The degree of polymerisation, N, is the number of monomer units in an average chain. It is given by the molar mass divided by the mass of the monomer, m. The value obtained depends on which of the various molar mass values are chosen:

$$N_n = \frac{M_n}{m}$$

$$N_w = \frac{M_w}{m}$$

$$N_v = \frac{M_v}{m}$$

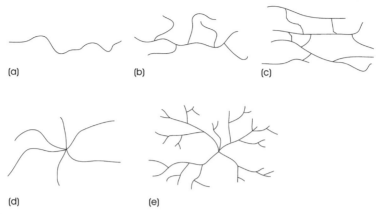

Figure 6.22   Polymer chain geometries: (a) linear; (b) branched; (c) cross-linked; (d) star; (e) dendrimer

6.4.2.2   Chain structure

Polymerisation involves the breaking and reforming of large numbers of chemical bonds. This energetic process results in a number of different molecular structures. Frequently, chains are not simply linear but also have side-branches, which can also cross-link molecules. Growth of several chains can also start from a small nucleation centre, and the resulting chains can branch to give a dendritic structure (Figure 6.22). Each of these molecular geometries imparts different physical properties to the resultant polymer.

Polyethylene (polythene) is a long-chain polymer 10^4 ethene (ethylene) units or more long. If these chains are relatively short and highly branched, the material has a low density, a low refractive index and is very flexible but weak. It is referred to as low-density polyethylene (LDPE). High-density polyethylene (HDPE) consists of linear molecules and has a molecular weight of between 200 000 and 500 000. It is much stronger than LDPE. Ultrahigh-molecular-weight polyethylene (UHMWPE), with a molecular weight of the order of 5 000 000, is stronger still. The mechanical properties of polyethylene are influenced further by the degree of crystallisation that occurs. On cooling slowly from the melt, some chains can order into crystalline regions 10–20 nm thick. These crystalline regions are of high density and of high refractive index.

Most polyethylene is a mixture of crystalline and amorphous regions, which is why it appears milky.

6.4.2.3   Crystal structure

Straight chains rarely occur in polymers. More often, as in solid polyethylene and many similar polymers, the chains fold back on themselves with a characteristic fold length (Figure 6.23a). The folded chains aggregate into blocks that have a regular structure similar to a crystal. This unit of microstructure is called a lamella (Figure 6.23b). The lamellae are not usually made from a single folded chain but are formed by a variety of neighbouring chains (Figure 6.23c). The parts of the chains not incorporated into the lamellae then link one lamella to another in the partly crystalline material. During crystallisation of the melt, lamellae form in three dimensions, from a nucleation site, to form a spherulite. This feature consists of a set of spokes, called lamellar fibrils, radiating out from a common centre into the amorphous interspoke regions. A two-dimensional cross-section of such a region is sketched in Figure 6.24.

The degree to which a polymer can crystallise depends on the details of the chain. For example, the amide group is a polar unit and forms hydrogen bonds with the carboxyl oxygen in nylons (Scheme 6.13, p. 186). These intermolecular forces hold the

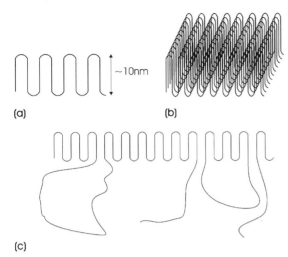

(a)   (b)

(c)

Figure 6.23 (a) The structure of polymer chains in polyethylene fold back on themselves approximately every 10 nm; (b) folded chains aggregate to form a lamella; (c) lamellae can contain more than one polymer chain, or a chain folded back into the arrangement

chains together, which produces a highly crystalline polymer with excellent strength.

The degree of crystallinity of a polymer can be determined if the density of polymer crystals and purely amorphous material is known. The reasoning is exactly the same as used in Vegard's law, for the determination of the lattice parameter of a solid

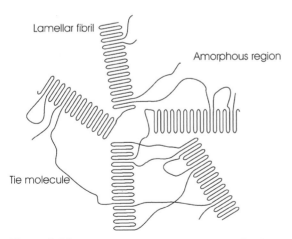

Figure 6.24 Schematic structure of a spherulite in a polymer such as polyethylene

solution (see Section 5.2.2). The fraction of crystalline polymer, x_c, is:

$$x_c = \frac{\rho_s - \rho_a}{\rho_c - \rho_a}$$

where ρ_s is the density of the sample, ρ_c is the density of the crystals, and ρ_a is the density of completely amorphous polymer.

6.4.2.4 Tacticity

The disposition of side-groups on a polymer chain has a great influence on properties, especially the flexibility of the chain, thus changing the melting point of the polymer, and the ability of the chains to pack together, changing strength and optical properties. Three different arrangements have been characterised: isotactic polymers have the side-groups all on one side of the chain (Figures 6.25a and 6.25b), syndiotactic polymers have the side-groups alternating (Figure 6.25c), and atactic polymers have a random arrangement of side-groups (Figure 6.25d). For example, atactic polypropylene is largely amorphous and weak. The stereoregular material syndiotactic polypropylene is crystalline, transparent and hard, whereas isotactic polypropylene crystallises and readily forms fibres. Polystyrene is similar. The atactic material is amorphous whereas syndiotactic polystyrene is crystalline. Poly(methyl methacrylate), used as a replacement for glass, is almost completely amorphous.

6.4.2.5 Cross-linking

The degree of cross-linking between chains changes properties dramatically. Weak cross-links tend to soften materials, whereas extensive cross-linking turns the material into a hard resin. Most thermoplastics, such the epoxy resins, are heavily cross-linked into a hard mass. Cross-linking is the key to elastomeric properties and is considered again below, in Section 6.4.4.

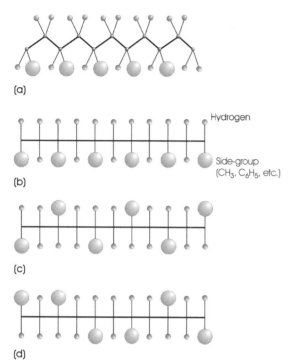

(a)

(b)

Hydrogen

Side-group
(CH₃, C₆H₅, etc.)

(c)

(d)

Figure 6.25   (a) Perspective view of an isotactic polymer chain; (b) a projection of part (a) from above to give a schematic plan of an isotactic polymer chain; (c) a syndiotactic polymer chain, depicted as per the polymer in part (b); (d) an atactic polymer chain, depicted as per the polymer in part (b)

6.4.2.6   Copolymers

Besides these different chain geometries, polymer properties are considerably changed by the polymerisation of two different monomers together. These materials are called copolymers and have different names according to the arrangement of the monomers (Figure 6.26). The physical and chemical properties of these materials can be regarded as a combination of the properties of the two polymers in the 'mixture'. For example, pure atactic polystyrene is transparent and brittle. Polybutadiene (synthetic rubber) is resilient but soft. High-impact polystyrene, a graft copolymer of these two materials, is durable, strong and transparent.

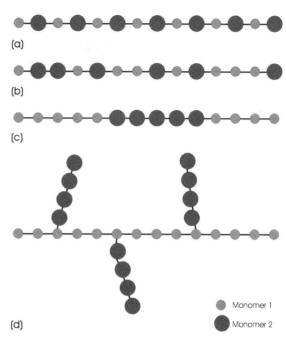

(a)

(b)

(c)

(d)

Monomer 1
Monomer 2

Figure 6.26   Schematic polymer geometry: (a) an alternating copolymer; (b) a random copolymer; (c) a block copolymer; and (d) a graft copolymer

6.4.3   Production of polymers

Because the properties of polymers depend so critically on microstructure, the manufacture of polymers is a highly skilled undertaking. Much of the evolution in the properties of plastics can be attributed to improvements in preparation methods.

The division of polymers into two categories – addition polymers and condensation polymers – is not very useful when formation reactions are concerned. Here the emphasis is better placed on the growth mechanism. There are two principal growth mechanisms – step growth and chain growth. In step growth, growing chains can link together to form longer chains (Scheme 6.6a). In chain growth, monomers are added, one by one, to the growing end of the chain, and different chains do not link together (Scheme 6.6b). Nylons, for example, are condensation polymers that form by a step-growth mechanism. In addition, monomer molecules or chains can link head to tail (the most common

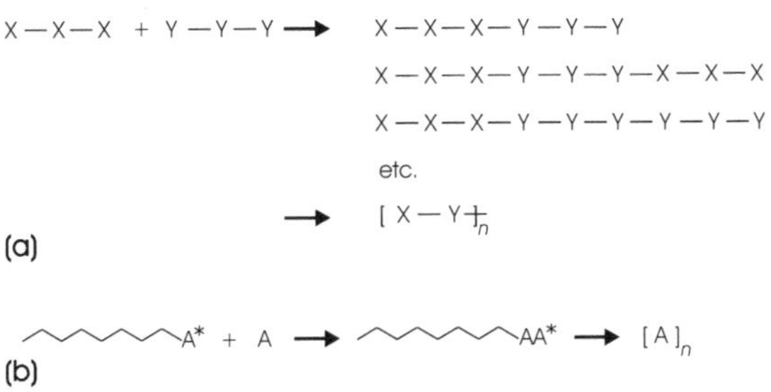

X—X—X + Y—Y—Y →   X—X—X—Y—Y—Y

X—X—X—Y—Y—Y—X—X—X

X—X—X—Y—Y—Y—Y—Y—Y

etc.

→   [X—Y ⁺]ₙ

(a)

~~~~~~A* + A → ~~~~~~AA* → [A]ₙ

(b)

Scheme 6.6 (a) Step-growth polymerisation. Short chains can link to give a variety of sequences. (b) Chain-growth polymerisation involves the addition of new molecules to one end of a growing chain. A^* represents a free radical or similar active centre where addition of new material can occur

way), head to head, or tail to tail. Random linkage can also occur, so increasing the complexity of the reaction.

6.4.3.1   Initiation

To form a polymer, the initial monomers must be activated in some way to start the reaction, a step called initiation. This can be accomplished by heat or high-energy radiation such as ultraviolet light. These processes, which are not reproducible enough for industrial production, contribute to the degradation of polymers in use. Industrially, the initiation stage is achieved by mixing a wide variety of active molecules with the monomers.

6.4.3.2   Propagation

The production of the polymer chains by linkage of the monomers, the second stage of the reaction, is called propagation. The mechanisms of propagation are complex and not all of the reaction steps are fully understood for all reactions. Nevertheless, the propagation stage is of key importance in the production of special polymers and, for this, catalysts are usually employed. In several cases, catalysts not only increase the rate of reaction but also ensure that the addition of the monomers to the growing polymer chain takes place in a constrained

way. That is, the reacting molecules approach the growing chain in only one direction. This leads to the production of polymers with a single tacticity, such as isotactic polystyrene. Ziegler–Natta and metallocene catalysts fall into this group and are used to prepare polymers with controlled structures. The production of polypropylene using a metallocene catalyst is described in Sub-section 6.4.3.4.

6.4.3.3   Termination

Growing polymer chains must ultimately stop growing. This is brought about by the process of termination. Chain termination can come about in two ways. The simplest is for two growing chains to meet and join. In such cases, the reactive end of the chain is usually a free radical, which is a molecule or a fragment of a molecule that has an unpaired electron. The unpaired electrons, which are extremely reactive, unite to form a pair bond joining the chains (Scheme 6.7a). A second mechanism, called disproportionation, allows one growing chain to take a hydrogen atom from another chain. The resultant is that one chain terminates with a $-CH_3$ group and one in a double bond (Scheme 6.7b). Further reaction at this double bond can continue to lengthen the chain. The hydrogen can also be extracted from the middle of a chain. This will terminate one chain in a $-CH_3$ group and create a free radical in the interior of the chain. This reactive region can act as a centre for new chain growth (described in Sub-section 6.4.3.5), and continued polymerisation will produce a branch (Scheme 6.7c). This is a very common occurrence in polyethylene polymerisation and results in the highly branched low-density polythene (LDPE) described in Sub-section 6.4.2.2. In order to make much stronger nonbranched high-density polythene, different preparation methods need to be used.

6.4.3.4   Metallocene catalysis

Metallocenes are derived from the cyclopentadiene anion, $(C_5H_5)^-$, shown in Scheme 6.8(a). These are stable units, in which a delocalised orbital, similar

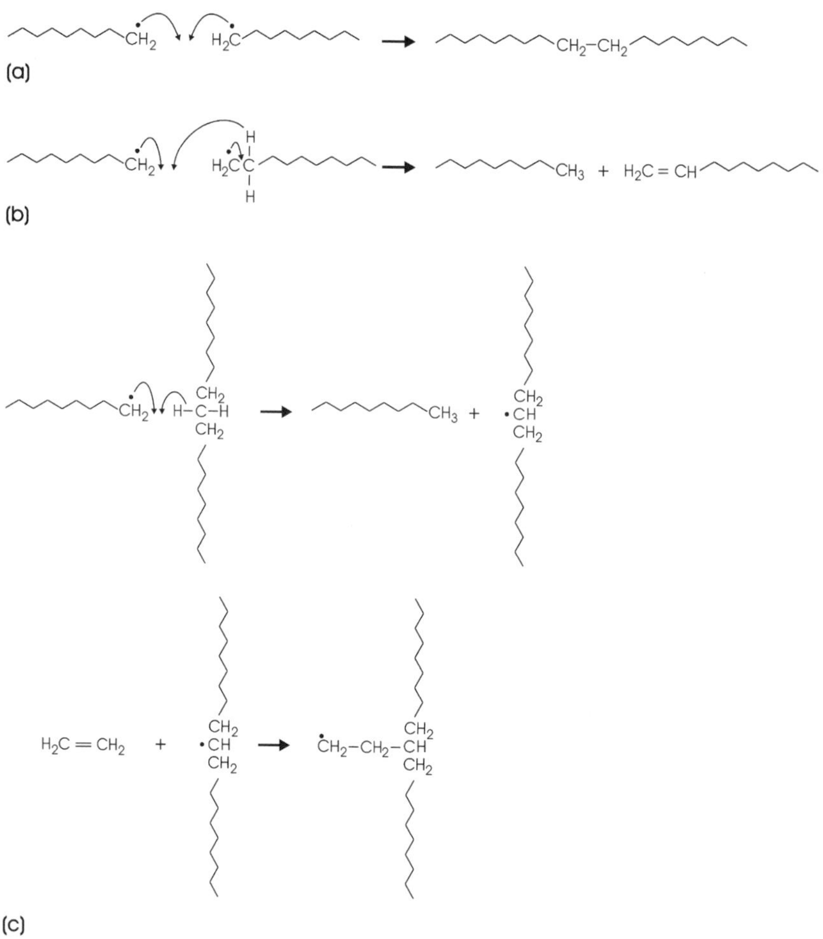

(a)

(b)

(c)

Scheme 6.7 (a) Chain termination as a result of two chains meeting; (b) chain termination as a result of disproportionation; and (c) disproportionation, leading to branching

to that in benzene, lies above and below the plane of the pentagon of carbon atoms. This orbital allows strong chemical bonds to form with metal cations, and the first metalocene investigated, ferrocene, contained Fe^{2+} sandwiched between the anions (Scheme 6.8b). At present, catalysts for polymer production are derived from a molecule called zirconocene, containing Zr^{4+} cations (Scheme 6.8c). In order to produce polymers with a precise structure, the components surrounding the Zr^{4+} ion are carefully modified. A change in the cyclopentadiene anions alters the geometry of the approach of monomers to the cation. The point where the poly-

mer chain grows is controlled by replacement of the Cl^- ions with methyl (CH_3) groups, as the methyl group acts as the point of attachment of successive monomer molecules.

In the molecule used to produce polypropylene, a complex structure called an indentyl group (a benzene ring linked to a cyclodiene), replaces each of the cyclopentadiene groups. The indentyl groups are actually linked by a short chain of two CH_2 groups (Scheme 6.8d). When these are opposed (Scheme 6.8e) the incoming monomers are guided into a position where only isotactic polypropylene can form. In molecules in which the bulky groups are

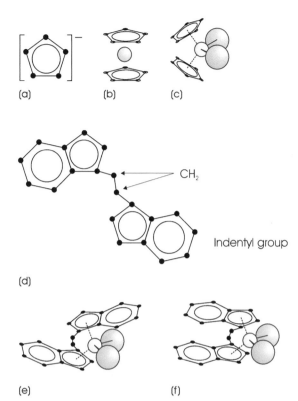

(a)   (b)   (c)

CH₂

Indentyl group

(d)

(e)   (f)

Scheme 6.8 (a) The cyclopentadiene anion, $[C_5H_5]^-$ (only the carbon atom framework is shown; the delocalised π molecular orbitals above and below the plane of the carbon atoms are represented by the circle). (b) The structure of a molecule of ferrocene, in which the Fe^{2+} cation is sandwiched between two $[C_5H_5]^-$ anions. (c) The structure of a molecule of zirconocene; the cation Zr^{4+} is bonded to two $[C_5H_5]^-$ anions and two Cl^- anions. (d) Two identyl groups linked by two CH_2 groups (the hydrogen atoms are omitted from this representation). (e) The molecular geometry required to produce isotactic polypropylene. (f) The molecular geometry required to produce atactic polypropylene

6.4.3.5 Free-radical polymerisation

Free-radical polymerisation combines initiation and propagation into one process. This method is used

to produce low-density branched polyethylene, poly(methyl methacrylate) and poly(vinyl acetate). Taking ethene (ethylene) as an example, the gas is pressurised to 100 atm. at 100 °C, and a small amount of an unstable initiator molecule, typically an organic peroxide or azide, is added (Schemes 6.9a and 6.9b). These decompose to form free radicals, which are molecules or molecular fragments that have an unpaired electron present. Because of this, they are extremely reactive chemically.

Free radicals are reactive enough to attack double bonds, thus initiating polymerisation (Scheme 6.9c). The reaction is able to satisfy the bonding requirements of the initial free radical but creates a new free radical in the process, which can attack another ethene molecule in turn, leading to continued chain growth (Schemes 6.9d and 6.9e). The reactions can be written as follows:

$$R^\bullet + CH_2=CHX \rightarrow R-CH_2-CHX^\bullet,$$
$$R-CH_2-CHX^\bullet + CH_2=CHX \rightarrow$$
$$R-CH_2-CHX-CH_2-CHX^\bullet$$

where $R^\bullet$ represents an organic free radical. Free-radical polymerisation can occur with many, but not all, ethene derivatives. Because there are no constraints on how the chain grows, the polymers are atactic.

6.4.4 Elastomers

Elastomers are materials that behave like rubber. They can be stretched to many times their original length and, when the force is released, they spring back to their original size and shape.

Elastomers are a subgroup of amorphous thermoplastics. The distinction depends on the glass transition temperature, T_g, of the polymer. Normal thermoplastics have a glass transition temperature well above room temperature and behave as hard, brittle solids. Those with a glass transition temperature below room temperature are soft and can be deformed easily at room temperature. These are elastomers. If an elastomer is cooled to well

on the same side of the Zr^{4+} cation (Scheme 6.8f) the approach of the monomers is variable, and atactic polypropylene is produced.

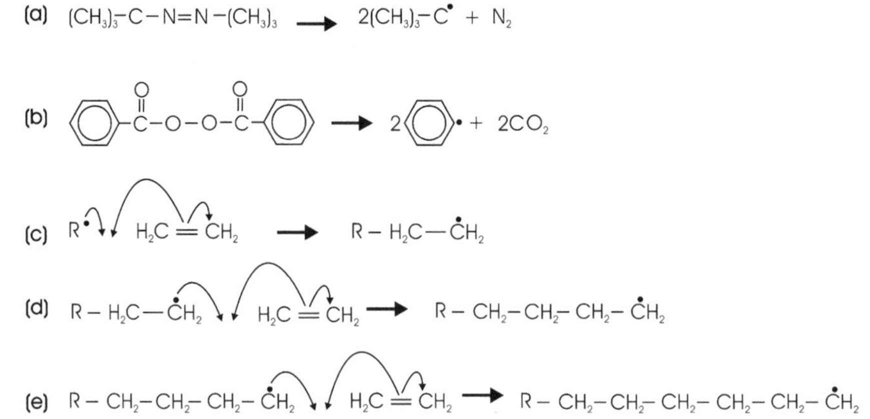

(a) $(CH_3)_3{-}C{-}N{=}N{-}(CH_3)_3 \longrightarrow 2(CH_3)_3{-}C^\bullet + N_2$

(b) ⟨O⟩$-\overset{\overset{O}{\|}}{C}$$-O-O-$$\overset{\overset{O}{\|}}{C}$⟨O⟩ $\longrightarrow$ 2⟨O⟩$^\bullet + 2CO_2$

(c) $R^\bullet \quad H_2C{=}CH_2 \longrightarrow R-H_2C{-}\overset{\bullet}{C}H_2$

(d) $R-H_2C{-}\overset{\bullet}{C}H_2 \quad H_2C{=}CH_2 \longrightarrow R-CH_2{-}CH_2{-}CH_2{-}\overset{\bullet}{C}H_2$

(e) $R-CH_2{-}CH_2{-}CH_2{-}\overset{\bullet}{C}H_2 \quad H_2C{=}CH_2 \longrightarrow R-CH_2{-}CH_2{-}CH_2{-}CH_2{-}CH_2{-}\overset{\bullet}{C}H_2$

Scheme 6.9 (a, b) The generation of free radicals, which contain an unpaired electron on one atom (marked ●), by bond breaking in an azide (part a) and a peroxide (part b). (c) Reactions of a free radical, $R^\bullet$, with ethene (ethylene) to extend the polymer chain. (d, e). Continued reaction and chain growth

below its glass transition temperature it will become hard and brittle. For example, a piece of rubber tubing dipped in liquid nitrogen becomes brittle and can easily be shattered with a hammer blow.

In its normal form, the microstructure of an elastomer is a mess of jumbled coiled polymer chains (Figure 6.27a). On being stretched, the chains tend to partly align parallel to one another (Figure 6.27b). The stretched state is thermodynamically less stable than the coiled state, and the material will revert to the coiled state when the deforming stress is removed. The entropy of the coiled state is greater than that of the stretched state and provides the driving force for the material to return to the coiled configuration. However, entropy alone does not control the key 'snap-back' property of elastomers. This is provided by cross-

linking *a few* of the elastomer molecules using other molecules (Figure 6.27). When the elastomer is now stretched, some of the bonds in the cross-links are stretched, and these spring back rapidly when the tension is released.

The best-known elastomer is natural rubber, polyisoprene (Scheme 6.10). Isoprene (Scheme 6.10a) is a liquid at room temperature, which polymerises readily to give the elastomer polyisoprene (Scheme 6.10b). The polymerisation produces two main geometrical isomers (see Section S2.1.) Natural rubber is the all-*cis* form of polyisoprene (Scheme 6.10c), in which the methyl (—CH_3) groups and hydrogen (H) atoms are on the same side of the carbon–carbon double bond. Rubber latex, a milky liquid, is a suspension of rubber in water. It is found in many plants (e.g. in dandelions) as well as rubber

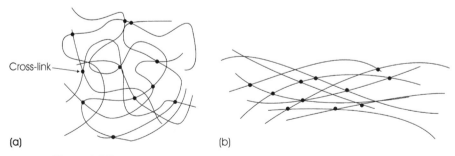

Cross-link

(a)   (b)

Figure 6.27 An elastomer in (a) an unstretched and (b) a stretched state

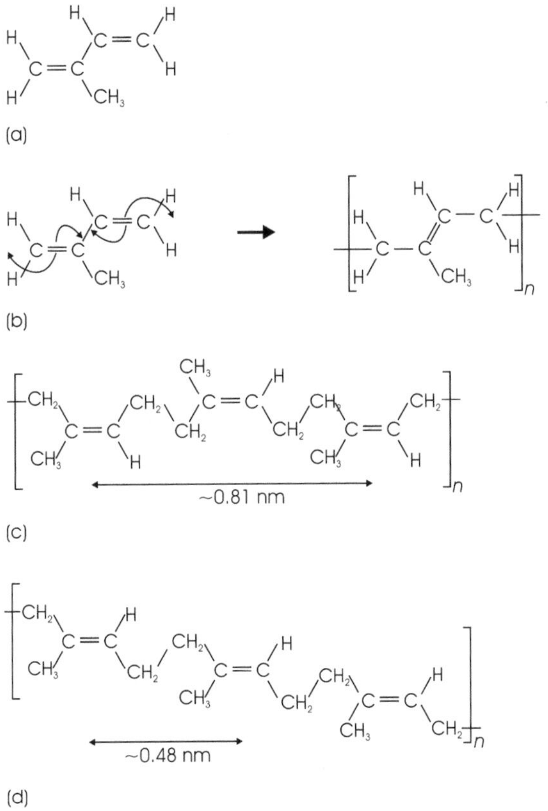

(a)

(b)

(c)

~0.81 nm

(d)

~0.48 nm

Scheme 6.10 (a) The structure of an isoprene molecule. (b) Bond redistribution and subsequent polymerisation to form poly(isoprene). (c) The structure of natural rubber, all-*cis*-poly(isoprene). (d) The structure of gutta-percha, all-*trans*-poly(isoprene)

trees. The rubber in latex can be coagulated by the addition of acetic acid, to give a soft, easily oxidised material called crepe or gum rubber. In its natural state rubber, like most amorphous thermoplastics, is sticky at a temperature above T_g and does not posses the typical elastomer properties of snap when stretched. The all-*trans* form, called gutta-percha (Scheme 6.10d) is also found in nature. It is harder and finds use in golf balls and dentistry.

The practice of cross-linking the polyisoprene chains in natural rubber to form a usable elastomer was discovered by Goodyear, in 1839. He heated natural rubber latex with sulphur, a process called vulcanisation. This transforms the sticky runny natural latex into a product in which the elastic

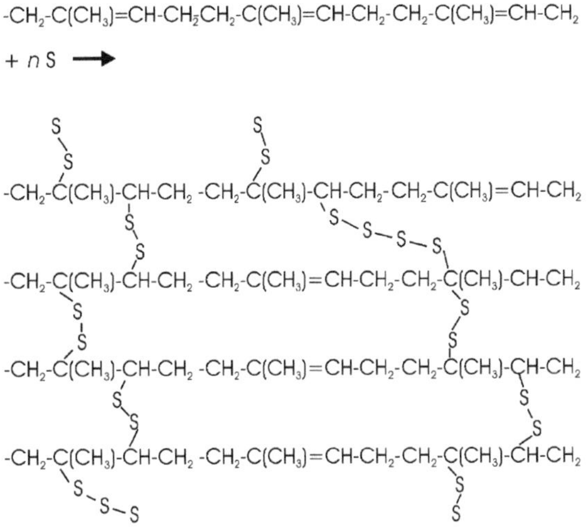

-CH₂-C(CH₃)=CH-CH₂CH₂-C(CH₃)=CH-CH₂-CH₂-C(CH₃)=CH-CH₂

+ *n* S ⟶

Scheme 6.11 The formation of sulphur cross-links between rubber molecules, produced during vulcanisation

properties are retained yet the stickiness is lost. The cross-linking (Scheme 6.11) utilises the remaining double bonds in the elastomer chains. These are opened by the cross-linking molecules to form the ties.

Cross-linking makes the polymer more rigid. For a soft rubber, 1–2 % of sulphur is added. If this process is carried too far, the whole mass of polymer turns into a solid block. Hard rubbers contain up to 35 % sulphur and, in essence, are transformed into thermosetting polymers similar to epoxy resins by this extensive cross-linking.

A large family of artificial rubbers related to natural rubber is now produced. One of the earliest examples was obtained by the polymerisation of butadiene in the presence of sodium (Na), to give buna (*Bu*tadiene + *Na*) rubber. Two other widely used rubbers are neoprene (Scheme 6.12a), a rubber resistant to organic solvents, and nitrile rubber (Scheme 6.12b), which is a copolymer of butadiene (CH_2=CH–CH=CH_2) and propenenitrile (CH_2=CH–CN).

6.4.5 The principal properties of polymers

Polymers consist mainly of carbon and hydrogen, and one of the principal properties of this group of

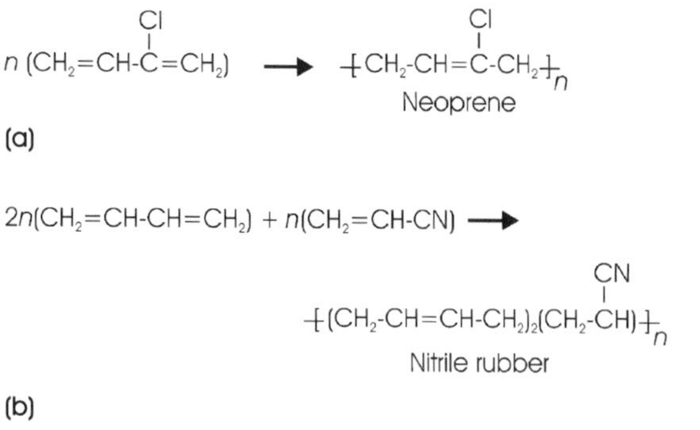

$$n\ (CH_2=CH-\overset{\overset{\textstyle Cl}{|}}{C}=CH_2) \longrightarrow \left.\!\!+\!CH_2\text{-}CH=\overset{\overset{\textstyle Cl}{|}}{C}\text{-}CH_2\!\!+\!\right._n$$

Neoprene

(a)

$$2n(CH_2=CH-CH=CH_2) + n(CH_2=CH-CN) \longrightarrow$$

$$\left.\!\!+\!(CH_2\text{-}CH=CH\text{-}CH_2)_2(CH_2\text{-}\overset{\overset{\textstyle CN}{|}}{CH})\!\!+\!\right._n$$

Nitrile rubber

(b)

Scheme 6.12 (a) The structural formula of neoprene, prepared from a chlorinated hydrocarbon precursor. (b) The structural formula of nitrile rubber, a copolymer of butadiene and a hydrocarbon containing a nitrile (cyanide, —CN) group

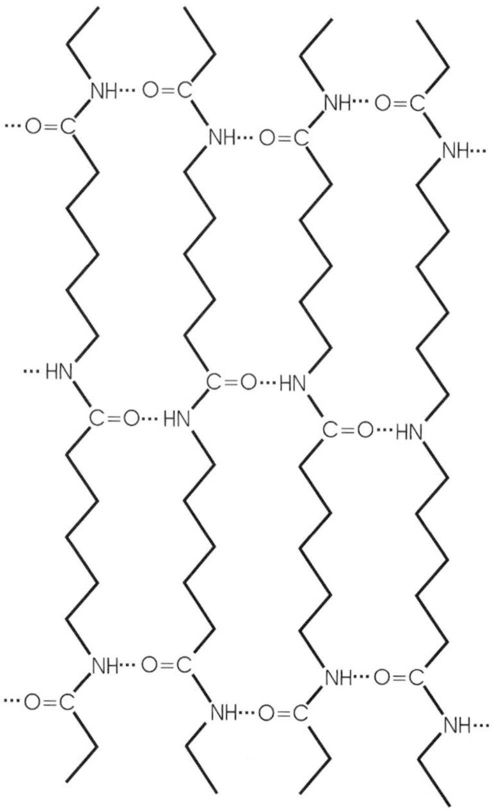

Scheme 6.13 Hydrogen bonding between chains of nylon 6. The hydrogen bonds (broken lines) between H and O draw the chains together and give a nylon fabric enhanced properties

materials is that they are low-density solids. This is enhanced by the fact that the polymer chains often do not pack together very closely, although the crystalline regions in polymers are denser than the overall solid.

Many properties of polymers depend on the functional groups that occur on the chains. Nylon fibres gain strength from >NH···O=C< hydrogen bonding (Scheme 6.13). This also allows them to absorb water and adds to the comfortable feel of these materials when used in clothing. The —NH— group in the polymers can pick up charge from the water to give $-NH_2^+-$. This is noticeable as electrostatic charge buildup on carpets and clothing.

Other properties of polymers can be varied widely by making changes to processing and additives. As we have seen, introducing a small number of cross-links into a thermoplastic elastomer can change it from a sticky substance to a useful rubber. Similarly, although polymers are inherently insulators, doping can turn them into good conductors of electricity, as explained in Section 13.2.8. Moreover, they can be made into electronic and ionic conductors, allowing these materials to be used as electrolytes and current collectors in lightweight batteries.

The behaviour of polymers under stress is very variable and depends on the polymer structure and microstructure. Thermoplastic polymers are brittle at low temperatures and are easily deformed and

plastic at higher temperatures. This is because the polymer chains are not linked to each other. In the brittle state, cracks can easily pass right through the solid, between the chains. At higher temperatures, the molecules can easily slip past each other. Crystalline regions in thermoplastics oppose these tendencies and add appreciably to the strength of the solid. Cross-linking, characteristic of thermosetting polymers, prevents molecular movement and results in solids that combine good mechanical strength with chemical stability.

One of the principal properties shown by polymers relates to this aspect of chemical stability. Polymers do not degrade rapidly and present an eyesore in many parts of the world. The ease of degradation of a polymer rapidly decreases as the degree of cross-linking of the polymer chains increases, as the many large dumps of used tyres indicate. The heart of the problem is that the carbon–carbon backbone of polymers is extremely resistant to chemical attack as the bonds are so strong. The same is true of the carbon–hydrogen bonds that make up much of the rest of the molecules. In order to make a polymer more degradable, weak links must be introduced into the chain. To some extent, double bonds are more susceptible to attack than are single bonds, and elastomers with only one double bond are less likely to be attacked and degraded than are those rubbers with two double bonds available. Degradability can be enhanced by the introduction of deliberate points of attack into the polymer. For example, inclusion of oxygen atoms, or hydroxyl or acid groups, allow for water penetration and aid bacterial attack.

6.5 Composite materials

Composites are solids made up of more than one material, designed to have enhanced properties compared with the separate materials themselves.

Composites are widely used in nature. For example, wood is composed of strong flexible cellulose plus stiff lignin, and bone consists of strong soft collagen (a protein) plus apatite (a brittle hard ceramic). Composites are also are widely used for advanced engineering applications, from use in aircraft, to use in high-technology leisure items such as skis and sails. One of the earliest applications of a composite was the Macintosh raincoat. The fabric for this garment consisted of a sandwich of natural rubber between two sheets of a woven natural polymer, cotton.

Man-made composites fall into three broad classes, depending on whether the main part of the composite, the matrix, is a polymer, a metal or a ceramic. Often, but not always, the composite combines materials from two classes, as in glass-fibre-reinforced plastics. However, the most widely used composite material, concrete, is a ceramic – ceramic composite. The most important classes of artificial composite are described below. The mechanical properties of composites are discussed in Section 10.4. Biological composites are very varied and will not be considered here.

6.5.1 Fibre-reinforced plastics

The main polymers (plastics) used as matrices in polymer composites are thermosetting resins, especially polyester and epoxy resins. Polyester resins are relatively inexpensive but tend to shrink during curing and tend to absorb water. Epoxy resins are more expensive but do not shrink on curing and are fairly resistant to water penetration. In principle, any highly cross-linked polymer will form a matrix. The resins are reinforced by filling with fibreglass, carbon fibre or strong polymer fibres such Kevlar[®], an aramid fibre.

The purpose of the matrix is to hold the fibres together in the desired orientation. Fibres alone tend to be brittle and, although they have good tensile strength, they cannot sustain compression readily. The purpose of the fibres is to add strength. The resultant strength depends on the type of fibre utilised and geometrical factors. These include the amount of fibre added and the length of the fibres (Figure 6.28), as well as the bonding between the matrix and the fibre inserts. The orientation of the fibres is also important. Composites are strong in

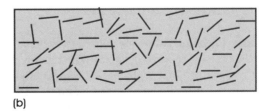

(a)   (b)

Figure 6.28 (a) Aligned fibre-reinforced composite; (b) random fibre-reinforced composite

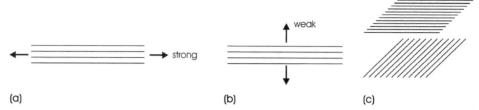

Figure 6.29   Reinforcing fibres tend to be strong in tension (part a) but are weak when subjected to a transverse force (part b); laminates in which the fibres are aligned in differing orientations offset this disadvantage (part c)

the direction of alignment and weaker normal to this direction. To overcome this, the orientation of successive layers of fibres is often changed to form a laminate (Figure 6.29).

6.5.2   Metal-matrix composites

Metals are frequently reinforced with continuous fibres to give improvements to strength. The fibres used are ceramic [e.g. silicon carbide (SiC) or alumina (Al_2O_3)] or metallic [e.g. boron (B) or tungsten (W)]. Fibres of these materials are difficult to fabricate and, unless superior performance is vital, it is easier to make composites by using small particles of a hard material such as alumina or silicon carbide. These ceramics are usually mixed with the molten alloy, which is then formed into the intended shape. Among the most widespread of metal–particle composites are the cemented carbide materials used as cutting tools for machining steel. The first of these was made from a matrix of the metal cobalt, containing hard tungsten carbide particles. Interestingly, this was the first combination tried by the producer and, although many other metal–metal carbide combinations were tried later, the cobalt–tungsten carbide composite remains the best for most purposes.

6.5.3   Ceramic-matrix composites

Ceramic-matrix composites are utilised to overcome the inherent brittleness of ceramics. The reinforcement consists of fibres or particles. The materials used include silicon carbide and alumina. The toughening comes about because the fibres or particles deflect or bridge cracks in the matrix.

Naturally occurring ceramic–ceramic composites include granite and marble. These are imitated in concrete, discussed below.

Although glass in composite materials is associated mostly with the strengthening component in the form of fibres, laminated glass is a widely used composite. This material consists of a thin plastic film sandwiched between two or more sheets of glass. The purpose of the polymer is to prevent the glass splintering on impact. It is widely used as bullet-proof glass.

6.5.4   Cement and concrete

Concrete is a composite material made from cement paste and aggregate (a coarse, stony material). The composition of concrete varies widely and depends on the intended application but always contains cement, water and aggregate. The mixture of cement, water and sand is called mortar, whereas cement and water alone constitute cement paste.

Early forms of cement, which were composed mainly of calcium hydroxide [$Ca(OH)_2$] partly transformed to calcium carbonate ($CaCO_3$) predate Roman times. The material was made from limestone, [impure calcium carbonate ($CaCO_3$)] heated or 'burned' to give quicklime or burnt lime, calcium oxide (CaO).

$$CaCO_3 (s) \rightarrow CaO (s) + CO_2 (g)$$

Quicklime reacts with water to release considerable heat, a process called slaking, to give slaked lime, calcium hydroxide [$Ca(OH)_2$]:

$$CaO (s) + H_2O (l) \rightarrow Ca(OH)_2 (s).$$

Table 6.6 Constituents of Portland cement

Chemical name	Mineral name	Chemical formula	Shorthand notation	Typical composition/wt%
Tricalcium silicate	Alite	Ca_3SiO_5	C_3S	40–65
Dicalcium silicate	Belite	Ca_2SiO_4	C_2S	10–20
Tricalcium aluminate		$Ca_3Al_2O_6$	C_3A	10
Tetracalcium aluminoferrite		$Ca_4Al_2Fe_2O_{10}$	C_4AF	10
Calcium sulphate dihydrate	Gypsum	$CaSO_4 \cdot 2H_2O$	CSH_2	2–5

The slaked lime was mixed to a paste with water and used to cement sand or stone. The paste slowly reacted with carbon dioxide in the air to give calcium carbonate again:

$$Ca(OH)_2\,(s) + CO_2\,(g) \rightarrow CaCO_3\,(s) + H_2O\,(l).$$

The Romans improved the process by adding volcanic ash to the limestone, to produce a far more durable cement, which is still seen today.

6.5.4.1 Portland cement

Portland cement, invented in 1892 by Joseph Aspdin of Leeds, was so called because it resembled expensive Portland stone (at least to the eye of the inventor). It is made from about 80 % limestone and about 20 % clay. It was widely adopted because it possessed superior qualities to the older quicklime-based material, including the especially important property of being able to harden in damp conditions. This latter property was especially valuable at a time when tunnel construction was widespread, including amongst other projects, the London underground system.

To make cement powder, the raw materials are ground with water to form a slurry. This is heated in a kiln at gradually increasing temperatures, initially to drive off water and then to decompose the calcium carbonate:

$$CaCO_3\,(s) \rightarrow CaO\,(s) + CO_2\,(g).$$

As the temperature increases, other reactions take place, and the reaction products partly melt and sinter, to produce 'clinker'. In the final stage of manufacture, the cooled clinker is ground, and

about 2–5 % gypsum, calcium sulphate hydrate ($CaSO_4 \cdot 2H_2O$) is added to produce cement powder.

Portland cement powder contains five major constituents. These are complex minerals and are known by their chemical names, mineral names and by a shorthand notation, listed in Table 6.6. There are also traces of other impurities in ordinary cement powder, which may have a large effect on the final strength and durability of the concrete. The composition of typical Portland cement is included in Table 6.6.

6.5.4.2 Hardening of cement

Cement hardens when the constituents react with water to produce an interlocking array of hydrated silicate crystals. The main reactions are rather indeterminate, because of the variable quantities of water and cement powder involved and because the reactions that take place are extremely complex. Broadly, the sequence of reactions that take place are (Figure 6.30):

- reaction of alite and water;
- reaction of belite and water;
- reaction of aluminate and water;
- reaction of ferrite and water;
- reaction of gypsum with aluminate and water.

Reaction of alite and water

$$\text{alite} + \text{water} \rightarrow \text{calcium hydroxide}$$
$$+ \text{calcium silicate hydrate}$$
$$Ca_3SiO_4 + H_2O \rightarrow Ca(OH)_2 + (CaO)_x(SiO_2)_y \cdot aq$$

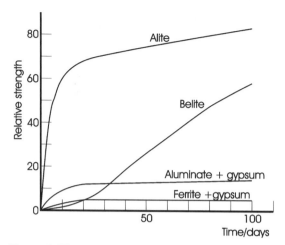

Figure 6.30 The approximate relative strengths of the components of Portland cement after hydration as a function of time elapsed

This chemical equation is not balanced because, in practice, x varies between the approximate limits of 1.8–2.2, y varies around a mean value of 1.0 and 'aq' means that water is also combined in the material in an indeterminate amount. The idealised composition of the calcium silicate hydrate phase is $Ca_2SiO_4 \cdot aq$. The reaction is rapid and continues for up to approximately 20 days. Considerable heat is evolved – in the order of 500 J per gram of powder – and care must be taken to remove this heat when forming large masses of concrete into structures. The reaction gives a high-strength product.

Reaction of belite and water

$$belite + water \rightarrow calcium\ hydroxide$$
$$+ calcium\ silicate\ hydrate$$
$$Ca_2SiO_5 + H_2O \rightarrow Ca(OH)_2$$
$$+ (CaO)_x(SiO_2)_y \cdot aq$$

The reaction, like that of alite, is imprecise, and the chemical equation is not balanced. Belite reacts slowly, taking about a year to harden and is responsible for the long-term strength of concrete. The heat liberated, approximately 250 J per gram of powder, is not as great as that liberated in the reaction of alite and, because the reaction is slower,

does not have such immediate consequences during construction.

Reaction of aluminate and water

$$aluminate + water \rightarrow calcium\ aluminate\ hydrate$$
$$Ca_3Al_2O_6 + 6H_2O \rightarrow Ca_3Al_2O_6 \cdot 6H_2O$$

This is a very rapid reaction and is completed in minutes. It is also a very energetic reaction, with about 900 J per gram released during the hydration. The product is of very low strength and is attacked by sulphate, a common impurity in many soils and rocks.

Reaction of ferrite and water   Ferrite and water react very slowly. Moderate amounts of heat are evolved – about 300 J per gram. The product is not attacked by sulphate and is beneficial mainly to the strength of the cement. However, the main importance of ferrite is cosmetic, as it influences the colour of the product.

Reaction of gypsum with aluminate and water

$$gypsum + aluminate + water \rightarrow etringite$$
$$Ca_3Al_2O_6 + 3CaSO_4 \cdot 2H_2O + 30H_2O$$
$$\rightarrow Ca_6Al_2S_3O_{18} \cdot 32H_2O$$

Gypsum is important as it reacts with aluminate to give etringite – calcium aluminium sulphate hydrate, $Ca_6Al_2S_3O_{18} \cdot 32H_2O$ – on the surface of the aluminate grains. This slows the reaction of aluminate with water and allows the wet cement paste to be worked for longer.

6.5.4.3   Heat of hydration

The hydration of cement involves a number of exothermic reactions which liberate a great deal of heat and, when building substantial concrete structures, this must be removed to prevent cracking or other deterioration of the structure. The evolution of heat takes place over a period, and the rate of heat evolution is as important as the total amount of heat given out. Several empirical relationships between the composition of the cement, the heat of hydration and the time elapsed have been developed. These take the typical form:

$$\text{heat of hydration} = A\,x_{C_3S} + B\,x_{C_2S} + C\,x_{C_3A} + D\,x_{C_4AF}$$

where x_i is the weight fraction of constituent i (see Table 6.6), and A, B, C and D are empirical constants that vary over time, reflecting the changes in the composition of the cement as it hardens; the heat of hydration is measured in joules per gram of cement. For example, the heats of hydration after three days, $H(3\,\mathrm{d})$, and after a year, $H(1\,\mathrm{yr})$, are given as follows:

$$H(3\,\mathrm{d}) = 240x_{C_3S} + 50x_{C_2S} + 880x_{C_3A} + 290x_{C_4AF}$$
$$H(1\,\mathrm{yr}) = 490x_{C_3S} + 225x_{C_2S} + 1160x_{C_3A} + 375x_{C_4AF}$$

6.5.4.4 Microstructures of cement and concrete

The microstructures that form as the paste reacts with water are important in controlling the final strength of the concrete. Initially, water reacts to give a silicate gel. This material is amorphous and produces a glutinous coating on the powder particles, holding the particles together and causing a certain amount of swelling to occur. The gel slowly crystallises, to give a mass of interpenetrating needles and plates. Gypsum reacts slowly to form hexagonal needle-like prisms of the silicate ettringite – hydrated calcium aluminium sulphate hydroxide, $Ca_6Al_2(SO_4)_3(OH)_{12}\cdot 26H_2O$ – which further interlock the mass. In addition, the material contains free water, at least in the early stages of reaction, and pores. The details of these reactions still remain to be completely worked out for all the constituents present.

The microstructure of concrete is further complicated by the presence of the aggregate. Although this is often supposed to be inert chemically it can also react with the other constituents and with water, particularly if this contains acidic or alkaline impurities.

Answers to introductory questions

Are hydrides alloys or ceramics?

Hydrides are composed of metals (elements or alloys) and hydrogen. The answer to the question will depend on the composition of the hydride. Broadly speaking, when only small amount of hydrogen is present, it occupies interstitial positions in the metallic crystal structure and the material behaves as a metal. With more hydrogen, the electrons dispersed throughout the metallic crystal begin to localise, resulting in a more brittle solid, although the electronic properties remain metallic. Eventually, chemical compounds form. These tend to resemble ceramics in that the electrons in the solid are localised and the materials are insulators. MgH_2, for example, is transparent and resembles an insulating ceramic such as MgO.

Are glasses liquids?

Glasses are not liquids; they are obviously solids. However, the structure of a glass is similar to the structure of the liquid phase that exists at high temperatures. Glasses are called supercooled liquids because they do not crystallise on cooling from this state. The liquid structure is frozen in and for kinetic reasons does not rearrange into a crystal.

Are polymers glasses?

Polymers are not glasses, as glasses are defined as inorganic materials formed by a high-temperature process. Polymers are (mostly) carbon compounds and are made by low-temperature routes from organic precursors, derived mainly from oil. Structurally, however, polymers often resemble glasses. Polymers that are largely amorphous in structure show a glass transition temperature. The glass transition temperature in a polymer gives an idea of how easily the polymer chains slip past each other. It is easily changed by adding small molecules, called plasticisers, to the polymer mix.

Why are plastic bags difficult to degrade?

Plastic bags are made of polymers that are formed from chains of carbon atoms linked mainly by strong sp^3 hybrid bonds. The majority of the other bonds in the molecules are the equally strong bonds between carbon and hydrogen, which are also mainly sp^3 hybrid bonds. These strong bonds resist chemical attack by water, oxygen and bacteria, the major forces for degradation in the natural world. They do slowly degrade when exposed to high-energy radiation, such as the ultraviolet radiation present in sunlight, but the process is slow and often affects the pigments in the plastic before having an effect on the polymer chains, so that although the bag appears to degrade, the polymer remains mostly intact. Plastic bags and other polymers can be made to degrade more rapidly by introducing weak links into the carbon–carbon chains similar to those found in naturally occurring polymers such as proteins.

Further reading

J.C. Anderson, K.D. Leaver, R.D. Rawlings, J.M. Alexander, 1998, *Materials Science*, 4th edn, Stanley Thornes, Cheltenham, Glos.

G.H. Beall, 1992, 'Synthesis and Design of Glass-ceramics', *Journal of Materials Education* **14** 315.

D.C. Boyd, D.A. Thompson, 1980, 'Glass', in *Kirk-Othmer Encyclopedia of Chemical Technology: Volume II*, 3rd edn, John Wiley & Sons, New York, p. 807.

W.D. Callister, 2000, *Materials Science and Engineering: An Introduction*, 5th edn, John Wiley & Sons, New York.

J.A. Ewen, 1997, 'New Chemical Tools to Create Plastics', *Scientific American* **276** (May) 60.

W.L. Johnson, 1999, 'Bulk Glass-forming Metallic Alloys: Science and Technology', *Materials Research Society Bulletin* **24** (October) 42.

S. Mindess, 1982, 'Concrete Materials', *Journal of Materials Education* **5** 983.

W.F. Smith, 1993, *Foundations of Materials Science and Engineering*, 2nd edn, McGraw-Hill, New York.

X. Xia, and P.G. Wolynes, 2000, 'Fragilities of Liquids Predicted from the Random First Order Transition Theory of Glasses', *Proceedings of the National Academy of Science* **97** 2990.

Problems and exercises

Quick quiz

1   An allotrope of a metallic element is:
(a) A high-temperature form
(b) A form with a different crystal structure
(c) A compound of the element

2   An interstitial alloy is an alloy in which:
(a) Impurity atoms occupy normally empty sites in the crystal structure
(b) Impurity atoms replace atoms in normally filled sites in the crystal structure
(c) Impurity atoms alloy and form a new crystal structure

3   In the A1, copper structure, face-centred cubic: the atoms are in contact along:
(a) A cube body diagonal
(b) A cube face diagonal
(c) A cube edge

4   Ordered alloy structures such as Cu_3Au are formed by:
(a) Rapidly cooling the alloy
(b) Slowly cooling the alloy
(c) Annealling the alloy

5   Steel is an alloy of iron and carbon in which the carbon occupies:
(a) Substitutional sites
(b) Interstitial sites
(c) Substitutional and interstitial sites

6   In each unit cell of the A1 (face-centred cubic) structure there are:
(a) 4 octahedral sites
(b) 8 octahedral sites
(c) 12 octahedral sites

7   In each unit cell of the A2 (body-centred cubic) structure there are:
(a) 4 tetrahedral sites
(b) 6 tetrahedral sites
(c) 8 tetrahedral sites

8 The tetrahedral sites in interstitial alloys can be occupied by:
(a) Hydrogen
(b) Nitrogen
(c) Both hydrogen and nitrogen

9 The inclusion of semimetals in a mixture:
(a) Hinders the formation of metallic glasses
(b) Aids the formation of metallic glasses
(c) Has no effect on the formation of metallic glasses

10 Crystallisation of metals is hindered by:
(a) The addition of glass to the melt
(b) The addition of other metals to the melt
(c) The addition of nonmetallic elements to the melt

11 The high electrical conductivity and reflectivity of metals is attributed to:
(a) The nature of the metallic bond
(b) The crystal structure of metals
(c) The defects present in metals

12 The ductility of metals is attributed to:
(a) The metallic bond
(b) Impurities in metals
(c) The crystal structure of metals

13 Metals can be hardened by:
(a) Adding impurities
(b) Removing impurities
(c) Removing defects

14 Silicon carbide is regarded as:
(a) An electroceramic
(b) An engineering ceramic
(c) A glass

15 Isomorphous replacement in silicate ceramics creates:
(a) Substitutional defects
(b) Interstitial defects
(c) No defects

16 Silicates are stable because of strong bonds between:

(a) Oxygen and oxygen
(b) Metal atoms and silicon
(c) Silicon and oxygen

17 Ionic silicates contain isolated:
(a) Silicate groups
(b) Silicate chains
(c) Silicate sheets

18 Clays are silicates containing:
(a) Chains of $[SiO_4]$ plus hydroxyl
(b) Networks of $[SiO_4]$ plus hydroxyl
(c) Sheets of $[SiO_4]$ plus hydroxyl

19 Ruby is a gemstone consisting of aluminium oxide containing small amounts of:
(a) Titanium impurity
(b) Chromium impurity
(c) Titanium and iron impurities

20 The ceramic zirconia, ZrO_2,
(a) Adopts the *fluorite* structure
(b) Adopts the *corundum* structure
(c) Adopts several structures, depending on temperature

21 The formation of a glass during ceramic production is called:
(a) Vitrification
(b) Sintering
(c) Glass – ceramic formation

22 A refractory ceramic is one that is:
(a) Difficult to process
(b) Particularly hard
(c) Able to withstand high temperatures

23 Pyrex® glass is also known as:
(a) Flint glass
(b) Borosilicate glass
(c) Soda-lime glass

24 The glass transition temperature marks the point at which:
(a) A glass transforms from a solid to a viscous liquid

(b) The glass can be moulded and blown

(c) The glass becomes stable

25 In glass technology, small ions such as those of phosphorus (P) and boron (B) are known as:
(a) Network modifiers

(b) Network formers

(c) Intermediates

26 A glass ceramic is:
(a) A glass processed at high temperatures

(b) A transparent ceramic

(c) A ceramic containing crystals and glass

27 Polymers are:
(a) Small molecules that can join together

(b) Very large molecules made of small units

(c) Molecules also called oligomers

28 Thermosetting plastics:
(a) Can be reformed repeatedly

(b) Can be formed only once

(c) Set at high temperatures

29 Polystyrene is an example of:
(a) An addition polymer

(b) A condensation polymer

(c) An elastomer

30 The formula of the monomer of poly(acrylonitrile) is
(a) $CH_2=CH_2 CN$

(b) $CH_3-CH=CN$

(c) $CH_2=CH CN$

31 Nylon 6, 6 is an example of:
(a) An addition polymer

(b) A condensation polymer

(c) An elastomer

32 Nylons are:
(a) Polyamides

(b) Polyesters

(c) Polycarbonates

33 The molar mass of a specific polymer:
(a) Is a fixed number

(b) Varies between narrow limits

(c) Varies between wide limits

34 The molecular mass of a monomer:
(a) Is a fixed number

(b) Varies between narrow limits

(c) Varies between wide limits

35 A lamella is:
(a) A spherulite in a polymer

(b) A chain in a polymer

(c) A crystalline region in a polymer

36 A polymer in which the side-groups lie on alternate sides of the polymer chain backbone is:
(a) Isotactic

(b) Syndiotactic

(c) Atactic

37 Copolymers are formed from:
(a) Two different monomers

(b) Two different polymers

(c) Mixed polymers

38 The growth of polymer chains by the joining of existing chains is called:
(a) Chain growth

(b) Step growth

(c) Link growth

39 The 'snap-back' property of elastomers is due to:
(a) A few weak cross-links between chains

(b) A few strong cross-links between chains

(c) Large numbers of cross-links between chains

40 A solid made by combining several different types of material is called:
(a) A combination material

(b) A consolidation material

(c) A composite material

41 Ceramic matrix composites are designed to overcome:
(a) The weight of ceramics

(b) The brittle nature of ceramics

(c) The inertness of ceramics

42 The initial hardening of Portland cement is attributed to:
   (a) Alite
   (b) Belite
   (c) Gypsum

43 The main source of heat when Portland cement hardens is due to the reaction of:
   (a) Tricalcium silicate
   (b) Dicalcium silicate
   (c) Tricalcium aluminate

44 The long-term hardening of Portland cement is attributed to the presence of:
   (a) Tricalcium silicate
   (b) Dicalcium silicate
   (c) Tricalcium aluminate

45 Which of the following is *not* used to make Portland cement:
   (a) Clay
   (b) Limestone
   (c) Portland stone

Calculations and questions

6.1 The radius of a gold atom is 0.144 nm. Gold adopts the A1 structure. Estimate the unit cell parameter of gold crystals.

6.2 The radius of a lead atom is 0.175 nm. Lead adopts the A1 structure. Estimate the unit cell parameter of lead crystals.

6.3 The radius of a palladium atom is 0.138 nm. Palladium adopts the A1 structure. Estimate the unit cell parameter of palladium crystals.

6.4 The radius of an iridium atom is 0.136 nm. Iridium adopts the A1 structure. Estimate the unit cell parameter of iridium crystals.

6.5 Nickel adopts the A1 structure, with a lattice parameter of 0.3524 nm. Estimate the metallic radius of nickel in this structure.

6.6 Rhodium adopts the A1 structure, with a lattice parameter of 0.3803 nm. Estimate the metallic radius of rhodium in this structure.

6.7 The radius of a barium atom is 0.224 nm. Barium adopts the A2 structure. Estimate the unit cell parameter of barium crystals.

6.8 The radius of a niobium atom is 0.147 nm. Niobium adopts the A2 structure. Estimate the unit cell parameter of niobium crystals.

6.9 Vanadium adopts the A2 structure, with a lattice parameter of 0.3024 nm. Estimate the metallic radius of vanadium in this structure.

6.10 Potassium adopts the A2 structure, with a lattice parameter of 0.5321 nm. Estimate the metallic radius of potassium in this structure.

6.11 At room temperature, iron adopts the A2 structure with a lattice parameter of 0.28665 nm.
   (a) Estimate the metallic radius of iron in this structure.
   (b) Ignoring thermal expansion, estimate the lattice parameter of the high-temperature A1 structure that exists above 912 °C.

6.12 At room temperature, calcium adopts the A1 structure with a lattice parameter of 0.5588 nm.
   (a) Estimate the metallic radius of calcium in this structure.
   (b) Ignoring thermal expansion, estimate the lattice parameter of the high-temperature A2 structure that exists above 445 °C.

6.13 At room temperature, strontium adopts the A1 structure. The metallic radius of strontium in this structure is 0.215 nm.
   (a) Estimate the lattice parameter of the A1 structure of strontium.
   (b) Ignoring thermal expansion, estimate the lattice parameter of the high-temperature A2 structure that exists above 527 °C.

6.14 The metallic radius of magnesium, which adopts the A3 structure, is 0.160 nm. Estimate the value of the lattice parameters a_0 and c_0 (ideal).

6.15 The metallic radius of rhenium, which adopts the A3 structure, is 0.138 nm. Estimate the value of the lattice parameters a_0 and c_0 (ideal).

6.16 The a_0 lattice parameter of titanium, which adopts the A3 structure, is 0.2951 nm. Estimate the radius r of titanium in this structure and the ideal value of the parameter c_0.

6.17 The a_0 lattice parameter of beryllium, which adopts the A3 structure, is 0.2286 nm. Estimate the radius r of beryllium in this structure and the ideal value of the parameter c_0.

6.18 At room temperature, hafnium adopts the A3 structure, with lattice parameters $a_0 = 0.3195$ nm, and $c_0 = 0.5051$ nm.
 (a) Estimate the metallic radius of hafnium in this structure.
 (b) Ignoring thermal expansion, estimate the lattice parameter of the high-temperature A2 structure that exists above 1742 °C.

6.19 At room temperature, yttrium adopts the A3 structure, with lattice parameters $a_0 = 0.3648$ nm, and $c_0 = 0.5732$ nm.
 (a) Estimate the metallic radius of yttrium in this structure.
 (b) Ignoring thermal expansion, estimate the lattice parameter of the high-temperature A2 structure that exists above 1481 °C.

6.20 The rapidly cooled form of the alloy CuAu has the A1 structure, in which the metals atoms are distributed at random over the available sites. The unit cell parameter is 0.436 nm. Estimate the density of the alloy.

6.21 Cartridge brass is a composition in the phase range of the alloy α-brass, which is made up with 30 wt% zinc and 70 wt% copper. The alloy has the A1 structure, in which the metal atoms are disordered over the available sites. The density is 8470 kg m^{-3}. Estimate the lattice parameter of the alloy.

6.22 Which of the following metals would be expected to form the most extensive substitutional solid solution with nickel, A1 structure, metallic radius 0.1246 nm, electronegativity 1.8?
 (a) Cobalt, A3 structure, metallic radius 0.125 nm, electronegativity 1.7.

(b) Chromium, A2 structure, metallic radius 0.128 nm, electronegativity 1.4.
 (c) Platinum, A1 structure, metallic radius 0.139 nm, electronegativity 2.1.
 (d) Silver, A1 structure, metallic radius 0.145 nm, electronegativity 1.8.

6.23 Which of the following metals would be expected to form the most extensive substitutional solid solution with copper, A1 structure, metallic radius 0.1278 nm, electronegativity 1.8?
 (a) Aluminium, A1 structure, metallic radius 0.143 nm, electronegativity 1.5.
 (b) Palladium, A1 structure, metallic radius 0.138 nm, electronegativity 2.0.
 (c) Vanadium, A2 structure, metallic radius 0.135 nm, electronegativity 1.4.
 (d) Titanium, A3 structure, metallic radius 0.146 nm, electronegativity 1.6.

6.24 Ceramics for use in technical applications are often divided into the groups 'traditional', 'engineering', 'electroceramics' and 'glass'. Define each of these categories and give a typical use for a material from each group. [Note: answer is not provided at the end of this book.]

6.25 Why does the refractory zirconia, ZrO_2, not find more high-temperature uses when pure? Explain how the problem is overcome by doping. [Note: answer is not provided at the end of this book.]

6.26 It is necessary to identify some plate-shaped crystals that have appeared in a glass melt. X-ray diffraction shows that they are single phase and of a structure related to muscovite mica, talc or pyrophyllite. the chemical analysis gives an empirical formula of $KF \cdot AlF \cdot BaO \cdot MgO \cdot Al_2O_3 \cdot 5MgSiO_3$. Suggest substitutions that could have been made in the muscovite mica, talc and pyrophyllite structures to produce the crystals. [Note: answer is not provided at the end of this book.]

6.27 The mineral pyrophyllite, $Al_2(OH)_2Si_4O_{10}$, has a Mohs hardness of 1; muscovite mica, $KAl_2(OH)_2Si_3AlO_{10}$, has a Mohs hardness of 2.5; and the mica margite, $CaAl_2(OH)_2$-$Si_2Al_2O_{10}$, has a Mohs hardness of 5. Explain this in terms of the structure and bonding in these compounds. [Note: answer is not provided at the end of this book.]

6.28 The viscosity of a soda-lime glass is given in Table 6.7. Estimate the glass transition temperature.

Table 6.7 Soda-lime: viscosity, with temperature; for Question 6.28

Viscosity/dPa s	Temperature/°C
5×10^{14}	450
5×10^{7}	700
1×10^{4}	1050
1×10^{2}	1450

6.29 Calculate the viscosity at (a) 940 °C and (b) 1400 °C of a high-silica glass for which η_0 is 3.5×10^{-5} Pa s and the activation energy, E, is 382 kJ mol^{-1}.

6.30 The viscosity parameters for a clear float glass are: softening point, 720 °C; annealing point, 535 °C; strain point, 504 °C. Estimate the activation energy for the viscosity.

6.31 The viscosity parameters for a glass are: softening point, 677 °C; annealing point, 532 °C; strain point, 493 °C. Estimate the activation energy for the viscosity.

6.32 The viscosity parameters for a borosilicate glass are: softening point, 794 °C; annealing point, 574 °C; strain point, 530 °C. Estimate the activation energy for viscosity.

6.33 The viscosity of a borosilicate glass is drawn in Figure 6.31(a). Estimate the activation energy of the viscosity and comment on the form of the Arrhenius plot.

6.34 The viscosity of a high-silica glass is drawn in Figure 6.31(b). Estimate the activation energy of the viscosity and comment on the form of the Arrhenius plot.

6.35 Write the reaction equation for the reaction of two monomer molecules of styrene to produce a dimer. [Note: equation is not shown in the answers at the end of this book.]
   (a) What weight of monomer is needed to produce 100 kg of polymer?
   (b) How many monomer molecules is this?

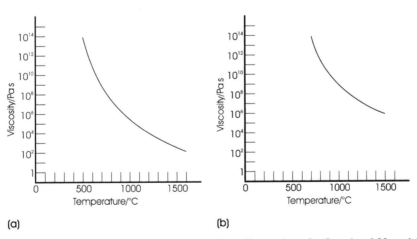

(a)   (b)

Figure 6.31 Plot of viscosity against temperature for (a) a borosilicate glass, for Question 6.33, and (b) a high-silica glass, for Question 6.34

(c) The number average molar mass of the polymer is 250 000 g mol^{-1}. What is the degree of polymerisation?

6.36 Write the reaction equation for the reaction of two monomer molecules of methyl methacrylate to produce a dimer. [Note: equation is not shown in the answers at the end of this book.]

(a) What weight of monomer is needed to produce 100 kg of polymer?

(b) How many monomer molecules is this?

(c) The number average molar mass of the polymer is 200 000 g mol^{-1}. What is the degree of polymerisation?

6.37 Nitrile rubber is a copolymer of butadiene, $(CH_2=CH-CH=CH_2)$ and acrylonitrile (propenenitrile), $(CH_2=CH-CN)$. Write the reaction equation between these two molecules to produce a dimer. [Note: equation is not shown in the answers at the end of this book.]

(a) What masses of the reactants are needed to give 100 kg of polymer?

(b) How many molecules of each reactant is this?

6.38 PET is a polymer of terephthalic acid and ethylene glycol (see Scheme 6.2). Write the hypothetical reaction equation between these two molecules to produce a dimer. [Note: equation is not shown in the answers at the end of this book.]

(a) What masses of the reactants are needed to give 100 kg of polymer?

(b) How many molecules of each reactant is this?

6.39 The structural formula of the aramid polymer Kevlar®, used in bullet-proof vests, is shown in Scheme 6.14. Write the formulae of the

Scheme 6.14   Kevlar®

two monomers used. [Note: the full version of Scheme 6.14 is shown at in the answers, at the end of this book.]

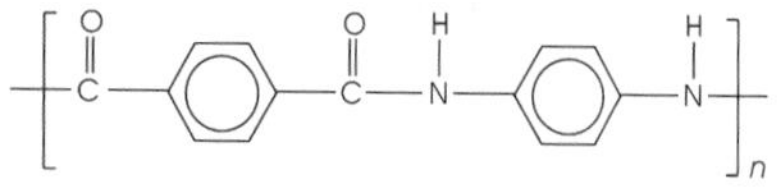

6.40 The structural formula of the polycarbonate PEN [poly(ethylene naphthalate)], used in recyclable jars and bottles, is shown in Scheme 6.15. Write the formulae of the two monomers used. [Note: the full version of Scheme 6.15 is shown in the answers, at the end of this book.]

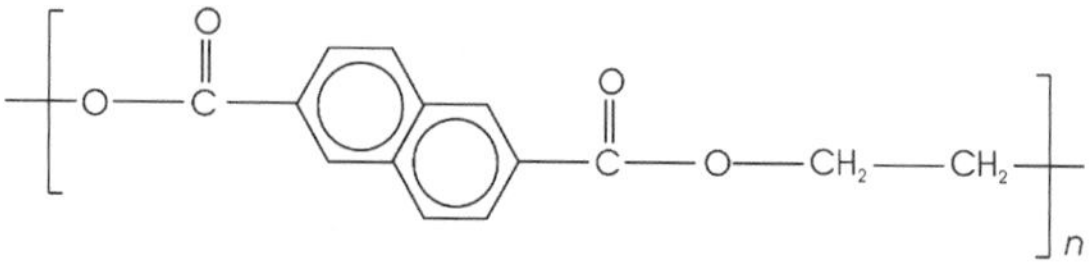

Scheme 6.15   Poly(ethylene naphthalate) (PEN)

6.41 Compute (a) the number-average molar mass and (b) the degree of polymerisation for polypropylene from the data given in Table 6.8.

Table 6.8   Polypropylene: data for Question 6.41

Molar mass/ g mol^{-1} Range	Mean in range	Fraction of molecules in range
5 000–10,000	7 500	0.01
10 000–15 000	12 500	0.09
15 000–20 000	17 500	0.17
20 000–25 000	22 500	0.18
25 000–30 000	27 500	0.20
30 000–35 000	32 500	0.17
35 000–40 000	37 500	0.09
40 000–45 000	42 500	0.06
45 000–50 000	47 500	0.03

6.42 Compute (a) the weight-average molar mass and (b) the degree of polymerisation for polypropylene from the data in Table 6.9.

6.43 The density of polyethylene crystals is 998 kg m^{-3}, and the unit cell has dimensions

Table 6.9   Polypropylene: data for Queston 6.42

Molar mass/g Range	mol^{-1} Mean in range	Weight fraction of molecules in range
5 000–10 000	7 500	0.01
10 000–15 000	12 500	0.07
15 000–20 000	17 500	0.16
20 000–25 000	22 500	0.20
25 000–30 000	27 500	0.23
30 000–35 000	32 500	0.18
35 000–40 000	37 500	0.08
40 000–45 000	42 500	0.05
45 000–50 000	47 500	0.02

$a = 0.741$ nm, $b = 0.494$ nm,   and $c = 0.255$ nm.

(a) How many CH_2 units are there in a unit cell?

(b) How many 'monomer' (CH_2–CH_2) units are there?

6.44   The density of amorphous polythene is approximately $810 \, \text{kg m}^{-3}$. Estimate the crystallinity of: low-density polyethylene, density $920 \, \text{kg m}^{-3}$; medium-density polyethylene, density $933 \, \text{kg m}^{-3}$; and high-density polyethylene, density $950 \, \text{kg m}^{-3}$. [Note: the density of crystalline polyethylene is given in Question 6.43.]

6.45   Calculate the heats of hydration after three days, $H(3 \, \text{d})$, and one year, $H(1 \, \text{yr})$, for the following Portland cement compositions, (wt%):

(a) 50 % C_3S; 25 % C_2S; 12 % C_3A; 8 % C_4AF;

(b) 45 % C_3S; 30 % C_2S; 7 % C_3A; 12 % C_4AF;

(c) 60 % C_3S; 15 % C_2S; 10 % C_3A; 8 % C_4AF.

PART 3

Reactions and transformations

7

Diffusion

- What is steady-state diffusion?

- How does one obtain a quick estimate of the distance moved by diffusing atoms?

- How does the energy barrier for ionic diffusion change when an electric field is present?

Diffusion originally described the way in which heat (believed to be a fluid) flowed through a solid. Later the same ideas were applied to describe the way in which a gas would spread out to fill the available volume. In solids, diffusion is the transport of atoms, ions or molecules under the influence of a driving force that is usually a concentration gradient. Diffusion takes place in solids at a much slower rate than in gases or liquids and, in the main, it is a high-temperature process. However, this is not always so, and in some solids the rate of diffusion at room temperature is considerable.

Movement through the body of a solid is called volume, lattice or bulk diffusion. In a gas or liquid, the diffusion is usually the same in all directions and the material is described as isotropic. This is also true in amorphous or glassy solids and in cubic crystals. In all other crystals, the rate of diffusion depends on the direction taken, and is anisotropic. Moreover, atoms can also diffuse along surfaces,

between crystallites or along dislocations. As the regular crystal geometry is disrupted in these regions, atom movement is often much faster than for volume diffusion. Diffusion by way of these pathways is often collectively referred to as short-circuit diffusion. In the following sections, volume diffusion along a single direction in an isotropic medium will be described.

Diffusion is studied by measuring the concentration of the atoms at different distances from the release point after a given time has elapsed. Raw experimental data thus consist of concentration and distance values. The speed at which atoms or ions move through a solid is usually expressed in terms of a diffusion coefficient, D, with units of $m^2 \, s^{-1}$. In general, it has been found that D depends on position and concentration, and hence D varies throughout the solid. In this chapter, attention is confined to diffusion when the concentration of the diffusing species is very small, so that concentration effects are not important, or where the diffusion coefficient does not depend on concentration or position. This is equivalent to stipulating that the diffusion coefficient is a constant.

7.1 Self-diffusion, tracer diffusion and tracer impurity diffusion

When atoms in a pure crystal diffuse under no concentration gradient or other driving force, the process is called self-diffusion. In such a case, the

Understanding solids: the science of materials.   Richard J. D. Tilley
© 2004 John Wiley & Sons, Ltd   ISBNs: 0 470 85275 5 (Hbk) 0 470 85276 3 (Pbk)

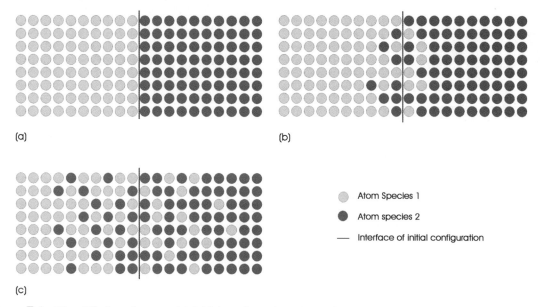

Figure 7.1   The diffusion of atoms: (a) initial configuration; (b) configuration after heating for shorter times; (c) configuration after heating for longer times; the interface becomes blurred as a result of interdiffusion of the two atom species

atomic movements are random, with motion in one direction just as likely as in another (Figure 7.1). The relevant diffusion coefficient is called the self-diffusion coefficient and is given the symbol D_{self}.

It is by no means easy to measure the self-diffusion coefficient of an atom because it is not possible to keep track of the movements of one atom in a crystal composed of many identical atoms. However, it is possible to measure something that is a very good approximation to the self-diffusion coefficient if some of the atoms can be uniquely labelled and their movement tracked. In this case, the diffusion coefficient that is measured is called the tracer diffusion coefficient, written D^*.

The experiment can be repeated with impurity tracer atoms, A, to yield the tracer impurity diffusion coefficient, D_A^*.

To illustrate this, consider the determination of the tracer diffusion coefficient of magnesium atoms in magnesium oxide, MgO, which crystallises with the *halite* (NaCl) structure. Initially, a thin layer of radioactive magnesium, which consists of the tracer atoms, is evaporated onto the surface of a carefully polished single crystal of MgO. This layer is oxidised to MgO by exposing the layers to oxygen gas, after which another carefully polished single-crystal slice of MgO is placed on top to form a diffusion couple (Figure 7.2).

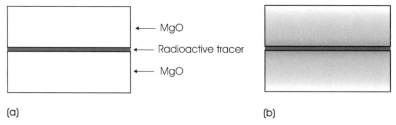

Figure 7.2   A diffusion couple formed by two crystals of MgO separated by radioactive material: (a) initially and (b) after heating

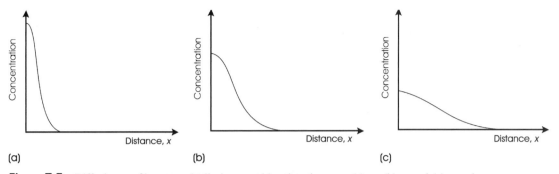

Figure 7.3   Diffusion profiles at gradually increased heating times, t_i: (a) t_1, (b) t_2 and (c) t_3, where $t_1 < t_2 < t_3$

The crystal sandwich is heated for a known time at the temperature for which the diffusion coefficient is required. The whole slab is then carefully sliced parallel to the original interface containing the radioactive MgO layer and the radioactivity of each slice, which is a measure of the concentration of radioactive magnesium in each section, is determined. A graph of the concentration of the radioactive component is then plotted against the distance from the interface to give a diffusion profile, or concentration profile (Figure 7.3). Note that the concentration of the tracer at any point, x, changes over time.

In order to obtain the diffusion coefficient from such profiles, we need to use the diffusion equation, also called Fick's second law:

$$\frac{d c_x}{d t} = D^* \frac{d^2 c_x}{d x^2} \qquad (7.1)$$

where c_x is the concentration of the diffusing radioactive ions (atoms m^{-3}) at a distance x from the original interface after time t has elapsed, and D^* is the tracer diffusion coefficient. When D^* is a constant, and does not depend on either concentration or position, the equation can be solved analytically to give an expression for c_x in terms of x. For the experimental arrangement in Figure 7.2, the solution is:

$$c_x = \frac{c_0}{2(\pi D^* t)^{1/2}} \exp\left(\frac{-x^2}{4 D^* t}\right) \qquad (7.2)$$

where c_0 is the initial concentration on the surface, usually measured in mol m^{-2}. A value for the tracer diffusion coefficient is obtained by taking logarithms of both sides of this equation, to give:

$$\ln c_x = \ln\left[\frac{c_0}{2(\pi D^* t)^{1/2}}\right] - \frac{x^2}{4 D^* t} \qquad (7.3)$$

This has the form:

$$\ln c_x = \text{constant} - \frac{x^2}{4 D^* t}$$

and a plot of $\ln c_x$ versus x^2 will have a gradient of $[-1/(4 D^* t)]$ (Figure 7.4). A measurement of the gradient gives a value for the tracer diffusion

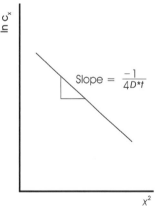

Figure 7.4   A straight-line graph of $\ln c_x$ versus x^2 from a diffusion experiment; the slope of the graph allows a value for the diffusion coefficient, D^*, to be determined

coefficient at the temperature at which the diffusion couple was heated. To obtain the diffusion coefficient over a variety of temperatures the experiments must be repeated.

In this experiment, there is a concentration gradient, because the concentration of the radioactive isotopes in the coating will be different from the concentration of radioactive isotopes, if any, in the original crystal pieces. This is why the term 'tracer diffusion coefficient' is used. However, if the layer of tracer atoms is very thin, the concentration gradient will be small and will rapidly become smaller as diffusion takes place and, in these circumstances, D^*, the tracer diffusion coefficient, will be very similar to the self-diffusion coefficient, D_{self}.

7.2 Nonsteady-state diffusion

The normal state of affairs during a diffusion experiment is one in which the concentration at any point in the solid changes over time, as in the example of tracer diffusion described in Section 7.1. This situation is called nonsteady-state diffusion, and diffusion coefficients are found by solving the

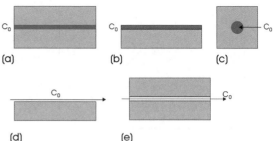

Figure 7.5 Common geometries for nonsteady-state diffusion: (a) thin-film planar sandwich; (b) open planar thin film; (c) small spherical precipitate; (d) open plate; and (e) sandwich plate. In parts (a)–(c) the concentration of the diffusant is unreplenished; in parts (d) and (e) the concentration of diffusant is maintained at a constant value, c_0, by gas or liquid flow

diffusion equation. Provided the diffusion coefficient, D, is not dependent on composition and position, analytical solutions can be found. Solutions for some diffusion experiment geometries, sketched in Figure 7.5, are summarised in Table 7.1.

The solution given as Equation (7.2), relevant to the sandwich arrangement described in Section 7.1, is illustrated in Figure 7.5(a). The solution to the

Table 7.1 Solutions of the diffusion equation

Experimental arrangement	Solution
Initial concentration c_0: Thin-film planar sandwich	$c_x = \dfrac{c_0}{2(\pi D t)^{1/2}} \exp\left(\dfrac{-x^2}{4Dt}\right)$
Open planar thin film	$c_x = \dfrac{c_0}{(\pi D t)^{1/2}} \exp\left(\dfrac{-x^2}{4Dt}\right)$
Small spherical precipitate	$c_r = \dfrac{c_0}{8(\pi D t)^{3/2}} \exp\left(\dfrac{-r^2}{4Dt}\right)$
Initial concentration, c_0, maintained constant: Open plate	$\dfrac{c_x - c_0}{c_s - c_0} = 1 - \mathrm{erf}\left[\dfrac{x}{2(Dt)^{1/2}}\right]$
Sandwich plate	$\dfrac{c_x - c_0}{c_s - c_0} = \dfrac{1}{2}\left\{1 - \mathrm{erf}\left[\dfrac{x}{2(Dt)^{1/2}}\right]\right\}$

Note: the geometries are shown in Figure 7.5; c_x and c_r, concentration of diffusing species at distance x or r (radial distance); D, diffusion coefficient; t, time; c_s, concentration of diffusing species at the surface

diffusion equation for the experimental situation in which the coated surface is uncovered (Figure 7.5b), is:

$$c_x = \frac{c_0}{(\pi D t)^{1/2}} \exp\left(\frac{-x^2}{4Dt}\right)$$

where c_x is the concentration of the diffusing species at a distance of x from the original surface after time t has elapsed, D is the diffusion coefficient, and c_0 is the initial concentration on the surface, usually measured in mol m^{-2}.

A commonly encountered problem concerns the rate at which a precipitate can dissolve in the surrounding medium. In the case of a small spherical precipitate containing c_0 moles of diffusant (Figure 7.5c), in an isotropic matrix, the solution of the diffusion equation is:

$$c_r = \frac{c_0}{8(\pi D t)^{3/2}} \exp[-r^2/4Dt]$$

where c_r is the concentration at a radial distance r from the precipitate as the precipitate dissolves in the surroundings.

In these examples, the initial concentration of the tracer is fixed and the amount remaining as the diffusion progresses will diminish over the course of the experiment. An important case of nonsteady-state diffusion occurs when the initial concentration at the surface is maintained as a constant throughout the experiment. This happens when gas molecules diffuse into a solid and the gas supply is constantly replenished. The solution for diffusion into a plate (Figure 7.5d), provided that no atoms ever reach the opposite side of the solid (i.e. the solid is regarded as infinitely thick) and the diffusion coefficient is a constant, is given by:

$$\frac{c_x - c_0}{c_s - c_0} = 1 - \mathrm{erf}\left[\frac{x}{2(Dt)^{1/2}}\right]$$

where c_s is the constant concentration of the diffusing species at the surface, c_0 is the uniform concentration of the diffusing species already present in the solid before the experiment, and c_x is the concentration of the diffusing species at a position x from the surface after time t has elapsed, and D is the (constant) diffusion coefficient of the diffusing species. The function erf $[x/2(Dt)^{1/2}]$ is frequently encountered in mathematics and is called the error function. It cannot generally be evaluated analytically, but numerical values are to be found in probability tables. An abbreviated list is given in Table 7.2.

In a pure material in which c_0 is considered to be zero,

$$c_x = c_s\left\{1 - \mathrm{erf}\left[\frac{x}{2(Dt)^{1/2}}\right]\right\}$$

If the same constant surface concentration holds, but with a sandwich geometry (Figure 7.4e), the solution is:

$$\frac{c_x - c_0}{c_s - c_0} = \frac{1}{2}\left\{1 - \mathrm{erf}\left[\frac{x}{2(Dt)^{1/2}}\right]\right\}$$

Table 7.2 Values of the error function, erf(z)

z	Erf(z)	z	Erf(z)	z	Erf(z)	z	Erf(z)
0	0	0.40	0.4284	0.85	0.7707	1.60	0.9763
0.025	0.0282	0.45	0.4755	0.90	0.7970	1.70	0.9838
0.05	0.0564	0.50	0.5205	0.95	0.8209	1.80	0.9891
0.10	0.1125	0.55	0.5633	1.00	0.8427	1.90	0.9928
0.15	0.1680	0.60	0.6039	1.10	0.8802	2.00	0.9953
0.20	0.2227	0.65	0.6420	1.20	0.9103	2.20	0.9981
0.25	0.2763	0.70	0.6778	1.30	0.9340	2.40	0.9993
0.30	0.3286	0.75	0.7112	1.40	0.9523	2.60	0.9998
0.35	0.3794	0.80	0.7421	1.50	0.9661	2.80	0.9999

Note: erf(z) is equal to 0 when z is equal to 0 and is equal to 1 when z is equal to 2.8; erf($-z$) = $-$erf(z).

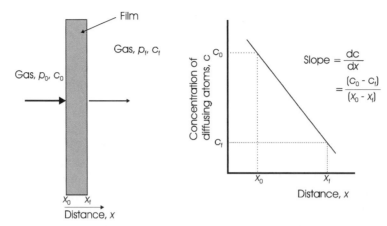

Figure 7.6   Steady-state diffusion. (a) a gas diffusing through a thin film and (b) the resultant diffusion profile. Note: c_0 and c_f, initial and final concentration of gas; p_0 and p_f, initial and final pressure of gas

7.3   Steady-state diffusion

Steady-state diffusion differs from nonsteady-state diffusion in that the concentration of the diffusing atoms at any point, x, and hence the concentration gradient at x, in the solid remains constant (Figure 7.6). Steady-state diffusion can occur when a gas permeates through a metal foil or thin-walled tube. The same situation can occur when oxygen diffuses through a plastic wrapping film. Hydrogen gas, for example, can be purified by allowing it to diffuse through a palladium 'thimble'.

Under steady-state conditions, the diffusion coefficient is obtained by using Fick's first law. This is written:

$$J_i = -D_i \frac{d c_i}{d x} \qquad (7.4)$$

where D_i is the diffusion coefficient of atom of type i, c_i is the concentration of atoms i, and x is the position in the solid. J_i is called the flux of atoms of type i, that is, the net flow of these atoms through the solid. It is measured in atoms (or a related unit such as grams or moles), per metre squared per second. When the steady state has been reached, the diffusion coefficient across the foil, D, will be given by:

$$D = \frac{J\, l}{c_0 - c_f}$$

where the concentrations on each side of the foil are c_0 and c_f and the foil thickness is l.

7.4   Temperature variation of the diffusion coefficient

Diffusion coefficients are usually found to vary considerably with temperature. This variation can often be expressed in terms of the Arrhenius equation:

$$D = D_0 \exp\left(\frac{-E}{R T}\right) \qquad (7.5)$$

where R is the gas constant, T is the temperature (in kelvin), and D_0 is a constant term referred to as the pre-exponential factor or frequency factor. The term E is called the activation energy of diffusion. Taking logarithms of both sides of this equation gives:

$$\ln D = \ln D_0 - \frac{E}{R T}$$

The activation energy can be determined from the gradient of a plot of $\ln D$ *versus* $1/T$ (Figure 7.7).

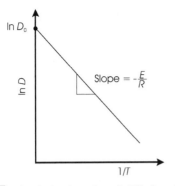

Figure 7.7   An Arrhenius plot of diffusion data: $\ln D$ versus $1/T$. Note: D, diffusion coefficient; T, temperature (in kelvin); D_0, pre-exponential, or frequency, factor

Table 7.3   Some representative values for self-diffusion coefficients

Atom	Matrix	Structure type	$D_0/\mathrm{m^2\,s^{-1}}$	E/kJ $\mathrm{mol^{-1}}$
Metals and semiconductors				
Cu	Cu	A1	2.0×10^{-5}	200
γ-Fe	Fe	A1	2.0×10^{-5}	269
α-Fe	Fe	A2	1.9×10^{-4}	240
Na	Na	A2	2.4×10^{-5}	44
Si	Si	Diamond	1.5×10^{-1}	455
Ge	Ge	Diamond	2.5×10^{-3}	303
Compounds				
Na^+	NaCl	Halite	8.4×10^{-8}	189
Cl^-	NaCl	Halite	0.17×10^{-4}	245
K^+	KCl	Halite	0.55×10^{-4}	256
Cl^-	KCl	Halite	1.3×10^{-6}	231
Mg^{2+}	MgO	Halite	2.5×10^{-9}	330
O^{2-}	MgO	Halite	4.3×10^{-13}	343
Ni^{2+}	NiO	Halite	4.8×10^{-10}	254
O^{2-}	NiO	Halite	6.2×10^{-12}	241
Pb^{2+}	PbS	Halite	8.6×10^{-13}	147
S^{2-}	PbS	Halite	6.8×10^{-13}	133
Ga	GaAs	Sphalerite	2.0×10^{-10}	400
As	GaAs	Sphalerite	7.0×10^{-5}	309
Zn^{2+}	ZnS	Sphalerite	3.0×10^{-12}	145
S^{2-}	ZnS	Sphalerite	2.1×10^{-4}	304
Cd^{2+}	CdS	Sphalerite	3.4×10^{-8}	193
S^{2-}	CdS	Sphalerite	1.6×10^{-10}	198

Note: literature values for self-diffusion coefficients vary widely, indicating the difficulty of making reliable measurements. The values in this table are intended to be representative only. The values of diffusion coefficients in the literature are mostly given in $\mathrm{cm^2\,s^{-1}}$; to convert the values given here to $\mathrm{cm^2\,s^{-1}}$, multiply by 10^4. D_0, pre-experimental, or frequency, factor; E, activation energy of diffusion.

Such graphs are known as Arrhenius plots. Diffusion coefficients found in the literature are usually expressed in terms of the Arrhenius equation D_0 and E values. Some representative values are given in Table 7.3.

7.5   The effect of impurities

Arrhenius plots for the majority of materials resemble Figure 7.7. However, Arrhenius plots obtained from very pure materials often consist of two straight-line parts with differing slopes (Figure 7.8). The region corresponding to diffusion at lower temperatures has a lower activation energy than the high-temperature region. The point where the two straight lines intersect is called a knee. If a number of different crystals of the same compound are studied, it is found that the position of the knee varies from one crystal to another, and depends on the impurity content. In Figure 7.8, for instance, crystal 1 would have a higher impurity concentration than crystal 2. The part of the plot sensitive to impurity content is called the impurity or extrinsic region. The high-temperature part of the plot is unaffected by the impurities present and is called the intrinsic region. The effect is explained in Section 7.11.

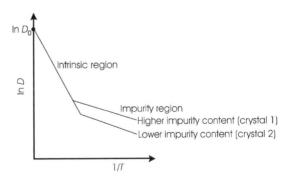

Figure 7.8   A form of Arrhenius plot found for almost-pure crystals with low impurity concentrations. Note: D, diffusion coefficient; T, temeprature (in kelvin); D_0, pre-exponential, or frequency, factor

7.6   The penetration depth

It is of considerable practical importance to have some idea of how far an atom or ion will diffuse into a solid during a diffusion experiment. For example, the electronic properties of integrated circuits are created by the careful diffusion of selected dopants into single crystals of very pure silicon, and metallic machine components are hardened by the diffusion of carbon or nitrogen from the surface into the bulk. An approximate estimate of the depth to which diffusion is significant is given by the penetration depth, x_P, which is the depth at which an appreciable change in the concentration of the tracer can be said to have occurred after a diffusion time t. It is obtained by using the equation

$$x_P = (D^* t)^{1/2} \tag{7.6}$$

A value for the distance moved during diffusion by a sequence of random jumps can be calculated. It is found that (Section S3.1.1):

$$\sqrt{(\langle x^2 \rangle)} = \sqrt{(n D t)}$$

where $\sqrt{(\langle x^2 \rangle)}$ is a quantity called the root mean square displacement of x, D is the diffusion coefficient, t is the diffusion time, and the factor n is of the order of 2 to 3. Real systems are more complex, and the approximation given in Equation (7.6) is adequate for practical purposes.

7.7   Self-diffusion mechanisms

During volume diffusion, an individual atom jumps from one stable position to another. If vacancies are present (Figure 7.9) atoms or ions can jump from a normal site into a neighbouring vacancy and so gradually move through the crystal. Movement of a diffusing atom into a vacant site corresponds to movement of the vacancy in the other direction. This process is therefore frequently referred to as vacancy diffusion. In practice, it is often very convenient, in problems where vacancy diffusion occurs, to ignore atom movement and to focus

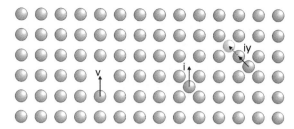

Figure 7.9   Self-diffusion mechanisms. Note: v, vacancy; i, interstitial; iy, interstitialcy

attention on the diffusion of the vacancies as if they were real particles.

In the case of interstitials, two diffusion mechanisms can be envisaged (Figure 7.9). An interstitial can jump to a neighbouring interstitial position. This is called interstitial diffusion and is the mechanism by which tool steels are hardened by incorporation of nitrogen or carbon. Alternatively, an interstitial can jump to a filled site and knock the occupant into a neighbouring interstitial site. This 'knock-on' process is called interstitialcy diffusion.

Substitutional impurities can also move by way of three mechanisms (Figure 7.10). As well as vacancy diffusion, an impurity can swap places with a neighbouring normal atom, exchange diffusion, and in ring diffusion co-operation between several atoms is needed to make the exchange. These processes have been found to take place during the doping of semiconductor crystals. Interstitial impurities can move by interstitial and interstitialcy jumps similar to those described above (Figure 7.10).

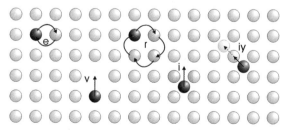

Figure 7.10   Impurity diffusion mechanisms. Note: v, vacancy; e, exchange; r, ring; i, interstitial; iy, interstitialcy

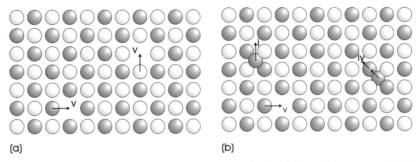

(a)   (b)

Figure 7.11   (a) Vacancy diffusion, v, in a crystal containing Schottky defects; (b) vacancy diffusion, v, interstitial diffusion, i and interstitialcy diffusion, iy, in crystals containing Frenkel defects

When Schottky defects are present in a crystal, vacancies are found on the cation and the anion sublattices, allowing both cation and anion diffusion to occur (Figure 7.11a). In the case of Frenkel defects, interstitial sites are occupied and vacancies occur, allowing for interstitial, interstitialcy and vacancy diffusion to take place in the same crystal (Figure 7.11b) so that three migration routes are possible.

7.8   Atomic movement during diffusion

The process of self-diffusion involves the movement of atoms in a random fashion through a crystal. No strong driving force, such as a concentration gradient, is present. Nevertheless, each time an atom moves it will have to overcome an energy barrier. This is because the migrating atoms have to leave normally occupied positions, which are, by definition, the most stable positions for atoms in the crystal, to pass through less stable positions not normally occupied by atoms. Often atoms may be required to squeeze through a bottleneck of surrounding atoms in order to move at all.

For one-dimensional diffusion the energy barrier can be supposed to take the form shown in Figure 7.12, where the height of the barrier is E. Obviously, the larger the magnitude of E the less chance there is that the atom has the necessary energy to make a successful jump.

We can gain an estimate of the probability that an atom will successfully move by using Maxwell–Boltzmann statistics. The probability, p, that a single atom will move from one position of minimum energy to an adjacent position will be given by the equation:

$$p = \exp\left(\frac{-E}{kT}\right) \qquad (7.7)$$

where E represents the energy barrier, k is the Boltzmann constant, and T is the absolute temperature. Equation (7.7) indicates that if E is very small

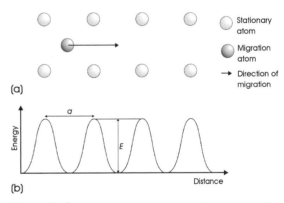

(a)

(b)

Figure 7.12   (a) An atom migrating from one stable position to another, separated by a distance a. (b) The energy barrier encountered has a periodicity equal to a and is a maximum when the diffusing atom has to pass a 'bottleneck' between two stationary atoms

the probability that the atom will clear the barrier approaches 1.0; if E is equal to kT the probability for a successful jump is about one third; and if E increases above kT the probability that the atom could jump the barrier rapidly becomes negligible.

The atoms in a crystal are vibrating continually with a frequency, ν, which is usually taken to have a value of about 10^{13} Hz at room temperature. It is reasonable to suppose that the number of attempts at a jump, sometimes called the attempt frequency, will be equal to the frequency with which the atom is vibrating. The number of successful jumps that an atom will make per second, Γ, will be equal to the attempt frequency, ν, multiplied by the probability of a successful move, that is,

$$\Gamma = \nu \exp\left(\frac{-E}{kT}\right) \qquad (7.8)$$

7.9   Atomic migration and diffusion coefficients

A flow of atoms along the x direction is related to the diffusion coefficient by Fick's first law:

$$J = -D\frac{dc}{dx}$$

where J is the number of atoms crossing a unit area in the solid each second (the atom flux), D is the diffusion coefficient, and c is the concentration of the diffusing species at point x after time t has elapsed. An expression for J in terms of atomic jumps is given in Section S3.1.2. The analysis shows that:

$$J = -\frac{1}{2}\Gamma a^2 \frac{dc}{dx}$$

where Γ is the number of successful jumps that an atom makes per second, and a is the separation of the stable positions. If we now compare this equation with Fick's first law:

$$D = \frac{1}{2}\Gamma a^2$$

Replacing Γ by Equation (7.8) gives:

$$D = \frac{1}{2}a^2 \nu \exp\left(\frac{-E}{kT}\right)$$

7.10   Self-diffusion in crystals

In real crystals it is necessary to take some account of the three-dimensional nature of the diffusion process. An easy way of doing this is to add a geometrical factor, g, into the equation for D so that it becomes:

$$D = g\,a^2 \nu \exp\left(\frac{-E}{kT}\right)$$

In the one-dimensional case, the factor 1/2 is a geometrical term to account for the fact that an atom jump can be in one of two directions:

$$D = \frac{1}{2}a^2 \nu \exp\left(\frac{-E}{kT}\right)$$

In a cubic structure diffusion can occur along six equivalent directions, and a value of g of 1/6 is appropriate:

$$D = \frac{1}{6}a^2 \nu \exp\left(\frac{-E}{kT}\right)$$

In the foregoing discussion, every possible atom jump is allowed. This may not be true in real crystals. For example, in the case of vacancy diffusion, no movement is possible if the vacancy population is zero. Equation (7.8) for the number of successful jumps ignores this, and should contain a term p_J, that expresses the probability that the jump is possible from a structural point of view:

$$\Gamma = p_J\nu \exp\left(\frac{-E}{kT}\right) \qquad (7.9)$$

For example, in the A1 (face-centred cubic) structure of magnesium, each metal atom is surrounded by 12 nearest neighbours. If on average throughout the crystal two of these sites are empty, the probability of a successful jump will be $p_J = 2/12 = 1/6$. The diffusion coefficient is then given by:

$$D = \frac{1}{6}a^2 \nu n \exp\left(\frac{-E}{kT}\right) \qquad (7.10)$$

7.11   The Arrhenius equation and the effect of temperature

A comparison of Equations (7.9) and (7.10) with the Arrhenius equation, Equation (7.5), which can be written in terms of the energy per defect rather than molar quantities as:

$$D = D_0 \exp\left(\frac{-E}{kT}\right)$$

shows that the pre-exponential factor D_0 is equivalent to $p_J g\, a^2 \nu$:

$$D_0 = p_J g\, a^2 \nu$$

and the activation energy, E, is equivalent to the height of the energy barrier to be surmounted.

The term p_J, which is the probability that a jump can take place for structural reasons, is closely related to the number of defects present. In most ordinary solids, the value of p_J is fixed by the impurity content. Any variation in D_0 from one sample of a material to another is accounted for by the variation of the impurity content. However, the value of p_J does not affect the energy of migration, E, so that Arrhenius plots for crystals will consist of a series of parallel lines (Figure 7.13).

In very pure crystals, another feature becomes important. In this case, the number of intrinsic defects may be greater than the number of defects due to impurities, especially at high temperatures. Under these circumstances, the value of p_J will be influenced by the intrinsic defect population, and can contribute to the observed value of E.

To illustrate this, suppose that cation vacancy diffusion is the predominant migration mechanism

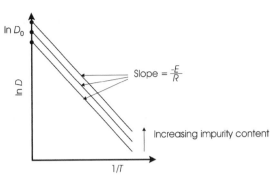

Figure 7.13   The variation of Arrhenius plots with impurity content. Note: D, diffusion coefficient; T, temperature (in kelvin); D_0, pre-exponential, or frequency, factor

and that the crystal, of formula *MX*, contains Schottky defects as the major type of intrinsic defect. The probability of a jump taking place is equal to the fraction of cation vacancies in the crystal, given by Equation (3.3):

$$\frac{n_S}{N} = \exp\left(\frac{-\Delta H_S}{2kT}\right) \qquad (7.11)$$

At high temperatures the appropriate form of Equation (7.10) becomes:

$$D = g \nu a^2 \exp\left(\frac{-E_v}{kT}\right) \exp\left(\frac{-\Delta H_S}{2kT}\right)$$

where E_v represents the height of the energy barrier to be overcome in vacancy diffusion, and ΔH_S is the enthalpy of formation of a Schottky defect.

Similarly, at high-temperatures, diffusion in a crystal of formula *MX* by interstitials will reflect the population of Frenkel defects present, given by Equation (3.6):

$$n_F = (N N^*)^{1/2} \exp\left(\frac{-\Delta H_F}{2kT}\right) \qquad (7.12)$$

In these circumstances, the probability factor, p_J, will be proportional to the number of Frenkel defects present, and it is possible to write Equation (7.10) in a form analogous to Equation (7.12):

$$D = g \nu a^2 \exp\left(\frac{-E_i}{kT}\right) \exp\left(\frac{-\Delta H_F}{2kT}\right)$$

where E_i represents the potential barrier to be surmounted by an interstitial atom and ΔH_F is the enthalpy of formation of a Frenkel defect.

Both of these equations retain the form:

$$D = D_0 \exp\left(\frac{-E}{RT}\right)$$

However, the activation energy, E, will consist of the energy required for migration, E_i or E_v, plus the energy of defect formation. For Schottky defects

$$E = E_v + \frac{\Delta H_S}{2}$$

and for Frenkel defects

$$E = E_i + \frac{\Delta H_F}{2}$$

This means that an Arrhenius plot will have a steeper slope at high temperatures where the point defect equilibria are significant, than at low temperatures, where the impurity content dominates. The plot will show a knee between the high-temperature and low-temperature regimes (Figure 7.8). A comparison of the two slopes will allow an estimate both of the energy barrier to migration and of the relevant defect formation energy to be made. Some values found in this way are listed in Table 7.4.

In the foregoing discussion, it has been supposed that the height of the potential barrier will be the same at all temperatures. This is probably not so. As the temperature increases the lattice will expand and, in general, E would be expected to decrease. Moreover, some of the other constant terms in the preceding equations may vary with temperature. For example, an expansion of the crystal is likely to lead to an increase in the vibration frequency, ν. The Arrhenius plots reveal this by being slightly curved.

Table 7.4 Approximate enthalpy values for the formation and movement of vacancies in alkali halide crystals

Material	$\Delta H/kJ$ mol^{-1}	E_1/kJ mol^{-1}	E_2/kJ mol^{-1}
(a) Schottky defects			
NaCl	192	84	109
NaBr	163	84	113
KCl	230	75	172
KBr	192	64	46
(b) Frenkel defects			
AgCl	155	13	36
AgBr	117	11	23

Note: for Schottky defects, ΔH is ΔH_S, the enthalpy of formation of a Schottky defect, and E_1 and E_2 are the energy barriers to be surmounted for vacancy diffusion by cations and anions, respectively; for Frenkel defects, ΔH is ΔH_F, the enthalpy of formation of a Frenkel defect, and E_1 and E_2 are the energy barriers to be surmounted for interstitial and vacancy diffusion, respectively.

7.12   Correlation factors for self-diffusion

So far, the diffusion of atoms has been considered to occur in a random fashion throughout the crystal structure. Each step was unrelated to the one before and the atoms were supposed to be jostled solely by thermal energy. However, diffusion of an atom in a solid may *not* be a truly random process and in some circumstances a given jump direction may depend on the direction of the previous jump.

Consider the vacancy diffusion of an atom in a crystal (Figure 7.14). The diffusing atom can be

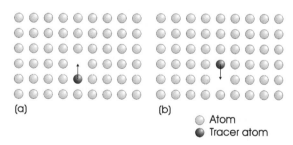

(a)   (b)

○ Atom
● Tracer atom

Figure 7.14 Vacancy diffusion of a tracer atom; in both part (a) and part (b) the most probable jump for the tracer is into the adjacent vacancy

regarded as a tracer atom. It is situated next to a vacant site, so that diffusion can take place by jumping into the vacant site (Figure 7.14a). The next jump of the tracer is not, now, an entirely random process. It is still next to the vacancy and it is more likely that the tracer will move back to the vacancy (Figure 7.14b), recreating the original situation. Hence, of the choices available to the tracer in Figure 7.14(b), a jump back to the situation shown in Figure 7.14(a) is of highest probability.

If attention is focused on the vacancy, a different result is found. Considering the situation in Figure 7.14(a), diffusion can occur by way of any of the atoms around the vacancy moving into the empty site. The vacancy, of course, does not prefer any of its neighbours so that its first jump is entirely random. The same is true of the succeding situation (Figure 7.14b). The vacancy will have no need to prefer a jump to the tracer position. Thus, the vacancy can always move to an adjacent cation site and hence can follow a truly random path.

When these processes are considered over many jumps, the mean square displacement of the tracer will be less that that of the vacancy, even though both have taken the same number of jumps. Therefore, it is expected that the observed diffusion coefficient of the tracer will be less than that of the vacancy. In these circumstances, the random-walk diffusion equations need to be modified by the introduction of a correlation factor, f. The correlation factor is given by the ratio of the values of the mean square displacement of the tracer, $\langle x^2 \rangle_{tracer}$, to that of the vacancy, $\langle x^2 \rangle_{vacancy}$, if the number of jumps considered is large:

$$f = \frac{\langle x^2 \rangle_{tracer}}{\langle x^2 \rangle_{vacancy}}$$

Correlation factors for vacancy diffusion generally take values of between 0.5 and 0.8.

In the case of interstitial diffusion, in which we have only a few diffusing atoms and many available empty interstitial sites, a correlation factor close to 1.0 would be expected. In effect, the interstitial atom moves in a 'sea of vacancies'. In the case of interstitialcy diffusion, this will not be true because

Table 7.5 Correlation factors for self-diffusion

Structure	Correlation factor (f)
Vacancy mechanism:	
Diamond	0.50
A2, body-centred cubic	0.7272
A1, face-centred cubic	0.7815
A3, hexagonal close packed	0.7812 (f_x, f_y)
A3, hexagonal close packed	0.7815 (f_z)
Intersticialcy mechanism:	
A1, face-centred cubic	0.80
Fluorite (cations)	1.00
A2, body-centred cubic	0.666
CsCl (cations)	0.832
AgBr (cations)	0.666

the number of vacancies will be equal to the number of interstitials present, which will always be rather small in proportion to the number of filled sites.

Table 7.5 lists some values for a variety of diffusion mechanisms in some common crystal structure types.

The diffusion of an impurity atom in a crystal, say K in NaCl, involves other considerations apart from those of a statistical nature. In such cases, the probability that the impurity will exchange with the vacancy will depend on factors such as the relative sizes of the impurity compared with the host atoms. In the case of ionic movement, the charge on the diffusing species will also play a part. These factors can can also be expressed in terms of the jump frequencies of the host and impurity atoms, in which case one is likely to be greater than the other.

7.13 Ionic conductivity

During ionic conductivity, ions jump from one stable site to another. Hence, the process can be described by equations similar to those for diffusion. The movement of the ions, however, is not random, but is influenced by the presence of an electric field, V, so that positive and negative ions move in opposite directions. The electric field

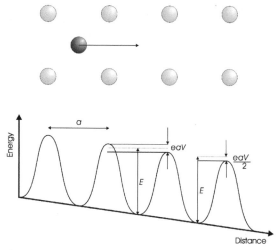

Figure 7.15 The energy barrier to ionic diffusion is higher against the direction of the electric field and lower in the direction of the electric field. The barrier difffference, eaV, is for a monovalent ion

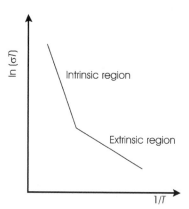

Figure 7.16 Arrhenius-type plot of $\ln(\sigma T)$ versus $1/T$ for ionic conductivity in a crystal containing only a small concentration of impurities. Note: σ, ionic conductivity; T, temperature (in kelvin)

changes the potential barrier encountered by a diffusing ion from that in Figure 7.12 to that in Figure 7.15. For monovalent ions, the barrier will be reduced by $\frac{1}{2} e a V$ in one direction and raised by the same amount in the other.

Following the methods set out for random diffusion, it is possible to caluclate the relative number of jumps that an ion will make with and against the field and hence obtain the ionic conductivity, σ (see Section S3.1.3). For low field strengths,

$$\sigma = \frac{\sigma_0}{T} \exp\left(\frac{-E}{kT}\right) \qquad (7.13)$$

where σ_0 is a constant, E is the mean height of the diffusion barrier, k is the Boltzmann constant, and T is the temperature (in kelvin). The constant, σ_0, is:

$$\sigma_0 = \frac{n \nu a^2 e^2}{k} \qquad (7.14)$$

where a is the jump distance, e the charge on the ions, ν is the vibration frequency or jump frequency, k is the Boltmann constant, and n is the number of mobile species present in the crystal.

Generally, n will be fixed and controlled by the impurity population. When ion migration takes place via a vacancy diffusion mechanism we can write:

$$\sigma_v = \frac{n_v \nu a^2 e}{kT} \exp\left(\frac{-E_v}{kT}\right)$$

and when it takes place by the migration of interstitials we can write:

$$\sigma_i = \frac{n_i \nu a^2 e}{kT} \exp\left(\frac{-E_i}{kT}\right)$$

In this regime, we can obtain measures of E_v or E_i directly from the Arrhenius-like plots of $\ln(\sigma T)$ versus $1/T$ (Figure 7.16).

In the case of very pure solids, it is necessary to take into account the number of intrinsic defects, as discussed above. The value of n must then reflect the type of intrinsic defect present. For example, should Schottky defects predominate, substitution of Equation (7.11) [Equation (3.3)] for n in Equations (7.13) and (7.14) gives:

$$\sigma_S = \frac{\nu a^2 e^2 N}{kT} \exp\left(\frac{-E_v}{kT}\right) \exp\left(\frac{-\Delta H_S}{2kT}\right)$$

Should Frenkel defects predominate, substituting Equation (7.12) [Equation (3.6)] for n in Equations (7.13) and (7.14) gives:

$$\sigma_F = \frac{\nu a^2 e^2 (N N^*)^{1/2}}{kT} \exp\left(\frac{-E_i}{kT}\right) \exp\left(\frac{-\Delta H_F}{2kT}\right)$$

In these cases, Arrhenius plots of $\ln \sigma T$ versus $1/T$ will show a knee between the low-temperature and high-temperature regions (Figure 7.16). The high-temperature value for E will be composed of two terms; namely,

$$E_S = E_v + \frac{\Delta H_S}{2}$$

for Schottky defects, and

$$E_F = E_i + \frac{\Delta H_F}{2}$$

for Frenkel defects.

7.14 The relationship between ionic conductivity and the diffusion coefficient

If both ionic conductivity and ionic diffusion occur by the same random-walk mechanism, a relationship between the self-diffusion coefficient, D, and the ionic conductivity, σ, can be derived. In the simplest case, assume that processes involve the same energy barrier, E, and jump distance, a (Figures 7.12 and 7.15). Further, if the diffusion is restricted to only one direction, the $+x$ direction, and each jump is allowed, the diffusion coefficient is

$$D = \nu a^2 \exp\left(\frac{-E}{kT}\right)$$

and the ionic conductivity of a monovalent ion is

$$\sigma = \frac{n \nu a^2 e^2}{kT} \exp\left(\frac{-E}{kT}\right)$$

Combining these two equation gives

$$\frac{\sigma}{D} = \frac{n e^2}{kT} \tag{7.15}$$

where n is the number of mobile ions per unit volume, e is the electronic charge, k is the Boltzmann constant, and T is the temperature (in kelvin). This equation is a simplified form of the Nernst–Einstein equation. For an ion of charge $+z$, the equation becomes

$$\frac{\sigma}{D} = \frac{n z^2 e^2}{kT} \tag{7.16}$$

The implication is that it is possible to determine the diffusion coefficient from the easier measurement of ionic conductivity. However, the assumption that both processes utilise exactly the same mechanism is important. In general, this is not true. In such a case, the relationship is slightly different from that in Equations (7.15) and (7.16) and depends on the details of the diffusion mechanism. For vacancy diffusion in a cubic structure,

$$\frac{\sigma}{D} = \frac{n z^2 e^2}{f_v kT} \tag{7.17}$$

where f_v is the correlation factor for vacancy self-diffusion, and n is the number of ions per unit volume on the relevent sublattice. In the case of interstitial diffusion in a cubic crystal, Equation (7.17) applies, with a correlation factor of 1:

$$\frac{\sigma}{D} = \frac{n z^2 e^2}{kT}$$

For interstitialcy diffusion,

$$\frac{\sigma}{D} = \frac{2 n z^2 e^2}{f_{iy} kT}$$

where f_{iy} is the appropriate intersticialcy correlation factor.

Answers to introductory questions

What is steady-state diffusion?

Steady-state diffusion differs from nonsteady-state diffusion in that the concentration of the diffusing atoms at any point, x, and hence the concentration gradient at x, in the solid, remains constant. Steady-state diffusion may be achieved when air diffuses through plastic food wrapping film.

Under steady-state conditions, the diffusion coefficient is obtained by using Fick's first law:

$$J_i = -D_i \frac{d c_i}{d x}$$

where D_i is the diffusion coefficient of atom of type i, c_i is the concentration of these atoms, and x is the position in the solid. J_i is the flux of atoms of type i; that is, it is the net flow of these atoms through the solid.

How does one obtain a quick estimate of the distance moved by diffusing atoms?

It is of considerable practical importance to have some idea of how far an atom or ion will diffuse into a solid during a diffusion experiment. An approximate estimate of the depth to which diffusion is significant is given by the penetration depth, x_P, which is the depth at which an appreciable change in the concentration of the tracer can be said to have occurred after a diffusion time t. It is obtained by using the equation

$$x_P = (D^* t)^{1/2}$$

This approximation is adequate for practical purposes.

How does the energy barrier for ionic diffusion change when an electric field is present?

During ionic conductivity, ions jump from one stable site to another. Hence, the process can be described by equations similar to those for diffusion. However, the movement of the ions is influenced by the presence of an electric field, V, so that positive and negative ions move in opposite directions. The electric field changes the potential barrier encountered by a diffusing ion from that encountered during diffusion in the following way. For ions with a charge of ze, the barrier will be reduced by $\frac{1}{2}zeaV$ in one direction and raised by the same amount in the other. When the mechanism of migration is the same, the relationship between the ionic conductivity and the diffusion coefficient for an ion of charge ze is given by a simplified form of the Nernst–Einstein equation:

$$\frac{\sigma}{D} = \frac{n z^2 e^2}{k T}$$

Further reading

W.D. Kingery, H.K. Bowen, and D.R. Uhlmann, 1976, *Introduction to Ceramics*, 2nd Edition, Wiley-Interscience, New York.

J.S. Kirkaldy, D.J. Young, 1987, *Diffusion in the Condensed State*, Institute of Metals, London.

A.D. LeClaire, 1976, 'Diffusion' in *Treatise on Solid State Chemistry, Volume 4: Reactivity of Solids* Ed. N.B. Hannay, Plenum, New York, Ch. 1.

R. Metselaar, 1984, 'Diffusion in Solids: Part 1', *Journal of Materials Education* **6** 229.

R. Metselaar, 1985, 'Diffusion in Solids: Part 2', *Journal of Materials Education* **7** 653.

R. Metselaar, 1988, 'Diffusion in Solids: Part 3', *Journal of Materials Education* **10** 621.

P.G. Shewman, 1963, *Diffusion in Solids*, McGraw Hill, New York.

Problems and exercises

Quick quiz

1 Diffusion through a crystalline structure is called:
   (a) Tracer diffusion
   (b) Volume diffusion
   (c) Self-diffusion

2 Short-circuit diffusion refers to diffusion along:
(a) Defects
(b) Lattices
(c) Planes

3 Self-diffusion is diffusion of:
(a) Radioactive atoms in a crystal
(b) Impurity atoms in a crystal
(c) Native atoms in a pure crystal

4 Tracer diffusion refers to the diffusion of:
(a) Marked atoms
(b) Traces of impurities
(c) Single atoms

5 A radioactive layer between two crystal slabs forms a:
(a) Difusion pair
(b) Diffusion couple
(c) Diffusion probe

6 A diffusion profile is a graph of:
(a) Concentration versus time
(b) Concentration versus distance
(c) Distance versus time

7 In order to obtain the diffusion coefficient from a diffusion profile, use:
(a) Fick's first law
(b) The Arrhenius equation
(c) The diffusion equation

8 What graph would you plot to determine the tracer diffusion coefficient in a diffusion couple:
(a) c versus x?
(b) $\ln c$ versus x^2?
(c) $\ln c$ versus $\ln x$?

9 Nonsteady-state diffusion is characterised by the fact that the concentration at:
(a) Any point in the solid changes over time
(b) Any point in the solid is constant
(c) The surface of the solid changes over time

10 Steady-state diffusion is characterised by the fact that the concentration at:
(a) Any point in the solid changes over time
(b) Any point in the solid is constant
(c) The surface of the solid changes over time

11 In steady-state diffusion the diffusion coefficient is obtained via:
(a) The diffusion equation
(b) The Arrhenius equation
(c) Fick's first law

12 The variation of a diffusion coefficient with temperature is given by:
(a) The diffusion equation
(b) Fick's first law
(c) The Arrhenius equation

13 The activation energy for diffusion can be determined by a plot of:
(a) $\ln D$ versus $1/T$.
(b) $\ln c$ versus $1/T$.
(c) $\ln D$ versus T.

14 The part of an Arrhenius plot above a knee (i.e. the high-temperature part) is called the:
(a) Extrinsic region
(b) Intrinsic region
(c) Impurity region

15 The penetration depth is:
(a) The distance that an atom can diffuse in a given time
(b) An estimate of the depth to which diffusion is significant
(c) The thickness of the surface layer on a crystal before diffusion

16 The mechanism of vacancy diffusion refers to atoms moving into:
(a) Vacant sites in the crystal structure
(b) Nonoccupied sites in the crystal structure
(c) Adjacent atom sites in the crystal structure

17 The mechanism of interstitial diffusion refers to atoms moving into:

(a) Vacant sites in the crystal structure

(b) Nonoccupied sites in the crystal structure

(c) Adjacent atom sites in the crystal structure

18 The diffusion mechanism of exchange involves:
(a) Two impurity atoms

(b) An impurity atom and a vacancy

(c) An impurity atom and a normal atom

19 In crystals containing Frenkel defects:
(a) One diffusion mechanism is possible

(b) Two diffusion mechanisms are possible

(c) Three diffusion mechanisms are possible

20 The geometrical factor in diffusion varies according to:
(a) The crystal structure

(b) The number of defects present

(c) The size of the diffusing atoms

21 Correlation factors take into account:
(a) Nonrandom diffusion

(b) Impurity content

(c) Frenkel and Schottky defects

22 An electric field changes the rate of diffusion of ions in a direction:
(a) Perpendicular to the applied electric field

(b) Parallel to the unit cell edge

(c) Parallel to the electric field

Calculations and questions

7.1 Show that the units of the diffusion coefficient are $m^2 \, s^{-1}$. [Note: answer is not provided at the end of this book.]

7.2 Show that the following two equations are equivalent:

$$\frac{c_x - c_0}{c_s - c_0} = 1 - \text{erf}\left(\frac{x}{2\sqrt{Dt}}\right)$$

$$\frac{c_s - c_x}{c_s - c_0} = \text{erf}\left(\frac{x}{2\sqrt{Dt}}\right)$$

[Note: answer is not provided at the end of this book.]

7.3 Radioactive nickel-63 was coated onto a crystal of CoO and made into a diffusion couple. The sample was heated for 30 min at 953 °C. The radioactivity perpendicular to the surface is given in Table 7.6(a). Calculate the impurity tracer diffusion coefficient of nickel-63 in CoO.

7.4 Radioactive cobalt-60 was coated onto a crystal of CoO and made into a diffusion couple. The sample was heated for 30 min at 953 °C. The radioactivity perpendicular to the surface is given in Table 7.6(b). Calculate the tracer diffusion coefficient of cobalt-60 in CoO.

Table 7.6 Diffusion couples: radioactivity perpendicular to the surface for (a) nickel-63 as a crystal of CoO, for Question 7.3; (b) cobalt-60 on a crystal of CoO, for Question 7.4; and (c) iron-59 on $(0\,0\,1)$ face of a single crystal of TiO$_2$, for Question 7.5

(a) Question 7.3		(b) Question 7.4		(c) Question 7.5	
activity[a]	distance/μm	activity[a]	distance/μm	activity[a]	distance/μm
80	6	110	10	520	300
50	10	70	20	400	400
20	14	39	30	270	500
6	18	23	40	185	600
5	20	9	50	130	650
				90	700
				53	800

[a] In counts per second.

7.5   Radioactive iron-59 was coated onto the $(0\,0\,1)$ face of a single crystal of TiO_2 (rutile) (tetragonal) and made into a diffusion couple. The sample was heated for $300\,s$ at $800\,°C$. The radioactivity perpendicular to the surface is given in Table 7.6(c). Calculate the impurity tracer diffusion coefficient of iron-59 parallel to the c axis in rutile.

7.6   Carbon-14 is diffused into pure α-iron from a gas atmosphere of $CO + CO_2$. The gas pressures are arranged to give a constant surface concentration of $0.75\,wt\%$ C. The diffusivity of ^{14}C into α-iron is $9.5 \times 10^{-11}\,m^2\,s^{-1}$ at $827\,°C$. Calculate the concentration of ^{14}C 1 mm below the surface after a heating time of 2 hours.

7.7   Using the data in Question 7.6, how long would it take to make the carbon content $0.40\,wt\%$ 1 mm below the surface?

7.8   An ingot of pure titanium metal is heated at $1000\,°C$ in an atmosphere of ammonia, so that nitrogen atoms diffuse into the bulk. The diffusivity of nitrogen in β-titanium, the stable structure at $1000\,°C$, is 5.51×10^{-12} $m^2\,s^{-1}$ at this temperature. What is the thickness of the surface layer of titanium that contains a concentration of nitrogen atoms greater than $0.25\,at\%$ after heating for 1 hour?

7.9   Zircalloy is a zirconium alloy used to clad nuclear fuel. How much oxygen will diffuse through each square metre of casing in a day under steady-state diffusion conditions, at $1000\,°C$, if the following apply: diffusivity of oxygen in zircalloy at $1000\,°C$, $9.89 \times 10^{-13}\,m^2\,s^{-1}$; concentration of oxygen on the inside of the container, $0.5\,kg\,m^{-3}$; oxygen concentration on the outside of the container, $0.01\,kg\,m^{-3}$; container thickness, 1 cm

7.10   Pure hydrogen is made by diffusion through a Pd-20\,%Ag alloy 'thimble'. What is the mass of hydrogen prepared per hour if: the total area of the thimbles used is $10\,m^2$; the thickness of each is 5 mm; and the operating temperature is $500\,°C$? The equili-

brium alloy in the surface of the thimble on the hydrogen–rich side has a composition $PdH_{0.05}$, and, on the further side, the hydrogen is swept away rapidly so that the surface is essentially pure metal. The diffusion coefficient of hydrogen in the alloy at $500\,°C$ is $1.3 \times 10^{-8}\,m^2\,s^{-1}$. Pd has the A1 (face-centred cubic) structure, with lattice parameter $a_0 = 0.389\,nm$.

7.11   The radioactive tracer diffusion coefficient of silicon atoms in silicon single crystals is given in Table 7.7(a). Estimate the activation energy for diffusion.

7.12   Using the data for Question 7.11, listed in Table 7.7(a), at what temperature in (°C) will the penetration depth be $2\,\mu m$ after 10 hours of heating?

7.13   The diffusion coefficient of carbon impurities in a silicon single crystal is given in Table 7.7(b). Estimate the activation energy for diffusion.

7.14   Using the data in Question 7.13, listed in Table 7.7(b), at what temperature in (°C) will the penetration depth be $10^{-4}\,m$ after 20 hours of heating?

7.15   The diffusion coefficient of radioactive Co^{2+} tracers in a single crystal of cobalt oxide, CoO, is given in the Table 7.7(c). Estimate the activation energy for diffusion.

7.16   The diffusion coefficient of radioactive Cr^{3+} tracers in the refractory oxide Cr_2O_3 is given in the Table 7.7(d). Estimate the activation energy for diffusion.

7.17   The impurity diffusion coefficient of Fe^{2+} impurities in a magnesium oxide, MgO, single crystal is given in Table 7.7(e). Estimate the activation energy for diffusion.

7.18   Calculate the diffusivity of ^{51}Cr in titanium metal at $1000\,°C$: $D_0 = 1 \times 10^{-7}\,m^2\,s^{-1}$; $E = 158\,kJ\,mol^{-1}$.

7.19   Calculate the diffusivity of ^{51}Cr in a titanium: 18 wt% Cr alloy at $1000\,°C$; $D_0 = 9 \times 10^{-2}\,m^2\,s^{-1}$; $E = 186\,kJ\,mol^{-1}$.

Table 7.7 Radioactive tracer diffusion coefficients, D^* with temperature, T, for: (a) silicon atoms in silicon single crystals, for Questions 7.11 and 7.12; (b) carbon impurities in a silicon single crystal, for Questions 7.13 and 7.14; (c) radioactive Co^{2+} tracers in a single crystal of CoO, for Question 7.15; (d) radioactive Cr^{3+} tracers in Cr_2O_3, for Question 7.16; and (e) Fe^{2+} impurities in a single crystal of MgO

$T/°C$	$D^*/m^2\,s^{-1}$
(a) Questions 7.11 and 7.12	
1150	8.82×10^{-19}
1200	3.40×10^{-18}
1250	1.20×10^{-17}
1300	3.90×10^{-17}
1350	1.18×10^{-16}
1400	3.35×10^{-16}
(b) Questions 7.13 and 7.14	
900	1.0×10^{-17}
1000	3.0×10^{-16}
1100	4.0×10^{-15}
1200	5.0×10^{-14}
1300	9.0×10^{-13}
1400	5.0×10^{-12}
(c) Question 7.15	
1000	1.0×10^{-13}
1100	3.5×10^{-13}
1200	9.0×10^{-13}
1300	2.0×10^{-12}
1400	4.0×10^{-12}
1500	8.0×10^{-12}
1600	1.5×10^{-11}
(d) Question 7.16	
1050	1.0×10^{-15}
1100	4.6×10^{-15}
1200	1.05×10^{-14}
1300	6.2×10^{-14}
1400	2.7×10^{-13}
1500	6.5×10^{-13}
(e) Question 7.17	
1150	2.0×10^{-14}
1200	3.2×10^{-14}
1250	5.0×10^{-14}
1300	7.5×10^{-14}
1350	1.0×10^{-13}

7.20 Calculate the diffusivity of ^{55}Fe in forsterite, Mg_2SiO_4, at 1150 °C: $D_0 = 4.17 \times 10^{-10}$ m^2 s^{-1}, $E = 162.2$ kJ mol^{-1}.

7.21 Calculate the diffusivity of ^{18}O in Co_2SiO_4 at 1250 °C: $D_0 = 8.5 \times 10^{-3}$ m^2 s^{-1}; $E = 456$ kJ mol^{-1}.

7.22 Calculate the diffusivity of Li in quartz, SiO_2, parallel to the c axis, at 500 °C: $D_0 = 6.9 \times 10^{-7}$ m^2 s^{-1}; $E = 85.7$ kJ mol^{-1}.

7.23 The diffusion coefficient of Ni^{2+} tracers in NiO is 1×10^{-15} m^2 s^{-1} at 1100 °C. Estimate the penetration depth of the radioactive Ni^{2+} ions into a crystal of NiO after heating for 1 hour at 1100 °C

7.24 Ge is diffused into silica glass for fibre optic light guides. How long should a fibre of 0.1 mm diameter be annealed at 1000 °C to be sure that Ge has diffused into the centre of the fibre? The diffusion coefficient of Ge in SiO_2 glass is 1×10^{-11} m^2 s^{-1}.

7.25 What is the probability of a diffusing atom jumping from one site to another at 500 °C and 1000 °C? The activation energy for diffusion is 127 kJ mol^{-1}.

7.26 Estimate the ratio of the ionic conductivity to diffusion coefficient for a monovalent ion diffusing in an ionic solid. Take a typical value for the number of mobile diffusing ions as the number of vacancies present, approximately 10^{22} defects per metre cubed, and T as 1000 K.

7.27 The ionic conductivity of F^- ions in the fast ionic conductor $Pb_{0.9}In_{0.1}F_{2.1}$ at 423 K is 1×10^{-4} Ω^{-1} m^{-1}. The cubic unit cell (*fluorite* type) has a cell parameter 0.625 nm, and there are on average, 0.4 mobile F^- anions per unit cell. Estimate the diffusion coefficient of F^- at 423 K.

7.28 The conductivity of SrO, an oxide with the *halite* (B1) structure, $a_0 = 0.5160$ nm, depends upon oxygen partial pressure. The value of the ionic conductivity is 2×10^{-3} Ω^{-1} m^{-1} at 900 °C under 0.1 atm O_2. Assuming the ionic conductivity is due to vacancy diffusion of Sr^{2+} ions, estimate the diffusion coefficient of Sr^{2+} at 900 °C.

7.29 The diffusion coefficient is often found to obey an equation of the type:

$$D = D_0 \exp\left(\frac{-E}{RT}\right)$$

Explain each of the terms D, D_0 and E in terms of atomic diffusion. [Note: answer is not provided at the end of this book.]

7.30 A plot of $\ln D$ versus $1/T$ is sometimes found to be composed of two straight-line sections, with differing slopes. Explain how this form of plot arises. [Note: answer is not provided at the end of this book.]

8

Reactions and transformations

- What is dynamic equilibrium?

- What defines a martensitic transformation?

- What is the main driving force for sintering?

The direction of a chemical reaction or a phase transformation can be determined from the equilibrium thermodynamic properties of the phases involved. Note, though, that the speed of any transformation is not accessible from thermodynamics. Thermodynamics clearly states that diamond will transform into graphite at room temperature, but the rate of the reaction is insignificant. This chapter is concerned mainly with the kinetics of reactions, the speed at which they occur. Marrying this aspect with thermodynamics lies outside the scope of this chapter, but some introductory notes are given in Section S3.2.

8.1 Dynamic equilibrium

Dynamic equilibrium is an important and powerful concept in many areas of science. It is useful when discussing matters as diverse as the population of plants or animals in a region, the ozone layer in the upper atmosphere or the number of dust particles at a particular height in a column of air.

8.1.1 Reversible reactions and equilibrium

Chemical equations are usually written to represent a change in one direction only, from reactants to products:

$$Si + O_2 \rightarrow SiO_2$$

The tendency for silica, SiO_2, once formed, to disproportionate into elemental silicon and oxygen is very small indeed. However, there are a number of important chemical reactions where the reverse tendency of a reaction is similar to the forward tendency. These can be represented by an equation:

$$a\,A + b\,B \rightleftharpoons x\,X + y\,Y$$

where the double arrows signify that the reaction can take place in either direction. In such a reversible reaction, the reactants and products are not at all clearly defined, and either A and B or X and Y fulfil both roles. A feature of reversible reactions is that they often appear to be static, as if there were no change occurring. This state of affairs is illusory. Reactants are being consumed and regenerated by the forward and backward reactions continuously. If no change seems to be taking place it is because the rates of the reactions in each direction

Understanding solids: the science of materials. Richard J. D. Tilley
© 2004 John Wiley & Sons, Ltd ISBNs: 0 470 85275 5 (Hbk) 0 470 85276 3 (Pbk)

are balanced. In this case, the system is said to be in equilibrium.

These concepts can be illustrated by reference to a simple reversible reaction involving the dimerisation of nitrogen dioxide, NO_2. The reaction is written:

$$N_2O_4 \rightleftharpoons 2\,NO_2$$

If a quantity of N_2O_4 is introduced into an evacuated vessel at a fixed temperature, ultimately the vessel will contain a mixture of both N_2O_4 and NO_2 molecules. Similarly, if a quantity of NO_2 is introduced into an identical evacuated vessel at the same temperature, after a time the vessel will again contain a mixture of the two molecules. After a sufficient length of time, the ratio of the two species present will be the same in both cases (Figure 8.1). The system is then at equilibrium. At equilibrium, N_2O_4 and NO_2 molecules are continuously changing from one to another, and the rate of the forward reaction will be the same as the rate of the reverse reaction. The important feature is that the equilibrium is a dynamic state of affairs.

However, the ratio of the two species in the container is constant when a time average is taken. The equilibrium position will be obtained no matter where we start. That is, it does not matter whether we take a vessel that, at the outset, contains only N_2O_4 or only NO_2. In either case the same mixture will eventually form. The appearance of stasis is an illusion created by the large number of reacting molecules involved.

A change of conditions may change the apparent equilibrium because it may effect one of the reactions, say the forward reaction, more than the other, the reverse reaction. The equilibrium state that will finally prevail is found to be dependent on external conditions of temperature, pressure (for gas reactions) and the concentration of reactants.

8.1.2 Equilibrium constants

In any reversible reaction involving gases or solutions that can be written

$$a\,A + b\,B \rightleftharpoons x\,X + y\,Y$$

the equilibrium constant, K_c, is given by:

$$K_c = \frac{[X]^x [Y]^y}{[A]^a [B]^b} \qquad (8.1)$$

where [A] denotes the concentration of compound A, and so on when the reaction has come to equilibrium. The units of K_c are concentration units. The value of the equilibrium constant, K_c,

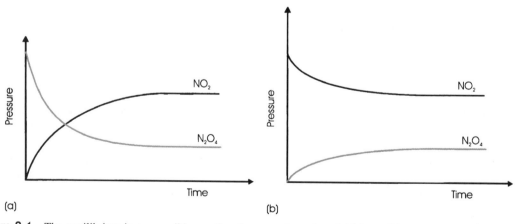

(a)   (b)

Figure 8.1 The equilibrium in a reversible reaction does not depend on initial conditions: (a) starting point 100 % N_2O_4; (b) starting point 100 % NO_2

depends only on temperature. It does not change when concentrations, volumes or pressures change. Pure liquids or pure solids that appear in the reaction equation (such as water for reactions in solution) do not figure in the equilibrium constant equation.

In reactions involving gaseous species, it is more convenient for the equilibrium constant to be expressed in terms of the partial pressures of the components:

$$K_p = \frac{p_X^x p_Y^y}{p_A^a p_B^b} \quad \text{at equilibrium,}$$

where (p_A) is the partial pressure of compound A, and so on.

For the reaction:

$$N_2O_4(g) \rightleftharpoons 2\,NO_2(g)$$

$$K_p = \frac{p_{NO_2}^2}{p_{N_2O_4}}$$

The units of K_p are pressure units.

8.1.3   Combining equilibrium constants

In the environment, or in many industrial processes, it is often necessary to work with several successive reversible reactions. The combination of equilibrium constants is straightforward. For any reversible reaction:

$$K(\text{reverse reaction}) = \frac{1}{K(\text{forward reaction})}$$

If several reactions can be added to produce an overall reaction, the equilibrium constants can be multiplied together to produce the overall equilibrium constant. For example:

$$C(s) + H_2O \rightleftharpoons CO(g) + H_2(g)$$

$$K_1 = \frac{p_{CO}p_{H_2}}{p_{H_2O}}$$

Note: Pure solids do not figure in equilibrium constants.

$$CO_2(g) + 2\,H_2(g) \rightleftharpoons 2\,H_2O(g) + C(s)$$

$$K_2 = \frac{p_{H_2O}^2}{p_{CO_2}p_{H_2}^2}$$

The overall reversible reaction is

$$CO_2(g) + H_2(g) \rightleftharpoons H_2O(g) + CO(g)$$

for which the equilibrium constant, K_3, is given by

$$K_3 = \frac{p_{H_2O}p_{CO}}{p_{CO_2}p_{H_2}}$$

$$= K_1 K_2$$

8.1.4   Equilibrium conditions

It is useful, and often vital, to know how a change in conditions will affect an equilibrium. For example, there is considerable debate at present about how a change in global temperature, global warming, will affect the (equilibrium) climate of the Earth. Le Chatelier's principle describes what will happen when a reaction at equilibrium is subjected to a change in conditions. It is found that the reaction proceeds towards a new equilibrium state that (at least partially) offsets the change in conditions.

The conditions of greatest importance are the concentrations of reactants or products, the volume or pressure and the temperature. Remember that the equilibrium constant depends only on temperature. It does not change when concentrations, volumes or pressures change. The following examples illustrate le Chatelier's principle.

8.1.4.1   Concentration

Addition of a reactant appearing in the equilibrium constant equation changes the equilibrium to remove as much as possible of the added component. For example, for

$$C(s) + CO_2(g) \rightleftharpoons 2\,CO(g)$$

if $CO_2(g)$ is added to the mixture, the equilibrium will shift from the left-hand side to the right-hand side to reduce the amount of $CO_2(g)$ present. In the new equilibrium state the amount of $CO_2(g)$ and CO will both be greater than in the original state. If $CO(g)$ is added, the equilibrium will shift from the right-hand side to the left-hand side to decrease the amount of CO present. The addition of C(s) produces no change, as the concentration of a solid does not influence the equilibrium constant.

8.1.4.2   Volume and pressure

A decrease in volume or an increase in applied pressure shifts the reaction equilibrium towards the side with fewer moles of gas. For example, with the reaction

$$N_2O_4(g) \rightleftharpoons 2\,NO_2(g)$$

an increase in pressure or decrease in volume will cause the equilibrium to move from the right-hand side to the left-hand side, which helps to offset the change applied. If the number of molecules is constant then volume and pressure changes do not affect the equilibrium.

8.1.4.3   Temperature

An increase in temperature shifts the reaction equilibrium in the direction in which heat will be absorbed. A decrease in temperature shifts the equilibrium in the direction in which heat will be liberated.

$$CaCO_3(s) \rightleftharpoons CaO(s) + CO_2(g) \quad \Delta H_R^\circ = +158\,kJ$$

is an endothermic reaction in which 158 kJ are consumed for each mole of CaO produced (see Section S3.2). An increase in temperature will shift the reaction in the direction that causes more heat to be consumed. The reaction will shift from left-hand side to the right-hand side, and there will be an increase in the amount of $CO_2(g)$ present at a higher temperature.

8.1.5   Pseudochemical equilibrium

Many nonchemical reactions can be treated by using similar techniques. One such is the creation of mobile electrons and holes in an intrinsic semiconductor. In this case, thermal energy excites some electrons from a filled valence band to the energetically nearby empty conduction band, the result of which is a population of mobile holes in the valence band and a population of mobile electrons in the conduction band. At ordinary temperatures, this is a dynamic equilibrium. There is a continuous excitation of electrons, and these are continuously falling back to the valence band and recombining with holes. The appearance of stasis is illusory.

This situation can be represented by a pseudochemical equation. As the electrons and holes are generated by thermal energy, the equation can be written:

$$\text{null} \rightleftharpoons [n] + [p]$$

where [n] is the concentration of electrons in the conduction band and [p] is the concentration of holes in the valence band. The equilibrium concentration is then

$$K = [n][p]$$

K will depend on temperature, in the expected way, but not on pressure or how much semiconductor is in the sample.

In an intrinsic semiconductor, the number of holes will equal the number of electrons. However, the equilibrium equation also applies to doped semiconductors, and the equilibrium constant derived for a pure intrinsic semiconductor is valid for a doped sample of the same semiconductor. This is an extremely useful finding because it means that as the concentration of electrons in a semiconductor is increased by doping the concentration of holes decreases, and vice versa. Thus an n-type semiconductor can be changed to a p-type semiconductor simply by increasing the number of holes present, by appropriate doping, and vice versa. This possibility underlies the fabrication of semiconductor devices. This information is used in Sections 13.2.2 and 13.2.4.

The equilibrium population of point defects is another example of a pseudochemical equilibrium. For the creation of a Frenkel defect on a cation array:

$$\text{null} \rightleftharpoons [\text{cation interstitial}] + [\text{cation vacancy}]$$

with an equilibrium constant

$$K = [\text{cation interstitial}][\text{cation vacancy}]$$

where square brackets indicate the concentrations of the defects. Other intrinsic point defect equilibria can be treated in the same way.

8.2 Phase diagrams and microstructures

Many of the processes that are used to make a useful material are grounded in reactions that involve a change of chemical composition or a change in the microstructure or nanostructure of the solid. The processing of steel is aimed largely at obtaining the correct microstructure, and the preparation of semiconductors involves altering the populations of defects present. This section outlines some of the ways in which these changes are accomplished.

8.2.1 Equilibrium solidification of simple binary alloys

One of the most important phase transitions occurs when a liquid transforms into a solid. A great deal of information concerning the microstructure of the solid can be obtained from a consideration of the phase diagram of the material, even though phase diagrams refer to equilibrium conditions and solidification is rarely carried out so slowly as to be an equilibrium process. For example, consider the solidification of a simple nickel–copper alloy, from the point of view of the phase diagram, reproduced in Figure 8.2.

In the liquid state, any nickel–copper alloy is homogeneous. Solid will begin to form as soon as

the temperature reaches the liquidus, at T_1 (Figure 8.2a). The initial composition of the liquid is l_1, (virtually equal to c) and that of the solid is s_1. The solid is rich in nickel compared with the original liquid composition. If the material is held at a temperature of T_1 for long enough, a dynamic equilibrium will be achieved. In this state, although the system appears to be static, the solid is continually dissolving and reforming. The atoms that are in the liquid and solid phases are continually being exchanged.

As the mixture is slowly cooled, this exchange leads to a continuous change in the composition of the solid and liquid phases. Consider the situation that will hold when the temperature drops slightly to T_2 (Figure 8.2b). In this case, the original solid of composition s_1 has been replaced by a solid of composition s_2 (much exaggerated in the figure). The new composition of the liquid in equilibrium with the solid is l_2. At all times, the atoms in the solid are dissolving in the liquid, and other atoms in the liquid are crystallising to form solid. Over a period of time the crystal present will always have the composition s_2, although the actual atoms that comprise the solid are forever changing. The same is true of the liquid. Further slow cooling, to temperature T_3, will cause the composition of the solid to change gradually to s_3, in equilibrium with liquid of composition l_3 (Figure 8.2c). This imaginary process can be continued until all of the original composition c is identical to the point on the solidus s_4, at temperature T_4 (Figure 8.2d). The final trace of liquid in composition with this solid has a composition l_4.

Thus, during equilibrium cooling, the composition of the solid will run down the solidus line, s_1 to s_2 to s_3, and so on, and the composition of the liquid in equilibrium with the solid runs down the liquidus from l_1 to l_2 to l_3, and so on, as the liquid cools. The composition of the solid phase when all of the liquid has solidified will be equal to that of the original liquid phase. Not only does the composition of the solid and liquid phases change continuously as the temperature falls through the two-phase region, but the number of small crystals present also increases. When temperature T_4 is reached, the microstructure of the solid consists of crystallites or grains

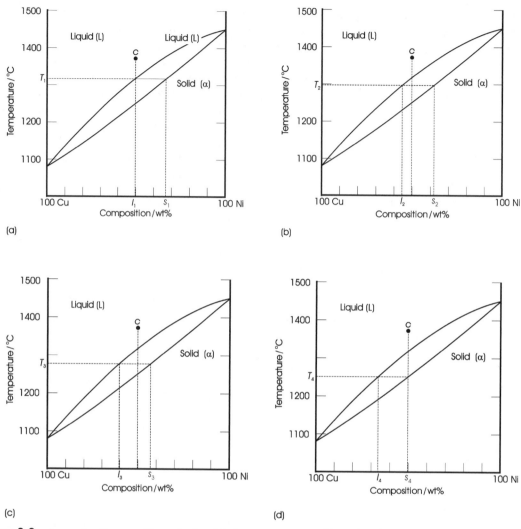

Figure 8.2 A sample of composition c in the nickel–copper system will consist of liquid, liquid plus solid, or solid, depending on the temperature T_i: (a) at T_1, the liquidus, the liquid has a composition l_1, virtually equal to c, and the infinitesimal amount of solid has a composition s_1; (b) at T_2, the liquid has a composition l_2 and the solid a composition s_2; (c) at T_3, the liquid has a composition l_3 and the solid a composition s_3; (d) at T_4, the solidus, the infinitesimal amount of liquid has a composition l_4 and the solid a composition s_4, equal to c

(Figure 8.3a). Further cooling in the solid will not initiate change of composition.

8.2.2 Nonequilibrium solidification and coring

During normal processing, cooling is usually rather fast, and solidification is rarely an equilibrium process. Solidification in these conditions is extremely complex. If cooling is not too fast, the first material to precipitate will still have a composition s_1 (Figure 8.2a). However, there will be insufficient time on further cooling for the original solid to equilibrate, and new material of composition s_2 (Figure 8.2b) will start to form on the nucleus of composition s_1. Ultimately, the solid will consist of

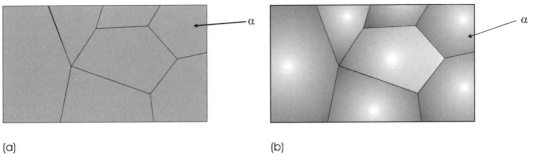

(a)   (b)

Figure 8.3 Microstructures of a solidified nickel–copper alloy: (a) after slow cooling the grains are homogeneous; (b) after rapid cooling, the grains are richer in one component (Ni) at the grain centres (paler shading) and richer in the other (Cu, darker shading) at the grain surfaces

a core that is richer in nickel than the outer regions, and there will be a gradation of composition of the alloy from the inside to the outside (Figure 8.3b). This is called coring. Coring can occur in all crystallites formed rapidly, including dendrites.

These nonequilibrium structures can be removed by heating the solid for an appropriate amount of time at a temperature below that of the solidus, a process called annealing. Annealing is effective only if the atoms can diffuse in the solid to correct the compositional differences generated during the cooling.

8.2.3 Solidification in systems containing a eutectic point

The composition and microstructure of a solid formed in a system showing a eutectic point depends critically on the composition of the liquid with respect to the eutectic composition. The situation will be explained by using the lead–tin (Pb–Sn) phase diagram described in Section 4.2.3.

In a liquid lead–tin alloy the two atom species are mixed at random. A liquid alloy with the same composition as that of the eutectic point, c_e, called the eutectic composition, is unique. It will pass directly into the solid state at a temperature 183 °C, without traversing a two-phase solid plus liquid region as it cools (Figure 8.4a). The solid that

forms must contain two phases, solid α and solid β. Thus, the random arrangement of atoms in the liquid must separate into the appropriate solid compositions on solidification. It is found that each grain contains a characteristic microstructure that consists of thin alternating lamellae of the two phases, α and β, called eutectic α and eutectic β (Figure 8.4b). The actual thickness of the lamellae and their shapes will depend on the relative diffusion coefficients of the two species. Note, though, that the *average* composition of the solid will be the same as that of the eutectic point.

The idealised microstructure of a solid formed when other liquid compositions cool is derived in a similar way. Suppose that the liquid with composition c_1, richer in lead than the eutectic composition (Figure 8.5a), is slowly cooled. At temperature T_1, below the liquidus, some solid α phase will have nucleated (Figure 8.5b). The composition of the solid phase, given by the tie line, is s_1. Similarly, the composition of the liquid phase is l_1. As slow cooling continues, the composition of the solid α crystallites will move along the solidus, as described in Section 8.2.1 for the nickel–copper alloys. At the same time, the composition of the liquid phase in contact with the crystallites will move along the liquidus. For example, at temperature T_2 the solid has a composition of s_2 and the liquid a composition of l_2 (Figure 8.5a). Ultimately, the horizontal solidus line will be reached at the eutectic temperature. At this point, any further drop

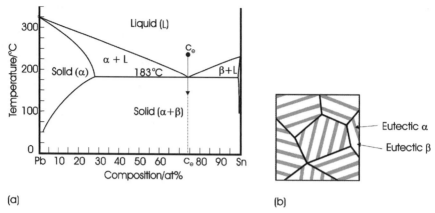

(a)   (b)

Figure 8.4   (a) The binary lead–tin phase diagram; c_e is the eutectic composition. (b) The microstructure of an alloy of composition c_e; each grain consists of alternating lamellae of eutectic α and eutectic β

in temperature will cause the remaining liquid to solidify. The microstructure of this latter material will be the same characteristic eutectic structure described above. The solid is a mixture of eutectic α and β, and crystallites of the α phase that formed in contact with the liquid, called primary α (Figure 8.5c).

A similar situation will occur for compositions on the tin-rich side of the eutectic point. In this case, the microstructure will consist of precipitates of primary β in a eutectic matrix.

In the case of a composition such as c_2 (Figure 8.6a), the first phase to form as the temperature

passes the liquidus is solid α in liquid. Ultimately, the temperature will fall below the solidus and, at a temperature T_1, for example, the solid will consist of grains of α phase (Figure 8.6b). On cooling further, at temperature T_2, for example (Figure 8.6a), the temperature will fall below the solvus, and the solid will consist of precipitates of the β phase in grains of the α phase (Figure 8.6c).

The compositions and amounts of the phases present at all times can be calculated by use of tie lines and the lever rule, as explained in Section 4.2.2.

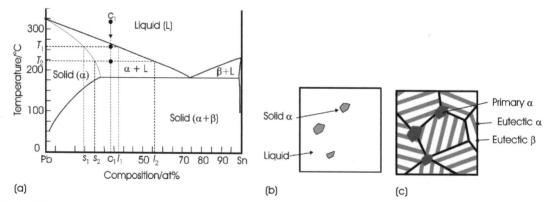

(a)   (b)   (c)

Figure 8.5   (a) The binary lead–tin (Pb–Sn) phase diagram; c_1 is a composition on the lead-rich side of the eutectic composition. (b) Cooling into the $(\alpha + L)$ two-phase region results in the formation of crystallites of solid α in a liquid matrix. (c) Cooling to below the solidus causes the remaining liquid to solidify into alternating lamellae of eutectic α and eutectic β. The grains of primary α are located in the grain boundaries

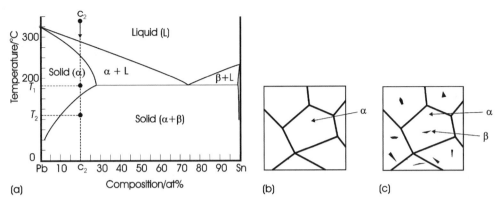

(a)   (b)   (c)

Figure 8.6 (a) The binary lead–tin (Pb–Sn) phase diagram; c_2 is a very lead-rich composition. (b) On cooling to a temperature T_1 the solid consists of homogeneous grains of composition c_2. (c) Cooling to T_2 results in the formation of small precipitates of β in each grain of α

8.2.4 Equilibrium heat treatment of steels

Steel is an alloy containing mainly iron, together with small amounts of carbon and other metals. Much of the usefulness of steel centres on the fact that the properties of the alloy, especially the mechanical properties, can be modified by changing the microstructures present in the finished product.

This amounts to the fact that there is not just one steel but a vast range of steels, each suited to a particular function. The changes in properties are brought about by different heating and cooling regimes and exploit the phases present in the iron–carbon (Fe–C) system. The principle microstructures found in steels can be explained by reference to the partial Fe–C phase diagram (Figure 8.7).

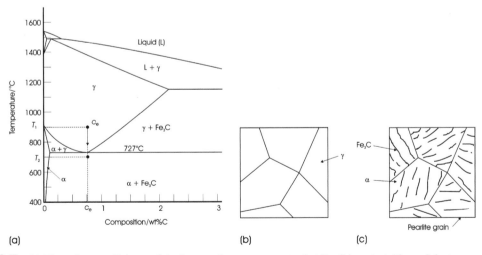

(a)   (b)   (c)

Figure 8.7 (a) The existence diagram of the iron–carbon system; note that Fe_3C is metastable, and the true equilibrium is between iron and graphite. (b) The microstructure of the solid at temperature T_1 consists of γ (astenite) with composition c_e, the eutectoid composition. (c) The microstructure of a solid of composition c_e at a temperature T_2 below the eutectoid composition consists of grains of pearlite, composed of alternating lamellae of Fe_3C and α

The most important transformation point in the Fe–C phase diagram from the point of view of steel is the eutectoid point. This point is at 0.76 wt% carbon (0.034 at% carbon or $FeC_{0.036}$) and a temperature of 727 °C. Above this temperature (T_1, Figure 8.7a), an alloy with the eutectoid composition, c_e, is a single phase, composed of grains of austenite (Figure 8.7b). On cooling through a eutectoid point, a single solid phase transforms to two solid phases. In the present case, on slow cooling, the single homogeneous phase austenite, which consists of face-centred cubic (fcc) iron containing 0.76 wt% carbon as interstitial carbon atoms, transforms into a two-phase mixture of ferrite (α-ferrite) and cementite (Fe_3C) at a temperature of 727 °C (T_2, Figure 8.7a). The driving force for the transformation into ferrite and cementite is likely to be internal strain. The fcc structure of austenite is strained by the interstitial carbon atoms, and this increases substantially as the temperature falls and the fcc unit cell contracts.

The transformation is complex. The phase diagram shows that the body-centred cubic (bcc) ferrite, one of the phases existing below the eutectoid temperature, is hardly able to dissolve any carbon. The transformation requires diffusion of the carbon to create carbon-poor regions that become ferrite, and carbon-rich regions that become cementite. At the same time the iron atoms must rearrange considerably. The transformation of the fcc austenite array into the bcc ferrite array requires considerable shuffling of the iron atoms. The rearrangement of the iron atoms in austenite to that found in cementite is much more complex. The structure of cementite is made up of hexagonal close-packed (hcp) layers of iron atoms twinned on every third $(1\,1\,\bar{2}\,2)$ plane with respect to the hcp unit cell. The transformation thus involves changing the ABC packing of austenite into multiply twinned ABAB packing. This twinning is a method of minimising the internal strain in the structure while maintaining the overall shape and volume of the solid (further information is found Section 8.3.3 in the discussion of shape-memory alloys). The carbon atoms lie in the twin planes of the cementite structure, where most room occurs.

The newly formed cementite nucleates at many sites simultaneously. The resulting solid, which is a mixture of ferrite and cementite, is called pearlite because it has a lustrous appearance in an optical microscope. Pearlite is not a compound or single phase but is a microstructure, made up of thin lamellae of cementite and α-ferrite side by side (Figure 8.7c). In this context, these phases are called eutectoid ferrite and eutectoid cementite.

The microstructures formed at compositions away from the eutectoid point mirror those previously described for eutectic transformations. Compositions to the iron-rich side of the eutectoid are called hypoeutectoid alloys. For example, a hypoeutectic composition c_1, at temperature T_1 (Figure 8.8a), consists of homogeneous grains of austenite (Figure 8.8b). As these cool slowly, the austenite region is exchanged for a two-phase region (temperature T_2, Figure 8.8a). This material is transformed to ferrite plus austenite. The ferrite often forms at grain boundaries, as the reaction is kinetically favoured in the disordered regions at grain boundaries (Figure 8.8c). The ferrite has an almost constant composition and holds only a very small amount of dissolved carbon in its bcc structure. The remaining austenite thus becomes carbon-rich. As cooling continues, the compositions of the two phases, determined by using tie lines, run down the phase boundaries. Ultimately, the temperature reaches the eutectoid temperature, 727 °C. Further cooling causes the remaining austenite to transform to pearlite (temperature T_3, Figure 8.8a). The microstructure now consists of pearlite (eutectoid ferrite and eutectoid cementite), together with the precipitates of the ferrite formed earlier, called proeutectoid ferrite (Figure 8.8d).

Compositions on the carbon-rich side of the eutectoid are called hypereutectoid alloys. Once again, as such an alloy is cooled slowly it will pass from single-phase austenite into a two-phase region. Consider a hypereutectic composition c_2 at temperature T_1 (Figure 8.9a). The microstructure will consist of grains of austenite (Figure 8.9b). On cooling to temperature T_2 (Figure 8.9a), cementite separates from the austenite, forming preferentially at the grain boundaries (Figure 8.9c). One reason

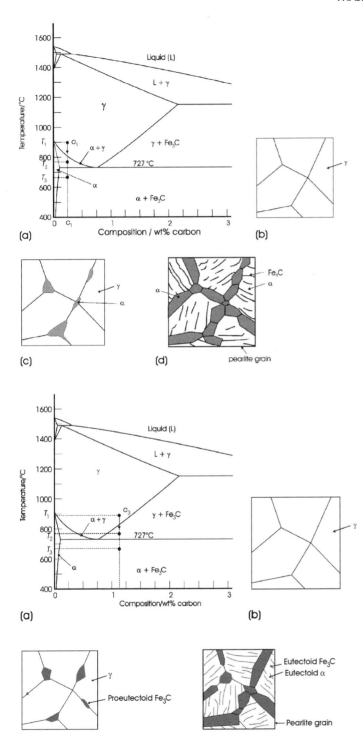

Figure 8.8 (a) The existence diagram of the iron–carbon system. (b) The microstructure of the solid with composition c_1, a hypoeutectoid composition, at a temperature T_1. (c) On cooling to a temperature T_2 the microstructure consists of grains of γ with precipitates of proeutectoid α at the grain boundaries. (d) At a temperature T_3 the microstructure consists of grains of pearlite with proeutectoid α at the grain boundaries

Figure 8.9 (a) The existence diagram of the iron–carbon system. (b) The microstructure of the solid with composition c_2, a hypereutectoid composition, at a temperature T_1. (c) At a temperature T_2 the microstructure consists of grains of γ with precipitates of proeutectoid Fe_3C at the grain boundaries. (d) At a temperature T_3 the microstructure consists of grains of pearlite with regions of proeutectoid Fe_3C at the grain boundaries

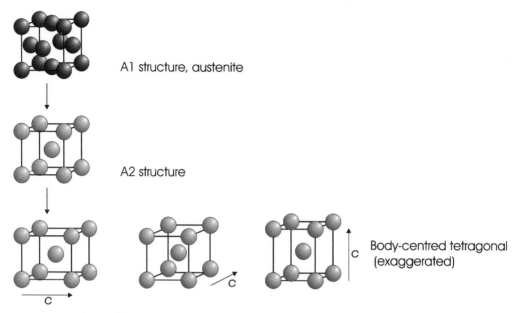

A1 structure, austenite

A2 structure

Body-centred tetragonal
(exaggerated)

Figure 8.10   Transformations occuring during the formation of martensite

for this is that the strain generated in the complex transformations taking place is more easily relieved at grain boundaries. This material is called proeutectoid cementite. As cooling continues, more proeutectoid cementite will form in the grain boundaries, and the composition of the remaining austenite will run down the phase boundary to the eutectoid point. Further cooling, below the eutectoid temperature, will cause any remaining austenite to transform into pearlite (temperature T_3, Figure 8.9a). The microstructure of the final solid will consist of proeutectoid cementite and pearlite, which itself consists of eutectoid cementite and eutectoid ferrite (Figure 8.9d).

8.2.5   Rapid cooling of steels

Rapid cooling will produce different microstructures from those obtained by slow cooling, only a few examples of which are described here. Rapidly cooling the homogeneous alloy austenite, with the fcc structure, to room temperature (a process called quenching), leads to a metastable phase called martensite. The fcc structure of the iron atoms in austenite transforms to a bcc structure of iron atoms in the process (Figure 8.10). The carbon atoms present, normally up to about 1.5 wt%, are trapped in the new structure, which causes a deformation such that one of the cubic cell edges is elongated. The phase is now body-centred tetragonal $(a = b \neq c, \alpha = \beta = \gamma = 90°)$, with the c axis longer than the a and b axes. The cubic phase has a lattice parameter of about 0.286 nm. The lattice parameters of the tetragonal phase vary with carbon content, a_0 taking a value of approximately 0.285 nm, and c_0 taking values between 0.292 nm and 0.300 nm. The ratio of c_0 to a_0 increases linearly with carbon content to a maximum of 1.08 in the highest carbon steels.

The formation of martensite is extremely rapid and only a slight displacement of atoms takes place. In particular, the carbon atoms have no time to diffuse any great distance, and the iron atoms do not have time to shear to form cementite. Once a nucleus of a martensite crystal forms, it will grow to its final size in a time of the order of 10^{-7} s. This is as fast as the speed of sound in iron. The rate of formation seems to be constant down to liquid helium temperatures. The change is called a martensitic transformation. Martensitic transformations are a general class of reactions that are diffusionless.

The temperature at which the transformation to martensite takes place is found to be composition-dependent. Martensite starts to form when the temperature reaches about 700 °C for the lowest-carbon-content steels but not until a temperature of about 200 °C for austenite with a carbon content of 1.2 wt%. The temperature at which the martensite starts to form is usually labelled M_s, the martensite start temperature, and the temperature at which the transformation is complete is labelled M_f. the martensite finish temperature.

The actual microstructure of the martensite that forms is also composition-dependent. Below about 0.6 wt% carbon the martensite forms in long blades and is called lath martensite. At compositions above about 1.0 wt% carbon the form is more lens-shaped in form, and is called plate martensite. At compositions between 0.6 wt% and 1.0 wt% carbon, a mixture of the two forms is found. In addition, most rapidly-quenched steels contain some austenite that has not transformed, intergrown with the laths or plates of martensite.

During the transformation, any one of the original bcc axes might elongate. In practice, all three possibilities occur at random. The martensite grows rapidly to form a grain of body-centred tetragonal structure within the surrounding cubic matrix. The elongation of the unit cell causes a stress around the crystal. Ultimately, the solid is made up of a set of interlocking martensite grains with the c axes aligned at random along the original cubic axes, in a matrix of untransformed austenite (Figure 8.11). This creates a strong internal stress field. The result is that martensite is both the hardest and the most brittle constituent of quenched steels. This effect is easily demonstrated. Heat an initially flexible steel piano wire to redness and then rapidly cool it by plunging it into cold water. The cold wire will be brittle and can be snapped by hand. The broken ends will scratch glass.

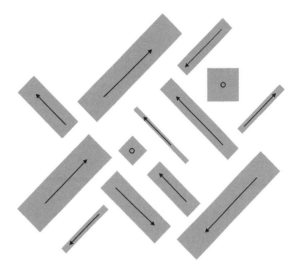

Figure 8.11   The microstructure of martensite regions in austenite. The arrows represent the direction of the elongated c axis; the c axis is normal to the figure in the domains marked ○

Martensite reverts to ferrite and cementite when the steel is heated above a temperature at which carbon diffusion becomes possible. This process is called tempering. During tempering, the components present revert to those described in Section 8.2.4. However, the microstructure of the steel will be different from that achieved by slower cooling. The properties of steels are controlled by choosing a cycle of slow cooling, quenching and tempering to optimise the microstructure formed with respect to the end use.

8.3   Martensitic transformations

8.3.1   Displacive transitions

Phase transitions of solids that do not involve a change in chemical composition are often divided into two types, reconstructive and displacive. In a reconstructive transition the parent crystal structure is broken apart and a new structure is formed from the constituents. There is no relationship between the unit cell of the parent form and that of the product. Single crystals undergoing a reconstructive transition usually fragment, and the reactions are usually rather slow. Displacive transitions, however, involve only small, coordinated movements of atoms and are usually rapid. They are often triggered by temperature and are reversible, with the crystal structure of the reactant phase closely related to that of the product.

There are a number of displacive transitions mentioned in this book. The order–disorder transformation of hydrogen atoms in hydrogen bonds in ferroelectric ceramics (Section 11.3.5) is one example. Displacive transitions that involve a change from an ordered arrangement of atoms to a random arrangement are commonly found in alloys. A subgroup of such order–disorder transitions, martensitic transitions, which can be used to produce shape-memory alloys, are considered in Sections 8.3.2 and 8.3.3.

8.3.2   Martensitic transitions in alloys

Martensitic transformations in alloys are essentially order–disorder displacive transitions that take place very rapidly, because atomic diffusion does not occur. The discussion of the formation of martensite in the Fe–C system, in Section 8.2.5, is an example. This transition is the transformation of a cubic phase containing excess carbon in interstitial sites into a tetragonal phase. As any one of three cubic axes can be elongated, three orientations of the martensite c axis can occur. This is a general feature of martensitic transformations and the different orientations that can arise are called variants or domains of the martensitic phase. These variants are simply twins (see Section 3.4.10).

An important martensitic transformation occurs in the titanium–nickel (Ti–Ni) system, as it is used in shape-memory alloys, described in Section 8.3.3. The phase in question is TiNi (Figure 8.12), called Nitinol. At temperatures above 1090 °C, TiNi has a bcc structure in which the atoms are distributed at random over the available sites in the crystal. Below

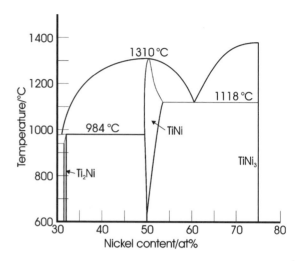

Figure 8.12   Part of the phase diagram of the titanium–nickel system

1090 °C, this structure orders to form the B2 (CsCl) structure (Figure 8.13). If this latter phase is quenched (cooled rapidly) to room temperature the structure transforms via a martensitic transformation into a monoclinic B19′ type. On cooling, the transformation starts at a temperature designated M_s, the martensite start temperature, and is complete by a temperature M_f, the martensite finish temperature. For the alloy NiTi, M_s is 60 °C and M_f is 52 °C.

The displacement of the atoms is a shear in the $\{1\,0\,1\}$ planes of the cubic structure (Figure 8.14). The shearing process nucleates at a number of points within the crystal as it cools, and each

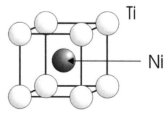

Figure 8.13   The B2 (CsCl) structure of the high-temperature form of TiNi

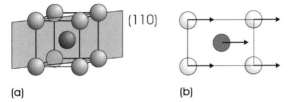

Figure 8.14 (a) The $(1\,1\,0)$ plane in TiNi; (b) the atoms in the $(1\,1\,0)$ plane; the shear displacement that occurs on transformation to martensite is indicted by arrows

grows to form a domain or variant. The shear operations across each side of a twin boundary are opposed to each other in direction and help to minimise the strain in the crystal (Figure 8.15). When the transformation is completed the material is highly twinned but maintains the same bulk shape as the original material (Figure 8.15c).

The formation of a heavily twinned material on cooling can be reversed by an increase in temperature, which causes the material to transform to the untwinned pre-martensite state. The transformation starts at a temperature, usually called A_s, the austenite start temperature, and is complete at a temperature A_f, the austenite finish temperature (Figure 8.16). (These terms are related to the fact that the best-known martensitic transformation is that of austenite to martensite, in steels.) For the alloy NiTi, A_s is 71 °C, and A_f is 77 °C. It is seen that M_s and M_f differ from A_s and A_f. This is a hysteresis phenomenon, commonly found in solid-state transformations.

The transformation can be greatly modified by the addition of other alloying elements or by variation of the heat treatment to which the alloy is subjected. For example, TiNi has an extension in composition from $Ni_{1.0}Ti_{1.0}$ to about $Ni_{0.86}Ti_{1.14}$ at 1118 °C (Figure 8.12). This phase range narrows at lower temperatures, so that cooling an alloy with a composition slightly richer in nickel than NiTi results in the formation of precipitates of $TiNi_3$ in the TiNi matrix. These considerably modify the mechanical properties of the martensitic phase. Moreover, this modification is dependent on precipitate morphology and size, and therefore on the rate of cooling. The formation of precipitates also influences M_s, and it has been found that a change in composition

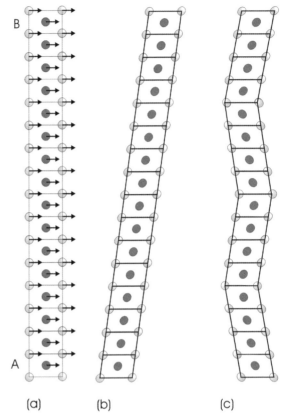

Figure 8.15 (a) A stack of unit cells of NiTi in the $(1\,1\,0)$ projection; shear displacement represented by arrows is applied to each cell in sequence, starting at A and ending at B, to produce a martensitic structure. (b) The deformed structure resulting from shear. (c) The overall shape of the original stack is more nearly maintained by repeated twinning

of as little as 1 at% can move M_s by more than 100 K.

The martensitic transformation is thus easily open to modification by heat treatment and alloying, which gives these microstructures great flexibility from the point of view of engineering design.

8.3.3 Shape-memory alloys

Shape-memory alloys are a group of metallic materials that can regain their original shape after

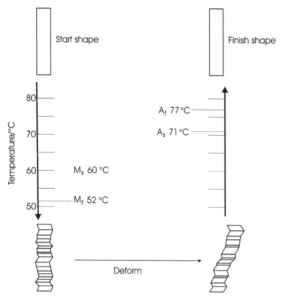

Figure 8.16 The sequence of events taking place during the deformation and recovery of a shape using a shape-memory alloy. Cooling the high-temperature shape below M_f transforms it into a multiply twinned form with the same overall shape. Deformation alters the distribution of the twin boundaries. Reheating the sample above A_f causes the material to revert to the high-temperature form. This removes the twins and allows the original shape to be recovered. The temperatures are appropriate to TiNi

deformation. This is rather remarkable and has been used in a wide variety of devices. These range from antennae on spacecraft, that can be crumpled into a small volume for launch and then unravel into a dish form on deployment, to spectacle frames that can be returned to their original shape after being sat on!

Shape-memory alloys show a thermoelastic martensitic transformation. This is a martensitic transformation, as described above, but which, in addition, must have only a small temperature hysteresis, some 10s of degrees at most, and mobile twin boundaries, that is, ones that move easily. Additionally, the transition must be crystallographically reversible. The importance of these characteristics will be clear when the mechanism of the shape-memory effect is described.

Initially, a shape-memory alloy is formed into the desired geometry at temperatures above M_s (Figure 8.16). On cooling, this original form is maintained, but the material transforms to a heavily twinned state. If this shape is now deformed, the twin boundaries move to accommodate the stress (Figure 8.16). It is for this reason that the twin boundaries must be mobile. In effect, the size of the individual twins increases, and the number of twins decreases. Above A_f the structure returns to the high-temperature form, which is untwined and no longer deformed (Figure 8.17), which is why the crystallographic transition must be reversible. The material has 'remembered' its original shape.

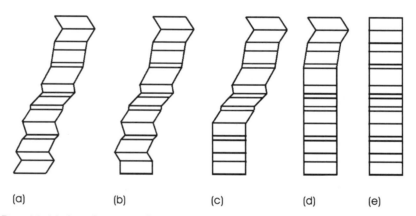

Figure 8.17 Parts (a)–(e) show the progressive recovery of the original shape of a deformed rod by reversion to the original crystal structure, which also removes twins

The process by which the stress is relieved by twin boundary migration leads to another important property, superelasticity, described in Section 10.1.3.

In the shape-memory transformation described, only the shape of the parent phase is 'remembered'. It is called the one-way shape-memory effect. It is also possible to produce alloys that display two-way shape-memory effects. In these materials, both the shape of the parent phase and the martensitic phase is 'remembered'. This reversible effect is caused by the fact that the nucleation of the martensite is very sensitive to the stress field. Introduction of lattice defects such as precipitates can restrict the number of variants that form and the positions where they nucleate. Such materials generate the martensitic shape on cooling below the temperature M_f. Cycling between higher and lower temperatures causes the alloy to switch alternately between the two shapes. There is considerable research interest in developing and exploiting two-way shape-memory effect alloys at present.

8.4   Sintering

8.4.1   Sintering and reaction

Sintering is the process by which a compacted powder is converted into a solid body by heating below the melting point of the main constituents, so that the object essentially remains a solid throughout the process. The first stage that occurs during sintering is an initial reduction in the surface roughness of the individual particles. This is followed by a period in which the particles start to join together. Finally, the solid becomes denser by the elimination of internal pores and voids (Figure 8.18).

Sintering is widely used in the ceramics and powder metallurgy industries. In general, the procedure involves the careful preparation of powders, pressing these powders into the desired shape and then heating to sinter the powder particles together into a strong solid. Ideally, there should be little change of overall shape during sintering, and the resulting object should be strong and pore-free. Although the initial powder is frequently a single phase, small amounts of additives are often included to achieve strength and to eliminate porosity. This desirable result is sometimes obtained by an additive that melts below the sintering temperature to form a small amount of liquid between the grains of the major component a process called liquid-phase sintering. Another variation is the technique of reaction sintering. This is exemplified by the production of silicon nitride objects. The desired shape is pressed from silicon powder, and this is heated in an atmosphere of nitrogen gas. The silicon sinters and reacts simultaneously to form a compact silicon nitride part.

Sintering can be brought about by a variety of reactions. Of these, material transport by viscous flow is important in glasses but less so in metals or ceramics. Evaporation and condensation is important in rather volatile compounds such halides and some oxides. Diffusion – bulk, grain boundary and surface diffusion – is important for refractory materials and, for these materials, the presence of traces

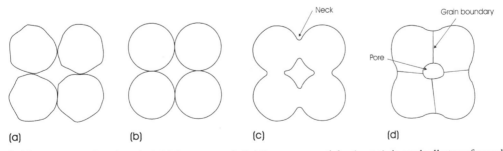

Figure 8.18   Stages in sintering. An initial compact of slightly uneven particles (part a), is gradually transformed into a solid (parts b–d). Usually, this involves shrinkage, change of shape and the formation of pores

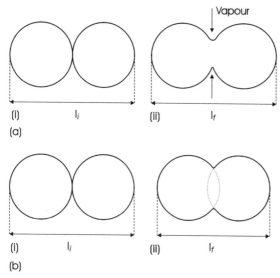

(a)

(b)

Figure 8.19   (a) Sintering via vapour transport: (i) initial configuration and (ii) final configuration. (no shrinkage). (b) Sintering via solid-state diffusion: (i) initial configuration and (ii) final configuration (shrinkage). Note: l_i and l_f, initial and final lengths, respectively

of liquid phase are also greatly beneficial in speeding up the reaction. In practice, the mechanism that operates has an influence on the final shape of the sintered object. Sintering that takes place by vapour phase transport of material gives a solid with little shrinkage, whereas sintering by way of solid-state diffusion decreases the separation between the constituent particles, often leading to significant shrinkage (Figure 8.19).

8.4.2   The driving force for sintering

As sintering does not usually involve chemical reactions, the driving force is not a reduction in Gibbs energy of reaction but a reduction in the surface area and the associated reduction in surface energy. This driving force can be illustrated for a flat surface that contains a spherical protuberance and a similar spherical depression, both of radius r (Figure 8.20a).

The vapour pressure over a curved surface, p, is related to the vapour pressure over a flat surface, p_0, by the Kelvin equation:

$$RT \ln\left(\frac{p}{p_0}\right) = \frac{2V\gamma}{r} = \Delta G$$

where V is the molar volume of the substance, γ is the surface energy of the solid–vapour interface, and ΔG is the difference in Gibbs energy between a flat surface and the curved surface. Thus the vapour pressure over a protuberance will be greater than the vapour pressure over the flat surface and will increase as the radius of the curved surface decreases. In the case of a depression, the radius r is now negative and it is necessary to write:

$$RT \ln\left(\frac{p}{p_0}\right) = \frac{2V\gamma}{-r} = \Delta G$$

Thus, the vapour pressure over the depression will be less than the vapour pressure over a flat surface. There will be a transfer of matter via the vapour

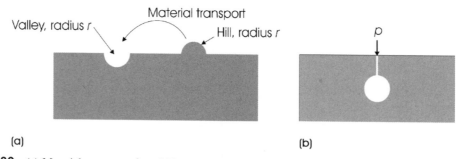

(a)   (b)

Figure 8.20   (a) Material transports from hills to valleys during vapour-phase sintering, because the vapour pressure over a hill is higher than that over a valley; the surface tends to become flat. (b) The excess pressure p inside a void balances the surface energy of the void

phase from a protuberance to a depression and the surface will tend to become flat.

It is important that articles fabricated by sintering do not contain large pores, because these lead to weakness and dimensional changes. The way in which powder granules link up during sintering will inevitably lead to pores forming unless the powders are carefully prepared. The size of the pores is dependent on the surface energy of the material in a similar way to that described above. In the case of a spherical pore (Figure 8.20b), the excess pressure in the pore, p, needed to balance the surface energy is given by:

$$p = \frac{2\gamma}{r}$$

where r is the radius of the sphere and γ is the surface energy. If the pressure is greater than that given, the pore will expand, whereas if it is smaller the pore will shrink, until equilibrium is reached. The ratio of the initial radius, r_i, to the final radius, r_f, of the pore is given by:

$$\frac{r_i}{r_f} = \left(\frac{2\gamma}{p_i\, r_i}\right)^{1/2}$$

where p_i is the initial pressure in the pore, and γ is the surface energy of the material. This equation indicates that small pores will tend to shrink and large pores will grow, with no change taking place when $r_i = r_f$, that is, when

$$p_i\, r_i = 2\gamma$$

Material transport also occurs (for a similar reason) when two spheres touch to form a neck (Figure 8.21). The spheres will have a positive and relatively large radius of curvature and the neck region a smaller negative radius of curvature. Matter will thus tend to be transported via the vapour phase from the larger spheres into the neck region, causing the particles to join. The transport of material to achieve this is not restricted to transport via the vapour phase. Bulk or surface diffusion can also be called into play to achieve the same result.

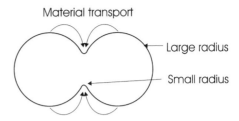

Figure 8.21 The transport of material from a surface of large radius, to a region of small radius (a neck) by vapour transport during sintering

8.4.3 The kinetics of neck growth

The rate of flow of material into the neck region between two spheres will depend on the mechanism of atomic transport. Four different mechanisms were investigated by Kuczynski, some 50 years ago. These were viscous flow, surface diffusion, bulk diffusion and vapour transport by evaporation and condensation. The rate of neck growth was found to be quite different from one mechanism to another. They are usually expressed in terms of the ratio of the neck radius x to the sphere radius r (defined in Figure 8.22):

For viscous flow,

$$\frac{x^2}{r} = k_1\, t$$

For surface diffusion,

$$\frac{x^5}{r^2} = k_2\, t$$

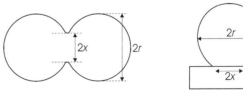

Figure 8.22 Definition of neck radius, x, and sphere radius, r, used in the kinetics of sintering

For bulk diffusion,

$$\frac{x^3}{r} = k_3 t$$

For vapour transport,

$$\frac{x^7}{r^3} = k_4 t$$

In these equations, k_1, k_2, k_3 and k_4 are constants which vary with temperature, and t is the time of reaction.

Experiments have shown that sodium chloride (NaCl) and other volatile materials sinter by a predominantly vapour transport mechanism whereas refractory oxides and metals sinter mainly by way of bulk diffusion.

8.5 High-temperature oxidation of metals

Corrosion of metals takes place by way of a variety of important chemical reactions. At ordinary temperatures, this process is often called tarnishing, and at high temperatures, scale formation. In this section, the reaction of metals with dry gases at relatively high temperatures is considered. This is referred to as direct corrosion, to distinguish it from many common forms of corrosion, including rust formation on iron, which need the presence of water. These latter reactions are considered in Section 9.4.

8.5.1 The driving force for oxidation

Direct corrosion can occur when a gas in the environment reacts directly with a metal. The discussion that follows applies to all gases, but reaction with oxygen will be used to illustrate this phenomenon. For example, both aluminium and calcium react rapidly with oxygen to form the appropriate oxide:

$$4\,Al + 3\,O_2 \rightarrow 2\,Al_2O_3,$$
$$2\,Ca + O_2 \rightarrow 2\,CaO.$$

The driving force for these reactions is the Gibbs energy of formation of the oxide (see Section S3.2). In favourable cases, such as that of aluminium, the product of corrosion may provide a protective coating, preventing further reaction from taking place. In less favourable circumstances, such as the corrosion of iron, the product may be loose or flake off, continually exposing new surfaces for corrosion.

The oxidation reaction will be influenced by the equilibrium oxygen pressure in the surroundings (Sections S1.5 and S3.2). Calculation of the equilibrium partial pressures over metal oxides shows that values lie between approximately 10^{-7} atmospheres to 10^{-40} atmospheres. As the oxygen partial pressure in air is about a fifth of an atmosphere it is clear that metals will have a tendency to oxidise. From the point of view of thermodynamics, there is always a considerable driving force for reaction.

8.5.2 The rate of oxidation

The rates of formation of oxide films vary widely. For thin layers, nominally less than approximately 100 nm in thickness, four rate laws have been established experimentally:
The cubic law,

$$x = (k_c\,t)^{1/3}$$

The logarithmic law,

$$x = k_1 - k_2 \ln t$$

The reciprocal logarithmic law,

$$\frac{1}{x} = k_1' - k_2' \ln t$$

The parabolic law,

$$x = (k_p\,t)^{1/2}$$

In each equation, x is the thickness of the film, t is the time of reaction, and k_c, k_p, etc., are experimentally determined constants.

In the case of thick layers, nominally over 100 nm, two laws have been observed:

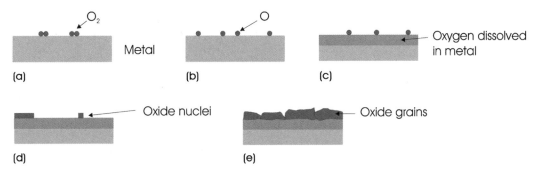

Figure 8.23 Oxide formation on a metal surface: (a) physical adsorption of oxygen (O_2) molecules from the air; (b) chemical adsorption of separated oxygen atoms (O) strongly bound to the surface; (c) penetration of some oxygen atoms into the metal to form a subsurface layer, as more oxygen arrives at the surface; (d) saturation of the surface and subsurface with oxygen, leading to formation of oxide nuclei on the surface; (e) surface layer of oxide grains

The linear law,

$$x = k\,t$$

The parabolic law,

$$x = (k_p\,t)^{1/2}$$

In both of these equations, x is the film thickness, t is the time of the oxidation, and k and k_p are experimentally determined constants. The constant k_p is called the parabolic rate constant. A linear rate is usually found when the film is porous or cracked. The parabolic equation is found when the film forms a coherent, impenetrable layer. As the rate of film growth, dx/dt, diminishes with time for the parabolic rate law, this equation is associated with protective kinetics. The parabolic rate law arises when the reaction is controlled by diffusion. The species with the lowest diffusion coefficient plays the most important role in this case.

8.5.3 Mechanisms of oxidation

The initial step of an oxidation reaction is usually the adsorption of oxygen onto the metal surface. Initially, the adsorbate will consist of oxygen molecules that are weakly bound to the surface (Figure 8.23a). This is called physical adsorption or physisorption. On most metals, the oxygen molecules rapidly dissociate into oxygen atoms, and the resulting layer of atoms is more strongly bound to the metal surface (Figure 8.23b). This stage is called chemical adsorption or chemisorption.

The oxidation begins by the diffusion of oxygen atoms into surface layers of the metal (Figure 8.23c). These do not form recognised oxides but a dilute solid solution that exists only in the surface regions. Nucleation of chemically recognisable oxides then occurs. At high temperatures and low partial pressures of oxygen, this takes place at random (Figure 8.23d). The oxide formed might be the normal oxide (e.g. MgO, on magnesium) or, if a number of oxides form, it might be a 'lower' oxide (e.g. FeO on iron). Sometimes, metastable oxides are also found that do not occur normally in the bulk. Continued growth proceeds so that the nuclei enlarge to form islands of oxide. Under some conditions, growth perpendicular to the surface might be much greater than lateral growth, in which case whiskers can form. Lateral growth is most frequent in dry conditions and, as such growth proceeds, the islands grow together to give a surface layer consisting of randomly oriented grains of oxide (Figure 8.23e).

This is an idealised picture and individual metals vary in their response to oxygen attack. Moreover, the overall reaction path is sometimes dependent on the temperature. Many other factors are important when the stability and further growth of this film is considered, including the electronic conductivity of the film and its thermal expansion coefficient compared with that of the parent metal.

One factor that is readily assessed is the volume of oxide produced by oxidation of a given volume of metal. This is called the Pilling–Bedworth ratio, X_{PB}, and is most readily expressed in terms of molar volumes:

$$X_{PB} = \frac{\text{molar volume of oxide}}{\text{molar volume of metal}}$$

For the reaction

$$m\,M + \frac{n}{2}\,O_2 \rightarrow M_m\,O_n$$

it is necessary to compute the Pilling–Bedworth ratio for the same amount of metal consumed as appears in the oxide. The result is:

$$
\begin{aligned}
X_{PB} \\
= \frac{(\text{molar mass of oxide } M_m O_n) \times (\text{density of the metal})}{m \times (\text{molar mass of the metal}) \times (\text{density of the oxide } M_m O_n)} \\
= \frac{M_o d_m}{m M_M d_o}
\end{aligned}
$$

where M_o is the molar mass of the oxide, d_m the density of the metal, M_m the molar mass of the metal, and d_o the density of the oxide.

If the Pilling–Bedworth ratio is less than 1 the oxide cannot cover the metal completely and the oxide film has an open or porous structure. Oxidation takes place continuously, and the oxidation kinetics tend to be linear. This type of behaviour is found for the alkali and alkaline earth metals. In the rare cases where the Pilling–Bedworth ratio is equal to 1, a closed layer can form which is stress-free. When the Pilling–Bedworth ratio is greater than 1, a closed layer forms with a certain amount of internal compressive stress present.

The location of this stress depends on the mechanism of the reaction, as discussed below. In cases where further formation of oxide is on the outer side of the layer, the stresses are easily relieved and the layer remains coherent. This results in protective oxidation, with parabolic reaction kinetics. The oxide film is called protective scale. If the new oxide film forms between the metal and an outer oxide layer, the stresses cannot be easily relieved, the oxide layer can crack, and spalling (fragmentation) can occur. Spalling is also possible when the value of X_{PB} is significantly greater than 1, because the additional volume that has to be accommodated generates stresses that lead to cracking. In general, cracks lead to faster oxidation, called accelerating or breakaway oxidation. The kinetics typical of breakaway oxidation initially follows a parabolic rate law, but this changes to a linear rate law when the film begins to crack.

In cases where a continuous and coherent layer of oxide film is present, further reaction can proceed only by diffusion of some of the reactants across the film. There are several possible mechanisms for this transport of material. In many solids, the passage of neutral atoms is less likely than the transport of charged particles, ions and electrons. In such cases, called ambipolar diffusion, the concentration gradient is not the only constraint on the system. In addition, and at all times, overall charge neutrality needs to be maintained.

Perhaps the most obvious mechanism for oxide formation (Figure 8.24a) is the diffusion of M^{n+} cations outward from the metal towards the gas atmosphere. If this occurs, a large negative charge would remain at the metal–metal oxide interface. This negative charge would act to slow the diffusing positively charged cations, and this would bring the reaction to a halt. To maintain electrical neutrality in the system, and to allow the reaction to continue, cation diffusion must be accompanied by a parallel diffusion of an appropriate number of electrons. When the electrons arrive at the surface, they react with oxygen molecules arriving on the oxide surface, to form O^{2-} ions. These are incorporated into the oxide film and, together with the arriving M^{n+} cations, allow the film to grow at the outer surface of the film. Note that film growth cannot continue if the oxide is an electrical insulator.

A similar mechanism envisages that oxygen ions diffuse across the film from the outer surface towards the metal (Figure 8.24b). The oxygen ions cannot be generated spontaneously and, once again, it is necessary for electrons to move through the

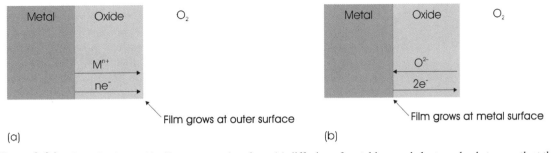

Figure 8.24   Growth of an oxide film on a metal surface: (a) diffusion of metal ions and electrons leads to growth at the outer (oxide/gas) side of the oxide film; (b) counter diffusion of electrons and oxide ions leads to growth at the inner (metal/oxide) side of the oxide film

oxide layer from the metal to make the ionisation possible. This leaves M^{n+} cations behind at the metal–metal oxide boundary. These cations are able to combine with the arriving O^{2-} anions to extend the oxide film at the metal–metal oxide inner boundary. Film growth will be curtailed if the film is an electrical insulator.

A third possibility, in which counter-diffusion of M^{n+} cations and O^{2-} anions occurs, is rare in oxide film formation and will be discussed in detail in Section 8.6.1.

The oxidation of copper metal in a low partial pressure of oxygen produces cuprous oxide, Cu_2O, by a mechanism involving diffusion of Cu^+ cations and electrons. The reaction is described by the chemical equation:

$$2\,Cu + \frac{1}{2}\,O_2 \rightarrow Cu_2O$$

The initial reaction results in the formation of a continuous film of oxide that is firmly attached to the metal surface. The rate of growth of the film is controlled by the slow diffusion of the Cu^+ ions. However, no corrosion could occur without the transport of electrons, as the mechanism depends on electron transport. The electronic conductivity of the film is therefore of major importance. The reason why both aluminium and chromium appear to be corrosion-resistant lies in the fact that, although oxide films form very rapidly in air, the films are insulators and prevent reaction from continuing. As the thin films are also transparent, the metals do not lose their shiny appearance.

8.6   Solid-state reactions

Reactions between two solids are analogous to the oxidation of a metal, because the product of the reaction separates the two reactants. Further reaction is dependent on the transport of material across this barrier. As with oxidation, cracking, porosity and volume mismatch can all help in this. In this section, the case when a coherent layer forms between the two reactants will be considered. The mechanism of the reaction may depend on whether electron transport is possible in the intermediate phase, and the rate of reaction will be controlled by the rate of diffusion of the slowest species. To illustrate the problems encountered a typical solid-state reaction, the formation of oxide *spinels*, is described.

8.6.1   Spinel formation

Spinel is a mineral with a composition $MgAl_2O_4$ (see Section 5.3.10). A large number of other oxides crystallise with the same structure, and these are collectively referred to as *spinels*. The formula of *spinels* is AB_2O_4, where A is most often a divalent cation, and B a trivalent cation, as in $MgAl_2O_4$ itself.

The spinel formation reaction can be represented by the chemical equation

$$MgO + Al_2O_3 \rightarrow MgAl_2O_4$$

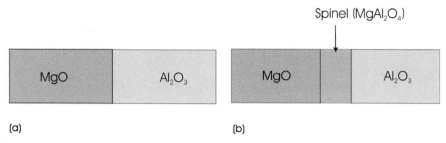

Figure 8.25 Growth of a layer of spinel (MgAl$_2$O$_4$) between crystals of magnesium oxide (MgO) and alumina (Al$_2$O$_3$): (a) MgO and Al$_2$O$_3$ crystals in contact; (b) separation of the reacting oxides by a layer of spinel

Suppose that a crystal of aluminium oxide, Al$_2$O$_3$, is placed in close contact with a crystal of magnesium oxide, MgO (Figure 8.25a). This arrangement, which resembles a diffusion couple (Section 7.1), is called a reaction couple. Initial reaction will result in the separation of the two reacting oxides MgO and Al$_2$O$_3$ by a layer of spinel, MgAl$_2$O$_4$ (Figure 8.25b). Continued reaction will depend on transport of reactants across the spinel layer. As in the case of metal oxidation, a number of mechanisms can be suggested but, because MgAl$_2$O$_4$ is an insulator, electron transport is not possible and so only mechanisms involving ions are permitted.

One such mechanism requires counter-diffusion of equal numbers O^{2-} anions and Mg^{2+} cations (Figure 8.26b). The electrical charges on the ions are equal and opposite, so no charge-balance problems arise. New spinel growth will take place at the Al$_2$O$_3$ interface. The diffusion of O^{2-} anions accompanied by an antiparallel diffusion of Al^{3+} cations (Figure 8.26c) is equally possible. Because of the difference in the ionic charges, the diffusion of two Al^{3+} cations needs to be balanced by the transport of three O^{2-} anions to maintain charge neutrality. Spinel growth will now take place at the MgO boundary. The counter-diffusion of Mg^{2+} and Al^{3+} is also possible (Figure 8.26d). To maintain charge neutrality, the diffusion of three Mg^{2+} cations must be balanced by the diffusion of two Al^{3+} cations. In this case, the spinel layer forms on either side of the initial boundary.

It has been found that the reaction between MgO and Al$_2$O$_3$ follows this latter mechanism. The

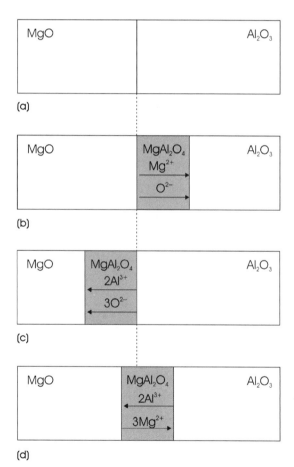

Figure 8.26 Mechanisms of spinel formation; (a) initial state; (b) diffusion of equal numbers of Mg^{2+} and O^{2-} ions; (c) diffusion of $2n$ Al^{3+} ions and $3n$ O^{2-} ions; (d) counter diffusion of $2n$ Al^{3+} ions and $3n$ Mg^{2+} ions

reactions at the boundary between Al_2O_3 and spinel are:

$$Al_2O_3 \rightarrow 2\,Al^{3+} + 3\,O^{2-}$$

and

$$3\,Mg^{2+} + 3\,O^{2-} + 3\,Al_2O_3 \rightarrow 3\,MgAl_2O_4$$

The reactions at the boundary between MgO and spinel are:

$$3\,MgO \rightarrow 3\,Mg^{2+} + 3\,O^{2-}$$

and

$$3\,O^{2-} + 2\,Al^{3+} + MgO \rightarrow MgAl_2O_4$$

These equations indicate that the spinel layer grows in an asymmetrical fashion. For every three Mg^{2+} ions that arrive at the Al_2O_3 boundary three $MgAl_2O_4$ molecules form, and for every two Al^{3+} ions which arrive at the MgO boundary only one $MgAl_2O_4$ molecule forms. Thus, the spinel layer will form in a ratio of 1:3 on either side of the initial boundary, with the thicker part on the Al_2O_3 side (Figure 8.27).

8.6.2   The kinetics of spinel formation

The rate at which the total thickness of the spinel layer grows is controlled by the speed of diffusion of the slowest cation. In reactions of this sort, if the spinel layer increases by an amount Δx in a period of time Δt, the rate of film growth, $\mathrm{d}(\Delta x)/\mathrm{d}(\Delta t)$, is given by:

$$\frac{\mathrm{d}(\Delta x)}{\mathrm{d}(\Delta t)} = \frac{k}{x}$$

where k is a constant and x is the film thickness at time t. Integration and rearrangement of this equation leads to the following equation:

$$x^2 = 2\,k\,t$$

where x is the thickness of the spinel layer, k is called the practical reaction rate constant, and t is the reaction time. This will be recognised as the

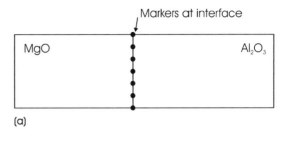

(a)

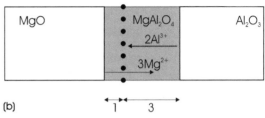

(b)

Figure 8.27   Markers used to determine the mechanism of spinel formation from magnesium oxide and alumina (a) inert markers at the interface between MgO and Al_2O_3 crystals; (b) after reaction the marker will appear to be within the $MgAl_2O_4$ layer when a cation counter-diffusion is in operation. The ratio of the layer thickness on each side of the boundary will depend on the charges on the ions

parabolic rate law described in Section 8.5.2 for metal oxidation, and k is also called the parabolic rate constant, k_p.

Many solid-state reactions give a parabolic rate law for the growth of an internal phase, and such a parabolic rate law is taken as evidence that the reaction is diffusion-controlled. The units of k are the same as the units of the diffusion coefficient, $m^2\,s^{-1}$. Generally, one can write:

$$k \propto D_A$$

where D_A is the diffusion coefficient of the ionic species that diffuses at the slowest speed.

Answers to introductory questions

What is dynamic equilibrium?

Dynamic equilibrium is an important and powerful concept in many areas of science. Dynamic equilibrium generally involves a reversible reaction, such as that represented by a chemical equation:

$$a\,A + b\,B \rightleftharpoons x\,X + y\,Y$$

or a physical equation:

$$\text{null} \rightleftharpoons [n] + [p]$$

where [n] is the concentration of electrons in the conduction band and [p] is the concentration of holes in the valence band of a semiconductor. The double arrows signify that the reaction can take place in either direction.

A feature of reversible reactions is that they often appear to be static, as if there were no change occurring. This state of affairs is illusory. Reactants are continuously being consumed and regenerated by the forward and backward reactions. If no change seems to be taking place, it is because the rates of the reactions in each direction are balanced. In this case, the system is said to be in equilibrium. At equilibrium, reactant and product are continuously being formed and consumed, and the rate of the forward reaction is the same as the rate of the reverse reaction. This feature is the key to the importance of dynamic equilibrium.

What defines a martensitic transformation?

Rapidly cooling the homogeneous alloy austenite, with the face-centred cubic structure, to room temperature leads to the formation of a metastable phase called martensite. The face-centred cubic structure of the iron atoms in austenite transforms to a body-centred cubic structure of iron atoms in the process. The carbon atoms present are trapped in the new structure, which causes a deformation such that one of the cubic cell edges is elongated. The formation of martensite is extremely rapid, and only a slight displacement of atoms takes place. In particular, neither the carbon nor the iron atoms have time to diffuse any great distance. Once a nucleus of a martensite crystal forms, it will grow to its final size in a time of the order of 10^{-7} s. The change is called the martensitic transformation.

Martensitic transformations are a general class of reactions that are similar to that found in steel. The main characteristic of this class of reactions is that they are very rapid, because no atom diffusion occurs. Martensitic transformations are defined by being diffusionless reactions.

What is the main driving force for sintering?

Sintering does not usually involve chemical reactions, and the driving force is a reduction in the surface area and the associated reduction in surface energy. This driving force can be illustrated for a flat surface that contains a spherical protuberance and a similar spherical depression. The vapour pressure over a curved surface is related to the vapour pressure over a flat surface by the Kelvin equation. This shows that the vapour pressure over a protuberance will be greater than the vapour pressure over the flat surface and will increase as the radius of the curved surface decreases. Similarly, the vapour pressure over a depression will be less than the vapour pressure over a flat surface. When a solid is heated, vapour transfer of matter will take place from a protuberance to a depression, and the surface will tend to become flat.

A body to be sintered consists of many small approximately spherical grains. Material transport occurs, for a similar reason, when two spheres touch to form a neck. The spheres will have a positive and relatively large radius of curvature, and the neck region a smaller negative radius of curvature. Matter will thus tend to be transported via the vapour phase from the larger spheres into the neck region, causing the particles to join.

The transport of material to achieve reduction in surface area is not restricted to transport via the vapour phase. Bulk or surface diffusion can also be called into play to achieve the same result.

Further reading

On dynamic equilibrium, see:

P. Atkins, L. Jones, 1997, *Chemistry. Molecules, Matter and Change*, 3rd edn, Freeman, New York, Ch. 13.

D.A. McQuarrie, D.A. Rock, 1991, *General Chemistry*, 3rd edn, Freeman, New York, Ch. 17.

On phase diagrams and microstructures, see:

J.C. Anderson, K.D. Leaver, R.D. Rawlings, J.M. Alexander, 1998, *Materials Science*, 4th edn, Stanley Thornes, Cheltenham, Glosc.

W.D. Callister, 2000, *Materials Science and Engineering: An Introduction*, 5th edn, John Wiley & Sons, New York.

W.F. Smith, 1993, *Foundations of Materials Science and Engineering*, 2nd edn, McGraw-Hill, New York.

On shape-memory alloys, see:

L. McDonald Schetky, 1979, 'Shape-memory Alloys', *Scientific American* **241** (November) 68–76.

Materials Research Society Bulletin **27** (February 2002) 91–127 (various authors).

On sintering, see:

G. Bickley Remmey, 1994, *Firing Ceramics*, World Scientific, Singapore.

On high-temperature oxidation, see:

J.H.W. de Witt, 1981, 'High Temperature Oxidation of Metals', *Journal of Materials Education (JEMMSE)* **3** 343.

On solid-state reactions, see:

R.J.D. Tilley, 1998, *Principles and Applications of Chemical Defects*, Stanley Thornes, Cheltenham, Glos.

Problems and exercises

Quick quiz

1 Dynamic equilibrium is attained when:
   (a) The forward and reverse reactions come to a standstill
   (b) The forward reaction slows to zero
   (c) The forward and reverse reactions proceed at the same speed

2 The equilibrium constant depends on which one of the following:
   (a) Pressure?
   (b) Temperature?
   (c) Concentration?

3 The equilibrium constant for a reaction equation derived by adding simpler reactions is given by:
   (a) The sum of the equilibrium constants of the individual reactions
   (b) The product of the equilibrium constants of the separate reactions
   (c) The sum of the reciprocals of the equilibrium constants of the individual reactions

4 If the equilibrium constant for the forward reaction is K, that for the reverse reaction is:
   (a) $1/K$
   (b) $-K$
   (c) $K/2$

5 If the concentration of a reactant is changed in an equilibrium situation:
   (a) The equilibrium constantly changes to offset the change
   (b) The reaction proceeds to minimise the change
   (c) The reaction reverses

6 If the volume of the vessel containing gaseous reactants at equilibrium is decreased:
   (a) The equilibrium constant changes to offset the change
   (b) The reaction proceeds to change the amount of reactants present
   (c) The reaction proceeds to minimise the number of molecules present

7 If the number of electrons in a semiconductor is increased at a constant temperature, the number of holes:
   (a) Decreases
   (b) Increases
   (c) Stays the same

8 During slow cooling of a Ni-Cu alloy, the composition of the crystallites:
   (a) Is constant
   (b) Varies continuously
   (c) Has the composition of the liquid

9 When a Ni-Cu alloy is cooled fairly rapidly, the core of the crystallites will have a composition:
   (a) Richer in the lower-melting-point parent phase
   (b) Richer in the higher-melting-point parent phase
   (c) Identical to that of the liquid phase

10 When a liquid alloy of lead and tin is cooled through a eutectic point, the microstructure of the solid produced contains
   (a) Eutectic α and eutectic β
   (b) Primary α and eutectic β
   (c) Eutectic α and primary β

11 When a liquid alloy of lead and tin, with a composition between the eutectic point and the α phase is cooled, the microstructure of the solid produced contains
   (a) Primary α and eutectic β
   (b) Primary α, eutectic α and primary β
   (c) Primary α, eutectic α and eutectic β

12 When austenite is cooled slowly through the eutectoid point, the material that forms is called:
   (a) Cementite
   (b) Ferrite
   (c) Pearlite

13 The microstructure of pearlite is composed of:
   (a) Eutectoid ferrite plus austenite
   (b) Eutectoid ferrite plus eutectoid cementite
   (c) Eutectoid cementite plus austenite

14 A hypoeutectoid steel has a composition that lies to the:
   (a) Iron-rich side of the eutectoid
   (b) Carbon-rich side of the eutectoid
   (c) Carbon-rich side of austenite

15 Slow cooling of a hypereutectoid steel composition produces a microstructure composed of:
   (a) Proeutectoid ferrite plus eutectoid cementite plus eutectoid ferrite
   (b) Proeutectoid cementite plus proeutectoid ferrite plus pearlite

   (c) Proeutectoid cementite plus eutectoid cementite plus eutectoid ferrite

16 Martensite is produced from austenite by:
   (a) Quenching the austenite
   (b) Annealing the austenite
   (c) Tempering the austenite

17 A martensitic transformation is described as:
   (a) A reconstructive transformation
   (b) A diffusionless transformation
   (c) An equilibrium transformation

18 Shape-memory alloys utilise:
   (a) Martensitic transformations
   (b) Eutectoid transformations
   (c) Reconstructive transformations

19 The shape-memory effect in shape-memory alloys requires that the solid contain:
   (a) Fixed twin boundaries
   (b) Mobile twin boundaries
   (c) No twin boundaries

20 The shape-memory alloy Nitinol has an approximate formula:
   (a) $NiTi_3$
   (b) $NiTi$
   (c) Ni_3Ti

21 The transformation of a compacted powder into a solid by heating is called:
   (a) Annealing
   (b) Tempering
   (c) Sintering

22 Liquid-phase sintering involves using:
   (a) Only liquids for the reaction
   (b) Solids mixed with liquids for the reaction
   (c) An additive which melts during the reaction

23 The driving force for sintering is:
   (a) Reduction in surface area
   (b) Reduction in total volume
   (c) Reduction in Gibbs energy of reaction

24 Direct corrosion of a metal requires the presence of:
 (a) A gas plus water
 (b) A dry gas
 (c) Air

25 The parabolic rate law for oxidation arises when:
 (a) The oxide film is cracked
 (b) The oxide film is thin
 (c) Diffusion controls the reaction

26 The Pilling–Bedworth ratio is used to predict the:
 (a) Likelihood of corrosion of a metal
 (b) Reactivity of a metal with a gas
 (c) Rate of corrosion of a metal

27 The counter-diffusion of ions is sometimes called:
 (a) Reaction diffusion
 (b) Ambipolar diffusion
 (c) Solid-state diffusion

28 The formation of spinel, $MgAl_2O_4$, involves the diffusion of:
 (a) Mg^{2+} and O^{2-} ions
 (b) Al^{3+} and O^{2-} ions
 (c) Mg^{2+} and Al^{3+} ions

29 A parabolic rate constant is characteristic of a reaction between two solids which is controlled by:
 (a) Sintering
 (b) Ionic diffusion
 (c) Vapour transport.

Calculations and questions

8.1 Write down the expression for the equilibrium constants K_c and K_p of the following reactions:
 (a) $CaCO_3$ (s) $\rightleftharpoons$ CaO (s) $+ CO_2$ (g).
 (b) C (s) $+ 2H_2$ (g) $\rightleftharpoons CH_4$ (g).

 (c) $2SO_2$ (g) $+ O_2$ (g) $\rightleftharpoons 2SO_3$ (g).
 (d) Zn(s) $+ CO_2$ (g) $\rightleftharpoons$ ZnO(s) $+$ CO(g).

8.2 $CaCO_3$ (limestone, chalk) is heated in a sealed container, initially in a vacuum, at 1000 K.
 (a) What will happen to the pressure in the vessel?
 (b) The equilibrium constant for the reaction (see Question 8.1) is $K_p = 0.039$ atm. What will be the pressure in the vessel at 1000 K?
 (c) The size of the vessel is doubled and equilibrium reestablished. What will be the new pressure?
 (d) A quantity of CO_2(g) prepared from radioactive carbon-14 is admitted to the vessel at equilibrium. What will occur initially?
 (e) What will the final distribution of radioactive carbon-14 be in the vessel?
 (f) During cement production $CaCO_3$ is heated in a kiln. Why does all of the limestone transform to CaO?

8.3 Deduce the relationship between K_c and K_p for the reaction:

$$N_2O_4(g) \rightleftharpoons 2NO_2(g)$$

8.4 Copper sulphate, $CuSO_4$, exists as hydrates $CuSO_4 \cdot 5H_2O$, $CuSO_4 \cdot 3H_2O$ and $CuSO_4 \cdot H_2O$.
 (a) Write equations for the equilibria between these forms.
 (b) Sketch the variation of water vapour pressure over a sample of copper sulphate as the average water content varies between $5H_2O$ and zero. [Note: answer is not shown at the end of this book.]
 (c) The equilibrium water vapour pressure at 25 °C (i) over a mixture of 50:50 $CuSO_4 \cdot 5H_2O{:}CuSO_4 \cdot 3H_2O$ is 6.266×10^3 Pa, (ii) over a 50:50 mixture of $CuSO_4.3H_2O{:}CuSO_4 \cdot H_2O$ is 4×10^3

Pa, and (iii) over a 50:50 mixture of $CuSO_4 \cdot H_2O$ and $CuSO_4$ (anhydrous) is 600 Pa. Calculate the equilibrium constants of the reactions involved in the dehydration of $CuSO_4$.

8.5 The number of intrinsic electrons in pure germanium at 300 K is 2.4×10^{19} m^{-3}. The material is doped with a concentration of phosphorus donors of 10^{22} phosphorus atoms per metre cubed. What is the equilibrium concentration of electrons and holes?

8.6 The intrinsic hole concentration in gallium arsenide at 200 °C is 4.1×10^{17} m^{-3}. The material is doped with a concentration of zinc acceptors of 10^{19} zinc atoms per metre cubed. What is the equilibrium concentration of electrons and holes?

8.7 The intrinsic electron concentration in a silicon sample at 500 °C is 6.5×10^{20} m^{-3}. (a) What quantity of aluminium dopant needs to be added to reduce the electron concentration to 6.6×10^{-19} m^{-3}? (b) The unit cell of silicon is cubic with a cell edge of 0.5431 nm and contains eight atoms of silicon. What percentage of silicon atoms will be replaced by aluminium atoms?

8.8 The gold–silver system forms a complete solid solution. The melting point of gold is 1064.43 °C and that of silver is 961.93 °C.
(a) During rapid cooling of a 50 at% gold:50 at% silver mixture, which phase will be richest in the core of a grain?
(b) Sketch the microstructure of the solid at a temperature of 800 °C if the melt is cooled very slowly. [Note: answer is not shown at the end of this book.]
(c) Sketch the microstructure of the solid at a temperature of 800 °C if the melt is cooled quickly. [Note: answer is not shown at the end of this book.]

8.9 The ruthenium–rhenium (Ru–Re) system forms a complete solid solution (for the phase diagram, see Figure 4.24, page 111). The melting point of ruthenium is 2334 °C and that of rhenium is 3186 °C.

(a) During rapid cooling of a 45 at% Re: 55 at% Ru mixture, which phase will be richest in the core of a grain?
(b) Sketch the microstructure of the solid at a temperature of 2000 °C if the melt is cooled very slowly. [Note: answer is not shown at the end of this book.]
(c) Sketch the microstructure of the solid at a temperature of 2000 °C if the melt is cooled quickly. [Note: answer is not shown at the end of this book.]

8.10 The corundum–chromia (Al_2O_3–Cr_2O_3) system forms a complete solid solution (for the phase diagram, see Figure 4.23, page 110). The melting point of corundum is 2035 °C and that of chromia is 2330 °C.
(a) During rapid cooling of a 33 mol% Al_2O_3: 67 mol% Cr_2O_3 melt, which phase will be richest in the core of a grain?
(b) Sketch the microstructure of the solid at a temperature of 2000 °C if the melt is cooled very slowly. [Note: answer is not shown at the end of this book.]
(c) Sketch the microstructure of the solid at a temperature of 2000 °C if the melt is cooled quickly. [Note: answer is not shown at the end of this book.]

8.11 With respect to the phase diagram of the copper–silver system (Figure 8.28):

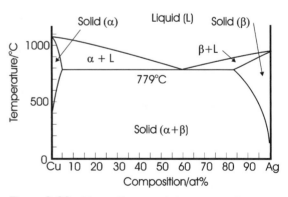

Figure 8.28 Phase diagram of the copper–silver system, for Question 8.11

(a) Sketch and label the microstructure of a solid containing 2.5 at% silver (about half-way across the α phase field at 800 °C) when the melt is slowly cooled to 800 °C and then to 400 °C. [Note: answer is not shown at the end of this book.]

(b) Sketch and label the microstructure of a solid containing 30 at% silver when the melt is slowly cooled to 700 °C. [Note: answer is not shown at the end of this book.]

(c) Sketch and label the microstructure of a solid formed by slowly cooling the eutectic composition, containing 60 at% Ag, to 700 °C. [Note: answer is not shown at the end of this book.]

8.12 With respect to the iron–carbon phase diagram (Figure 4.16; page 102):

(a) Sketch and label the microstructure of a solid containing 3.44 wt% carbon when the melt is slowly cooled to 800 °C, and then to 600 °C. [Note: answer is not shown at the end of this book.]

(b) Sketch and label the microstructure of a solid containing 62 wt% carbon when the melt is slowly cooled to 800 °C. How much of each phase is present? [Note: answer is not shown at the end of this book.]

(c) The alloy in part (b) is further cooled to 600 °C. What phases will now be present? [Note: answer is not shown at the end of this book.]

8.13 With respect to the iron–carbon phase diagram (Figure 4.16, page 102):

(a) Sketch and label the microstructure of a solid containing 1.0 wt% carbon when the melt is slowly cooled to 1000 °C, and then to 750 °C and, finally, to 700 °C. [Note: answer is not shown at the end of this book.]

(b) Sketch and label the microstructure of a solid containing 0.3 wt% carbon when the melt is slowly cooled to 1000 °C, and then to 750 °C and, finally, to 700 °C.

[Note: answer is not shown at the end of this book.]

(c) Sketch and label the microstructure of a solid containing 0.76 wt% carbon when the melt is slowly cooled to 1000 °C, and then to 750 °C and, finally, to 700 °C. [Note: answer is not shown at the end of this book.]

8.14 The surface energy of solid corundum, Al_2O_3, at 1850 °C is 0.905 J m^{-2}. Calculate the relative pressure, p/p_0, over a hemispherical 'hill' of diameter 1×10^{-7} m on the surface of a flat plate of corundum at this temperature.

8.15 The surface energy of solid magnesia, MgO, at 1500 °C is 1.2 J m^{-2}. Estimate the pressure differential between a spherical pit of diameter of 5×10^{-7} m and that over the surface of a flat plate of magnesia at this temperature.

8.16 Assuming that the ideal gas law ($pV = nRT$) holds, deduce the formula

$$\frac{r_i}{r_f} = \left(\frac{2\gamma}{p_i\, r_i}\right)^{1/2}$$

where r_i and r_f are the initial and final radii of the pore in the material, respectively, p_i is the initial gas pressure in the pore, and γ is the surface energy of the material. [Note: answer is not provided at the end of this book.]

8.17 A spherical pore in a soda-lime glass at 1200 K contains trapped gas. Assuming the gas trapped in the pore is at atmospheric pressure, and the surface energy of the glass at 1200 K is 0.350 J m^{-2}, calculate the equilibrium pore size when the initial pore is of diameter:

(a) 0.4 μm.

(b) 4 μm.

(c) 40 μm.

8.18 Using the data in Question 8.17, what is the equilibrium size of a pore that will neither shrink nor expand?

8.19 Solid titanium nitride, TiN, has a surface energy of 1.19 J m^{-2} at 1200 °C. If the voids in this material contain nitrogen gas at the same partial pressure as found in the atmosphere, approximately 8×10^4 Pa, estimate the maximum void size to ensure that voids in a sintered ceramic will shrink.

Table 8.1 Data for Question 8.20

Metal/metal oxide	Density/kg m^{-3}	
	Metal	Oxide
Cu/CuO	8933	6315
Fe/Fe$_2$O$_3$	7873	5240
K/K$_2$O	862	2320
Ti/TiO$_2$	4508	4260
Al/Al$_2$O$_3$	2698	3970
Na/Na$_2$O	966	2270

8.20 Which of the metals listed in Table 8.1 are likely to be protected from corrosion by the formation of a protective oxide film?

Table 8.2 Data for Question 8.21

Time of heating/h	Film thickness/ nm
1	4.74×10^{-7}
2	0.67×10^{-6}
3	0.82×10^{-6}
4	0.95×10^{-6}
5	1.06×10^{-6}
6	1.16×10^{-6}

8.21 A nickel foil is oxidised at $1000\,^\circ$C. The film thickness as a function of time is given in Table 8.2. Confirm that the rate is parabolic and calculate the parabolic rate constant.

8.22 When copper is oxidised under low partial pressures of oxygen gas, it forms Cu$_2$O via a parabolic rate law. The rate constant is 5.38×10^{-10} m^2 s^{-1} at 0.05 atm pressure and $900\,^\circ$C.

(a) What will the film thickness be after oxidation of copper foil for 10 h at this temperature?

(b) What will the weight of the copper oxide film be?

(c) Experimentally, it is easier to measure the weight gain as a function of time rather than film thickness. What will the weight gain of the film be?

8.23 The thickness of the layer of the *spinel* NiAl$_2$O$_4$ formed between NiO and Al$_2$O$_3$ when reacted at $1350\,^\circ$C is given in Table 8.3. Check that the reaction is diffusion con-

Table 8.3 Data for Question 8.23

Layer thickness/μm	time/h
1.0	20
1.4	40
1.8	60
2.0	80
2.3	100

trolled and calculate the rate constant. The reaction equation is:

$$NiO + Al_2O_3 \rightarrow NiAl_2O_4$$

9

Oxidation and reduction

- What is an electrochemical cell?

- What are the electrode materials in nickel–metal-hydride batteries?

- What information is contained in a Pourbaix diagram?

Oxidation and reduction involve the transfer of electrons from one compound to another. Reactions involving oxidation and reduction are called redox reactions. The first important application of a redox reaction was the construction of the first battery, reported by Alessandro Volta in 1800. Volta's battery simply consisted of a stack of alternating silver and zinc discs. A sheet of porous material, parchment, cloth or hide, saturated with a salt solution, was inserted between each silver and zinc disc. A current was found to flow when a wire connected the bottom silver disc to the top zinc disc. The voltage increased as the number of layers in the stack increased. What happens, in fact, is that one component is being oxidised, and one is being reduced. The subsequent electron transfer takes place in the external circuit, and is observed as an electric current.

One of the major problems encountered with the Volta pile was severe corrosion of the metals, and many early experiments were directed towards solving this problem. Thus it is apparent that batteries and corrosion are closely linked and, indeed, both are oxidation and reduction reactions. Oxidation and reduction is also involved in the related process of electrolysis, which underlies electroplating, a method of preventing corrosion. Finally, oxidation and reduction reactions underpin all life processes, although this aspect is not covered here.

9.1 Redox reactions

9.1.1 Oxidation and reduction

Oxidation and reduction reactions involve the transfer of electrons from one chemical compound to another. Oxidation is electron loss. It is characterised by an increase in the oxidation number of the species oxidised (see Section S3.3). For example, when solid zinc metal in contact with water is oxidised it forms zinc ions in solution and releases two electrons:

$$Zn(s) \rightarrow Zn^{2+}(aq) + 2e^- \quad [0 \rightarrow +2]$$

The change in oxidation number is shown in brackets, $[M \rightarrow N]$.

Understanding solids: the science of materials. Richard J. D. Tilley
© 2004 John Wiley & Sons, Ltd ISBNs: 0 470 85275 5 (Hbk) 0 470 85276 3 (Pbk)

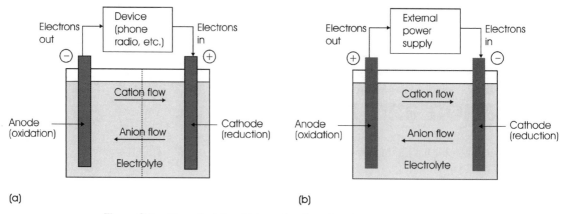

Figure 9.1 The principle of (a) a galvanic cell and (b) an electrolytic cell

An oxidising agent causes oxidation. It takes electrons from the species being oxidised and becomes reduced in the process.

Reduction is the opposite of oxidation and is equivalent to electron gain. The oxidation number of a species decreases on reduction. For example, copper ions in water solution can gain two electrons and form metallic solid copper in the process:

$$Cu^{2+}(aq) + 2e^- \rightarrow Cu(s) \quad [+2 \rightarrow 0]$$

A reducing agent causes reduction. It gives up electrons to the species being reduced and becomes oxidised itself.

These reactions, which cannot occur in isolation, are called half-reactions. The reduced and oxidised pair of species found in a half-reaction is called a redox couple. The redox couple is written: oxidised species/reduced species, for example, Cu^{2+}/Cu or Zn^{2+}/Zn. A reaction involving oxidation and reduction, two half-reactions, is called a redox reaction. The redox reaction in the above examples is:

$$Cu^{2+}(aq) + Zn(s) \rightarrow Cu(s) + Zn^{2+}(aq)$$

9.2 Galvanic cells

An electrochemical cell is a device for the conversion of electrical to chemical energy, and vice versa,

by way of redox reactions. Electrochemical cells consist of two metal electrodes (an anode and a cathode) in contact with an electrolyte that is able to conduct ions but not electrons (Figure 9.1). A galvanic cell (Figure 9.1a) uses a spontaneous chemical reaction to produce an external electric current. Galvanic cells are called batteries in colloquial speech. Electrolytic cells (Figure 9.1b), employed for electrolysis and electroplating, use external electrical power to force nonspontaneous chemical reactions to take place.

9.2.1 The Daniel cell

The principle behind all galvanic cells can be explained with reference to one of the simplest, the Daniell cell, which was invented in 1836. It consists of a zinc rod in a solution of zinc sulphate, and a copper rod in a solution of copper sulphate. To complete the circuit, a porous solid layer, which allows ions to pass between the sulphate electrolytes, and an external metallic conductor between the zinc and copper, are needed. In this case, electrons then pass around the external circuit and ions travel through the electrolyte solutions (Figure 9.2).

Experiment will show that the zinc rod is negative and is called the anode, whereas the copper rod is positive and is called the cathode. When the external connection is completed, a current of electrons

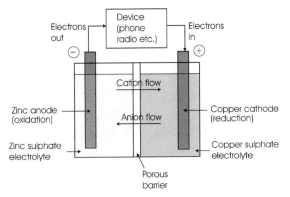

Figure 9.2 A Daniell cell

flows from the zinc anode to the copper cathode. The reaction is spontaneous and needs no external assistance. The current will persist until one of the reactants is consumed.

The electrons are generated by a spontaneous redox reaction in the cell. Oxidation occurs at the zinc anode and reduction takes place at the copper cathode. The reactions taking place are as follows. Anode half-reaction (oxidation, electron generation):

$$Zn(s) \rightarrow Zn^{2+}(aq) + 2e^-$$

Cathode half-reaction (reduction, electron consumption):

$$Cu^{2+}(aq) + 2e^- \rightarrow Cu(s)$$

Zinc is oxidised to Zn^{2+} ions on one side of the porous barrier and Cu^{2+} ions are reduced to metallic copper on the other. Electrons pass via the external circuit. Ions pass through the electrolytes to maintain charge balance in the cell. The chemical reaction being used to generate electricity, called the cell reaction, is obtained by adding the anode and cathode half-reactions, ensuring, by appropriate multiplication, that the number of electrons cancels out, as follows.
Cell reaction (redox reaction):

$$Zn(s) + Cu^{2+}(aq) \rightarrow Zn^{2+}(aq) + Cu(s)$$

There are a number of other processes occurring as well as the generation of an electric current. The

zinc anode is being dissolved. If left for sufficient time the rod degrades and is noticeably corroded. At the same time the copper rod gains weight and a layer of new copper metal forms on the surface. Electroplating is occurring.

9.2.2 Standard electrode potentials

A combination of *any* two dissimilar metallic conductors can be used to construct a galvanic cell. The cell potential defines the measure of the energy available in a cell. A high cell potential signifies a vigorous spontaneous redox reaction. The unit of potential is the volt, V. A Daniell cell, for example, has a potential of 1.1 V.

Because of the multiplicity of possible cells it is more convenient to consider the cell potential as being made up from separate voltage contributions from anode and cathode half-reactions. We can then write:

$$E_{cell} = E_{cathode}(\text{reduction half-reaction})$$
$$+ E_{anode}(\text{oxidation half-reaction})$$

In general, only the reduction half-reaction potentials are listed in tables, as in Table 9.1. The potential of an oxidation half-reaction is the negative of the value of the reduction half-reaction. Moreover, it is convenient to standardise the concentrations of the components of the cells. If the cell components are in their standard states, standard electrode potentials, $E°$, are recorded:

$$E_{cell}° = E_{cathode}° + E_{anode}°$$

For cells involving ionic solutions the standard state is a solution of 1 molar concentration, and for cells involving gases these are at 1 atmosphere pressure.[1]

It is not possible to measure the voltage generated by half a cell, and it has been agreed that the voltage should be measured with respect to a reference electrode using hydrogen gas, called the standard hydrogen electrode. The reference electrode is a

[1] The recommended standard state pressure is 10^5 Pa, but all tabulated data, including that listed here, refer to a standard state of 1 atmosphere (101325 Pa).

Table 9.1 Standard reduction potentials, $E°$, at 25 °C

Half-reaction	$E°$/V
$F_2 + 2e^- \rightarrow 2F^-$	+2.87
$Au^+ + e^- \rightarrow Au$	+1.69
$Ce^{4+} + e^- \rightarrow Ce^{3+}$	+1.61
$Cl_2 + 2e^- \rightarrow 2Cl^-$	+1.36
$O_2 + 4H^+ + 4e^- \rightarrow 2H_2O$	$+1.23^a$
$Br_2 + 2e^- \rightarrow 2Br^-$	+1.09
$Hg^{2+} + 2e^- \rightarrow Hg$	+0.85
$Ag^+ + e^- \rightarrow Ag$	+0.80
$Fe^{3+} + e^- \rightarrow Fe^{2+}$	+0.77
$I_2 + 2e^- \rightarrow 2I^-$	+0.54
$O_2 + 2H_2O + 4e^- \rightarrow 4OH^-$	$+0.40^a$
$Cu^{2+} + 2e^- \rightarrow Cu$	+0.34
$AgCl + e^- \rightarrow Ag + Cl^-$	+0.22
$2H^+ + 2e^- \rightarrow H_2$	0^b
$Fe^{3+} + 3e^- \rightarrow Fe$	-0.04
$O_2 + H_2O + 2e^- \rightarrow OH^- + HO_2^-$	-0.08
$Pb^{2+} + 2e^- \rightarrow Pb$	-0.13
$Sn^{2+} + 2e^- \rightarrow Sn$	-0.14
$Ni^{2+} + 2e^- \rightarrow Ni$	-0.25
$Fe^{2+} + 2e^- \rightarrow Fe$	-0.44
$Fe(OH)_3 + e^- \rightarrow Fe(OH)_2 + OH^-$	-0.56
$Zn^{2+} + 2e^- \rightarrow Zn$	-0.76
$2H_2O + 2e^- \rightarrow 2OH^- + H_2$	-0.83^c
$Ti^{2+} + 2e^- \rightarrow Ti$	-1.63
$Al^{3+} + 3e^- \rightarrow Al$	-1.66
$Be^{2+} + 2e^- \rightarrow Be$	-1.85
$Sc^{3+} + 3e^- \rightarrow Sc$	-2.10
$Mg^{2+} + 2e^- \rightarrow Mg$	-2.36
$La^{3+} + 3e^- \rightarrow La$	-2.52
$Na^+ + e^- \rightarrow Na$	-2.71
$K^+ + e^- \rightarrow K$	-2.93
$Li^+ + e^- \rightarrow Li$	-3.05

^{a} +0.81 at pH 7.
^{b} Zero, by definition.
^{c} -0.42 at pH 7.
Standard conditions: concentration of each ion is 1 mol l^{-1} and all gases are at 1 atm pressure.

mixture of H^+ ions and H_2 gas in the standard state of $H^+(aq)$ at 1 mol l^{-1} and $H_2(g)$ at 1 atm. The standard electrode potential of this half-reaction is defined as zero:

$$2H^+(aq) + 2e^- \rightarrow H_2(g), \quad E° = 0 \text{ V}$$

When the hydrogen electrode is incorporated in a cell, it will form either the anode or the cathode,

depending on the other metal involved. For example, in a cell made with zinc, the zinc electrode is found to be the anode and the hydrogen electrode is the cathode, as follows.
Anode half-reaction (oxidation, electron generation):

$$Zn(s) \rightarrow Zn^{2+}(aq) + 2e^- \quad E° = +0.76 \text{ V}$$

Cathode half-reaction (reduction, electron consumption):

$$2H^+(aq) + 2e^- \rightarrow H_2(g) \quad E° = 0 \text{ V}$$

Cell reaction:

$$Zn(s) + 2H^+(aq) \rightarrow Zn^{2+}(aq) + H_2(g) \quad E_0 = +0.76 \text{ V}$$
$$E°(\text{cell}) = +0 \text{ V} + (+0.76 \text{ V}) = 0.76 \text{ V}$$

The Zn is oxidised to Zn^{2+}, and the $H^+(aq)$ is reduced to $H_2(g)$. Note that because zinc is the anode, the value of $E°$ for the anode half-reaction is the negative of that given in Table 9.1.

In a cell using copper, the hydrogen electrode is found to be the anode and the copper electrode the cathode, as follows.
Anode half-reaction (oxidation, electron generation):

$$H_2(g) \rightarrow 2H^+(aq) + 2e^- \quad E° = 0 \text{ V}$$

Cathode half-reaction (reduction, electron consumption):

$$Cu^{2+}(aq) + 2e^- \rightarrow Cu(s) \quad E_0 = +0.34 \text{ V}$$

Cell reaction:

$$H_2(g) + Cu^{2+}(aq) \rightarrow 2H^+(aq) + Cu(s)$$
$$E_0 = +0.34 \text{ V}$$
$$E°(\text{cell}) = +0.34 \text{ V} + 0 \text{ V} = +0.34 \text{ V}$$

When many cells are compared, it is found that the anode is always the material that has the lowest tendency to be reduced. From the two examples above, it is seen that the order of the electrodes with respect to this tendency is $Zn^{2+}/Zn < H^+/H_2(g) < Cu^{2+}/Cu$. Thus, in a Daniell cell, zinc forms the anode and copper the cathode, as follows.

Anode half-reaction (oxidation, electron generation):

$$Zn(s) \rightarrow Zn^{2+}(aq) + 2e^- \quad E^\circ = +0.76\,V$$

Cathode half-reaction (reduction, electron consumption):

$$Cu^{2+}(aq) + 2e^- \rightarrow Cu(s) \quad E^\circ = +0.34\,V$$

Cell reaction:

$$Zn(s) + Cu^{2+}(aq) \rightarrow Cu(s) + Zn^{2+}(aq)$$
$$E^\circ = +0.34\,V$$
$$E^\circ(cell) = +0.34\,V + (+0.76\,V) = 1.10\,V$$

As before, note that because zinc is the anode, the value of E° is the negative of that given in Table 9.1.

The comparison of each element to a hydrogen electrode in a standard galvanic cell allows the reduction tendency to be ranked. A table of these values arranged so that the elements with the greatest tendency to be reduced (the most strongly oxidising) are at the top is referred to as the electrochemical series. An abbreviated electrochemical series is given in Table 9.1. The oxidised species in a redox couple has the ability to oxidise the reduced species in any redox couple below it in the table. Moreover, such a reaction will be spontaneous. For example, fluorine gas, F_2, will have the highest tendency to be reduced, or gain electrons, and lithium metal, Li, has the highest tendency to be oxidised, or lose electrons. Mixing these elements will lead to a spontaneous, and very vigorous, reaction, leading to the production of Li^+ and F^- ions.

When forming a galvanic cell, the couple higher in the table forms the cathode and the couple below in the table forms the anode. This is written using a standard notation (see Section 53.4):

lower couple (anode) || higher couple (cathode)

When written out fully this means:

The lower couple, which forms the anode, undergoes the reaction

'reduced species $\rightarrow$ oxidised species + ne$^-$',

whereas the higher couple, which forms the cathode, undergoes the reaction

'oxidised species + ne$^-$ $\rightarrow$ reduced species'

The cell diagrams and cell voltages for the cells described earlier are written:

$$Zn(s)|Zn^{2+}(aq)||H^+(aq)|H_2(g)|Pt \quad E^\circ = 0.76\,V$$
$$Pt|H_2(g)|H^+(aq)||Cu^{2+}(aq)|Cu(s) \quad E^\circ = 0.34\,V$$
$$Zn(s)|Zn^{2+}(aq)||Cu^{2+}(aq)|Cu(s) \quad E^\circ = 1.10\,V$$

- E°(cathode) is obtained directly from the table of standard reduction potentials;

- E°(anode) is the *negative* of the standard reduction potential;

- E°(cell) is the cathode standard reduction potential plus the *negative* of the anode standard reduction potential.

The voltage of a cell when all of the reactants are in the standard conditions defined above is called the standard cell potential, E°. The standard cell potential of any cell can be derived from the appropriate standard reduction potentials.

9.2.3 Cell potential and free energy

The free energy of a reaction is a measure of the tendency for a chemical reaction to take place (see Section S3.2). The direction of spontaneous change in a process is that of decreasing free energy. If we write ΔG_r as the free energy change of a reaction, ΔG_r is *negative* for a spontaneous change. The cell potential is related to the Gibbs energy of the cell reaction by:

$$\Delta G_r = -nE_{cell}F$$

Where E_{cell} is the cell potential in volts, defined to be positive; F is the Faraday constant, 96485 C mol^{-1}; and n is the number of moles of electrons that migrate from anode to cathode in the cell reaction. Thus a galvanic cell is also a Gibbs energy meter.

When the electrodes are in their standard states, the free energy change is called the standard reaction free energy, ΔG°_r, and the cell voltage is just the standard cell potential, E°. In this case:

$$\Delta G^{\circ}_r = -nE^{\circ}F$$

For example, in the Daniell cell:

$$Zn(s) + Cu^{2+}(aq) \rightarrow Cu(s) + Zn^{2+}(aq)$$

Two electrons are transferred in the cell reaction, hence $n = 2$. [Note that the number of electrons taking part in the reaction is usually clear from the half-reactions rather than from the cell reaction.] Hence,

$$\Delta G_r = -2E_{cell}F$$
$$= -(1.93 \times 10^5)E_{cell}$$

When E_{cell} is measured in volts, the value of ΔG_r is in joules. When the concentrations of both Cu^{2+} and Zn^{2+} are 1 mol l^{-1} (the standard state),

$$\Delta G^{\circ}_r = -nE^{\circ}F$$

9.2.4   Concentration dependence

The potential generated by a cell is dependent on the concentration of the components present. The relationship is given by the equation:

$$\Delta G_r = -nE_{cell}F = RT \ln\left(\frac{Q}{K}\right)$$

where R is the gas constant, T is the temperature (in kelvin), Q is the reaction quotient, and K is the equilibrium constant (see Section 8.1).

The reaction quotient, Q_c, of a reaction

$$aA + bB \rightleftharpoons xX + yY$$

is given by:

$$Q_c = \frac{[X]^x[Y]^y}{[A]^a[B]^b}$$

where [A] denotes the concentration of compound A at any time, and so on. For reactions involving gases, the concentration term can be replaced by the partial pressure of the gaseous reactants, to give:

$$Q_p = \frac{p_X^x p_Y^y}{p_A^a p_B^b}$$

These expressions are identical to those for the equilibrium constant given in Section 8.1.2, except that they apply to any concentrations or partial pressures, not just those when the system is at equilibrium. Pure liquids or solids, or water in solutions, appearing in the reaction equation do not appear in the equations for Q. (The quantity of importance is the activity rather than the concentration. Activity and concentration are equal in dilute solutions; see Section S3.2).

At equilibrium:

$$Q = K$$

The cell voltage is then given by rearrangement of

$$-nE_{cell}F = RT \ln\left(\frac{Q}{K}\right)$$

to yield

$$E_{cell} = -\frac{RT}{nF} \ln\left(\frac{Q}{K}\right)$$

This equation for E_{cell} shows:

- if $Q/K < 1$, $E_{cell} > 0$, $\Delta G_r < 0$, and the cell reaction is spontaneous from left to right.

- if $Q/K > 1$, $E_{cell} < 0$, $\Delta G_r > 1$, and the cell reaction is spontaneous from right to left;

- if $Q/K = 1$, $E_{cell} = 0$, $\Delta G_r = 0$, and the cell reaction is at equilibrium.

It is convenient to separate the term relating to the equilibrium constant and write the equation for the cell voltage as

$$E_{cell} = \frac{RT}{nF} \ln K - \frac{RT}{nF} \ln Q \qquad (9.1)$$

When all species are in the standard state, $Q = 1$, $\ln Q = 0$, and the cell voltage is $E°$, hence:

$$E_{cell} = E° = \frac{RT}{nF} \ln K$$

substituting $E°$ into Equation (9.1) gives:

$$E_{cell} = E° - \frac{RT}{nF} \ln Q \qquad (9.2)$$

where E_{cell} is the cell voltage, $E°$ is the standard cell voltage, R is the gas constant, T is the absolute temperature, n is the number of electrons transferred in the cell reaction, F is the Faraday constant, and Q is the reaction quotient of the cell reaction. Equation (9.2) is the Nernst equation. Inserting values for the constants, the Nernst equation can be written:

$$E = E° - \frac{0.02569}{n} \ln Q$$

or

$$E = E° - \frac{0.05916}{n} \log Q$$

9.2.5 Chemical analysis using galvanic cells

The measurement of the potential of a galvanic cell can be used to determine the concentration of the ions in a solution via the Nernst equation. For example, the cell reaction of the Daniell cell is:

$$Zn(s) + Cu^{2+}(aq) \rightarrow Cu(s) + Zn^{2+}(aq)$$

The appropriate form of the Nernst equation is:

$$E = E° - \frac{0.02569}{n} \ln \left\{ \frac{[Zn^{2+}]}{[Cu^{2+}]} \right\}$$

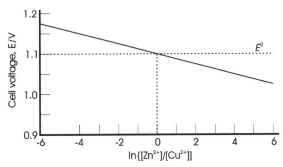

Figure 9.3 Plot of cell voltage, E, versus $\ln \{[Zn^{2+}]/[Cu^{2+}]\}$ for a Daniell cell

A plot of E versus $\ln \{[Zn^{2+}]/[Cu^{2+}]\}$ is (ideally) a straight line (Figure 9.3). Provided that the concentration of one of the ions is known, the concentration of the other can be determined from the experimental value of the cell potential.

The most widespread use of this analytical technique is the measurement of the concentration of hydrogen ions (pH). To illustrate this, consider how the combination of a standard electrode with a (nonstandard) hydrogen electrode can be used to measure the concentration of hydrogen ions present at the hydrogen electrode. The standard electrode is often a particularly stable electrode, called a calomel electrode, which uses the redox couple Hg_2Cl_2/Hg, Cl^-. The cell is:

$$Pt \mid H_2(g) \mid H^+(aq) \parallel Cl^-(aq) \mid Hg_2Cl_2 \mid Hg(l)$$

Anode reaction:

$$H_2(g) \rightarrow 2H^+(aq) + 2e^- \quad E° = 0\,V$$

Cathode reaction:

$$Hg_2Cl_2(s) + 2e^- \rightarrow 2Hg(l) + 2Cl^-(aq)$$
$$E° = +0.27\,V$$

Cell reaction:

$$Hg_2Cl_2(s) + H_2(g) \rightarrow 2H^+(aq) + 2Cl^-(aq) + 2Hg(l)$$

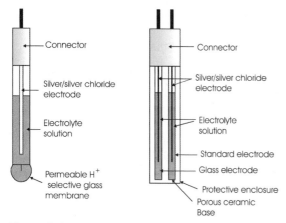

Figure 9.4 (a). A glass electrode for the measurement of pH and (b) experimental arrangement of glass electrode and standard electrode in a single cell

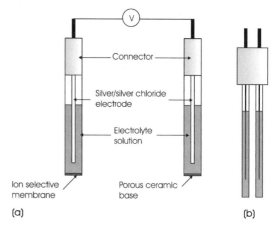

Figure 9.5 (a) Schematic arrangement of an ion selective electrode for the measurement of ion concentrations in solution and (b) experimental arrangement, where the electrodes are attached to a connecting unit

The cell voltage is found to be directly proportional to the pH:

$$E(\text{cell}) = E' + 0.0592 \times \text{pH}$$

Where E' is the cell constant, which depends on the Cl^- concentration. The pH can be read directly on a voltage scale.

The hydrogen electrode is not practical, and commercial pH meters are constructed with a glass electrode that has a sensing element made of a thin membrane of a special glass sensitive only to H^+ ions (Figure 9.4a). The potential of the electrode is found to be proportional to the pH of the surrounding solution, and the response of this electrode is similar to that of a hydrogen electrode.

Apart from the calomel electrode, one of the most common standard electrodes used in pH meters is a silver/silver chloride electrode. This consists of a silver wire coated with silver chloride immersed in a four-molar solution of potassium chloride (KCl), saturated with silver chloride (AgCl). The half-reaction is:

$$AgCl(s) + e^- \rightarrow Ag(s) + Cl^-$$
$$E^\circ = 0.2046 \, V \text{ at } 25\,^\circ C$$

In practice, the hydrogen ion selective electrode and the standard electrode are packaged together in a small plastic cylinder, which is easily transported (Figure 9.4b). Measurement is made by dipping the electrode into the solution to be checked, and the voltage is transformed into a pH reading electronically.

The same measurement principle can be used to determine the concentration of other ions in solution. An electrode that is (ideally) sensitive to one ion only, called an ion selective electrode, is paired with a standard electrode (Figure 9.5). The potential developed by such a combination is of the general form

$$E = E' + \frac{2.303RT}{nF} \log(c)$$

where E' is a constant characteristic of the ion selective electrode and the reference electrode, and c is the concentration of the ion. This is a linear dependence, in which the slope is $(2.303RT/nF)$. Experimentally, the value of the slope is found to lie between 50–60 mV for monovalent ions and 25–30 mV for divalent ions.

The critical component of an ion selective electrode is a membrane that acts to pass the selected ions into the interior of the electrode assembly (Figure 9.5a). These are of two principal types. Crystal membranes consist of a polycrystalline or single-crystal plate. For example, fluoride ion (F^-)

sensors are made from single-crystal lanthanum trifluoride (LaF_3) doped with europium difluoride (EuF_2). The Eu^{2+} ions substitute for La^{3+} in the LaF_3 matrix, and each substituted ion is accompanied by a vacancy on the F substructure to maintain charge neutrality. The large number of vacancies thus generated increases the diffusion coefficient of F^- in LaF_3 enormously. The membrane has a similar permeability to F^- as the surrounding liquid and is found to be highly selective for the passage of F^- ions.

The other type of membrane in use consists of a polyvinyl chloride (PVC) disc, impregnated with a large organic molecule that can react with the ion. The binding must be weak enough for the ion to be passed from one molecule to another across the membrane under the driving force of a concentration gradient. For example, K^+ ion selective membranes are made using the antibiotic valinomycin. This has a structure that accommodates the K^+ ions and can pass them on from one molecule to another. The operation of this material mimics the way in which living cells transfer ions across the cell membrane. Unfortunately, such molecules are not usually completely specific for a single ion and usually also channel chemically similar species. The potassium membrane, for example, can also pass lesser amounts of sodium ions.

9.3   Batteries

Batteries, which are galvanic cells, fall into one of three main types. A primary cell is a battery that cannot be recharged. A secondary cell is a battery that can be recharged and reused. A fuel cell has a continuous input of chemicals (fuel) to produce a continuous output of current.

In all batteries, oxidation occurs at the anode of the cell; the electrode removes electrons from the species in the electrolyte. It is given a negative sign in diagrams (Figures 9.1 and 9.2). It consists of a relatively easily oxidised metal such as zinc, cadmium or nickel, sometimes in contact with a current collector such as a graphite rod. At the cathode of a battery, reduction occurs; the electrode gives elec-

trons to species in the electrolyte. The cathode is marked as positive in a diagram (Figures 9.1 and 9.2). The active component in the cathode is often a metal oxide such as MnO_2 or PbO_2, which is capable of being reduced, sometimes in contact with a metallic current collector. Electrons flow from anode to cathode (negative to positive) via an external circuit. Anions (negative ions) travel towards the anode through the electrolyte. Cations (positive ions) travel towards the cathode through the electrolyte. Normally, a redox reaction in a battery will stop if the reactants are separated. However, if the electrons are allowed to travel via an external circuit from one electrode to the other the reaction can continue, even when the components are separated.

In a battery, the electrons moving around the external circuit are used to do useful work. The driving force for this is the energy of the cell reaction. Batteries are spent, or need recharging, when one or both of the components has been used up.

There are large numbers of different batteries manufactured, and in the descriptions that follow only a small selection of commonly encountered batteries is described.

9.3.1   'Dry' and alkaline primary batteries

The widely used primary 'dry-cell' or, Leclanché cell, which was invented in 1866, gives a voltage of about 1.5 V. It is especially used for intermittent applications, such as for flashlights. In this cell the redox couple used is Zn/MnO_2. The current collector is a graphite rod buried in the positive cathode, a mixture of MnO_2 and carbon. This is kept moist by the electrolyte, aqueous ammonium chloride, NH_4Cl. The anode (negative terminal) is the container itself, made from zinc (Figure 9.6).

As with all commercial batteries, the real cell reaction is complex, and the reactions are approximated as follows.

Anode reaction (oxidation, electron generation):

$$Zn(s) \rightarrow Zn^{2+}(aq) + 2e^-$$

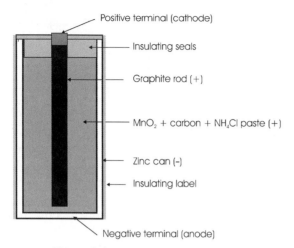

Figure 9.6   Section through a dry cell

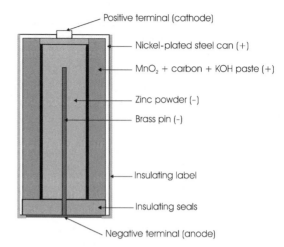

Figure 9.7   Section through an alkaline cell

Cathode reaction (reduction, electron consumption):

$$MnO_2(s) + H_2O(l) + e^- \rightarrow MnO(OH)(s) + OH^-(aq)$$

Cell reaction:

$$Zn(s) + 2MnO_2(s) + 2H_2O(l) \rightarrow 2MnO(OH)(s)$$
$$+ 2OH^-(aq) + Zn^{2+}(aq)$$

The main problem encountered with this cell is the buildup of Zn^{2+} and OH^- at the respective electrodes, which is why the battery is mainly used intermittently. When the battery is not being used the concentrations of these reaction products falls again. This is because the OH^- ions migrate to the Zn anode where they form ammonia with NH_4^+ ions in the electrolyte:

$$NH_4^+(aq) + OH^-(aq) \rightarrow H_2O(l) + NH_3(aq)$$

The concentration of the Zn^{2+} ions in the vicinity of the anode subsequently drops as a result of reaction with the NH_3:

$$Zn^{2+}(aq) + 4NH_3(aq) \rightarrow Zn(NH_3)_4^{2+}(aq)$$

The cell is spent when ionic conduction is no longer possible because of the buildup of $Zn(NH_3)_4Cl_2$.

Alkaline cells use the same zinc–manganese dioxide couple as Leclanché cells. However, the ammonium chloride electrolyte is replaced with a solution of about 30 wt% potassium hydroxide (KOH) to improve ionic conductivity. The cell reactions are identical to those above, but the battery construction is rather different (Figure 9.7). The negative material is zinc powder, and the anode (negative terminal) is a brass pin. The positive component is a mixture of MnO_2 and carbon powder that surrounds the anode. A porous cylindrical barrier separates these components. The positive terminal (cathode) is the container, which is a nickel-plated steel can.

9.3.2   Lithium-ion primary batteries

Lithium has a number of advantages over other materials for battery manufacture. It is the lightest true metal, and it also has a high electrochemical reduction potential, that is, it occurs at the bottom of Table 9.1. There is one disadvantage in using lithium in that it is very reactive, a feature that poses problems not only in manufacture but also in the selection of the other battery components. Despite this, there are a large number of lithium-based primary cells available, both in traditional cylindrical form and as button and flat coin cells.

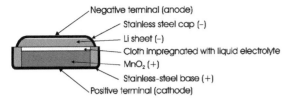

Figure 9.8   Section through a lithium coin cell

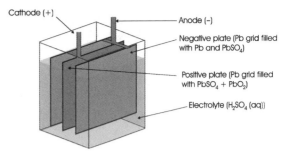

Figure 9.9   A single cell in a lead – acid battery

The lithium forms the anode in such cells, and a variety of compounds may form the anode. The most usual of these is manganese dioxide (MnO_2), giving a working potential of approximately 3 V. Because lithium reacts vigorously with water, the electrolyte must be nonaqueous and is frequently a solution of lithium salts in a polar organic liquid. The conductivity of such solutions is low compared with that of aqueous solutions of hydroxides, which means that the design of a cell is constrained by the need for large electrode areas separated by a thin electrolyte. The coin cell is a natural result of such considerations (Figure 9.8).

The cell reactions are poorly understood, but can be written schematically as follows.

Anode reaction (oxidation, electron generation):

$$Li(s) \rightarrow Li^+(aq) + e^-$$

Cathode reaction (reduction, electron consumption):

$$MnO_2(s) + e^- \rightarrow MnO_2^-(s)$$

Cell reaction:

$$Li(s) + MnO_2(s) \rightarrow LiMnO_2(s) \quad E^\circ = 3.2\,V$$

9.3.3   The lead–acid secondary battery

Plante discovered the basic technology of the rechargeable lead–acid battery in 1859. Since then there have been many refinements to the materials used, but the operating principles remain the same. This battery is the widely used car battery in use to this day. The anode is lead (Pb), the cathode lead dioxide (PbO_2) and the electrolyte is dilute sulphuric acid [$H_2SO_4(aq)$] (Figure 9.9). As with all batteries, the chemical reactions taking place are complex, but schematically the processes are as follows.

Anode reaction (oxidation, electron generation):

$$Pb(s) + HSO_4^-(aq) \rightarrow PbSO_4(s) + H^+(aq) + 2e^-$$

Cathode reaction (reduction, electron consumption):

$$PbO_2(s) + 3H^+(aq) + HSO_4^-(aq) + 2e^- \rightarrow$$
$$PbSO_4(s) + 2H_2O(l)$$

Cell discharge reaction:

$$PbO_2(s) + Pb(s) + 2H^+(aq) + HSO_4^-(aq) \rightarrow$$
$$2\,PbSO_4(s) + 2H_2O(l)$$

The cell potential is about 2 V. A car battery consists of (usually) six cells in series (a battery of cells), to give 12 V. Sulphuric acid is used up during operation, so the state of charge of the battery can be estimated by measuring the concentration of the acid, usually via density.

The cell-charging reaction is the reverse of the discharge reaction, and to charge a battery the cell reaction is simply driven backwards by an imposed external voltage.

9.3.4   Nickel–cadmium (Ni–Cd, nicad) rechargable batteries

These were, until recently, the most widespread rechargeable batteries available for home use.

However, concerns about the toxicity of cadmium have accelerated the replacement of these batteries by nickel–metal hydride batteries, described in Section 9.3.5. In nickel–cadmium (nicad) batteries, the anode is cadmium and the cathode is an unstable nickel oxyhydroxide, formed in the unusual conditions found in the cell, and written variously as $Ni(OH)_3$ or $NiO(OH)$. It is generally formed together with stable nickel hydroxide, $Ni(OH)_2$. The electrolyte is NaOH or KOH. The anode and cathode are assembled in a roll separated by a cellulose separator containing the electrolyte. The cathode/separator/anode roll is contained in a nickel-plated stainless steel can (Figure 9.10). The cell voltage is 1.3 V but the working voltage is usually nearer to 1.2 V. The schematic cell reactions are as follows.

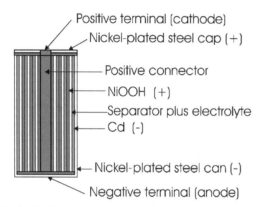

Positive terminal (cathode)
Nickel-plated steel cap (+)

Positive connector
NiOOH (+)
Separator plus electrolyte
Cd (-)

Nickel-plated steel can (-)

Negative terminal (anode)

Figure 9.10   Section through a nichel–cadmium (nicad, Ni – Cd) cell

Anode reaction (oxidation, electron generation):

$$Cd(s) + 2OH^-(aq) \rightarrow Cd(OH)_2(aq) + 2e^-$$

Cathode reaction (reduction, electron consumption):

$$2NiO(OH)(s) + 2H_2O(l) + 2e^- \rightarrow$$
$$2Ni(OH)_2(s) + 2OH^-(aq)$$

Cell discharge reaction:

$$Cd(s) + 2NiO(OH)(s) + 2H_2O(l) \rightarrow Cd(OH)_2(s)$$
$$+ 2Ni(OH)_2(s)$$

The charging reaction is the reverse of the discharge reaction, driven by an external voltage.

9.3.5   Nickel-metal-hydride rechargeable batteries

These batteries, frequently called Ni–MH batteries, are now replacing nicad batteries (Section 9.3.4) for many uses. They rely on hydrogen storage in a metal alloy for the reversible operation.

Many metals absorb large quantities of hydrogen to form, initially, nonstoichiometric interstitial alloys and, at greater concentrations, various alloy phases with definite compositions M_aH_b (see also Section 6.1.3.2). In many of these materials the hydrogen incorporation is reversible, so that hydrogen is taken up at high hydrogen pressure and released at lower pressure. This reversible uptake of hydrogen is exploited in Ni–MH batteries.

The positive electrode, the cathode, is similar to that in nicad cells and consists of a mixture of $NiO(OH)/Ni(OH)_3$ and $Ni(OH)_2$. An alloy that supports hydride formation replaces the cadmium as the negative anode. The alloy most commonly used is derived from $LaNi_5$, in which a mixture of other lanthanides replaces the lanthanum, and a nickel-rich alloy replaces the nickel, to give a general formula LnM_5. The anode is composed of an agglomeration of alloy powder. A small amount of potassium hydroxide is added as an electrolyte. The cell voltage is 1.3 V, making these cells suitable for the direct replacement of nicad batteries. The cell construction is identical to that of the nicad cell (Figure 9.10), with the cadmium replaced by metal hydride. The approximate cell reactions are as follows.

Anode reaction (oxidation, electron generation):

$$MH_x(s) + OH^-(aq) \rightarrow MH_{x-1}(s) + H_2O(l) + e^-$$

Cathode reaction (reduction, electron consumption):

$$2\,NiO(OH)(s) + H_2O(l) + e^- \rightarrow Ni(OH)_2(s)$$
$$+ OH^-(aq)$$

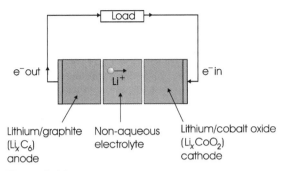

Figure 9.11 A lithium rechargable 'Sony' cell, in discharge operation

Cell discharge reaction:

$$MH_x(s) + 2\,NiO(OH)(s) \rightarrow MH_{x-1}(s)$$
$$+ Ni(OH)_2(s)$$

The charging reaction is the reverse of the discharge reaction, driven by an external voltage.

9.3.6 Lithium-ion rechargeable batteries

The advantages of lithium primary cells extend to secondary cells. In particular, the high power available and the lightness make them ideal for portable electronic devices. The first successful lithium-ion rechargeable battery was introduced by Sony in 1991 and is often called the Sony cell. The principle of the cell is shown in Figure 9.11. The difficulties of working with lithium metal are overcome by using nonstoichiometric intercalation compounds (see Section 3.4.5). The electrolyte is, as with the lithium primary cells, a nonaqueous solution of lithium salts in a polar organic liquid.

The active component of the anode is lithium metal contained in graphite. The structure of graphite (Figure 5.21, page 132), consists of strongly linked layers of carbon atoms, arranged in a hexagonal array, linked by weak van der Waals bonds. It has long been known that graphite can take in alkali metal atoms between the weakly linked layers of carbon hexagons to form intercalation compounds. The nominal composition of the lithium–

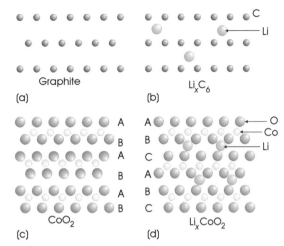

Figure 9.12 Transformations in electrode materials; graphite (part a) changes layer stacking when lithium atoms are intercalated (part b); CoO_2 changes from hexagonal closest packing of the oxygen anions to cubic closest packing when lithium atoms are intercalated (part d)

graphite intercalation material is Li_xC_6, with x varying from 0 to approaching 1.0. The stacking of the hexagonal carbon layers is staggered in the pure compound (Figure 9.12a), but in the lithium-containing phase they are directly over each other (Figure 9.12b). The stacking, therefore, alters with the degree of lithium incorporation, limiting the performance of the electrode. The cathode material is the nonstoichiometric oxide Li_xCoO_2. This material is also an intercalation compound, in which the Li^+ ions lie between layers of composition CoO_2 (see Figures 9.12c and 9.12d). In theory, the composition range of the cathode material is from $LiCoO_2$ to CoO_2 (i.e., x varying from 1 to 0) but in battery operation the degree of nonstoichiometry is restricted, and x generally takes values in the range of 0 to about 0.45. As with graphite, intercalation of lithium changes the layer stacking. In this case, CoO_2 is built of hexagonal close-packed (ABAB) layers of oxygen atoms (Figure 9.12c), whereas $LiCoO_2$ has cubic (ABCABC) close packing (Figure 9.12d). This change degrades the oxide during use and is the major reason why the useful composition range is limited.

At present there is much research work on improving the cathode and the anode materials in these cells, and a number of alternative layer compounds are actively being tested.

The cell reactions are similar to those utilised in the lithium primary cell. During discharge, Li^+ ions are transported from the anode to the cathode via the following reactions.

Anode reaction (oxygen, electron generation):

$$Li_xC_6(s) \rightarrow 6C(s) + xLi^+(s) + xe^-$$

Cathode reaction (reduction, electron consumption):

$$Li_{0.55}CoO_2(s) + xLi^+(s) + xe^- \rightarrow Li_{0.55+x}CoO_2(s)$$

Cell discharge reaction:

$$Li_xC_6(s) + Li_{0.55}CoO_2(s) \rightarrow 6C(s) + Li_{0.55+x}CoO_2(s)$$

These reactions can also be written as:

Anode reaction (oxidation, electron generation):

$$Li(s) \rightarrow Li^+(s) + e^-$$

Cathode reaction (reduction, electron consumption):

$$Co^{4+}(s) + e^- \rightarrow Co^{3+}(s)$$

Cell discharge reaction:

$$Li(s) + Co^{4+}(s) \rightarrow Li^+(s) + Co^{3+}(s)$$

The charging reaction is the reverse of the discharge reaction, driven by an external voltage.

9.3.7 Fuel cells

Batteries have a fixed amount of reactants present, stored in the battery casing. Fuel cells are primary cells with a continuous input of chemical reactants and a continuous output of power. The reactants are stored separately from the electrodes and electrolyte and can be replenished when necessary. There is much research at present on fuel cells as a source of clean electricity. The reaction chosen is the produc-

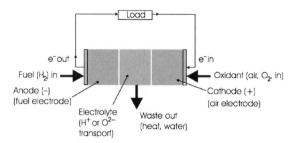

Figure 9.13   A fuel cell: fuel is added continuously to the electrodes, and electricity is produced continuously; waste products are heat and water

tion of water from hydrogen gas, H_2, and oxygen gas, O_2, giving a cell voltage of about 1.2 V. The concept is simple (Figure 9.13), and early fuel cells, containing an alkaline solution, typically potassium hydroxide solution, KOH(aq), as electrolyte, demonstrate the principles involved.

Anode reaction (oxidation of H_2; electron generation):

$$H_2(g) + 2OH^-(aq) \rightarrow 2H_2O(l) + 2e^-$$
$$E^\circ = -0.83\,V$$

Cathode reaction (reduction of O_2; electron consumption):

$$O_2(g) + 2H_2O(l) + 4e^- \rightarrow 4OH^-(aq) \quad E^\circ = 0.40\,V$$

Cell discharge reaction:

$$2H_2(g) + O_2(g) \rightarrow 2H_2O(l)$$

Current research is centred on making compact cells of high efficiency. They are described in terms of the electrolyte that is used. The principle types are alkali fuel cells, described above, with aqueous KOH as electrolyte, MCFCs (molten carbonate fuel cells), with a molten alkali metal or alkaline earth carbonate electrolyte, PAFCs (phosphoric acid fuel cells), PEMs (proton exchange membranes), using a solid polymer electrolyte that conducts H^+ ions, and SOFCs, (solid oxide fuel cells), with solid electrolytes that allow oxide ion, O^{2-}, transport. The majority of cells use a catalyst to speed up the reactions taking place. These are mostly platinum-based and are deposited on the electrodes as small, highly dispersed particles, but the MCFC has an advantage in that it uses a nickel catalyst, which is much cheaper than platinum.

As the cell reaction is between hydrogen and oxygen, the continuous supply of these gases is vital. Oxygen poses no problem, as it is freely available from the air. The best way to provide hydrogen fuel has not yet been resolved. Hydrogen reservoirs containing pressurised gas, or liquid hydrogen, have been considered. Another approach envisages the formation of hydrogen directly on demand from a hydrocarbon store such as methane, petrol or oil, using a catalyst. One promising avenue seems to be the use of hydrides for this purpose. Nonstoichiometric metal hydrides, similar to those used in Ni–MH batteries, which can reversibly store hydrogen as atoms, are contenders for this purpose.

Of the cells available, SOFCs appear to be nearest to commercial use for large-scale electricity generation. A variety of designs are being explored, including planar and tubular geometries (Figure 9.14). In all cases, single cells are linked to give a fuel-cell stack by an interconnect material. The electrolyte in these cells is derived from calcia-stabilised zirconia (see Section 3.4.5). This nonstoichiometric oxide has approximately 15 % of the oxygen sites unoccupied. Because of this, the oxygen ion diffusion coefficient in the solid is very high. The cell operating principle is drawn in Figure 9.14. Oxygen in the form of air is supplied to the cathode, often called the air electrode. The oxygen gas is ionised and oxygen ion transport across the electrolyte ensues. Hydrogen fuel is supplied to the anode or fuel electrode. Here it reacts with the oxide ions to form water.

Anode reaction (oxidation of H_2; electron generation):

$$H_2(g) + O^{2-}(s) \rightarrow H_2O(l) + 2e^-.$$

Cathode reaction (reduction of O_2; electron consumption):

$$\tfrac{1}{2}O_2(g) + 2e^- \rightarrow O^{2-}(s).$$

Cell discharge reaction:

$$H_2(g) + \tfrac{1}{2}O_2(g) \rightarrow H_2O(l).$$

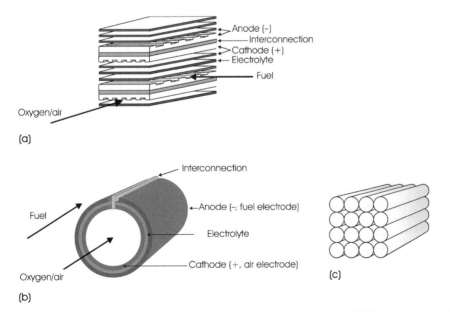

Figure 9.14   (a) An expanded view of a stack of planar design solid-oxide fuel cells (SOFCs); (b) tubular design of an SOFC; (c) a stack of tubular SOFCs

A problem with SOFCs is that diffusion of oxygen ions in the electrolyte is too slow at room temperature to make the cells viable. At present, satisfactory cell operation is accomplished only when the electrolyte is held at temperatures in excess of 650 °C, although intensive research is continually lowering this temperature.

9.4   Corrosion

Corrosion refers to the degradation of a metal by electrochemical reaction with the environment. At room temperature, the most important corrosion reactions involve water, and the process is known as aqueous corrosion. (Corrosion at high temperatures in dry air, called oxidation tarnishing, or direct corrosion, is considered in Section 8.5.) Aqueous corrosion involves a set of complex electrochemical reactions in which the metal reverts to a more stable condition, usually an oxide or mixture of oxides and hydroxides (Figure 9.15). In many cases the products are not crystalline and are frequently mixtures of compounds. Aside from the loss of metal, the corrosion products may be voluminous. In this case, they force overlying protective layers away from the metal and so allow corrosion to proceed unchecked, which exacerbates the damage.

The extent of aqueous corrosion often depends on the presence of impurities and trace contaminants in the water present. For example, carbon-steel reinforcing bars in concrete corrode more severely in acidic conditions and in the presence of chloride ions, a process called electrochemical attack. On the other hand, alkaline conditions inhibit the rate of corrosion.

In principle, corrosion is easily prevented by one of two methods – by modifying the environment (which includes coating the metal) or by replacing the corrodible metal with a corrosion-resistant metal. However, these simple remedies are not always possible, and corrosion is a major economic factor across the world.

9.4.1   The reaction of metals with water and aqueous acids

The reaction of a metal with an aqueous acid to yield hydrogen, a severe form of corrosion, involves oxidation of the metal and reduction of hydrogen ions in solution, $H^+(aq)$, to H_2 gas, and so can be thought of in terms of an electrochemical cell. The tendency for a reaction to occur follows the order of the electrochemical series (Table 9.1). Metals below H_2 in the electrochemical series – those with a negative standard reduction potential – will react with aqueous acids to release hydrogen gas. Those above it will not react with acid. Thus, zinc will dissolve in acid to give hydrogen, whereas copper will not.

Such reactions are generally written in terms of the overall reaction:

$$Zn + 2HCl \rightarrow ZnCl_2 + H_2(g)$$

However, more insight into the process is given by using the two half-reactions:

$$2H^+(aq) + 2e^- \rightarrow H_2(g)$$
$$Zn(s) \rightarrow Zn^{2+}(aq) + 2e^-$$

that is,

$$Zn(s) + 2H^+(aq) \rightarrow Zn^{2+}(aq) + H_2(g)$$

The same information can also be used to predict which metals may dissolve in acid rain. For example, both lead and tin can enter the water supply in acid rain areas, and, for the same reason, acidic water will react with lead water pipes.

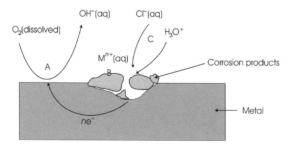

Figure 9.15   Aqueous corrosion reactions: A, electrode reactions; B, deposition reactions; C, impurity reactions

The reaction of metals with water can be examined by using identical principles. Two reactions are important: oxidation and reduction. When water acts as an oxidising agent it is reduced to H_2. This is similar to the oxidation of a metal by an acid (H^+) to give H_2. Metals in the electrochemical series *below* the couple,

$$2H_2O(l) + 2e^- \rightarrow H_2(g) + 2OH^-(aq)$$
$$E^\circ = -0.83 \text{ V at pH } 14$$

will react with water and produce hydrogen gas. These metals (Al, Mg, Na, K and Li) are the most reactive. For example, the reaction of magnesium with water is written:

$$Mg(s) + 2H_2O(l) \rightarrow Mg(OH)_2 + H_2(g)$$

and, in terms of half reactions,

$$2H_2O(l) + 2e^- \rightarrow H_2(g) + 2OH^-(aq)$$
$$Mg(s) \rightarrow Mg^{2+}(aq) + 2e^-$$
$$Mg(s) + 2H_2O(l) \rightarrow H_2(g) + 2OH^-(aq) + Mg^{2+}(aq)$$

In practice, the reaction products from these low-temperature processes are invariably ill-defined amorphous materials consisting of poorly soluble oxyhydroxides.

Under standard conditions ($E^\circ = -0.83$ V), the concentration of the OH^-(aq) ions is 1 molar and the pH is 14 – very alkaline conditions indeed. In order to determine whether neutral water, at pH 7, will react, it is necessary to use the Nernst equation to redefine the reaction voltage:

$$2H_2O(l) + 2e^- \rightarrow H_2(g) + 2OH^-(aq)$$
$$E = -0.42 \text{ V at pH } 7$$

The revised reduction potential is −0.42 V. It is of interest to compare this with the reduction potential for iron. The Fe^{2+}/Fe couple has $E^\circ = -0.44$ V, which is almost equal to that of neutral water. Thus, surprisingly, iron has no (or at best only a slight tendency) to be corroded by pure water. Corrosion of iron only takes place in water containing dissolved oxygen, discussed in Sections 9.4.3 and S3.6.

When water acts as a reducing agent it is oxidised to O_2. The relevant equation is:

$$2H_2O(l) \rightarrow 4H^+(aq) + O_2(g) + 4e^-$$

This is the reverse of the reduction half-reaction

$$O_2(g) + 4H^+(aq) + 4e^- \rightarrow 2H_2O(l)$$
$$E^\circ = +1.23 \text{ V at pH } 0$$

The presence of the H^+(aq) ions indicates that the water is acidic. The concentration of H^+(aq) in the standard state is 1 molar, and the pH will be 0, equivalent to very acidic conditions. In these conditions, water will be able to reduce redox couples above this reaction in the electrochemical series and liberate O_2. For example, the couple Co^{3+}/Co^{2+} has $E^\circ = +1.82$ V, so that Co^{3+} is reduced by acidified water to give O_2 thus:

$$Co^{3+}(aq) + e^- \rightarrow Co^{2+}(aq)$$
$$O_2(g) + 4H^+(aq) + 4e^- \rightarrow 2H_2O(l)$$

that is,

$$4Co^{3+}(aq) + 2H_2O(l) \rightarrow 4Co^{2+}(aq) + O_2(g)$$
$$+ 4H^+(aq)$$

The reduction potential for neutral water can be calculated via the Nernst equation:

$$O_2(g) + 4H^+(aq) + 4e^- \rightarrow 2H_2O(l),$$
$$E = +0.81 \text{ V at pH } 7$$

There are a number of half-reactions involving nonmetals that lie above this value in the electrochemical series, and these will be reduced. In all of these cases, reaction will produce oxygen gas. For example, the dissolution of chlorine, Cl_2, in water will produce oxygen, although the unstable oxyacid, HOCl, forms as an intermediate. The reaction is:

$$2H_2O(l) \rightarrow 4H^+(aq) + O_2(g) + 4e^-$$
$$2Cl_2(g) + 4e^- \rightarrow 4Cl^-(aq)$$
$$2Cl_2 + 2H_2O \rightarrow 4HCl + O_2(g)$$

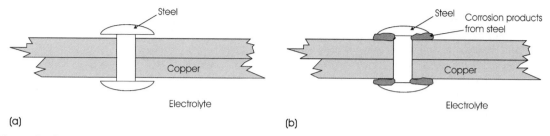

Figure 9.16 Dissimilar metal corrosion of a steel rivet in contact with copper and an electrolyte: (a) before and (b) after corrosion

9.4.2 Dissimilar-metal corrosion

Two different metals that are connected and immersed in an electrolyte form an electrochemical cell. If a current is allowed to flow, one metal will be consumed and one will remain the same or be increased in some way. These processes lead to dissimilar metal corrosion. In order for dissimilar metal corrosion to occur, it is necessary to have an anode, a cathode, an electrolyte and a connection from anode to cathode. The anode component corrodes whereas the cathode remains unattacked. The tendency for such reactions to take place spontaneously can be judged from the electrochemical series. Three examples follow.

9.4.2.1 Copper and iron/steel

Copper and iron or steel in juxtaposition can form a cell in which the copper becomes the cathode and the iron the anode. Several reactions are possible. One of these is as follows.
Anode reaction:

$$Fe(s) \rightarrow Fe^{2+}(aq) + 2e^-$$

Cathode reaction:

$$O_2 + 2H_2O(l) + 4e^- \rightarrow 4(OH^-)$$

Cell reaction:

$$2\,Fe(s) + O_2 + 2H_2O(l) \rightarrow 4(OH^-) + 2Fe^{2+}(aq)$$

The Fe^{2+} is soluble and the iron will gradually dissolve (Figure 9.16). The copper is not attacked and serves only to complete the cell. If there is a small anode area, such as an exposed nail head, the attack is more pronounced.

These cells have had considerable influence historically. Wooden sailing ships were attacked below the waterline by wood-boring barnacles. Severe infestation could ultimately lead to the destruction of the bottom of the hull and catastrophic loss of the vessel. To prevent this, ships were sheathed in copper, a practice that gave rise to the expression 'copper-bottomed', meaning sound or reliable. Unfortunately, iron or steel nails were often used to secure the copper sheathing. In the presence of alkaline water and oxygen (i.e. surface seawater, which has a pH of about 8.5 and a high content of dissolved oxygen) the nails corroded and the copper sheathing was lost.

9.4.2.2 Galvanization

Coating steel sheet with zinc, a procedure called galvanizing, is widely used to prevent corrosion of the steel. The zinc coating does not corrode in air because initial reaction produces a dense layer of insulating zinc oxide that protects the surface from further reaction (as described for aluminium and chromium in Section 8.5.3). Should the zinc coating become penetrated, so that both zinc and steel (iron) are exposed to the air, corrosion is inhibited by the formation of a galvanic cell (Figure 9.17). The reason is that in the presence of water and oxygen the zinc will become the anode in the cell formed and will corrode in preference to the exposed iron. Several reactions are possible, including the following.

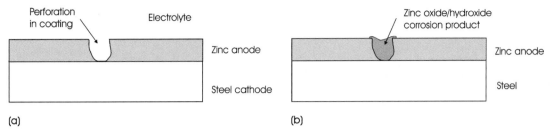

Figure 9.17   The protective coating formed by zinc on steel: galvanisation. (a) Before and (b) after formation of the protective corrosion product

Anode reaction:

$$Zn(s) \rightarrow Zn^{2+} + 2e^-$$

Cathode reaction:

$$O_2 + 2H_2O(l) + 4e^- \rightarrow 4(OH^-)$$

Cell reaction:

$$2\,Zn(s) + O_2 + 2\,H_2O(l) \rightarrow 4(OH^-) + 2\,Zn^{2+}(aq)$$

The Zn^{2+} ions react in ordinary conditions to produce ZnO or a zinc oxyhydroxide. These are inert and form insoluble deposits that help to prevent further corrosion. Overall, steel coated with zinc corrodes far more slowly than does bare steel.

9.4.2.3   Tin plate

Steel coated with tin was widely used on food cans, or 'tins', until replaced by aluminium or plastic coatings. Steel coated with tin corrodes faster than steel alone. Unlike the situation with zinc, a scratch in the coating allows an electrochemical cell to form in which the steel (iron) forms the anode and corrodes as follows.
Anode reaction:

$$Fe(s) \rightarrow Fe^{2+} + 2e^-$$

Cathode reaction:

$$O_2 + 2H_2O(l) + 4e^- \rightarrow 4(OH^-)$$

Cell reaction:

$$2Fe(s) + O_2 + 2H_2O(l) \rightarrow 4(OH^-) + 2Fe^{2+}(aq)$$

No protective oxide forms and the corrosion is enhanced compared with uncoated steel. This accounts for the fact that old 'tin cans' on rubbish tips are always badly corroded.

9.4.3   Single-metal electrochemical corrosion

Two subtle corrosion effects can occur when a single metal is in contact with an electrolyte – differential aeration and crevice corrosion. Differential aeration can cause corrosion when no obvious galvanic cells are in evidence. To illustrate this effect, suppose we have a cell with a copper anode and cathode. If the concentration of the electrolyte and the temperature of each cell compartment is the same, no potential is generated and no corrosion occurs. However, bubble O_2 into the one compartment, which becomes the cathode compartment, and corrosion will occur in the other, which forms the anode compartment. Differential aeration is, in fact, a concentration effect, and can be understood by using the Nernst equation. Electrons will flow from anode to cathode and the anode will corrode.
Anode reaction:

$$Cu(s) \rightarrow Cu^{2+} + 2e^-$$

Cathode reaction:

$$O_2 + 2H_2O + 4e^- \rightarrow 4(OH^-)$$

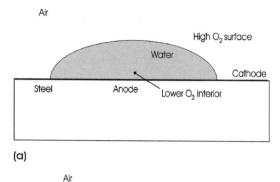

(a)

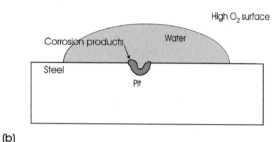

(b)

Figure 9.18 Differential aeration leading to pitting in steel: (a) before and (b) after corrosion

Cell reaction:

$$2Cu(s) + O_2 + 2H_2O(l) \rightarrow 4(OH^-) + 2Cu^{2+}(aq)$$

This type of corrosion can happen within a water drop on steel. The surface contains more dissolved oxygen than the interior of the drop and creates a circular cathode (Figure 9.18). The less-aerated centre forms an anode, and corrosion produces a pit at the centre of the drop.

Anode reaction:

$$Fe(s) \rightarrow Fe^{2+} + 2e^-$$

Cathode reaction:

$$O_2(g) + H_2O(e) + 4e^- \rightarrow 4(OH^-)$$

Cell reaction:

$$2Fe(s) + O_2 + 2H_2O(l) \rightarrow 4(OH^-) + 2Fe^{2+}(aq)$$

Similar corrosion effects can be seen in narrow crevices. For example, narrow channels between a damp steel rivet and a damp plate can receive less oxygen than the surface of either. The crevice becomes anodic, and corrosion may occur (Figure 9.19). This problem is often enhanced when the corrosion product has a high volume. The resulting stress may lever the rivet head off. This effect is termed crevice corrosion.

Corrosion of a single metal can also occur even in the absence of significant differential aeration. This puzzling occurrence is due to the presence of anodic and cathodic regions on the metal. These can be generated during heat treatment and cold working of metals. For example, the regions of a metal subjected to cold working are often anodic compared with the remainder of the material. In contact with an electrolyte these areas will tend to corrode as a result of the formation of a galvanic cell, even though no concentration effects exist.

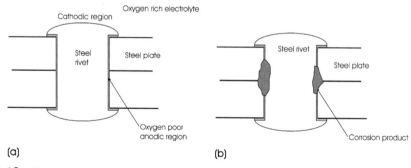

(a)   (b)

Figure 9.19 Corrosion in a crevice as a result of differential aeration: (a) before and (b) after corrosion

9.5   Electrolysis

Whereas a galvanic cell uses a spontaneous chemical reaction to produce an electric current, an electrolytic cell uses an electric current to drive a nonspontaneous chemical reaction. A rechargeable battery thus operates as a galvanic cell when being used and as an electrochemical cell when being charged. The process occurring in an electrochemical cell is called electrolysis. Electrolytic cells are widely used in the preparation of chemicals such as magnesium and aluminium and in electroplating.

9.5.1   Electrolytic cells

Electrolytic cells do not, in general, require the electrodes to be in separate compartments, and so are simpler in construction than are galvanic cells (compare Figures 9.1a,b and 9.20). However, the reactions at the anode (oxidation) and at the cathode (reduction) are identical to those in a galvanic cell. Similarly, during operation, electrons from the external supply enter the cell via the cathode, and leave it via the anode, as in a galvanic cell. Cations in the electrolyte move away from the anode and towards the cathode, whereas anions in the electrolyte move away from the cathode and towards the anode. Unfortunately, the anode of an electrolytic

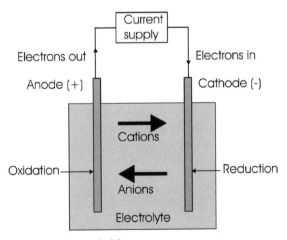

Figure 9.20 An electrolysis cell

cell is labelled + and the cathode −, whereas in a galvanic cell the reverse is true, and the anode is labelled − and the cathode +.

The potential that has to be supplied to make the nonspontaneous reactions occur must be (in principle) the reverse of the potential generated by the spontaneous reaction. That is, the potential must be the potential encountered in the cell reaction, but reversed. In practice, this minimum amount of potential has to be exceeded. The actual amount of extra potential to be supplied, the overpotential, is a function of the electrode materials and surface conditions.

In a situation where the fluid contains several species that could be oxidised or reduced, those requiring the least input of energy will preferentially react. This aspect is explored in more detail below.

9.5.2   Electrolysis of fused salts

Davy electrolysed fused hydroxides of potassium and sodium in 1807, and in this way he discovered the metals potassium and sodium. A number of important metals are still produced by a similar route, the electrolysis of fused salts, including sodium (Na), magnesium (Mg), calcium (Ca) and aluminium (Al). Sodium, magnesium and calcium are all produced via electrolysis of the molten chloride. Taking molten sodium chloride (NaCl), as an example, the reactions are as follows.
Anode reaction:

$$2Cl^-(l) \rightarrow Cl_2(g) + 2e^-$$

Cathode reaction:

$$Na^+(l) + e^- \rightarrow Na(l)$$

Electrolysis reaction:

$$2NaCl(l) \rightarrow 2\,Na(l) + Cl_2(g)$$

In order to lower the temperature of the process, a eutectic mixture of $CaCl_2$ and NaCl is used, which

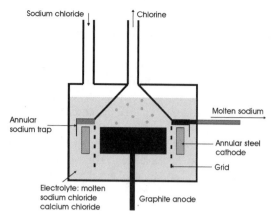

Figure 9.21 The Downs cell for the production of sodium; the cathode is steel and the anode is graphite

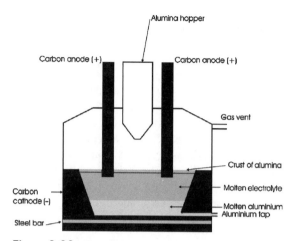

Figure 9.22 The Hall – Hérault cell for aluminium production; both electrodes are made of carbon

considerably reduces the melting point of the solid. The cells employed, Downs cells, use a steel cathode and inert carbon anodes (Figure 9.21). Because of the inclusion of $CaCl_2$, some calcium metal is also released during the operation of the cell. This is returned to the melt, whereas the sodium is recovered.

The electrolytic production of aluminium is more complex. Hall and Héroult developed the current process independently in 1886. In this process, alumina, Al_2O_3, is dissolved in molten sodium aluminium fluoride (Na_3AlF_6) and electrolysed (Figure 9.22). Originally, the aluminium oxide was derived directly from the mineral bauxite, and the sodium aluminium fluoride was used in the form of the naturally occurring mineral cryolite. These days, synthetic cryolite is used, and the aluminium oxide is produced from a wider variety of mineral sources, consisting of aluminium oxide-hydroxides.

The composition of the electrolyte is approximately 80–90 % cryolite. The amount of Al_2O_3 is continuously replenished as the electrolysis continues. The anode, which is consumed in the reaction, is carbon, and the cathode is carbon reinforced with steel bars, contained in a steel 'pot'. Although the fine details of the reactions occurring are still not completely known, the framework of the process is believed to be as follows.

Anode reaction:

$$2Al_2O_3(l) + 3C(s) + 24F^-(l) \rightarrow 4AlF_6^{3-}(l) + 3CO_2(g) + 12e^-$$

Cathode reaction:

$$4AlF_6^{3-}(l) + 12e^- \rightarrow 4Al(l) + 24F^-(l)$$

Electrolysis reaction:

$$2Al_2O_3(l) + 3C(s) \rightarrow 4Al(l) + 3CO_2(g)$$

A lesser reaction,

$$2Al(l) + 3CO_2(g) \rightarrow Al_2O_3(s) + 3CO(g)$$

produces alumina, which forms a protective crust on the molten cryolite.

9.5.3 The electrolytic preparation of titanium by the Fray–Farthing–Chen Cambridge process

The preparation of titanium metal uses a batch process, the Kroll process, which involves the reduction of titanium tetrachloride with magnesium metal. The starting material is usually the ore ilmenite, $FeTiO_3$. This is heated with chlorine and carbon at $900\,°C$ to form titanium tetrachloride, $TiCl_4$:

$$2\,FeTiO_3 + 7Cl_2 + 6C \rightarrow 2TiCl_4 + 2FeCl_2 + 6CO$$

The chloride mixture must then be distilled to separate the $TiCl_4$ from the $FeCl_2$. Titanium tetrachloride is a reactive and corrosive liquid that requires very careful handling at all times. This chemical is reduced with molten magnesium metal in a sealed vessel under argon at $950–1150\,°C$:

$$TiCl_4 + 2Mg \rightarrow Ti + 2MgCl_2$$

The resulting titanium is purified by reaction with acids, including aqua regia (a 1:3 mixture of concentrated nitric and hydrochloric acids). This final product can then be cast into ingots under vacuum. The final titanium metal product is very expensive, not only because of the chemicals involved but also because production is a batch process and not continuous.

Because of the expense of this process, schemes using electrolysis of molten titanium salts, similar to the production of aluminium, have been widely investigated. To date, none of these has worked well. Although metal can be produced in this way, it is often dendritic in form, and very reactive, oxidising on contact with air. The various valence states of titanium found in melts (4+, 3+ and 2+) lower the efficiency of the methods and contribute to unreliable results.

A new process, called the FCC Cambridge Process, uses a slightly different electrochemical approach. The method is named after its discoverers, Fray, Farthing and Chen, working in the University of Cambridge, England. The key to the method, and what distinguishes it from earlier electrolysis attempts, lies in the use of slightly nonstoichiometric titanium dioxide, TiO_2, as the cathode in an electrochemical cell. Titanium dioxide itself is an insulator with a very high relative permittivity. For the rutile form of the oxide, the values are approximately 80 parallel to the a axis and about 137 parallel to the c axis of the tetragonal unit cell. However, removal of oxygen from rutile to

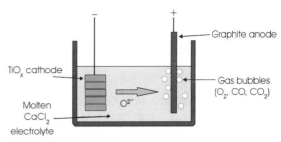

Figure 9.23   The design of the FFC Cambridge cell for the production of titanium metal by the electrolysis of slightly reduced titanium dioxide, TiO_x. Oxygen is transported through the electrolyte, molten $CaCl_2$, and liberated at the graphite anode, as oxygen gas, carbon monoxide and carbon dioxide

give a nonstoichiometric oxide is easy. Moreover, these nonstoichiometric materials are good electronic conductors, and even a composition as close to TiO_2 as $TiO_{1.995}$ conducts electricity well. Thus, the trick is not to dissolve the oxide in a flux, and transport titanium ions, but to make the cathode of pellets of slightly reduced rutile, and transport oxygen ions. The electrolyte is molten $CaCl_2$, and graphite is used as the anode (Figure 9.23). During electrolysis, oxygen is pulled out of the titanium oxide cathode and transported to the graphite anode, where some of it reacts to form carbon oxides and some is released as oxygen. The oxide cathode is gradually converted to pellets of titanium metal in a sponge-like form. This material does not oxidise easily and can be melted and turned into ingots with minimum additional processing.

The cell reaction can be written in a simplified form as follows.

Anode reaction (oxidation, electron generation):

$$xC(s) + xO^{2-}(l) \rightarrow xCO(g) + 2xe^-$$

Cathode reaction (reduction, electron consumption):

$$TiO_x(s) + 2xe^- \rightarrow xO^{2-}(l) + Ti(l)$$

Electrolysis reaction:

$$TiO_x + xC(s) \rightarrow Ti(s) + xCO(g)$$

In reality, carbon monoxide, carbon dioxide and oxygen are produced at the anode. These anode reactions can be approximated by:

$$aC(s) + aO^{2-}(l) \rightarrow aCO(g) + 2ae^-$$
$$bC(s) + 2bO^{2-}(l) \rightarrow bCO_2(g) + 2be^-$$
$$cO^{2-}(l) \rightarrow \frac{c}{2}O_2(g) + 2ce^-$$

where $(a + 2b + c)$ is equal to x in TiO_x; that is, the cell reaction is:

$$TiO_{a+2b+c} + (a+b)C(s) \rightarrow Ti + aCO(g)$$
$$+ bCO_2(g) + \frac{c}{2}O_2(g)$$

This new process works well with a number of other oxides that are difficult to convert into metals by conventional redox methods, including Cr_2O_3, ZrO_2, Nb_2O_5 Ta_2O_5 and WO_3, which can all be produced in an electronically conducting form. Moreover, if the cathode is made of solid solutions or mixed oxides, alloys can be produced directly. Using this technique, simple alloys such as $TiAl_3$ and Ni_3Ti and more complex compounds such as Ti_6Al_4V have been synthesised.

9.5.4 Electrolysis of aqueous solutions

The central equation in the electrolysis of water is

$$2H_2O(l) \rightarrow 2H_2(g) + O_2(g) \quad E = -1.23\,V \text{ at pH}\,7$$

This nonspontaneous reaction is the reverse of that used in the fuel cell. There are two aspects of the reaction to note. First, pure water is a poor conductor of electricity, and dilute solutions, often of acids such as sulphuric acid, H_2SO_4, are used. Second, because of overpotentials, it is necessary for a voltage greater than 2 V to be applied to the electrodes for the reactions to take place. These are as follows.
Anode reaction (oxidation of H_2O; electron generation):

$$2H_2O(l) \rightarrow O_2(g) + 4H^+(aq) + 4e^-$$
$$E = +0.81\,V \text{ at pH}\,7$$

Cathode reaction (reduction of H_2O, electron consumption):

$$2H_2O(l) + 2e^- \rightarrow H_2(g) + 2OH^-(aq)$$
$$E = -0.42\,V \text{ at pH}\,7$$

Electrolysis reaction:

$$2H_2O(l) \rightarrow 2H_2(g) + O_2(g)$$

When other anions and cations are present they can be discharged in preference if the energy requisite is less than that of the reactions above. For example, in the presence of sulphate ions, $SO_4^{-2}(aq)$, these may also give up electrons at the anode and be oxidised. However, this needs a potential of 2.05 V, which is more than the competing water reaction and so does not occur to any extent.

The same principles apply to all other aqueous electrolysis reactions. The two reactions involving the oxidation and reduction of water will always be competitors, and the oxidation or reduction reaction taking place will be that requiring the lowest potential. For example, consider the reduction of brine (concentrated salt solution) (Figure 9.24). The solution contains $Na^+(aq)$ and $Cl^-(aq)$. The two competing cathode reactions are:

$$Na^+(aq) + e^- \rightarrow Na(s) \quad E° = -2.17\,V$$
$$2H_2O(l) + 2e^- \rightarrow H_2(g) + 2OH^-(aq)$$
$$E° = -0.83\,V$$

This indicates that hydrogen gas will be evolved at the cathode rather than sodium metal. The two anode reactions are:

$$2Cl^-(aq) \rightarrow Cl_2(g) + 2e^- \quad E° = +1.36\,V$$
$$2H_2O(l) \rightarrow O_2(g) + 4H^+(aq) + 4e^-$$
$$E° = +1.23\,V$$

It would be expected, therefore, that oxygen gas would be evolved at the anode. However, the overpotential for oxygen production is substantial and in

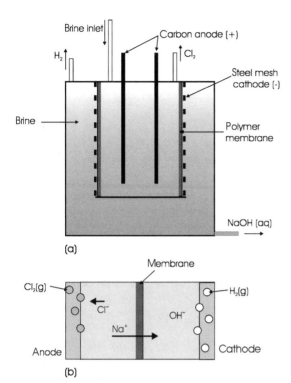

(a)

(b)

Figure 9.24 The chloralkali process: (a) membrane cell: (b) the membrane is permeable only to Na^+ ions, leading to segregation of Cl^- and OH^- in the anode and cathode compartments

fact chlorine production occurs experimentally. The overall electrolysis reaction is:

$$2\,NaCl(aq) + 2\,H_2O(l) \rightarrow 2\,NaOH(aq)$$
$$+ Cl_2(g) + H_2(g)$$

The chloralkali process, which involves the electrolysis of brine, is widely used for the production of sodium hydroxide and chlorine gas. During electrolysis it is necessary to keep the sodium hydroxide separate from the chlorine, to prevent the formation of sodium hypochlorite, NaOCl, and this determines cell design. In older processes, the cathode used was flowing mercury. At this electrode, sodium is formed, and this dissolves in the mercury to form a sodium amalgam. The sodium amalgam is removed continually from the cell and reacted with water to produce hydrogen gas and

caustic soda in a separate reactor. Pollution from the mercury has made this cell unviable, and the electrolysis of brine is now carried out by means of membrane cells.

In a membrane cell, a polymer membrane separates the anode and cathode compartments. This allows the passage of Na^+ ions but blocks the Cl^- and OH^- ions (Figure 9.24). The concentration of $Na^+(aq)$ and $OH^-(aq)$ gradually rises in the cathode region and is continuously removed.

9.5.5 The amount of product produced during electrolysis

The chemical nature of the products of electrolysis is determined by the reduction potential of the appropriate redox couple. The amount of product formed depends only on the amount of electricity that has passed through the cell. This fact was first recognised by Faraday, who formulated what are now known as Faradays laws of electrolysis:

1. The mass of substance produced at an electrode is directly proportional to the quantity of electricity that has passed through the cell;

2. The mass of a substance produced by a given quantity of electricity is directly proportional to the molar mass of the substance and inversely proportional to the numbers of electrons transferred per molecule of the substance.

The amount of electricity that is provided in an electrolysis experiment, Q, is given by:

$$Q = I\,t$$

where the amount of electricity is measured in coulombs, the current, I, is in amperes, and the time, t, in seconds. One mole of electrons has a charge

$$(1.6022 \times 10^{-19}\,C) \times (6.0222 \times 10^{23}\,mol^{-1})$$
$$= 9.6485 \times 10^4\,C\,mol^{-1}$$

This is called the Faraday constant, F. Thus, the quantity of electricity needed to produce one mole of a monovalent element is 9.6485×10^4 C. Double this amount is needed for a divalent element, and so on. The number of moles of electrons provided in any electrolysis experiment, Q_m, is given by

$$Q_m = \frac{I\,t}{F}$$

To obtain the mass of an element that is formed, it is necessary to multiply by the molar mass produced by 1 mole of electrons,

$$m = \left(\frac{I\,t}{F}\right)\left(\frac{M}{z}\right)$$

where m is the mass produced, I is the current (A), t is the time (s), M is the molar mass of the element (g mol^{-1}), and z is the charge on the ion involved.

9.5.6 Electroplating

Electroplating is the deposition of a metallic coating onto a metal object by means of electrolysis. It is widely carried out for decorative purposes and for corrosion prevention. The principles of electroplating do not differ from those given for electrolysis. However, in practice, the production of a high-quality film is critically dependent on a large number of factors, especially the cleanliness of the surface to be plated. In addition, commercial plating solutions also contain organic additives to enhance film adherence.

To illustrate the process, consider a (schematic) description of nickel plating (Figure 9.25). The metal object to be plated is connected to negative input from a direct current (dc) source, so as to form the cathode of the cell. The electrolyte is a solution of a soluble nickel salt in water, $NiCl_2$ for example. The anode of the cell is a rod of nickel metal. The current drives the nickel ions in solution towards the cathode, where they are deposited via the following reaction:

$$Ni^{2+}(aq) + 2e^- \rightarrow Ni(s) \quad E^\circ = +0.23\,V$$

Two moles of electrons must be supplied to deposit 1 mole of nickel. At the anode, the chloride ions are

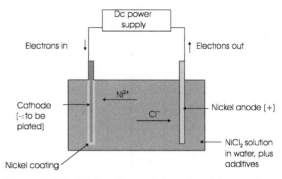

Figure 9.25 Nickel electroplating: the nickel anode is dissolved and transported to the object to be plated, which is the cathode, under the driving force of the external power supply

discharged, as in the chloralkali process:

$$2Cl^-(aq) \rightarrow Cl_2(g) + 2e^- \quad E^\circ = +1.36\,V$$

Chlorine gas is evolved, and the electrons released oxidise the nickel anode, which dissolves in the process:

$$Ni(s) \rightarrow Ni^{2+}(aq) + 2e^-$$

The overall result is the transfer of nickel from the anode to the cathode:

$$Ni(s)\ anode \rightarrow Ni(s)\ cathode$$

It is this same process that causes the corrosion of anodes in all batteries. As mentioned above, the real reactions that take place during electroplating are far more complex than this. As a first approximation these can be written as follows.
Anode reaction:

$$Ni(s) + 2Cl^-(aq) \rightarrow Ni^{2+}(aq) + Cl_2(g) + 4e^-$$

Cathode reaction:

$$2Ni^{2+}(aq) + 4e^- \rightarrow 2Ni(s)$$

Plating reaction:

$$Ni(s) + 2Cl^-(aq) + Ni^{2+}(aq) \rightarrow 2Ni(s) + Cl_2(g)$$

9.6 Pourbaix diagrams

9.6.1 Passivation and corrosion

Many reactions that occur in water are sensitive to acidity (pH), concentration and to the relative oxidising or reducing conditions in the neighbourhood. This is especially true of corrosion. However, corrosion does not always occur, and under certain combinations of acidity and reduction potential iron, copper, zinc and other metals can resist corrosion. This feature is called passivation.

In order to determine whether a metal will corrode, it necessary to write down all the possible half-reactions that can be envisaged and then determine how these will vary with acidity, concentration and oxidation potential. It is tedious to carry out these calculations, and the results are often not especially lucid. The overall scheme of reactivity can, however, be represented graphically. Such diagrams are called Pourbaix diagrams. Although initially formulated to assist in demarcation of corrosion-resistant conditions for metals, they have found applicability in other areas, including electrochemistry, the earth sciences, chemical engineering and metallurgy and in the disposal of hazardous or radioactive waste.

A Pourbaix diagram uses the oxidising/reducing potential and the acidity of the environment as parameters to quantify the reactivity of the system under consideration, usually for aqueous environments. The oxidising capability is plotted on the ordinate (y axis) as a voltage. The use of voltage to express oxidation and reduction is simply an adaptation of the electrochemical series, and the voltage used is that of the half-reaction measured against a standard hydrogen electrode. A table of reduction half-reactions and associated voltages is also a table of relative oxidising and reducing capabilities. The acidity is plotted on the abscissa (x axis) as pH. The area of the diagram is divided up into stability fields, which show where a certain species is stable.

Most corrosion reactions of interest take place in the presence of water, and the area of the diagram in which water is present is indicated. This region, the stability field of water, is defined as the range of pH

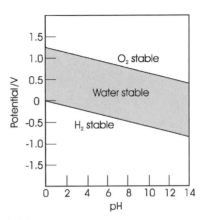

Figure 9.26 A Pourbaix diagram showing the stability field of water (shaded)

and oxidation/reduction potential over which water is stable both to oxidation and to reduction at 25 C and 1 atm pressure (Figure 9.26). Above the upper boundary, water is oxidised to O_2 gas, and below the lower line it is reduced to H_2 gas. The calculation of the stability field of water is given in Section S3.5.

9.6.2 Variable valence states

Transition metals display several valence states. Generally, the valence state that is stable depends on the acidity and the oxidation potential of the environment. Iron (Fe), and its compounds, illustrate these possibilities. Iron is present in the Earth's core as liquid metal, Fe^0. In the mantle, or in reducing conditions in sediments, Fe is present as Fe^{2+} [Fe(II), ferrous]. In oxidising conditions, Fe exists as Fe^{3+} [Fe(III), ferric]. The commonest iron-containing minerals, Fe_2O_3 (haematite, Fe^{3+}), Fe_3O_4 (magnetite, lodestone, Fe^{2+}, Fe^{3+}) and $FeCO_3$ (siderite, Fe^{2+}), reflect the different formation conditions of the minerals in the Earth's crust.

Irrespective of origin, all iron compounds tend to the stable Fe^{3+} state in air. Iron itself corrodes in moist air and reacts with nonoxidising acids to yield H_2 and Fe(II) salts. Fe(II) salts are subsequently oxidised to Fe(III) salts in air. This is a slow reaction in acidic solution and is rapid in a basic

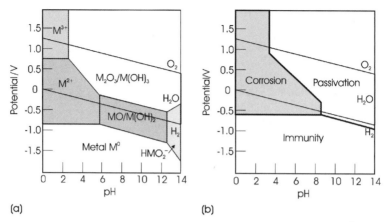

Figure 9.27   (a) Pourbaix diagram for an element, M, with two valence states, M^{2+} and M^{3+}, in water; (b) simplified version of the diagram in part (a), showing the range of conditions over which corrosion, passivation and immunity are likely to occur

solution, when insoluble oxyhydroxides, typically labelled Fe(OH)$_3$, are precipitated.

These confusing relations occur for most transition metals, as well as for the lanthanides and actinides. The various stability regions are most easily understood via the appropriate Pourbaix diagram.

9.6.3   Pourbaix diagram for a metal showing two valence states, M^{2+} and M^{3+}

A schematic Pourbaix diagram for a typical transition metal such as iron, showing two valence states, M^{2+} and M^{3+}, is drawn in Figure 9.27(a). Remember that the boundaries are concentration dependent, and the figure is representative of a typical concentration. [The method of construction of the diagram for the real iron–water–air system is given in Section S3.6.]

The upper left-hand corner of the diagram represents conditions that are oxidising and acidic. Under these conditions the higher valence state, M^{3+}, is the stable form. The region over which this remains true is called the stability field of the M^{3+} ions. [Note that, in reality, the stable species is generally a hydrated ion such as $M(H_2O)_6^{3+}(aq)$.]

As the pH increases, that is, as the acidity decreases, one moves towards the right-hand side

of the diagram. Ultimately, the stable ion M^{3+} is replaced by the oxide, M_2O_3, or the hydroxide, $M(OH)_3$, as the stable species, and a new stability field is entered. The half-reaction is:

$$M^{3+}(aq) + 3H_2O(l) \rightarrow M(OH)_3(s) + 3H^+(aq)$$

The change in acidity is revealed by the formation of $H^+(aq)$, but the oxidation state of the M^{3+} valence has not changed during this transformation, and the metal is trivalent in both M_2O_3 and $M(OH)_3$. [Once again, note that in real systems the material that forms in solution is often an ill-defined oxyhydroxide and not a simple compound.] Somewhere between the high-pH and low-pH regions a boundary exists that separates the two stability fields. The boundary on Figure 9.27 (a) is a vertical line. The boundary therefore indicates the pH at which a precipitate of hydroxide would be expected to form as the acidity of the environment changes. *Vertical boundaries separate stability fields that involve a change in acidity (H^+ or OH^- concentration) and no change in oxidation state (e^- transfer).*

Return to the upper left-hand corner of Figure 9.27(a), the M^{3+} stability field. As the oxidising power of the environment decreases, that is, as the voltage on the ordinate decreases, one moves towards the bottom left-hand side of the diagram, representing more reducing conditions. Ultimately, the lower valence state, M^{2+}, becomes the stable species. The half-reaction is:

$$M^{3+} + e^- \rightarrow M^{2+}$$

The change in oxidation state is revealed by the half-reaction, which also confirms that there is no change in acidity. At some point, a boundary between the stability field of M^{3+} and M^{2+} is crossed. The boundary between the two stability fields is horizontal. *Horizontal boundaries separate stability fields that involve a change in oxidation state (e^- transfer) and no change in acidity (H^+ or OH^- concentration).*

A continued reduction in the oxidising potential, moving further towards the lower left-hand side of the diagram, causes the M^{2+} ion to be replaced by more stable metal, M^0. The boundary between the stability fields for M^{2+} and M^0 is horizontal, for the reason given above.

Return to the lower part of the M^{2+} stability field and consider the consequence of decreasing the acidity, that is, increasing the pH. Ultimately, the M^{2+} stability field gives way to one in which the oxide MO or the hydroxide $M(OH)_2$ are preferred. The boundary between the stability fields is vertical as no change in oxidation state is involved.

Repeat this in the upper part of the M^{2+} stability field. The conditions then correspond to a decrease in acidity under oxidising conditions. The M^{2+} stability field will now give way to a field in which either oxide, M_2O_3, or hydroxide, $M(OH)_3$, are preferred. In this case, the stability field boundary marks a change in both oxidation state and acidity, and the stability field boundary is sloping. For example, when M^{2+} produces a trivalent hydroxide, $M(OH)_3$, the half-reaction is:

$$2M^{2+}(aq) + 6H_2O \rightarrow 2M(OH)_3 + 6H^+(aq) + 2e^-$$

The redox nature of the reaction is revealed by the production of electrons, and the acid–base nature of the reaction by the production of H^+. *Sloping boundaries separate stability fields that involve a change in acidity (H^+ or OH^- concentration) and a change in oxidation state (e^- transfer).*

The sloping nature of the boundary shows that the formation of a precipitate will depend on the local pH and oxidising power. Small changes can make a large difference in whether an ion will stay in solution or transform into a solid. These considerations are important in whether dangerous metals can dissolve and spread out from spoil tips or radioactive waste dumps.

One other stability field is mapped out, the acid ion HMO_2^-. This is included as illustrative of the complex species that can form. Although they do not usually play a part in normal environmental concerns, they are relevant to reactions in unusual or extreme conditions.

9.6.4 Pourbaix diagram displaying tendency for corrosion

Figure 9.27(a) is redrawn in Figure 9.27(b) with the stability fields labelled differently. There are no chemical species indicated but instead the tendency for corrosion is indicated. Corrosion is likely when the metal is in an environment corresponding to the stability fields in which aqueous M^{3+} and M^{2+} ions are stable. The regions in which a solid occurs are less likely to corrode extensively as the initial formation of a precipitate will prevent further corrosion from taking place. In these stability fields the metal is passivated. Finally, the region in which metal is stable is labelled as immune to corrosion. The range of oxidation and pH conditions under which the metal would be immune to corrosion, passivated or corrode are clearly distinguished.

9.6.5 Limitations of Pourbaix diagrams

The positions of the boundaries to the phase fields are concentration-dependent, and diagrams are usually constructed with concentrations that are relevant to the problems under consideration. It is important to keep this in mind when using these diagrams. Moreover, they are derived by means of equilibrium thermodynamic data and are only as accurate as the available data. This is of high quality for well-known systems such as iron–water–air, but for some systems involving radioactive materials the data are less accurate. In addition, these diagrams do not consider any kinetic or crystallographic aspects. For example, both glass and diamonds are thermodynamically unstable but are of importance for all that. Unstable reaction products may be of considerable importance in corrosion. Moreover, kinetic factors such as the flow rate of any solutions, or changes in temperature, are not catered for. Similarly, the definition of passivation must be treated with caution. An important feature of a passive film is that it is coherent with the metal and does not crack or contain pores. The Pilling–Bedworth ratio (Section 8.5.3) is an attempt to determine if a film is likely to fulfil the requirements of passivation.

Answers to introductory questions

What is an electrochemical cell?

An electrochemical cell is a device for the conversion of electrical to chemical energy, and vice versa, by way of redox reactions. Electrochemical cells consist of two metal electrodes (an anode and a cathode) in contact with an electrolyte that is able to conduct ions but not electrons.

There are two types of electrochemical cell. A galvanic cell uses a spontaneous chemical reaction to produce an external electric current. Galvanic cells are called 'batteries' in colloquial speech. Electrolytic cells, employed for electrolysis and electroplating, use external electrical power to force nonspontaneous chemical reactions to take place.

What are the electrode materials in nickel–metal-hydride batteries?

These batteries, frequently called Ni–MH batteries, rely on hydrogen storage in a metal alloy for the reversible operation. The positive electrode, the cathode, does not consist of nickel. It consists of a complex mixture of $NiO(OH)/Ni(OH)_3$ and $Ni(OH)_2$.

The negative electrode, the anode, consists of an alloy that supports metal hydride formation. Many metals absorb large quantities of hydrogen to form nonstoichiometric interstitial alloys. In many of these materials the hydrogen incorporation is reversible, so that hydrogen is taken up at high hydrogen pressure and released at lower pressure. This reversible uptake of hydrogen is exploited in nickel–metal-hydride batteries. The alloy most commonly used is derived from $LaNi_5$, in which a mixture of other lanthanides replaces the lanthanum, and a nickel-rich alloy replaces the nickel, to give a general formula LnM_5. The anode is composed of an agglomeration of alloy powder. A small amount of potassium hydroxide is added as an electrolyte. The cell voltage is 1.3 V, and the approximate cell reactions are as follows.

Anode reaction (oxidation, electron generation):

$$MH_x(s) + OH^-(aq) \rightarrow MH_{x-1}(s) + H_2O(aq) + e^-$$

Cathode reaction (reduction, electron consumption):

$$2NiO(OH)(s) + H_2O(l) + e^- \rightarrow Ni(OH)_2(s) + OH^-(aq)$$

Cell discharge reaction:

$$MH_x(s) + 2NiO(OH)(s) \rightarrow MH_{x-1}(s) + Ni(OH)_2(s)$$

The charging reaction is the reverse of the discharge reaction, driven by an external voltage.

What information is contained in a Pourbaix diagram?

Many reactions that occur in water are sensitive to acidity (pH), and to the relative oxidising or reducing conditions in the neighbourhood. This is especially true of corrosion.

In order to determine whether a metal will corrode, it necessary to write down all of the possible half-reactions that can be envisaged and then determine how these will vary with acidity, concentration and oxidation potential. It is tedious to carry out these calculations and the results are often not especially lucid. The overall scheme of reactivity can be represented graphically on Pourbaix diagrams. Although initially formulated to assist in demarcation of corrosion-resistant conditions for metals, they have found applicability in other areas, including electrochemistry, the earth sciences, chemical engineering and metallurgy and in the disposal of hazardous or radioactive wastes.

The raw information on a Pourbaix diagram is essentially the oxidising/reducing potential, plotted along the y axis, and the acidity of the environment, plotted along the x axis. The area of the diagram is divided up into stability fields, which show the oxidation and acidity limits between which a particular chemical species is stable. Because most of the reactions of interest take place in the presence of water, the area of the diagram in which water remains stable is indicated. This region, the stability field of water, is defined as the range of pH and oxidation/reduction potential over which water is stable to both oxidation and reduction. Above the upper boundary water is oxidised to O_2 gas, and below the lower line it is reduced to H_2 gas.

The stability fields are labelled in a fashion determined by the application of the diagram. For geochemical applications, these are usually ionic species or minerals. For the purposes of corrosion science, the fields are labelled corrosion, passivation or inert, depending on which of these possibilities is judged as the most likely to occur.

Further reading

D.A. Brookins, 1998, *Eh – pH Diagrams for Geochemistry*, Springer-Berlin.

R.M. Dell, D.A.J. Rand, 2001, *Understanding Batteries*, Royal Society of Chemistry, Cambridge.

D.A. McQuarrie, D.A. Rock, 1991, *General Chemistry*, 3rd edn, Freeman, New York, Ch. 21.

D.R. Sadoway, A.M. Mayes, 2002, 'Portable Power: Advanced Rechargable Lithium Batteries', *Materials Research Society Bulletin* **27** 590.

D.F Shriver, P.W. Atkins, C.H. Langford, 1994, *Inorganic Chemistry*, 2nd edn, Oxford University Press, Oxford, Ch. 7.

K.R. Trethewey, J. Chamberlain, 1995, *Corrosion*, 2nd edn, Longman, Harlow, Essex.

S.C. Singhal, 2002, 'Science and Technology of Solid-oxide Fuel Cells', *Materials Research Society Bulletin* **25** (March), 16.

Problems and exercises

Quick quiz

1 Oxidation is equivalent to:
   (a) Electron gain
   (b) Electron loss
   (c) Electron transfer

2 During reduction, the oxidation number of the species being reduced
   (a) Increases
   (b) Decreases
   (c) Does not change

3 A redox reaction is one in which:
   (a) Oxidation *and* reduction occurs
   (b) Oxidation *or* reduction occurs
   (c) Oxygen takes part

4 A galvanic cell uses:
   (a) An external power supply to cause a chemical change
   (b) A battery to cause chemical change
   (c) A spontaneous chemical reaction to produce an external electric current

5 An electrolytic cell uses:
   (a) A spontaneous chemical reaction to produce an external electric current
   (b) An external power supply to cause a chemical change
   (c) A battery to cause chemical change

6 In a galvanic cell the anode:
   (a) Is the negative terminal
   (b) Is the positive terminal
   (c) Links the cell compartments

7 Batteries are examples of:
   (a) Electrolytic cells

(b) Corrosion cells

(c) Galvanic cells

8 In a battery, oxidation takes place at the:
(a) Cathode

(b) Anode

(c) Neither

9 A hydrogen electrode can be:
(a) The anode of a cell

(b) The cathode of a cell

(c) Either the cathode or the anode

10 The tendency for reduction of an element is given by:
(a) The electrochemical series

(b) The standard cell potential

(c) The hydrogen electrode

11 The couple *higher* in the electrochemical series:
(a) Forms the cathode

(b) Forms the anode

(c) Sometimes forms the cathode and sometimes the anode

12 The cell potential is a measure of:
(a) The free energy change of the cell reaction compared with the hydrogen electrode

(b) The free energy of the anode reaction

(c) The free energy of the cell reaction

13 The Nernst equation describes:
(a) The free energy of a galvanic cell

(b) The variation of the potential of a galvanic cell with concentration

(c) The reaction equation of a galvanic cell

14 pH meters are based on the operating principles of:
(a) Batteries

(b) Galvanic cells

(c) Electrolytic cells

15 A rechargable battery is called:
(a) A primary cell

(b) A secondary cell

(c) A fuel cell

16 In an alkaline dry cell the zinc is:
(a) Oxidised and forms the anode

(b) Reduced and forms the cathode

(c) Is neutral and forms the container

17 In a lithium primary cell the lithium is:
(a) Oxidised and forms the anode

(b) Reduced and forms the cathode

(c) Is neutral and forms the container

18 In a Ni–MH battery the anode is formed by:
(a) Nickel

(b) Nickel hydroxides

(c) A metal alloy hydride

19 When a metal reacts with an acid it is:
(a) Reduced

(b) Oxidised

(c) Neither oxidised nor reduced, simply dissolved

20 When a metal reacts with water it is:
(a) Oxidised

(b) Reduced

(c) Neither oxidised nor reduced, simply corroded

21 During dissimilar metal corrosion, the metal that corrodes is:
(a) The anode

(b) The cathode

(c) Neither

22 During electrolysis for the production of a metal, the metal is produced:
(a) At the anode

(b) At the cathode

(c) Between the anode and cathode

23 The amount of chemical produced during electrolysis is governed by:
(a) The voltage applied

(b) The concentration of the reactants

(c) The amount of electricity passed

24 A Pourbaix diagram plots:
(a) Oxidation potential against free energy

(b) Oxidation potential against temperature

(c) Oxidation potential against pH

25 A Pourbaix diagram does *not* give information about:

(a) The corrosion resistance of a metal

(b) The rate of corrosion of a metal

(c) The solubility of a metal

Calculations and Questions

9.1 Classify the following reactions as oxidation (ox), reduction (red.) or redox reactions. The equations are representative and not balanced.
(a) $Fe_3O_4 \rightarrow Fe_2O_3$.
(b) $O_2(g) \rightarrow 2O^{-2}$.
(c) $Zn(s) + Cu^{2+}(aq) \rightarrow Zn^{2+}(aq) + Cu(s)$.
(d) $Br_2(l) \rightarrow Br^-(aq)$.
(e) $Mg(s) + Cl_2(g) \rightarrow MgCl_2(s)$.
(f) $Fe_2O_3(s) \rightarrow 2Fe(l)$.

9.2 Classify the following reactions as oxidation (ox.), reduction (red.) or redox reactions. The equations are representative and not balanced.
(a) $Co_3O_4 \rightarrow Co_2O_3$.
(b) $CuO \rightarrow Cu$.
(c) $Zn(s) + H^{2+}(aq) \rightarrow Zn^{2+}(aq) + H_2(g)$.
(d) $N_2(l) \rightarrow 2N^{3-}(aq)$.
(e) $2Ca(s) + O_2(g) \rightarrow 2CaO(s)$.
(f) $Al_2O_3(s) \rightarrow 2Al(l)$.

9.3 A Volta pile is made with six silver and six zinc discs.
(a) Which metal forms the anode and which the cathode?
(b) Write the anode reaction, the cathode reaction and the cell reaction.
(c) Determine the voltage of the pile.

9.4 What is the standard reaction free energy for the cell reaction of a Daniel cell?

9.5 Estimate the value of RT/nF for monovalent, divalent and trivalent ions at 27 °C.

9.6 (a) Write the cathode and anode reactions and the overall cell reaction for a cell with nickel and zinc electrodes.

(b) Determine the standard cell voltage.

(c) Calculate the cell voltage if the concentrations of the ions in solution are: Zn^{2+}, $0.016 \, mol \, dm^{-3}$, Ni^{2+}, $0.087 \, mol \, dm^{-3}$.

9.7 Determine the voltage of the cell in Question 9.6 if the cell is operated at 50 °C.

9.8 A cell constructed with a hydrogen electrode is represented by:

$Pt|H_2(g)|H^+(aq)||Cu^{2+}(aq)|Cu \quad E^\circ = 0.34 \, V$

(a) Write the anode reaction, the cathode reaction and the overall cell reaction.

(b) Derive an expression for the variation of the cell voltage with the pH of the acid solution and the Cu^{2+} concentration.

(c) A cell constructed with a Cu^{2+} concentration of $1 \, mol \, dm^{-3}$ and a hydrogen pressure of 1 atm has a voltage of 0.855 V. Estimate the pH of the acid solution.

9.9 For the cell with an overall cell reaction:

$$Zn(s) + Fe^{2+}(aq) \rightarrow Zn^{2+}(aq) + Fe(s)$$

(a) Write the anode and cathode reactions.

(b) Determine the standard cell potential of the cell reaction.

(c) Determine the reaction free energy of the cell reaction.

9.10 What will the voltage of the cell in the Question 9.9 be (a) if the Fe^{2+} ion concentration is changed to $0.35 \, mol \, dm^{-3}$ and (b), if the temperature of the cell is subsequently raised to 35 °C?

9.11 Derive the equation for the pH meter using a calomel electrode. This is a particularly stable electrode, which uses the redox couple Hg_2Cl_2/Hg, Cl^-. The cell is:

$Pt|H_2(g)|H^+(aq)||Cl^-(aq)|Hg_2Cl_2|Hg(l)$
$E^\circ = +0.27 \, V$

The cell reaction is:

$$Hg_2Cl_2(s) + H_2(g) \rightarrow 2H^+(aq) + 2Cl^-(aq) + 2Hg(l)$$

The calomel half-reaction is:

$$Hg_2Cl_2(s) + 2e^- \rightarrow 2Hg(l) + 2Cl^-(aq)$$
$$E^\circ = 0.27\,V$$

9.12   A voltammeter connected to the cell in Question 9.11 was calibrated with a buffer solution of pH 7.0 and showed a voltage of 0.12 V. What is the pH of a solution that gives a voltage of (a) 0.195 V, (b) 0.48 V?

9.13   A cell of the type in Question 9.11 is made up with a $Cl^-(aq)$ concentration of 0.5 mol dm^{-3}. The cell voltage is 0.48 V. Determine the pH of the $H^+(aq)$ component.

9.14 E° for for the couple NiO(OH)/Ni(OH)$_2$ is +0.49 V. Determine the standard reduction potential for the anode reaction

$$MH_x(s) + OH^-(aq) \rightarrow MH_{x-1}$$

in a Ni–MH cell, knowing that the cell voltage is 1.3 V.

9.15   Three clean steel nails are treated in the following ways: (a) completely immersed in tap water; (b) completely immersed in boiled distilled water; (c) partly immersed in tap water. The results found after several days are: for treatment (a), rust present, especially on the head and point; for treatment (b), little rust present; for treatment (c), dense ring of rust at the waterline and pitting just below the surface of the water. Explain these findings [see P.J. Guichelaar and M.W. Williams, 1990. 'A Simple Demonstration of Corrosion Cells', *Journal of Material Education* **12** 331].

9.16   A titanium container is to be used to store nuclear waste in mine conditions that are damp and broadly reducing so that corrosion will tend to produce $Ti^{2+}(aq)$. If the container lid is fixed with steel bolts, should these be plated and, if so, what metals might be considered?

9.17   Buried metal pipes can be protected from corrosion by connecting them to blocks of metal such as magnesium, called sacrificial anodes, which corrode in preference to the pipe. A cell made from the redox couples Fe/Fe^{2+} and Mg/Mg^{2+} is a laboratory representation of a sacrificial anode. (a) Draw the cell diagram [not shown in the answers at the end of this book], (b) write the cell half-reactions and the cell reaction and (c) calculate the standard cell voltage.

9.18   Explain how pitting can occur under a water droplet on a steel plate. [Note: answer is not provided at the end of this book.]

9.19   Water coming from a waste tip contains (a) Cu^{2+}; (b) Zn^{2+}; (c) Pb^{2+} and (d) Sn^{2+} ions in solution. What will happen when these come into contact with a steel pipe? Write the half-reactions and the overall reactions expected.

9.20   (a) Write an equation for the electrolysis of molten MgCl$_2$ to produce Mg metal and Cl$_2$ gas. (b) What mass of each of these elements is produced if a current of 30 A is passed for 3 hours through the cell?

9.21   The simplified equation for the production of aluminium by the Hall–Héroult process is:

$$2Al_2O_3(l) + 3C(s) \rightarrow 4Al(l) + 3CO_2(g)$$

What quantity of electricity must be passed to produce (a) 2 moles and (b) 2 kg of aluminium? (c) If a current of 25 A is used, how long will it take to produce these amounts?

9.22   A solution of a metal sulphate, MSO$_4$, is used to electroplate an object. The plating bath is operated for 2 hours at a current of 0.5 A. If 1.095 g of metal plates out, determine the molar mass of the metal and identify it.

9.23   On a Pourbaix diagram,
(a) What general types of equilibria apply to horizontal lines?
(b) What general types of equilibria apply to vertical lines?
(c) What general types of equilibria apply to sloping lines?

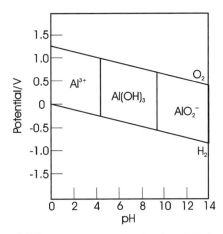

Figure 9.28 Pourbaix diagram for the Al–H$_2$O system within the water stability field, for Question 9.24

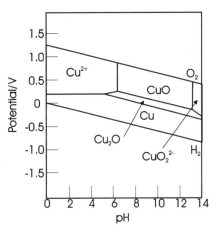

Figure 9.29 Pourbaix diagram for the Cu–H$_2$O system within the water stability field, for Question 9.25

[Note: answer is not provided at the end of this book.]

9.24 Figure 9.28 shows the Pourbaix diagram for the Al – H$_2$O system within the water stability field, for an Al^{3+} concentration of 1×10^{-6} mol dm^{-3}.
  (a) What oxidation states of Al exist?
  (b) Why are there no sloping boundaries on this diagram?
  (c) Will Al be soluble in water with an acidity of pH 3?
  (d) What species will exist in the surface waters of lakes and streams (pH $\sim$ 7, $E \sim 0.6$ V)?
  (e) Over what pH range is Al passivated?

9.25 Figure 9.29 shows the Pourbaix diagram for the Cu – H$_2$O system within the water stability field, for a Cu^{2+} concentration of 1×10^{-6} mol dm^{-3}.
  (a) What oxidation states of Cu exist?
  (b) Under what conditions will copper be soluble in water of pH 3?
  (c) Label the diagram to show the regions where copper corrodes, is passivated and is immune to corrosion.
  (d) What equilibria do the two vertical lines on the diagram represent?

  (e) What equilibrium does the horizontal line represent?
  (f) What equilibrium does the sloping line between the copper and Cu$_2$O stability fields represent?
  (g) What equilibrium does the sloping line between the Cu$_2$O and CuO stability fields represent?

9.26 The water types met with in nature are:
  (a) fresh surface waters, pH $\sim$ 7, $E \sim 0.6$ V;
  (b) organic-rich fresh water, pH $\sim$ 5, $E \sim 0$ V;
  (c) organic-rich waterlogged soil, pH $\sim$ 4.5, $E \sim -0.1$ V;
  (d) acid bog water, pH $\sim$ 3, $E \sim 0.1$ V;
  (e) fresh water polluted by mine drainage, pH $\sim$ 3, $E \sim 0.8$ V;
  (f) surface ocean water, pH $\sim$ 8, $E \sim 0.55$ V;
  (g) organic-rich ocean water, pH $\sim$ 9, $E \sim -0.4$ V

Sketch these onto the stability field of water on the Pourbaix diagram of Figure 9.29 [sketch not shown at the end of this book.]. What copper species will be present in each of these?

PART 4

Physical properties

10

Mechanical properties of solids

- How are stress and strain defined?

- Why are alloys stronger than pure metals?

- What are solid lubricants?

Cars are not made of glass, and bottles are not made of steel, although cars and bottles may be made from polymers. The reasons for this are mostly bound up with the mechanical properties of the materials. These are controlled by two interacting features of the solid, the strength of the chemical bonds in the material, and the defects present.

The effects of variation in chemical bond strength are displayed by the mechanical properties of a crystal structure. For example, the clays and the micas are sheet silicates. The structures are built from sheets of atoms linked by strong chemical bonds. These sheets are linked into stacks by weak chemical bonds. This causes a marked anisotropy in behaviour, in which the strength of the solid in the plane of the sheets is quite different from the strength perpendicular to the sheets. The slightly greasy feel of many of these types of mineral is due to the fact that when some of the solid is rubbed between finger and thumb the weakly bound sheets slide over each other, creating stacking defects. Such materials are exploited as solid lubricants.

The strongest minerals are those with no weak bonds, but contain infinite three-dimensional networks of $[SiO_4]$ tetrahedra, linked to each other by all four vertices. Despite this strength, they often shatter and are brittle. Metals, in contrast, although weaker in some ways, can be hammered into sheets (are malleable) or drawn into wires (are ductile) without breaking.

Chemical bonding, however, is rarely directly responsible for the observed mechanical properties of a material. In fact, most solids are far weaker than the strength predicted on the basis of the chemical bonds present. In these cases, the defects present, especially dislocations, stacking defects and grain boundaries, control matters.

Materials reveal their mechanical properties when subjected to forces. The application of a force results in a deformation. The amount of deformation will depend on the magnitude of the force and its direction measured with respect to the crystallographic axes. Both force and deformation are vector quantities. In the discussion below, it will be assumed that all materials are isotropic in this respect and that there is no crystallographic relationship between force and deformation, which are both presumed to be scalars (numbers, See section S4.13). In fact, in much of the discussion, especially of the elastic properties of solids, the atomic structure is ignored, and the solids are treated as if they were continuous. This viewpoint cannot explain plastic deformation, and knowledge of the crystal structure of the solid is needed to understand the

Understanding solids: the science of materials. Richard J. D. Tilley
© 2004 John Wiley & Sons, Ltd ISBNs: 0 470 85275 5 (Hbk) 0 470 85276 3 (Pbk)

role of dislocations. In the final section, concerned with properties on a nanoscale, knowledge of the microstructure of the solid at atomic dimensions is essential.

10.1 Deformation

10.1.1 Strength

Everyone has a subjective idea of strength, and some materials, such as steel, are universally regarded as strong, whereas others, such as plastics, are considered weak. However, the strength of a material will depend exactly on how it is evaluated. A reliable measure of the strength of a solid is the amount of force that can be applied to it before it breaks (Figure 10.1).

- A material that is stretched is in tension, and suffers tensile forces (Figure 10.1a). The tensile strength of metals and polymer fibres is usually high.

- A material that is squeezed is in compression, and suffers compressive forces (Figure 10.1b). Compressive strength needs to be high in building materials that have to bear heavy loads.

- A material that is subjected to opposed forces is said to be sheared, and suffers shear forces

(Fig. 10.1c). Many polymers behave like very viscous liquids and have very low shear strength.

- A material that is twisted is subjected to a torsional load (Figure 10.1d). Torsional strength is important for shafts that transmit rotation.

- A solid that is flexed or bent; (Figure 10.1e) is subject to both tension and compression, and the flexural strength gives a measure of the amount of bending that an object can sustain without fracture.

- Finally, the impact strength of a solid measures the resistance to a sudden blow (Figure 10.1f). Glass has low impact strength, whereas wooden bats have high impact strength.

A component may experience all of these forces and resultant deformations at the same time, but here each will be treated separately.

The interdependence between the forces applied and the deformations that are produced are summarised by a number of moduli, of which Young's modulus, also called the elastic modulus, is the best known (see Section 10.1.4 and Section S4.1).

10.1.2 Stress and strain

The force (often called the load) applied to an object is defined in terms of the stress on the object. Stress is measured as the force applied to a unit area of the

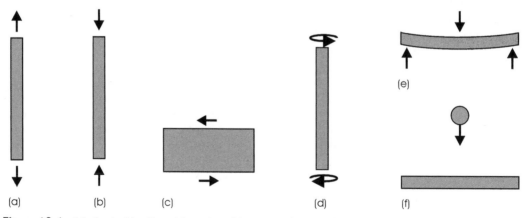

Figure 10.1 Mechanical loading: (a) tension, (b) compression, (c) shear, (d) torsion, (e) bend and (f) impact

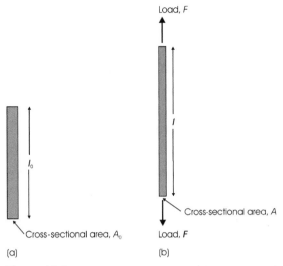

Load, F

l_0

Cross-sectional area, A_0

Cross-sectional area, A

Load, F

l

(a)   (b)

Figure 10.2 A rod in tension: (a) initial state and (b) final state

specimen. The application of a stress results in a dimensional change, which is called the strain.

For a rod-shaped specimen (Figure 10.2), used for the evaluation of metal or polymer samples:

$$\sigma_T = \frac{F}{A}$$

where σ_T is the true stress, F is the force (or load) applied to the rod, and A is the cross-sectional area subjected to the force. The force (or load) is measured in newtons, N, and the area is measured in m^2. Stress is measured in pascals, Pa (N m^{-2}), commonly cited as MPa $= 10^6$ N m^{-2}, or GPa $= 10^9$ N m^{-2}.

For practical purposes, it is often adequate to ignore the continuous change in cross-sectional area that occurs when a force is applied. The stress so defined is called the engineering (or nominal) stress, σ:

$$\sigma = \frac{F}{A_0}$$

where F is the average force (or load) applied, and A_0 is the initial cross-sectional area of the sample.

The elongation of the rod when subjected to a force is equivalent to the strain. The increment in

tensile strain experienced, $\Delta \varepsilon_T$, when a rod is extended, is defined as the ratio of the increase in length, Δl, to the total length:

$$\Delta \varepsilon_T = \frac{\Delta l}{l}$$

The total true strain is then given by:

$$\varepsilon_T = \int_{l_0}^{l} \frac{\mathrm{d}l}{l_0}$$
$$= \ln\left[\frac{l}{l_0}\right]$$

where l is the final length of the specimen, and l_0 is the original length. As strain is a ratio, it has no units. If the incremental changes are ignored, the engineering (or nominal) strain is:

$$\varepsilon = \frac{l - l_0}{l_0}$$

where l is the final length of the specimen, and l_0 is the original length of the specimen. In the stress–strain diagrams for metals and polymers that follow, engineering stress and engineering strain are plotted.

Many materials are used under compression rather than tension. At low loads, the compressed material behaves in a similar way to materials tested under tension. In compression tests, the value of the force is taken as negative and hence we have negative values of stress and strain compared with those obtained in tension.

10.1.3 Stress–strain curves

A great deal can be learned about the mechanical properties of materials by stressing them until they fracture or break. The most common mechanical test involving metals or polymers is the tensile test, in which a sample of the solid is stretched. The test uses a standard test piece with a shape dependent on the material to be tested. Metals usually have a central cylindrical section, of known gauge length,

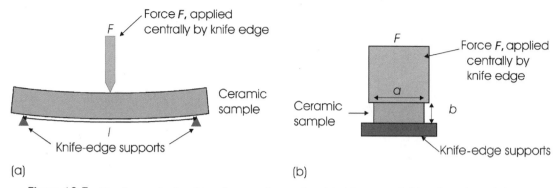

Figure 10.3   The three-point bend test for ceramic samples: (a) sideways and (b) end-on view of the test

usually 50 mm, on which the measurements are made. Polymer samples tend to be sections cut from plates and are larger in dimension. During the test, the instrument applies a force to the sample at a constant rate and simultaneously records the change in dimensions of the test piece.

The stress and strain relationships for a ceramic specimen are more often determined by bending a bar, plate or cylinder of material (Figure 10.3). In this test, the lower part of the ceramic is under tension, and the upper surface is under compression. As ceramic materials are generally much stronger in compression, failure is initiated on the surface under tension. The maximum stress in the upper surface of a deformed sample, σ^{max}, is given by:

$$\sigma_T^{max} = \frac{3\,F\,l}{2\,a\,b^2}$$

where F is the applied force and l is the arc length of the deformed sample between the supports, with a bar with width a normal to the force and thickness b parallel to the force (Figure 10.3). The strain corresponding to the maximum stress is related to the maximum deflection, δ, at the centre of the bar by

$$\varepsilon_T^{max} = \frac{6\,\delta\,b}{l}$$

The cross-sectional area of the ceramic specimen is not greatly altered during the test, so that the true stress is measured. In testing a ceramic, the force or load is slowly increased until fracture.

A plot of load against extension, stress against strain or, more commonly, engineering stress against engineering strain, gives a good picture of the mechanical behaviour of the solid in question (Figure 10.4). The behaviour of brittle materials such as ceramics, brittle metals such as cast iron, or polymers that are chilled to well below the glass transition temperature is drawn in Figure 10.4(a). In these materials, the stress is usually directly proportional to the strain over all or most of the range up to fracture. Metals initially show a similar linear relationship, but the plot ultimately curves and extensive deformation occurs before fracture (Figure 10.4b). This type of curve is typical of a ductile solid. The curves for most polymers are very temperature-sensitive. Thermoplastic polymers above the glass transition temperature give rise to a plot that curves in the opposite way from that of a ductile metal (Figure 10.4c). Elastomers (Figure 10.4d) deform at far lower stress levels than other materials.

The linear part of the stress–strain curve is the elastic region. Here, removal of the load will allow the solid to return to its original dimensions, quite reversibly. In the case of elastomers, this reversibility is maintained over the whole of the stress–strain curve.

For all other solids, once the elastic region is passed, the deformation of the solid is not reversed when the stress is removed, and some degree of permanent deformation remains. This is called plastic deformation. For metals, the point at which elastic behaviour changes to plastic behaviour is

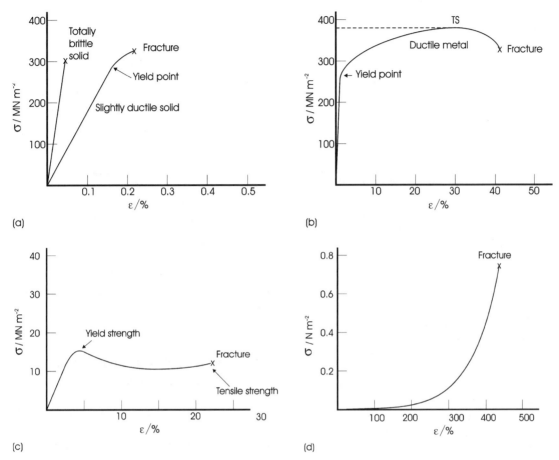

Figure 10.4 Schematic engineering stress–engineering strain (σ–ε) curves for (a) brittle and slightly ductile solids, (b) ductile metals, (c) a typical polymer, and (d) rubber, an elastomer. Note the different stress scale in part (d); point x represents fracture of the specimen; point TS is the ultimate tensile strength

called the yield point. This occurs at a value of stress called the yield stress. In the case of slightly ductile materials (Figure 10.4a), only a small amount of plastic deformation occurs before the material breaks into two. For a ductile metal, a large amount of deformation is possible before fracture. The maximum load that can be sustained, corresponding to point TS in Figure 10.4b, is called the tensile strength (or ultimate tensile strength) of the metal. For a polymer, once the elastic region is passed, almost no increase in stress will bring about a large amount of plastic deformation (Figure 10.4c). Anyone who has carried an overloaded

plastic bag for any distance will have practical experience of this. The bag will support the load for a period, and then suddenly start to stretch until it breaks. Elastomers show extensive plastic deformation under any load, but this is always reversible, and so this behaviour differs from the plastic deformation found in the other materials.

The ultimate tensile strength of a material with respect to its weight is an important engineering parameter. This is termed the specific strength:

$$\text{specific strength} = \frac{\text{tensile strength}}{\text{specific gravity}}$$

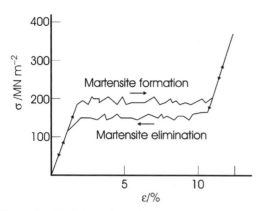

Figure 10.5 Schematic engineering stress–engineering strain (σ–ε) curve for a shape-memory alloy showing superelasticity

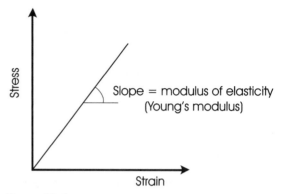

Figure 10.6 The modulus of elasticity, Young's modulus, is defined as the slope of the stress–strain curve in the linear (elastic) region

The major properties obtained from the engineering stress–engineering strain tensile curve are modulus of elasticity, yield strength, ultimate tensile strength and amount of elongation at fracture.

The stress–strain behaviour is different again in many shape-memory alloys (Section 8.3.3). In these solids, the stress–strain plot shows a region of extreme deformation (Figure 10.5). The curve returns to the original pathway on release of the stress, although rarely following exactly the same pathway, a hysteresis effect. This is called super-elasticity. In the super-elastic region, martensite is forming (even at temperatures above A_f, the austenite finish temperature), under the influence of the applied stress. For example, the martensitic transformation in the shape-memory alloy Nitinol is caused by a shear of the atoms in $\{101\}$ planes (see Section 8.3.2). Stress has a similar shearing effect on the structure and can cause the martensitic transformation to occur above A_f. This is called a stress-induced martensitic transformation. The application of further stress causes more martensite to form and, in so doing, the stress is released. When the transformation is complete and the variant boundaries are fixed in place, the material reverts to normal behaviour. However, above A_f the martensitic form is unstable in the absence of stress. Thus, as the stress is released the martensite plates revert to the parent structure, but normally the start of the reverse transformation is initiated at a lower stress than the forward transformation, resulting in hysteresis. In practice, the degree of hysteresis is found to depend on the composition of the alloy and prior heat treatment.

10.1.4 Elastic deformation: the elastic (Young's) modulus

Elastic deformation is reversible deformation. The slope of the elastic region is a measure of the elastic modulus (Young's modulus) of the material. When the force applied to a material is relatively small and the material is subject only to elastic deformation, the stress is related to the strain by Hooke's law:

$$\sigma = E\,\varepsilon \qquad (10.1)$$

The constant of proportionality, E, is Young's modulus (Figure 10.6). As weight is an important consideration in many applications, the specific modulus is often quoted as a material parameter:

$$\text{specific modulus} = \frac{\text{modulus of elasticity}}{\text{specific gravity}}$$

The elastic deformation experienced is a result of pulling atoms apart or pushing atoms together and so is directly related to interatomic bonding. If the chemical bonds between the atoms can be accurately described in terms of energies, then the amount of deformation that will result from a given applied force can be calculated (Section S4.1.7).

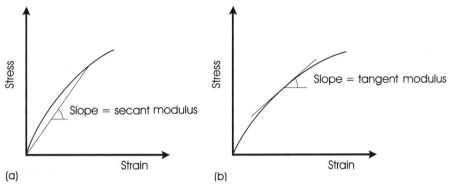

Figure 10.7   The (a) secant modulus and (b) tangent modulus of a material

In many materials, especially those in which one (or more) component(s) of the bonding is (are) relatively weak, Hooke's Law is not obeyed exactly. Instead, the stress–strain curve shows a distinct curve (Figure 10.7). This is termed nonlinear behaviour. In such cases, a single value for Young's modulus cannot be obtained. As an approximation, it is possible to measure the tangent to the curve at any point, to obtain the tangent modulus or, alternatively, the secant modulus can be used (Figure 10.7). This is the slope of the line drawn from the origin to intersect the stress–strain curve at a specified value of the strain.

Representative values of the modulus of elasticity are given in Table 10.1.

10.1.5   Poisson's ratio

Although measures of engineering stress and engineering strain assume constant cross-sectional area of the rod being stressed, a material deformed elastically longitudinally (in compression or tension) has an accompanying lateral dimensional change. This is described by Poisson's ratio, ν. If a tensile stress σ_z produces an axial strain $+\varepsilon_z$ and lateral contractions $-\varepsilon_x$ and $-\varepsilon_y$, (in isotropic materials $-\varepsilon_x = -\varepsilon_y$),

$$\nu = \frac{-\varepsilon(\text{lateral})}{\varepsilon(\text{longitudinal})}$$
$$= \frac{-\varepsilon_x}{\varepsilon_z} \qquad (10.2)$$

Table 10.1   Representative values of the modulus of elasticity, E, and of the Poisson ratio, ν

Material	E/GPa	ν
Aluminium	70.3	0.345
Copper	129.8	0.343
Iron (cast)	≈152	≈0.27
Magnesium	44.7	0.291
Nickel	219.2	0.306
Titanium	115.7	0.321
Tungsten	411.0	0.280
Brass	≈100	≈0.35
Bronze	≈105	≈0.34
Steel, mild	≈212	≈0.29
Alumina	379.2	0.22
Magnesium oxide	210.3	0.23
Silicon carbide	468.9	0.17
Silica glass	72.4	0.17
Epoxy resin	≈3.2	≈0.35
Nylon 6,6	≈2.0	≈0.39
Polycarbonate	≈2.4	≈0.36
Polystyrene	≈3.5	≈0.33

The negative sign is to ensure that the numerical value of Poisson's ratio is positive. For isotropic materials, the theoretical value of ν is $\frac{1}{2}$. Most metals show values in the range 0.25–0.35. Some materials, counterintuitively, have a negative Poisson's ratio, and get thicker when under tension. They are discussed later in this chapter. Representative values of Poisson's ratio are given in Table 10.1.

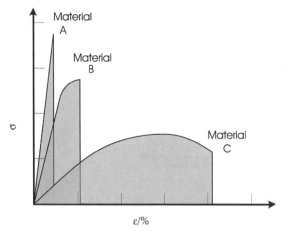

Figure 10.8 The toughness of a solid can be represented by the area under a engineering stress–engineering strain (σ–ε) curve to fracture. Material A has the highest elastic modulus, but material C has the greatest toughness

10.1.6 Toughness and stiffness

Toughness is often a much more important material property than strength. Toughness can loosely be defined as the amount of energy absorbed by a material before it fractures. A tough material has a high resistance to the propagation of cracks, and so tough materials are both strong and ductile. The toughness can be estimated by the area under the stress–strain curve (Figure 10.8).

The ease with which a material can be extended, the stiffness of the solid, is represented by the slope of the stress–strain curve in the initial, elastic region. A stiff material shows only a small strain for a given stress. The stiffness is thus equivalent to the modulus of elasticity. It is seen from Figure 10.8 that the material with the highest modulus of elasticity is not necessarily the toughest. In this case, material C is the toughest of the three, although it has the lowest stiffness. The toughness of a specimen will depend on specimen geometry and the way in which the stress is applied.

10.1.7 Brittle fracture

Many materials are elastic and brittle, especially at lower temperatures. Single crystals frequently fracture by cleavage along crystal planes in which the bonding is relatively weak (Figure 10.9a). A polycrystalline material can fracture in two ways. Fracture across the constituent crystallites is akin to crystal cleavage and is called transgranular or transcrystalline fracture. [Single crystals can fracture only in a transgranular fashion.] In some materials, the weakest part is the region between crystallites, and so the fracture surface runs along the boundaries between the constituent crystallites. This is called intergranular fracture (Figure 10.9b). Amorphous materials such as glass or brittle polymers fracture to produce a smooth surface resembling the inside of a seashell. This is called conchiodal fracture (Figure 10.9c). Materials containing voids or several phases, such as porcelain, which contains glass, crystals, and voids, frequently fracture in the neighbourhood of these defects (Figure 10.9d).

Polymer fibres are rather different in that they can be thought of as aligned molecules. Many polymer fibres, such as Kevlar[®], carbon fibre and ultrahigh-molecular-weight polythene have tensile strengths comparable to or better than steel. Rigid polymers, such as methyl methacrylate, polycarbonate, or many thermoplastics and elastomers at low temperatures, behave in a brittle fashion.

The elastic deformation in all these materials can be successfully explained by considering the strength of the chemical bonds between the constituent atoms of the solid or the fibre molecule, and it is logical to conclude that fracture takes place when the tensile stress is greater than the chemical bond strength (Section S4.1.7). Rapid failure occurs when a crack propagates through the solid by the breaking of successive chemical bonds. There is a fundamental difference between metals and ceramics, on the one hand, and polymers, on the other, in this respect. In metals and ceramics, the bonding within the solid is more or less uniform and consists of fairly strong bonds, metallic, ionic or covalent. In a brittle polymer, the material is made up of crystals joined by amorphous regions in which the bonding is weak, often hydrogen bonds. Although the fracture of the material proceeds in the same way as in a metal or ceramic, the weakness of the bonds that have to be broken makes these materials much less

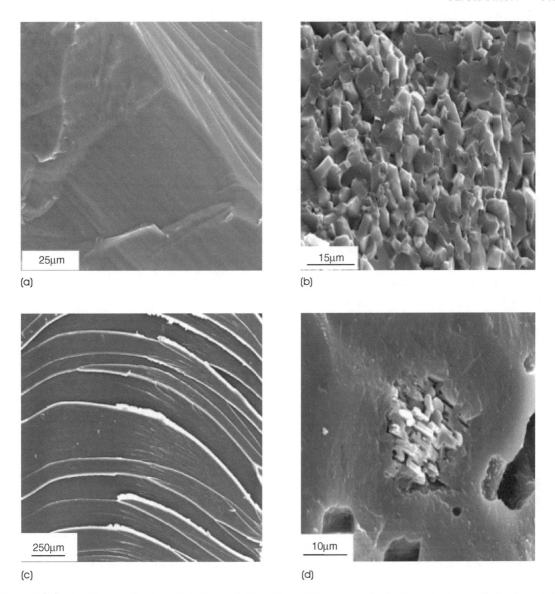

Figure 10.9 (a) Cleavage fracture of single crystalline silicon, (b) transgranular fracture of a fine-grained polycrystal-line alumina ceramic crucible. (c) Conchoidal fracture of brittle poly(methyl methacrylate). (d) Fracture of porcelain showing a cluster of crystallites and pores in a glass-like matrix

strong. In the case of polymer fibres, of course, the bonds that have to break are strong covalent bonds, and thus these materials are very strong, at least in the fibre direction.

It is possible to estimate the critical stress required for fracture from the interatomic forces (Section S.4.2.1). The critical stress, σ_c, is given by:

$$\sigma_c = \left(\frac{\gamma E}{r_0}\right)^{1/2} \qquad (10.3)$$

where r_0 is the equilibrium spacing of the atoms in the material, E is the modulus of elasticity, and γ is the surface energy of the solid. The theoretical strength, determined from Equation (10.3), is calculated to be about $E/6$.

Measurements consistently reveal that the real strength of solids is much less than this value, lying somewhere in the region of $E/100$ to $E/1000$. In order to resolve the anomaly, it is necessary to consider the distribution of stress in a solid in the region of a crack tip. This was determined in the early years of the 20th century for elliptical cracks, a geometry chosen to make calculations feasible in the period before computers were available. It was found that the stress in the region of a crack tip (or, in fact, at many other objects such as voids, sharp corners or sharp grain boundaries) is much greater than the applied stress. Such flaws are called stress raisers. The amount that the stress is increased at a crack tip with an elliptical cross-section, the stress concentration factor, K_t, is given by:

$$K_t = 1 + 2\left(\frac{a}{\rho}\right)^{1/2}$$

where a is the length of a surface crack (Figure 10.10a) or half the length of an internal crack (Figure 10.10b), and ρ is the radius of curvature of the crack tip. For a relatively long crack with a sharp tip (Figure 10.10c).

$$K_t \approx 2\left(\frac{a}{\rho}\right)^{1/2} \tag{10.4}$$

Clearly, for a long crack with a sharp tip, the stress at the tip can be many times that of the nominal applied stress.

When the stress at the crack tip reaches the strength of the material as calculated from chemical bond strength, bonds will break and the crack will lengthen, leading to fracture. The critical stress for a material with a crack is found to be (Section S4.2.2):

$$\sigma_c = \left(\frac{E\gamma}{2a}\right)^{1/2} \tag{10.5}$$

This indicates that the strength of the solid, as measured by the average applied stress, decreases as the length of the crack in the material increases.

The role of cracks in the process of brittle fracture was first investigated quantitatively by Griffith some 90 years ago. In this theory, called the Griffith theory of brittle fracture, fracture was considered to be due to the presence of microscopic cracks or flaws, called Griffith flaws, distributed throughout the solid. Griffith suggested that failure occurred when the energy introduced into the solid as a result

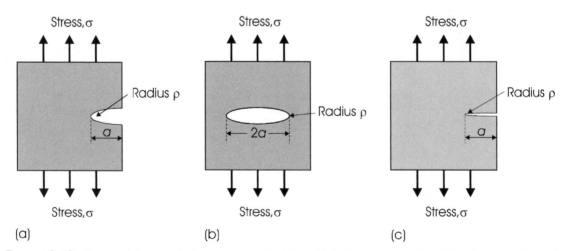

Figure 10.10   Stress at (a) an elliptical groove in a solid, (b) an elliptical pore in a solid, and (c) a sharp crack in a solid

of the tension was equal to the energy needed to create two new surfaces on either side of a growing crack. [Remember that at this time the concept of chemical bonds was still being worked out.] Using the energy approach, the critical stress, σ_c, to fracture the solid was determined to be

$$\sigma_c = \left(\frac{2\gamma E}{\pi a}\right)^{1/2} \qquad (10.6)$$

where γ is the surface energy of the solid (per m²), E is the modulus of elasticity, and a is the length of a surface crack or half the length of an internal crack (Figure 10.10). Equation (10.6) is known as the Griffith equation. This equation gives the maximum stress that a solid containing an internal crack of length $2a$, or a surface crack of length a, that the solid can sustain without the crack growing. The longer the crack, the lower the stress needed to cause fracture.

Equations (10.5) and (10.6) can be made identical by setting

$$\frac{\rho}{4 r_0} = \frac{2}{\pi}$$

This gives an approximate value for the crack radius, ρ:

$$\rho \approx 2.5 \, r_0$$

This gives a rule-of-thumb measure for the separation of the two surfaces. The surfaces that are formed as the crack opens can be considered separate when the crack has opened to about $5 \, r_0$, or about 1 nm.

10.1.8 Plastic deformation of metals and ceramics

Metals are normally ductile, and plastic deformation can be considerable. This is a valuable property of metals and is used, for example, to fabricate dish shapes of one sort or another by impressing a die into a sheet of metal. The metal deforms plastically and retains the dished shape when the load on the die is removed. Ceramics are regarded as brittle, and normally the same fabrication process applied to a ceramic would shatter it. However, at high temperatures, many ceramic materials can deform plastically, and at low temperatures many metals become brittle and lose ductility. Thus, it is convenient to treat these two apparently dissimilar crystalline materials together. Polymers, which contain crystalline and amorphous regions, are considered in Section 10.1.11.

In the case of an ordinary metal, once the loading in a tensile test is continued past the yield stress (point A, Figure 10.11a), the material will be permanently

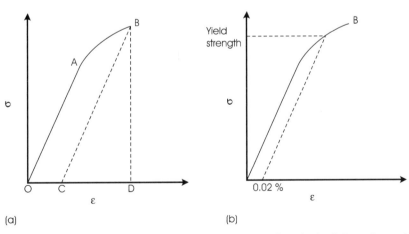

(a) (b)

Figure 10.11 (a) Engineering stress–engineering strain (σ–ε) curve for plastic deformation, which results in a permanent distortion of the material, represented by OC. (b) The yield strength, determined by the intersection of an offset line, here at 0.02 %, parallel to the σ–ε curve Note: A, yield stress; B, load release point; D, strain under load; C strain at release of load

deformed. As the load is released, the plot of stress against strain will follow the line BC, which is parallel to AO. When the stress falls to zero, CD is the amount of elastic strain recovered, and OC represents the plastic or permanent deformation that has occurred. A rod so stressed will not return to its original thickness, but remains thinner than at the start. This property, used in drawing rods down into wires, is called plastic deformation.

The yield strength (a load) or yield stress (a stress value) is the point at which plastic deformation occurs (point A, Figure 10.11a). For most materials, the transition from elastic to plastic behaviour is rarely abrupt, and a single 'point' does not mark the boundary between elastic and plastic deformation. In order to obtain a guide as to when a stress value passes that required for plastic deformation to occur, it is usual to select a value of the stress that leads to 0.2 % plastic strain (0.002 strain). This is also called the 0.2 % offset yield strength. [This value is arbi-

trary, and any value could be chosen, such as 0.1% offset yield strength.] The yield strength is determined by drawing a line from the 0.02 % strain point parallel to the elastic section of the curve, to intersect the stress–strain curve (Figure 10.11b).

10.1.9 Dislocation movement and plastic deformation

From an atomic point of view, the origin of plastic deformation is most readily understood by studying single crystals. The deformation is revealed as a series of steps or lines, where parts of the crystal have moved relative to each other. The process can be one of slip, in which atom planes have slid sideways, or one of twinning, known as mechanical twinning. In slip, a small slice of crystal is moved sideways (Figure 10.12a). The slices are usually the order of a few hundred atoms wide and, during slip,

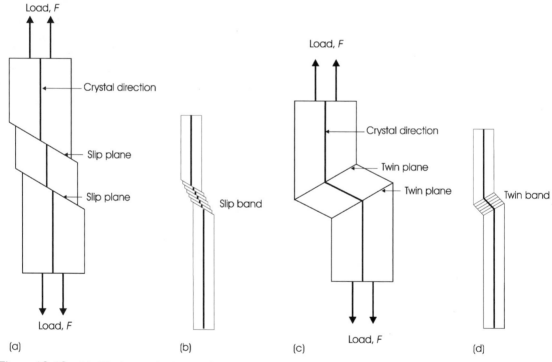

Figure 10.12 (a) Slip in a rod, characterised by diagonal planes across which atoms in the crystal have sheared because of an applied load; (b) slip band, in which slip planes are aggregated into narrow regions; (c) mechanical twin planes, across which the atoms in the crystal are reflected because of an applied load; (d) twin band, in which twin planes are aggregated. Note: slip and twinning are both caused by stress and are difficult to distinguish in macroscopic samples

a number of these translated slices group together to form a slip band (Figure 10.12b), which has the appearance of a line on the crystal face. The crystal direction along the deformed crystal does not change during slip.

The process of twinning gives a crystal in which the atom positions are reflected across the twin plane (Figure 10.12c). Twins also tend to occur in bands (Figure 10.12d). The distinction between slip and twinning is best determined by means of X-ray crystallography or transmission electron micro-scopy, as these techniques are able to reveal the crystallographic relations on each side of a planar boundary.

Slip is generally easier than twinning in metals, but the reverse is true for many ceramic materials. In polycrystalline solids, both twinning and slip may operate.

The amount of energy required to move a plane of atoms from one stable position to another, from A to B in Figure 10.13, can be calculated from the strength of the chemical bonds in the solid, in much the same way as the modulus of elasticity and strength of a solid is estimated (Section S4.2). It is found that the calculated energy is greater than that observed in practice. The discrepancy was resolved by a realisation that slip is achieved by the movement of dislocations through the structure. The motion of an edge dislocation through a metal-lic crystal is illustrated in Figure 10.14. In fact, slip is the process by which plastic deformation is produced by dislocation motion. The plane on which the dislocation moves is called the slip plane. The result of the dislocation movement is

slip across a single atom plane (Figures 10.15a and 10.15b). A similar situation occurs with screw dislocation movement (Figures 10.15c and 10.15d). Mixed dislocations move by a combination of both mechanisms.

The movement of dislocations is constrained by crystallography. Some planes allow movement to take place more easily than do others, and these preferred planes are called slip planes. Similarly, dislocation movement is easier in some directions than it is in others. These preferred directions are called slip directions. The combination of a slip plane and a slip direction is called a slip system.

In general, slip planes are planes with the highest area density of atoms, and slip directions tend to be directions corresponding to the direction in the slip plane with the highest linear density of atoms. For example, the planes containing most atoms in metals that adopt the face-centred cubic (A1) struc-ture are $\{1\,1\,1\}$ planes (Figure 10.16). In these planes, the greatest linear density of atoms occurs along the $\langle 1\,1\,0 \rangle$ directions. In such a metal there are four different $\{1\,1\,1\}$ planes and three different $\langle 1\,1\,0 \rangle$ directions. The number of slip systems is thus equal to 12. The possibilities are conveniently written as $\{1\,1\,1\}\langle 1\,\bar{1}\,0 \rangle$.

Slip in hexagonal metal crystals occurs mainly parallel to the basal plane of the unit cell, normal to the c axis. The slip systems can be described as $\{0\,0\,0\,1\}\langle 1\,1\,\bar{2}\,0 \rangle$, of which there are three. Body-centred cubic metals have slip described by $\{1\,1\,0\}\langle \bar{1}\,1\,1 \rangle$, giving 12 combinations in all. Other slip systems also occur in metals, but those described operate at lowest energies.

10.1.10 Brittle and ductile materials

Although ceramics are regarded as being predomi-nantly brittle, many show ductility at high tempera-tures. This is a general phenomenon, not limited to ceramics, and these ductile-to-brittle transitions are of great importance in engineering. It is generally found that materials are brittle at low temperatures and ductile at high temperatures, although the definition of 'high' and 'low' is very sensitive to the material under consideration. For an elastomer

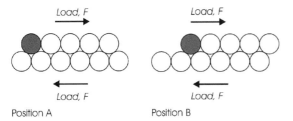

Figure 10.13 Slip, caused by a shear stress displacing atoms planes with respect to each other; it becomes increasingly difficult to move the atoms as the length of the atom rows increases

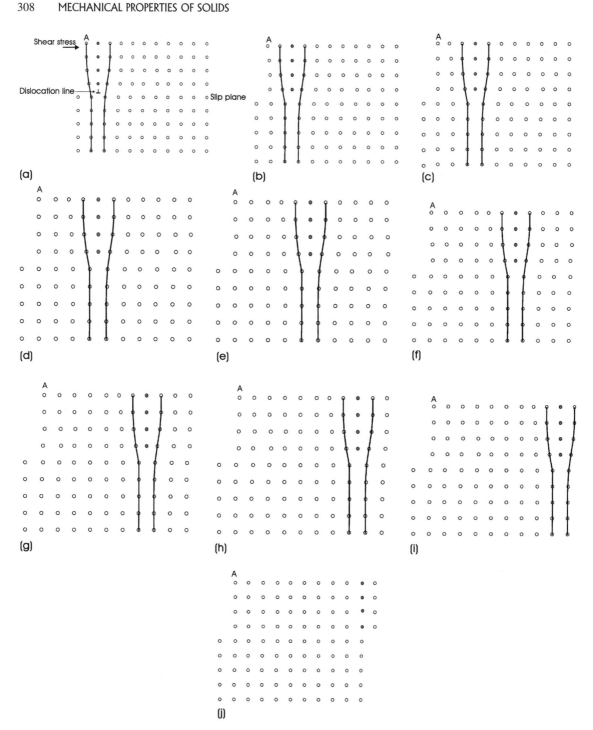

Figure 10.14 Slip caused by the movement of an edge dislocation under a shear stress. At each step (from part a to part b, part b to part c, etc.) only a small number of atomic bonds must be broken, and the process occurs at much lower shear stress levels than are needed in perfect crystals (Figure 10.13)

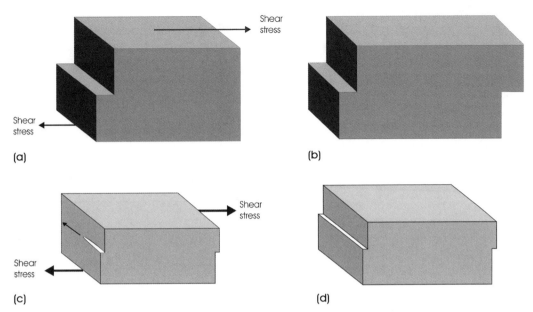

Figure 10.15 (a) Initial configuration in a crystal containing an edge dislocation; (b) final configuration after the dislocation has moved through the solid. (c) Initial configuration of a crystal containing a screw dislocation; (d) final configuration after the dislocation has moved through the solid

such as rubber, room temperature is already 'high'. In the case of polymeric materials, the glass transition temperature, T_g, gives a good indication of the boundary between the brittle and ductile regions.

In the case of ceramics, ductility begins to become important at higher temperatures. As an empirical guide, this is above the Tamman temperature, which is about half of the melting temperature in kelvin. In this regime, ceramics can deform via slip. However, dislocation motion is much harder in ceramics than it is in metals, partly because of the stronger bonding but also because of electrostatic

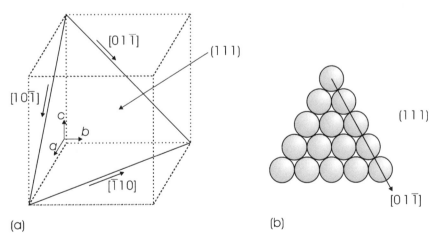

Figure 10.16 (a) The (1 1 1) plane in a crystal of a metal with the face-centred cubic (A1, copper) structure (shaded). The directions (arrows) along each of the edges are given. (b) The same (1 1 1) plane represented as a packing of atoms. One direction, [0 1 $\bar{1}$], is marked

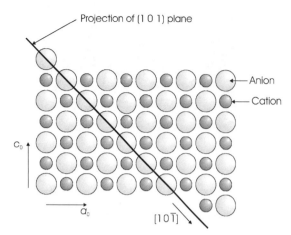

Figure 10.17 A $\{1\,1\,0\}$ slip plane in the *halite* (NaCl, B1) structure

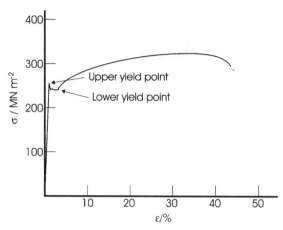

Figure 10.18 Engineering stress–engineering strain (σ–ε) curve for a typical carbon steel

repulsion between similarly charged ions, and twinning is often the favoured mechanism of releasing stress. The reason why ceramics tend to be brittle and metals ductile is therefore not due to the presence of dislocations in metals and their absence in ceramics but because of the greater difficulty of slip in ceramics at normal temperatures.

The preferred slip plane in ionic crystals with the *halite* (NaCl) structure, such as NaCl or LiF, is $\{1\,1\,0\}$, and the slip direction used is $\langle 1\,\bar{1}\,0 \rangle$. This slip system is sketched in Figure 10.17. For the more metallic *halite* structure solids such as titanium carbide (TiC), the slip system is similar to that in face-centred cubic metals, $\{1\,1\,1\}\langle 1\,\bar{1}\,0 \rangle$.

For some metals, notably steels, there is an abrupt break in the stress–strain plot at the upper yield point (Figure 10.18). This is followed by continued deformation at a lower yield stress, at the lower yield point, before the curve rises again. Between the upper and lower yield points, deformation occurs in localised regions that have the form of bands, rather than across the specimen in a uniform manner. The reason for this is that the dislocations, which would move during plastic deformation, are pinned in the steel, mainly by the interstitial carbon atoms present. At the upper yield stress, these become mobile and, once released, they can move and multiply at a lower stress value. This is analogous to sticking and slipping when a body over-

comes friction. The region in which dislocation movement starts forms a band. Ultimately, the whole of the dislocation network is mobile, at which point the stress–strain curve begins to rise again, as in a ductile material.

10.1.11 Plastic deformation of polymers

Polymers that show considerable amounts of plastic deformation are usually partly crystalline thermoplastics or elastomers. The yield strength is taken as the initial maximum of the curve, and the tensile strength as the fracture point (Figure 10.4c). The tensile strength can be less than the yield point, especially at higher temperatures. In partly crystalline polymers, the stress is imposed on both the amorphous region and the crystalline region. The weakest links are those between the coiled chains in the amorphous regions, and these slip past each other as the sample elongates (Figure 10.19). Ultimately, these molecules become more or less aligned, and the stress now acts on the molecular chains, in which strong covalent bonding occurs. At this point, the crystalline regions can begin to slip, and, ultimately, fracture will occur. Elastomers behave differently in that the coiled regions predominate and can be uncoiled to give enormous extension before the molecules are all more or less

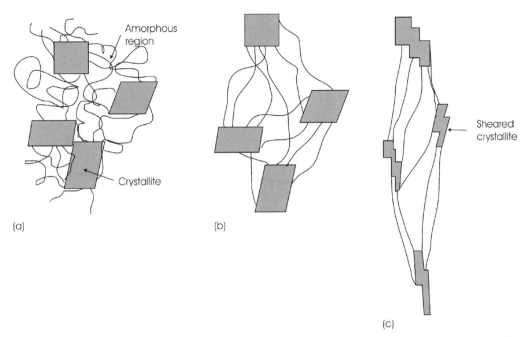

Figure 10.19 Deformation of a semicrystalline polymer (the crystalline regions are represented by shaded quadrilaterals). (a) The unstressed state. (b) On loading the polymer the molecules in the amorphous regions elongate initially. (c) When the polymer chains can no longer accommodate the stress, the crystalline regions deform

parallel and before the strong covalent bonds of the molecules are stressed (see Section 6.4.4).

10.1.12 Fracture following plastic deformation

Solids that show considerable plastic deformation before fracture generally fail in a different way from brittle solids, discussed above. An important parameter that characterises this type of failure is the ductility of the material. Ductility is a measure of the degree of plastic deformation that can be sustained by a material at fracture. The ductility of metals can be estimated by measuring the percentage elongation of a sample after fracture, where:

$$\text{elongation} = \frac{\text{final length }(l) - \text{initial length }(l_o)}{\text{initial length }(l_o)}$$

$$\text{percentage elongation} = \frac{l - l_o}{l_o} \times 100$$

During the early stages of deformation of a metal, the test sample retains its original shape, although when the stress is released the original dimensions are not recovered. When a metal fractures after considerable plastic deformation, the central part of the sample is found to have formed a neck. This occurs after the highest point in the stress–strain curve (Figure 10.20). Because a large amount of the final deformation occurs in a rather small neck region, the percentage elongation measured will depend on the value taken for l_0. It is for this reason that standard gauge lengths, usually 5 cm for metal samples, are used for specimens that are to undergo tensile testing.

The ductility can also be estimated by a measurement of the reduction in the cross-sectional area of the sample at the break. The percentage reduction in area of a sample after fracture is:

$$\text{reduction in area} = \frac{\text{initial area }(A_o) - \text{final area }(A_f)}{\text{initial area }(A_o)}$$

$$\text{percentage reduction in area} = \frac{A_o - A_f}{A_o} \times 100$$

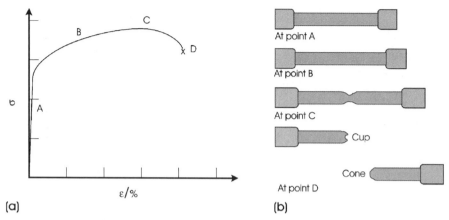

Figure 10.20 (a) Tensile engineering stress–engineering strain (σ–ε) curve for a ductile metal. (b) The corresponding shape of the test piece during the test

The fracture surface after ductile fracture has a characteristic shape, one side resembling a cup and the other a cone (Figure 10.20b, for point D; Figure 10.21).

Although most metals are ductile, some metals are brittle. Tungsten, a particularly brittle metal, is of considerable importance in daily life. It is the best material for use in incandescent lightbulbs, the normal lightbulbs that contain a white-hot filament. In the earliest years of the electric light industry, carbon filaments were used, but these were clearly seen to be a poor choice from the outset. Tungsten, an ideal choice, was too brittle to be formed into filaments or even coiled as wire. The process for making tungsten ductile involves drawing the wire

at temperatures above the ductile-to-brittle transformation temperature, but at a temperature too low for the metal to recrystallise, and adding alloying elements, notably potassium, to modify the resulting grain structure of the wire.

The term ductility is generally reserved for metals and in the case of polymers ductility is replaced by elongation. The elongation is simply measured as the length of the polymer after stretching, l, divided by the original length, l_0, often given as a percentage:

$$\text{percentage elongation} = \frac{l}{l_0} \times 100$$

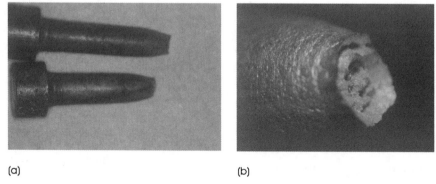

Figure 10.21 (a) A fractured semi-ductile steel tensile test specimen, showing necking and cup and cone fracture. (b) Detail of cup and cone fracture surface from a copper tensile test specimen. The diameter of the neck is approximately 2 mm

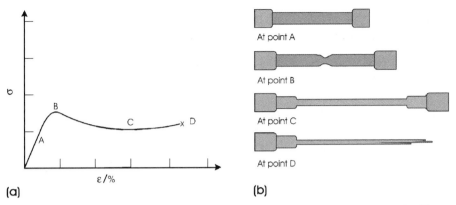

Figure 10.22 (a) Tensile engineering stress–engineering strain (σ–ε) curve for a polymer. (b) The corresponding shape of the test piece during the test

The deformation of a polymer tensile specimen is rather different from that of a metal. Polymers neck, and the neck region itself will extend in a ribbon until the material ultimately tears (Figure 10.22). The term 'percentage elongation at fracture' used for metals is generally replaced by 'percentage elongation at break' for polymers.

In both metals and polymers, fracture is preceded by the formation of voids or holes. In a metal, these voids tend to segregate at the central region of the neck, and coalesce into a large crack. This crack ultimately spreads to the edge of the neck. The fracture surface has a pulled-out, fibrous texture exhibited in the 'cup' and the 'cone' regions of the break. In the case of a polymer sheet, the holes are formed by the continual alignment of the polymer chains. Ultimately, the separate holes coalesce to form a long tear. As with metals, the edges of the tear have a pulled-out, fibrous texture.

10.1.13 Strengthening

Knowledge of the ways in which solids fail can be exploited to improve the mechanical properties, particularly the strength, of solids. Pure metals tend to be soft and relatively weak. Ductility can be reduced and the metal strengthened by restricting dislocation movement. However, if this is continued too far the metal will become brittle. A compromise is often required.

Historically, three methods have been used to strengthen metals. Two of these, reduction in grain size and work hardening, are brought about by heat treatment and mechanical deformation, for example hammering. The third method, alloying, was for most of historical times based on empirical observation. As technology advanced, these methods have been supplemented by precipitation strengthening, in which careful heat treatment is used to initiate the formation of precipitates within the metal matrix. In all cases, the metal is strengthened because dislocation movement is restricted.

Alloying, first widely used to transform soft copper into much stronger bronze, relies on the strain set up in the crystal structure by the impurity or dopant atoms. This strain field impedes dislocation movement because dislocations also generate a strain in the structure. The two strain components mutually repel each other to hinder slip. If sufficient of the second component is added, precipitates of a second phase can form in the crystal matrix. These hinder dislocation movement simply by blocking slip planes. Reduction of grain size has a similar effect. Dislocations can cross from one grain to another only by expending energy, and often grain boundaries block almost all dislocation movement. Work hardening, which is also called cold working or strain hardening, is the result of repeated deformation. This causes dislocation numbers to increase. Initially, this can lower the strength but beyond a certain dislocation density dislocations become tangled and movement becomes impeded. Again, this is due to interaction of the strain fields around the dislocations.

Ceramics normally fail in a brittle mode. They can be strengthened by the removal of the Griffith flaws. Although this is not always possible, careful protection of the surface from contamination, chemical reaction or mechanical damage will help. Optical fibres, when freshly drawn, are extremely strong, but reaction of the surface with water vapour present in the air rapidly degrades the fibre strength. For this reason, freshly drawn fibres are immediately coated with a polymer to maintain strength. Similarly, single-crystal ceramics can have surfaces carefully polished to increase strength. Polycrystalline ceramics need to be fabricated in such a way as to minimise the amount of internal porosity.

Polymers are somewhat different. The strength of a polymer is generally greatly enhanced by increasing the amount of crystalline phase present with respect to the amount of amorphous phase present. Similarly, cross-linking will transform a soft thermoplastic into a hard and brittle solid. However, a strengthening route available to polymer science that is not available for metals and ceramics is the formation of block copolymers. Copolymers of an elastomer and a fibre will give a strong but elastic fibre, and copolymers of elastomers with brittle polymers will give a tough and flexible solid.

For many engineering (and biological) applications, solids are strengthened by mixing two materials together. In the case of polymers, these can be blended, but for metals and ceramics the two phases remain separate, and the material is called a composite (see Section 10.4).

10.1.14 Hardness

Hardness is a property that is intuitively understood but difficult to define. It is usually taken to be a measure of the resistance of a material to permanent (plastic) local deformation and is often measured by forcing a chosen solid into the surface of the material to be tested. Hardness therefore measures a compressive property rather than a tensile property.

The first use of hardness was in the characterisation of minerals. The hardness of a mineral was determined by observation of whether it could scratch or be scratched by another mineral. This is sometimes called the scratch hardness of a solid. The *ad hoc* system in use in medieval times was quantified by Mohs, who listed 10 minerals that, as far as possible, were equally spaced on a hardness scale of 1 to 10. Even today, the commonest description of hardness is still in terms of this scale, called the Mohs scale of hardness. The minerals chosen, and their hardness are: 1, talc $[Mg_3Si_4O_{10}(OH)_2]$; 2, gypsum $(CaSO_4 \cdot 2H_2O)$; 3, calcite $(CaCO_3)$; 4, fluorite (CaF_2); 5, apatite $[Ca_5(PO_4)_3(OH)]$; 6, orthoclase $(KAlSi_3O_8)$; 7, quartz (SiO_2); 8, topaz $[Al_2SiO_4(OH)_2]$; 9, corundum (Al_2O_3); and 10, diamond (C). In addition, Mohs suggested the following values: fingernail, less than 2; copper coin, 3; pocket knife blade, just over 5; window glass, 5.5; and a steel file, 6.5. Each of these materials will scratch the surface of the solid immediately below it in the list. The softest material is talc, which will not scratch any of the others. Diamond is the hardest material known.

To measure hardness more precisely a known load is applied slowly to a hard indenter that is placed onto the smooth surface to be tested. The surface is deformed plastically, and the indent size or depth after the indenter is removed is taken as the measure of the indentation hardness of the material. The hardness is often recorded as an empirical hardness number, related to the size of the indentation.

There are four major indentation hardness tests, which differ from each other in the shape of the indenter (Figure 10.23). The first of these, described in 1900, was the Brinell test, using a 10 mm steel ball indenter (Figure 10.23a), giving the Brinell hardness number, BHN. This was suitable only for metals softer than steel. In 1920 Rockwell developed a number of tests, including the B, E, F and G scales, in which the indenter is steel, and the A, C and D scales, using a conical diamond indenter with a spherical tip (Figures 10.23b and 10.23c). In the Rockwell test the difference in size between the

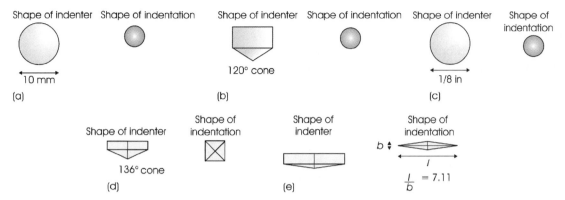

Figure 10.23 Hardness indenters: (a) Brinell steel or tungsten carbide sphere; (b) Rockwell A, C, D diamond cone; (c) Rockwell B, F, G steel sphere; (d) Vickers diamond pyramid; and (e) Knoop diamond pyramid

indentations caused by a small and a large load are compared to give the Rockwell hardness number, RHN. The Vickers test, introduced in 1924, uses a pyramidal diamond indenter (Figure 10.23d) and a standard load time of 15 s, to give the Vickers hardness number, VHN. The Knoop test, widely used for brittle materials such as minerals and glass, was introduced in 1939. This method uses an elongated pyramidal diamond indenter (Figure 10.23e) and gives the Knoop hardness number, KHN. Both the Vickers test and the Knoop test use small loads, of the order of grams, and measure what is generally called microhardness.

When hardness was first measured, it was reported simply as a series of internally consistent 'hardness numbers', as Mohs hardness still is. Vickers hardness can readily be converted into conventional units. The Vickers hardness is given by the applied load (kg) divided by the projected area of the indentation (mm^2). The units of kg mm^{-2} is readily converted into SI units by converting the load in kg to N, and the area into m^2, so that 1 kg mm^{-2} is equal to 9.81 MPa.

The development of thin-film engineering and the increasing use of nanomaterials have led to the development of nano-indenters. These are described in more detail in Section 10.3.3.

The hardness of a material is related to the strength of the chemical bonds within the solid. As with other mechanical properties, however, the direct correlation is masked by the defect structure of the solid. The deformation, that acts as a measure of hardness, is produced by compression and subsequent plastic deformation. The compression is a temporary elastic displacement operating while the load is imposed. Plastic deformation is due to dislocation slip, initiated by shear stress. It has recently been found there is a good correlation between shear modulus and hardness (Figure 10.24). This allows the hardness of new materials

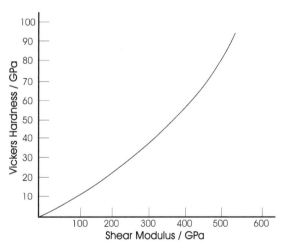

Figure 10.24 Schematic relationship between Vickers hardness and shear modulus. Adapted from D.M. Teter, 1998, 'Computational Alchemy: The Search for New Superhard Materials', *Materials Research Society Bulletin* **23** (January) 22

to be predicted from knowledge of the shear modulus.

In recent years, efforts have been made to try to make a harder material than diamond. The hardness of diamond is attributed to the strong sp³ hybrid bonds linking the crystal into one giant molecule, and most attempts at a synthetic alternative have focused on iso-electronic crystals. These are crystals with the same average number of bonding electrons available as in diamond, four per atom. The idea is to force the atoms in the new material to bond via strong sp³ tetrahedral hybrids. The first such material made was cubic boron nitride, BN. It is seen from the periodic table that boron lies one place to the left of carbon, and has three valence electrons per atom, whereas nitrogen, one place to the right, has five. The combination of equal numbers of boron and nitrogen atoms gives an average of four valence electrons per atom. Cubic boron nitride is the second hardest material, after diamond. [A form of boron nitride resembling graphite, with a low hardness, is also known (Section 5.3.8).]

Compounds intermediate in composition between diamond and cubic boron nitride, such as BC$_2$N, have recently been synthesised, as well as other iso-electronic compounds such as B$_2$CO and B$_5$NO$_2$, but none, so far, has the hardness of diamond.

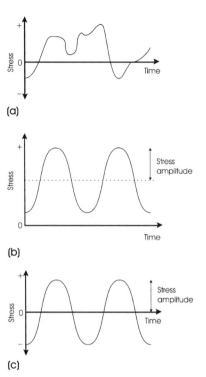

Figure 10.25 Stress cycles: (a) irregular, (b) positive sinusoidal cycle and (c) sinusoidal with alternating positive and negative stress

10.2   Time-dependent properties

A number of mechanical properties are a function of time. Although these are not so amenable to theoretical analysis as the elastic properties of solids, they are of enormous practical importance. Two of these properties, fatigue and creep, will be described in this section.

10.2.1   Fatigue

When placed under a cyclical or varying load a material may fail as a result of fatigue. This is invariably at a much lower stress than the part could withstand during a single application of the stress. Failure under repeated cycling is called fatigue failure. Fatigue affects moving parts in machinery

but is also of importance in components that are slightly flexed in a repetitive fashion, such as an aircraft fuselage under the varying pressure regimes in the atmosphere.

The stress pattern imposed on a solid can vary greatly, from sinusoidal changes, often used in testing laboratories, to completely irregular patterns (Figure 10.25). In the laboratory, a sample is tested by imposing a cyclic strain and testing until the part fails. The results are displayed as the stress amplitude, S, plotted against the number of cycles that the sample can tolerate, N, before failure (Figure 10.26). Alloys of iron (ferrous alloys) have a behaviour represented by curve A, and most nonferrous alloys and pure metals such as copper or aluminium have a behaviour represented by curve B. In curves of type A, the abrupt change in slope is called a knee, and the part of the curve parallel to the N axis is called the endurance limit or fatigue limit. At stress levels

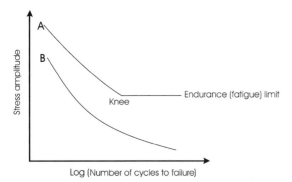

Figure 10.26 S–N curves (stress amplitude plotted against the logarithm of the number of cycles to failure): curve A, typical steel; curve B, many nonferrous metals and alloys, such as copper, aluminium, brasses, etc.

below the endurance limit, fatigue failure will never occur.

The process of fatigue is empirically divided into a number of stages. The initial damage occurs in stage 1. An originally perfect specimen will crack at an angle of approximately 45° to the direction of the cyclic stress (Figure 10.27). The surface also becomes distorted at the crack, giving rise to intrusions and extrusions, called persistent slip bands. The crack propagates across a small number of grains at this time. Stage 2 of the failure is indicated

by a change in crack direction to approximately normal to the cyclic stress direction. During this stage, the crack opens a little during each cycle, gradually spreading across much of the specimen. In practice, it is not easy to separate stages 1 and 2 of the process. Stage 3 represents the final failure of the component. This occurs when the crack in stage 2 has grown to such an extent that the whole part fails catastrophically, by ductile or brittle failure, depending on the nature of the material.

As might be anticipated, an important factor in fatigue is the state of the surface. Surface flaws and roughness can increase the stress locally by a large amount, thus initiating stage 1 of the process. Additionally, notches, sharp edges, holes or changes in cross-sectional area, called stress raisers, should also be avoided. (Aircraft have rounded windows, rather than square ones, for this reason!) The local environment is also relevant, and fatigue initiated by corrosion, called corrosion fatigue, is often important in chemical plants.

10.2.2 Creep

Creep is the gradual elongation of a material under a constant load or stress. That is, under a constant stress, a gradual increase in strain (creep) can be

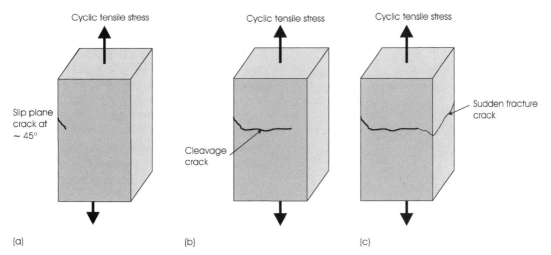

Figure 10.27 Fatigue failure: (a) stage 1, consisting of initial crack formation on a slip plane; (b) stage 2, consisting of crack growth along cleavage planes; and (c) stage 3, consisting of sudden failure as a result of rapid crack propagation

observed. Although creep is not very important at normal temperatures for most metals and ceramics, creep of polymers can be extensive. It is especially important in high-temperature applications such as in turbine blades, where even the smallest change in dimensions can lead to catastrophic failure. Creep is mainly due to progressive plastic deformation of the solid under the constant load. Because the strain in the material increases, creep must be caused by the movement of material. It can be movement of crystallites at grain boundaries, movement of dislocations or diffusion of atoms. The mechanism of creep will often depend on the temperature as well as the solid in question. For example, diffusion is more likely at high temperatures in metals and ceramics. Low-temperature creep in plastics is due to movement between the coiled polymer chains in the material.

Creep is usually displayed as a graph of strain against time, in a creep curve (Figure 10.28). The section of the curve OA is the extension that occurs when the specimen is first stressed. This is elastic deformation for most ordinary loads and corresponds to the straight-line part of a stress–strain curve. Creep as such is indicated by the remaining parts of the curve.

10.2.2.1 Primary creep

The initial section is the primary stage (Figure 10.28), also called primary creep. In this regime,

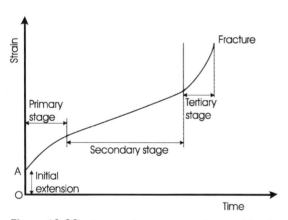

Figure 10.28 A normal creep curve for a solid subjected to a constant load at constant temperature

the creep rate is continually decreasing. The strain is largely believed to be by dislocation movement in the case of metals and ceramics, and by polymer chains sliding past each other in plastics. This part of the creep curve is described by one of two equations:

$$\varepsilon = \alpha \, \log t$$

where ε is the strain, t is the time, and α is a constant. This equation is applicable at lower temperatures, fitting well with polymers such as rubber. At higher temperatures the equation

$$\varepsilon = \beta t^{m}$$

is applicable, where ε is the strain, β and m are constants, and t is the time. The value of m can vary between about 0.03 to 1.0, a range of about 10^{2}.

The decreasing rate of creep is attributed to cold working in the case of metals. Here, the dislocations move until they begin to impede each other. In the case of polymers, uncoiling of tangled polymer chains proceeds at slower rates as the energetically easiest rearrangements give way to rearrangement involving cross-linking or crystal movement.

10.2.2.2 Secondary creep

The middle, secondary, stage of the curve (Figure 10.28) is linear and is called secondary or steady-state creep. In this case, the internal relaxation in the solid is balanced by the strain induced by the load. This linear portion of the curve is described approximately by an equation

$$\varepsilon = K t$$

where ε is the strain, and K is a constant, the slope of the linear portion of the curve.

A number of mechanisms have been proposed to explain secondary creep. At lower temperatures, the rearrangements of dislocations into lower-energy configurations, by dislocation glide, is thought to be the most important process. As the temperature increases, dislocation climb is thought to become increasingly important. These processes lead to the reduction in the number of dislocations present, a process called recovery. This mechanism of creep, in which dislocation movement controls the creep rate, is called dislocation creep or power law creep, as the slope of the creep curve, K, is given by an equation of the general type

$$K = \frac{d\varepsilon}{dt} = \frac{\pi^2 D \sigma^n}{(bN)^{1/2} G^n k T} \qquad (10.7)$$

where D is the diffusion coefficient of the rate-limiting species, G is the shear modulus of the solid, b is the Burgers vector of the rate-limiting dislocations, N is the number of dislocations, k is Boltzmann's constant, T is the absolute temperature, and σ is the stress. The exponent n has a magnitude of about 5–6 for metals and ceramics and about 2 for plastics.

At higher temperatures, diffusion is widely believed to control the rate of creep in ceramics and metals, and this process is called diffusion creep. Two mechanisms have been suggested. At relatively lower temperatures in the diffusion creep regime, creep rate is limited by diffusion along the grain boundaries, that is, short-circuit diffusion is the rate-limiting step. This is referred to as Coble creep (Figure 10.29a). The slope of the secondary creep curve for Coble creep is given by:

$$\frac{d\varepsilon}{dt} = \frac{47 \sigma \Omega D_b \delta}{k T d^3} \qquad (10.8)$$

where σ is the stress, Ω is the atomic volume, D_b is the short-circuit diffusion coefficient, δ is the grain boundary width, d is the grain diameter, k is the Boltzmann constant, and T is the absolute temperature.

At relatively higher temperatures, diffusion of atoms within the grains, that is, bulk diffusion, becomes more important in metals and ceramics. This is called Herring–Nabarro creep (Figure 10.29b). In this model, the slope of the secondary creep curve is given by:

$$\frac{d\varepsilon}{dt} = \frac{13.3 \sigma \Omega D}{k T d^2} \qquad (10.9)$$

where Ω is the volume of the vacancy, D is the vacancy diffusion coefficient for diffusion within the grains, d is the grain diameter, k is the Boltzmann constant, T is the absolute temperature, and σ is the stress.

In both of these processes, atom diffusion is away from boundaries under compression towards boundaries under tension, leading to relief of the stress at the grain boundaries generated by the load on the sample. This causes the grains to elongate along the direction of the stress. [Note that diagrams often show the flow of vacancies rather than the flow of atoms. Vacancy flow is opposite to atom flow, and is from boundaries more or less parallel to the compressive stress to boundaries more or less perpendicular to the compressive stress.]

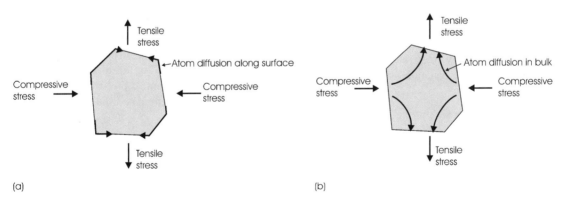

(a)   (b)

Figure 10.29 (a) Coble creep, in which atoms diffuse along grain boundary surfaces to grain boundaries parallel to the compressive stress; (b) Herring–Nabarro creep, in which atoms diffuse within grains towards boundaries parallel to the compressive stress. Note that vacancy diffusion will be in the opposite direction from that of atom diffusion (shown)

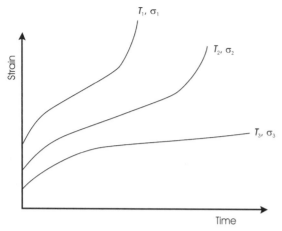

Figure 10.30 The effect of temperature on creep curves. When the stresses σ_1, σ_2 and σ_3 are equal, the temperature values T_1, T_2 and T_3 lie in the sequence $T_1 > T_2 > T_3$. When the temperatures T_1, T_2 and T_3 are equal, the stress values σ_1, σ_2 and σ_3 lie in the sequence $\sigma_1 > \sigma_2 > \sigma_3$

Equations (10.8) and (10.9) reveal that the rate of diffusion creep increases rapidly as the grain size decreases. This has implications for the creep of thin films and nanomaterials. In such fine-grained solids, creep is often surprisingly large, compared with the bulk phase, because of the grain boundary effect.

The various mechanisms proposed for secondary creep do not operate in totally separate temperature regimes. All tend to overlap, and the rate of creep can often best be generalised as a complex function of the many material parameters used in Equations (10.8) and (10.9).

Despite this complexity, the temperature dependence of the secondary creep rate can often be given by an Arrhenius-type equation (see Section 7.4):

$$K = \frac{d\varepsilon}{dt} = A\,\sigma^n \exp\left[\frac{-Q}{RT}\right]$$

where A, Q and n are constants. Q is called the activation energy for creep and is found to have a similar magnitude to the activation energy for self-diffusion in the material. The constant n varies from about 2 to about 6 and is dependent on the material in question.

The whole of the creep curve is enhanced at higher temperatures or at greater stress loading (Figure 10.30).

10.2.7.3   Tertiary Creep

The tertiary stage, or tertiary creep, is characterised by a rapid increase in the strain, leading to failure or creep rupture (Figure 10.28). At this stage, voids form in the region of the fracture and there is considerable grain boundary movement. As would be anticipated, this part of the curve is difficult to analyse theoretically.

10.3   Nanoscale properties

In general, one should not suppose that the properties of bulk materials will apply to materials at the nanoscale level. With respect to the mechanical properties of small-scale solids, it is known that the elastic behaviour, due to bond stretching and twisting, does not vary significantly in nanoparticles compared with that in the bulk. Other properties are more sensitive. For example, the rate of diffusion creep (Nabarro–Herring and Coble creep) is dependent on grain size. Hence, creep will be enhanced in compacts of nanoparticles and in thin films.

In this section, three illustrative examples of the impact of scale on mechanical properties are outlined. First, solid lubricants are discussed, underlining the connection between crystal structure and the observed mechanical properties. Second, auxetic materials, in which crystal structure and microstructure combine to produce materials with negative values of Poisson's ratio, are discussed. Last, thin films, in which mechanical properties are measured by methods similar to that used in the bulk, are considered.

10.3.1   Solid lubricants

Solid lubricants embody the opposite properties to those of hardness. Solid lubricants are soft and feel greasy to the touch. As with all lubricants, solid lubricants reduce friction and wear and prevent damage between surfaces in relative motion. Solid lubricants have advantages over normal liquid lubricants in certain conditions, including: at high temperatures, where liquids can decompose or boil; at low temperatures, where liquids can freeze; or in a vacuum, where liquids can evaporate or contaminate the vacuum. With respect to use in a vacuum, space applications are becoming more important. Originally, space programmes were frequently limited by electronic reliability. Now that these problems are largely solved, the mechanical wear of moving parts has become of prime importance, and solid lubricants that can operate under the harsh conditions imposed by space exploration are being sought continually.

Solid lubricants need low shear strength in at least one dimension. Solid lubricants fall into three main classes – inorganic solids with a lamellar (layer-like) crystal structure, solids that suffer plastic deformation easily, and polymers in which the constituent chains can slip past each other in an unrestricted way. Although soft metals such as tin and lead have been used for many years as bearings, and polymers, especially Teflon® (PTFE, polytetrafluoroethylene) are used as a coating to create 'nonstick' surfaces, the two categories of most importance are layer structures and soft inorganic compounds.

The most familiar solid lubricant is graphite. This has a layer structure (Figure 10.31a). The layers are about 0.335 nm apart and are linked by weak van der Waals bonds. However, dry graphite is not a perfect lubricant, and much of the lubricating action seems to stem from adsorbed water vapour on and between the layers. This prevents pure graphite from being used for vacuum applications. The lubrication properties are greatly enhanced by forcing the layers apart by inserting atoms into the van der Waals gap. Of these materials, fluorinated graphite, CF_x, in which x can take values between 0.3 and 1.1, is one of the most successful. This material, originally developed for use in lithium batteries (see Section 9.3.6), has a structure in which fluorine atoms covalently bond to the carbon atoms. This destroys the sp^2-bonded skeleton of the planar carbon layers, which become puckered as in diamond. (Figure 10.31b). The increase in spacing between the layers, from 0.335 nm in graphite to 0.68 nm in CF_x, weakens the interlayer interaction so much that lubricating properties are greatly enhanced.

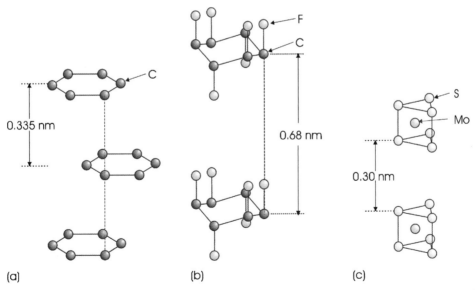

Figure 10.31 The structures of solid lubricants: (a) graphite; (b) graphite fluoride, CF_x; and (c) molybdenum disulphide, MoS_2

Molybdenum disulphide, MoS_2, is another layer structure widely used as a solid lubricant (Figure 10.31c). It is composed of MoS_2 layers, weakly linked by van der Waals bonds. The Mo atoms are surrounded by six sulphur atoms in the form of a trigonal prism. Although the MoS_2 layers are closer together than are the carbon layers in graphite, MoS_2 acts as a much better lubricant for most purposes.

Both graphite and molybdenum disulphide oxidise at higher temperatures in air, and the main alternatives used are soft inorganic fluorides. Like many ceramics, although they show brittleness at room temperature, they display ductility at temperatures of about half the melting point. A solid lubricant that is widely used in the temperature range from approximately 540 °C to 900 °C is the eutectic mixture of calcium fluoride, CaF_2, and barium fluoride, BaF_2. The melting point of the eutectic is 1022 °C.

At present, there is much research on solid lubricants, driven by the requirements of space exploration, and of high temperature engines as well as of a large range of devices in between.

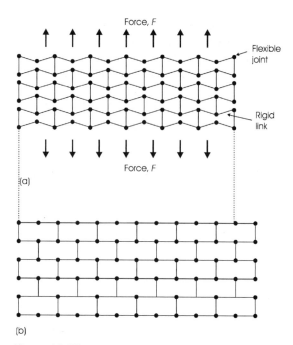

Figure 10.32   An auxetic network. The bonds (drawn as lines) between elements of the structure are of fixed length, but the links (circles) are flexible. (a) Initial state. (b) Configuration under an applied force, F; the material expands both parallel and perpendicular to the direction of the force

10.3.2   Auxetic materials

Auxetic materials expand laterally when subjected to a tensile strain. That is, unlike elastic, which gets thinner when pulled, auxetic substances get fatter when pulled. They have a negative Poisson's ratio. This counterintuitive mechanical property was first noticed in foam-like structures (Figure 10.32). Since then the property has been found in other materials

One method of producing auxetic structures is to utilise the microstructures of semicrystalline polymers. These solids are tailored so that rigid blocks of structure are linked by strong bonds. In the usual geometry (left-hand side of Figure 10.33a) the solid behaves normally and becomes thinner as the material is stretched (right-hand side of Figure 10.33a). However, if the bonds have a re-entrant geometry (left-hand side of Figure 10.33b), a tensile force will cause the blocks to move apart, and the material will become fatter (right-hand side of Figure 10.33b). In semicrystalline polymers,

the crystallites form the rigid blocks, and the disordered polymer chains between them form the bonds. This type of structure has been made in highly crystalline polyethylene [ultra high-molecular-weight polyethylene (UHMWPE)] and polypropylene fibres. The transformation from a normal polymer to an auxetic solid is achieved by making the crystalline portion the major component and reducing the polymer chains linking the crystallites to such an extent that they become short noncoiled links that act as the rigid bonds depicted in Figure 10.33b.

Surprisingly, many cubic metals behave in a similar fashion, although the effect has been masked by the fact that the mechanical properties are most often measured on polycrystalline solids. For example, the commonplace metallic alloy β-brass, CuZn, which has the CsCl structure, is noticeably auxetic. A tensile stress applied along the $[0\,0\,1]$ direction

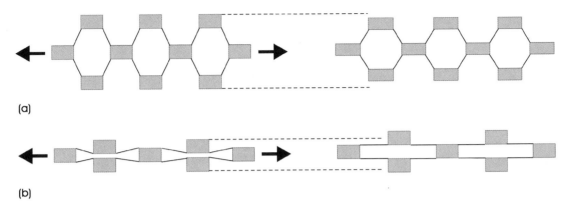

Figure 10.33 A solid composed of rigid blocks of crystal linked by bonds of constant length, but with flexible links to the crystallites: (a) A normal solid becomes thinner under tension (right-hand side) (c) A material with re-entrant bonds (left-hand side) is auxetic and becomes thicker under tension (right-hand side)

will result in a contraction along the lateral direction, perpendicular to the $(1\,0\,0)$ and $(0\,1\,0)$ planes. Poisson's ratio has a normal value of $+0.39$. However, a tensile stress applied along the $[1\,1\,1]$ direction will result in an expansion in the lateral direction, with a Poisson's ratio of -0.39.

The reason for this behaviour can be anticipated from the bonding and structure in the material. The main bonds in β-brass are between the copper and zinc atoms and are directed along the $\langle 1\,1\,1 \rangle$ directions, towards the corners of the unit cell from the central atom. Weaker bonds lying along the cell edges link the copper atoms (Figure 10.34a). In mechanical terms, it is as if the central atom is linked to the atoms at the unit cell corners by strong springs, and the atoms at the unit cell corners are linked to each other by weaker springs aligned along the unit cell edges. Tension along the $[0\,0\,1]$ direction will have a minimal effect on the strong bonds and be taken up by the weaker bonds (Figure 10.34b). In this case, the $\langle 1\,1\,1 \rangle$ strong bonds are not stretched or compressed but simply bend a little. As a result, the atoms at the cell corners move in slightly and the unit cell contracts perpendicular to the $(1\,0\,0)$ and $(0\,1\,0)$ faces, similar to a slight folding of a four-spoke umbrella (Figure 10.34c).

The arrangement of the atoms perpendicular to the $(1\,1\,1)$ plane is drawn in Figure 10.34d. Between the two apical zinc atoms, A and B, are two triangles of zinc atoms, with the copper atom sandwiched in the centre. Tension along the $[1\,1\,1]$ direction is along a line AB, and hence directly along a strong bond and strongly resisted. As the atoms A and B move apart, the triangles of zinc atoms move slightly closer together, rather like a three-spoke umbrella opening slightly. This causes an expansion of the structure in the direction perpendicular to line AB (Figure 10.34e).

It has been found that many cubic metals with structures related to the body-centred cubic structure are, in fact, auxetic. A number of silicates also show this unusual property. In each case, the relationship between the bonding and structure controls the response to the tensile stress.

10.3.3 Thin films

Films with a thickness of the order of nanometres are at the heart of microelectronics, and it has become apparent that the mechanical properties of these films must be known in order to ensure complete control over device manufacture and operation. There are a number of microscopic methods available for the determination of the mechanical properties of such films (Figure 10.35). The technique most widely used is the small-scale equivalent to the hardness test, called

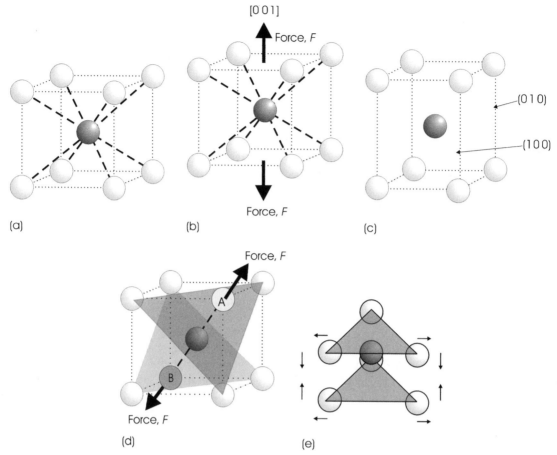

Figure 10.34 (a) The CsCl structure of β-brass (strong bonds are drawn as heavy broken lines, and weak bonds as light broken lines). The application of a force along the $[0\,0\,1]$ direction to the unit cell of β-brass (part b), results in contraction perpendicular to the $(1\,0\,0)$ and $(0\,1\,0)$ faces (part c). (d) The $(1\,1\,1)$ planes in β-brass; a force along the $[1\,1\,1]$ direction is directed along strong bonds, causing the triangles of zinc atoms above and below the central copper atom to come together slightly, which causes the structure to expand perpendicular to the $[1\,1\,1]$ direction (part e). Note: A and B are apical zinc atoms

nano-indentation. It has the advantage that the properties can be measured while the film is attached to a substrate and the surface can be tested in a large number of different places. It is also useful for measuring the properties of surfaces that have been modified by, for example, laser irradiation or optical coatings.

In a nano-indentation test, the load applied to the indenter and the resultant displacement are measured as a function of time. In this sense, the test mirrors the conventional tensile or compression test. The parameters obtained are the elastic modulus, Poisson's ratio and the hardness of the film. However, there are considerable differences between nano-indentation and conventional tests. For example, there is no clearly distinguished elastic region, and the deformation produced during nano-indentation involves both elastic and plastic deformation. Additionally, the area over which the load is applied changes continually during indentation and, because

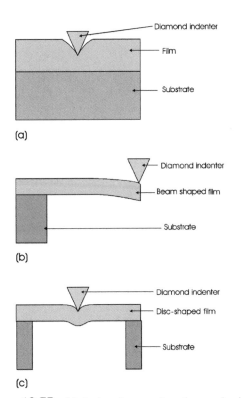

(a)

(b)

(c)

Figure 10.35 Methods of measuring the mechanical properties of thin films: (a) Nano-indentation, (b) beam deflection and (c) disc deflection. (Adapted from G.M. Pharr and W.C. Oliver, 1992, 'Measurement of Thin Film Mechanical Properties Using Nanoindentation', *Materials Research Society Bulletin* **XVII** (July) 28

of the small scale of the measurements, the interaction between the indenter, the film and the substrate cannot be ignored.

The indenter geometry is different from those used in large-scale testing (Figure 10.23). A diamond indenter with a triangular pyramidal shape, called a Berkovich indenter, is used (Figure 10.36a). The loads used can be as small as 0.01 μN, and displacements as small as 0.1 nm can be measured. The displacement of the indenter tip follows a different path on unloading compared with loading, and considerable hysteresis is found. This means that interpretation of the results is less obvious than for large-scale experiments.

A representative load–displacement curve for unloading is drawn in Figure 10.36b. The maximum load applied is F_{max} and the corresponding maximum displacement is h_{max}. The initial slope of the unloading curve, dF/dh, is a measure of the stiffness, $S_{initial}$, and is given by:

$$S_{initial} = \frac{dF}{dh} = \frac{F_{max}}{h_{max} - h_0}$$

where h_0 is the extrapolation of the initial slope to $F = 0$. The slope is related to the elastic modulus of the indenter, the film and the substrate.

If the indentation does not exceed approximately 20 % of the film thickness the substrate does not

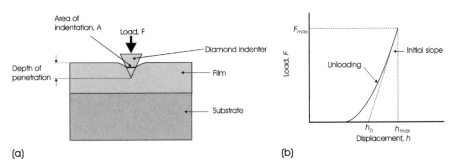

(a)   **(b)**

Figure 10.36 (a) The indentation of a thin film. The indentation area and depth will depend on the properties of the indenter and the substrate as well as the film being tested. (b) Schematic unloading curve for a nano-indentation experiment. Adapted from G.M. Pharr and W.C. Oliver, 1992, 'Measurement of Thin Film Mechanical Properties Using Nano-indentation', *Materials Research Society Bulletin* **XVII** (July) 28

need to be taken into account. In this case, the hardness of the film is given by:

$$H = \frac{F_{max}}{A}$$

where A is the area of the indentation. The value of the contact area, A, is a function of the depth of penetration, and must often be determined experimentally. For an ideally sharp indenter:

$$A = 24.5 \, h_0^2$$

The slope is related to the elastic modulus of the indenter, the film and the substrate, and when the substrate is ignored the initial slope is given by:

$$S_{initial} = \frac{2}{\sqrt{\pi}} E_r \sqrt{A}$$

where A is the area of indentation at maximum load, F_{max}, and E_r is the reduced elastic modulus, given by:

$$\frac{1}{E_r} = \frac{1 - \nu_f^2}{E_f} - \frac{1 - \nu_i^2}{E_i}$$

where the subscripts i and f refer to indenter and film, respectively, E is the relevant elastic modulus, and ν is the relevant value of Poisson's ratio.

10.4   Composite materials

Composites are solids made of more than one material type, designed to have enhanced properties compared with those of the separate materials themselves (see Sections 6.1.5, 6.2.4 and 6.4.5). The mechanical properties of composite materials are often difficult to obtain because of the complex microstructures found, especially in biological structures. However, in simple cases these can be modelled.

10.4.1   Elastic modulus of large-particle composites

The elastic modulus of a composite containing large particles depends on the volume fraction of the constituent phases. The elastic modulus falls between two limits, an upper and a lower limit:

$$E_c(\text{upper limit}) = E_m V_m + E_p V_p$$
$$E_c(\text{lower limit}) = \frac{E_m E_p}{E_m V_p + E_p V_m}$$

where E_c, E_m and E_p are the elastic moduli of the composite, matrix and particles, respectively, and V_c (equal to 1.0), V_m and V_p are the corresponding volume fractions.

The cemented carbides are large-particle composites. These materials are used as cutting tools to machine hard steels. The first cemented carbide made was formed of tungsten carbide particles embedded in a matrix of cobalt metal. The hard carbide cuts the steel but is brittle. The toughness comes from the cobalt matrix.

The large-particle composite in greatest use is concrete.

10.4.2   Elastic modulus of fibre-reinforced composites

The objective of fibre reinforcement is to endow a lightweight matrix with high strength and stiffness. A critical fibre length, l_c, is necessary to achieve this:

$$l_c = \frac{\sigma_f d}{2 \, \tau_c}$$

where d is the fibre diameter, σ_f is the (ultimate) tensile strength of the fibre, and τ_c is the shear yield strength of the matrix. The fibre length relative to the critical fibre length gives rise to two terminologies:

- if $l > 15 \, l_c$ the material is termed a continuous fibre composite;

- if $l < 15 \, l_c$ the material is termed a discontinuous fibre composite.

In practice, the fibre orientation, concentration and distribution all contribute to the final properties of

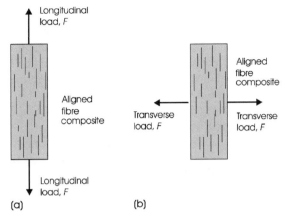

Figure 10.37 (a) A longitudinal load applied to an aligned fibre composite and (b) a transverse load applied to an aligned fibre composite

the composite. Moreover, the mechanical properties of the composite depend on whether the load is applied along the fibre direction or normal to it.

10.4.2.1 Longitudinal load on a continuous and aligned fibre composite

To obtain a value of the elastic modulus of a composite in which the fibres are both continuous and aligned parallel to each other, it is simplest to assume that the deformation of the fibres and the matrix is the same, that is, that they are firmly bonded together (Figure 10.37a). This is termed the iso-strain condition, in which case:

$$\varepsilon_c = \varepsilon_m = \varepsilon_f$$

where ε_c is the strain experienced by the composite as a whole, ε_m is the strain experienced by the matrix, and ε_f is the strain experienced by the fibre. In this case it is reasonable to suppose that the tensile stress, σ, is divided into parts acting on the fibre and on the matrix in proportion to the volume fraction of each. With these assumptions:

$$E_{cl} = E_m V_m + E_f V_f$$

where E_{cl} is the elastic modulus of the composite under a longitudinal load, E_m is the elastic modulus

of the matrix, E_f is the elastic modulus of the fibre, V_f is the volume fraction of fibre, given by

$$V_f = \frac{\text{volume of fibre}}{\text{total volume}}$$

and V_m is the volume fraction of the matrix, given by

$$V_m = \frac{\text{volume of matrix}}{\text{total volume}}$$

For two components:

$$V_m = 1 - V_f$$
$$E_{cl} = E_m(1 - V_f) + E_f V_f$$

This is the same equation as the upper bound for large-particle composites.

The fraction of the load carried by each component is given by:

$$\frac{F_f}{F_m} = \frac{E_f V_f}{E_m V_m}$$

where F_f is the load carried by fibre, and F_m is the load carried by the matrix.

10.4.2.2 Transverse load on a continuous and aligned fibre composite

When a load is applied in a transverse direction to a composite which contains continuous and aligned fibres (Figure 10.37b) the iso-strain condition is unreasonable and it is more appropriate to assume an iso-stress condition applies:

$$\sigma_c = \sigma_m = \sigma_f = \sigma$$

where σ represents the stress and the subscripts have the same meaning as in Sub-section 10.4.2.1. In this case:

$$E_{ct} = \frac{E_m E_f}{E_f V_m + E_m V_f}$$

$$= \frac{E_m E_f}{(1 - V_f)E_f + E_m V_f}$$

where E_{ct} is the modulus of elasticity in a transverse direction for the composite. This is the same equation as the lower bound for large-particle composites.

10.4.3 Elastic modulus of a two-phase system

Many ordinary solids, such as ceramics, are made up of several phases. Strictly speaking, these are not composite materials, but similar reasoning can be applied to obtain the elastic modulus and other mechanical properties of such systems. Although the equations for a solid composed of several phases with a complex microstructure are frequently unwieldy, simpler equations exist for well-defined geometries.

A ceramic body composed of two phases, one of which is distributed as particles within the matrix of the other, has a modulus of elasticity given by:

$$E_c(\text{upper limit}) = E_m V_m + E_p V_p$$

$$E_c(\text{lower limit}) = \frac{E_m E_p}{E_m V_p + E_p V_m}$$

where E_c, E_m and E_p are the elastic moduli of the ceramic, matrix and particles, respectively, and V_c (equal to 1.0), V_m and V_p are the corresponding volume fractions. These equations are identical to those for large-particle composites given in Section 10.4.1.

A ceramic body composed of layers aligned parallel to a uniaxial stress, in which the strain is shared equally by the two phases, the iso-strain condition, has an elastic modulus:

$$E_c(\text{upper limit or Voigt bound}) = E_m V_m + E_p V_p$$

This approximation is called the Voigt model, and the value of the elastic modulus is often known as the Voigt bound. The expression is identical to that for a continuous aligned fibre composite under a longitudinal load, and gives the elastic modulus when the load is applied parallel to the sheets. Similarly, if the stress is applied perpendicular to the layers, and an iso-stress condition applies (the Reuss model), the elastic modulus is:

$$E_c(\text{lower limit or Reuss bound}) = \frac{E_m E_p}{E_m V_p + E_p V_p}$$

The value of the elastic modulus, often called the Reuss bound, is identical to that for transverse loading on a fibre composite, and gives a value for the elastic modulus normal to the layers. In both of these equations, E_c, E_m and E_p are the elastic moduli of the ceramic, matrix and particles, respectively, and V_c (equal to 1.0), V_m and V_p are the corresponding volume fractions.

The presence of pores in ceramics usually leads to weakness. The elastic modulus of a body with a Poisson's ratio of 0.3, containing isolated closed pores, is described by the equation:

$$E_c = E_0(1 - 1.9 V_p + 0.9 V_p^2)$$

where E_c is the elastic modulus of the porous ceramic, E_0 is the elastic modulus of the nonporous ceramic, and V_p is the volume fraction of pores.

Answers to introductory questions

How are stress and strain defined?

Stress is defined with respect to the force applied to an object and is measured as the force applied to a unit area of the specimen. The application of a stress results in a dimensional change, which is called the strain.

For a rod-shaped specimen,

$$\text{stress}(\sigma_T) = \frac{F}{A}$$

where F is the force (or load) applied to the rod, and A is the cross-sectional area subjected to the force.

For practical purposes, it is often adequate to ignore the continuous change in cross-sectional area that occurs when a force is applied. The stress so defined is called the engineering (or nominal) stress.

$$\sigma = \frac{F}{A_0}$$

where F is the average force (or load) applied, and A_0 is the initial cross-sectional area of the sample.

The strain that is found on application of a load to a rod is equal to the elongation of the rod. The increment in tensile strain experienced, $\Delta \varepsilon$, when a rod is extended, is defined as the ratio of the increase in length, Δl, to the total length:

$$\text{strain}(\Delta \varepsilon_T) = \frac{\Delta l}{l}$$

The total strain is then given by:

$$\text{strain}(\varepsilon_T) = \int_{l_0}^{l} \frac{\mathrm{d}l}{l_0}$$

$$= \ln\left(\frac{l}{l_0}\right)$$

where l is the final length of the specimen, and l_0 is the original length. As strain is a ratio, it has no units. If the incremental changes are ignored, the engineering (or nominal) strain is:

$$\varepsilon = \frac{(l - l_0)}{l_0}$$

where l is the final length of the specimen, and l_0 is the original length of the specimen.

Many materials are used under compression rather than tension. At low loads, the compressed material behaves in a similar way to materials tested under tension. In compression tests, the value of the force is taken as negative and hence we have negative values of stress and strain compared with those obtained in tension.

Why are alloys stronger than pure metals?

Four principal methods have been used to strengthen metals. Three of these – reduction in grain size, work hardening and precipitation strengthening – are brought about by heat treatment and mechanical deformation (e.g. hammering). The fourth method – alloying – was for most of historical times based on empirical observation. In all cases, the metal is strengthened because dislocation movement is restricted.

Alloying, first widely used to transform soft copper into much stronger bronze, relies on the strain set up in the crystal structure by the impurity or dopant atoms. This strain field impedes dislocation movement because dislocations also generate a strain in the structure. The two strain components mutually repel each other to hinder slip. If sufficient of the second component is added, precipitates of a second phase can form in the crystal matrix. These hinder dislocation movement simply by blocking slip planes.

What are solid lubricants?

Solid lubricants, which are greasy to the touch, are highly anisotropic solids with a low shear strength in at least one dimension. Solid lubricants fall into three main classes – inorganic solids with a lamellar (layer-like) crystal structure, solids that suffer plastic deformation easily and polymers in which the constituent chains can slip past each other in an unrestricted way. The categories of most importance are layer structures and soft inorganic compounds.

The most familiar solid lubricant is graphite, which has a layer structure. The layers are about 0.335 nm apart and are linked by weak van der Waals bonds. Dry graphite is found to be a poor lubricant and, in reality, the lubrication properties are brought about by forcing the layers apart by inserting atoms or molecules into the van der Waals gap. In ordinary circumstances this is water, and much of the lubricating action of graphite seems to stem from adsorbed water vapour on and between the layers. Fluorinated graphite, CF_x, in which fluorine is inserted between the layers, is another very successful lubricant.

Molybdenum disulphide, MoS_2, is another layer structure widely used as a solid lubricant. It is composed of MoS_2 layers, weakly linked by van der Waals bonds. The Mo atoms are surrounded by six sulphur atoms in the form of a trigonal prism. Although the MoS_2 layers are closer together than the carbon layers in graphite, MoS_2 acts as a much better lubricant for most purposes.

Both graphite and molybdenum disulphide oxidise at higher temperatures in air, and the main alternatives used are soft inorganic fluorides. Like many ceramics, although they show brittleness at room temperature, they display ductility at temperatures of about half the melting point. A solid lubricant that is widely used in the temperature range from approximately 540 °C to 900 °C is the eutectic mixture of calcium fluoride, CaF_2, and barium fluoride, BaF_2. The melting point of the eutectic is 1022 °C.

Further reading

L.C. Bryant, B.P. Bewlay, 1995, 'The Coolidge Process for Making Ductile Tungsten', *Materials Research Society Bulletin* **XX** (August) 67.

M.E. Eberhart, 1999, 'Why Things Break', *Scientific American* **281** (October) 44.

S.S. Hecker, A.K. Ghosh, 1976, 'The Forming of Sheet Metal', *Scientific American* **235** (November) 100.

H.J. McQueen, W.J.M. Tegart, 1995, 'The Deformation of Metals at High Temperatures', *Scientific American* **232** (April) 116.

G.M. Pharr, W.C. Oliver, 1992, 'Measurement of Thin Film Mechanical Properties Using Nanoindentation', *Materials Research Society Bulletin* **XVII** (July) 28.

S.L. Semiatin, G.D. Lahoti, 1981, 'The Forging of Metals', *Scientific American* **245** (August) 82.

D.M. Teter, 1998, '*Computational Alchemy: The Search for New Superhard Materials*', *Materials Research Society Bulletin* **23** (January) 22.

Problems and exercises

Quick quiz

1 *A* material under a tensile force is:
   (a) Stretched
   (b) Twisted
   (c) Compressed

2 Torsional forces occur in a material that is:
   (a) Sheared
   (b) Twisted
   (c) Compressed

3 Both tensile and compressive forces occur in a rod that is:

   (a) Bent
   (b) Twisted
   (c) Stretched

4 For a rod-shaped specimen of a metal, the stress is defined as:
   (a) The change in length per unit length
   (b) The change in length per unit force
   (c) Force per unit area

5 For a rod-shaped specimen of a metal, the strain is defined as:
   (a) The change in length per unit length
   (b) The change in length per unit force
   (c) Force per unit area

6 The engineering strain is given by:
   (a) Force divided by the original cross-sectional area
   (b) The change in length divided by the original length
   (c) Force divided by the original length

7 The strength of a ceramic sample is usually tested by:
   (a) Stretching a bar
   (b) Bending a bar
   (c) Twisting a bar

8 The engineering stress versus engineering strain curve for a solid indicates:
   (a) Only elastic behaviour
   (b) Only ductile behaviour
   (c) Both elastic and ductile behaviour

9 The initial (linear) part of an engineering stress–engineering strain curve represents:
   (a) Plastic deformation
   (b) The tensile strength
   (c) Elastic deformation

10 In comparison with a strongly cross-linked polymer, a weakly cross-linked form of the polymer will have:
   (a) A higher Young's modulus
   (b) A lower Young's modulus
   (c) The same Young's modulus

11 Permanent deformation of a solid when all stress is removed is a sign of:
   (a) Plastic deformation
   (b) Elastic deformation
   (c) Tensile deformation

12 The defects mainly held responsible for plastic deformation are:
   (a) Point defects
   (b) Precipitates
   (c) Dislocations

13 The tensile strength of a solid is:
   (a) The maximum load that can be carried before fracture
   (b) The maximum load before plastic deformation occurs
   (c) The load when fracture occurs

14 The yield stress indicates the point at which:
   (a) The solid starts to thin down (neck)
   (b) The elastic behaviour changes to plastic behaviour
   (c) The solid starts to fracture

15 The specific strength of a material is:
   (a) The tensile strength divided by the specific gravity
   (b) The tensile strength divided by the volume
   (c) The tensile strength divided by the weight

16 An engineering stress–engineering strain curve does *not* give information on:
   (a) The modulus of elasticity of the material
   (b) The yield strength of the material
   (c) The hardness of the material

17 Superelasticity is caused by:
   (a) Hysteresis
   (b) Martensite formation
   (c) Deformation

18 The slope of the stress–strain curve in the elastic region is a measure of:
   (a) Young's modulus
   (b) The tangent modulus
   (c) The secant modulus

19 The lateral dimensional change that accompanies a longitudinal strain is:
   (a) Poisson's modulus
   (b) Poisson's ratio
   (c) Poisson's strain

20 Fracture of a polycrystalline solid that takes place between the crystallites is:
   (a) Intergranular fracture
   (b) Transgranular fracture
   (c) Cleavage

21 Conchoidal fracture is displayed by brittle solids that are:
   (a) Polycrystalline
   (b) Amorphous
   (c) Natural (biomaterials)

22 The Griffith theory of brittle fracture postulates that the fracture is due to:
   (a) Dislocations
   (b) Surface defects
   (c) Small cracks

23 During plastic deformation, dislocations move preferentially on:
   (a) Slip planes
   (b) Twin planes
   (c) Slip directions

24 Ceramics are more brittle than metals because:
   (a) They contain fewer dislocations than metals
   (b) Dislocation movement is simpler than in metals
   (c) Dislocation movement is more difficult than in metals

25 The initial process of plastic deformation in semicrystalline polymers is a result of deformation in:
   (a) The crystalline regions
   (b) The amorphous regions
   (c) The crystalline and the amorphous regions

26 The ductility of a metal can be estimated from:
   (a) The elongation at fracture

(b) The stress applied at fracture

(c) The strain at the yield point

27 Ceramics can be strengthened by:
(a) Immobilising dislocations

(b) Removing Griffith flaws

(c) Increasing the degree of crystallinity

28 One of the following does *not* describe hardness:
(a) Knoop

(b) Brinell

(c) Poise

29 Materials that fail after repeated cycles of stress are said to suffer:
(a) Fatigue failure

(b) Creep failure

(c) Catastrophic failure

30 Fatigue failure does not occur at stress levels below:
(a) The persistence limit

(b) The fatigue limit

(c) The endurance limit

31 The gradual elongation of a material under a constant load is called:
(a) Fatigue

(b) Creep

(c) Yield

32 Steady-state creep is defined by:
(a) The initial part of a creep curve

(b) The middle part of the creep curve

(c) The final part of the creep curve

33 Both Coble creep and Herring–Nabarro creep describe:
(a) Power law creep

(b) Creep due to dislocation movement

(c) Creep due to atomic diffusion

34 The temperature dependence of creep is often described by:
(a) An Arrhenius law

(b) A linear law

(c) A parabolic rate law

35 Solid lubricants often have:
(a) Layer structures

(b) Amorphous structures

(c) Liquid crystal structures

36 Materials that expand when under tension are called:
(a) Nanomaterials

(b) Eutactic materials

(c) Auxetic materials

37 The hardness of thin films is measured by using:
(a) A Vickers indenter

(b) A Knoop indenter

(c) A Berkovich indenter

38 Cemented carbides that are used in cutting tools are:
(a) Longitudinally reinforced fibre composites

(b) Large-particle composites

(c) Ceramic composites

Calculations and questions

10.1 A weight of 500 kg is hung from a 2 cm diameter rod of brass. What is the engineering stress?

10.2 A weight of 3500 kg load is applied to a 1.5 cm diameter rod of nickel. What is the engineering stress?

10.3 A steel wire 75 cm long and 1 mm in diameter is subjected to a load of 22 kN. The elastic modulus of the steel 201.9 GPa. Calculate the new length.

10.4 A rod of bronze 150 cm long and 3 mm in diameter is subjected to a load of 30 kN. The elastic modulus of the bronze is 105.3 GPa. Calculate the new length.

10.5 A rod of copper 60 cm long is subjected to a tensile stress of 300 MPa. The elastic modulus of copper is 129.8 GPa. Calculate the new length.

10.6 A rod of aluminium 100 cm long is subjected to a tensile stress of 250 MPa. The elastic modulus of aluminium is 70.3 GPa. Calculate the new length.

10.7 A cast-iron rod of length 200 mm and dimensions 10 mm × 20 mm is subjected to a load of 70 kN. An extension of 0.46 mm is observed. Calculate the Young's modulus of the cast iron.

10.8 A zinc bar of length 125 mm and dimensions 5 mm × 7.5 mm is subjected to a load of 40 kN. An extension of 1.23 mm is observed. Calculate the Young's modulus of zinc.

10.9 Calculate Poisson's ratio for a bar of metal originally 10 mm × 10 mm × 100 mm, which is extended to 101 mm, if there is no change in the overall volume of the sample.

10.10 A copper bar of 10-mm square section is subjected to a tensile load that increases its length from 100 mm to 102 mm. The value of Poisson's ratio for copper is 0.343. Calculate the new dimensions of the bar.

10.11 A brass rod of 12.5 mm diameter is subjected to a tensile load that increases its length from 150 mm to 151.5 mm. The value of Poisson's ratio for brass is 0.350. Calculate the new diameter.

10.12 A cylindrical titanium rod of diameter 15 mm is subjected to a tensile load applied along the long axis. The modulus of elasticity of the metal is 115.7 GPa, and Poisson's ratio is 0.321. Determine the magnitude of the load needed to produce a contraction in diameter of 5×10^{-3} mm if the deformation is elastic.

10.13 A steel rod of diameter 16.2 mm and length 25 cm is subjected to a force of 50 000 N in tension along the long axis. The modulus of elasticity is 210 GPa, and Poisson's ratio is 0.293. Determine (a) the amount that the specimen will elongate in the direction of the applied force and (b) the change in diameter of the rod.

10.14 A niobium bar of dimensions 15-mm square and of length 300 mm is subjected to a tensile force of 25 000 N. The elastic modulus of niobium is 104.9 GPa, and Poisson's ratio is 0.397. Determine (a) the engineering stress, (b) the elongation, (c) the engineering strain and (d) the change in the cross-section of the bar.

10.15 A tungsten rod of 12.5 mm diameter and of length 350 mm is subjected to a tensile force of 90 000 N. The elastic modulus of tungsten is 411.0 GPa, and Poisson's ratio is 0.280. Determine (a) the engineering stress, (b) the elongation, (c) the engineering strain and (d) the change in the diameter of the rod.

10.16 A tensile test specimen of magnesium has a gauge length of 5 cm. The metal is subjected to a tensile loading until the gauge markings are 5.63 cm apart. Calculate (a) the engineering stress and (b) the percentage elongation.

10.17 A tensile test specimen of brass has a gauge length of 5 cm. The metal is subjected to a tensile loading until the gauge markings are 6.05 cm apart. Calculate (a) the engineering stress and (b) the percentage elongation.

10.18 An aluminium alloy specimen of 3 mm diameter with 50 mm gauge length was tested to destruction in a tensile test. The results are given in Table 10.2. The maximum load applied was 8100 N and the final length between the gauge marks was 53.2 mm.

Table 10.2 Data for Question 10.18

Load/N	Extension/mm
1000	0.06
2000	0.180
3000	0.290
4000	0.402
5000	0.504
6000	0.697
7000	0.900
7500	1.297
8000	2.204
7150 (fracture)	3.200

(a) Plot an engineering stress versus engineering strain curve [Note: not shown in the answers at the end of this book.]

(b) Determine the modulus of elasticity of the alloy

(c) Determine the tensile strength of the alloy

(d) Determine the 0.2 % offset yield strength of the alloy

(e) Determine the percentage elongation at fracture

10.19 A steel specimen of 12 mm diameter with 50 mm gauge length was tested to destruction in a tensile test. The results are given in Table 10.3. The maximum load applied was 152 kN.

(a) Plot an engineering stress versus engineering strain curve [Note: not shown in the answers at the end of this book.]

(b) Determine the modulus of elasticity of the alloy

Table 10.3 Data for Question 10.19

Load/kN	extension/mm
10	0.030
20	0.064
30	0.098
40	0.130
50	0.170
60	0.195
70	0.218
80	0.256
90	0.294
100	0.335
110	0.400
120	0.505
130	0.660
140	0.898
150	1.300
152 (maximum)	1.500
150	1.700
140	1.960
133 (fracture)	2.070

(c) Determine the tensile strength

(d) Determine the 0.1 % offset yield stress of the alloy

(e) Determine the percentage elongation at fracture

10.20 Figure 10.38 shows the engineering stress–engineering strain behaviour of a carbon steel. Determine:

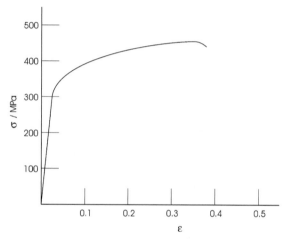

Figure 10.38 Engineering stress–engineering strain (σ–ε) curve of a carbon steel, for Question 10.20

(a) The modulus of elasticity

(b) The stress at 0.2 % offset strain (proof stress)

(c) The maximum load that can be sustained by a rod of diameter 12.5 mm

(d) The change in length of a rod originally 250 mm long subjected to an axial stress of 400 MPa

10.21 A tensile test carried out on a sample of polypropylene of dimensions 12.5 mm width, 3.5 mm thick and gauge length 50 mm gave the data in Table 10.4. Estimate:

Table 10.4 Data for Question 10.21

Force/N	Extension/mm
25	0.018
50	0.042
75	0.071
100	0.115
125	0.145
150	0.187
175	0.230
200	0.285
225	0.345
250	0.387
275	0.460
300	0.543
286 (break)	0.720

(a) the initial modulus;

(b) the secant modulus at 0.2 % strain;

(c) the tangent modulus at 0.2 % strain;

(d) the secant modulus at 0.4% strain;

(e) the tangent modulus at 0.4% strain;

(f) the percentage elongation at break.

10.22 A copper–nickel alloy has a 1 % offset yield strength of 350 MPa and a modulus of elasticity of 130 GPa.

(a) Determine the maximum load that may be applied to a specimen of cross-section 10 mm × 13 mm without significant plastic deformation occurring.

(b) If the original specimen length is 100 mm, what is the maximum length to which it can be stretched elastically?

10.23 Using the Griffith criterion, Equation (10.6), estimate the stress at which a glass plate containing a surface crack of 1.2 μm will fracture as a result of a force applied perpendicular to the length of the crack. The elastic modulus of the glass is 71.3 GPa and the surface energy of the glass is 0.360 J m^{-2}.

10.24 A glass plate has to withstand a stress of 10^8 N m^{-2}. Using the data Question 10.23, what will be the critical crack size for this to be achieved?

10.25 A plate of high-density polyethylene has a surface crack 7.5 μm in one face. The plate fractures in a brittle fashion when a force of 6 × 10^6 N m^{-2} is applied in a direction perpendicular to the crack. The elastic modulus of the polyethylene is 0.95 GPa. Estimate the surface energy of the material.

10.26 Determine (a) the upper-bound and (b) the lower-bound elastic modulus of an ingot of magnesium metal containing 30 vol% magnesia (MgO) particles. The modulus of elasticity of magnesium is 44.7 GPa and that of magnesia is 210.3 GPa.

10.27 An aluminium alloy is to be strengthened by the incorporation of beryllium oxide (BeO) particles. Calculate (a) the upper-bound and (b) the lower-bound elastic moduli of a composite consisting of 40 wt% alloy and 60 wt% BeO. The elastic modulus of the alloy is 70.3 GPa and its density is 2698 kg m^{-3}. The elastic modulus of BeO is 301.3 GPa and its density is 3010 kg m^{-3}.

10.28 Compute the modulus of elasticity of a composite consisting of continuous and aligned glass fibres of 50 % volume fraction in an epoxy resin matrix under (a) longitudinal and (b) transverse loading. The modulus of elasticity of the glass fibres is 76 GPa and that of the resin is 3 GPa.

10.29 Compute the modulus of elasticity of a composite consisting of continuous and aligned carbon fibres of 60 % weight fraction in an epoxy resin matrix under (a) longitudinal and (b) transverse loading. The modulus of elasticity of the carbon fibres is 290 GPa and the density is 1785 kg m^{-3}. The elastic modulus of the resin is 3.2 GPa and its density is 1350 kg m^{-3}.

10.30 Determine (a) the Voigt and (b) the Reuss bounds to the modulus of elasticity of a ceramic material consisting of layers of alumina and a high-silica glass. The modulus of elasticity of the alumina is 380 GPa and that of the glass is 72.4 GPa, and the glass represents 30 vol% of the solid.

10.31   A mineral with an approximate formula $Mg_7Si_8O_{23}$ has a structure that is made up of alternating layers with compositions of $7MgO$ and $8SiO_2$. Estimate the elastic modulus of the material when stressed (a) parallel and (b) perpendicular to the layers. The elastic modulus and density of MgO are 210.3 GPa and 3580 kg m^{-3}, respectively,

and for silica are 72.4 GPa and 2650 kg m^{-3}m, respectively.

10.32   The elastic modulus of sintered calcia stabilised zirconia with a porosity of 5 % is 151.7 GPa. Estimate the elastic modulus of completely-pore-free material.

10.33   The elastic modulus of sintered silicon carbide with 5 % porosity is 468.9 GPa. What is the porosity of a specimen with an elastic modulus of 350 GPa?

11

Insulating solids

- How are the relative permittivity and refractive index of a transparent solid related?

- What is the relationship between ferroelectric and pyroelectric crystals?

- How can a ferroelectric solid be made from a polycrystalline aggregate?

Solids have traditionally been divided into three classes when the electrical properties are described. Those that conduct electricity well are called conductors. This group is typified by metals. Those solids that conduct poorly are called semiconductors. This group contains elements such as silicon and germanium, and large numbers of minerals such as the iron sulphide fool's gold, FeS_2. Solids that do not conduct electricity are called insulators or dielectrics. Many oxides, such as magnesium oxide, MgO, and most polymers, such as polyethylene, are insulators.

This division is far too coarse to encompass the wide range of electrical properties that are now known. It is quite feasible to turn an insulating oxide into a very good 'metallic' conductor, and

'metallic' polymers are well known. However, all of the various possibilities can be understood in terms of the themes already presented – chemical bonding, crystal structure and microstructure. In this chapter the insulators are described. In Chapter 13 those materials that are conductors of electricity are discussed.

11.1 Dielectrics

11.1.1 Relative permittivity and polarisation

Insulators are explained in terms of chemical bonding as those solids in which the outer electrons are unable to move through the structure. They are localised in strong bonds if the material is considered to be a covalent compound, or else are restricted to the region close to an atomic nucleus if the compound is supposed to be ionic. In either case, these electrons are trapped and cannot move from one region to another.

Insulating materials are often referred to as dielectrics. One of the most important parameters used to describe an insulator is its dielectric constant, properly called the relative permittivity, ε_r.

Dielectrics form the working material in capacitors. A capacitor consisting of two parallel metal plates separated by a dielectric (including air) has a

Understanding solids: the science of materials. Richard J. D. Tilley
© 2004 John Wiley & Sons, Ltd   ISBNs: 0 470 85275 5 (Hbk) 0 470 85276 3 (Pbk)

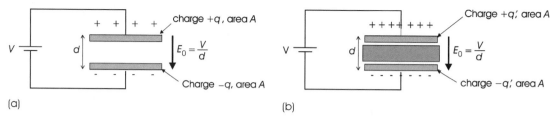

Figure 11.1 (a) Charges will accumulate on metal plates as a result of an applied voltage, V; the field between the plates, E_0, is equal to V/d. (b) A slab of dielectric (insulator) inserted between the plates will cause the charges on the plates to increase in proportion to its relative permittivity

capacitance, c, equal to the ratio of the charge on either of the metallic foils, q, to the potential difference, V, between them:

$$c = \frac{q}{V} \qquad (11.1)$$

The relative permittivity of a material may be determined by making it a part of such a capacitor. If we arrange for two parallel metallic plates to be connected to a battery, a certain amount of charge will accumulate on the plates (Figure 11.1). In a vacuum (or air, in practice), we find

$$q = \frac{\varepsilon_0 A V}{d}$$

where q is the charge on the capacitor, ε_0 is a constant, the permittivity of free space, A is the area of the plates, and d their separation. A comparison with Equation (11.1) shows that the term $\varepsilon_0 A/d$ is the capacitance, c, of the device:

$$c = \frac{\varepsilon_0 A}{d}$$

If the region between the plates is now filled with a dielectric the charge and the capacitance increase by an amount ε_r:

$$q' = \frac{\varepsilon_0 \varepsilon_r A V}{d}$$

where ε_r is the relative permittivity of the material and has no units. The new capacitance, c', is:

$$c' = \frac{\varepsilon_0 \varepsilon_r A}{d}$$

The relative permittivity describes the response of a solid to an electric field. The electric field is a vector quantity, E, which has a direction pointing from positive to negative. When an insulating material is exposed to an external electric field, E_0, arising from, for example, two charged metallic plates as in a capacitor, the negatively charged electrons and the positively charged nuclei will experience forces acting in opposite directions. The positive nuclei tend to move in the direction of the imposed electric field and the negative electrons in the opposite direction. The extent of this relative displacement will depend on how strongly the electrons are bound, the masses of the nuclei and the strength of the applied electric field. The displacement results in the formation of induced internal electric dipoles (Figure 11.2).

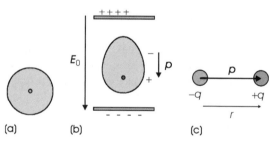

Figure 11.2 (a) For an atom in the absence of an electric field, the centres of gravity of the positive (nucleus) and negative (electrons) charges are coincident. (b) In an electric field these become separated to create a dipole. (c) An electric dipole consists of two equal and opposite charges separated by a distance r. The dipole moment is given by the vector p, which points from negative to positive. The dipole gives rise to an electric field, E, in the surrounding volume

Other constituents of the solid can also become polarised, and add to the effect, as described below.

Electric dipoles are vectors that run from the negative charge to the positive (Figure 11.2c). The magnitude of the dipole moment of a dipole, a vector p, is defined as:

$$p = q \times r$$

where q is the charge on the dipole, and r is the charge separation (see also Section S4.3). These induced dipoles add together with a result that opposite surfaces of the solid become positively and negatively charged (Figure 11.3). The solid becomes polarised. The polarisation of the dielectric, P, is a vector quantity, defined as the electric dipole moment per unit volume. The polarisation vector, P, points from the negative surface to the positive surface. In the case of a solid that is uniform in all directions (e.g. a glass or cubic crystal) it is found that at ordinary field strengths P is proportional and parallel to the applied electric field, E_0, and we can write:

$$P = \varepsilon_0 \chi E_0 \qquad (11.2)$$

where χ is called the dielectric susceptibility of the material, and ε_0 is the permittivity of free space. At higher electric field strengths, such as those found in laser beams, this relation can break down, and in this case it is better to replace the right-hand side of Equation (11.2) with a series, with $\varepsilon_0 \chi E$ as the first term (see Section 14.9.1). At normal field strengths, the electric susceptibility is related to the relative permittivity by the equation

$$\chi = (\varepsilon_r - 1)$$
$$P = (\varepsilon_r - 1)\varepsilon_0 E_0 \qquad (11.3)$$

11.1.2 Polarisability

The bulk polarisation observed when a material is placed in an electric field is a result of the polarisability of the constituent parts of the solid, which can be the atoms, ions and so on that make up the solid. For ordinary electric field strengths, the electric dipole moment, p, induced in a constituent, which might be an atom, for example, is proportional to the polarisability, α, of the constituent and the *local* electric field, E_{loc}, acting on the constituent, thus:

$$p = \alpha E_{loc}$$

Note that the local field, E_{loc}, is not the same as the applied field, E_0, as it will include contributions from other dipoles present in the structure. Some of these may be permanent dipoles, but others will be temporary dipoles, induced by the applied electric field itself. When the dipole moment is in units of C m and the electric field strength is in $V\,m^{-1}$, the units of α are $C\,m^2\,V^{-1}$. [These SI units are still rarely encountered in the literature, which mostly quotes values of the polarisability volume α', given in m^3, cm^3 or $\mathring{A}^3$. See Section S4.3 for the conversion between the two units.] If there are N dipoles per unit volume:

$$P = N \alpha E_{loc} \qquad (11.4)$$

In principle, all constituents of a solid, including defects and internal surfaces, contribute to the polarisability. Thus, a solid with two components, A and B, containing 10 units of A and 15 units of B, of polarisability α_A and α_B, would have a polarisability:

$$P = 10p_A + 15p_B = 10\,\alpha_A\,E_{loc} + 15\,\alpha_B\,E_{loc}$$

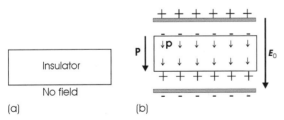

Figure 11.3 (a) In the absence of an applied electric field a dielectric has no surface charge. (b) In an electric field, E_0, the material has a surface charge as a result of the formation of internal dipoles, p, that induce an observable polarisation, P

Similarly, if there are N_j types of constituent j of polarisability α_j, in a solid, the observed polarisation, P, is:

$$P = \sum_j N_j p_j = \sum_j N_j \alpha_j E_{\text{loc}}$$

where the local electric field acting on the constituent, E_{loc}, may vary from site to site in the crystal. The most important sources of polarisation in insulating solids are derived from the atomic constituents that make up the material, as well as defects that may be present (Figure 11.4).

11.1.2.1   Electronic polarisability α_e

In the absence of an electric field, the electronic charge cloud surrounding an atom (at a little distance from the atom) is symmetrically disposed around the nucleus. In an electric field this charge cloud becomes deformed and the centre of the electronic negative charge is no longer coincident with the positive nuclear charge (Figure 11.4a), and a dipole will arise.

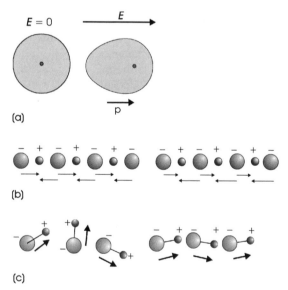

(a)

(b)

(c)

Figure 11.4   The effects of an electric field: (a) Electronic polarisation, (b) ionic polarisation and (c) orientational polarisation; dipoles are shown as arrows

11.1.2.2   Ionic polarisability α_i

Charged ions in a solid will suffer a displacement in an electric field. Similarly, the charge distribution in bonds between atoms can be altered by an electric field. This is illustrated for a linear chain of anions and cations in Figure 11.4(b).

11.1.2.3   Orientational polarisability α_d

A number of common molecules, including water, carry a permanent dipole. If such molecules are exposed to an electric field they will try to orient the dipole along the field (Figure 11.4c). As the movement of molecules in solids is restricted, orientational polarisability is more often noticed in gases and liquids.

11.1.2.4   Space charge polarisability α_s

If a material has mobile charges present they will move under the influence of the electric field, with positive charges moving towards one electrode and negative charges towards the other. These will tend to build up until the charge in the electrode regions, the space charge, inhibits further movement and equilibrium is reached. Good ionic conductors often show pronounced space charge effects.

11.1.2.5   Bulk polarisability α_{tot}

The observed bulk polarisability α_{tot} will arise from the sum of all the separate terms defined above. The total polarisability can be written as:

$$\alpha_{\text{tot}} = \alpha_e + \alpha_i + \alpha_d + \alpha_s$$

Note that other contributions can arise in some solids. In particular, if a solid contains a considerable number of defects these can make a significant contribution to the observed polarisation.

11.1.3   Polarisability and relative permittivity

To relate polarisability, α, to the relative permittivity, ε_r, it is necessary to remember that each constituent of the solid is polarised by a *local* electric field. This local field, E_{loc}, is not the same as the applied field, E_0, but will also include contributions from internal fields E_1, E_2, E_3 and so on arising from the induced and permanent dipoles in the structure:

$$E_{loc} = E_0 + E_1 + E_2 + E_3 + \cdots$$

Lorentz, using classical electrostatic theory, showed that the local field in an isotropic insulator such as a gas, a glass or a crystal with cubic symmetry is uniform everywhere and given by:

$$E_{loc} = E_0 + \frac{P}{3\,\varepsilon_0} \qquad (11.5)$$

Using Equations (11.3)–(11.5) it is possible to derive the most widely used relationship between relative permittivity and polarisability, the Clausius–Mossotti relation, Equation (11.6), usually written:

$$\frac{\varepsilon_r - 1}{\varepsilon_r + 2} = \frac{N\,\alpha}{3\,\varepsilon_0} \qquad (11.6)$$

where α is the polarisability of the material, and N is the number of atoms or formula units of structure per unit volume. If there are j types of atom or structural unit, all with differing polarisabilities, the sum of $N_j\,\alpha_j$ is needed. Remember that this equation is applicable only to homogeneous isotropic materials that do not contain permanent dipoles or dipolar molecules. In fact, this means that its use is restricted to glasses, amorphous solids and cubic crystals that show only electronic and ionic polarisability. However, it is often taken to be approximately true for crystals of lower symmetry, provided that they do not contain permanent dipolar molecules.

Several alternative forms of the Clausius–Mossotti equation are encountered. Frequently, the term N is replaced by its reciprocal, the volume of one atom or one formula unit of structure, $V_m = 1/N$ and is set out in terms of α, thus:

$$\alpha = 3\,\varepsilon_0\,V_m\,\frac{\varepsilon_r - 1}{\varepsilon_r + 2} \qquad (11.7)$$

Quite often, the equation is expressed in terms of the molar polarisability, P_m. This form is obtained by multiplying both sides of Equation (11.6) by M/ρ, where M is the molar mass of the material and ρ is its density, to obtain:

$$\frac{(\varepsilon_r - 1)M}{(\varepsilon_r + 2)\rho} = \frac{N_A\,\alpha}{3\,\varepsilon_0} = P_m, \qquad (11.8)$$

where $P_m = N_A\,\alpha/3\,\varepsilon_0$ and N_A is the Avagadro constant, given by MN/ρ.

As mentioned in Section 11.1.2, the literature mainly quotes the polarisability volume, α', rather than the SI polarisability, α. In this case, the common form of the Clausius–Mossotti equation encountered is:

$$\alpha' = \left(\frac{3\,V_m}{4\,\pi}\right)\left(\frac{\varepsilon_r - 1}{\varepsilon_r + 2}\right) \qquad (11.9)$$

where V_m is the volume of one formula unit of structure. The units of α' will be the same as those of V_m.

11.1.4 The frequency dependence of polarisability and relative permittivity

As the total polarisability of a material, α, is made up several contributions, the relative permittivity, ε_r, can also be thought of as made up from the same contributions. In a static electric field, all the various contributions will be important, and both α and ε_r will arise from electrons, ions, dipoles, defects and surfaces. However, if a variable, especially alternating, electric field acts on the solid the situation changes.

At low enough frequencies the value of the relative permittivity measured will be identical to the static value, and all polarisability terms will contribute to ε_r. However, space charge polarisation is usually unable to follow changes in electric field that occur much faster than that of radio frequencies, about 10^6 Hz, and this contribution will no longer be registered at frequencies higher than this value. Similarly, any dipoles present are usually unable to rotate to and fro in time with the alternations of the electric field when frequencies reach the microwave region, about 10^9 Hz, and at higher

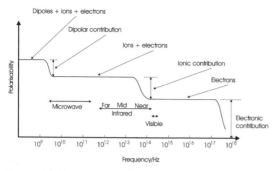

Figure 11.5 The contribution of electronic, dipole, ionic and orientation polarisability to the overall polarisability of a solid

frequencies this contribution will be lost. Ionic polarisability, involving the movement of atomic nuclei, is no longer registered when the frequency of the field approaches that of the infrared range, 10^{12} Hz. Electrons, being the lightest components of matter, still respond to an alternating electric field at frequencies corresponding to the visible region, 10^{14} Hz, but even the contribution of electronic polarisability 'drops out' at ultraviolet frequencies (Figure 11.5).

The interaction of an alternating electric field with a solid in the frequency range between the infrared and ultraviolet – the optical range – is more commonly expressed as the refractive index, because a lightwave consists of oscillating electric and magnetic fields. The relative permittivity in the optical region is related to the refractive index n by:

$$n^2 = \varepsilon_r \qquad (11.10)$$

This relationship as such is not well obeyed for most compounds if the static or low-frequency relative permittivity is used, as can be judged from Table 11.1. The relationship can be correctly interpreted by using the relative permittivity due to electronic polarisation in the equation. With this in mind, substitution of the relationship given in Equation (11.10) into the Clausius–Mossotti equation yields the Lorentz–Lorenz equation:

$$\frac{n^2 - 1}{n^2 + 2} = \frac{N\,\alpha_e}{3\,\varepsilon_0} \qquad (11.11)$$

The to-and-fro interaction of the components of a solid with an alternating electric field dissipates energy and results in heating of the dielectric. It also causes a lag between the phase of the input field and the phase of the output field. The action of an alternating electric field in best described by using the complex dielectric constant

$$\varepsilon_r = \varepsilon_0(\varepsilon' - i\,\varepsilon'')$$

Table 11.1   The relative permittivity and refractive index of some crystals

Compound	Symmetry	Relative permittivity, ε_r	Frequency Hz	Refractive index, n	n^2
Diamond	Cubic	5.66	10^3	2.418	5.85
Periclase, MgO	Cubic	9.65	10^2–10^8	1.735	3.010
Spinel, MgAl$_2$O$_4$	Cubic	8.6	–	1.719	2.955
Fluorite, CaF$_2$	Cubic	6.81	10^2–10^{11}	1.434	2.056
Corundum, Al$_2$O$_3$	Hexagonal		10^2–10^9		
perpendicular to c		9.34		1.761	3.101
along c		11.54		1.769	3.129
Beryl, Be$_3$Al$_2$Si$_6$O$_{18}$	Hexagonal		10^3		
perpendicular to c		6.86		1.589	2.525
along c		5.95		1.582	2.503
Rutile TiO$_2$	Tetragonal		10^4–10^6		
along a and b		86		2.609	6.807
along c		170		2.900	8.410

where $i = \sqrt{-1}$. The loss tangent, $\tan \delta$, then specifies the phase lag, where:

$$\tan \delta = \frac{\varepsilon''}{\varepsilon'}$$

The loss tangent is a measure of the energy loss in a capacitor. For good dielectrics, $\tan \delta$ is about 10^{-4} and is relatively insensitive to the frequency of the applied field.

11.1.5 Polarisation in nonisotropic crystals

In the preceding discussion it has not, strictly speaking, been necessary to regard the electric field and polarisation as vectors. However, in most crystalline solids the direction of an applied electric field is unlikely to be parallel to the induced polarisation. When the shapes of groups of atoms in a crystal are considered it is reasonable to think that polarisation will be easier in some directions than in others. To treat this it is usual to define a set of orthogonal axes in the crystal and refer the applied electric field to these axes. These axes are conveniently taken to coincide with the crystallographic axes for tetragonal and orthorhombic systems. In hexagonal systems one axis is taken to coincide with the crystallographic c axis, and the other two are normal to the c axis. In monoclinic and triclinic crystals it is still possible to define three Cartesian axes, although the relationship between these and the crystallographic axes is not so simple.

The relative permittivity (as well as the refractive index) for such crystals is quoted as three values corresponding to the polarisations projected onto the axes. Some representative values are given in Table 11.1.

It may sometimes be necessary to estimate the polarisability of a solid in the absence of experimental data. Polarisability is not particularly easy to measure, but the relative permittivity is. The Clausius–Mossotti equation, Equation (11.6), is generally used to obtain polarisability from relative permittivity. The equation gives reasonable values for isotropic solids showing only ionic and electronic polarisation. If the refractive index is known, the Lorentz–Lorenz equation, Equation (11.11), will yield the electronic polarisability of the material. Hence, by difference, the ionic polarisability can be estimated.

In the absence of relative permittivity data for the solid under consideration it is possible to make use of the additivity rule. In its simplest form, we can write:

$$\alpha(\text{compound}) = \sum \alpha(\text{components})$$

For example, for an oxide mineral:

$$\alpha(\text{mineral}) = \Sigma \, \alpha(\text{component oxides})$$
$$\alpha(\text{Mg}_2\text{SiO}_4) = 2 \, \alpha(\text{MgO}) + \alpha(\text{SiO}_2)$$

A more extended form of the additivity rule is obtained if the 'components' are actual ions or atoms. In the example above, we would then write:

$$\alpha(\text{Mg}_2\text{SiO}_4) = 2 \, \alpha(\text{Mg}^{2+}) + \alpha(\text{Si}^{4+}) + 4 \, \alpha(\text{O}^{2-})$$

11.2 Piezoelectrics, pyroelectrics and ferroelectrics

11.2.1 The piezoelectric and pyroelectric effects

In a normal dielectric, the observed polarisation of the material is zero in the absence of an electric field, and this does not change if the material is heated or subjected to mechanical deformation. In a piezoelectric solid a surface electric charge develops when the solid is subjected to a mechanical stress such as pressure, even in the absence of an external electric field. This is called the direct piezoelectric effect. The effect is reversible and the inverse (or converse) piezoelectric effect, in which a voltage applied to a crystal causes a change in shape, also occurs in piezoelectric crystals. The piezoelectric effect generally varies from one direction to another in a crystal, and in some directions a crystal may show no piezoelectric effect at all whereas in other directions it is pronounced.

Piezoelectric solids are a subset of dielectrics. All piezoelectrics are dielectrics, but only some dielectrics are piezoelectrics.

In the case of a pyroelectric solid a change of temperature induces a polarisation change. The change in polarisation found on heating is reversed on cooling. Pyroelectric crystals are a subset of piezoelectrics. All pyroelectric crystals are piezoelectrics, but not all piezoelectrics demonstrate pyroelectricity. A material that is a pyroelectric is found to possess a spontaneous polarisation, P_s. This means that a pyroelectric crystal shows a permanent polarisation that is present both in the absence of an electric field and in the absence of mechanical stress.

The relative permittivity values normally encountered in crystals are rather small (Table 11.1). Some crystals, however, exhibit relative permittivity values many orders of magnitude higher than those in normal dielectrics. For example, one crystallographic polymorph of barium titanate, $BaTiO_3$, has a relative permittivity, ε_r, of the order of 20 000 (more values are given in Table 11.2.). By analogy

to magnetic behaviour, this behaviour is called ferroelectricity, and the materials are called ferroelectrics. Ferroelectrics also possess a spontaneous polarisation, P_s, in the absence of an electric field and a mechanical distortion. They are, therefore, a subset of pyroelectrics and, as such, all ferroelectrics are also pyroelectrics and piezoelectrics. The feature that distinguishes ferroelectrics from pyroelectrics is that the direction of the spontaneus polarisation, P_s, can be switched (changed) in an applied electric field, as described below.

The hierarchy of insulating properties can be summarised thus (Figure 11.6):

- If polarisation, P, changes with applied electric field, E, we have a dielectric.

- In some dielectrics the polarisation, P, can arise from mechanical stress, σ, to give a piezoelectric.

- In some piezoelectrics, there is a spontaneous polarisation, P_s, when the applied electric field,

Table 11.2 Ferroelectrics and antiferroelectrics

Compound	Formula	T_C/K	$P_s/C\,m^{-2}$	Approximate ε_r
Hydrogen-bonded compounds:				
Rochelle salt	$NaK(COO \cdot CHOH)_2 \cdot 4\,H_2O$	298	0.01	5×10^3
Triglycine sulphate	$(NH_2CH_2COOH)_3 \cdot H_2SO_4$	322	0.03	2×10^3
Potassium dihydrogen sulphate	KH_2PO_4	123	0.05	6×10^5
Polar groups:				
Sodium nitrite	$NaNO_2$	436	0.08	1.1×10^3
Perovskites:				
Barium titanate	$BaTiO_3$	403	0.26	1×10^4
Lead titanate	$PbTiO_3$	763	0.80	9×10^3
Potassium niobate	$KNbO_3$	691	0.30	4.5×10^3
Tungsten bronzes:				
Sodium barium niobate	$Ba_2NaNb_5O_{15}$	833	0.40	6×10^4
Antiferroelectrics:				
Tungsten trioxide	WO_3	1010	0	300*
Ammonium dihydrogen phosphate	$NH_4H_2PO_4$	148	0	57, 10*
Lead hafnate	$PbHfO_3$	476	0	200*
Lead zirconate	$PbZrO_3$	503	0	150*
Sodium niobate	$NaNbO_3$	627	0	700, 70*

*Varies with crystal direction. Upper and lower values are given when these differ substantially.
Note: T_C, Curie temperature; P_s, spontaneous polarisation; ε_r, relative permittivity.

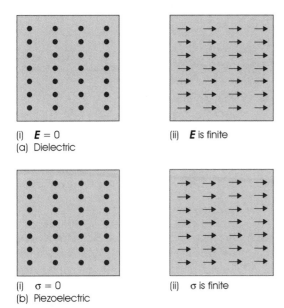

(i) **E** = 0   (ii) **E** is finite
(a) Dielectric

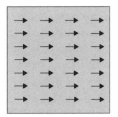

(i) σ = 0   (ii) σ is finite
(b) Piezoelectric

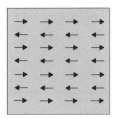

(c) Pyroelectric and ferroelectric, **E** = 0, σ = 0

(d) Antiferroelectric: **E** = 0, σ = 0

Figure 11.6 Schematic relationship between dielectric solids (**E** is an applied electric field, and σ is an applied stress). (a) Dielectric: (i) **E** = 0, (ii) **E** is finite; a dielectric, normally unpolarised, becomes polarised is an electric field). (b) Piezoelectric: (i) σ = 0, (ii) σ is finite; (a piezoelectric, normally unpolarised, develops a polarisation when subjected to stress, even is no electric field). (c) Pyroelectric and ferroelectric: **E** = 0, σ = 0. (d) Antiferroelectric: **E** = 0, σ = 0 (pyroelectric, ferroelectric and antiferroelectric solids contain dipoles when both electric field and stress are zero)

E, and the stress, σ, are zero, which changes with temperature, T, to give pyroelectrics.

- In some pyroelectrics the spontaneous polarisation, P_s, is easily switched in an electric field, to give a ferroelectric.

11.2.2   Piezoelectric mechanisms

A piezoelectric crystal develops surface charges as a result of bulk polarisation due to the formation of internal dipoles or to the rearrangement of existing dipoles. To give an idea of how polarisation can be produced on the application of pressure, two examples are described below.

11.2.2.1   Example 1: metal–oxygen tetrahedra

In the first example, suppose that a crystal is built up of metal–oxygen MO_4 tetrahedra. (Note: piezoelectricity is not confined solely to crystals containing tetrahedral groups.) In an ideal MO_4 tetrahedron the centre of gravity of the negative charges, arising from the combined effects of the oxygen atoms and the chemical bonds, will coincide with the centre of gravity of the positive charges arising in the metal atom, M (Figure 11.7a). A force applied to the top of a tetrahedron will cause a deformation. The oxygen–metal bond in line with the force will resist deformation most, as the positive metal and negative oxygen atoms are being forced together. The basal triangle of oxygen atoms will be flattened and as there are no metal atoms to oppose this change directly, it will occur to a greater degree than will the other deformation. The centre of gravity of the negative charges will no longer coincide with the centre of gravity of the positive charge, and a dipole will result.

The direction of the force is important and in some directions stress will not cause polarisation to occur. A force directed perpendicular to a tetrahedron edge (Figure 11.7b) will deform all bonds equally, and not give rise to any dipoles.

When this idea is applied to a crystal, the results for all the tetrahedra need to be added together. If

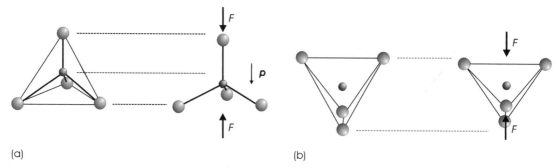

Figure 11.7 (a) A load, F, applied to a tetrahedron along a bond gives rise to a dipole as a result of distortion. (b) A load, F, applied perpendicular to a tetrahedron edge does not

the unit cell in the crystal has a centre of symmetry the overall polarisation of adjacent tetrahedra must add to zero. If the structure lacks a centre of symmetry the dipoles will add to give the unit cell an overall dipole moment. In this case an external polarisation will appear. Thus, piezoelectric materials are characterised by a lack of a centre of symmetry. In fact, there are 21 crystal classes that lack a centre of symmetry and, of these, 20 are piezoelectric classes.

11.2.2.2   Example 2: dipole-containing crystals

The piezoelectric effect can also be generated in a crystal already containing dipoles. In some materi-

als the elementary dipoles add to zero in the absence of stress. When the crystal is deformed the dipole directions rotate slightly, so that an overall polarisation is observed. This happens in quartz, SiO_2, one of the best-known piezoelectric solids. Although frequently drawn in a idealised form, with regular $[SiO_4]$ tetrahedra (Figure 11.8a), at room temperature the tetrahedra are considerably distorted, and each gives rise to a small permanent dipole (Figure 11.8b). In the unstressed structure these cancel, to give no net polarisation. When stress is applied, the tetrahedra distort slightly (Figure 11.8c), with a consequence that the dipoles no longer cancel and an overall polarisation is produced.

In the case of crystals showing spontaneous polarisation, the elementary dipoles are already

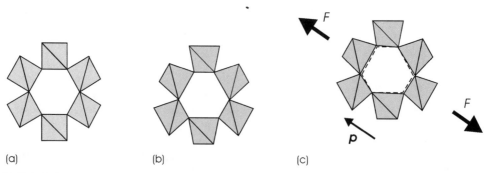

Figure 11.8 (a) Part of the idealised structure of high-temperature (β) quartz drawn as corner-connected $[SiO_4]$ tetrahedra, projected down the c axis (note that the tetrahedra are arranged in a helix, not in rings). (b) Part of the structure of room-temperature (α) quartz; the tetrahedra are distorted and each gives rise to an electric dipole, but these add to zero in the unit shown and over a unit cell. (c) Application of a load, F, to the structure in the direction drawn distorts the structure from the unstressed form, shown as dotted lines, so that the dipoles no longer cancel. This leads to the overall dipole, p. Similar diagrams can be drawn for other directions; not all give rise to observable dipoles

ordered. In this case the application of stress will cause a change in the relative dispositions of the dipoles. This will lead to a change in the observed polarisation. In these solids the piezoelectric effect is a measurement of the change in polarisation that has accompanied the stress.

The requirement that the piezoelectric effect is restricted to noncentrosymmetric crystals implies that piezoelectricity should not be observed in a polycrystalline solid. This is because the individual grains will polarise in random directions that will cancel overall. It is possible to get around this problem in some piezoelectric materials, as described in Section 11.3.8.

11.2.3   Piezoelectric polymers

At first sight it might seem surprising that polymers can exhibit piezoelectricity, but it is so. Indeed, the requirements to produce piezoelectricity are the same as those just given. That is, the material should contain pressure-induced or pressure-sensitive elementary dipoles, and these should be incorporated into a crystalline matrix that lacks a centre of symmetry. Piezoelectric polymers generally rely on permanent dipoles on the polymer chains. There are two main sources of these dipoles in polymers: strongly polar bonds such as carbon–fluorine (C–F), carbon–chlorine (C–Cl), carbon–nitrogen (C–N) and hydrogen bonds. Polar carbon–fluorine bonds are found in polymers such as poly(vinyl fluoride), $[CH_2-CHF]_n$, known as PVF (Figure 11.9a), and poly(vinylidene fluoride), $[CH_2-CF_2]_n$, known as PVF_2 (Figure 11.9b). The negative end of the dipole is located on the fluorine atom, $C \leftarrow F$, and a smaller dipole is found on the carbon–hydrogen bond, $C \rightarrow H$ (Figure 11.9a.i). In PVF, the overall dipole moment is greatest in the isotactic form of the polymer, in which all of the fluorine atoms are on the same side of the carbon–carbon backbone (Figure 11.9a.ii). As would be expected, atactic polymers, in which the fluorine atoms have a random distribution do not show a significant piezoelectric effect. The polymer PVF_2 also has an overall dipole composed of two $C \leftarrow F$ dipoles (Figure 11.9b.i) opposed by two $C \rightarrow H$

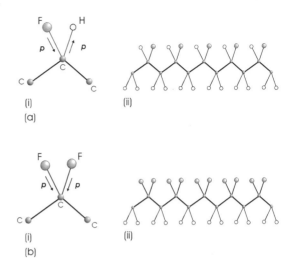

Figure 11.9 (a) Poly(vinyl fluoride), PVF, $[CH_2-CHF]_n$: (i) dipoles present in a tetrahedral unit of PVF and (ii) isotactic structure of a polymer chain of PVF. (b) Poly(vinylidene fluoride), PVF_2, $[CH_2-CF_2]_n$: (i) dipoles present in a tetrahedral unit of PVF_2 and (ii) isotactic structure of a polymer chain of PVF_2

dipoles. The isotactic form of the polymer (Figure 11.9b.ii) has the highest net dipole moment. Defects in the chain, especially caused by irregular linking of the monomer units during polymerisation, reduce the overall dipole moment of the chains.

Hydrogen bonding produces the polarisation in polyamides, better known as nylons. The relative configurations of the hydrogen-bond dipoles depend on the spacing between the amide groups along the polymer chain (Figure 11.10). In the case of even polymers, such as nylon 6 (Figure 11.10a), dipoles are opposed along the chain. In odd polymers, such as nylon 5 (Figure 11.10b), the dipoles are aligned to give an observable polarisation.

In addition to the presence of elementary dipoles, it is important for the polymer to crystallise or partly crystallise into noncentrosymmetric structures. The polymer chains can usually pack together in several different ways. For example, poly(vinylidene fluoride), PVF_2, can crystallise in four forms. The arrangement of the chains in one nonpolar and one polar form is drawn schematically in Figure 11.11. Naturally, the degree of crystallinity of the

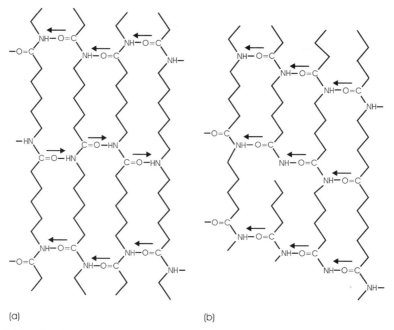

Figure 11.10 (a) The electric dipoles present in chains of an even nylon, nylon 6; no overall dipole moment is observed. (b) The electric dipoles present in chains of an odd nylon, nylon 5; the dipoles add to produce an observed dipole moment

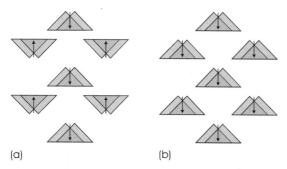

Figure 11.11 The schematic crystal structure of two forms of poly(vinylidene fluoride), PVF$_2$, viewed down the polymer chains, shown as double triangles; the electric dipoles in the chains are drawn as arrows. (a) The dipoles cancel in a centrosymmetric structure and the material is a nonpiezoelectric. (b) The dipoles add together in a non-centrosymmetric structure and the material is piezoelectric

polymer strongly influences the magnitude of the observed piezoelectric effect. Careful processing is important in the production of good piezoelectric films.

Piezoelectric plastic sheets can also be fabricated. These materials are known as electrets. Electrets are thin polymer films of high resistance that are polarised in a high field or by having an electric charge 'sprayed' onto the surface from a discharge. The resistance is so high that they retain the polarisation so induced permanently. In these materials, the dipole moment is very large, because of the large separation (in atomic terms) of the charges on the opposed faces of the plastic.

Although polymer piezoelectrics do not generally have as high piezoelectric coefficients as ceramic materials, they have some important advantages. Among other things, polymer films are of low density and are flexible, which makes them suitable for use in sensors and transducers in microphones, keyboards and flat-panel speakers.

11.2.4 The pyroelectric effect

As noted briefly in Section 11.2.1, a pyroelectric crystal possesses a spontaneous polarisation, P_s. The polarisation is a result of elementary dipoles in the crystal that are aligned to give an observable external bulk polarisation at all times (Figure 11.6c). Despite this, it is a matter of common observation that a pyroelectric crystal does not usually show an external charge. This is because the surface charges are neutralised by ions or other charged particles picked up from the air. Nevertheless, when a pyroelectric crystal is heated or cooled the spontaneous polarisation will change as a result of the thermal expansion of the solid, but the collection of neutralising particles will take time to arrive, and a pyroelectric effect will be seen. Pyroelectric crystals kept clean and in a vacuum maintain their surface charges for many days.

The relationship between the change in spontaneous polarisation, ΔP_s and the change in temperature, ΔT, can be written as:

$$(\Delta P_s)_i = \pi_i \, \Delta T$$

where π_i is the pyroelectric coefficient, with units of $C\,m^{-2}\,K^{-1}$, and i takes values of 1, 2 or 3 and refers to the (unique) x, y or z axis. Typical values of π are of the order of $10^{-5}\,C\,m^{-2}\,K^{-1}$.

As in the case of piezoelectrics, the elementary dipoles will cancel out if the crystallographic unit cell has a centre of symmetry. However, another condition is also needed to produce a spontaneous polarisation, the presence of a unique polar axis, which is a direction in the crystal unrelated by symmetry to any other direction, not even the antiparallel direction. The dipoles lie parallel to the polar axis of the crystal (see Section 5.1.3). Of the 20 piezoelectric crystal classes, only 10 fulfil this criterion and give rise to the pyroelectric effect. The relationship between the appearance of piezoelectricity and pyroelectricity and the symmetry of the crystal is set out in Figure 11.12.

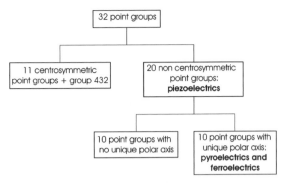

Figure 11.12 The relationship between point group and piezoelectric and pyroelectric properties

The pyroelectric effect that is normally observed in a crystal is, in fact, composed of two separate effects called the primary (or true) pyroelectric effect and the secondary pyroelectric effect. If a crystal is fixed so that its size is constant as the temperature changes, the primary effect is measured. Normally, though, a crystal is unconstrained. An additional pyroelectric effect will now be measured, the secondary pyroelectric effect, caused by strains in the crystal produced by the thermal change. In general, the secondary effect is much greater than the primary effect, but both are utilised in devices.

Among the structurally simplest pyroelectrics are hexagonal ZnO (zincite) and the isostructural hexagonal ZnS (wurtzite). In these crystals, the structure is built of layers of metal and nonmetal atoms, with the metals surrounded by a tetrahedron of nonmetals (Figure 11.13). The tetrahedra are

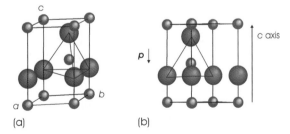

Figure 11.13 (a) The structure of hexagonal ZnO (zincite); a ZnO_4 tetrahedron is outlined. (b) An electric dipole, p, parallel to the c axis, arises in the unsymmetrical ZnO_4 tetrahedron; the wurtzite form of ZnS is isostructural

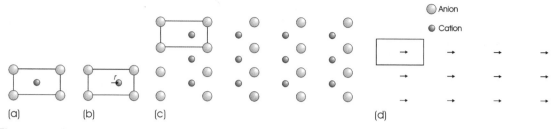

Figure 11.14   (a) A cation-centred unit cell; (b) cation displacement, *r*, creating an electric dipole in each cell; (c) an array of unit cells; and (d) an array of electric dipoles

slightly flattened, which gives rise to an electric dipole lying parallel to the polar axis, the *c* axis. Materials showing the pyroelectric effect are used as infrared radiation detectors.

11.3   Ferroelectrics

11.3.1   Ferroelectric crystals

Ferroelectrics are distinguished from pyroelectrics by virtue of the fact that the spontaneous polarisation, P_s, can be switched in direction. A variety of crystallographic features can result in ferroelectric behaviour, and many different chemical compounds are classified as ferroelectrics. Some details are given below for three important classes of ferroelectrics – those involving hydrogen bonds, those involving polar groups and those involving medium-sized transition-metal cations.

Because of this wide variation, it is preferable to explain the process by way of a simple model. Suppose we have a rectangular array of anions and that the structure gains stability when the cations are displaced slightly from the centre of the surrounding anion coordination polyhedron (i.e. if the cations are placed as illustrated in Figure 11.14b rather than as in Figure 11.14a). (This frequently happens for relatively small cations.) The centre of gravity of the anion array will not now coincide with the positive cation, and each 'rectangle' in the structure now contains a dipole (Figure 11.14c). Repetition of this motif results in an aligned dipole array characteristic of a ferroelectric (Figure 11.14d).

The cation displacement can take place in one of two directions. A plot of the potential energy of the cation against position will have two minima, separated by a potential energy barrier, ΔU, corresponding to the two alternative displacements (Figure 11.15). During crystallisation a cation in a crystal nucleus might occupy either of these positions at random. Thereafter, local interactions tend to make cations in adjoining 'rectangles' line up so as to form a parallel set of dipoles. Crystal nuclei in different parts of the crystal will take any of the two possible orientations. As growth continues the crystallites ultimately touch and the crystal contains domains of differing polarisation. For example, in Figure 11.14(d) the cation displacement could also occur in the opposite direction to that drawn, resulting in the domain boundary shown in

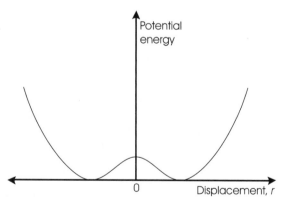

Figure 11.15   The variation of potential energy with cation displacement, $\pm r$, from the centre of a surrounding anion polyhedron

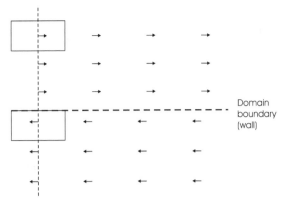

Figure 11.16 Domains due to the differing alignment of dipoles in adjacent regions of a crystal. The regions are separated by a domain wall, which extends over several tens of nanometres in practice

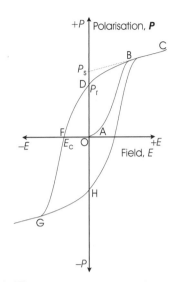

Figure 11.17 Hysteresis behaviour of the polarisation, *P*, in relation to the applied electric field, *E*, for a ferromagnetic crystal. As the field takes values between $+E$ and $-E$, the polarisation, *P*, takes values between $+P$ and $-P$

Figure 11.16. The number of different domains that are found in an actual crystal will depend on the number of different displacement directions that are possible. This is generally greater than two, and domain structures can be complicated. In general, crystals that exhibit a domain structure are called ferroic materials.

11.3.2   Hysteresis in ferroelectric crystals

In general, a ferroelectric crystal will be composed of an equal number of domains oriented in all the equivalent directions allowed by the crystal symmetry. The overall polarisation of the crystal will be zero. If we now apply a small electric field, *E*, in a nominally positive direction, the crystal will behave like a normal dielectric, as the value of *E* is not great enough to overcome the energy barrier ΔU (Figure 11.15). In Figure 11.17 this corresponds to the segment O–A. As *E* increases, cations will start to gain sufficient energy to overcome the energy barrier and will be able jump from one potential well to the other. The elementary dipole direction will switch. Gradually, all of the domains will change orientation and the observed polarisation will now increase rapidly, corresponding to section A–B of Figure 11.17. Ultimately, all of the dipoles will be aligned parallel and the crystal will, in

principle, consist of a single domain. This is the state of saturation, B–C (Figure 11.17). On reducing and then reversing the applied electric field the converse takes place. Gradually, dipoles switch direction, following path C–D–F–G, to reach saturation, with dipoles pointing in the opposite direction, at G (Figure 11.17). Reversal of the electric field again causes a reversal of dipole direction, and the curve will follow the path G–H–C. This closed circuit is called a hysteresis loop. The value OD is called the remanent polarisation, P_r, and OF is called the coercive field, E_c. Extrapolation of the linear portion of the curve, B–C, to $E = 0$ gives the value of the spontaneous polarisation P_s. The most important characteristic of a ferroelectric is that the spontaneous polarisation can be reversed by the application of a suitably oriented electric field.

11.3.3   Antiferroelectrics

Ferroelectricity is governed by two types of factors: (a) chemical bonds, which are short-range forces, and (b) dipolar interactions, which are long-range forces (Section 3.1.1). Calculations of the energy of ferroelectric crystals indicate that a minimum energy results when all the elementary dipoles are parallel or all the dipoles are in an antiparallel arrangement (Figure 11.6). The antiparallel arrangement is found in antiferroelectrics. Dielectric measurements are needed to establish the antiferroelectric behaviour of a material.

The balance between ferroelectric and antiferroelectric states is delicately poised, and some antiferroelectrics readily transform to ferroelectric states. This transformation is often accompanied by a change in the crystal structure of the solid. For example, orthorhombic lead zirconate, $PbZrO_3$, which is antiferroelectric, can transform to rhombohedral lead zirconate, $PbZrO_3$, which is ferroelectric. In the system $PbZrO_3$–$PbTiO_3$, as the smaller Ti^{4+} ion replaces the larger Zr^{4+} ion, the antiferroelectric phase is replaced by a ferroelectric state. Some ferroelectrics and antiferroelectrics are listed in Table 11.2.

11.3.4 The temperature dependence of ferroelectricity and antiferroelectricity

The fact that an applied field can cause the polarisation to alter its direction implies that the atoms involved make only small movements and that the energy barrier between the different states is low. With increasing temperature the thermal motion of the atoms will increase, and eventually they can overcome the energy barrier separating the various orientations. Thus at high temperatures the distribution of atoms becomes statistical and the crystal behaves as a normal dielectric and no longer as a polar material. This is referred to as the paraelectric state. The temperature at which this occurs is known as the Curie temperature, T_c, or the transition temperature. The relative permittivity often rises to a sharp peak in the neighbourhood of T_c.

The temperature dependence of the relative permittivity of many ferroelectric crystals in the paraelectric state can be described fairly accurately by a relationship called the Curie–Weiss law:

$$\varepsilon_r = \frac{C}{T - T_c} \qquad (11.12)$$

where ε_r is the relative permittivity, C is a constant, T_c is the Curie temperature, and T is the absolute temperature. The value of the constant, C, is determined by a plot of $1/\varepsilon_r$ versus T:

$$\frac{1}{\varepsilon_r} = \frac{T}{C} - \frac{T_c}{C}$$

Ideally the graph is linear, with a slope of $1/C$ and an intercept on the T axis of T_c (Figure 11.18). Frequently, the point of intercept, T_0, is slightly different from the measured value of T_c, and the Curie–Weiss equation is often written in the form

$$\varepsilon_r = \frac{C}{T - T_0}$$

where T_0 is the extrapolated Curie temperature (Figure 11.18).

11.3.5 Ferroelectricity due to hydrogen bonds

Hydrogen bonds are formed when a hydrogen atom sits between two electronegative atoms in an off-centre position (Section 3.1.1). At temperatures below the Curie temperature the hydrogen atoms

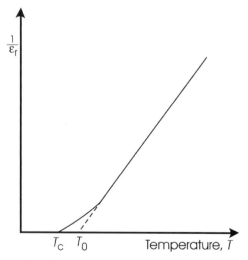

Figure 11.18 The Curie–Weiss behaviour of a ferroelectric solid above the Curie temperature T_c. Note: ε_r, relative permittivity; T, absolute temperature

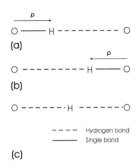

(a)

(b)

- - - - Hydrogen bond
——— Single bond

(c)

Figure 11.19 (a) and (b) Below the ferroelectric transition temperature, T_c, hydrogen atoms in hydrogen bonds lie to one side or the other of the centre. (c) Above T_c the hydrogen atoms are, on average, central. Hydrogen bonds are shown as broken lines, and normal covalent bonds are shown as continuous lines

are ordered on one side of the hydrogen bond or the other (Figure 11.19a). As the hydrogen in a hydrogen bond can occupy two equally stable positions, it is not difficult to see a possible origin for the switching. A sufficiently high electric field will swap the dipole direction by causing the hydrogen ion to jump to the alternative position (Figure 11.19b). At temperatures higher than the Curie temperature, atomic vibrations induced by thermal energy overcome the barrier between the two alter-

native positions and the hydrogen atoms will occupy an average position between the two adjacent electronegative atoms (Figure 11.19c). The polar nature of the solid is lost. The compounds in this group are ordered at lower temperatures and become disordered at higher temperatures. This type of change is called an order–disorder transition.

Hydrogen bonding is the origin of ferroelectricity in potassium dihydrogen phosphate (KH_2PO_4), Rochelle salt [sodium potassium tartrate; Na(COO.-CHOH $\cdot$ CHOH $\cdot$ COO)K $\cdot$ 4H$_2$O] and triglycine sulphate [(NH$_2$CH$_2$COOH)$_3$ $\cdot$ H$_2$SO$_4$]. However, the interaction of the hydrogen bonds with other features of the crystal structure usually makes each compound unique, and the dipoles giving rise to ferroelectricity may lie in other parts of the structure.

This feature is well illustrated by the transition in KH_2PO_4. At a temperature above 123 K KH_2PO_4 is paraelectric. The skeleton of the structure (Figure 11.20a) is made up of regular (PO_4) tetrahedra connected by hydrogen bonds (broken lines in Figure 11.20a). On average, the hydrogen atoms are found at the centres of the hydrogen bonds. The low-temperature form of KH_2PO_4 exists below 123 K. In this structure, the hydrogen atoms (shown as black circles) order (Figure 11.20b) so that each (PO_4) tetrahedron in the high-temperature form is converted into a [PO$_2$(OH)$_2$] tetrahedron.

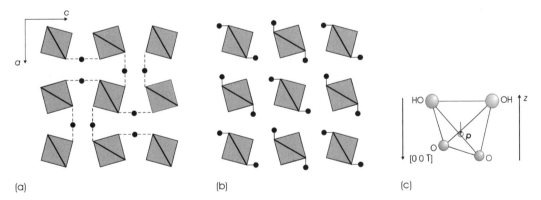

(a)   (b)   (c)

Figure 11.20 Potassium dihydrogen phosphate, KH_2PO_4: (a) skeleton of the structure of KH_2PO_4 projected down the [0 0 1] direction (the PO$_4$ tetrahedra are shown in projection as squares; the hydrogen bonds are shown as broken lines; the potassium atoms are omitted); (b) low-temperature structure showing ordered H atoms (black circles). (c) The displacement of the P atoms in the PO$_4$ tetrahedra as a result of the H-atom ordering, and the subsequent formation of (OH) groups, induces an electric dipole, p, parallel to the c axis

The phosphorus atoms in the tetrahedra are off-centre, pushed away by the hydrogen atoms. The dipoles responsible for ferroelectricity arise in these tetrahedra. They lie along the $[00\bar{1}]$ direction and the z axis is the polar axis. The O—H····O bonds are almost perpendicular to these dipoles, and, although hydrogen bonding is the prime cause of ferroelectricity, the hydrogen bonds themselves are not the seat of the dipoles. However, the off-centre positions of the phosphorous and hydrogen atoms are closely linked. When an external field is applied the hydrogen and the phosphorous atoms switch in concert.

Ammonium dihydrogen phosphate, $NH_4H_2PO_4$, is structurally and chemically very similar to the ferroelectric KH_4PO_4, but the proton ordering is different, and $NH_4H_2PO_4$ is an antiferroelectric below the transition temperature.

11.3.6 Ferroelectricity due to polar groups

Compounds with dipolar groups such as $(NO_3)^-$, which is pyramidal, and nitrite, $(NO_2)^-$, which is shaped like an arrowhead, can also produce ferroelectric phases. At temperatures below the Curie temperature these angular units are locked into one position in the solid in an ordered array. In cases where the geometry of the crystal structure will allow, a sufficiently high electric field can reorient such groups, thus causing the dipole to point in a different direction. At temperatures above the Curie temperature these groups disorder, and ferroelectric behaviour is lost. The compounds in these two groups are ordered at lower temperatures and become disordered at higher temperatures. This type of change is an order–disorder transition.

As an example, consider sodium nitrite, $NaNO_2$. The structure is similar to that of halite, NaCl, and, if the NO_2 groups were spherical instead of shaped like blunt arrowheads, it would be identical (Figure 11.21). These groups point along the b axis with their planes perpendicular to the a axis. As each NO_2 group is polar, the structure is polar, with the dipoles pointing along the b axis. In an applied electric field the NO_2 groups can be made to reverse, and the material is a ferroelectric. There are two ways in which the dipoles could change

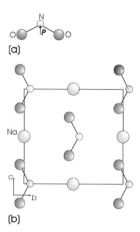

Figure 11.21 (a) The structure of a planar nitrite, (NO_2^-) group; the electric dipole, p, in each unit points towards the nitrogen (N) atom. (b) The low-temperature structure of sodium nitrite, $NaNO_2$, projected down the $[1\,0\,0]$ direction; the dipoles are aligned along the b axis. In the high-temperature paraelectric form the dipoles are arranged at random along $+b$ and $-b$

direction. The nitrogen atom could flip between the oxygen atoms, but this does not occur, as the NO_2 ion is fairly rigid. Instead, the NO_2 groups rotate in their own plane.

The ferroelectric-to-paraelectric phase transition occurs at 165 °C, and in the paraelectric phase the net dipole moment has been lost. Although this could be due to free rotation of the NO_2 groups, this, in fact, does not happen. In reality, the high-temperature structure is disordered. Half of the NO_2 dipoles point along $+b$ and half point along $-b$. This transition is an order–disorder transition and its onset is gradual. At 150 °C, 15 °C below the transition temperature, some 10 % of the NO_2 groups have reversed their orientation.

11.3.7 Ferroelectricity due to medium-sized transition-metal cations

Ferroelectric properties can be attributed to the presence of medium-sized cations in many oxides with structures related to that of perovskite, for example $BaTiO_3$ and $KNbO_3$. These contain ions such as Ti^{4+} and Nb^{5+}, surrounded by six O^{2-} ions in an octahedral geometry. At lower temperatures, these cations are usually displaced from the centre of the surrounding oxygen coordination polyhedron. These ions can jump from one off-centre position to another under the influence of an electric field. Above the Curie temperature the ions occupy an average position in the coordination polyhedron, and a net dipole is lost. This type of transformation is a displacive transition.

As an example, consider barium titanate, $BaTiO_3$. Above 120 °C the paraelectric form of $BaTiO_3$ has the cubic perovskite structure, with $a_0 = 0.4018$ nm (Figure 11.22a). The large Ba^{2+} cations are surrounded by 12 oxygen ions, and the medium-sized Ti^{4+} ions are situated at the centre of an octahedron

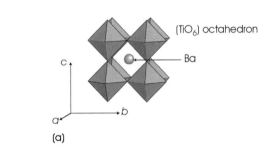

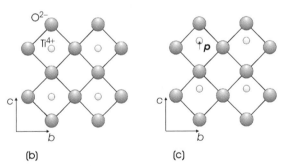

Figure 11.22 (a) The cubic perovskite structure of high-temperature cubic barium titanate, $BaTiO_3$, drawn as corner-shared TiO_6 octahedra; the Ba^{2+} ions occupy the unit cell centre. (b) Projection of the structure down the [1 0 0] direction, with the Ba^{2+} ions omitted. (c) Schematic view of the low-temperature tetragonal form, with the displacements of the Ti^{4+} ions exaggerated and the Ba^{2+} ions omitted. The electric dipoles, p, generated by the off-centre displacement of the Ti^{4+} ions, point along the c axis of the tetragonal unit cell

of oxygen ions. Below 120 °C the phase becomes ferroelectric. Between 120 °C and above 5 °C the unit cell is tetragonal (Figure 11.22b), with $a_0 = 0.3997$ nm, and $c_0 = 0.4031$ nm. (There are other structures with lower symmetries found at temperatures below 5 °C that will not be considered here.) As the crystal cools through the cubic–tetragonal transition temperature, the cubic cell expands slightly along one edge to produce the tetragonal c axis and is slightly compressed along the other two edges to form the tetragonal a and b-axes. The change from cubic to tetragonal is accompanied by an off-centre movement of the octahedrally coordinated Ti^{4+} ions, accompanied by a slight change in octahedron dimensions. This results in the formation of a dipole pointing along the c axis (Figure 11.22c). The change in the Ba^{2+} positions is almost negligible. The polar axis is the c axis. The off-centre Ti^{4+} position in an octahedron can be changed in an electric field and hence tetragonal $BaTiO_3$ is a ferroelectric.

There is no preference as to which of the original cubic axes becomes the polar direction, and so this can take one of six equivalent directions, parallel to $\pm x$, $\pm y$ or $\pm z$. On cooling, a ferroelectric domain pattern forms, reflecting that the transformation takes place along any of the allowed cubic directions in various parts of the phase.

11.3.8 Poling and polycrystalline ferroelectric solids

A ferroelectric crystal does not normally show any observable polarisation, because the domain structure leads to overall cancellation of the effect. Polycrystalline ceramics would be expected to be similar. In order to form a material with an observable polarisation the crystals are poled. This process involves heating the crystals above the Curie point, T_c, and then cooling them in a strong electric field. The effect of this is to favourably orient dipoles so that the crystal or polycrystalline ceramic shows a strong ferroelectric effect. The majority of ferroelectric materials used are, in fact, polycrystalline.

The same is true of polymer piezoelectrics. In these materials, the crystallites that give rise to piezoelectricity are oriented at random within the polymer matrix. Poling can give the dipoles an overall preferred orientation. Naturally, poling will not affect a crystallite that does not show a permanent dipole, and so poling applies only to pyro-electric and ferroelectric materials.

11.3.9 Doping and modification of properties

Many ferroelectric materials show interesting and potentially useful properties, but not at the temperature or pressure required for a particular application. It is then necessary to change or 'tune' the property to the fit the application. This change is frequently brought about by the replacement of one or more of the constituents of the compound, or the deliberate addition of impurities. As in the case of semiconductors and other materials, the deliberate addition of impurities to change the physical properties of a solid is known as doping.

Ferroelectric oxides, for example, are of interest as capacitor materials because of their high permittivity levels, but usually the sharp maximum in dielectric constant at the Curie point must be broadened and moved to room temperature. Consider $BaTiO_3$, which has a high dielectric constant at the Curie temperature, about 393 K. The Curie temperature can be increased by the replacement of some of the Ba^{2+} ions by Pb^{2+} ions. These ions are 'softer' (i.e. more easily polarised) than the Ba^{2+} ions, as they have a lone pair of electrons and so are more easily affected by an applied electric field. The resultant compound retains the crystal structure of $BaTiO_3$ but has a formula $Ba_{1-x}Pb_xTiO_3$. The compound $Ba_{0.6}Pb_{0.4}TiO_3$ has a Curie temperature of approximately 573 K, an increase of 200 K. In a similar way, the Curie temperature can be lowered by the substitution of Ba^{2+} ions by Sr^{2+}. These ions are smaller than Ba^{2+} ions and can be considered to be 'harder' and more difficult to polarise. The compound $Ba_{0.6}Sr_{0.4}TiO_3$ has a Curie temperature of 0 °C. The Curie temperature can also be lowered by the replacement of some of the Ti^{4+} ions by Zr^{4+} or Sn^{4+} ions. In this type of doping, the normal ferroelectric behaviour, due to off-centre Ti^{4+} ions in TiO_6 octahedra, is simply modified.

Answers to introductory questions

How are the relative permittivity and refractive index of a transparent solid related?

The observed bulk polarisability of a solid, α, will arise from the sum of a number of separate terms such as electronic polarisability, ionic polarisability and so on. As the total polarisability of a material is made up several contributions, the relative permittivity, ε_r, can also be thought of as made up from the same contributions. In a static electric field, all the various contributions will be important and both α and ε_r will arise from electrons, ions, dipoles, defects and surfaces.

An alternating electric field acts on these components in different ways. At low frequencies the value of the relative permittivity measured will be identical to the static value, and all polarisability terms will contribute to ε_r. However, space charge polarisation is usually unable to follow changes in electric field that occur much faster than that of radio frequencies, about 10^6 Hz, and this contribution will no longer be registered at frequencies higher than this value. Similarly, any dipoles present are usually unable to rotate to and fro in time with the alternations of the electric field when frequencies reach the microwave region, about 10^9 Hz, and at higher frequencies this contribution will be lost. Ionic polarisability, involving the movement of atomic nuclei, is no longer registered when the frequency of the field approaches that of the infrared range, 10^{12} Hz. Electrons, being the lightest components of matter, still respond to an alternating electric field at frequencies corresponding to the visible region, 10^{14} Hz.

The interaction of an alternating electric field with the electrons in a solid in the frequency range between the infrared and ultraviolet–the optical range–is more commonly expressed as the refractive index, because a lightwave consists of oscillating electric and magnetic fields. The relative permittivity in the optical region is related to the refractive index by:

$$n^2 = \varepsilon_r$$

What is the relationship between ferroelectric and pyroelectric crystals?

In a piezoelectric solid a surface electric charge develops when the solid is subjected to a mechanical stress such as pressure, even in the absence of an external electric field. This is called the piezoelectric effect, and crystals that exhibit this behaviour are called piezoelectrics. Piezoelectric solids are a subset of dielectrics. All piezoelectrics are dielectrics, but only some dielectrics are piezoelectrics.

In the case of a pyroelectric solid a change of temperature induces a polarisation change. Pyroelectric crystals are a subset of piezoelectrics. All pyroelectric crystals are piezoelectrics, but not all piezoelectrics demonstrate pyroelectricity. A material that is a pyroelectric is found to possess a spontaneous polarisation, P_s. This means that a pyroelectric crystal shows a permanent polarisation that is present both in the absence of an electric field and in the absence of mechanical stress.

Some crystals, however, exhibit relative permittivity values many orders of magnitude higher than found in normal dielectrics. By analogy with magnetic behaviour, this behaviour is called ferroelectricity, and the materials are called ferroelectrics. Ferroelectrics also possess a spontaneous polarisation, P_s, in the absence of an electric field and a mechanical distortion. They are, therefore, a subset of pyroelectrics and, as such, all ferroelectrics are also pyroelectrics and piezoelectrics. The feature that distinguishes ferroelectrics from pyroelectrics is that the direction of the spontaneous polarisation, P_s, can be switched (changed) in an applied electric field.

How can a ferroelectric solid be made from a polycrystalline aggregate?

A ferroelectric crystal does not normally show any observable polarisation because the domain struc-

ture leads to overall cancellation of the effect. Polycrystalline ceramics would be expected to be similar. In order to form a material with an observable polarisation the crystals are poled. This process involves heating the crystals above the Curie point, T_c, and then cooling them in a strong electric field. The effect of this is to favourably orient dipoles so that the crystal or polycrystalline ceramic shows a strong ferroelectric effect. The majority of ferroelectric materials used are, in fact, polycrystalline.

The same is true of polymer piezoelectrics. In these materials, the crystallites that give rise to piezoelectricity are oriented at random within the polymer matrix. Poling can give the dipoles an overall preferred orientation. Naturally, poling will not affect a crystallite that does not show a permanent dipole, and so poling applies only to pyroelectric and ferroelectric materials.

Further reading

P.E. Dunn, S.H. Carr, 1989, 'A Historical Perspective on the Occurrence of Piezoelectricity in Materials', *Materials Research Society Bulletin* **XIV** (February) 22.

G.H. Haertling, 1999, 'Ferroelectric Ceramics: History and Technology', *Journal of the American Ceramics Society* **82**.

W.D. Kingery, H.K. Bowen, D.R. Uhlmann, 1976, *Introduction to Ceramics*, 2nd edn., John Wiley & Sons, New York, Ch. 18.

M.E. Lines, A.M. Glass, 2001, *Principles and Applications of Ferroelectrics and Related Materials*, Oxford Classics, Oxford University Press, Oxford.

H.D. Megaw, 1973, *Crystal Structures*, Saunders, Philadelphia, PA.

R.E. Newnham, 1975, *Structure-Property Relations*, Springer, Berlin, Ch. 4.

R.E. Newnham, 1997, 'Molecular Mechanisms in Smart Materials', *Materials Research Society Bulletin* **22** 20.

R.D. Shannon, R.A. Oswald, 1991, 'Dielectric Constants of YVO_4, Fe-, Ge- and V-containing Garnets, the Polarizabilities of Fe_2O_3, GeO_2, and V_2O_5 and the Oxide Additivity Rule', *Journal of Solid State Chemistry* **95** 313.

R.L. Withers, J.G. Thompson, A.D. Rae, 1991, 'The Crystal Chemistry Underlying Ferroelectricity in $Bi_4Ti_3O_{12}$, Bi_3TiNbO_9 and Bi_2WO_6', *Journal of Solid State Chemistry* **94** 404.

Problems and exercises

Quick quiz

1 The relative permittivity of a material is also called the:
(a) Dielectric constant
(b) Dielectric permittivity
(c) Dielectric susceptibility

2 The polarisation of a solid in an electric field is due to:
(a) The formation of charges on the surface
(b) The flow of charges from one surface to the other
(c) The formation of electric dipoles

3 The dielectric susceptibility relates the:
(a) Polarisation of a solid to the charges in a solid
(b) Polarisation to the capacitance of a solid
(c) Polarisation of a solid to the electric field

4 The polarisability is *not* a property of:
(a) Surfaces
(b) Atoms
(c) Ions

5 One of the following will *not* contribute to the polarisability of a nonmolecular solid in a static electric field:
(a) Ionic polarisability
(b) Orientational polarisability
(c) Electronic polarisability

6 Which of the following contributes to the polarisability of a solid in a very-high-frequency electric field:
(a) Ionic polarisability?
(b) Electronic polarisability?
(c) Space charge polarisability?

7 The relative permittivity of a crystal with cubic symmetry is:
(a) The same in all directions
(b) Different along each crystallographic axis

(c) Different along two out of the three axes

8 The relative permittivity of a crystal with hexagonal symmetry is:
(a) The same in all directions
(b) The same along the *a* and *b* axes, and different along the *c* axis
(c) Different along the *a*, *b* and *c* axes

9 The relative permittivity of a crystal with tetragonal symmetry is:
(a) The same along the *a*, *b* and *c* axes
(b) The same along the *a* and *b* axes, and different along the *c* axis
(c) Different along the *a*, *b* and *c* axes

10 The relative permittivity of a crystal with orthorhombic symmetry is:
(a) The same along the *a*, *b* and *c* axes
(b) The same along the *a* and *b* axes, and different along the *c* axis
(c) Different along the *a*, *b* and *c* axes

11 In the direct piezoelectric effect:
(a) An applied voltage causes a dimensional change
(b) A dimensional change produces a voltage
(c) An applied voltage produces a temperature change

12 A spontaneous polarisation is *not* found in:
(a) Piezoelectric solids
(b) Ferroelectric solids
(c) Pyroelectric solids

13 A solid that has a switchable spontaneous polarisation is:
(a) A pyroelectric
(b) A piezoelectric
(c) A ferroelectric

14 A piezoelectric must contain:
(a) Tetrahedral groups
(b) Octahedral groups
(c) No centre of symmetry

15 For a polymer to be potentially piezoelectric it needs:
(a) An amorphous structure
(b) Permanent dipoles
(c) A crystalline structure

16 Electrets are:
(a) Polymer sheets with a permanent surface charge
(b) Polymer sheets with permanent internal dipoles
(c) Polymer sheets that are charged easily

17 The crystal structure of pyroelectric crystals must contain:
(a) A polar axis
(b) A centre of symmetry
(c) Switchable permanent dipoles

18 Ferroelectric crystals must posses:
(a) Hydrogen bonds
(b) Switchable dipoles
(c) Polar groups

19 Ferroics are solids that contain:
(a) Domain structures
(b) Ferroelectric domains
(c) Magnetic domains and ferroelectric domains

20 Hysteresis is characteristic of:
(a) Pyroelectric crystals
(b) Piezoelectric crystals
(c) Ferroelectric crystals

21 The dipoles in an antiferroelectric crystal are:
(a) In parallel rows
(b) In antiparallel rows
(c) Partly aligned

22 The paraelectric state of a ferroelectric is:
(a) The high-temperature phase
(b) The low-temperature phase
(c) The antiferroelectric phase

23 The Curie temperature is *not* the temperature at which:

(a) A ferroelectric transforms to a paraelectric state
(b) A paraelectric transforms to a ferroelectric state
(c) A ferroelectric transforms to an antiferroelectric state

24 The temperature dependence of the relative permittivity of many ferroelectric crystals obeys the Curie–Weiss Law:
(a) In the ferroelectric state
(b) In the paraelectric state
(c) At low temperatures

25 Hydrogen bonding is *not* the cause of ferroelectricity in:
(a) Triglycine sulphate
(b) Rochelle salt
(c) Sodium nitrite

26 The cause of ferroelectricity in perovskite materials is often due to:
(a) The presence of medium-sized cations in octahedral coordination
(b) The presence of hydrogen bonds
(c) The presence of polar groups

27 The process by which polycrystalline solids can be made ferroelectric is:
(a) Annealing
(b) Sintering
(c) Poling

Calculations and questions

11.1 The plates on a parallel plate capacitor are separated by 0.1 mm and filled with air.
(a) What is the capacitance if the plates have an area of 1 cm^2?
(a) If the space between the plates is filled with a polyethylene sheet, with a relative permittivity of 2.3, what is the new capacitance?

11.2 A parallel plate capacitor is connected to a battery and acquires a charge of 200 μC on each plate. A polymer is inserted and the charge on the plates is now found to be 750 μC. What is the relative permittivity of the polymer?

11.3   A parallel plate capacitor has a capacitance of 8 nF and is to be operated under a voltage of 80 V. What is the relative permittivity of the dielectric if the maximum dimensions of the capacitor are 1 mm × 1 cm × 1 cm?

11.4   The dipole moment of the molecule nitric oxide, NO, is 0.5×10^{-30} C m. The N–O bond length is 0.115 nm.
(a) What is the charge on the atoms?
(b) Which atom is more positive?

11.5   The dipole moment of a water molecule is 6.2×10^{-30} C m, the H–O bond length is 95.8 pm and the H–O–H bond angle is 104.5°. Determine the charge on each atom.

11.6   The dipole moment of the molecule HCN is 9.8×10^{-30} C m and the charges, which reside on the H and N atoms (with carbon central), are measured to be 3.83×10^{-20} C.
(a) What is the dipole length and hence the approximate length of the molecule?
(b) The molecules are packed into a cubic structure, with a lattice parameter of 0.512 nm, each unit cell containing one molecule at $(0, 0, 0)$. Determine the bulk polarisation of the solid when all dipoles are aligned.

11.7   The dipole moments of the following molecules are:
• carbon monoxide, CO (linear), 0.334×10^{-30} C m;
• nitrous oxide, N_2O (linear), 0.567×10^{-30} C m;
• ammonia, NH_3 (tetrahedral, N at one vertex, H–N–H bond angle = 106.6°), 4.837×10^{-30} C m;
• sulphur dioxide, SO_2 (angular, O–S–O bond angle = 119.5°), 5.304×10^{-30} C m.
The covalent radii of the atoms involved are: C, 0.077 nm; O, 0.074 nm; N, 0.074 nm; H, 0.037 nm; S, 0.104 nm. Calculate the nom-

inal charges on the atoms as a fraction of the electron charge.

11.8   Derive the Clausius–Mosotti equation using text Equations (11.3), (11.4) and (11.5). [Note: derivation is not given in the answers at the end of this book.]

11.9   The following data are for a single crystal of periclase, MgO: relative permittivity, ε_r, 9.65; refractive index, n, 1.736; cubic unit cell, a_0, 0.4207 nm, $Z = 4$ formula units of MgO.
(a) Estimate (i) the electronic polarisability and (ii) the corresponding polarisability volume.
(b) Estimate (i) the ionic polarisability and (ii) the corresponding polarisability volume.

11.10 The following data are for a single crystal of α-quartz, one of the mineral forms of SiO_2: average relative permittivity, ε_r, 4.477; average refractive index, n, 1.5485; hexagonal unit cell, a_0, 0.49136 nm, c_0, 0.54051 nm, $Z = 3$ formula units of SiO_2. Quartz is hexagonal, which means that the refractive indices and relative permittivity depend on crystallographic direction (see Table 14.1) and the Clausius–Mossotti and Lorentz–Lorentz relations apply only approximately. Nevertheless, from the data given:
(a) Estimate (i) the electronic polarisability and (ii) the corresponding polarisability volume.
(b) Estimate (i) the ionic polarisability and (ii) the corresponding polarisability volume.

11.11 Use the additivity rule to estimate (a) the ionic, (b) the electronic and (c) the total polarisability of forsterite, Mg_2SiO_4, given the information in Questions 11.9 and 11.10, assuming that only ionic and electronic polarisations are important.

11.12 Use the following crystallographic and optical data for forsterite to estimate the electronic polarisability of the mineral: forsterite is orthorhombic, $a_0 = 0.4758$ nm, $b_0 = 1.0214$ nm, $c_0 = 0.5984$ nm, $Z = 4$ formula units of Mg_2SiO_4. The three principle refractive indices are 1.635, 1.651 and 1.670.

11.13 Estimate (a) the polarisability and (b) the relative permittivity of the garnet $Ca_3Ga_2Ge_3O_{12}$, using the following data: $Ca_3Ga_2Ge_3O_{12}$, cubic, $a_0 = 1.2252$ nm, $Z = 8$; CaO, polarisability, α, 5.22×10^{-30} m^3; Ga_2O_3, polarisability, α, 8.80×10^{-30} m^3; GeO_2, polarisability, α, 5.50×10^{-30} m^3.

11.14 The relative permittivity of the garnet $Y_3Fe_5O_{12}$, a magnetic oxide, was measured as 15.7. Calculate (a) the polarisability and (b) the volume polarisability of Y_2O_3 if the volume polarisability given for Fe_2O_3 is 10.5×10^{-30} m^3. The unit cell of $Y_3Fe_5O_{12}$ is cubic, $a_0 = 1.2376$ nm, $Z = 8$ formula units of $Y_3Fe_5O_{12}$.

11.15 Use the additivity rule to estimate (a) the polarisability and (b) the volume polarisability of mullite, Al_2SiO_5. The following volume polarisabilities were found in the literature: SiO_2, 4.84×10^{-30} m^3; Al_2O_3, 7.70×10^{-30} m^3. The experimental value is 15.22×10^{-30} m^3; (c) comment on the accuracy of the method.

11.16 Use the additivity rule to estimate (a) the polarisability and (b) the volume polarisability of diopside, $CaMgSi_2O_6$. The following volume polarisabilities were found in the literature. Magnesium oxide, MgO, 3.32×10^{-30} m^3, SiO_2, 4.84×10^{-30} m^3; CaO, 5.22×10^{-30} m^3. The experimental value of the volume polarisability is 18.78×10^{-30} m^3; (c) comment on the accuracy of the method.

11.17 The polarisability for the oxide $Mn_3Al_2Si_3O_{10}$ was estimated to be 35.83×10^{-30} C m^2 V^{-1}. Determine the value of the polarisability for Mn^{2+} ions in this oxide given the following data: α (Al^{3+}), 0.32×10^{-30} C m^2 V^{-1}; α (Si^{4+}), $0.11 \times$ 10^{-30} C m^2 V^{-1}; α (O^{2-}), 2.64×10^{-30} C m^2 V^{-1}.

11.18 Show that the units for the direct piezoelectric coefficient, d, of C N^{-1} and m V^{-1} are equivalent. [Note: answer is not shown at the end of this book.]

11.19 The relevant value of the piezoelectric coefficient, d, for a quartz is given as 2.3 pC N^{-1}. Calculate the polarisation of a plate of dimensions 10 cm $\times$ 5 cm $\times$ 0.5 mm when a mass of 0.5 kg is placed on it.

11.20 An electret film with the very high piezoelectric coefficient of 170 pC N^{-1} is used in speakers, microphones and keyboards and related devices where mechanical and electrical signals are coupled. Calculate the change in thickness when 500 V are applied across a film 0.1 mm thick.

11.21 The semiprecious gemstone tourmaline, with an approximate formula $CaLi_2Al_7(OH)_4(BO_3)_3Si_6O_{18}$, has a pyroelectric coefficient, π_i, of 4×10^{-6} C m^{-2} K^{-1}. The unique polar axis is the crystallographic c axis. What is the change in polarisation caused by a change of temperature of 100 °C?

11.22 The measured relative permittivity, ε_r, of a ceramic sample of $PbZrO_3$ as a function of temperature, T, is given in Table 11.3. Determine (a) the Curie temperature, T_c, and (b) the Curie constant, C, for this sample.

Table 11.3 Data for question 11.22

ε_r	130	142	166	222	360	420	472	556
$T/°C$	50	100	150	200	225	230	234	235
ε_r	775	3200	3000	2840	2440	1620	1240	840
$T/°C$	236	238	240	242	250	275	300	350

11.23 The measured relative permittivity, ε_r, of a ceramic sample of $Cd_2Nb_2O_7$ as a function of temperature, T, is given in Table 11.4. Determine (a) the Curie temperature, T_c and (b) the Curie constant, C, for this sample.

Table 11.4 Data for question 11.23

ε_r	4500	4125	3750	3500	3225	3000	2800	2600
$T/°C$	−80	−75	−70	−65	−60	−55	−50	−45
ε_r	2465	2280	2115	2000	1860	1750	1630	1560
$T/°C$	−40	−35	−30	−25	−20	−15	−10	−5

11.24 The measured relative permittivity, ε_r, of a crystal of triglycine sulphate as a function of temperature, T, is given in Table 11.5. Determine (a) the Curie temperature, T_c, and (b) the Curie constant, C, for triglycine sulphate.

Table 11.5

ε_r	120	190	280	400	540	730	1300	~7000
$T/°C$	40	41	42	43	44	45	46	47
ε_r	1100	830	700	590	520	460	420	380
$T/°C$	48	49	50	51	52	53	54	55
ε_r	250	180	130	110				
$T/°C$	60	65	70	75				

11.25 Zincite (zinc oxide, ZnO) has a hexagonal unit cell, with $a_0 = 0.3250$ nm, $c_0 = 0.5207$ nm and a unit cell volume of 47.63×10^{-27} m^3. The atom positions are Zn, $(\frac{1}{3}, \frac{2}{3}, 0)$, $(\frac{2}{3}, \frac{1}{3}, \frac{1}{2})$; O, $(\frac{1}{3}, \frac{2}{3}, 0.38)$, $(\frac{2}{3}, \frac{1}{3}, 0.58)$. There are two formula units of ZnO in the unit cell. (a) Sketch the unit cell [sketch not shown at the end of this book] and (b) estimate the maximum spontaneous polarisation of ZnO, assuming that the structure is ionic.

11.26 Calculate the maximum spontaneous polarisation of a crystal of sodium nitrite, NaNO$_2$, given that the unit cell is orthorhombic, with $a_0 = 0.360$ nm, $b_0 = 0.575$ nm, $c_0 = 0.535$ nm, $Z =$ two formula units of NaNO$_2$. The dipole moment of each N–O bond is 0.5×10^{-30} C m^{-1}, and the O–N–O angle is 115°.

11.27 (a) Calculate the dipole moment of a TiO$_6$ octahedron in BaTiO$_3$, with a very slightly distorted perovskite structure, in which $a_0 = 0.3997$ nm, $c_0 = 0.4031$ nm, tetragonal, assuming that the compound is fully ionic and that the Ti^{4+} ions are displaced by 0.012 nm along the c axis of the unit cell. (b) Determine the maximum spontaneous polarisation under these conditions.

11.28 (a) Calculate the dipole moment of an NbO$_6$ octahedron in KNbO$_3$, with a very slightly distorted perovskite structure, in which $a_0 = 0.4002$ nm, $c_0 = 0.4064$ nm, tetragonal, assuming that the compound is fully ionic and that the Nb^{5+} ions are displaced by 0.017 nm along the c axis of the unit cell. (b) Determine the maximum spontaneous polarisation under these conditions.

11.29 (a) Calculate the dipole moment of a TiO$_6$ octahedron in PbTiO$_3$, with a very slightly distorted perovskite structure, in which $a_0 = 0.3899$ nm, $c_0 = 0.4153$ nm, tetragonal, assuming that the compound is fully ionic and that the Ti^{4+} ions are displaced by 0.030 nm along the c axis of the unit cell. (b) Determine the maximum spontaneous polarisation under these conditions.

11.30 Most ceramics are electrical insulators. Describe the combination of factors that would allow a ceramic to be classified as a ferroelectric rather than just an insulator. [Note: answer is not provided at the end of this book.]

11.31 Both silica glass and quartz, SiO$_2$, are composed of SiO$_4$ tetrahedra and neither material posses a centre of symmetry. Why is silica glass not a piezoelectric, whereas quartz is? [Note: answer is not provided at the end of this book.]

12

Magnetic solids

- What atomic feature renders a material para-magnetic?

- Why do ferromagnetic solids show a domain structure?

- What is a ferrimagnetic material?

Magnets pervade everyday life. In reality, these magnets are examples of a group of materials called ferromagnetic compounds. If a small rod magnet is freely suspended, it will align (approximately) north–south. The end pointing north is called the north pole of the magnet and the end pointing south, the south pole of the magnet. This fact proves that the Earth also acts as a magnet. It is found that opposite magnetic poles attract each other and similar magnetic poles repel each other. [Because of this, the end of a freely suspended magnet that points towards the north should be labelled as a south pole, but it is too late to change things now!] A ferromagnetic solid behaves as if surrounded by a magnetic field, and will attract or repel other ferro-magnets via the interactions of the magnetic fields.

Not only solids but also wires carrying an electric current give rise to magnetic fields, the strength of which is proportional to the current flowing. Most solids, however, including wires not carrying a current, and ferromagnetic materials above a certain temperature, are loosely termed 'nonmagnetic'. Strictly speaking, this is inaccurate, as these materials simply exhibit extremely weak magnetic effects.

12.1 Magnetic materials

12.1.1 Characterisation of magnetic materials

The weak magnetic properties of most solids can be measured using a Gouy balance (Figure 12.1a). In this equipment, the sample is suspended between the poles of an electromagnet from a sensitive balance. The vast majority of solids show only miniscule magnetic effects. Of these, most weigh slightly less when the electromagnet is on than when the electromagnet is turned off (Figure 12.1b). The materials, which are weakly repelled by the magnetic field, are the diamagnetic materials. The rest of the 'nonmagnetic' group weigh slightly more when the electromagnet is on than when it is off (Figure 12.1c). These substances are drawn weakly into a magnetic field and are called para-magnetic materials. Ferromagnetic materials are strongly attracted to one or other of the pole pieces of the magnet and the technique does not give a result.

Diamagnetic and paramagnetic substances are characterised by their 'susceptibility' to the magnetic

Understanding solids: the science of materials. Richard J. D. Tilley
© 2004 John Wiley & Sons, Ltd ISBNs: 0 470 85275 5 (Hbk) 0 470 85276 3 (Pbk)

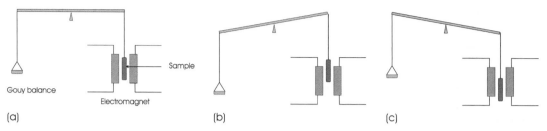

Figure 12.1 The weak magnetic properties of a solid can be determined by its behaviour in the field generated by an electromagnet, in a Gouy balance: (a) sample with electromagnet turned off; (b) diamagnetic sample with electromagnet turned on (the sample appears to weigh less); and (c) paramagnetic sample with the electromagnet turned on (the sample appears to weigh more)

field. For materials that are isotropic, a group that includes gases and liquids as well as glasses, cubic crystals and polycrystalline solids, the magnetic susceptibility, χ, is defined by:

$$M = \chi H \qquad (12.1)$$

where M is the magnetisation of the sample and H is the magnetic field, which are usually defined as vectors. The magnetic susceptibility is a dimensionless quantity. In nonisotropic solids, M and H are not necessarily parallel. The value of the magnetic susceptibility will then vary with direction in the crystal. The units used in magnetism are given in Section S4.4.

These weak magnetic materials can also be characterised by the extent to which the magnetic field is able to penetrate into the sample, the magnetic permeability. The magnetic permeability of a compound, μ, is defined by:

$$\mu = \mu_0(1 + \chi) \qquad (12.2)$$

where μ_0 is a fundamental constant, the permeability of free space. Diamagnetic materials have a value of μ less than μ_0, whereas paramagnetic materials have value of μ greater than μ_0. These values are equivalent to field enhancement in a paramagnetic substance, and field reduction in a diamagnetic material (Figure 12.2).

12.1.2 Types of magnetic material

Magnetic materials can be classified in terms of the arrangements of magnetic dipoles in the solid.

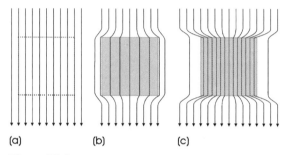

Figure 12.2 Weak magnetic materials differ in their response to an external magnetic field: (a) no solid present; (b) a diamagnetic solid repels the external field and the density of field lines in the solid decreases; (c) a paramagnetic solid attracts the external field and the density of the magnetic field lines in the sample increases. B (induction) and H (magnetic field) decrease inside a diamagnetic solid and increase inside a paramagnetic solid

These dipoles can be thought of, a little imprecisely, as microscopic bar magnets attached to the various atoms present.

Materials with no elementary magnetic dipoles at all are diamagnetic (Figures 12.3a and 12.3b). The imposition of a magnetic induction generates weak magnetic dipoles that oppose the applied induction. The magnetic susceptibility, χ, of a diamagnetic substance is negative and very slightly less than 1. There is no appreciable variation of diamagnetism with temperature.

Paramagnetic solids are those in which some of the atoms, ions or molecules making up the solid possess a permanent magnetic dipole moment. These dipoles are isolated from one another. The

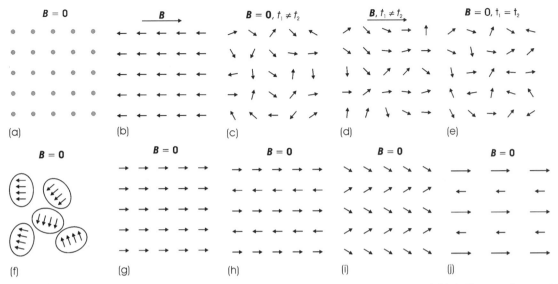

Figure 12.3   The effect of an applied magnetic field, H, or induction, B, on a solid. (a) and (b) A diamagnetic material has no dipoles present in the absence of magnetic induction; weak dipoles are induced that oppose the field to give a weak repulsion. (c) and (d) In the absence of magnetic induction, the elementary magnetic dipoles present are random; in a field, there is a tendency to align in the induction direction to give a weak attraction. The diagrams apply only for an instant, at time t_1. The dipoles continually change orientation and will be differently disposed at another time t_2. (e), A spin glass is similar to (c), but below a temperature T_f, the orientation of the dipoles changes slowly. (f) A cluster glass has oriented dipoles in small volumes below a temperature T_f. (g) A ferromagnetic solid has aligned dipoles in the absence of induction. (h) An antiferromagnetic solid has dipoles in an antiparallel arrangement in the absence of magnetic induction. (i) A canted magnetic material has dipoles arranged in a canted fashion in the absence of a field. (j) A ferrimagnetic solid has two opposed dipole arrays

solid, in effect, contains small, noninteracting atomic magnets. In the absence of a magnetic field, these are arranged at random and the solid shows no net magnetic moment. In a magnetic field, the elementary dipoles will attempt to orient themselves parallel to the magnetic induction in the solid, and this will enhance the internal field within the solid and give rise to the observed paramagnetic effect (Figures 12.3c and 12.3d). The alignment of dipoles will not usually be complete, because of thermal effects and interaction with the surrounding atoms in the structure. Thus the situation shown in Figures 12.3(c) and 12.3(d) is a snapshot at any instant, t_1. At any other instant, t_2, the orientation of the spins would still be random but it would be different from that drawn. The magnetic effect is much greater than diamagnetism, and the magnetic susceptibility, χ, of a paramagnetic solid is positive and slightly greater than 1.

The partial orientation of the elementary dipoles in a paramagnetic solid will be counteracted by thermal agitation, and it would be expected that at high temperatures the random motion of the atoms in the solid would cancel the alignment resulting from the magnetic field. The paramagnetic susceptibility would therefore be expected to vary with temperature. The temperature dependence is given by the Curie law:

$$\chi = \frac{C}{T} \qquad (12.3)$$

where χ is the magnetic susceptibility, T is the absolute temperature, and C is the Curie constant. Curie law dependence in a solid (Figure 12.4a), is indicative of the presence of isolated paramagnetic ions or atoms in the material.

Interacting magnetic dipoles can produce a variety of magnetic properties in a solid. The

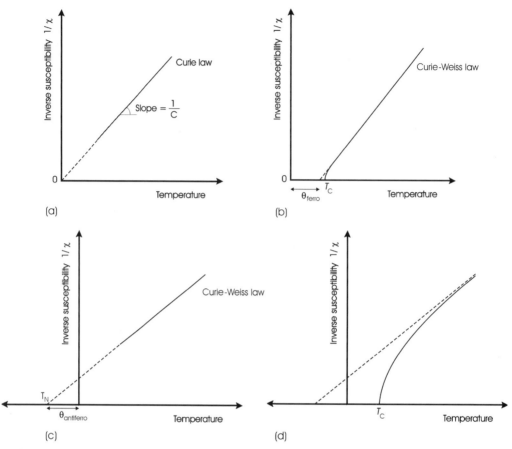

Figure 12.4 The temperature dependence of the reciprocal magnetic susceptibility: (a) Curie law behaviour of a paramagnetic solid; (b) Curie–Weiss law behaviour of a ferromagnetic solid above the Curie temperature, in the paramagnetic state; (c) Curie–Weiss law behaviour of an antiferromagnetic solid; and (d) behaviour of a ferrimagnetic solid

interactions can increase sufficiently that at low temperatures the random reorientation of the dipoles is restricted and changes only slowly with time. The directions of the spins are said to become frozen below a freezing temperature, T_f, to produce a spin glass (Figure 12.3e). In other materials, the interactions below T_f are strong enough for local ordering but because of the crystal structure, the localised regions are restricted and no long-range order occurs (Figure 12.3f). This arrangement is called a cluster glass, and is found at low temperatures in compounds such as Prussian blue, $KFe_2(CN)_6$, which has a magnetic freezing point of approximately 25 K.

Ferromagnetic materials are those in which the magnetic moments align parallel to each other over considerable distances in the solid (Figure 12.3g). An intense external magnetic field is produced by this alignment. Ferromagnetism is associated with the transition elements, with unpaired d electrons, and the lanthanides and actinides with unpaired f electrons.

Above a temperature called the Curie temperature, T_C, all ferromagnetic materials become paramagnetic. The transition to a paramagnetic state comes about when thermal energy is greater than the magnetic interactions, and causes the dipoles to disorder. *Well above* the Curie temperature,

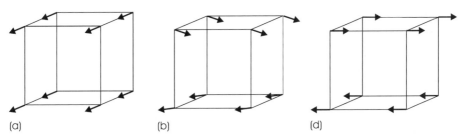

Figure 12.5 The relationship between the ordered arrays of magnetic dipoles arranged on a square lattice: (a) ferromagnetic configuration, (b) canted antiferromagnetic configuration and (c) antiferromagnetic configuration

ferromagnetic materials obey the Curie–Weiss law:

$$\chi = \frac{C}{T - \theta} \qquad (12.4)$$

The Curie–Weiss constant, θ, is positive, has the dimensions of temperature, and a value usually close to, but not quite identical to, the Curie temperature, T_C, (Figures 12.4b). The transition is reversible, and on cooling, ferromagnetism returns when the magnetic dipoles align parallel to one another as the temperature drops through the Curie temperature.

It is energetically favourable in some materials for the elementary magnetic moments to align in an antiparallel fashion (Figure 12.3h). These are called antiferromagnetic compounds. Above a temperature called the Néel temperature, T_N, this arrangement disorders and the materials revert to paramagnetic behaviour. Well above the Néel temperature, antiferromagnetic materials obey the Curie–Weiss law:

$$\chi = \frac{C}{T - \theta} \qquad (12.5)$$

In this case, the constant, θ, is negative (Figure 12.4c). Cooling the sample through the Néel temperature causes the antiferromagnetic ordering to reappear.

Ferromagnetic ordering and antiferromagnetic ordering represent the extremes of dipole orientation. In a number of solids, neighbouring magnetic dipoles are not aligned parallel to one another, but at an angle, referred to as a canted arrangement (Figure 12.3i). Such canted arrangements can be thought of as an intermediate configuration between ferromagnetic and antiferromagnetic states. For example, a ferromagnetic ordering of magnetic dipoles arranged on a square lattice (Figure 12.5a) can be transformed by canting of alternate layers (Figure 12.5b) into an antiferromagnetic configuration (Figure 12.5c).

An important group of solids has two different magnetic dipoles present, one of greater magnitude than the other. When these line up in an antiparallel arrangement they behave rather like ferromagnetic materials (Figure 12.3j). They are called ferrimagnetic materials. A ferrimagnetic solid shows a more complex temperature dependence. This is because the distribution of the magnetic ions over the available sites is sensitive both to temperature and to the spin interactions. The behaviour can be approximated by the equation

$$\frac{1}{\chi} = \frac{T}{C} + \frac{1}{\chi_0} + \frac{\xi}{T - \theta}$$

where the parameters χ_0, θ, and ξ depend on the population of the available cation sites and the spin interactions. These are often taken as constants and a graph of $1/\chi$ against temperature is a straight line except near the Curie temperature, T_C (Figure 12.4d).

Although magnetic solids are generally thought to be inorganic compounds or metals, there is much current interest in organic magnetic materials.

12.1.3   Atomic magnetism

Magnetic properties reside in the subatomic particles that make up atoms. Of these, electrons make the biggest contribution, and only these will be considered in this chapter. Each electron has a magnetic moment due to the existence of a magnetic dipole, which can be thought of as a minute bar magnet linked to the electron. There are two contributions to the magnetic dipole moment of an electron bound to an atomic nucleus, which, in semiclassical models, are attributed to orbital motion and spin, related to the two quantum numbers l and s, introduced in Chapter 1.

The magnetic properties of electrons are described in terms of an atomic unit of magnetic dipole called the Bohr magneton, μ_B. It is given by:

$$\mu_B = \frac{eh}{4\pi m_e} \qquad (12.6)$$

where e is the electron charge, m_e is the electron mass, and h is Planck's constant. The magnitude of the total magnetic dipole moment of a single electron associated with the orbital quantum number, l, the orbital component, is given by:

$$m_{\text{orbital}} = \mu_B[l(l+1)]^{1/2}$$

The magnitude of the total contribution to the magnetic dipole moment of a single electron associated with the spin, s, the spin component, is given by:

$$m_{\text{spin}} = g\mu_B[s(s+1)]^{1/2}$$

where g is called the free electron g value, equal to 2.0023, and s can take a value of $\frac{1}{2}$.

The orbital and spin components are linked, or coupled, on isolated atoms or ions to give an overall magnetic dipole moment for the atom, as explained in Section 1.4.2. The commonest procedure for calculating the resultant magnetic dipole moment is called Russell–Saunders coupling (see Section S1.3). In summary, individual electron spin quantum numbers, s, are added to give a many-electron total spin quantum number, S, and the individual orbital quantum numbers, l, are added to give a many-electron total orbital angular momentum quantum number, L. The many-electron quantum numbers L and S are further combined to give a total angular momentum quantum number, J. More detail

is given in Section 12.2.2 for the lanthanides and transition metals of interest here.

The magnetic dipole moment of the atom is given by Equation (12.7):

$$m_{\text{atom}} = g_J\mu_B[J(J+1)]^{1/2} \qquad (12.7)$$

where g_J is called the Landé g factor, given by:

$$g_J = 1 + \frac{J(J+1) - L(L+1) + S(S+1)}{2J(J+1)} \qquad (12.8)$$

The method of calculating S, L and J is described in Section S1.3.

A completely filled orbital contains electrons with opposed spins and, in this case, the value of S is zero. Similarly, a completely filled s, p, d or f orbital set has L equal to zero. This means that atoms with filled closed shells have no magnetic moment. The only atoms that display a magnetic moment are those with incompletely filled shells. These are found particularly in the transition metals, with incompletely filled d shells, and the lanthanides and actinides, which have incompletely filled f shells.

12.2   Weak magnetic materials

12.2.1   Diamagnetic materials

None of the atoms in a diamagnetic material has magnetic dipoles. A magnetic field, however, will induce a small magnetic dipole that is present only for as long as the magnetic field persists. The induced dipole is opposed to the magnetic induction, B, as sketched in Figure 12.3. The effect is analogous to the effect on a macroscopic conductor brought into a magnetic field, described by Lenz's Law. (A conducting electrical circuit brought into a magnetic field will contain an induced current such as to yield a magnetic field opposed to the inducing field.) This very small effect gives rise to the typical diamagnetic response.

12.2.2   Paramagnetic materials

The magnetic dipole moment of a paramagnetic solid containing paramagnetic atoms or ions will be given by Equations (12.7) and (12.8). To obtain values for S, L and J that are of interest in understanding the magnetic behaviour (following Section 1.4.3 and Section S1.3.2):

1. Draw a set of boxes corresponding to the number of orbitals available, seven for the lanthanides and five for the transition metals.

2. Label each box with the value of m_l, highest on the left and lowest on the right. For lanthanides these run 3, 2, 1, 0, −1, −2, −3. For transition metals these run 2, 1, 0, −1, −2.

3. Fill the boxes with electrons, from left to right. When all boxes contain one electron, start again at the left.

4. Sum the m_s values of each electron, $+\frac{1}{2}$ or $-\frac{1}{2}$. This is equal to the maximum value of S.

5. Sum the m_l values of each electron to give a maximum value of L.

6. Calculate the J values, $J = L + S$, to $J = L - S$. The ground state is the lower value if the shell is

up to half full and the higher value if the shell is over half full.

Equations (12.7) and (12.8) work well for many compounds, such as the salts of lanthanide ions, as can be judged from Table 12.1. Lanthanide ions have a partly filled 4f shell, and these orbitals are well shielded from any interaction with the surrounding atoms by filled 5s, 5p and 6s orbitals, so that, with the notable exceptions Eu^{3+} and Sm^{3+}, they behave like isolated ions.

For the transition metals, especially those of the 3d series, interaction with the surroundings is considerable. This has two important consequences. One is the curious fact that 3d transition metal ions in paramagnetic solids often have magnetic dipole moments corresponding only to the electron spin contribution, given by the quantum number S. The orbital moment, L, is said to be 'quenched'. In such materials, Equation (12.8) reduces to

$$g_J = 1 + \frac{S(S+1) + S(S+1)}{2S(S+1)} = 2$$

Table 12.1 Calculated and observed magnetic dipole moments for the lanthanides

Ion	Configuration[a]	S	L	J	Magnetic dipole moment[b]	
					calc.	meas.
La^{3+}, Ce^{4+}	f^0	0	0	0	0	Dia.[c]
Ce^{3+}, Pr^{4+}	f^1	$\frac{1}{2}$	3	$\frac{5}{2}$	2.54	~2.5
Pr^{3+}	f^2	1	5	4	3.58	3.5
Nd^{3+}	f^3	$\frac{3}{2}$	6	$\frac{9}{2}$	3.62	3.5
Pm^{3+}	f^4	2	6	4	2.68	–
Sm^{3+}	f^5	$\frac{5}{2}$	5	$\frac{5}{2}$	0.84	1.5
Sm^{2+}, Eu^{3+}	f^6	3	3	0	0	3.4[d]
Eu^{2+}, Gd^{3+}, Tb^{4+}	d^7	$\frac{7}{2}$	0	$\frac{7}{2}$	7.94	~8.0
Tb^{3+}	f^8	3	3	6	9.72	9.5
Dy^{3+}	f^9	$\frac{5}{2}$	5	$\frac{15}{2}$	10.63	10.6
Ho^{3+}	f^{10}	2	6	8	10.61	10.4
Er^{3+}	f^{11}	$\frac{3}{2}$	6	$\frac{15}{2}$	9.59	9.5
Tm^{3+}	f^{12}	1	5	6	7.57	7.3
Yb^{3+}	f^{13}	$\frac{1}{2}$	3	$\frac{7}{2}$	4.54	4.5
Yb^{2+}, Lu^{3+}	f^{14}	0	0	0	0	Dia.[c]

[a] The configurations of the lanthanide metals are given in Section S1.2.3.
[b] Units: Bohr magnetons; calc., calculated from Equation (12.7) in text; meas., measured value.
[c] Diamagnetic.
[d] Eu^{3+}

To a good approximation, Equations (12.7) and (12.8) can then be replaced by a spin only formula:

$$m = g_J[S(S+1)]^{1/2}\mu_B \qquad (12.9)$$

where $g_J = 2$, S is the total spin quantum number, equal to $s_1 + s_2 + s_3 + \cdots$ for the unpaired electrons, and μ_B is the Bohr magneton. In the case of n unpaired electrons on each ion, each of which has s equal to $\frac{1}{2}$,

$$S = s_1 + s_2 + s_3 \cdots s_n = \frac{1}{2} + \frac{1}{2} + \frac{1}{2} + \cdots = \frac{n}{2}$$

The spin-only formula can then be written:

$$m = [n(n+2)]^{1/2}\mu_B \qquad (12.10)$$

The magnetic dipole moments calculated in this way are compared with the experimental values in Table 12.2. It is seen that in many cases the agreement is very good. In cases where the experimental value is considerably different from the spin-only value, orbital effects are significant.

The second feature of the transition metal ions, especially noticeable in the 3d series, is that many ions have two apparent spin states: high spin and low spin. For example, Fe^{3+}, with a $3d^5$ electron configuration, sometimes appears to have five unpaired spins and sometimes only one. This behaviour is found for 3d ions with configurations between $3d^4$ and $3d^7$. The explanation of this phenomenon lies with crystal field splitting of the d-electron energy levels, explained in Section S4.5. In summary, when the ions are in octahedral surroundings, the d-electron levels are split into two groups, one of lower energy, containing three levels, labelled t_{2g}, and one of higher energy, containing two levels, called e_g. If the splitting is large, the d electrons preferentially occupy the lower, t_{2g}, group. Spin pairing occurs for more than three d electrons, giving rise to a low spin state. In the alternative case, when the splitting is small, the electrons occupy all of the d orbitals. Spin pairing is avoided, and the ion has a high spin state.

The crystal field splitting energy between the t_{2g} and e_g set of orbitals is generally similar to the energy of visible light. This means that the magnetic behaviour of a material can be changed by irradiation with light of a suitable wavelength, which will promote electrons from the lower to the upper energy state. Molecules in which this state of affairs can occur are known as spin-crossover complexes. In them, a magnetic state can be switched from, say, diamagnetic to paramagnetic, by irradiation,

Table 12.2 The magnetic properties of the 3d transition metal ions

Ion	Configuration[a]	S	Magnetic dipole moment[b]	
			calc.	meas.
Sc^{3+}, Ti^{4+}, V^{5+}	d^0	0	0	Dia.[c]
Ti^{3+}, V^{4+}	d^1	$\frac{1}{2}$	1.73	1.7–1.8
Ti^{2+}, V^{3+}	d^2	1	2.83	2.8–2.9
Cr^{3+}, Mn^{4+}, V^{2+}	d^3	$\frac{3}{2}$	3.87	3.7–4.0
Cr^{2+}, Mn^{3+}	d^4	2	4.9	4.8–5.0
Mn^{2+}, Fe^{3+}	d^5	$\frac{5}{2}$	5.92	5.7–6.1
Co^{3+}, Fe^{2+}	d^6	2	4.9	5.1–5.7
Co^{2+}	d^7	$\frac{3}{2}$	3.87	4.3–5.2
Ni^{2+}	d^8	1	2.83	2.8–3.5
Cu^{2+}	d^9	$\frac{1}{2}$	1.73	1.7–2.2
Cu^+, Zn^{2+}	d^{10}	0	0	Dia.[c]

[a] The configuration of the 3d transitions metals are given in Section S1.2.2.
[b] Units: Bohr magnetons; calc., calculated from Equation (12.9) in text; meas., measured value.
[c] Diamagnetic.

resulting in photoinduced magnetism. The application of this photoinduced spin-crossover is being studied in single molecules, with the aim of making ever-smaller data storage devices.

12.2.3 The temperature dependence of paramagnetic susceptibility

As described in Section 12.1.2, the temperature dependence of the susceptibility of paramagnetic compound is given by the Curie law, Equation (12.3). The Curie law is indicative of the presence of isolated paramagnetic ions or atoms in the material.

In classical physics, the magnetic dipole can lie in any direction with respect to the magnetic field. In real atoms this is not possible, and the direction of the magnetic moment vector can only take values such that the projection of the vector on the magnetic field direction, z, has values of M_J, where M_J is given by:

$$M_J = J, J - 1, J - 2, \cdots - J$$

There are $(2J + 1)$ values of M_J. These orientations can be represented as a set of energy levels, as sketched in Figure 12.6. The separation of the energy levels is:

$$\Delta E = g_J \mu_B B$$

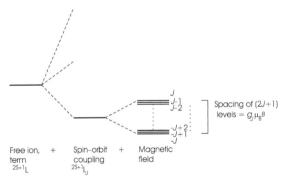

Figure 12.6 The splitting of atomic energy levels. The Russell–Saunders free ion term (left) becomes more complex when the spin and orbital contributions couple (centre). In a magnetic field each spin–orbit level splits into a further $(2J + 1)$ equally spaced levels (right)

where g_J is the Landé g factor, of Equation (12.8), and B is the magnitude of the magnetic induction. The average value of the magnetic dipole moment, $\langle m \rangle$, will depend on the population of these levels when at thermal equilibrium. The bulk magnetisation will then be given by:

$$M = N \langle m \rangle$$

where N is the number of magnetic dipoles per unit volume.

It is found that the value of M is:

$$M = N g_J \mu_B J B_J(x) \qquad (12.11)$$

where $B_J(x)$ is the Brillouin function:

$$B_J(x) = \frac{2J + 1}{2J} \coth \left\{ \frac{2J + 1}{2J} x \right\} - \frac{1}{2J} \coth \left\{ \frac{x}{2J} \right\} \qquad (12.12)$$

and

$$x = g_J \mu_B \frac{JB}{kT}$$

The classical case, in which the magnetic dipole moments can rotate freely, is equivalent to a continuum of energy levels with J tending to infinity. In this case, the magnetisation is given by:

$$M = N m L(x) = N m \left(\coth x - \frac{1}{x} \right) \qquad (12.13)$$

where

$$x = \frac{mB}{kT}$$

and the function $L(x)$ is the Langevin function.

The rather complex equation for the magnetisation of a paramagnetic substance can be simplified for the case of $x \ll 1$, which holds in the case of higher temperatures and lower magnetic fields. In fact, even when the magnetic induction is equal to 1 T, a temperature of 300 K is sufficient to make $x \ll 1$). In this case, the magnetisation is given by:

$$M = \frac{N g_J^2 \mu_B^2 B J (J + 1)}{3kT} \qquad (12.14)$$

Substituting $B = \mu_0 H$, for paramagnetic materials and rearranging, gives:

$$\frac{M}{H} = \chi = \frac{N\mu_0 g_J^2 \mu_B^2 J(J+1)}{3kT} \qquad (12.15)$$

This has the same form as the Curie equation, and the Curie constant is given by:

$$C = \frac{N\mu_0 g_J^2 \mu_B^2 J(J+1)}{3k} \qquad (12.16)$$

The value of χ is proportional to the number of magnetic dipoles present. The susceptibility is thus given in several forms. In the derivation above, N is the number of dipoles per unit volume, and the susceptibility is called the volume susceptibility, χ_v in m^{-3}. The mass susceptibility, χ_m, in which N refers to the number of magnetic dipoles per unit mass, is given by χ_v/ρ, where ρ is the density in $kg\,m^{-3}$. The susceptibility of one mole of the substance, the molar susceptibility, χ_{molar}, is given by $(\chi_v V_m)$, where V_m is the molar volume of the material.

12.3   Ferromagnetic materials

12.3.1   Ferromagnetism

The most important temperature effect is shown by just a small number of paramagnetic materials. In these, below a temperature called the Curie temperature, T_C, the material takes on quite new magnetic properties and is no longer paramagnetic, but strongly ferromagnetic. The change is reversible, and heating above the Curie temperature causes the material to revert to paramagnetic behaviour. Below the Curie temperature, the magnetic susceptibility is found to be 10^4 or more times higher than that of a normal paramagnetic substance. This state of affairs is found in only four elements – Fe, Co, Ni and Gd – at room temperature. Two lanthanides, Tb and Dy, have a temperature range (well below room temperature) over which they are ferromagnetic, as given in Table 12.3. However, in the second half of the 20th century, many other ferromagnetic

Table 12.3 Ferromagnetic and antiferromagnetic compounds

Ferromagnetic compound	T_C/K	Antiferromagnetic compound	T_N/K
Fe	1043	Cr	310
Co	1388	α-Mn	100
Ni	627	α-Fe_2O_3	950
Gd	293	CuF_2	69
Tb	220–230	MnF_2	67
Dy	87–176	$CoCO_3$	18
CrO_2	386	NiO	523
$SmCo_5$	973	CoO	293
$Nd_2Fe_{14}B$	573	FeO	198
		MnO	116
		K_2NiF_4	97
		$LaFeO_3$	750

Note: T_C, Curie temperature; T_N, Néel temperature.

compounds have been synthesised, notably oxides and alloys of the 3d transition metals and lanthanides.

The magnetisation of a compound, M, is the dipole moment per unit volume. For ferromagnetic compounds, the saturation magnetisation, M_s, when all of the dipoles are aligned, can be measured and used to estimate the effective magnetic moment on the atoms in the structure when the unit cell is known.

$$M_s = \text{(number of dipoles per unit volume)}$$
$$\times \text{(effective magnetic moment on each dipole)}$$
$$= N_{unpaired}\, m_{eff}\, \mu_B$$

In addition, the equation for the inductance (see Section S4.4),

$$B = \mu_0 H + \mu_0 M$$

can be approximated to:

$$B = \mu_0 M$$

as B is much greater than H in most of these materials.

The first theory to account for the existence of ferromagnetic solids was proposed by Weiss. He suggested that an internal 'molecular' field existed inside the magnetic compound and that this acted to align the magnetic moments of neighbouring atoms. The result was that a magnetisation was present even when the magnetic field was zero. Equation (S4.11) in Section S4.4 is replaced by Equation (12.17):

$$\boldsymbol{B} = \mu_0(\boldsymbol{H} + \lambda\boldsymbol{M}) \qquad (12.17)$$

where λ is the molecular field constant that indicates the strength of the molecular field. The resulting magnetisation of the solid is then given by Equation (12.11).

The Brillouin function, $B_J(x)$, is now a function of M:

$$x = \frac{g_J\mu_B JB}{kT} = \frac{g_J\mu_B J(H + \lambda M)}{kT}$$

Equation (12.11) cannot be solved analytically and, in the past, graphical solutions have been utilised. Nowadays, it is easier to use a computer to give numerical results. The way in which the value of M_s varies with temperature is shown in Figure 12.7. There are three special cases to note.

- Case 1: $T = 0$ K, $M = M_{sat} = Ng_J\mu_B J$

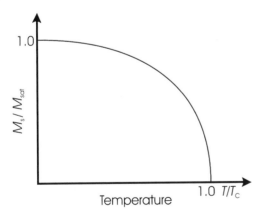

Figure 12.7 Variation of the relative spontaneous magnetisation, M_s/M_{sat}, as a function of relative temperature, T/T_C. When $T = 0$ the magnetisation is equal to the saturation magnetisation, M_{sat}. When $T = T_C$, the Curie temperature, the spontaneous magnetisation is zero

- Case 2: $T = T_C$, $M = 0$,

$$T_C = \frac{N\mu_0 g_J^2\mu_B^2 J(J+1)\lambda}{3k} = \lambda C \qquad (12.18)$$

- Case 3: $T \gg T_C$,

$$\chi = \frac{N\mu_0 g_J^2\mu_B^2 J(J+1)}{3k(T - \lambda C)} \qquad (12.19)$$

$$= \frac{C}{T - T_C}$$

The theory gives good agreement with the observations. Equation (12.18) shows that a high molecular field corresponds to a high Curie temperature, as one would expect. When there is a high interaction between the magnetic dipoles, it will be harder to disrupt the ordering by temperature effects alone. Equation (12.19) shows that ferromagnetic compounds obey the Curie–Weiss law well above a transition temperature, T_C, at which the material loses its ferromagnetic properties.

12.3.2 Exchange energy

The interaction that leads to the molecular field is quantum mechanical in origin and is related to the electron interactions that lead to chemical bonding. It is called the exchange interaction, giving rise to the exchange energy. To illustrate this concept we will focus on the important 3d orbitals. The electron distribution in an atom or an ion with several d electrons results from electrostatic repulsion between these electrons. This interaction is equivalent to the classical Coulomb repulsion between like charges, and is called the Coulomb repulsion. The total contribution to the energy is obtained by summing all the various electron–electron interactions, and is summarised as the Coulomb integral. This energy term is decreased if the electrons avoid each other as much as possible. Now, electrons with opposite spins tend to occupy near regions of space, whereas electrons with parallel spins tend to avoid the same regions of space. Thus, the electrostatic energy is decreased if the electrons all have parallel spins. In this case, as we know, they occupy different d orbitals, as far as possible. The energy decrease due to the preference for parallel spins is called the exchange energy. This idea has been encountered as Hund's first rule (Section S1.3.2) which says that in a free atom the state of lowest energy, the ground state, has as many electrons as possible with parallel spins. This is the same as saying that the ground state of a free atom has a maximum value of the spin multiplicity, or the equivalent quantum number S.

However, when the atom is introduced into a solid or a molecule another interaction, chemical bonding, is important. Chemical bonding is the result of placing electrons from neighbouring atoms into bonding orbitals. For bonding to occur, and the energy of the pair of atoms to be lowered, the electron spins must be antiparallel. However, the exchange interaction between electrons on neighbouring atoms is still present, tending to make the electron spins adopt a parallel orientation. In general, bonding energy is greater than the exchange energy, and paired electrons are more often found in solids and molecules. However, if the bonding forces are weak, the exchange energy can dominate, and unpaired electrons with parallel spins are favoured.

The 3d transition metals are notable in that the d orbitals do not extend far from the atomic nucleus, and so bonding between d orbitals is weak and the exchange energy is of greater importance. As one moves along the 3d-elements, it is found the d orbitals become more compact, making bonding less favourable and increasing the exchange energy. The interplay between bonding energy and exchange energy can be pictured in terms of the ratio of D/d, where D is atomic separation of the interacting atoms, and d is the diameter of the interacting d orbitals. A plot of D/d, against exchange interaction, a Bethe–Slater diagram, is shown in Figure 12.8. If the value of the exchange interaction is positive then ferromagnetism is to be expected. Of all of the 3d transition metals, only Fe, Co and Ni have positive values of D/d, and these are the only ferromagnetic 3d transition elements. The lanthanides have slightly positive values of D/d and are expected to be weakly ferromagnetic.

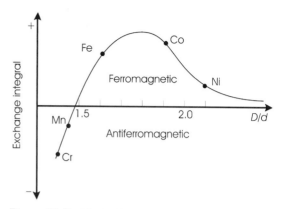

Figure 12.8 The Bethe–Slater curve for the magnitude of the exchange integral as a function of D/d, where D is the separation of the atoms in a crystal, and d is the diameter of the 3d orbital

12.3.3 Antiferromagnetism and superexchange

The interaction between the d orbitals giving rise to the conflict between the exchange energy and the bonding energy does not operate so simply in many compounds. In these, the transition metal ions are separated by a nonmetal such as oxygen, and the compounds are antiferromagnetic. The interaction of the d orbitals on the cations via the intermediate anion is called superexchange. It occurs in transition metal oxides such as nickel oxide, NiO, which is a typical antiferromagnetic.

In nickel oxide, the Ni^{2+} ions have eight d electrons. The oxide would be expected to be paramagnetic, as the Ni^{2+} ions are separated from each other by nonmagnetic oxygen ions. Instead, antiferromagnetic ordering occurs, and the magnetic susceptibility increases with temperature up to 250 °C before showing paramagnetic behaviour. The coupling between the ions leading to an antiferromagnetic ordering is related to a degree of covalent bonding between the Ni^{2+} and O^{2-} ions. The nickel atoms in NiO have an octahedral coordination, and the d orbitals are split by the crystal field (Section S4.5), so that the $d_{x^2-y^2}$ and d_{z^2} orbitals each contain one unpaired electron. As drawn schematically in Figure 12.9, an unpaired d electron in

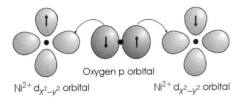

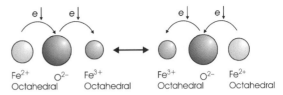

Figure 12.9 Superexchange leading to antiferromagnetic alignment of spins on cations

Figure 12.10 Double exchange, possible only with ferromagnetic alignment of spins on cations

the $3d_{x^2-y^2}$ orbital can have a covalent interaction with an electron in a filled p orbital, but only if the electrons have opposed spins. The oxygen 'spin down' electron will interact with a d orbital in which the electron is 'spin up'. This is true for both Ni^{2+} ions either side of an oxygen ion. In this case, as shown, the arrangement leads to an antiferromagnetic alignment of the unpaired d electrons on the Ni^{2+} ions. This is generally true, and superexchange tends to lead to an antiferromagnetic ordering in a solid.

It would be expected that the Néel temperature would increase with the degree of covalency, and this is found to be so. Table 12.3 shows that the Néel temperature for the series of 3d transition metal oxides MnO, FeO, CoO and NiO increases in the direction from MnO to NiO, as the covalent nature of the solids increase. Additionally, the dumbbell shape of the p orbitals suggests that the interaction should be greatest for a linear *M–O–M* configuration and minimum for a 90° angular *M–O–M* configuration. The oxides MnO, FeO, CoO and NiO all have the *halite* structure in which the linear *M–O–M* configuration holds, with metal ions at the corners of the cubic unit cell and oxygen ions at the centre of the cell edges. Thus, it becomes apparent that both chemical bonding and crystallography have a part to play in deciding on the effectiveness of magnetic interactions. This aspect is marked in the ferrimagnetic compounds.

12.3.4 Ferrimagnetism and double exchange

In ferrimagnetic materials two subsets of magnetic moments exist, with one of the subsets in an antiparallel arrangement with respect to the first,

as illustrated in Figure 12.3(j). The observed magnetic moment is due to the difference between the two sets of magnetic dipoles. Many of the magnetic ceramics are ferrimagnetic materials. The ancient 'lodestone', Fe_3O_4, is a ferrimagnetic. Above the Néel temperature (858 °C for Fe_3O_4), the material reverts to paramagnetic behaviour.

In the ferrimagnetic state, there are several interactions between the spin subsets. As with the antiferromagnetic compounds, they can only be understood in terms of the crystal structures and the chemical bonding in the compounds. The most important feature of ferrimagnetic materials is that two valence states must exist for the cations. In the case of Fe_3O_4, these are Fe^{2+} and Fe^{3+}. Other ion pairs, such as Mn^{2+} and Mn^{3+}, can also behave in the same way. The coupling between these ions in called double exchange. In this mechanism, the two ions are separated by an anion such as oxygen (Figure 12.10). An Fe^{2+} ion, for example, transfers an electron to an adjacent O^{2-} ion which simultaneously transfers one of its electrons to a neighbouring Fe^{3+} ion. In essence, the electron hops from Fe^{2+} to Fe^{3+} via the intermediate oxygen ion. The charges on the Fe ions are reversed in the process. Because all of the orbitals on the oxygen ion are full, and electrons are spin paired, a 'spin-down' electron moving from Fe^{2+} will displace a 'spin-down' electron from O^{2-} onto Fe^{3+}. This double exchange is favourable only if there are parallel spins on the two cations involved. Thus, double exchange leads to ferromagnetic ordering.

The other important factor in ferrimagnetic compounds is the crystal structure of the compound. This can also be illustrated by the oxide Fe_3O_4, which adopts the inverse *spinel* structure (see Section 5.3.10). In this structure the cations are distributed between tetrahedral sites and octahedral sites in the following way, $(Fe^{3+})[Fe^{2+}Fe^{3+}]O_4$, where (Fe^{3+}) represents cations in tetrahedral sites, and $[Fe^{2+}Fe^{3+}]$ represents cations in octahedral sites. In Fe_3O_4, the Fe^{2+} and Fe^{3+} cations that take part in double exchange are restricted to those that occupy the octahedral sites. The population of Fe^{3+} ions in the tetrahedral sites is not involved. The reason for this separation lies with the d-orbital energies, which are controlled by the crystal field splitting (see Section S4.5). The d-orbital energy of a cation in an octahedral site is quite different from that of a cation in a tetrahedral site. This difference in energy prevents double exchange between Fe^{2+} ions in octahedral sites and Fe^{3+} ions in tetrahedral sites. However, superexchange is possible between the occupants of the octahedral and tetrahedral sites, which tends to give an antiferromagnetic ordering between these cations.

The outcome of these two interactions is the ferrimagnetic structure. In general, therefore, ferrimagnetic compounds have a common cation in two valence states and two sets of different sites over which these cations are ordered. Many crystalline structures potentially fulfil these requirements, including the cubic and hexagonal ferrites, described in Sections 12.3.5 and 12.3.6.

The saturation magnetisation, M_s, of ferrites, can be estimated by assuming the spin-only magnetic moments listed in Table 12.2 apply. It is necessary to determine, from the crystal structure, which spins cancel, and then assuming all the excess moments are completely aligned:

M_s = (number of unpaired dipoles per unit volume)

$\times$ (magnetic moment on each dipole)

$= N_{unpaired}\, m_{ion}\, \mu_B$

where m_{ion} is taken from Table 12.2. The equation for the inductance (see Section S4.4),

$$B = \mu_0 H + \mu_0 M$$

can be approximated to

$$B = \mu_0 M$$

as B is much greater than H in most of these materials.

12.3.5 Cubic spinel ferrites

Ferrites with the cubic *spinel* structure (see Section 5.3.10) are soft magnetic materials (see Section 12.4.3), widely used in electronic circuitry. The formula of all ferrites can be written as $A^{2+}Fe_2^{3+}O_4$, where A^{2+} can be chosen from a large number of medium-sized cations, for example Ni^{2+} or Zn^{2+}. The majority of the important ferrites are inverse *spinels*, in which the A^{2+} cations occupy the octahedral sites, together with half of the Fe^{3+} ions. The other half of the Fe^{3+} cations is found in the tetrahedral sites. Thus nickel ferrite would be written $(Fe^{3+})[Ni^{2+}Fe^{3+}]O_4$, where the cations in octahedral sites are enclosed in square brackets and those in tetrahedral sites in parentheses. Lodestone, or magnetite, Fe_3O_4, described in Section 12.3.4, is an example in which the cations are Fe^{2+} and Fe^{3+}, and the cation distribution is $(Fe^{3+})[Fe^{2+}Fe^{3+}]_2O_4$.

The magnetic properties of the ferrites can be tailored by changing the A^{2+} cations present. This chemical manipulation of physical properties is widely used and is the cornerstone of silicon electronics. In the ferrites, dopant cations substitute for the A^{2+} and Fe^{3+} ions present. There is great flexibility in this, as not only can the ions be changed but also the distribution over the octahedral and tetrahedral sites. Some ions, such as Cr^{3+}, have a strong preference for octahedral sites and are rarely found in tetrahedral sites. Other ions, such as Fe^{3+}, can adapt to either geometry more easily. Complex mixtures of cations, for example $Ni_aMn_bZn_cFe_dO_4$, are often used to tailor quite specific magnetic behaviour. In all of these complex formulae, remember that the total of the cations must be in accord with the *spinel* formula, $A^{2+}B_2^{3+}O_4$, both with respect to the total numbers but also with respect to the charges.

In reality, the distribution of the cations is rarely perfectly normal or inverse, and the distribution tends to vary with temperature. As the magnetic properties depend sensitively on the cation distributions, processing conditions are important if the desired magnetic properties are to be obtained.

12.3.6 Hexagonal ferrites

An important group of ferrites is known that have hard magnetic properties (see Section 12.4.3). As in the case of the cubic ferrites, they are derived from a single structure type, this time that of magnetoplumbite, $Pb^{2+}Fe_{12}^{3+}O_{19}$. As can be seen, all of the Fe ions are trivalent. The unit cells of these phases are hexagonal, hence the general name of hexagonal ferrites. The variety of ferrites derived from magnetoplumbite is achieved in the same way as the cubic ferrites, by substitution of the cations present by others. The magnetic properties of the hexagonal ferrites can be changed by fine-tuning the amounts and types of M^{2+} and M^{3+} ions in the structure, which can substitute for the Pb^{2+} and the Fe^{3+} cations present. However, the system is more complex than that of the cubic *spinels*, in which only one parent structure forms. In the hexagonal ferrites hundreds of structures occur.

The two simplest hexagonal ferrites, $BaFe_{12}O_{19}$, known as ferroxdure, and $SrFe_{12}O_{19}$, are widely used as permanent magnets in many applications, especially in electric motors. The structure of both is the magnetoplumbite type. The structure can be illustrated by describing $BaFe_{12}O_{19}$, which is built from a close packing of oxygen and oxygen-plus-barium layers (Figure 12.11). There are 10 of these layers in the unit cell, which contains two $BaFe_{12}O_{19}$ formula units. Using the description of close-packed layers described in Section 5.4.1, the relative positions are:

$$A\ B'\ A\ B\ C\ A\ C'\ A\ C\ B$$

where the symbols B' and C' represent mixed barium and oxygen, BaO_3, layers, and the other symbols represent oxygen-only, O_4 layers. These 10 layers are thought of as being arranged in a sequence of four alternating slices, two that resemble spinel, called S blocks, and two containing the

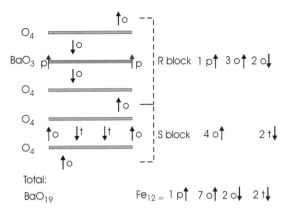

Figure 12.11 The idealised structure and spin arrangement of $BaFe_{12}O_{19}$. Half of the unit cell is drawn, composed of one R block and one S block. The composition of the cubic close-packed layers (wide lines) is indicated on the left-hand side. The spin directions on the cations are represented by arrows, and the coordination of the cations by letters; o = octahedral, t = tetrahedral, p = trigonal prismatic. The overall composition and spin status is summarised at the bottom of the diagram

BaO_3 layer, called R blocks. Half a unit cell, consisting of one R block and one S block, is drawn in Figure 12.11. [In the lower half of the cell the R and S blocks are inverted with respect to those shown, but the site geometries remain the same.] The Fe^{3+} cations occupy nine octahedral sites, two tetrahedral sites and one trigonal pyramidal (five-coordinate) site per $BaFe_{12}O_{19}$ unit. The arrangement of the spins on the Fe^{3+} ions in these sites is such that four point in one direction, parallel to the hexagonal c axis, and the other eight point in the opposite direction. The arrangement is different from that shown in Figure 12.3(j), because the spins are all of the same magnitude and, in this case, the numbers of spins are unbalanced. Because the spins are aligned along the crystallographic c axis, the materials show a high degree of anisotropy. The hexagonal ferrites are widely used in permanent ceramic magnets.

12.4   Microstructures of ferromagnetic solids[1]

12.4.1   Domains

So far, in this discussion, emphasis has been laid on the atomic interactions that lead to magnetic properties. Especially in ferromagnetic solids, the microstructure of the solid is of importance and dominates applications. The microstructures found are the result of long-range interactions between the elementary magnets. For an isolated atom or ion with unpaired electron spins, no interaction occurs between the elementary magnets, and magnetic properties are reflected in a Curie law type of dependence.

When the magnetic atoms are not isolated, as in ferromagnetic solids, interactions that are more complex arise. These are revealed by the way in which such materials respond to being magnetised in an external magnetic field. All ferromagnetic substances remain magnetised to some extent after the external field is removed, and this feature characterises ferromagnetism. Some materials retain a state of magnetisation almost indefinitely. These are called magnetically hard materials and are used to make permanent magnets. However, many ferromagnetic materials appear to lose most of their magnetisation rather easily. These are called magnetically soft materials.

If the ordering forces between the atomic magnetic moments are strong enough to lead to ferromagnetism, it is reasonable to ask why ferromagnetic materials often show no obvious magnetic properties. The answer lies in the microstructure of hard and soft magnetic solids. On a microscopic scale, all such crystals are permanently magnetised but are composed of magnetic domains or 'Weiss domains'. Domains are regions that have all of the elementary magnetic dipoles aligned

(Figure 12.12a), but the directions in neighbouring domains are different, so that the net result is that the external magnetic field is very small. Domains are distinct from the grain structure of a metal and occur in single crystals as well as in polycrystalline ferromagnetic solids, and the geometry of the domains varies from one ferromagnetic solid to another. Ferromagnetic materials, like ferroelectrics, belong to the group of ferroic materials.

Two main interactions need to be considered in order to understand domain formation, an electrostatic interaction and a dipole–dipole interaction. The parallel ordering, as we know, is caused by the exchange energy. This is effectively electrostatic and short range, and is equivalent to the ionic bonding forces described in Section 2.1.2. The external magnetic field arising from the parallel dipoles gives rise to magnetostatic energy. This is reduced if the external field is reduced. One way of achieving this is to form the aligned dipoles into *closure domains* with antiparallel orientations (Figure 12.12b). This is achieved by the dipole–dipole interaction, which is one of the weak bonding forces described in Section 3.1.1. This interaction between the magnetic dipoles, which is long range, falls approximately as the cube of the distance between atoms. It leads to a preferred antiparallel arrangement of the dipoles.

When the two interactions are compared, it is found that at short ranges the electrostatic force is the most important, and an arrangement of parallel spins is of lowest energy. As the distance from any dipole increases, the short-range electrostatic interaction falls below that of the dipole–dipole interaction. At this distance, the system can lower its energy by reversing the spins (Figure 12.12c). The domains form as a balance between these two competing effects, short-range electrostatic and long-range dipolar. The balance is achieved for domains of 1–100 micrometers in size.

The alignment from one domain to the adjoining one is not abrupt but depends on the balance between the electrostatic and dipole interactions. The dipole orientation is found to change gradually over a distance of several hundred atomic diameters. This structure is called a domain or Bloch wall (Figure 12.13).

[1]Ferromagnetic and ferrimagnetic materials behave in the same way with respect to the subject matter of this section. To avoid repetition, only the term ferromagnetic will be used, but it should be understood that ferrimagnetic materials are identical.

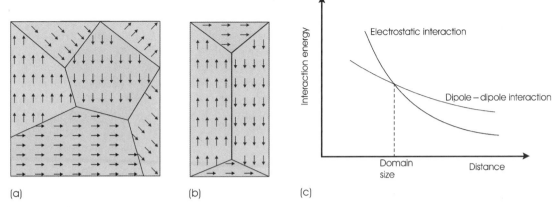

Figure 12.12 (a) Representation of Weiss domains; the magnetic dipoles are represented by arrows; (b) Domain closure; and (c) graph of interaction energy as a function of distance. The dipole–dipole interactions in the solid tend to produce antiparallel alignment of magnetic dipoles, and the electrostatic interactions a parallel alignment of magnetic dipoles. The domain size is a reflection of the point where these interactions balance

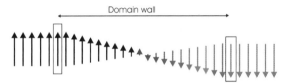

Figure 12.13 Schematic representation of a Bloch wall between two magnetic domains. On the far left the magnetic dipoles (arrows) point upwards. Moving the right they spiral through 180° until, at the for right, they point downwards. The extent of the wall is indicated by the magnetic dipoles enclosed in rectangles

The geometry of domains also depends on crystal structure. Most ferromagnetic solids are magnetically anisotropic. This means that the magnetic moments align more readily in some crystallographic directions than in others. These are known as 'easy' and 'hard' directions respectively. The domain geometry will reflect this crystallographic preference.

12.4.2   Hysteresis

In essence, the magnetic state of a ferromagnetic solid will depend on its history, a phenomenon called hysteresis. This is because the domain structure reflects the way in which the solid has previously been exposed to magnetic fields and to high temperatures. However, the state of magnetisation is reproducible and cycles in a constant fashion as a function of the direction and strength of an applied external magnetic field. Hysteresis is perhaps the most characteristic feature of strongly magnetic materials.

On placing a ferromagnetic material into a uniform external magnetic field, the dipoles tend to align themselves along the field direction. In a typical ferromagnetic solid, the alignment of some domains is favourable in that the magnetic moments of the constituents are already close to that of the external field. In other domains, the magnetic moments will be at an angle to the field or even opposed to it. The tendency to align with the field will be opposed by local effects in the solid, and at low field strengths little will happen. However, as the external field strength increases, some dipoles will rotate to align with the external field. The effect of this is that domains that are arranged so that the dipoles they contain are more or less aligned in the field direction will grow at the expense of those that are poorly aligned. Ultimately, all of the domains that remain will have parallel magnetic moments

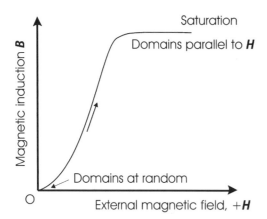

Figure 12.14 The initial part of a magnetic induction–external magnetic field (**B**–**H**) loop for a ferromagnetic solid

(Figure 12.14). At this point, the magnetisation of the sample will be a maximum and has reached saturation.

Reversal of the applied field will cause the magnetic moments to align in the opposite direction. Once again, this will be opposed by the internal energy of the solid, as new domain walls have to be created and domain walls must move. The **B**–**H** path will therefore not follow the original path, but will trace a new path that 'lags behind' (the definition of hysteresis) the old one (Figure 12.15). When

the applied field reaches zero, the value of **B** is still positive, and this is called the remanence, B_r. The magnetic flux, **B**, will be reduced to zero at a value of the applied field called the coercivity, H_c. Ultimately, domain alteration will cease when all the magnetic moments are now parallel to the reversed applied field and saturation is again reached. Reversal of the field once again will cause the value of **B** to trace a new path, but one that eventually reaches saturation in the initial direction. Repeated cycles now follow this outer pathway.

12.4.3 Hysteresis loops: hard and soft magnetic materials

The shape of the hysteresis loop for a ferromagnetic material is of importance. The area of the loop is a measure of the energy required to complete a hysteresis cycle. A soft magnetic material is one that is easily magnetised and demagnetised. The energy changes must be low, and the hysteresis loop has a small area (Figure 12.16). Soft magnetic materials are used in applications such as transformers, where the magnetic behaviour must mirror the variations of an electric current without large energy losses. Ferrites with the cubic *spinel* structure are generally soft magnetic

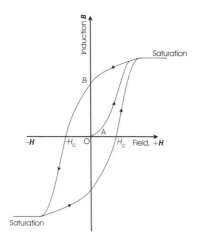

Figure 12.15 The complete **B**–**H** (hysteresis) loop of a ferromagnetic solid

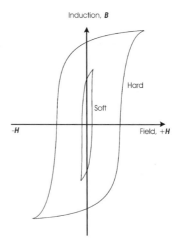

Figure 12.16 Schematic **B**–**H** loops for soft and hard magnetic solids

materials. The value of the saturation magnetisation is sensitive to the composition of the material, and the formulae of cubic ferrites are carefully tailored to give an appropriate value. However, the shape of the hysteresis loop is strongly influenced by the microstructure of the solid. Grain boundaries and impurities hinder domain wall movement and so change the coercivity and remanence of a sample. Ferrites for commercial applications not only have to have carefully chosen compositions but also have to be carefully fabricated in order to reproduce the desired magnetic performance.

Hard magnetic materials have rather rectangular, broad, hysteresis curves (Figure 12.16). These materials are used in permanent magnets, with applications from door catches to electric motors. The hexagonal ferrites are hard magnetic materials, as are alloys such as $SmCo_5$ and $Nd_2Fe_{14}B$.

Naturally, there is a continuum of magnetic materials between soft and hard magnetic materials, and the characteristics of any particular solid are tailored to the application. Of these, magnetic storage of information is of great importance. The data are stored by magnetising small volumes of the storage material, often a thin film deposited on a solid surface or a flexible tape. The magnetic material that receives the signal to be stored must be soft enough magnetically to respond quickly to small energising electric signals but be hard enough magnetically to retain the information for long periods.

12.5 Free electrons

So far, in this chapter, the band model of the solid has been ignored. This is because magnetism is associated with the d and f orbitals. These orbitals are not broadened greatly by interactions with the surroundings and even in a solid remain rather narrow. The resulting situation is quite well described in terms of localised electrons placed in d or f orbitals on a particular atom. However, some aspects of the magnetic properties of solids can be explained only by band theory concepts.

12.5.1 Pauli paramagnetism

Most metals show paramagnetic behaviour. It is rather small and independent of temperature and is quite different from the Curie law behaviour of a 'normal' paramagnetic ion. The cause of the weak paramagnetism was explained by Pauli in terms of the Fermi–Dirac statistics of electrons in solids. It is known as Pauli paramagnetism.

As outlined in Section 2.3, electrons in a normal metal occupy the (partly filled) uppermost band. All of the electrons are spin-paired, with two electrons being allocated to each energy state. In this state, the metal would be diamagnetic. A small change takes place when the metal is placed in a magnetic field. Those electrons with a spin parallel to the magnetic field have a slightly lower energy than those with a spin opposed to the magnetic field. Those electrons near the very top of the (partly filled) conduction band (i.e. those at the Fermi surface) can reorient themselves in the applied magnetic field. This results in an imbalance in the numbers of 'spin-up' and 'spin-down' electrons, and the metal becomes paramagnetic. Calculation shows that the number influenced is very small (for the same reason that the specific heat contribution of the electrons is very small) and so the magnetic susceptibility is small. The position is drawn in Figure 12.17. The paramagnetic susceptibility χ is

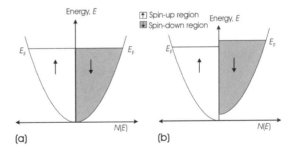

Figure 12.17 (a) The density of states, $N(E)$, for electrons in a metal is made up of equal numbers of spin-up and spin-down electrons. In the absence of a magnetic field, these are of equal energy. (b) The application of a magnetic field causes these to separate in energy. The electrons opposed to the field have a slightly higher energy than those parallel to the field

proportional to the density of states at the Fermi level, $N(E_F)$, and it is found that

$$\chi \propto \mu_0 \mu_B N(E_F),$$

where μ_0 is the permittivity of free space and μ_B the Bohr magneton. Thus, a measurement of the Pauli paramagnetism of a normal metal provides a method of assessing the density of states at the Fermi level. Temperature does not change the numbers of electrons at the Fermi surface appreciably, and so the paramagnetism is virtually temperature-independent.

All electrons not in the conduction band will be in filled shells, paired, and not magnetically active.

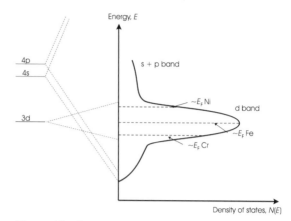

Figure 12.18 Energy band overlap of 3d, 4s and 4p orbitals for the 3d transition metals

12.5.2 Transition metals

The band model just outlined would lead to the expectation that all metals would show Pauli paramagnetism. However, three 3d transition metals, iron, cobalt and nickel, together with a few lanthanides, are ferromagnetic metals. The simple band picture must be expanded to account for this complication.

To illustrate this, consider the situation in the 3d transition metals. The electron configuration and measured magnetic moments of these metals is given in Table 12.4, and the band picture is illustrated in Figure 12.18. The d band is narrow and is overlapped by broad outer bands from the s and p orbitals. The s and d electrons will be allocated to this composite band. [However, it is still to be expected that the number of 'spin-up' and 'spin-down' electrons will be identical.] Now, a stable ferromagnetic state can arise only if there is

an excess of spins that are parallel and maintained in this state by the exchange energy of the system. To obtain this, energy must be supplied to promote electrons at the Fermi energy so that they can reverse spins. The rather small energy needed to produce Pauli paramagnetism is donated by the magnetic field. When this is removed, the separation of the spin states vanishes. In a ferromagnetic metal, the exchange energy must be sufficient to counterbalance the increase in energy of the electrons at the Fermi level. The balance depends on the form of the $N(E)$ curve, especially in the region of the Fermi surface. The 4s and 4p bands are broad, with a low density of states at the Fermi level, so that relatively small numbers of electrons are promoted. The gain in exchange energy is not sufficient to balance the energy outlay, and the metal remains paramagnetic. In contrast, the 3d band is narrow, with a high density of states at the Fermi level. Larger numbers of electrons can be promoted and reverse spin, leading to greater exchange energy. This is sufficient to stabilise the ferromagnetic ordering.

When the density of states at the Fermi surface is calculated with a high precision, it is found that only the three metals iron, cobalt and nickel will have sufficient exchange energy to retain a ferromagnetic state. The experimental value of the magnetic moment (Table 12.4) is also calculated to be equal to the number of holes in the d band. Thus, the d band of iron contains 7.78 electrons, (2.22 holes),

Table 12.4 Magnetic moments and electron configurations of the ferromagnetic 3d transition metals

Element	Electron configuration	Magnetic moment/μ_B
Iron, Fe	[Ar] $3d^6$ $4s^2$	2.22
Cobalt, Co	[Ar] $3d^7$ $4s^2$	1.72
Nickel, Ni	[Ar] $3d^8$ $4s^2$	0.6

that of cobalt contains 8.28 electrons (1.72 holes), and that of nickel contains 9.4 electrons (0.6 holes).

The magnetic properties of alloys of these ferromagnetic metals with nonmagnetic metals confirm this interpretation. For example, both copper and zinc have 10 d electrons. As they are alloyed with a ferromagnetic metal, the d band is filled and the ferromagnetic properties diminish. These vanish when the d band is just filled with electrons. A good example is provided by the nickel–copper system, because the alloys all have the same face-centred cubic crystal structure as the parent Ni and Cu phases (see Section 4.2.2). The electron configuration of copper is $[Ar]\, 3d^{10}\, 4s^1$, and in this metal the 3d band is filled. As copper is added to nickel to form the alloy, the excess copper electrons are added to the d band to gradually fill it. At 20 °C, the ferromagnetic–paramagnetic change occurs at a composition of 31.5 wt% copper, that is, approximately $Ni_{235}\, Cu_{100}$.

Exact details of the magnetic properties of most alloys depend on the crystal structures of the alloys and the form of the density of states curve at the Fermi level, as well as the relative proportions of the metals present.

12.6 Nanostructures

12.6.1 Small particles and data recording

The magnetic properties of very small particles have been of interest for a considerable period. The earliest applications of these properties arose in the use of magnetic particles for magnetic recording media. The essence of magnetic data storage is the existence of two easily distinguished magnetic states. Magnetic data storage uses a thin magnetisable layer laid down on a tape or disc. A 'write head' generates an intense magnetic field that changes the direction of magnetisation in the surface material. The induced magnetisation pattern in the layer is the stored data. The data are read by sensing the magnetic field changes by using a 'read head'.

The data storage layers are composed mostly of small magnetic particles in a polymer film. The

magnetic response of the film depends critically on the domain structure of the particles and the particle shape. Ideally, small single-domain particles are used, each of which has only two directions of magnetisation, directed along the + and − directions of a single crystallographic axis. The direction is switched by the 'write head' and sensed by the 'read head'.

The commonest magnetic particles in use at present are $\gamma\text{-}Fe_2O_3$, or maghemite, cobalt doped $\gamma\text{-}Fe_2O_3$, and chromium dioxide, CrO_2, all of which have acicular (needle-shaped) crystals. Chromium dioxide is a ferromagnetic oxide with the rutile structure. The crystal structure of $\gamma\text{-}Fe_2O_3$ is curious. Despite its formula, it has the *spinel* structure, and is closely related to the inverse *spinel* magnetite, $(Fe^{3+})[Fe^{2+}Fe^{3+}]O_4$, where round brackets enclose cations on tetrahedral sites and square brackets enclose cations on octahedral sites. $\gamma\text{-}Fe_2O_3$ is a ferrimagnetic *spinel* that has a formula $(Fe^{3+})[Fe^{3+}_{5/3}Va_{1/3}]O_4$, where Va represents vacancies on octahedral Fe^{2+} sites. It is thus a ferrite with an absence of M^{2+} ions. The direction of magnetisation in all of these compounds corresponds to the needle axis. The crystallites are aligned in a collinear array, with the needle axis parallel to the direction of motion of the tape or disc, by applying a magnetic field during the coating operation. This is called the longitudinal orientation or direction.

12.6.2 Superparamagnetism and thin films

Domain walls are created because of the competition between long-range dipole–dipole interactions, which lead to an antiparallel alignment of magnetic dipoles, and short-range electrostatic interactions, which tend to produce a parallel alignment of magnetic dipoles. When particles of a magnetic solid are below the domain size, the electrostatic interactions dominate (Figure 12.12c). The magnetic dipoles tend to align parallel to each other, and a *superparamagnetic* state results. This need not only occur in small particles. Isolated clusters of magnetic atoms or ions in a solid can also exhibit this feature. For example, nanoclusters of ferromagnetic particles embedded in polymers often show superparamagnetic behaviour, as can small magnetic precipitates in a nonmagnetic matrix. The same seems to be true of the magnetic particles that occur in certain bacteria that are able to navigate along the Earth's magnetic field gradient.

The same interactions lead to the observation that the magnetisation in thin films changes with film thickness. In thick films, of ordinary dimensions, domains form and the magnetic flux is trapped in the film (Figure 12.19a). When the thickness of a film is reduced to below single domain size, the magnetic dipoles align in a parallel direction in a longitudinal manner. The film has a north and a south pole at it extremities (Figure 12.19b). An external magnetic field is now detectable that runs more or less parallel to the film surface. Further reduction in film thickness to just a few atom layers results in the individual electron spins realigning normal to the film plane. North and south poles are now on the film surfaces, and the external magnetic field lies perpendicular to the film plane (Figure 12.19c). The implications of these scale effects are of importance in attempts to increase the density of magnetic recording media.

12.6.3 Molecular magnetism

The ultimate size limit for magnetic recording of data may well rest with magnetic molecules in which the cations show high-spin and low-spin states. For example, consider a single 3d transition metal ion such as Fe^{2+} in an octahedral crystal field (Section S4.5). There are two magnetic states, corresponding to the high-spin and the low-spin configurations (Figure 12.20). The low-spin state is diamagnetic, with the spin quantum number, S, zero, and the high-spin state is paramagnetic, with the spin quantum number, S, equal to two.

The crystal field-splitting energy between the t_{2g} and e_g set of orbitals is generally similar to the energy of visible light. This means that the magnetic

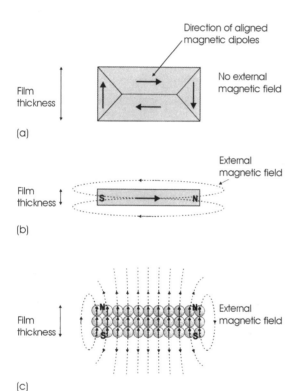

(a)

(b)

(c)

Figure 12.19 (a) Domain closure in a bulk film prevents magnetic flux from escaping. (b) A film less than a domain wide has magnetic dipoles aligned, so the flux escapes longitudinally; acicular crystals in magnetic recording media are ideally in this form. (c) A thin film of the order of one atomic thickness has elementary dipoles aligned perpendicular to the film, allowing flux to escape normal to the film

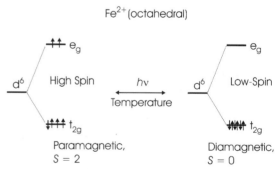

Figure 12.20 The Fe^{2+} ion in an octahedral crystal field can exist in either a high-spin or a low-spin state. These two states have different magnetic properties, and can be interconverted by radiation and heat energy

behaviour of a material can be changed by irradiation of light of a suitable wavelength, which will promote electrons from the lower-energy to the upper-energy state. Molecules in which this state of affairs can occur are known as spin-crossover complexes. In them, a magnetic state can be switched from, say, diamagnetic to paramagnetic, by irradiation, resulting in *photoinduced magnetism*. It is clearly possible to consider this as a potential data-storage unit.

In practice, Fe^{2+} ions are not isolated, but exist in compounds. The distribution of the ions between the high-spin and low-spin states will depend on molecular interactions and the temperature. If the transition-metal ions are linked by chemical bridges involving strong interactions, it is possible to induce a cooperative interaction between the cations, so that all cross over at once. There is still a long way to go before isolated ions are used to store bits of data, but research in this area is making rapid progress.

Answers to introductory questions

What atomic feature renders a material paramagnetic?

Paramagnetic solids are those in which some of the atoms, ions or molecules making up the solid possess a permanent magnetic dipole moment. These dipoles are isolated from one another. The solid, in effect, contains small, noninteracting atomic magnets. In the absence of a magnetic field, these are arranged at random and the solid shows no net magnetic moment. In a magnetic field, the elementary dipoles will attempt to orient themselves parallel to the magnetic induction in the solid, and this will enhance the internal field within the solid and give rise to the observed paramagnetic effect. The alignment of dipoles will not usually be complete, because of thermal effects and interaction with the surrounding atoms in the structure. Moreover, the orientation of the spins changes continuously. The magnetic susceptibility, χ, of a paramagnetic solid is positive and slightly greater than 1.

Why do ferromagnetic solids show a domain structure?

Two main interactions need to be considered in order to understand domain formation, a dipole–dipole interaction and an electrostatic interaction. The parallel arrangement of the spins is caused by the exchange energy. This is effectively electrostatic and short-range. The external magnetic field arising from the parallel dipoles gives rise to magnetostatic energy. This is reduced if the external field is reduced. One way of achieving this is to form the aligned dipoles into closure domains with antiparallel orientations. This is achieved by the dipole–dipole interaction, which is long-range, and leads to a preferred antiparallel arrangement of the dipoles. When the two interactions are compared, it is found that at short ranges the electrostatic force is the most important, and an arrangement of parallel spins is of lowest energy. As the distance from any dipole increases, the short-range electrostatic interaction falls below that of the dipole–dipole interaction. At this distance, the system can lower its energy by reversing the spins. The domains form as a balance between these two competing effects – short-range electrostatic and long-range dipolar. The balance is achieved for domains of 1–100 micrometers in size.

The alignment from one domain to the adjoining one is not abrupt but depends on the balance between the electrostatic and dipole interactions. The dipole orientation is found to change gradually over a distance of several hundred atomic diameters. This structure is called a domain or Bloch wall.

What is a ferrimagnetic material?

Ferrimagnetic materials behave rather like ferromagnetic materials. They have two different magnetic dipoles present, one of greater magnitude that the other. These line up in an antiparallel arrangement.

A ferrimagnetic solid shows complex temperature dependence because the distribution of the magnetic ions over the available sites is sensitive both to temperature and to the spin interactions.

Further reading

A.J. Epstein, 2003, 'Organic-based Magnets: Oportunities in Photoinduced Magnetism, Spintronics, Fractal Magnetism and Beyond', *Materials Research Society Bulletin* **28** 492.

J.H. Judy, 1990, 'Thin Film Recording Media', *Materials Research Society Bulletin* **XV** (March) 63.

J.S. Miller, A.J. Epstein, 2000, 'Molecule-based Magnets – An Overview', *Materials Research Society Bulletin* **25** (November) 21.

M.P. Sharrock, 1990, 'Particulate Recording Media', *Materials Research Society Bulletin* **XV** (March) 53.

N. Spaldin, 2003, *Magnetic Materials*, Cambridge University Press, Cambridge.

Problems and exercises

Quick quiz

1 A Gouy balance cannot be used to investigate:
   (a) Diamagnetic materials
   (b) Paramagnetic materials
   (c) Ferromagnetic materials

2 The material which is slightly attracted to a magnetic field is:
   (a) Diamagnetic
   (b) Paramagnetic
   (c) Ferromagnetic

3 A material that does not contain any magnetic dipoles on atoms or ions is:
   (a) A diamagnetic material
   (b) A paramagnetic material
   (c) An antiferromagnetic material

4 The Curie law describes the magnetic behaviour of:
   (a) Paramagnetic solids
   (b) Diamagnetic solids
   (c) Ferrimagnetic solids

5 Above the Curie temperature, ferromagnetic materials become:
   (a) Diamagnetic
   (b) Paramagnetic
   (c) Antiferromagnetic

6 The magnetic behaviour of ferromagnetic solids is described by the Curie–Weiss law:
   (a) Above the Curie temperature
   (b) Below the Curie temperature
   (c) At low temperatures

7 The Néel temperature is the temperature at which an antiferromagnetic material becomes:
   (a) Ferromagnetic
   (b) Ferrimagnetic
   (c) Paramagnetic

8 At high temperatures, the magnetic behaviour of ferrimagnetic materials can be described by:
   (a) The Curie–Weiss law
   (b) The Curie law
   (c) Neither of these

9 A ferrimagnetic compound contains:
   (a) All magnetic dipoles arranged parallel to one another
   (b) Two sets of magnetic dipoles arranged antiparallel to each other
   (c) One set of magnetic dipoles arranged in an antiparallel arrangement

10 The magnetic properties of electrons is due to:
   (a) The electron spin only
   (b) The orbital motion only
   (c) Both the orbital motion and spin

11 Atoms and ions that show magnetic properties have:
   (a) Only filled electron orbitals
   (b) Only partly filled electron orbitals
   (c) Some filled and some partly filled orbitals

12 The orbital contribution to the magnetic moment is quenched for:
   (a) Lanthanide atoms and ions
   (b) 3d transition metal atoms and ions
   (c) Both lanthanides and transition metals

13 The number of room-temperature ferromagnetic elements is:
   (a) Four
   (b) Three
   (c) Five

14 The Weiss molecular field constant describes:
   (a) Magnetic interactions between dipoles in a ferromagnetic material
   (b) Magnetic interactions between molecules in a paramagnetic material
   (c) Bonding interactions between molecules in a ferromagnetic material

15 The exchange energy in a magnetic material is lower for electrons with:
   (a) Parallel spins
   (b) Antiparallel spins
   (c) Paired electrons with no overall spin

16 The interaction of d orbitals and oxygen atoms that leads to antiferromagnetic ordering is called:
   (a) Exchange
   (b) Superexchange
   (c) Double exchange

17 Ferrimagnetic compounds contain metal ions in:
   (a) One valence state
   (b) Two valence states
   (c) Several valence states

18 The interaction between magnetic atoms in a ferrimagnetic material is called:
   (a) Double exchange
   (b) Superexchange
   (c) Interaction exchange

19 Ferrites with the *spinel* structure are:
   (a) Paramagnetic materials
   (b) Ferromagnetic materials
   (c) Ferrimagnetic materials

20 Most *spinel* structure ferrites have:
   (a) A normal *spinel* structure

   (b) An inverse *spinel* structure
   (c) A ferromagnetic *spinel* structure

21 The general formula of cubic ferrites can be written:
   (a) $A_2^{2+}Fe_3^{3+}O_4$
   (b) $A^{2+}Fe_2^{3+}O_4$
   (c) $A^{2+}Fe_3^{3+}O_4$

22 The general formula of hexagonal ferrites is:
   (a) $A^{2+}Fe_{12}O_{19}$
   (b) $A_2^{2+}Fe_{12}O_{19}$
   (c) $BaFe_{12}O_{19}$

23 Hexagonal ferrites are:
   (a) Paramagnetic materials
   (b) Ferromagnetic materials
   (c) Ferrimagnetic materials

24 Materials that lose their magnetism easily are called:
   (a) Soft magnetic materials
   (b) Hard magnetic materials
   (c) Impermanent magnetic materials

25 Hexagonal ferrites are:
   (a) Hard magnetic materials
   (b) Soft magnetic materials
   (c) Intermediate magnetic materials

26 The significant microstructure of a ferromagnetic solid is the presence of:
   (a) Magnetic grains
   (b) Magnetic domains
   (c) Hysteresis

27 A magnetic domain is a region within which:
   (a) All magnetic dipoles are parallel to the crystal axis
   (b) All magnetic dipoles are parallel to the magnetic field
   (c) All magnetic dipoles are aligned parallel to each other

28 A Bloch wall in a ferromagnetic solid:
   (a) Is the external surface of the ferromagnetic region

(b) Is the region between two domains

(c) Is another name for a grain boundary

29 The walls between magnetic domains are the result of:

(a) Competition between long-range interactions and thermal energy

(b) Competition between short-range interactions and chemical bonding

(c) Competition between long-range and short-range interactions

30 Hysteresis is caused by:

(a) Rearrangement of magnetic domains in an applied magnetic field

(b) Rearrangement of grains in an applied magnetic field

(c) Disordering of magnetic dipoles in an applied magnetic field

31 Pauli paramagnetism refers to:

(a) The paramagnetic behaviour of ions

(b) The paramagnetic behaviour of ferromagnetic metals above the Curie temperature

(c) The paramagnetic behaviour of metals

32 The ferromagnetic properties of the 3d transition metals are explained by:

(a) The overlap of 3s and 3d orbitals

(b) The overlap of 3p and 3d obitals

(c) The overlap of 3d and 4s orbitals

33 Superparamagnetism can be observed in:

(a) Very small ferromagnetic particles

(b) Very small paramagnetic particles

(c) Paramagnetic particles in a nonmagnetic matrix

Calculations and questions

12.1 Determine the ground-state values of S, J, L and m for the f^3 ion Nd^{3+}.

12.2 Determine the ground-state values of S, J, L and m for the f^7 ion Gd^{3+}.

12.3 Determine the ground-state values of S, J, L and m for the d^4 ion Mn^{3+} in (a) the high-spin state and (b) the low-spin state.

12.4 Determine the ground-state values of S, J, L and m for the d^7 ion Co^{2+} in (a) the high-spin state and (b) the low-spin state.

12.5 Why don't lanthanide ions possess high-spin and low-spin states? [Note: answer is not provided at the end of this book.]

12.6 Show that the 'spin-only' formula

$$m = g[S(S + 1)]^{1/2}$$

is equivalent to

$$m = [n(n + 2)]^{1/2},$$

where n is the number of unpaired spins. [Note: derivation is not given in the answers at the end of this book.]

12.7 Estimate the saturation magnetisation, M_s, for a sample of ferromagnetic nickel metal (a) assuming only the unpaired d electrons contribute to the magnetism and the spins can be added as if the material were paramagnetic and (b) using the measured magnetic moment per nickel atom of $0.58\ \mu_B$. The metal has an A1 structure with a cubic unit cell parameter of 0.3524 nm.

12.8 Estimate the saturation magnetisation, M_s, for a sample of ferromagnetic cobalt metal (a) assuming only the unpaired d electrons contribute to the magnetism and the spins can be added as if the material were paramagnetic and (b) using the measured magnetic moment per cobalt atom of $1.72\ \mu_B$. The metal has an A3 structure with a hexagonal unit cell parameter of $a_0 = 0.2507$ nm, and $c_0 = 0.4069$ nm.

12.9 Iron has a saturation magnetisation of $1.72 \times 10^6\ \mathrm{A\,m^{-1}}$. What is the measured magnetic moment, in Bohr magnetons, of an iron atom. Iron has an A2 structure, with a cubic unit cell parameter of 0.2867 nm.

12.10 Nickel has a saturation magnetisation of 0.489×10^6 A m^{-1}. What is the measured magnetic moment, in Bohr magnetons, of a nickel atom. Nickel has the A1 structure, with a cubic unit cell parameter of 0.3524 nm.

12.11 The magnetic moment of Fe^{3+} ions in the species [Fe(H$_2$O)$_6$]$^{3+}$ is 5.3 μ_B. What is the likely electron configuration of the Fe^{3+} ion?

12.12 The magnetic moment of Fe^{3+} ions in the species [Fe(CN)$_6$]$^{3-}$ is 2.3 μ_B. What is the likely electron configuration of the Fe^{3+} ion? What can you conclude about the orbital contribution to the magnetic moment for this species?

12.13 The species [Co(NH$_3$)$_6$]$^{3+}$, containing Co^{3+} ions, gives rise to diamagnetic solids, whereas the species [CoF$_6$]$^{3-}$, also containing Co^{3+} ions, has a strong magnetic moment and gives rise to paramagnetic solids.
(a) What is the likely electron configuration of the Co^{3+} ions in these two species?
(b) Calculate the expected magnetic moment of Co^{3+} in [CoF$_6$]$^{3-}$.

12.14 The species [Fe(CN)$_6$]$^{4-}$, containing Fe^{2+} ions, gives rise to diamagnetic solids, whereas the species [Fe(NH$_3$)$_6$]$^{2+}$, also containing Fe^{2+} ions, has a strong magnetic moment and gives rise to paramagnetic solids.
(a) What is the likely electron configuration of the Fe^{2+} ions in these two species?
(b) Calculate the expected magnetic moment of Fe^{2+} in [Fe(NH$_3$)$_6$]$^{2+}$.

12.15 The relationship between the energy of a light photon, E, and its frequency, ν, is $E = h\nu$ [see Section 1.2.2, Equation (1.3)]. The crystal field splitting of diamagnetic [Co(NH$_3$)$_6$]$^{3+}$, containing Co^{3+} ions (see Question 12.13) is 4.57×10^{-19} J. What wavelength light would produce photoinduced paramagnetism in this molecule?

12.16 The relationship between the energy of a light photon, E, and its frequency, ν, is

$E = h\nu$ [see Section 1.2.2, Equation (1.3)]. The crystal field splitting of diamagnetic [Fe(CN)$_6$]$^{4-}$, containing Fe^{2+} ions (Question 12.14) is 4.57×10^{-19} J. What wavelength light would produce photoinduced paramagnetism in this molecule?

12.17 Calculate (a) the paramagnetic energy level splitting for a Pr^{3+} ion in a magnetic induction of 0.5 T and (b) the corresponding wavelength of radiation for a transition between these energy levels.

12.18 Calculate (a) the paramagnetic energy level splitting for a Ho^{3+} ion in a magnetic induction of 0.75 T and (b) the corresponding wavelength of radiation for a transition between these energy levels.

12.19 (a) Calculate the paramagnetic energy level splitting for a V^{4+} ion in a magnetic induction of 0.25 T if the orbital contribution is quenched and (b) calculate the corresponding wavelength of radiation for a transition between these energy levels.

12.20 (a) Calculate the paramagnetic energy level splitting for a Ni^{2+} ion in a magnetic induction of 0.6 T if the orbital contribution is quenched and (b) calculate the corresponding wavelength of radiation for a transition between these energy levels.

12.21 Calculate the mass susceptibility of NiSO$_4 \cdot$7H$_2$O at 20 °C (in units of kg^{-1}), which contains isolated Ni^{2+} ions. The density of the compound is 1980 kg m^{-3}. Assume that the spin-only formula is adequate.

12.22 Calculate the mass susceptibility of CuSO$_4 \cdot$5H$_2$O at 20 °C, (in units of kg^{-1}), which contains isolated Cu^{2+} ions. The density of the compound is 2284 kg m^{-3}. Assume that the spin-only formula is adequate.

12.23 Calculate the volume susceptibility of MnSO$_4 \cdot$4H$_2$O at 20 °C, (in units of m^{-3}), which contains isolated Mn^{2+} ions. The density of the compound is 1980 kg m^{-3}.

Assume that the spin-only formula is adequate.

12.24 (a) Calculate the value of x in the Brillouin function $B_J(x)$, where $x = g_J \mu_B JB/kT$, for an Mn^{2+} ion with five unpaired electrons in an inductance of 0.5 T. Assume that the orbital angular momentum is quenched, so that $L = 0$. Take S as 5/2, from Table 12.2. The Curie law requires that $x \ll 1$. (b) Estimate the temperature at which x is 0.01.

12.25 (a) (i) Calculate the value of x in the Brillouin function $B_J(x)$, where $x = g_J \mu_B JB/kT$, for a Ti^{3+} ion with one unpaired electron in an inductance of 0.45 T. Assume that the orbital angular momentum is quenched, so that $L = 0$, and take S as 1/2, from Table 12.2. The Curie law requires that $x \ll 1$. (ii) Estimate the temperature at which x is 0.005. (b) Repeat the calculation assuming that the orbital angular momentum is not quenched, and $L = 2$, $J = 3/2$. (iii) Estimate the temperature at which x is 0.005.

12.26 The saturation magnetisation of iron is $1.752 \times 10^6 \, A \, m^{-1}$. Calculate the effective magnetic moment on an iron atom. Iron adopts the A2 structure with a unit cell edge of 0.2867 nm.

12.27 The saturation magnetisation of nickel is $0.510 \times 10^6 \, A \, m^{-1}$. Calculate the effective magnetic moment on a nickel atom. Nickel adopts the A1 structure with a unit cell edge of 0.3524 nm.

12.28 The saturation magnetisation of cobalt is $1.446 \times 10^6 \, A \, m^{-1}$. The crystal structure of cobalt is disordered in normal circumstances (see Section 6.1.1).
  (a) Calculate the number of magnetic dipoles per unit volume in cobalt, knowing the effective magnetic moment per atom is 1.72 μ_B.

(b) If it is assumed that there is one atom per primitive cubic unit cell, determine the length of the unit cell edge.

12.29 Estimate (a) the saturation magnetisation and (b) the magnetic inductance for the cubic ferrite $CoFe_2O_4$ with the inverse spinel structure. Co^{2+} is a d^7 ion. The cubic unit cell has a lattice parameter of 0.8443 nm and contains eight formula units. Assume that the orbital angular momentum is completely quenched in this material.

12.30 Estimate (a) the saturation magnetisation and (b) the magnetic inductance for the cubic ferrite $NiFe_2O_4$ with the inverse spinel structure. Ni^{2+} is a d^8 ion. The cubic unit cell has a lattice parameter of 0.8337 nm and contains eight formula units. Assume that the orbital angular momentum is completely quenched in this material.

12.31 Estimate (a) the saturation magnetisation and (b) the magnetic inductance for the hexagonal ferrite 'ferroxdur', $BaFe_{12}O_{19}$. The hexagonal unit cell has parameters $a_0 = 0.58836 \, nm$, $c_0 = 2.30376 \, nm$ and volume $7.128 \times 10^{-28} \, m^3$. There are two formula units in the unit cell. Assume that the orbital angular momentum is completely quenched in this material.

12.32 Estimate (a) the saturation magnetisation and (b) the magnetic inductance for the hexagonal ferrite $SrFe_{12}O_{19}$. The hexagonal unit cell contains two formula units and has a volume $7.975 \times 10^{-28} \, m^3$. Assume that the orbital angular momentum is completely quenched in this material.

12.33 Derive a formula for the saturation magnetisation of a cubic ferrite $A^{2+}Fe_2O_4$, which is only partly inverse. The fraction of Fe^{3+} ions on tetrahedral sites is given by λ, where $0 < \lambda < 0.5$ (see Section 5.3.10). [Note: derivation is not given in the answers at the end of this book.]

13

Electronic conductivity in solids

- How are donor atoms and acceptor atoms in semiconductors differentiated?

- What is a quantum well?

- What are Cooper pairs?

The electronic conductivity of solids varies widely (Figure 13.1). Solids that allow an electric current to flow when a small voltage is applied are called conductors. Solids that do not allow a current to flow are called insulators. Semiconductors fall between these two classes. As gauged from Figure 13.1, it is a matter of degree whether a solid is a conductor, semiconductor or an insulator but, irrespective of the magnitude of the effect, conductivity requires the presence of mobile charge carriers. In this chapter, solids that have reasonable numbers of mobile charge carriers present, either because of their native electronic properties or because they have been deliberately introduced by doping, are considered. In addition, superconductors, a group of materials that appear not to use 'normal' conductivity mechanisms, are described.

13.1 Metals

13.1.1 Metals, semiconductors and insulators

One of the defining physical properties of a metal is its electrical conductivity. (Section S4.6 gives definitions of this and related terms.) Electrical conductivity in a solid is attributable to electrons that are free to move (i.e. that gain energy) under the influence of an applied electric field. The metallic bonding model, described in Sections 2.3.2 and 2.3.3, allows conductivity to be understood most easily. In this model, the electrons on the atoms making up the solid are allocated to energy bands that run throughout the whole of the solid. A simple one-dimensional band-structure diagram, called a flat-band diagram (Figure 13.2), allows the broad distinction between conductors, semiconductors and insulators to be understood.

If the number of electrons available fills the energy band completely, and the energy gap between the top of the filled band and the bottom of the next higher (empty) energy band is large, the material is an insulator (Figure 13.2a). This is because the electrons have no means of taking up the additional energy needed to allow them to move under a low voltage, because all of the energy levels are filled or inaccessible. Only a considerable voltage will cause electrons to jump from the

Understanding solids: the science of materials. Richard J. D. Tilley
© 2004 John Wiley & Sons, Ltd ISBNs: 0 470 85275 5 (Hbk) 0 470 85276 3 (Pbk)

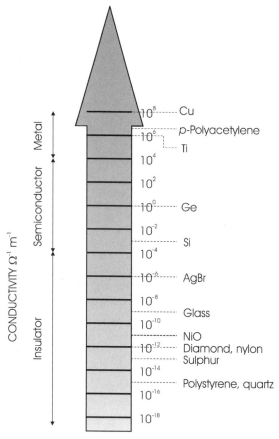

Figure 13.1 The range of electronic conductivity in solids

the conduction band. The energy gap is the band gap, E_g.

Although this picture is simple, it reveals an important feature. It is found that the 'vacancies' left in the valence band when electrons are promoted to the conduction band also contribute to the conduction process. To a good approximation, these vacancies can be equated to positive electrons, and move in the opposite direction to the electrons in an applied field. They are called positive holes, or just holes. Semiconductors are characterised by an increase in conductivity with temperature because the number of mobile charge carriers, electrons and holes, will increase as the temperature increases.

If the band gap is so small that thermal energy at normal temperatures is sufficient to generate a very high number of charge carriers in each band the material is classed as a degenerate semiconductor. At 0 K intrinsic semiconductors become insulators.

Semiconductivity can arise in an insulator if the material contains an appreciable number of impurities (added intentionally or not). Similarly, the conductivity of an intrinsic semiconductor can be manipulated by adding suitable impurities. The impurities can act as donors, donating electrons to the conduction band to form n-type semiconductors (Figure 13.2c), or as acceptors, accepting electrons from the valence band, and thus donating holes to the valence band, to form p-type semiconductors (Figure 13.2d). When all of the impurities are fully ionised (i.e. when all the donor levels have lost an electron or all the acceptor levels have gained an electron) the exhaustion range has been reached. If the donors and acceptors are present in equal numbers the material is said to be a compensated semiconductor. At 0 K these materials are also insulators. It is difficult in practice to distinguish between compensated extrinsic semiconductors and intrinsic semiconductors.

When there are insufficient electrons to fill the highest band (the valence band becomes indistinguishable from the conduction band in this case) even small amounts of energy will be able to move the topmost electrons into higher energy levels, and small voltages will produce significant conductivity and the solid is a metal (Figure 13.2e). The

completely filled band to the next-highest empty band, and when such a transfer does occur the insulator has broken down.

If the energy gap between the filled and empty band is small enough that thermal energy is sufficient to cause some electrons to jump from the lower filled band to the upper empty band, the electronic properties change. Such materials are called intrinsic semiconductors (Figure 13.2b). Once electrons arrive in the empty band, they can contribute to electrical conductivity, as there are empty energy levels around them and the solid is transformed from an insulator into a poor electronic conductor. The now almost filled band is called the valence band, and the almost empty band is called

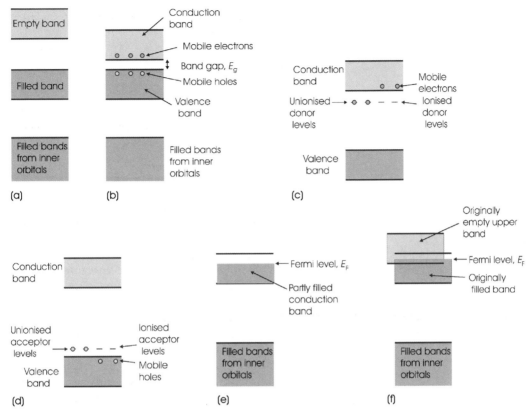

Figure 13.2 Energy-band representations of materials: (a) insulators, (b) intrinsic semiconductors, (c) n-type extrinsic semiconductors, (d) p-type extrinsic semiconductors, (e) metals and (f) semimetals. The innermost filled energy bands are omitted in parts (c) and (d)

uppermost filled energy levels form the Fermi surface. Should the bottom of the $(n + 1)$th band lie energetically lower than the top of a full nth band, electrons will spill over into the bottom of the empty band until the Fermi level intersects both sets of bands. Holes and electrons now exist, and coexist at 0 K. This type of material is called a semimetal (Figure 13.2f).

From the point of view of metallic conductivity, the nature of the atoms composing the structure is not important. The primary point is that there should be an upper band, a conduction band, which is party filled with electrons. This depends on crystal structure and the extent to which the outer electron orbitals overlap. A large number of compounds have sufficient overlap of the outer orbitals to generate bands of reasonable width, and can loosely

be classed as metals. Many oxides, including ReO_3 and TiO, as well as nitrides, phosphides and silicides have high conductivities and can often be considered as metals or semimetals. Large numbers of sulphides and selenides are considered to be best classified as high-conductivity semiconductors, whereas tellurides tend to be regarded as metallic.

13.1.2   Conductivity of metals and alloys

Although the conductivity of a metal is dependent on the shape of the Fermi surface, a good idea of the conduction properties of metals can be gained by considering the properties of a metal with a half-filled conduction band and a spherical Fermi

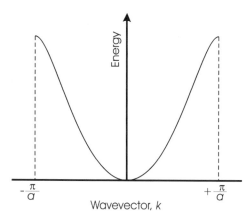

Figure 13.3 The energy versus wave vector curve for a free electron in a metal

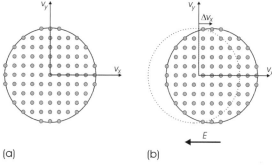

(a)   (b)

Figure 13.4 The Fermi surface (circled) marks the boundary up to which velocity states (shaded circles) of free electrons in a 'two-dimensional' metal are occupied. (a) In the absence of an electric field, equal numbers of electrons are moving in all directions, and the sum of the velocities is zero. (b) In an electric field, E, the distribution of velocities changes (solid outer boundary compared with the dotted boundary), causing more to move against the direction of the electric field. This overfall drift velocity is observed as electronic conductivity

surface. The ideas already described with respect to the metallic bond (Section 2.3) are able to account for this.

Return to the simple energy versus wave-vector diagram of a 'one-dimensional' metal (Figure 13.3). The wave vector, k, can be expressed in terms of the velocity or the momentum of the electrons:

$$k = \frac{2\pi}{\lambda} = \left(\frac{2\pi}{h}\right)p = \left(\frac{2\pi m}{h}\right)v \qquad (13.1)$$

where λ is the wavelength of the electron wave, p is the momentum of the electron, m is the mass of the electron, v is the velocity of the electron, and h is Planck's constant. Thus k is proportional to both momentum and velocity. Each energy level is then associated with an electron velocity, which increases as the Fermi surface is approached. The Fermi surface therefore divides the occupied velocity states from the unoccupied states.

In the case of a 'two-dimensional' metal, the Fermi surface can be represented as a circle, with velocity components along the x and y axes (Figure 13.4a). Because velocity states are filled up to the Fermi surface, when the metal is at normal equilibrium the velocities sum to zero. Thus, although the electrons are in motion, no current flows.

When an electric field, E, is applied to the metal, each electron experiences a force $-eE$ and a change in momentum, Δp, which is equivalent to a change in the value of the wave vector, Δk, and velocity,

Δv, of the electrons. An electric field applied along the x axis will translate the velocity distribution parallel to the x axis by a small amount opposite to the direction of the applied field, shown as the solid-line outer circle as compared with the dotted circle in Figure 13.4b. The velocities no longer sum to zero along x, and electrons drift in a direction opposite to that of the applied field. A current then flows. The effect is limited by collisions with atoms, impurities and defects and a steady current is reached.

The electronic conductivity of a metal can be estimated from this model. It is found (see Section S4.7) that the conductivity, σ, and its reciprocal, the resistivity, ρ, are given by:

$$\sigma = \frac{e^2 n\tau}{m_e^*}$$

$$\rho = \frac{m_e^*}{e^2 n\tau}$$

where e is the electron charge, n is the number of mobile electrons in the metal, τ is the average time an electron spends between two successive scattering events, and m_e^* is the effective mass of the electrons. This latter term is used to account for the fact that the dynamics of electrons in solids is quite different from that in a vacuum, and measurements show that the mass that applies to electrons in a vacuum (the rest mass) needs to be replaced by an effective mass. The effective mass is not a constant but depends on temperature and the direction in the crystal that the electron is travelling.

If it is assumed that the velocity of an electron is reduced to zero each time it is scattered, then it is seen that the current will decay to zero in a time τ after the voltage is turned off. The term τ, the time between successive scattering processes, is also called the relaxation time. The mean free path of the electron, which is the length of the path between successive scattering events, is given by:

$$\Lambda = \tau v_F$$

where v_F is the electron velocity at the Fermi surface.

The conductivity is often written in terms of another variable, the mobility of the electrons, μ_e:

$$\sigma = ne\mu_e$$

where the mobility is given by:

$$\mu_e = \frac{e\tau}{m_e^*}$$

Scattering is the main cause of resistivity. The electron wave can be scattered in a variety of ways, of which three are of most importance. The first is the interaction of the electron wave with lattice vibrations, called phonons. This is called thermal scattering. As the temperature increases so do the lattice vibrations, and the resistivity rises. At low temperatures the resistivity drops gradually to a finite value, maintained at absolute zero (Figure 13.5), except for the superconductors, described later in this chapter. This is an intrinsic property of a metal and cannot be altered. Structural imperfections present in the solid also contribute to resistivity. These are mainly defects such as dislocations and grain boundaries, or else impurities. As with lattice vibrations, they scatter the electron waves and so increase resistivity. Defects and impurities are extrinsic features that can be removed by careful processing.

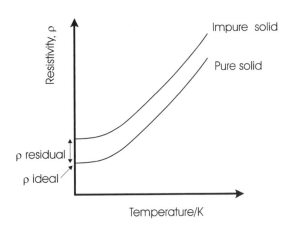

Figure 13.5 The variation of the resistivity of a metal with temperature; impure solids and alloys have a higher resistivity than pure metals, at all temperatures

The different scattering processes can be allocated relaxation times as follows. Suppose that the distance between thermal scattering events is Λ_{th}. The number of scattering events per second, $n_{thermal}$, will be given by:

$$n_{thermal} = \frac{v_F}{\Lambda_{th}}$$
$$= 1/\tau_{th}$$

where v_F is the electron velocity at the Fermi surface, and τ_{th} is the thermal relaxation time. Analogous equations can be written in respect of the distance between defect scattering, Λ_{def}, and between impurity scattering, Λ_{imp}. Thus the resistivity can be written:

$$\rho = \frac{m_e^*}{e^2 n_F \tau_{th}} + \frac{m_e^*}{e^2 n_F \tau_{def}} + \frac{m_e^*}{e^2 n_F \tau_{imp}}$$

Taking into account these features, the total resistivity can be written as:

$$\rho = \rho_{phonons} + \rho_{defects} + \rho_{impurities}$$

This is known as Mattiesen's rule, sometimes written as:

$$\rho = \rho_{ideal} + \rho_{residual}$$

where ρ_{ideal} is the intrinsic component due to phonon interactions and $\rho_{residual}$ is the extrinsic contribution.

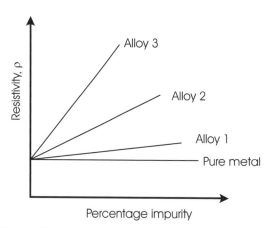

Figure 13.6 The variation of the resistivity of alloys with concentration of the alloying elements

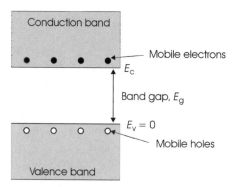

Figure 13.7 The simplest 'flat-band' energy-band description of an intrinsic semiconductor

The resistivity of a substitutional solid solution alloy will generally be greater than that of a pure metal because the elements added to form the alloy have the same effect as impurities (Figure 13.6). If the alloying atoms order to form a new crystal structure, the disruptive scattering of the electron wave is suppressed and the resistivity will drop.

A consequence of the electron collisions is a transfer of energy from the mobile electrons to the structure. This is revealed as heat energy and accounts for the fact that an electric current generates heat. This effect, which takes place uniformly along the length of the conductor, is called Joule heating. The amount of heat generated is:

$$P = I^2 R$$

where P is the power output (W), I is the current (A), and R is the resistance (Ω). This heating poses problems for closely packed electronic circuits, which have to be cooled to function correctly.

13.2 Semiconductors

13.2.1 Intrinsic semiconductors

The simplest band picture of a semiconductor is drawn in Figure 13.7. The energy gap between the top of the valence band, E_v, and the bottom of the conduction band, E_c, is called the band gap, E_g. Electron energy increases (and is defined to be positive) when measured *upwards* from the top of the valence band, which is usually taken as the energy zero. Hole energy increases (and is defined to be positive) when measured *downwards* from the top of the valence band. At absolute zero, the valance band will be full and the conduction band empty. As the temperature increases, some electrons will be promoted across the narrow band gap, and the material will show a small degree of conductivity. Because the number of electrons promoted will increase with temperature, the conductivity will rise with temperature. This increase is characteristic of a semiconductor and is of greater importance than the magnitude of the conductivity. Remember that for metals conductivity falls with increasing temperature.

The conductivity, σ, of a semiconductor is made up of two components, one due to electrons,

$$\sigma_e = ne\mu_e$$

and one due to holes,

$$\sigma_h = pe\mu_h$$

so that the overall conductivity will be given by:

$$\sigma(\text{total}) = ne\mu_e + pe\mu_h$$

where the subscripts e and h refer to electrons and holes, respectively; the number of electrons is given by n; the number of holes is given by p; and the mobility is given by μ.

It is possible to determine the number of electrons excited into the conduction band by thermal energy in an intrinsic semiconductor using Fermi–Dirac statistics (see Section 2.3.7 and Section S4.12). It is found that:

$$n \propto \exp\left(-\frac{E_g}{2kT}\right)$$

where n is the number of electrons in the conduction band, E_g is the band gap between the top of the valence band and the bottom of the conduction band, k is the Boltzmann constant, and T is the absolute temperature. As the number of holes is equal to the number of electrons in an intrinsic semiconductor,

$$p \propto \exp\left(-\frac{E_g}{2kT}\right)$$

The total conductivity, σ, which is proportional to the number of electrons and holes, can be expressed in the form:

$$\sigma = \sigma_0 \exp\left(-\frac{E_g}{2kT}\right)$$

where σ_0 is a constant, E_g is the band gap between the top of the valence band and the bottom of the conduction band, k is the Boltzmann constant, and T is the absolute temperature. Taking logarithms of each side of this equation, we obtain

$$\ln \sigma = \ln \sigma_0 - \frac{E_g}{2kT}$$

The gradient of a plot of conductivity versus $1/T$ will therefore yield a value for the (thermal) band gap (Figure 13.8).

An alternative method of obtaining the magnitude of the band gap is via the absorption of radiation. In a semiconductor, almost all of the energy levels below the conduction band are occupied. This means that low-energy radiation directed at a crystal will not interact with the electrons, and the crystal

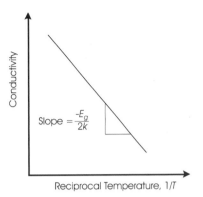

Figure 13.8   The variation of resistivity versus reciprocal temperature for an intrinsic semiconductor

will be transparent. As the energy gradually increases, eventually the energy will just be sufficient to promote an electron from the top of the valence band to the bottom of the conduction band. The radiation will now be absorbed and the crystal will become opaque. The (optical) band gap can be equated to the energy at which this change occurs. Thus,

$$E_g = h\nu_g$$

where $h\nu_g$ is the energy of the photon required to promote an electron from the valence band and create a hole in its place.

Note that the absorption of radiation is more complex than this simple model suggests, and a more complete description will be given in Section 14.1.3. Moreover, the measured value of the optical band gap is usually slightly different from the thermal band gap. This is simply a reflection of the fact that the bands in a semiconductor are not flat as drawn in Figure 13.7 but have a more complex curved shape.

Approximate values for the band gap in some semiconductors is given in Table 13.1. The band gap decreases as the atom size increases (i.e. as one moves down the relevant group in the periodic table). Thus, within the group of elements listed, diamond is best regarded as an insulator, whereas grey tin is regarded as a metal. The second group

Table 13.1 Approximate values for the band gap of some semiconductors

Compound	Symbol or formula	Band gap/eV
Elements:		
Diamond	C	5.47
Silicon	Si	1.12
Germanium	Ge	0.66
Grey tin	Sn	0.08
III–V semiconductors:		
Gallium nitride	GaN	3.36
Gallium phosphide	GaP	2.26
Gallium arsenide	GaAs	1.42
II–VI semiconductors:		
Cadmium sulphide	CdS	2.42
Cadmium selenide	CdSe	1.70
Cadmium telluride	CdTe	1.56

Note: the band gap is normally given in electron volts in most compilations; 1 eV is equal to 1.60219×10^{-19} J.

are called III–V semiconductors because they are compounds of elements in groups III and V (now groups 13 and 15) of the periodic table. The last compounds listed are called II–VI semiconductors because they are compounds of elements in groups II and VI (now groups 12 and 16) of the periodic table.

The decrease in band gap with size of atom is simply a consequence of the fact that the outer orbitals of larger atoms overlap more and give rise to wider bands. As a consequence, the gaps are narrower.

13.2.2 Carrier concentrations in intrinsic semiconductors

To determine more accurate values for the number of holes and electrons present in an intrinsic semi-conductor it is appropriate to use Fermi–Dirac statistics and the density of states at the bottom of the conduction band (see Section S4.8). To a good approximation, it is found that the number of electrons in the conduction band per unit volume,

n_i, which is equal to the number of intrinsic holes in the valence band, p_i, at an absolute temperature, T, is:

$$n_i = p_i = 4.826 \times 10^{21} \left(\frac{m_e^* m_h^*}{m_e^2} \right)^{3/4} T^{3/2} \exp\left(-\frac{E_g}{2kT} \right)$$

(13.2)

where m_e^* is the effective mass of the electron, m_h^* is the effective mass of a hole, and E_g is the band gap. In an intrinsic semiconductor

$$n = p = n_i = p_i$$

Equation (13.2) shows that, at a given temperature

$$np = \text{constant}$$

The temperature dependence of the equilibrium constant is given by:

$$np = 2.33 \times 10^{43} \left(\frac{m_e^* m_h^*}{m_e^2} \right)^{3/2} T^3 \exp\left(-\frac{E_g}{kT} \right)$$

Assuming that the effective mass of electrons and holes is independent of temperature, we obtain

$$np \propto T^3 \exp\left(-\frac{E_g}{kT} \right)$$

This equation applies to doped semiconductors as well as to intrinsic semiconductors, a finding of considerable practical importance. To a good approximation, the Fermi energy lies at the centre of the band gap (Section S4.8):

$$E_F = \frac{1}{2} E_g + \frac{3}{4} kT \ln\left(\frac{m_h^*}{m_e^*} \right) \approx \frac{1}{2} E_g$$

Writing the total conductivity, $\sigma(\text{total})$, as

$$\sigma(\text{total}) = ne\mu_e + pe\mu_h$$

we obtain

$$\sigma(\text{total}) = 773.1 \left(\frac{m_e^* m_h^*}{m_e^2} \right)^{3/4} T^{3/2} (\mu_e + \mu_h)$$
$$\times \exp\left(-\frac{E_g}{2kT} \right)$$

(13.3)

Although the mobility of electrons and holes decreases with temperature, exactly as in a metal, it is found that the conductivity of an intrinsic semiconductor increases with temperature, as the exponential term in Equations (13.2) and (13.3) dominates the other terms.

13.2.3 Extrinsic semiconductors

The deliberate addition of carefully chosen impurities to silicon, germanium and other semiconductors is called doping. It is carried out so as to modify the electronic conductivity, and the dopants are chosen so as to add either electrons or holes to the material. It is possible to gain a good idea of how this is achieved very simply. Suppose that an atom such as phosphorus, P, ends up in a silicon crystal. This can occur, for example, if a small amount of phosphorus impurity is added to molten silicon before the solid is crystallised. Experimentally the impurity atom is found to occupy a position in the crystal that would normally be occupied by a silicon atom, and so forms a substitutional defect.

Silicon adopts the diamond structure, in which each atom is linked to four tetrahedrally disposed neighbours by four sp^3 hybrid bonds (Section 5.3.6). These use all of the four (3s^2, 3p^2) valence electrons available. Phosphorus is found one place to the right of silicon in the periodic table, which indicates that the atom has one more electron in its complement. The outer electron structure of phosphorus is 3s^2, 3p^3, and, after forming four sp^3 hybrid bonds, one electron is spare and still associated with the phosphorus atom (Figure 13.9a.i). This electron is available to enhance the electrical conductivity if it can enter the conduction band. Atoms such as phosphorus are called donors when they are added to silicon as they can donate the unused valence electron to the conduction band.

The simplest model to employ for the estimation of the energy needed to free the electron uses the Bohr theory of the hydrogen atom. To recall this model, a single electron is attracted to a

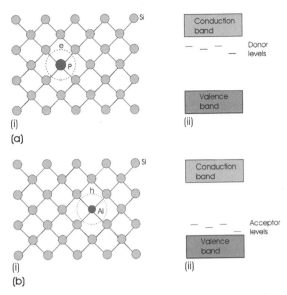

Figure 13.9 (a) (i) Donor impurity in a crystal of an extrinsic semiconductor and (ii) the associated energy-band diagram; donor impurities add donor energy levels below the conduction band. (b) (i) Acceptor impurity in a crystal of an extrinsic semiconductor and (ii) the associated energy-band diagram; acceptor impurities add acceptor energy levels above the valence band

positive nucleus consisting of a single proton. The energy needed to free this electron is given by:

$$E = -\frac{m_e e^4}{8\varepsilon_0^2 h^2}$$

where m_e is the mass of the electron, e is the electron charge, ϵ_0 is the permittivity of free space, and h is the Planck constant. The value of E is -2.18×10^{-18} J (-13.6 eV). The negative value reflects the fact that zero is taken as the energy of a completely free electron. To apply this to an electron located at a phosphorus atom, suppose that the attraction of the phosphorus nucleus is 'diluted' by the relative permittivity of silicon, ϵ_r, and the mass of the electron is replaced by the effective mass m_e^*. The energy to free the electron is now:

$$E(\text{P}) = -\frac{m_e^* e^4}{8\varepsilon_0^2 \varepsilon_r^2 h^2}$$

$$= \frac{Em_e^*}{m_e \varepsilon_r^2}$$

As the effective mass of an electron in silicon is approximately one tenth of the electron rest mass, and as the relative permittivity of silicon is about 10, the energy needed to free the electron is about one hundredth of the band gap, which suggests that the electron should be very easily liberated. Donor energies are often represented by an energy level, the donor level, drawn under the conduction band (Figure 13.9a.ii).

An analogous situation arises if silicon is doped with an element such as aluminium. Aluminium is also found to form a substitutional defect. However, aluminium is found one place to the left of silicon in the periodic table, with a valence electron configuration $3s^2$, $3p^1$, and has one valence electron less than silicon. One of the four resulting sp^3 hybrid bonds will be an electron short. This is equivalent to the introduction of a positive hole, which is localised on the aluminium impurity (Figure 13.9b.i). Providing that the energy needed is not too great, an electron from the full valance band can be promoted to fill the bond, leaving a hole in the valence band. The energy needed to free the hole, calculated using the Bohr model, is similar to that of an electron. Once in the valence band, the hole is free to move and to contribute to the conductivity. Dopant atoms from the left of silicon in the periodic table, such as aluminium, are called acceptors, because they can be imagined to accept an electron from the filled valence band and so create a hole that takes part in conductivity. These impurities can also be represented by energy levels, drawn just above the top of the valence band (Figure 13.9b.ii).

The energy to liberate donor electrons and acceptor holes is about 8×10^{-21} J (0.05 eV). These values are comparable to room temperature thermal energy, and most extrinsic electrons and holes should be free at room temperature. In this state, the donors and acceptors are said to be ionised, and

the semiconductor crystal will be a reasonable conductor. If donor atoms are present in great numbers, they will govern the conductivity, which will be by electrons. The material is said to be n-type. If acceptors are present in greatest quantities, then holes will control the conduction, and the material is said to be p-type. When both electrons and holes are present and both contribute to the conductivity we talk of majority and minority carriers.

The number of holes and electrons present is not solely dependent on the number of donors and acceptors present. The equilibrium equation,

$$np = \text{constant} = n_i^2 \qquad (13.4)$$

governs populations. If large numbers of donors are added, so that n increases, the number of holes decreases accordingly. Similarly, if large numbers of acceptors are added, the population of holes increases and the number of electrons decreases. This vital fact makes it possible to fabricate silicon-chip electronic circuits, founded on the ability to transform n-type to p-type silicon, and vice versa, at will. If the relationship given in Equation (13.4) did not hold, successive doping cycles would simply increase overall conductivity.

13.2.4   Carrier concentrations in extrinsic semiconductors

The electronic properties of an extrinsic semiconductor are determined by the number of mobile charge carriers, electrons and holes, both intrinsic and extrinsic, and by the position of the Fermi energy (Figure 13.10). The total numbers of mobile electrons and holes is related to n_i and p_i, the numbers of intrinsic electrons, and holes by:

$$np = \text{constant} = n_i^2 = p_i^2$$

The number of holes created by the addition of the acceptors is equal to N_a^-, where the number of acceptor atoms per unit volume is N_a, and the number of acceptors that have gained an electron,

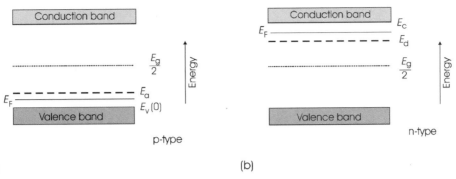

Figure 13.10   The position of the Fermi energy in (a) a p-type semiconductor and (b) an n-type semiconductor at low temperatures

ionised acceptors, is N_a^-. Similarly, the number of electrons created by the addition of the donors is equal to N_d^+, where the number of donor atoms per unit volume is N_d, and the number of donors that have lost an electron, ionised donors, is N_d^+. As a doped crystal must remain electrically neutral,

$$n + N_a^- = p + N_d^+ \qquad (13.5)$$

Equations (13.4) and (13.5) allow the numbers of mobile charge carriers and the position of the Fermi level to be found as a function of dopant concentration, using a similar approach to that outlined in Section S4.8 for intrinsic semiconductors. Unfortunately, the results are not easily expressed analytically or displayed graphically.

At elevated temperatures, most of the donors and acceptors will be ionised, and the electroneutrality condition, Equation (13.5), can be written as:

$$n + N_a = p + N_d \qquad (13.6)$$

In cases where an n-type semiconductor contains only donors, and at high temperatures, so that the donors are completely ionised, Equations (13.4) and (13.6) give the number of electrons and holes as

$$n(\text{n-type}) \approx N_d$$

$$p(\text{n-type}) \approx \frac{n_i^2}{N_d}$$

Similarly, for a p-type semiconductor crystal that contains only acceptors at high temperatures, the number of electrons and holes is:

$$n(p\text{-}type) \approx \frac{n_i^2}{N_a}$$

$$p(p\text{-}type) \approx N_a$$

The position of the Fermi level is a function of the temperature and the concentration of the electrons and holes (Figure 13.11). At low temperatures, the position of the Fermi level approaches the conduction band in n-type materials and the valence band in p-type materials. In effect, the Fermi level changes in order to maintain a balance between the charges (Figure 13.12). As the temperature increases, the contribution from the intrinsic electrons and holes increases. In essence, the semiconductor gradually changes from an extrinsic towards an intrinsic material. Because of this, the Fermi level approaches the position found in intrinsic semiconductors, near to the middle of the band gap. The rapidity of this change will depend on the concentrations of donors and acceptors in the semiconductor. A semiconductor will maintain its extrinsic character to higher temperatures when doped with higher concentrations of impurities.

13.2.5 Characterisation

Although the resistance of a metal can be measured easily by attaching two contacts, this gives

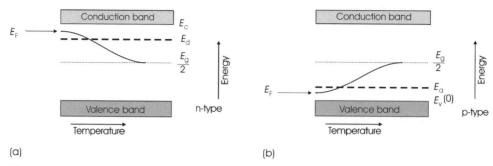

Figure 13.11   The variation of the position of the Fermi energy of (a) an n-type and (b) a p-type semiconductor with temperature

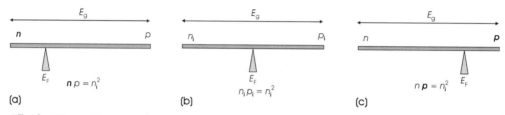

Figure 13.12   The position of the Fermi energy in semiconductors varies with temperature and dopant concentration so as always to maintain the relationship np = constant: (a) in an n-type semiconductor the concentration of electrons is much higher than the concentration of holes, and the Fermi energy is to the left-hand side of the 'balance'; (b) in an intrinsic semiconductor the concentration of holes and electrons is equal and the Fermi energy is central; (c) in an n-type semiconductor the concentration of holes is much higher than the concentration of electrons, and the Fermi energy is to the right-hand side of the 'balance'. As the temperature increases the intrinsic contribution to the hole and electron concentrations becomes more important, and the Fermi energy tends towards the centre

unreliable results for semiconductors. For these materials, the resistivity is most often determined by using a four-point probe on rectangular (Figure 13.13a) or disc-shaped (Figure 13.13b) samples. The probes are sharply pointed, equally spaced, needles and press down with a known force on the surface of the sample. The current passes between two outer probes, and the voltage drop is measured between the two inner probes. The relationship between the measured

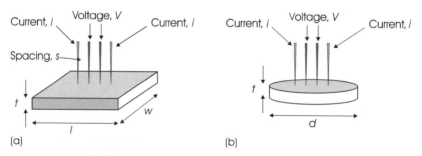

Figure 13.13   Arrangement in the four-point probe method of measurement of resistance of semiconductors for (a) rectangular specimen and (b) a disc. The spacing, s, of the probes is much smaller than the dimensions of the specimen. The current is measured between the outer probes, and the voltage drop is measured between the inner two probes

voltage and current and the resistivity depends on the geometry of the experimental setup. For a bulk specimen, in which the thickness of the sample is much greater than the spacing between the probes,

$$\rho = 2\pi s\left(\frac{V}{I}\right)$$

where ρ is the resistivity of the material, s is the distance between the probe needles, V is the voltage drop between the middle two needles, and I is the current between the outer two needles. In the case of a thin film, in which the probe spacing, s, is much greater than the film thickness, t, and much smaller than the distance to the edge of the film, that is $l, w \gg s \gg t$ (Figure 13.13a), or $d \gg s \gg t$ (Figure 13.13b), the expression for the resistivity is:

$$\rho = \frac{\pi t}{\ln 2}\frac{V}{I}$$

$$\approx 4.54t\frac{V}{I}$$

This formula is independent of the probe spacing, s.

The resistance of thin films is sometimes reported as the sheet resistance, R_s. This is defined in terms of the bulk resistance of a material, R, of resistivity ρ, and the dimensions of the sample. For a rectangular specimen (Figure 13.13a):

$$R_s = \frac{\rho}{t}$$

$$\rho = \frac{Rwt}{l}$$

hence

$$R_s = \frac{Rw}{l}$$

The sheet resistance is quoted in unit of ohms per square ($\Omega/\square$). Thus

$$\rho = R_s t \; (\Omega\,m)$$

The carrier type can be found in two ways, via the Hall effect or the Seebeck effect. The second technique is described in Section 15.2.2. The Hall effect, discovered in 1879, relies on the displacement of moving charges in a magnetic field to determine the nature of the mobile carriers. A slab of material is arranged so that the current flow is normal to a fairly strong (0.2 T), magnetic induction (Figure 13.14a). The moving charges will experience a force due to the magnetic induction in the solid that is normal both to current direction and to magnetic field. That is, if the current I is along the x axis and the magnetic induction is along the z axis the charge carriers will be deflected in the y-axis direction. This deflection will build up until the electric field density is just strong enough to oppose further charge displacement. The result is a voltage, the Hall voltage, along the y-axis. Perhaps counterintuitively the charge carriers will be deflected in the same direction, irrespective of the charge that they carry, and the sign of the voltage will give the sign of the charge carriers. The value of the Hall voltage, V, can be appreciable.

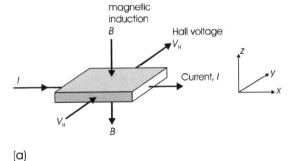

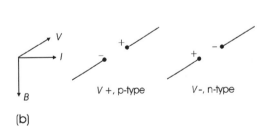

(a)   (b)

Figure 13.14   (a) Arrangement of the measurement of the Hall effect, showing the axes used; (b) the sign of the Hall voltage with respect to the direction of the current and magnetic induction

The relationship between the current, magnetic induction and electric field is:

$$E_y = \pm R_H J_x B_z \qquad (13.7)$$

where E_y is the electric field along the y axis, measured in $V\,m^{-1}$, J_x is the current density along the x axis, measured in $A\,m^{-2}$, and B_z is the magnetic induction along the z axis, measured in T. The constant of proportionality, the Hall coefficient, R_H, has units $m^3\,C^{-1} = V\,m\,A^{-1}\,T^{-1}$. The sign of the Hall coefficient differentiates n-type and p-type semiconductors. For example, if the magnetic induction is aligned along $-z$ (Figure 13.14a), a positive Hall voltage compared with the orthogonal V, B and I axes, yields a positive value for R_H and the material is p-type, whereas a negative voltage and negative R_H indicates that the material is n-type (Figure 13.14b).

The Hall coefficient is related in a simple way to the number of mobile charge carriers. Suppose that the current is made up of electrons flowing parallel to x with a drift velocity v_x. The magnetic field present will exert a force, F, on an electron:

$$F = e v_x B_z$$

The force exerted by an electric field, E_y, on an electron is given by

$$F = E_y e$$

When the two forces are equal and equilibrium is achieved,

$$E_y = v_x B_z$$

The current density is given by:

$$J = -n e v_x$$

where n is the number of mobile electrons per unit volume. Substituting the expressions for E_y and J into Equation (13.7), and rearranging, we obtain:

$$R_H = \frac{-1}{ne}$$

where n is the number of electrons per unit volume, each with a charge of $-e$. For a flow of positive holes:

$$R_H = \frac{1}{pe}$$

where p is the number of mobile holes per unit volume, each with a charge of $+e$.

In general, only one mobile charge carrier, either electrons or holes, is present. In this case, by measuring both the conductivity, σ, and the Hall coefficient, R_H, it is possible to determine the number of charge carriers and their mobility as:

$$\sigma_e = n e \mu_e$$
$$-R_H \sigma_e = \mu_e$$

or

$$\sigma_h = p e \mu_h$$
$$+R_H \sigma_h = \mu_h$$

If both electrons and holes are present:

$$\sigma(\text{total}) = n e \mu_e + p e \mu_h$$
$$= |e|(n \mu_e + p \mu_h)$$

and

$$R_H = \frac{n \mu_e^2 + p \mu_h^2}{(n \mu_e + p \mu_h)^2} \frac{1}{|e|}$$

The mobility derived from a Hall measurement is often called the Hall mobility to distinguish it from the drift mobility (see Section S4.7).

For noncubic single crystals the Hall coefficient varies with direction, although the effect is averaged when measurements are made on polycrystalline samples. A number of ordinary metals have positive Hall coefficients, including Be, Al, Cd, In, As and W. These results could not be explained in terms of the classical 'electron gas' model. In some metals, such as Er and Ho, the Hall coefficient varies from negative to positive as a function of crystallographic direction. The explanation of these findings requires a detailed knowledge of the three-dimensional shape of the Fermi surface in these metals.

13.2.6   The p–n junction diode

The electrical behaviour that emerges when a region of p-type semiconductor is adjacent to a region of n-type semiconductor, a p–n junction diode, often just called a diode, is different from the behaviour of the separate components. (Note that p–n junction diodes are fabricated by selective doping of different regions of a semiconductor crystal and not by joining separate crystals together.)

The simplest band picture of a p–n junction is shown in Figure 13.15. In separated materials, the Fermi energies are unequal (Figures 13.15a and 13.15b). When a p-type region abuts an n-type

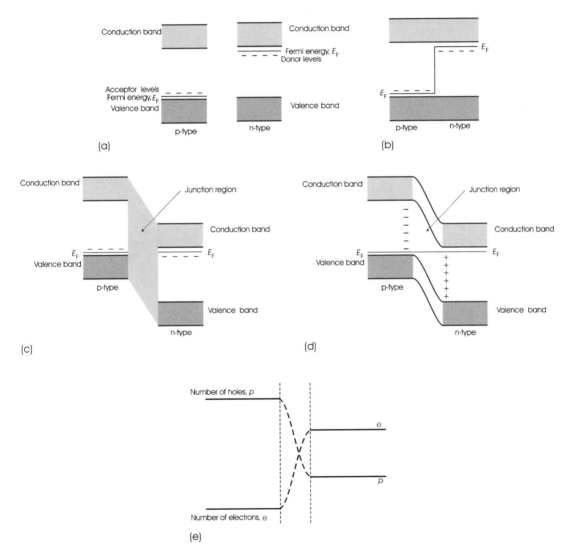

Figure 13.15   The p–n junction. (a) energy bands of separated p-type and n-type materials; (b) schematic energy bands for juxtaposed p-type and n-type materials. (c) energy bands on joining; (d) distorted energy bands in the junction region at equilibrium; and (e) the numbers of electrons and holes across the junction region at equilibrium

region, electrons move into the p-type region from the n-type side, and holes move into the n-type region from the p-type region, by diffusion. This can be thought of as electrons spilling *down* the Fermi surface, E_F, from p to n, and holes spilling *up* the Fermi surface, E_F, from p to n. Most of these moving displaced charge carriers will recombine, holes with electrons in the n region, and electrons with holes in the p region. As electrons leave the n-type material, positively charged donor atoms are left behind, and negatively charged acceptor atoms are left in the p-type material as holes leave. These charges will create an electric potential, the contact potential, of about 0.3 V. At equilibrium, the Fermi level must be the same on each side of the junction (Figure 13.15c). The energy levels have been shifted vertically with respect to each other, by the difference in E_F values, to give a distorted band structure in the junction region (Figure 13.15d), and the electron and hole populations change dramatically as the junction is traversed (Figure 13.15e).

The transition region has a width of about 1 μm. The density of mobile charge carriers in the transition region is low, and for this reason the transition region is also called the depletion region. At equilibrium (thermal and electrical) there will still be an exchange of carriers at the junction, but the current in each direction will be the same. Dynamic equilibrium holds.

The conductivity of the p–n junction in one direction is totally different from that in the other. An applied voltage, which will drop across the transition region, because of the absence of mobile charge carriers, can be applied with the positive side connected either to the p-type region or to the n-type region. The arrangement in which the positive voltage is connected to the p-type region is called forward bias. This causes the potential barrier to be reduced. The Fermi level changes to allow electrons to 'roll down' the Fermi level and the holes 'roll up' the Fermi level, causing a current to flow (Figure 13.16a). Under a forward bias there is a rapid increase in the current flowing across the junction. The situation when the negative voltage is applied to the p-type region is called reverse bias (Figure 13.16b). The effect of the reverse bias is to raise the potential barrier, so that electrons cannot

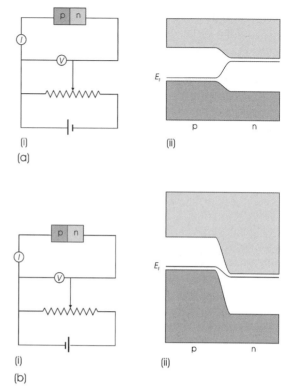

Figure 13.16   (a) (i) A p–n junction under forward bias and (ii) band structure of the junction; (b) (i) a p–n junction under reverse bias and (ii) band structure of the junction

pass up it or holes down it. Current flow now virtually ceases.

The change of current with applied voltage is given by the Shockley equation. In a simplified form this is:

$$I = I_0 \left[\exp\left(\frac{eV}{kT}\right) - 1 \right]$$

where I_0 is a constant term, the saturation current, determined by the junction geometry and the doping levels; e is the charge on the electrons and holes; V is the applied voltage; k is the Boltzmann constant; and T is the temperature (Figure 13.17).

The total current across the device, which is constant for any applied voltage, is made up of

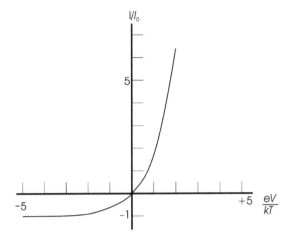

Figure 13.17 The ideal current–voltage characteristics for a p–n junction. Note: I and I_0, the current and the saturation current, respectively; e, charge on the electrons and holes; V, applied voltage; k, the Boltzmann constant; T, temperature

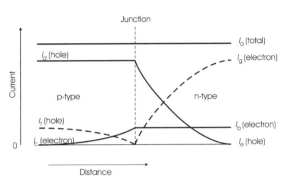

Figure 13.18 The currents I_a–I_g flowing across a p–n junction under forward bias

hole and electron flows in opposite directions. Consider the situation when a forward bias is applied (Figure 13.18). The number of electrons moving to the left will increase rapidly, by a factor of $\exp(V/kT)$. If V is 0.1 V, this is a factor of about $\times 55$ at room temperature. Thus the number of electrons appearing at the p-type boundary is about 55 times higher than the equilibrium concentration there. A similar situation describes the holes appearing at the n-type boundary. In general, the hole current will be greater than the electron current under forward bias, but the actual currents on each side will depend on the doping levels.

When the electrons reach the p-type region they are annihilated by combination with holes. The penetration depth depends on a number of factors, but can be taken to be about 1 mm. In order to maintain charge neutrality, the hole population must be replenished and holes must be move into the p-type region from the left to balance the electron density. Similarly, the holes that arrive in the n-type region are gradually annihilated by recombination with electrons. In order to maintain charge balance, an extra electron current must flow into the n-type region from the right.

The total current flowing will be made up of six components (Figure 13.18):

- I_a, total current (constant) $= I_b + I_e + I_g = I_c + I_d + I_f$;

- I_b, electron current flowing in the n-type region (constant);

- I_c, injected electron current in p-type region (decaying);

- I_d, hole current in p-type region (constant);

- I_e, injected hole current in n-type region (decaying);

- I_f, declining hole current in p-type region to balance and annihilate I_c;

- I_g, declining electron current in n-type region to balance and annihilate I_e.

This is quite different than in a metal, in which an applied voltage allows the mobile electrons to acquire a drift voltage. In a p–n junction, minority carriers are injected into both regions. They were not there originally and arise as a consequence of the applied voltage.

From what has been said, it can be seen that a p–n junction diode acts as a rectifier. That is, the device exhibits a very low resistance to current flow in one direction and a very high resistance in the other.

13.2.7 Modification of insulators

The properties of insulators can be modified in the same way as the properties of semiconductors, and an insulator can be transformed into a semiconductor by suitable doping. The impurities can act as donors, donating electrons to the conduction band, or as acceptors, accepting electrons from the valence band, and thus donating holes to the valence band. Nickel oxide, NiO, has the *halite* (B1) structure and can be considered to be an ionic oxide, containing equal numbers of Ni^{2+} and O^- ions. Green nickel oxide can be reacted with colourless lithium oxide, Li_2O, to give a black solid solution $Li_xNi_{1-x}O$. The Li^+ ions occupy Ni^{2+} sites in the structure to form substitutional defects. In order to maintain charge neutrality, every Li^+ ion in the crystal must be balanced by a Ni^{3+} ion. This can be regarded as a Ni^{2+} ion together with a trapped hole. The situation is thus analogous to that of Al^{3+} doped into silicon, and the defect can be regarded as an acceptor (Figure 13.19a). The process of creating electronic defects in a crystal in this way is called valence induction. Black $Li_xNi_{1-x}O$ is a p-type semiconductor, and as the holes are only weakly bound to the cations the material posses a high conductivity.

It is equally possible to impart n-type conductivity to an insulator by suitable doping. An example is provided by the reaction of small amounts of gallium oxide, Ga_2O_3, with zinc oxide, ZnO. In this case, Ga^{3+} ions substitute for Zn^{2+} ions in the zinc oxide structure to form $Ga_xZn_{1-x}O$. To maintain charge neutrality, one electron must be added to balance each Ga^{3+} ion in the structure. It is generally believed that these rest on Zn^{2+} ions to generate Zn^+ ions in the crystal. The electrons are not strongly attached to the Zn^{2+} ions, and each Zn^+ ion can be regarded as a donor (Figure 13.19b).

The key to valence induction is that the cation of the host structure can take two valence states, one of which is regarded as the host cation plus a weakly bound hole or electron.

The electronic properties of complex oxides can be changed in the same way, provided that one of

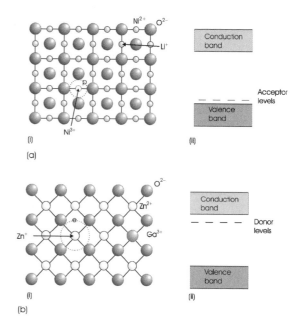

Figure 13.19 The conversion of insulating oxides into semiconductors. (a) (i) Nickel oxide (NiO) doped with lithium oxide (Li_2O), making it a p-type semiconductor, and (ii) the energy-band structure of Li^+-doped NiO. (b) (i) Zinc oxide (ZnO) doped with gallium oxide (Ga_2O_3), making it an n-type semiconductor, and (ii) the energy-band structure of Ga^{3+}-doped ZnO

the cations present can take part in the valence change. $SrVO_3$ provides an example. The structure of this phase is of the cubic *perovskite* type. This material, which is an insulator, contains Sr^{2+} ions in the large sites between the metal–oxygen octahedra and V^{4+} ions in the octahedra. If some of the Sr^{2+} ions are replaced by La^{3+} ions, charge neutrality is maintained by transforming some V^{4+} ions into V^{3+} ions. The V^{3+} ions can be regarded as V^{4+} ions plus a trapped electron, and are donors. Thus although $SrVO_3$ is a poor electronic conductor, $La_xSr_{1-x}VO_3$ is quite a good one.

13.2.8 Conducting polymers

Ordinary polymers are good insulators, and they are widely used in this capacity, as insulating covering

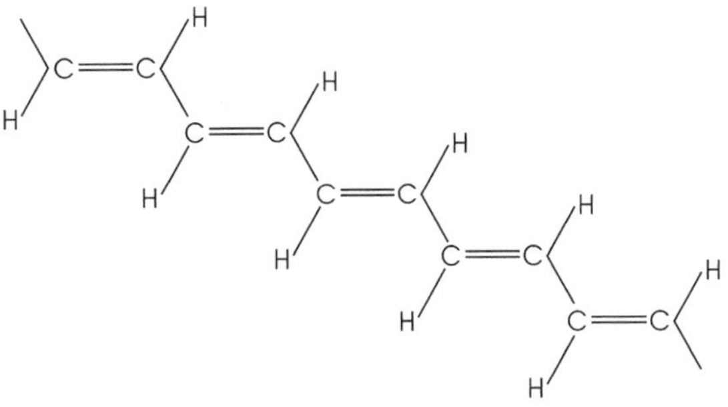

Scheme 13.1 Part of a conjugated hydrocarbon molecule, in which carbon atoms are linked alternately by single and double bonds. Note: C, carbon; H, hydrogen; bonds connecting the atoms are shown as lines

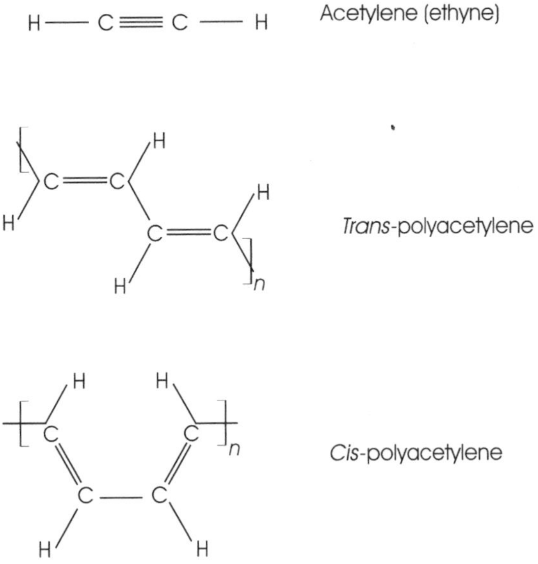

Scheme 13.2 Schematic structures of polyacetylenes: (a) acetylene (ethyne), (b) repeating unit of the *trans* polymer and (c) repeating unit of the *cis* polymer

on cables and other electrical conductors. The molecular feature that allows polymers to become electronically conducting is the presence of conjugated double and single bonds (Scheme 13.1). The framework of the molecule is composed of sp^2 hybrid σ bonds, at 120° to each other (see Section 2.2.6). The double bonds are formed by overlap of π orbitals above and below the plane of the carbon chains. These are not located between specific pairs of carbon atoms, as drawn, but are spread over all of the molecule as delocalised π orbitals, similar to those above and below the planes of the sheets of carbon atoms in graphite. A long molecule will have extensive delocalised π orbitals, and it could be anticipated that these would lead to high electronic conductivity along the backbone of the molecule, to give a 'one-dimensional' metal.

The first conducting polymer to be synthesised was polyacetylene. When polymerised, acetylene (ethyne) forms a silvery flexible film of polyacetylene. Acetylene (ethyne) has a formula C_2H_2. The carbon atoms are linked by a triple bond, consisting of 1 sp-hybrid σ bond and two π bonds (Scheme 13.2). Generally, polymerisation leads to the all-*cis* polymer. At room temperature this changes to the thermodynamically stable all-*trans* form. These two forms are geometrical isomers (see Section S2.1). Both are poor insulators, with the *trans* form having a conductivity similar to that of silicon (approximately $10^{-3}\,\mathrm{S\,m^{-1}}$), and the *cis* form with a conductivity similar to that of water (approximately $0.1\,\mathrm{S\,cm^{-1}}$).

This is rather surprising if the band structure of these materials is considered. The sp^2 hybrid orbitals are filled and would give lower-energy filled bands that do not contribute to the conductivity. The remaining p_z orbitals on each carbon contain one electron. A chain of CH units, each with one unpaired outer electron, is analogous to a chain of alkali metal atoms such as lithium. On the one hand, if the separation of the carbon atoms is rather large, each electron would be completely localised on each carbon and the polymer would be a magnetic insulator, possibly ferromagnetic or antiferromagnetic (Figure 13.20a; see Section 12.3). On the other hand, if the intercarbon distance is small, each electron is able to delocalise over all of the carbon atoms in a π band. Each atom would contribute just one electron, which would produce a half-full band and the material should show metallic conductivity (Figure 13.20b).

The discrepancy is resolved in the following way. A linear chain of equispaced atoms in a metal, such as a chain of sodium atoms, is found to be energetically unstable. Instead, the spacing between the atoms will adjust itself so that an energy gap opens

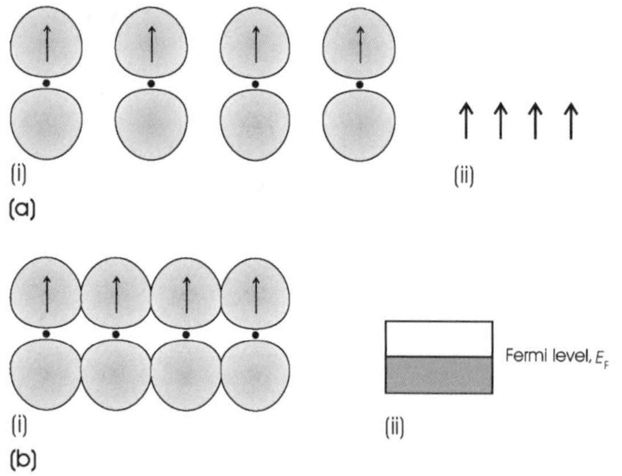

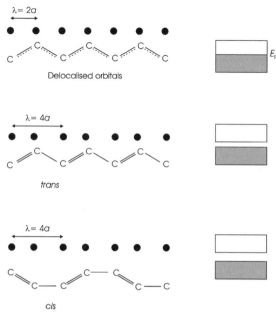

Figure 13.20 (a) (i) A chain of isolated half-occupied p orbitals, which will lead to a magnetic insulator and (ii) the magnetic dipoles in one possible orientation. (b) (i) A chain of overlapping p orbitals, which will lead to (ii) a half-filled conduction band and metallic conductivity

Figure 13.21 (a) Delocalised orbitals along a polymer chain leads to equally spaced atoms and a half-filled energy band. Peierls distortion in (b) *trans*-polyacetylene and (c) *cis*-polyacetylene, leads to alternating short and long bonds and a band structure similar to that of an intrinsic semiconductor

at the Fermi level. The variation in spacing is called a Peierls distortion. As a consequence, the half-filled band is transformed into two – a filled band and an empty band – and, as a result the solid becomes a semiconductor (Figure 13.21). In agreement with this, C–C bonds along the chain in *trans* and *cis* polyacetylene alternate between two lengths. The band gap is approximately 1.8 eV (2.88×10^{-19} J) in the *trans* form and about 2 eV (3.20×10^{-19} J) in the *cis* form of the polymer.

Polyacetylene is transformed into a metallic conductor by doping. This involves oxidation or reduction of the polymer. Electron acceptors such as halogens (chlorine, iodine, etc.) oxidise the polymer. In this process, electrons are taken from the filled lower band and used to form halide ions, leaving holes, which result in a 'p-type' material. A typical reaction is:

$$[CH]_n + \frac{3x}{2}I_2 \rightarrow [(CH)_n^{x+}x(I_3^-)]$$

The iodine enters the polymer between the molecular chains.

Doping with alkali metals (lithium, sodium, etc.) reduces the polymer. In this process, the alkali metal donates electrons to the empty band, forming an alkali metal ion and transforming the polymer into an 'n-type' material. A typical reaction can be written as:

$$[CH]_n + xLi \rightarrow [(CH)^{x-}xLi^+]$$

Conductivity of these doped materials is of the order of 10^8 S m^{-1}, similar to that of copper or silver. The variation of the conductivity with dopant concentration is illustrated in Figure 13.22. Notice that much greater concentrations of dopant are needed (at %) than those used in the traditional semiconductors such as silicon (parts per million). Heavy doping has transformed the insulating polymer into a material that seems to be metallic. This is an example of an insulator-to-metal transition.

However, the conductivity and other electrical properties of both the semiconducting and the

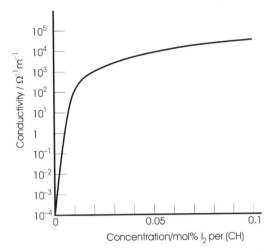

Figure 13.22 Schematic variation of the conductivity of polyacetylene doped with iodine. The concentration is in moles of I_2 per CH unit. The conductivity changes from that typical of an insulator to that associated with a metal

metallic region are not identical to those of the semiconductors and metals already discussed. For example, the conductivity in the metallic region of iodine-doped polyacetylene decreases as the temperature decreases (Figure 13.23) whereas in an

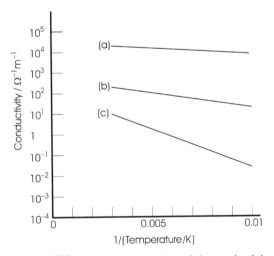

Figure 13.23 Schematic variation of the conductivity of polyacetylene doped with iodine versus reciprocal temperature. Dopant concentration is approximately (a) 0.15 mol%, (b) 0.01 mol% and (c) 0.005 mol% of I_2 per CH unit

ordinary metal the reverse is true. Moreover, the resistivity seems to have an Arrhenius-like temperature dependence:

$$\sigma = \sigma_0 \exp\left(-\frac{E_a}{kT}\right)$$

where E_a is the activation energy, k is Boltzmann's constant, T is the temperature, and σ_0 is a constant (Figure 13.23). This suggests that the dopant does not simply add mobile carriers to a conduction or valence band, as with silicon.

The explanation of this feature is bound up with both the microstructure and the nanostructure of the polymer. Polyacetylene is composed of ordered, fairly crystalline, regions linked by disordered regions in which the polymer chains are twisted and coiled, which indicates that the band picture is a severe approximation. This is why dopants do not simply add mobile charge carriers to bands. The dopants, in fact, modify the nanostructure of the polymer chains. Acetylene polymerises preferentially to form the *cis* isomer, and this transforms into the stable *trans* form. The transformation process is initiated at random, rather like the nucleation of a crystal in a liquid. In the same way that crystallites grow together to form a polycrystalline mass, when two regions of polymer chain that have transformed from *cis* to *trans* meet there is frequently a mismatch (Figure 13.24). This mismatch is called a soliton, which can be thought of as an interruption in the orderly pattern of conjugated double bonds, extending over several adjacent carbon atoms. In some cases there will be a π electron missing, which will give the soliton a positive charge. In other cases, the soliton can attract an unbound π electron to give a neutral soliton, or two

Figure 13.24 A soliton in *trans*-polyacetylene. The shaded ellipse in the centre of the soliton may represent either an electron hole (creating a positively charged soliton), a single electron (creating a neutral soliton) or two electrons (creating a negative soliton)

π electrons to give a negative soliton. Doping adds to the soliton density along the polymer chains by extracting or adding electrons.

Solitons give rise to two effects. First, at the soliton, the alternating bond lengths required for the Peierls transition are suppressed. Ultimately, when enough solitons are present, the Peierls transition fails and a half-filled band is reinstated, recreating the metallic state. Second, the solitons give rise to impurity energy levels in the band gap. When enough of these are present, they merge and bridge the band gap, again making high conductivity possible.

The conduction process is not simply the spread of electron waves throughout the solid, as in a crystalline metal. Instead, the charge on the soliton jumps from one location to a neighbouring one under the influence of an electric field. This is similar to ionic conductivity, and this produces the Arrhenius-like behaviour.

The conductivity of the polymer depends on the microstructure of the solid. Stretching the polymer sheet at moderate temperatures increases the alignment of the chains and leads to significant improvement in properties along the chain direction. In this form, polyactylene is widely used as an electrode material in lightweight batteries.

13.3 Nanostructures and quantum confinement of electrons

Section 3.2.5 introduced the ideas of quantum wells, quantum wires and quantum dots. These structures have unique electronic and optical properties because of the way in which electrons are localised, or confined. In bulk solids, electrons are located on atom cores in ionic solids, in localised bonds in normal covalent solids, delocalised over molecular orbitals, as in graphite, or completely delocalised, as in metals. Quantum nanostructures confine the electrons (and holes in semiconductors) at a scale different from any of these, and this gives rise to the novel properties that such structures possess. An electron or hole bonding energy much greater than thermal energy characterises strongly confined charge carriers.

13.3.1 Quantum wells

A quantum well is constructed by laying down a thin layer of a semiconductor with a smaller band gap within a semiconductor with a larger band gap. The most studied quantum well structures are those formed from a layer of gallium arsenide, GaAs, sandwiched in gallium aluminium arsenide, GaAlAs (Figure 13.25a). Gallium arsenide has a band gap of about 1.42 eV, whereas aluminium arsenide has a band gap of about 2.16 eV. Gallium

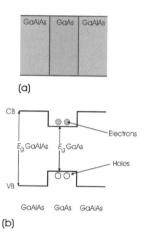

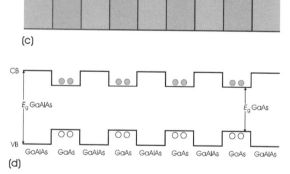

Figure 13.25 (a) A single quantum well of gallium arsenide, GaAs, formed from a thin layer in gallium aluminium arsenide, GaAlAs; (b) the energy band structure of the quantum well; (c) a multiple quantum well superlattice in gallium arsenide; and (d) the schematic energy band structure of the superlattice. Note: CB, conduction band; VB; valence band; E_g, band gap

aluminium arsenide alloys have band gaps between these values. The electrons in the thin GaAs layer are effectively trapped in the 'well' formed in the conduction band of the composite material (Figure 13.25b). Similarly, the holes in the thin layer of semiconductor are trapped at the 'hill' in the valence band of the composite material.

The properties of single quantum wells are enhanced when a number of these features are combined to form a multiple quantum well or superlattice (Figures 13.25c and 13.25d).

The dimension at which quantum confinement becomes important, Δx, is derived from the Heisenberg uncertainty principle (see Section 1.2.1). It is found that

$$\Delta x \approx \left(\frac{h^2}{m_e^* kT} \right)^{1/2}$$

for electrons, and

$$\Delta x \approx \left(\frac{h^2}{m_h^* kT} \right)^{1/2}$$

for holes, where h is the Planck constant, m_e^* is the effective mass of the electron in the semiconductor, m_h^* is the effective mass of the hole in the semiconductor, k is the Boltzmann constant, and T is the temperature (in kelvin). For quantum confinement to be important, quantum wells must be about 10 atom layers or less in thickness.

The energy of an electron in a quantum well can be calculated using the approach outlined in Section 2.3.6. If it is assumed that the electron is free, and trapped by an infinite boundary potential, the same equations for a free electron in a metal apply. Thus, the energy, E, of a free electron in a rectangular parallelepiped with edges a, b and c is given by Equation (2.15):

$$E(n_x, n_y, n_z) = \frac{h^2}{8m_e} \left(\frac{n_x^2}{a^2} + \frac{n_y^2}{b^2} + \frac{n_z^2}{c^2} \right)$$

where h is the Planck constant, m_e is the mass of the electron, and n_x, n_y and n_z are the quantum numbers along the three axes. Exactly the same equation will apply to a free electron confined to a slab of

material, although it is better to replace the electron mass with the effective mass, m_e^*. In the case of a quantum well, the electron is confined in one dimension, say x, and unconfined in two directions, which can be taken as y and z, so it is convenient to rewrite Equation (2.15) as:

$$E(n_x, n_y, n_z) = \left(\frac{h^2}{8m_e^*} \right) \left(\frac{n_x^2}{a^2} \right) + \left(\frac{h^2}{8m_e^*} \right) \left(\frac{n_y^2}{b^2} + \frac{n_z^2}{c^2} \right)$$

(13.7)

The values of b and c can be taken as about 1 cm, and the value of a is about 10^{-8} m. The energy is therefore dominated by the first term in Equation (13.7). This introduces a new set of energy levels, associated with electron waves trapped in the well (Figure 13.26). The electron energy level in the lowest, $n = 1$, state is raised by $h^2/8m_e^* a^2$ compared with the base of the well. These energy levels are called electron subbands, and when the energy levels trap the electron strongly the electrons are strongly confined.

Exactly the same equations apply to holes, where the effective mass m_h^* replaces m_e^*. The energy levels that arise from trapped holes are called hole subbands.

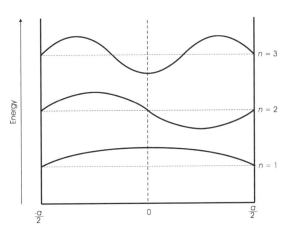

Figure 13.26 The first three energy levels for an electron trapped in a one-dimensional quantum well are the same as an electron trapped on a line (Section 2.3.6, Figure 2.27, page 48)

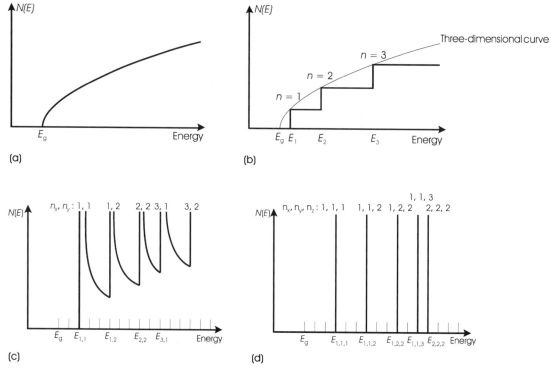

Figure 13.27   (a) The density-of-states function, $N(E)$, for a free electron in a metal; (b) the step-like density-of-states function for an electron trapped in a quantum well; (c) the density-of-states function of an electron trapped on a quantum wire; and (d) the density-of-states function of an electron trapped in a quantum dot

The density of states for a free electron is parabolic in shape (Figures 2.29, page 49, and 13.27a). The imposition of the quantum well changes the curve, and the smooth continuum is now broken into steps (Figure 13.27b). The height of the steps depends on the dimensions of the quantum well material.

The optical properties of quantum wells are described in Section 14.11.1.

13.3.2   Quantum wires and quantum dots

The above considerations can be applied equally well to confinement in two or three dimensions – so-called quantum wires and quantum dots (see Section 3.2.5). For a quantum wire with restricted dimensions along a and b, the free electron confined

in an infinite potential well will have energy levels given by:

$$E(n_x, n_y, n_z) = \left(\frac{h^2}{8m_e^*}\right)\left(\frac{n_x^2}{a^2}\right) + \left(\frac{h^2}{8m_e^*}\right)\left(\frac{n_y^2}{b^2}\right) + \left(\frac{h^2}{8m_e^*}\right)\left(\frac{n_z^2}{c^2}\right)$$

(13.8)

where a and b are small, and c is large. The density-of-states curve for this situation is drawn in Figure 13.27(c). In the case of the quantum dot, Equation (13.8) is retained, but the third dimension, c, is also small. The density-of-states curve for a quantum dot is shown in Figure 13.27(d).

The interesting optical properties of these structures are described in Section 14.11.2.

13.4   Superconductivity

13.4.1   Superconductors

In 1911 H. Kamerlingh Ohnes found that mercury lost all electrical resistance when cooled to the temperature of liquid helium (4.2 K) and reached the superconducting state. Subsequently, a large number of materials, including metallic elements, alloys, organic compounds, sulphides, oxides and nitrides have been found to exhibit superconductivity. In the superconducting state all electrical resistance is lost and electrical current, once started, will flow forever without diminishing (Figure 13.28). Superconductivity is a quantum mechanical feature, one of the few that are apparent in the 'macroscopic' world, and some of its features are puzzling when viewed from the standpoint of classical physics.

The temperature at which most materials become superconducting is called the superconducting transition temperature, T_c, (Figure 13.28). Most metallic elements have a T_c that is below 10 K. For many years the highest value of T_c recorded was close to 18 K, and this seemed to be a genuine limit, but new techniques of preparation pushed this up to 23.2 K, in the alloy Nb_3Ge, by 1970. This temperature stood as the record until the ceramic superconductors were discovered, in 1986. Although the T_c of the first ceramic compound recognised as a superconductor, $La_{1.85}Ba_{0.15}CuO_4$, was only about 30 K, the very existence of superconductivity in a ceramic led to an explosion of research that produced large numbers of new ceramic superconductors. The current record for T_c, 138 K, is held by the oxide $Hg_{0.8}Tl_{0.2}Ba_2Ca_2Cu_3O_{8.33}$.

13.4.2   The effect of magnetic fields

The defining characteristic of superconductors is the loss of all electrical resistivity. However, other features are important. When a superconductor is cooled in a magnetic field it expels the magnetic induction, B, in its interior. This is called the Meissner effect. Ideally, this expulsion is complete so that a superconductor behaves as perfectly diamagnetic (see Figures 13.29a and 13.29b). In effect, in the superconducting state a surface current is produced that is sufficient to generate an internal magnetic field that exactly cancels the magnetic induction. This current also distorts the external magnetic field so that it does not penetrate the superconductor, which is thus screened (Figure

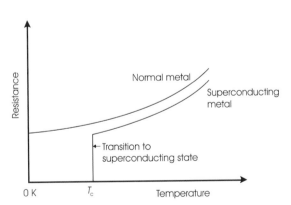

Figure 13.28 The variation of the resistance of a normal metal compared with that of a superconducting metal as the temperature approaches 0 K. In a superconducting metal all resistance is lost at the transition temperature, T_c, which is close to 0 K

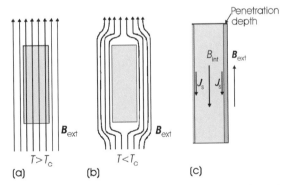

Figure 13.29 The Meissner effect; (a) a normal metal or a superconducting metal above the transition temperature, T_c, allows the penetration of magnetic flux into the bulk; (b) in a superconducting metal below the transition temperature the magnetic flux is 'expelled'; (c) a surface current, J_s, is induced in the superconductor below the transition temperature which creates an internal magnetic flux, B_{int}, that cancels the external flux, B_{ext}

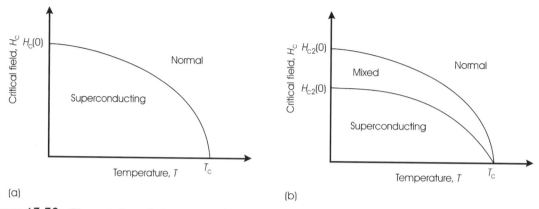

(a)   (b)

Figure 13.30   The variation of the superconducting properties with the external magnetic field: (a) a type-I superconductor $H_c(0)$ is the critical field at 0 K; (b) a type-II superconductor; $H_{c2}(0)$ and $H_{c1}(0)$ are, respectively, the upper and lower critical fields at 0 K

13.29c). The surface current is confined to a thin layer, called the penetration depth, which has a value of between 10 nm and 100 nm. The exclusion of the magnetic induction costs energy, and when that cost outweighs the energy gained by the formation of the superconducting state the material reverts to normal behaviour. This occurs at a critical field, H_c, in Type-I superconductors. The value of the critical field, H_c, is temperature-dependent so that a phase boundary can be mapped out in H–T phase space (Figure 13.30a). The relationship between H_c and the critical temperature, T_c, is given by the approximation:

$$H_c \approx H_c(0)\left[1 - \left(\frac{T}{T_c}\right)^2\right] \qquad (13.9)$$

where $H_c(0)$ is the critical magnetic field at 0 K, and T is the absolute temperature. [Note that values of the critical field are often quoted in tesla, the unit of magnetic induction, not the unit of magnetic field, $A\,m^{-1}$. In these cases, equate the 'field' in tesla to $\mu_0 H$, where H is the true field, measured in $A\,m^{-1}$.]

In many superconductors the transition between the superconducting and normal state is not sharp. When the external magnetic field reaches some lower critical magnetic field value, H_{c1}, magnetic flux starts to penetrate the material. The material becomes normal when the magnetic field reaches a

higher value, the upper critical field, H_{c2}. This behaviour characterises type-II superconductors (Figure 13.30b). The temperature dependence given in Equation (13.9) holds if H_c is relaced by $H_{c2}(0)$.

When the field reaches H_{c1}, filaments of magnetic flux cross the superconductor. These regions are no longer superconducting, so that some of the sample is in the superconducting state and some is normal (Figure 13.31). These small cylindrical regions of normal material are isolated from the superconducting matrix by surface currents forming a vortex.

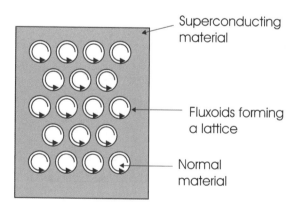

Superconducting material

Fluxoids forming a lattice

Normal material

Figure 13.31   A type-II superconductor in the mixed state. The solid contains small threads of normal material, fluxoids, which penetrate the superconducting bulk. These repel each other by virtue of the surface currents, and so form a regular array, a fluxoid lattice

The normal threads are called vortices or fluxoids. The currents act so as to repel each other, so that the flux lines form an ordered structure called a fluxon lattice, flux lattice or vortex lattice. At the core of each flux vortex the material is effectively normal, but is surrounded by a region of superconductor.

The magnetic flux enclosed by a loop of superconductor must be quantised in multiples of $h/2e$, where h is the Planck constant, and e is the electronic charge. The unit of flux is called the flux quantum, or fluxon, Φ_o, with a value of 2.07×10^{-15} Wb (T m^2). The flux enclosed in any circuit must then be $nh/2e$, with n taking integral values.

As the magnetic field increases, the amount of normal phase also increases relative to the superconducting part. Ultimately, the flux lines are so close together that no superconducting material exists between them, and the solid becomes normal. The field that finally destroys the superconductivity is the upper critical field, H_{c2}.

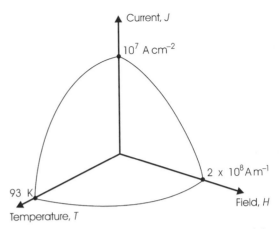

Figure 13.32   The region supporting superconductivity depends on the field, the temperature and the current in the solid. The values shown here for the critical field, temperature and current are typical of the high-temperature ceramic superconductor $YBa_2Cu_3O_{7-x}$

13.4.3   The effect of current

Although a superconducting solid exhibits no resistivity, at a certain current, the critical current, J_c, a superconductor reverts to a normal resistive state. The critical current is a function of the temperature and the external magnetic field, and hence the superconducting state of a solid can be mapped out as a volume with respect to current, field and temperature axes (Figure 13.32). As either temperature or magnetic field increase, the critical current decreases. The critical current is also greatly influenced by the microstructure of the solid, and careful processing is needed to obtain superconducting samples with high values of the critical current. For working devices a value of J_c of about 10^6 A cm^{-2} or greater needs to be achieved.

13.4.4   The nature of superconductivity

In a normal metal, resistivity is the result of interactions of the current-carrying electrons with the crystal structure. This clearly does not happen in the superconducting state and a completely different theory of electrical conduction is needed to account for the properties that superconductors possess.

Type-I superconductors are well explained by the Bardeen–Cooper–Schrieffer (BCS) theory. In this, the superconducting state is characterised by having the mobile electrons coupled in pairs. Each pair consists of two electrons with opposite spins, called Cooper pairs. At normal temperatures, electrons strongly repel one another. As the temperature falls and the lattice vibrations diminish, a weak attractive force between pairs of electrons becomes significant. In Type-I superconductors the 'glue' between the Cooper pairs are phonons (lattice vibrations).

The coupling can be envisaged in the following way. As an electron passes through a crystal, it interacts with the surrounding positively charged atomic cores, weakly attracting them (Figure 13.33). This leads to a slightly enhanced region of positive charge in the neighbourhood of the passing electron. Naturally, this weak attraction is swamped at high temperatures by thermal vibrations. At low temperatures another electron is able to feel the influence of the distortion. This second electron is weakly attracted to the slightly positive region

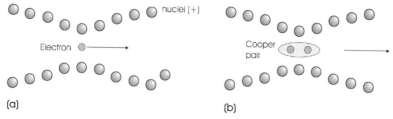

Figure 13.33 (a) An electron passing through a solid attracts the positively charged atomic nuclei slightly, creating a slightly enhanced region of positive charge; (b) at low temperatures, another electron can be attracted into this positive region to form a Cooper pair, which behaves as a single particle

generated by the first electron, and is carried along with it. The two electrons are thus linked.

Formally, this is described as linkage by the quanta of lattice vibration called phonons. When an electron in a Cooper pair passes through the crystal, the atom cores are attracted and then spring back as the electron passes. This causes a phonon to be emitted that is picked up by the other electron in the Cooper pair. At the same time, this electron is also causing phonons to be emitted, which are picked up by the first electron. This phonon exchange acts as the weak glue between the electron pairs. This coupling is easily destroyed even at low temperatures, and Cooper pairs are constantly forming and breaking apart. The pairs of electrons behave quite differently from single electrons. For example, they share the same wavefunction and are able to pass through the crystal unimpeded.

In the newly discovered high-temperature superconductors (described in Section 13.4.5) it has been shown that the electrons are also paired. Unfortunately, the BCS theory is not able to account for the much stronger coupling that must occur in these solids, and no satisfactory theory has yet been suggested.

13.4.5   Ceramic 'high-temperature' superconductors

High-temperature superconductors are copper oxides that maintain the superconducting state to temperatures above that of liquid nitrogen. One of the most interesting features of these new oxide superconductors is that they are all nonstoichiometric

oxides and the superconductivity is closely correlated to this feature. The stoichiometric variation of greatest significance with respect to superconductivity is the oxygen content. Some representative materials are listed in Table 13.2.

Crystallographically, the phases are all related to the *perovskite* structure type, ABO_3, where A is a

Table 13.2 Some high-temperature superconducting oxides

Compound[a]	T_c/K
$La_{1.85}Sr_{0.15}CuO_4$	34
$Nd_{2-x}Ce_xCuO_4$	20
$YBa_2Cu_3O_{6.95}$	93
Double Bi-O or Tl-O layers:	
$Bi_2Sr_2CuO_6$	10
$Bi_2CaSr_2Cu_2O_8$	92
$Bi_2Ca_2Sr_2Cu_3O_{10}$	110
$Tl_2Ba_2CuO_6$	92
$Tl_2CaBa_2Cu_2O_8$	119
$Tl_2Ca_2Ba_2Cu_3O_{10}$	128
$Tl_2Ca_3Ba_2Cu_4O_{12}$	119
Single Tl-O or Hg-O layers:	
$TlCaBa_2Cu_2O_7$	103
$TlCa_2Ba_2Cu_3O_9$	110
$HgBa_2CuO_4$	94
$HgCaBa_2Cu_2O_6$	127
$HgCa_2Ba_2Cu_3O_8$	133
$Hg_{0.8}Tl_{0.2}Ca_2Ba_2Cu_3O_{8.33}$	138
$HgCa_3Ba_2Cu_4O_{10}$	126

[a] The formulae are representative and do not always show the exact oxygen stoichiometry for the optimum T_c values given.

Note: T_c, superconducting transition temperature.

large cation and B is copper (see Section 5.4.3). The structures of the superconductors can be built up of slices of *perovskite* structure linked by slabs with structures mainly equivalent to slices of the *halite* and *fluorite* structures. The copper valence in most compounds lies between the formal values of Cu^{2+} and Cu^{3+}. Moreover, the appearance of superconductivity and the transition temperature are closely connected to the composition of these nonstoichiometric solids, which all exhibit considerable degrees of oxygen composition variation.

13.4.5.1 Lanthanum cuprate, La₂CuO₄

The phase La_2CuO_4 contains trivalent La and divalent Cu and adopts a slightly distorted version of the K_2NiF_4 structure in which the CuO_6 octahedra are lengthened along the c axis compared with the regular octahedra in the parent phase (Figure 13.34). The structure can be thought of as sheets of the *perovskite*-type one CuO_6 octahedron in thickness, and can also be described as built from CuO_2 and LaO layers, stacked in the sequence ... CuO_2, LaO, LaO, CuO_2 ... .

When prepared in air by heating the oxides CuO and La_2O_3, the compound is usually stoichiometric, with oxygen content close to 4.00. Electronically, the material is an insulator. The substance can be transformed into a superconductor via valence induction to generate Cu^{3+} ions. Replacing some of the La^{3+} cations with the alkaline earth cations Ba^{2+}, Sr^{2+} or Ca^{2+} achieves this. In the resulting compounds the dopant A^{2+} cations substitute for La^{3+}. Because these ions have a lower charge than the La^{3+}, charge neutrality can be maintained by creating one Cu^{3+} ion for each A^{2+}, to give a formula $La_{2-x}A_xCu^{2+}_{1-x}Cu^{3+}_xO_4$.

The first copper oxide superconductor, $La_{2-x}Ba_xCuO_4$, was produced in this way, although the isostructural material $La_{2-x}Sr_xCuO_4$ has been investigated in most detail. This latter compound shows a maximum T_c of 37 K at a composition near to $La_{1.85}Sr_{0.15}CuO_4$. Charge neutrality can be maintained in one of two ways in this compound. If valence induction occurs, one Cu^{3+} forms for each Sr^{2+} substituent, to give a formula $La_{2-x}Sr_xCu^{2+}_{1-x}Cu^{3+}_xO_4$. It is also possible to generate one oxygen vacancy for every two Sr^{2+} added, to give a charge-neutral formula $La_{2-x}Sr_xCu^{2+}O_{4-(x/2)}$.

The balance between these two alternatives is very delicately poised, and leads to a surprising situation. Initially, the Cu^{3+} option is preferred. Because Cu^{3+} can be looked on as Cu^{2+} together with a trapped hole, valence induction generates a hole population which leads to a superconducting state. As the Sr^{2+} concentration rises the Cu^{3+} population rises, peaking when x is approximately 0.2. As more Sr^{2+} is added, the preferred defect now becomes the oxygen vacancy and the oxygen content of the parent phase falls below 4.0. The number of Cu^{3+} ions decreases as oxygen vacancies form, and when the concentration of Sr^{2+} reaches approximately 0.32 all of the compensation is via vacancies and the material is no longer a superconductor (Figure 13.35). The nature of the defects in the phase is, therefore, directly related to the appearance of superconductivity.

13.4.5.2 Yttrium barium copper oxide (123), YBa₂Cu₃O₇

The compound $YBa_2Cu_3O_7$ has been widely studied because it is relatively easy to prepare and it was the first superconductor discovered with a T_c above the boiling point of liquid nitrogen. The crystal

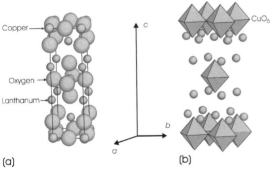

Copper

Oxygen

Lanthanum

c

b

a

CuO_6

(a)   (b)

Figure 13.34 The crystal structure of La_2CuO_4, shown as (a) atom packing and (b) CuO_6 octahedra. The dimensions of the room-temperature orthorhombic unit cell are $a_0 = 0.535$ nm, $b_0 = 0.540$ nm, and $c_0 = 1.314$ nm

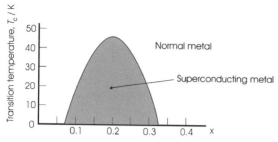

Figure 13.35 The variation of the superconducting transition temperature, T_c, with composition parameter x, for $La_{2-x}Sr_xCuO_4$. The superconducting state exists over only a small region of composition. Outside of this, the solid does not show superconducting behaviour

structure of $YBa_2Cu_3O_7$ consists of three perovskite-like unit cells stacked one on top of the other (Figure 13.36). The middle perovskite unit contains Y as the large A atom and Cu as the smaller B atom. The cells above and below this contain Ba as the A atom and Cu as the B atom, to give a metal formula of YBa_2Cu_3 as one would expect for a tripled perovskite cell, $A_3B_3O_9$. The unit cell of the superconductor should contain nine oxygen atoms.

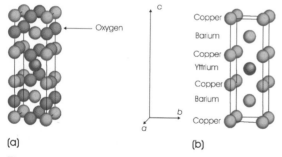

(a)   (b)

Figure 13.36 (a) The crystal structure of the high-temperature ceramic superconductor $YBa_2Cu_3O_7$; (b) the crystal structure showing cations only, revealing that the metal skeleton of the material is identical to a stack of three *perovskite*-like unit cells. The oxygen atoms are distributed so that no copper ions have six oxygen neighbours, as they would in a *perovskite*. Some copper ions have only four oxygen neighbours, thus changing the composition from $YBa_2Cu_3O_9$ to $YBa_2Cu_3O_7$. The dimensions of the orthorhombic unit cell are $a_0 = 0.381$ nm, $b_0 = 0.388$ nm, and $c_0 = 1.165$ nm

Instead the seven oxygen atoms present are arranged in such a way as to give the copper atoms square pyramidal and square planar coordination, rather than octahedral coordination as in the normal perovskites. If the ions are allocated the normal formal charges of Y^{3+}, Ba^{2+} and O^{2-}, the Cu must take an average charge of 2.33, which can be considered to arise from the presence of two Cu^{2+} ions and one Cu^{3+} ion per unit cell.

The appearance of superconductivity in this material is closely related to the oxygen content, and, like La_2CuO_4, it is a hole superconductor. At the exact composition $YBa_2Cu_3O_{7.0}$ the material is an insulator. Superconductivity appears when a small amount of oxygen is lost. The compound can readily lose oxygen down to a composition of $YBa_2Cu_3O_{6.0}$, and the superconducting transition temperature changes continuously over this composition range (Figure 13.37). The maximum value of T_c, close to 93 K, is found near to the composition $YBa_2Cu_3O_{6.95}$. As more oxygen is removed the value of T_c falls to a plateau of approximately 60 K, when the composition lies between the approximate limits of $YBa_2Cu_3O_{6.7}$ to $YBa_2Cu_3O_{6.5}$. Continued oxygen removal down to the phase limit of $YBa_2Cu_3O_{6.0}$ rapidly leads to a loss of superconductivity. The exact behaviour of any

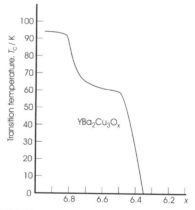

Figure 13.37 The variation of the superconducting transition temperature, T_c, with oxygen content, x, for $YBa_2Cu_3O_x$. Outside of the approximate range of x of 6.95–6.35 the material behaves as a typical insulating ceramic

sample depends on the defects present and this in turn depends on whether the samples are cooled quickly or slowly.

The oxygen atoms are not lost at random during reduction but come solely from the CuO_4 square planar units. This has the effect of converting the copper coordination from square planar to linear, resulting in CuO_2 chains running through the structure. This structural feature is of vital importance in allowing the superconducting transition to take place.

13.4.6 Josephson junctions

Cooper pairs can tunnel through a thin layer of insulator, called a weak link, separating two superconducting regions, without destroying the coupling between them or the superconductivity of the adjoining phases. In such cases a direct current (dc) flows across the insulating layer without the application of an accompanying voltage. This is called the dc Josephson effect. In essence, the insulator behaves as if it were a superconductor. The Josephson effect persists only for a certain range of currents and, eventually, above a critical current, i_c, a voltage develops across the junction.

When two Josephson junctions are connected in parallel (Figure 13.38a), the maximum current, i_c, that can flow across the device is function of the magnetic flux enclosed in the loop. As the magnetic flux penetrating the loop varies, the value of i_{cmax} varies from a maximum at zero flux or an integral number of flux quanta to a minimum for a half-integer number of flux quanta. The relationship is given by:

$$i_{cmax} = 2I_J \cos\left(\frac{\pi\Phi}{\Phi_0}\right)$$

where I_J is a constant depending on the junction geometry, Φ is the enclosed magnetic flux, and Φ_0 is the flux quantum.

A dc SQUID is a device based on this effect, used for the measurement of microscopic magnetic fields. The acronym SQUID stands for supercon-

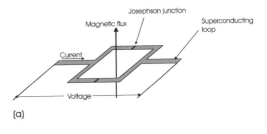

(a)

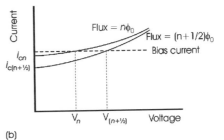

(b)

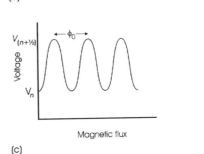

(c)

Figure 13.38 (a) A dc (direct-current) SQUID (superconducting quantum interference device) circuit, consisting of a loop of superconducting material containing two Josephson junctions, one in each arm. (b) With a bias current above the critical current, i_c, the voltage depends on the number of flux quanta that penetrate the loop. (c) In a varying magnetic field, the voltage cycles sinusoidally, with a period equal to the flux quantum

ducting quantum interference device. In operation, a bias current is used which is just above that needed to produce a voltage across the circuit. The critical current i_{cn} when a whole number of flux quanta penetrate the loop is higher than when an odd half number of flux quanta penetrate the loop, $i_{c[n+(1/2)]}$. Because of this, the voltage recorded, V_n, $V_{n+(1/2)}$, or an intermediate value, depends on the magnetic flux penetrating the loop (Figure 13.38b), and as the magnetic flux changes the voltage varies in a sinusoidal fashion (Figure 13.38c). A SQUID is a magnetic flux to voltage converter. Typically, the gain is about 1 volt per flux quantum. As fractions of a volt are easily measured, a SQUID has a sensitivity of approximately 10^{-4} to 10^{-6} of a flux quantum. This is sensitive enough to measure the magnetic fields produced by changes in the electrical activity of the brain.

Answers to introductory questions

How are donor atoms and acceptor atoms in semiconductors differentiated?

Donor atoms donate electrons to the conduction band in a semiconductor. This means that they have at least one extra valence electron over that required to fulfil the local bonding requirements of the structure. This electron can be promoted to the conduction band and enhance the number of mobile electrons present. For the elemental semiconductors silicon and germanium, this indicates atoms to the right of group 14 in the periodic table. The converse is true for acceptors. These have one less valence electron than required to fulfil the local bonding requirements in the structure. This shortfall is made up by appropriating an electron from the valence band, leaving behind a hole that adds to the existing hole population. For the elemental semiconductors silicon and germanium, this indicates atoms to the left of group 14 in the periodic table.

What is a quantum well?

A quantum well consists of a thin layer of a semiconductor with a smaller band gap within a semiconductor with a larger band gap. The most studied quantum well structures are those formed from a layer of gallium arsenide, GaAs, sandwiched in gallium aluminium arsenide, GaAlAs. The electrons in the thin GaAs layer are effectively trapped in the 'well' formed in the conduction band of the composite material. Similarly, the holes in the thin layer of semiconductor are trapped

at the 'hill' in the valence band of the composite material.

What are Cooper pairs?

Cooper pairs are pairs of mobile electrons that form at very low temperatures. Each pair consists of two electrons with opposite spin. The electrons share the same wavefunction and are able to pass through the crystal unimpeded.

At normal temperatures, electrons strongly repel one another. As the temperature falls and the lattice vibrations diminish, a weak attractive force between pairs of electrons becomes significant. In Type-I superconductors the 'glue' between the Cooper pairs are phonons (lattice vibrations). When an electron in a Cooper pair passes through the crystal, the atom cores are attracted and then spring back as the electron passes. This causes a phonon to be emitted that is picked up by the other electron in the Cooper pair. At the same time, this electron is also causing phonons to be emitted, which are picked up by the first electron. This coupling is easily destroyed even at low temperatures, and Cooper pairs are constantly forming and breaking apart.

Further reading

A. Cottrell, 1988, *Introduction to the Modern Theory of Metals*, Institute of Metals, London.

P.A. Cox, 1987, *The Electronic Structure and Chemistry of Solids*, Oxford University Press, Oxford.

W.B. Pearson, 1972, *The Crystal Chemistry and Physics of Metals and Alloys*, Wiley-Interscience, New York, especially Ch. 5.

M. Aldissi, 1987, 'Recent Advances in Inherently Conducting Polymers and Multicomponent Systems', *Journal of Materials Education* **9** 333.

A.P. Epstein, 1997, 'Eletrically Conducting Polymers: Science and Technology', *Materials Research Society Bulletin* **22** (June) 6.

A.J. Heeger, 2001, 'Semiconducting and Metallic Polymers: The Fourth Generation of Polymeric Materials', *Materials Research Society Bulletin* **22** (November) 900.

J.R. Waldron, 1996, *Superconductivity of Metals and Cuprates*, Institute of Physics, Bristol.

The following articles in *Scientific American* (in order of publication) give a good overview of superconductivity

R.M. Hazen, 1988, 'Perovskites', *Scientific American* **258**, (June) 52.

A.M. Wolsky, R.F. Giese, E.J. Daniels, 1989, 'The New Superconductors: Prospects for Applications', *Scientific American* **260** (February) 44.

R.J. Cava, 1990, 'Superconductors Beyond 1–2–3', *Scientific American* **263** (August) 24.

J. Clarke, 1994, 'SQUIDs', *Scientific American* **271** (August) 36.

R. de B. Ouboter, 1997, 'H.K. Oness's Discovery of Superconducivity', *Scientific American* **276** (March) 84.

A number of review articles, which give an indication of the rapid progress made in the science and engineering of high-temperature superconductor compounds, are to be found in

Materials Research Society Bulletin **XIV** (January 1989).

Materials Research Society Bulletin **XV** (June 1990).

Materials Research Society Bulletin **XVII** (August 1992).

Materials Research Society Bulletin **XIX** (September 1994).

Problems and exercises

Quick quiz

1 An insulator is a material with a full valence band, an empty conduction band and:
   (a) No band gap
   (b) A small band gap
   (c) A large band gap

2 An intrinsic semiconductor is a material with a full valence band, an empty conduction band and:
   (a) No band gap
   (b) A small band gap
   (c) A large band gap

3 In an intrinsic semiconductor, the current is carried by:
   (a) Electrons
   (b) Holes
   (c) Electrons and holes

4 A degenerate semiconductor has:
   (a) Few electrons and holes present
   (b) Only electrons or else holes present
   (c) Large numbers of electrons and holes present

5 Donors make a semiconductor:
   (a) p-type
   (b) n-type
   (c) Degenerate

6 Acceptors make a semiconductor:
   (a) p-type
   (b) n-type
   (c) Degenerate

7 A compensated semiconductor has:
   (a) Equal numbers of donors and acceptors present
   (b) Equal numbers of electrons and holes present
   (c) No intrinsic electrons and holes present

8 A metal has:
   (a) A partly filled uppermost band
   (b) Overlapping uppermost bands
   (c) Zero band gap between the two uppermost bands

9 A semimetal has:
   (a) A partly filled uppermost band
   (b) Overlapping uppermost bands
   (c) Zero band gap between the two uppermost bands

10 In a metal *not* carrying a current:
   (a) The electrons are stationary
   (b) The electrons at the Fermi surface are moving with random velocities
   (c) The average velocity of all electrons sums to zero

11 The conductivity of a metal:
    (a) Increases as the temperature increases
    (b) Falls as the temperature increases
    (c) Is insensitive to temperature

12 The conductivity of a semiconductor:
    (a) Increases as the temperature increases
    (b) Falls as the temperature increases
    (c) Is insensitive to temperature

13 The resistivity of a metal arises from:
    (a) Electron scattering from vibrations of the atoms in the material
    (b) Electron scattering from defects
    (c) Electron scattering both from defects and from atomic vibrations

14 The conductivity of a semiconductor increases as the energy band gap:
    (a) Decreases
    (b) Increases
    (c) Varies

15 A semiconductor crystal is transparent to radiation with energy:
    (a) Greater than the band gap
    (b) Less than the band gap
    (c) Exactly equal to the band gap

16 The band gap of a semiconductor:
    (a) Does not depend on atom size
    (b) Increases with increasing atom size
    (c) Decreases with increasing atom size

17 Atoms to the right of silicon and germanium in the periodic table act as:
    (a) Donors
    (b) Acceptors
    (c) Neither

18 Atoms to the left of silicon and germanium in the periodic table make the material:
    (a) n-type
    (b) p-type
    (c) Neither

19 Electrons give rise to:
    (a) A positive Hall coefficient
    (b) A negative Hall coefficient
    (c) A neutral Hall coefficient

20 When a current flows across a p–n junction under forward bias it is made up of:
    (a) Six components
    (b) Four components
    (c) Two components

21 Doping a transition metal oxide with a cation of lower valence will make it:
    (a) p-type
    (b) n-type
    (c) Cause no change

22 Conducting polymers contain:
    (a) Metal atoms in the structure
    (b) Conjugated double bonds
    (c) Conjugated triple bonds

23 Doping polyacetylene with sodium makes it:
    (a) A metallic conductor.
    (b) A p-type semiconductor
    (c) An n-type semiconductor

24 A semiconductor quantum well is:
    (a) A thin layer of a semiconductor on an insulator
    (b) Alternating layers of two semiconductors on an insulator
    (c) A thin layer of a semiconductor within a different semiconductor

25 Electrons in a quantum wire are strongly confined:
    (a) In one-dimension
    (b) In two dimensions
    (c) In three dimensions

26 A type-I superconductor:
    (a) Does not interact with magnetic fields
    (a) Draws an external magnetic field into itself
    (c) Expels internal magnetic fields

27 A material in the superconducting state can be thought of as:

(a) A perfect diamagnetic solid

(b) A perfect paramagnetic solid

(c) A perfect ferromagnetic solid

28 A superconductor carrying a current becomes a normal conductor:

(a) Below the critical current

(b) At the critical current

(c) Above the critical current

29 Superconductivity in conventional (low-temperature) superconductors is due to:

(a) Pairs of electrons

(b) Pairs of holes

(c) Electron–hole pairs

30 Ceramic superconductors mostly have structures closely related to:

(a) Spinel

(b) Perovskite

(c) Halite

31 A Josephson junction consists of:

(a) A very thin layer of insulator separating two superconducting regions

(b) A very thin layer of superconductor separating two insulating regions

(c) A thin layer of superconducting material separating two metallic regions

32 A SQUID measures:

(a) Resistivity

(b) Superconductivity

(c) Magnetic fields

Calculations and questions

13.1 The electrical resistivity of gold, at 273 K, is $2.05 \times 10^{-8}\,\Omega\,\mathrm{m}$. Gold adopts the A1 structure with a cubic lattice parameter, a_0, of 0.4078 nm. The velocity of electrons at the Fermi surface is $1.40 \times 10^6\,\mathrm{m\,s^{-1}}$. Each gold atom contributes one electron to the structure. Calculate the relaxation time, τ,

and the mean free path, Λ, of the electrons. Compare Λ with the interatomic spacing of gold atoms in the crystal.

13.2 The electrical resistivity of silver, at 273 K, is $1.47 \times 10^{-8}\,\Omega\,\mathrm{m}$. Silver adopts the A1 structure with a cubic lattice parameter, a_0, of 0.4086 nm. The velocity of electrons at the Fermi surface is $1.39 \times 10^6\,\mathrm{m\,s^{-1}}$. Each silver atom contributes one electron to the structure. Calculate the relaxation time, τ, and the mean free path, Λ, of the electrons. Compare Λ with the interatomic spacing of silver atoms in the crystal.

13.3 The electrical resistivity of rubidium, at 273 K, is $11.5 \times 10^{-8}\,\Omega\,\mathrm{m}$. Rubidium adopts the A2 structure with a cubic lattice parameter, a_0, of 0.5705 nm. The velocity of electrons at the Fermi surface is $8.1 \times 10^7\,\mathrm{m\,s^{-1}}$. Each rubidium atom contributes one electron to the structure. Calculate the relaxation time, τ, and the mean free path, Λ, of the electrons. Compare Λ with the interatomic spacing of rubidium atoms in the crystal.

13.4 The electrical resistivity of magnesium, at 273 K, is $4.05 \times 10^{-8}\,\Omega\,\mathrm{m}$. Magnesium adopts the A3 structure with hexagonal lattice parameters of $a_0 = 0.3209$ nm, and $c_0 = 0.5211$ nm. The velocity of electrons at the Fermi surface is $1.58 \times 10^6\,\mathrm{m\,s^{-1}}$. Each magnesium atom contributes two electrons to the structure. Calculate the relaxation time, τ, and the mean free path, Λ, of the electrons. Compare Λ with the interatomic spacing of magnesium atoms in the crystal.

13.5 The electrical resistivity of liquid mercury as a function of temperature is given in Table 13.3. The velocity of electrons at the Fermi

Table 13.3 Data for Question 13.5

$\rho/(10^{-8}\,\Omega\,\mathrm{m})$	94.1	103.5	128.0	214.0	630.0
$T/^{\circ}\mathrm{C}$	0	100	300	700	1200

surface is $1.52 \times 10^6 \, \text{m s}^{-1}$. Determine how the mean free path, Λ, varies with temperature. The density of liquid mercury is $13456 \, \text{kg m}^{-3}$. Assume that this does not vary with temperature and that each mercury atom contributes two mobile electrons to the liquid.

13.6   The resistivity of cadmium metal crystals at room temperature is $7.79 \times 10^{-8} \, \Omega \, \text{m}$ parallel to the c axis, and $6.54 \times 10^{-8} \, \Omega \, \text{m}$ parallel to the a axis. Cadmium adopts the A3 structure with hexagonal unit cell parameters of $a_0 = 0.2979 \, \text{nm}$, $c_0 = 0.5620 \, \text{nm}$ and each Cd atom contributes two mobile electrons to the crystal. Calculate the mobility of the electrons along the unit cell axes.

13.7   The resistivity of zinc metal crystals at room temperature is $6.05 \times 10^{-8} \, \Omega \, \text{m}$ parallel to the c axis, and $5.83 \times 10^{-8} \, \Omega \, \text{m}$ parallel to the a axis. Zinc adopts the A3 structure with hexagonal unit cell parameters of $a_0 = 0.2665 \, \text{nm}$, $c_0 = 0.4947 \, \text{nm}$ and each Zn atom contributes two mobile electrons to the crystal. Calculate the mobility of the electrons along the unit cell axes.

13.8   The resistivity of a sample of brass containing 70 wt% Cu and 30 wt% zinc is $6.3 \times 10^{-8} \, \Omega \, \text{m}$ at $0 \, °C$, compared with that of pure copper, which is $1.54 \times 10^{-8} \, \Omega \, \text{m}$ at

the same temperature. Determine the residual resistivity at this temperature.

13.9   The resistivity of a sample of bronze containing 90 wt% Cu and 10 wt% tin is $1.36 \times 10^{-7} \, \Omega \, \text{m}$ at $0 \, °C$, compared with that of pure copper, which is $1.54 \times 10^{-8} \, \Omega \, \text{m}$ at the same temperature. Determine the residual resistivity at this temperature.

13.10   The resistivity at $0 \, °C$ for nickel and for a number of nickel alloys is given in Table 13.4. Plot the resistivity against the amount of nickel in the alloy. Comment on the shape of the plot in terms of the possible structures of the alloys. [Note: graph is not shown in the answers at the end of this book.]

13.11   The resistivity at $0 \, °C$ aluminium and for a number of aluminium alloys is given in Table 13.5. Plot the resistivity against the amount of aluminium in the alloy. Comment on the shape of the plot in terms of the possible structures of the alloys. [Note: graph is not shown in the answers at the end of this book.]

13.12   The resistivity at $0 \, °C$ for copper and for a number of copper alloys is given in Table 13.6. Plot the resistivity against the amount of copper in the alloy. Comment on the shape of the plot in terms of the possible structures of the alloys. [Note: graph is not shown in the answers at the end of this book.]

Table 13.4   Data for Question 13.10

	Nickel	Alumel	Chromel P	Nichrome	Monel
Resistivity/$10^{-8} \, \Omega \, \text{m}$	6.16	28.1	70	107.3	42.9
Wt% Ni	100	95	90	77.3	67.1

Table 13.5   Data for Question 13.11

	Aluminium	RR59	RR57	Alpax gamma	Lo-Ex
Resistivity/$10^{-8} \, \Omega \, \text{m}$	2.42	3.5	3.95	3.5	3.95
Wt% Al	100	93	89	87	85

Table 13.6 Data for Question 13.12

	Copper	Bronze	Manganin	Brass	German silver	Constanin
Resistivity/$10^{-8}\,\Omega\,m$	1.54	19.8	41.5	6.3	40	49
Wt% Cu	100	90	84	70	62	60

13.13 Estimate the number of intrinsic electrons, n, and holes, p, and the product np, for a crystal of silicon at 300 K taking the effective mass of electrons and holes as equal to the electron rest mass, m_e. The band gap, E_g, is 1.12 eV.

13.14 Estimate the number of intrinsic electrons, n, and holes, p, and the product np, for a crystal of gallium arsenide, GaAs, at 300 K, taking the effective mass of electrons and holes as equal to the electron rest mass, m_e. The band gap, E_g, is 1.42 eV.

13.15 Estimate the number of intrinsic electrons, n, and holes, p, and the product np, for a crystal of gallium arsenide, GaAs, at 300 K, taking the effective mass of electrons to be $0.067m_e$ and holes as $0.082m_e$. The band gap, E_g, is 1.42 eV.

13.16 Estimate the number of intrinsic electrons, n, and holes, p, and the product np, for a crystal of cadmium selenide, CdSe, at 300 K, taking the effective mass of electrons to be $0.13m_e$ and holes as $0.45m_e$. The band gap, E_g, is 1.70 eV.

13.17 Determine the band gap, E_g, of silicon from the conductivity data given in Table 13.7.

What is the minimum frequency, ν, of light that will excite an electron across the band gap?

13.18 Determine the band gap, E_g, of germanium from the conductivity data given in Table 13.8. What is the minimum frequency, ν, of light that will excite an electron across the band gap?

13.19 Calculate the donor energy-level position in silicon doped with phosphorus using the 'Bohr model'. The effective mass of an electron is $0.33m_e$ and the relative permittivity of silicon is 11.7.

13.20 Calculate the acceptor energy-level position in germanium doped with aluminium using the 'Bohr model'. The effective mass of a hole is $0.16m_e$ and the relative permittivity of germanium is 16.0.

13.21 Calculate the donor energy-level position in indium phosphide, InP, doped with tin, using the 'Bohr model'. The effective mass of an electron is $0.067m_e$ and the relative permittivity of indium phosphide is 12.4.

13.22 Calculate the acceptor energy level position in gallium arsenide, GaAs, doped with zinc, using the 'Bohr model'. The effective mass

Table 13.7 Data for Question 13.17

Temperature /°C	227	277	327	377	427	477	527
Conductivity/$\Omega^{-1}\,m^{-1}$	3×10^{-4}	9×10^{-4}	3×10^{-3}	8×10^{-3}	2.5×10^{-2}	6×10^{-2}	8×10^{-2}

Table 13.8 Data for Question 13.18

Temperature /°C	5	47	82	122	162	240	344	441
Conductivity/$\Omega^{-1}\,m^{-1}$	0.0001	0.0008	0.004	0.009	0.05	0.1	0.6	1.0

Table 13.9 Data for Question 13.29

Current (along x)/mA	8	16	24	32	40
Voltage (along y)/V	0.081	0.159	0.233	0.318	0.401

of a hole is $0.082m_e$ and the relative permittivity of gallium arsenide is 13.2.

13.23 The energy gap for gallium arsenide, GaAs, is 1.4 eV at 300 K. The effective mass of electrons is $0.067m_e$ and of holes is $0.082m_e$. How near to the band gap centre is the Fermi level?

13.24 The energy gap for gallium phosphide, GaP, is 2.26 eV at 300 K. The effective mass of electrons is $0.82m_e$ and of holes is $0.60m_e$. How near to the band gap centre is the Fermi level?

13.25 A semiconductor containing 10^{20} holes m^{-3} and 10^{18} electrons m^{-3} has a conductivity of $0.455\,\Omega^{-1}\,\mathrm{m}^{-1}$. The ratio of the mobilities of electrons and holes, μ_e/μ_h, is 10. What are the hole and electron mobilities?

13.26 Gallium arsenide, GaAs, is doped with 10^{18} donor atoms. The hole mobility is $0.04\,\mathrm{m^2\,V^{-1}\,s^{-1}}$, and the electron mobility is $0.85\,\mathrm{m^2\,V^{-1}\,s^{-1}}$. Using the results of Question 13.15, determine the conductivity of the sample.

13.27 A 1-cm cube of n-type germanium supports a current of 6.4 mA when a voltage is applied across two parallel faces. The charge carriers have a mobility of $0.39\,\mathrm{m^2\,V^{-1}\,s^{-1}}$. Determine the Hall coefficient of the crystal assuming that only the majority charge carriers need be considered.

13.28 (a) Estimate the Hall coefficient, R_H, for intrinsic silicon, using the following values: $n_i = 1 \times 10^{16}\,\mathrm{m^{-3}}$; mobility of electrons = $0.15\,\mathrm{m^2\,V^{-1}\,s^{-1}}$; and mobility of holes =

$0.045\,\mathrm{m^2\,V^{-1}\,s^{-1}}$. (b) Calculate the Hall voltage, V, for a 1-cm cube of pure silicon at 20 °C, in a magnetic induction of 0.2 T, when a current of 10^{-3} A is applied to a cube face.

13.29 A single crystal of germanium doped with antimony to make it p-type is used in a Hall experiment. The crystal dimensions are 20 mm × 10 mm × 1 mm in the x, y, and z directions, respectively. In a constant induction of 0.9 T along the z direction, the current (along x) and voltage (along y) readings were obtained, as given in Table 13.9. Calculate the Hall coefficient, R_H, and the carrier density, ρ.

13.30 Using Figure 13.23, estimate the activation energy for the conductivity of polyacetylene doped with iodine for the three concentrations shown; that is, for (a) 0.15 mol%, (b) 0.01 mol% and (c) 0.005 mol% of I_2 per CH unit.

13.31 The transition metal titanium forms two slightly nonstoichiometric oxides, TiO_2 and Ti_2O_3. The oxide TiO_2 loses a small amount of oxygen to form TiO_{2-x}. Is it likely to show p-type or n-type semiconductivity? The oxide Ti_2O_3 gains a slight amount of oxygen to form Ti_2O_{3+x}. Is it likely to show n-type or p-type semiconductivity?

13.32 The oxide $LaCoO_3$ is an insulator. What are the charges on the cations present? The compound is doped with Sr^{2+} to form $La_{1-x}Sr_xCoO_3$, in which the Sr^{2+} substitutes for the La. What are the valences of the ions in this material? Is the doped material a p-type or an n-type semiconductor?

13.33 The compound Mg_2TiO_4 is an inverse *spinel*, which can be written as $(Mg)[Mg Ti]O_4$, where the round brackets indicate cations in tetrahedral sites and the square brackets indicate cations in octahedral sites. The compound $MgTi_2O_4$ is a normal *spinel*, $(Mg)[Ti_2]O_4$. (see Section 5.3.10). Both compounds are insulators. What are the charges on the ions? A small amount of $MgTi_2O_4$ is doped into Mg_2TiO_4 to form $Mg_{2-x}Ti_{1+x}O_4$. What are the charges on the ions and how are they distributed in the spinel structure? Will the material be a p-type or an n-type conductor?

13.34 Gallium arsenide has an electron effective mass of $0.067m_e$ and a hole effective mass of $0.082m_e$. Calculate the dimension at which quantum confinement becomes significant for (a) n-type and (b) p-type gallium arsenide at 300 K. Check the units of the calculation. (c) The crystal structure of gallium arsenide consists of layers of GaAs each 0.327 nm in thickness; how many layers are needed for quantum confinement of electrons or holes?

13.35 Gallium phosphide, GaP, has an electron effective mass of $0.82m_e$ and a hole effective mass of $0.60\ m_e$. Calculate the dimension at which quantum confinement becomes significant for (a) n-type and (b) p-type gallium phosphide. (c) The crystal structure of gallium phosphide consists of layers of GaP each 0.315 nm in thickness; how many layers are needed for quantum confinement of electrons or holes?

13.36 Compare the energy of the $n = 1$ energy level in the three directions for a quantum well of dimensions $1\ cm \times 1\ cm \times 10\ nm$, in a material in which the effective mass of the electron is $0.1m_e$.

13.37 Determine the energies of the lowest three states in a gallium arsenide–aluminium arsenide quantum well structure in which the potential well has a width of 9.8 nm, corresponding to 30 GaAs layers. The electron effective mass is $0.067m_e$.

13.38 Determine the energy of the $n = 1$ energy level in a fragment of n-type gallium nitride $20\ nm \times 393\ nm \times 1700\ nm$. The effective mass of electrons in this material is $0.19m_e$.

13.39 The critical field of the superconductor $PbMo_5S_6$ is given as 60 T. What is the critical field in $A\ m^{-1}$?

13.40 The critical field, H_{c2}, for the type-II superconductor Nb_3Sn at 15 K is given as 7 T. Estimate the value of the critical field at 0 K. The superconducting transition temperature, T_c, is 25 K.

13.41 What value of the critical field, B_{c2}, will cause the superconductivity of the type-II superconductor Nb_3Ge to be lost at 20 K. The value of the critical field at 0 K is given as 37 T, and the superconducting transition temperature is 23.6 K.

13.42 The superconductor $Nd_{1-x}Ce_xCuO_4$ is derived from the insulator Nd_2CuO_4 by substitution of some Nd^{3+} by Ce^{4+}. What are the charges on the Cu ions present? Will the superconductivity be via holes or electrons?

13.43 The compound $La_2SrCu_2O_6$ can take up oxygen to a composition $La_2SrCu_2O_{6.2}$. What are the charges on the Cu ions present in each of these phases? Are either potential high-temperature superconductors and, if so, would the superconductivity be via holes or electrons.

14

Optical aspects of solids

- What are lasers?

- Why are thin films often brightly coloured?

- What produces the colour in opal?

When light falls onto a solid, it is absorbed and/or scattered. Scattering is a broad term that has been subdivided into several categories, owing to the historical development of the subject. Thus, the topic 'scattering' tends to refer to the interaction of light with small particles. Other aspects of scattering include reflection and refraction, which is the scattering of light by a surface, and diffraction, the scattering of light by objects that are of similar dimensions to that of a light wave.

In this chapter, sources of light are considered before absorption and scattering are described, under the traditional headings. In addition, the appearance of objects, and fibre optics, both of interest in the context of optical properties of solids, are considered.

14.1 The electromagnetic spectrum

14.1.1 Light waves

Light is the form of energy detected by the eye. When light interacts with atoms and molecules it behaves as a stream of small particles, called photons. At larger scales, photons can be ignored and light can be treated as a wave. Light waves are part of the electromagnetic spectrum, with wavelengths ranging continuously from very long radiowaves, with wavelengths of gigametres (Gm), to high-energy 'cosmic rays', with wavelengths of the order of femtometres (fm); (see Figure 14.1). The wave and particle descriptions are linked by the fact that the wave equations describe the statistical behaviour of large numbers of photons.

A light wave has an electric and magnetic component, consisting of an oscillating electric and magnetic field, each described by a vector. As far as the topics in this chapter are concerned, the magnetic component need not be considered and only the electric component, specified by an electric field vector, needs to concern us. Light waves can

Understanding solids: the science of materials. Richard J. D. Tilley
© 2004 John Wiley & Sons, Ltd   ISBNs: 0 470 85275 5 (Hbk) 0 470 85276 3 (Pbk)

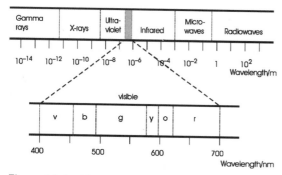

Figure 14.1 The electromagnetic spectrum. The visible region occupies only a small part of the whole, from approximately 400–700 nm. Note: v, violet; b, blue; g, green; y, yellow; o, orange; r, red

then be represented by the equation:

$$y = a_0 \sin\left[\left(\frac{2\pi}{\lambda}\right)(x + vt)\right]$$

where y is the magnitude of the electric field vector at position x and time t, a_0 is the amplitude of the wave, a constant; the wavelength of the light is λ, and the velocity of the wave is v. The peaks in the wave are referred to as crests and the valleys as troughs (Figure 14.2). Any point on the wave, a crest say, is moving in the x direction with a velocity v.

The velocity of a light wave, v, is related to the frequency, ν, of the wave by the equation:

$$v = \lambda\nu$$

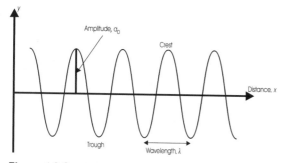

Figure 14.2 A light wave is characterised by an amplitude, a_0, and a wavelength, λ. For optical purposes, only the electric component, with an amplitude in the x–y plane, is needed

The frequency, ν, has units of Hz (Hertz) or s^{-1}. The velocity of light in a vacuum, $2.99792 \times 10^8 \, m \, s^{-1}$, which has a special significance in physics, is given the symbol c. The velocity of light in anything other than a vacuum is less than c. The units commonly used in discussions of light, optics and spectra are given in Section S4.9.

The part of the electromagnetic spectrum detected by human eyes is called the visible spectrum. Perception of the different wavelengths is called colour. The shortest wavelength of light that an average observer can perceive corresponds to the colour violet, $\lambda = 400$ nm, and the longest wavelength of light perceived by an average observer corresponds to the colour red, $\lambda = 700$ nm. Between these two limits, the other colours of the spectrum occur in the sequence from red, orange, green, blue, indigo and violet (Table 14.1).

Ultraviolet light has wavelengths shorter than violet. 'Ultraviolet A' is closest to the violet region and has the longest wavelengths. Although invisible to humans, many animals can detect ultraviolet A radiation. 'Ultraviolet B' and 'ultraviolet C' are at shorter wavelengths, and are energetic enough to damage biological cells. Radiation with wavelengths longer than red are referred to as infrared radiation. Although not visible, the longer wavelengths of infrared radiation, called thermal infrared, are detectable as the feeling of warmth on the skin.

A beam of light is said to be monochromatic when it is composed of only a very narrow range of wavelengths, and is said to be coherent when all of the waves that make up the beam are completely in phase, that is, when the crests and troughs of the waves are 'in step'. Normal light is incoherent, and laser light is coherent.

Many of the effects of the interaction of light with solids can be explained in terms of interference between light waves. If two light waves occupy the same region of space at the same time, they can add together, or interfere, to form a product wave (Figure 14.3). If two identical waves are exactly in step (Figure 14.3a.i) then they will add to produce a resultant wave with twice the amplitude by the process of constructive interference (Figure 14.3a.ii). If the two waves are out of step, the resultant amplitude will be less, as a result of

Table 14.1   The visible spectrum

Colour	λ/nm	ν/Hz	Energy J	eV
Deep red	700	4.29×10^{14}	2.84×10^{-19}	1.77
Orange-red	650	4.62×10^{14}	3.06×10^{-19}	1.91
Orange	600	5.00×10^{14}	3.31×10^{-19}	2.06
Yellow	580	5.17×10^{14}	3.43×10^{-19}	2.14
Yellow-green	550	5.45×10^{14}	3.61×10^{-19}	2.25
Green	525	5.71×10^{14}	3.78×10^{-19}	2.36
Blue-green	500	6.00×10^{14}	3.98×10^{-19}	2.48
Blue	450	6.66×10^{14}	4.42×10^{-19}	2.75
Violet	400	7.50×10^{14}	4.97×10^{-19}	3.10

destructive interference. If the waves are sufficiently out of step that the crests of one correspond to the troughs of the other (Figure 14.3b.i), the resulting amplitude will be zero (Figure 14.3b.ii).

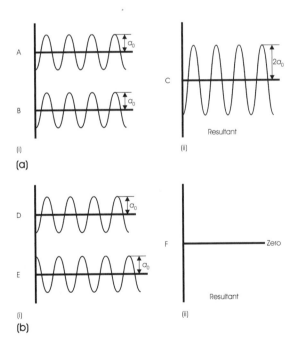

Figure 14.3   The interference of light waves: (a) two waves, A and B, in step (part i) add to give a resultant wave, C, with twice the amplitude of the original waves (part ii); (b) two waves, D and E, out of step (part i) add to zero, F (part ii)

14.1.2   Photons

When dealing with events at an atomic scale it is necessary to regard light as composed of particles, called photons. The energy of a photon is given by:

$$E = h\nu = \frac{hc}{\lambda}$$

where h is Planck's constant, and c is the velocity of light in a vacuum. The relationship between the wavelength and the frequency is:

$$\nu\lambda = c$$

Lattice vibrations in solids, which are at infrared frequencies, are often treated as particles rather than waves, and are called phonons.

14.1.3   The interaction of light with matter

When light interacts with individual atoms or molecules it is best regarded as a stream of photons. Isolated atoms and molecules have sharp energy levels. When a stream of photons encounters an atom or a molecule, the light will be absorbed only if the photon energy precisely matches the energy between the occupied energy level and one of the higher energy levels:

$$E(\text{photon})_{\text{abs}} = h\nu = E_2 - E_1$$

where E_1 represents the energy of the initial state, and E_2 represents the energy of the final (higher-energy) state. If the photon energy does not correspond exactly to the energy level separation between the occupied state and an upper energy state, it will not interact. Under normal circumstances, atoms or molecules are usually to be found in the lowest energy level, called the ground state, E_0, and the photon energy must be equal to the energy between the ground state and a higher energy state.

When an atom or molecule is in a high energy state, it may lose energy by dropping from the higher energy state to the lower state. At the same time, a photon is emitted, which carries off the excess energy. The energy of the photon is given by

$$E(\text{photon})_{\text{emit}} = h\nu = E_2 - E_1$$

where E_2 represents the energy of the initial high-energy state, and E_1 represents the energy of the final, lower-energy state. Ultimately, by emitting one or more photons, an atom or molecule will return to the ground state. Because the energy levels are sharp, the photons have precise energies, and the emission spectrum will consist of one or more sharp lines.

Although the photon energy may match the energy-level separation, photon absorption or emission need not necessarily occur. If the energy requirement is met, the actual probability of the transition occurring is governed by quantum mechanical selection rules. When the transition is allowed, which is equivalent to being highly probable, many photons will be absorbed or emitted. Colours caused by the absorption or emission of photons in allowed transitions are strong. The colours of dyes originate in allowed transitions. In cases where the transitions are forbidden – those that have a low probability of occurring – very few photons will be absorbed or emitted. Colours arising from forbidden transitions are weak and are typified by the colours displayed by 3d transition metal ions in solution.

As atoms or molecules are crowded together densely, the sharp energy levels give way to broader energy bands (see Section 2.3.1). In these cases, the

sharp lines in the emission spectrum change to broader lines. In solids and liquids the interactions between the atoms is so great that the sharp outer energy levels are transformed into broad energy bands. The emission and absorption of light then becomes more complex. Nevertheless, in favourable circumstances, the interaction of light with the solid is still understood in simple terms. For example, the energy gap between the valence and conduction band in semiconductors can be measured by observing the energy at which photons are just absorbed. When photons with an energy less than the semiconductor band gap, E_g, fall on a semiconductor, they are not absorbed, as electrons in the valance band do not gain enough energy to reach the valence band. When the photon energy is greater than E_g, the transition can occur, and the photons are absorbed. The energy at which absorption just starts is a measure of the optical band gap of the semiconductor.

14.2 Sources of light

14.2.1 Luminescence

The emission of radiation by solids at relatively low temperatures is called luminescence. Luminescence is subdivided into a number of categories, some of which are listed in Table 14.2. Fluorescence is

Table 14.2 Types of luminescence

Type	Source of energy
Fluorescence	Electronic excitation between allowed states
Phosphorescence	Electronic excitation between forbidden states
Triboluminescence	Mechanical bond breaking, fracture or friction
Chemiluminescence	Chemical reactions
Photoluminescence	Visible or ultraviolet light
Cathodoluminescence	Electron bombardment
Thermoluminescence	Increase of temperature
Electroluminescence	Applied electric field
Bioluminescence	Luminescence in a living organism

characterised by the immediate re-release of the exciting energy as light. Laser action is a form of fluorescence. Phosphorescence is typified by the slow conversion of the exciting energy into light, so that light emission is delayed, often by considerable lengths of time. Phosphors are materials that absorb radiation and re-emit a portion at a lower wavelength. They are widely used in fluorescent lamps, which convert ultraviolet radiation to visible light.

Fireflies and many other animals emit radiation by chemiluminescence. In this process, chemicals mix and produce light as one of the products of reaction. Many solids emit light when crushed, a manifestation of triboluminescence. There are no general mechanisms for luminescence, and each type needs to be treated independently.

14.2.2 Incandescence

Incandescence is the emission of light by a hot body. When light from an incandescent object is spread out according to wavelength by a prism the result is the continuous fan of colours listed in Table 14.1, called a continuous spectrum. However, the radiation emitted extends over a continuous range of wavelengths much broader than the visible spectrum and is both incoherent and unpolarised. For a solid body a little above room temperature all the wavelengths of the emitted energy lie in the infrared and are discernable as a sensation of warmth. As the temperature increases the overall energy of the radiation increases and the peak of the wavelength range moves towards shorter wavelengths. At a temperature of about 700 °C, the shortest wavelengths emitted creep into the red end of the visible spectrum. The colour of the emitter is seen as red and the object is said to become red hot. At higher temperatures, the wavelengths of the radiation given out extend increasingly into the visible region and the colour observed changes from red to orange and thence to yellow. When the temperature of the emitting object reaches about 2500 °C, all visible wavelengths are present and the body is said to be white hot.

The most important incandescent object for us is the Sun, which is the ultimate source of energy on Earth. The solar spectrum has a maximum near 560 nm. Light is perceived as white if it has a make-up like that of the solar spectrum. The human eye is most sensitive to the maximum in the solar spectrum, which corresponds to yellow-green, and is noticeably less sensitive to violet and red light.

The intensity of the radiation emitted by incandescent solids can be understood by considering a black body, which is an object that absorbs and emits all wavelengths perfectly. A graph of the intensity of the radiation issuing from a black body as a function of wavelength is called a black-body emission spectrum (Figure 14.4). The shape of the curve is found to be dependent only on the temperature of the body. As the temperature increases the peak in the curve moves to shorter wavelengths (higher energies). The solar spectrum has a form quite similar to the emission spectrum of a black body with temperature of about 5700 °C (about 6000 K).

The form of the black-body emission spectrum cannot be explained by classical physics, and its

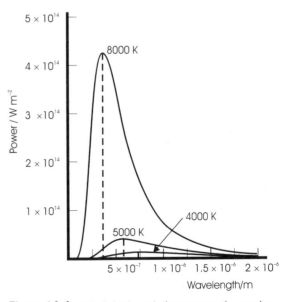

Figure 14.4 Black-body emission spectra: the maxima move to shorter wavelengths as the temperature of the emitter increases

successful theoretical explanation, by Planck in 1901, initiated quantum theory. In order to reproduce the form of the curve, Planck postulated that the energy absorbed or given out by the atoms and molecules in the black body had to be delivered in packets, or quanta. Energy emission was quantised. The relationship between the energy of a single quantum, E, and the frequency of the radiation, ν, was given by:

$$E = h\nu$$

The energy of the radiation emitted by a black body, E (in W m^{-2}), into a hemispherical region of space, in the wavelength interval, $\delta\lambda$, is given by:

$$E\mathrm{d}\lambda = \frac{2\pi hc^2 \mathrm{d}\lambda}{\lambda^5 \left[\exp\left(\frac{hc}{\lambda kT}\right) - 1\right]}$$

where T is the absolute temperature of the black body.

14.2.3 Fluorescence and solid-state lasers

The word laser is an acronym for the expression *l*ight *a*mplification by *s*timulated *e*mission of *r*adiation. The process of stimulated emission, described below, produces light that is coherent, giving it quite different properties from those of light from an incandescent source, which is incoherent.

When a photon of energy $h\nu$ is absorbed by an ion, atom or molecule it passes from a lower-energy state, the ground state, to an upper or excited state. The transition will take place if the frequency of the photon, ν, is given exactly by:

$$\nu = \frac{E_2 - E_0}{h}$$

where E_0 is the energy of the ground state, E_2 is the energy of the excited state, and h is Planck's constant. If an atom in the excited state, E_2, makes a transition to the ground state, E_0, energy will be emitted with the same frequency, given by the same equation. In 1917, Einstein suggested that

there should be *two* possible types of emission process rather than just one. The most obvious is that an atom in an excited state can randomly change to the ground state: a process called spontaneous emission. Alternatively, a photon having an energy equal to the energy difference between the two levels, that is, $E_2 - E_0$, can interact with the atom in the excited state causing it to fall to the lower state and emit a photon at the same time: a process called stimulated emission.

Spontaneous emission occurs in incandescence. The light photons all have the same frequency but the waves possess random phases and the light is incoherent. Stimulated emission occurs in lasers. The photons produced have the same energy and frequency as the one that caused the emission, and the light waves of all photons are coherent.

The key to laser action is to obtain atoms or molecules in an excited state and keep them there long enough for photons to pass and trigger stimulated emission. There are two theoretical difficulties to overcome. Under normal circumstances the number of atoms in the excited energy level, N_2, relative to those in the ground state, N_0 – the relative population – will be extremely small for energy levels that are sufficiently separated to give rise to visible light. This may be confirmed by the Boltzmann law, which gives the relative populations under conditions of thermal equilibrium by:

$$\frac{N_2}{N_0} = \exp\left[-\frac{(E_2 - E_0)}{kT}\right]$$

where k is the Boltzmann constant, T is the absolute temperature, and E_2 and E_0 are the energies of the excited state and the ground state, respectively. Under equilibrium conditions it is not possible to increase the population N_2 over N_0, a situation called a population inversion.

A second problem compounds the difficulty. The ratio R of the rate of spontaneous emission to stimulated emission under conditions of thermal equilibrium is given by:

$$R = \exp\left(\frac{h\nu}{kT}\right) - 1$$

At 300 K, at visible wavelengths, R is much greater than 1. Thus, stimulated emission will be negligible compared with spontaneous emission (see Section S4.10).

The solution to the difficulties lies in recognising that the transition from one energy level to another is associated with a transition probability. If atoms can be excited into energy levels from which the probability of a transition is small, the atoms will remain in this state for long enough to produce a population inversion and so be available for stimulated emission.

14.2.4   The ruby laser: three-level lasers

The ruby laser, invented in 1960, was the first device to put the ideas just described into practice. Rubies are crystals of alumina (aluminium oxide, corundum, Al_2O_3), containing about 0.5 % chromium ions, Cr^{3+}, in place of aluminium ions, Al^{3+}. Ruby is a dilute solid solution, and the Cr^{3+} ions form substitutional defects. The laser action involves only the Cr^{3+} ions and is due to the transition of electrons from the ground state to higher energy levels among the 3d orbitals.

The energy levels of the Cr^{3+} ions are derived from the term scheme of an isolated Cr^{3+} ion on which crystal field interactions have been imposed (see Sections S1.3.1 and S4.5). The three electrons on Cr^{3+} ions give rise to two sets of energy levels, one set associated with all electron spins parallel ($\uparrow\uparrow\uparrow$) and one set with one spin-paired couple ($\uparrow\downarrow\uparrow$) (Figure 14.5). The ground state, E_0, has all spins parallel. Electron transitions involved in colour (optical transitions) are only allowed between levels in which the total amount of electron spin does not change. As the ground state has electrons with parallel spins, ($\uparrow\uparrow\uparrow$), the allowed transitions are also to states with parallel spins. Such transitions are called spin-allowed transitions.

In the case of ruby there are two important spin-allowed transitions:

- $E_0 \rightarrow E_2$, at 556 nm, absorbs yellow-green;

- $E_0 \rightarrow E_3$, at 407 nm, absorbs violet.

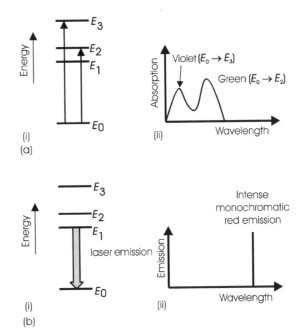

Figure 14.5   The energy levels involved in the ruby laser: (a) transitions from the ground state, E_0, to energy levels E_2 and E_3 (part i) produce the normal colour of ruby, revealed in the absorption spectrum (part ii); (b) the transition from the level E_1 to the ground state, E_0 (part i) is responsible for the laser emission (part ii)

Lying at a lower energy than E_2 is an energy level E_1. This energy level is due to an electron configuration in which two electrons are spin-paired, ($\uparrow\downarrow\uparrow$). A transition from E_0, E_2 or E_3 to E_1 is not allowed under the total electron spin rule. None of these transitions would normally be involved in transitions that produce colour. However, in ruby, excited Cr^{3+} ions in states E_2 or E_3 can lose energy to the crystal structure and drop down to level E_1. This process operates under different conditions from the optical transitions and is independent of spin. The energy is taken up in lattice vibrations and the ruby crystal warms up. This is called a radiationless or phonon-assisted transition. Typical rates of the transitions are:

- $E_2 \rightarrow E_0$, $3 \times 10^5 \ s^{-1}$;

- $E_2 \rightarrow E_1$, $2 \times 10^7 \ s^{-1}$.

The second of these two transitions is about 100 times faster than the first. The rates of the transitions from the E_3 energy level to E_1 and E_0 are of a similar magnitude. This means that on irradiating the ruby with white light, Cr^{3+} ions will be excited to energy levels E_2 and E_3, and then a significant number end up in the E_1 state rather than returning to the ground state. The transition from E_1 to the ground state is not allowed because of the spin rule and so atoms in the E_1 state have a long lifetime. (The spontaneous emission rate is $2 \times 10^2 \, s^{-1}$.) Thus, it is possible to build a population inversion between the E_1 and E_0 levels.

Laser operation takes place in the following way. An intense flash of white light is directed onto the crystal. This process is called optical pumping. This excites the Cr^{3+} ions into the E_2 and E_3 states. These then lose energy by radiationless transitions and 'flow over' into state E_1. An intense initial flash will cause a population inversion to form between E_1 and E_0. About 0.5 ms after the start of the pumping flash, some spontaneous emission will occur from E_1. In order to prevent these first photons from escaping from the crystal without causing stimulated emission from the other excited ions one end is coated with a mirror and the other with a partly reflecting mirror. In this arrangement, the photons are reflected back and forth, causing stimulated emission from the other populated E_1 levels. Once started, the stimulated emission rapidly depopulates these levels in an avalanche. There will be a burst of red laser light of wavelength 694.3 nm, which emerges from the partly reflecting surface.

Following the light burst, the upper levels will be empty and the process can be repeated. The ruby laser generally operates by emitting energy in short bursts, each of which lasts about 1 ms, a process referred to as pulsed operation. The ruby laser is called a three-level laser, because three energy levels are involved in the operation. These are the ground state (E_0), an excited state reached by optical absorption or pumping (E_2 or E_3), and an intermediate state of long lifetime (E_1) reached by radiationless transfer and from which stimulated emission (laser emission) occurs to the ground state.

It is energetically costly to obtain a population inversion in a three-level laser because one must pump more than half the population of the ground state to the middle level. Very little of the electrical energy supplied to the flash lamp ends up pumping photons, and carefully designed reflectors are essential. The energy lost in the transitions from E_3 and E_2 to E_1 ends up as lattice vibrations, which cause the crystal to heat up considerably. To make sure that the ruby does not overheat and shatter it is necessary to cool the crystal and to space the pulses to allow the heat to dissipate. Although the ruby laser was the first laser made, the three-level mode of operation makes it inefficient.

14.2.5 The neodymium (Nd^{3+}) solid-state laser: four-level lasers

A more energy-efficient device can be made employing four-level lasers. Laser operation takes place in the following sequence of steps (Figure 14.6). Atoms in the ground state, E_0, are excited to a rather high energy level, E_2, by optical pumping. This process needs to be fast and efficient. Subsequently, the atoms in E_2 lose energy without radiating light, to an intermediate state I_1. This step should also be fast and efficient. However, once in

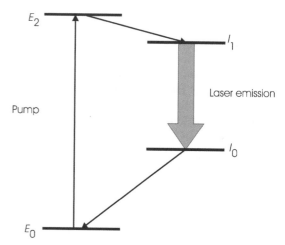

Figure 14.6 The principle transitions in a four-level laser: the laser transition occurs between two intermediate energy levels, I_1 and I_0; the transition that drives the laser, the pump transition, is between the ground state, E_0, and the energy level E_2

I_1, atoms should have a long lifetime and not lose energy quickly. When another intermediate state, I_0, is present and sufficiently high above the ground state to be effectively empty, a small population in I_1 gives a population inversion between I_1 and I_0. Ultimately, a few photons will be released as some atoms drop from I_1 to I_0. These can promote stimulated emission between I_1 and I_0, allowing laser action to take place. Atoms return from I_0 to E_0 by a step that needs to be rapid. If the energy corresponding to the transitions from E_2 to I_1 and from I_0 to E_0 can be easily dissipated, continuous operation rather than pulsed operation is possible.

The most important four-level solid-state laser uses neodymiumions ions (Nd^{3+}) as the active centres. The important transitions taking place in Nd^{3+} ion lasers are due to the transitions of f electrons (Section S1.2.3). The f-electron levels are rather sharp and can be approximated to free ion energy levels, because these levels are shielded from the effects of the surrounding crystal lattice by outer electron orbitals. Above the f-electron energy levels lie energy bands of considerable width derived from the interaction of the 5d and 6s orbitals (Figure 14.7). Optical pumping excites the ions from the ground state to these wide bands. This process is very efficient because broad energy bands allow a wide range of wavelengths to pump the laser and because the transitions are allowed. In addition, loss of energy from the excited state down to the f-electron energy levels is fast. The energy loss halts at the ^4F pair of levels. The principal laser transition is from these ^4F levels to a set of levels labelled $^4I_{11/2}$. The emission is at approximately 1060 nm, in the infrared. This is a useful wavelength as it coincides with a reasonably low-loss region of silica-based optical fibres.

Practical lasers contain about 1 % Nd^{3+} and can have quite high power outputs. The most common host materials are glass, yttrium aluminium garnet (YAG) and calcium tungstate, $CaWO_4$. They can be operated continuously or pulsed. At higher Nd^{3+} concentrations the lifetime of the ^4F upper state drops from about 200 μs in a typically 1 % doped material to about 5 μs at higher dopant concentrations. This is due to Nd–Nd interactions and associated changes in lattice vibration characteristics.

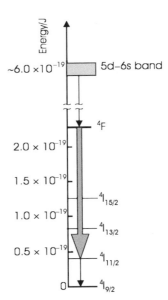

Figure 14.7 The energy levels of most importance in the neodymium laser. The pump transitions are from the ground state to a broad 5d – 6s band. The main laser transition occurs between the ^4F and $^4I_{11/2}$ levels

Under these conditions, laser operation is no longer possible.

14.2.6 Light-emitting diodes

The structure of a p–n junction diode was described in Section 13.2.6. A key feature of this device is the potential barrier that builds up in the junction region. Under equilibrium conditions, this serves to separate the electrons in the n-type region from the holes in the p-type region. When a positive voltage, called a forward bias, is applied to the p-type side of the junction, the equilibrium barrier height falls and the junction region broadens (Figure 14.8). Electrons and holes now enter the junction and recombine. The energy released is approximately equal to the band gap. If this appears as light, the diode acts as a lamp and is the familiar light-emitting diode, or LED.

There are a number of aspects of importance for a working device. Clearly, the band gap must be such as to give out visible wavelengths. The most widely

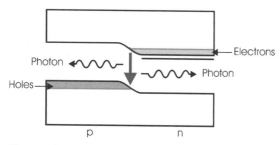

Figure 14.8 The principle of LED (light-emitting diode) operation: under a forward bias, electrons and holes recombine in the junction region and emit radiation

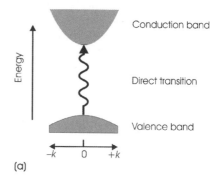

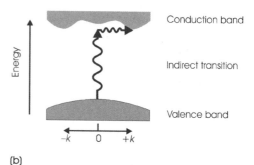

Figure 14.9 (a) In a direct transition between the valence band and the conduction band, the wave vector, k, of the excited electron does not change; (b) in an indirect transition, the wave vector, k, changes, making the transition much less probable than a direct transition

used LED materials are solid solutions based on the semiconductors gallium arsenide (GaAs), gallium phosphide (GaP), gallium nitride (GaN), aluminium arsenide (AlAs) and aluminium phosphide (AlP). By varying the amounts of each of these elements in the solid solution, light emission across the visible spectrum can be achieved.

A second consideration is vitally important. Ideally, LEDs must be highly efficient and give an adequate light output under a small voltage. To achieve this, the nature of the excitation of an electron from the valence band to the conduction band and the reverse process of annihilation must be efficient. This efficiency depends on the detailed band structure of the semiconductors, and the flat-band model used in Chapter 13 is no longer adequate.

There are two possibilities of importance. The first is that the lowest point of the valance band corresponds to the highest point of the conduction band; that is, they are at the same value of the wave vector, k (Figure 14.9a). In this case, an optical transition between the bands can take place without a change of k or electron momentum. Such a transition is called a direct transition. In terms of quantum mechanics, the transition has a high probability of occurring when a photon of the correct energy hits the semiconductor, and the efficiency of the process is high.

In the second case, the lowest point of the valence band is not at the same value of the wave vector k as the highest point in the conduction band (Figure 14.9b). In this case, the photon that is to promote the electron from the top of the valance band to the conduction band must also interact with the lattice to pick up (or lose) sufficient momentum to make the transition possible. That is, the photon must interact with a phonon to make the transition occur. Such a transition is called an indirect transition. The probability of an indirect transition occurring is quite low.

The indirect process is represented by the equation

$$h\nu = E_g \pm h\nu_{phonon}$$

where $h\nu$ is the energy of the photon needed to transfer the electron, E_g is the optical band gap, and $h\nu_{phonon}$ is the energy of the phonon involved. The $\pm$ term depends on whether the phonon is absorbed or emitted. In general, the phonon energy is small, from 0.01–0.03 eV (1.6×10^{-21}–4.8×10^{-21} J), so that this results in only a small error when optical absorption is used to measure the band gap.

The energy band structure of crystalline silicon gives rise to an indirect transition. The same is true of crystalline germanium, Ge, and for this reason neither of these materials is used in LEDs. Note, though, that amorphous silicon seems to show a direct transition. The energy band structure of gallium arsenide, GaAs, favours a direct transition, and for this reason GaAs is the preferred material for LEDs. The reason why gallium arsenide solid solutions are limited to the red and yellow colours is because of the fact that the transitions become indirect when the band gap becomes appropriate to green or blue emission. To overcome this problem gallium nitride (GaN) and related materials are being explored as blue and violet emitting LEDs.

14.2.7 Semiconductor lasers

Semiconductor lasers are, in essence, identical to LEDs, although the physical structure of lasers tends to be more complex. A semiconductor laser consists of a p–n junction in which one component has been heavily doped. The junction is placed under forward bias and a high current is passed across the device. Initially, the junction acts as an LED, and electrons recombine at random with holes to give out light by way of spontaneous emission. If the current is high enough, at some point in the junction region the number of electrons in the conduction band (from the n-type region) exceeds the number in the valence band (from the p-type region). When this population inversion is achieved, stimulated emission occurs and light emission from the junction region is coherent laser light (Figure 14.10). To increase the chance of stimulated emission occurring and to make the beam directional, two ends of the device are polished, and the whole is constructed of a number of carefully engineered layers with varying electronic characteristics.

An advantage of semiconductor lasers is that they are very efficient. Moreover, because the emission

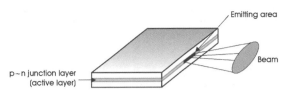

Figure 14.10 Schematic diagram of a semiconductor laser; the beam is emitted from a thin p–n junction active layer

comes from the host material itself (not a small quantity of dopant as in ruby or neodymium lasers) these lasers are very powerful for their size.

14.3 Colour and appearance

14.3.1 Luminous solids

The colour of a solid depends on the wavelengths of light that travel from the solid to the observer's eye. Luminous objects emit radiation directly and the colour of the object will be the overall perception of the wavelength range recorded in the eye. When all wavelengths are present at a reasonable intensity the object is regarded as white. Individual wavelengths are not perceived separately, and a mixture of wavelengths corresponding to, say, red and yellow, is seen as orange and not as a combination of colours. This combination of different wavelengths of light is called additive coloration. Mixing just three different wavelengths of light in various proportions can reproduce the perceived colours of all light sources. The three wavelengths are called additive primary colours and the process of mixing lights to obtain other colours is called additive mixing. There is no fixed set of primary colours, and any three colours loosely designated as red, blue and yellow will suffice for the purpose. Mixing the three additive primary colours in equal proportions will produce white light.

14.3.2 Nonluminous solids

Nonluminous solids interact with light passively. When light of a particular wavelength falls onto such a solid it might be absorbed, by absorption centres in the material, in which case the energy of the light may end up in the solid structure as heat, and the solid is described as opaque to the absorbed wavelength. Other absorption centres may re-emit light as luminescence. Alternatively, the light may pass through the solid, no energy transfer is made and the solid will be transparent to that wavelength. The light that leaves the material is the transmitted light. Finally, for the material to be visible, at least some of the incident light has to be scattered towards the observer. The commonest scattering process is reflection, which takes place at surfaces. Smooth surfaces reflect light uniformly, known as specular reflection. Rough surfaces will reflect incident light in all directions, called diffuse reflection. Scattering centres, for example crystallites, act to reflect the light internally, giving an opalescent appearance. The appearance of a solid will depend on which of these processes occur (Figure 14.11).

For example, a solid that has no internal surfaces that reflect the light and hinder its passage, and no internal absorption centres that subtract energy from the incident beam, will appear to be transparent. The same transparent material in powder form will

appear white. This is because no light is absorbed but each granule of the powder scatters light of all wavelengths. A proportion of the incident light is soon deflected in the direction of the observer by multiple reflections. Crystallites of transparent material deposited inside a glass also scatter light. If there are sufficient numbers, the glass will appear milky white. Opal glass is deliberately made to produce large concentrations of internal surfaces and appear uniformly white. Many thermoplastic materials consist of crystalline regions embedded in amorphous material. This behaviour is typified by polyethylene. The crystalline regions are of high density and of high refractive index (see Section 14.4.1) and so scatter light, which is why polyethylene appears milky.

Nontransparent objects become coloured by the selective absorption of radiation. If white light falls onto a material that absorbs blue and yellow light, the object will transmit and reflect red light. It will appear red. This process is called subtractive coloration. There are three subtractive primary colours which when blended produce the subtractive colour spectrum. These are cyan, which is red-absorbing, magenta, which is green-absorbing, and yellow, which is blue-absorbing. An object containing the appropriate amounts of the three subtractive primary colours will absorb all of the light falling onto its surface and appear black to the eye. Printing inks function by way of subtractive absorption.

Metals are a particular category of opaque solids. Metals are characterised by free electrons, and those at the Fermi surface have empty energy levels readily available. Thus, any light falling on the surface of a metal will be absorbed in the surface layers. However, the excited electrons rapidly fall back to lower energy levels. Most of the light is re-emitted, but a proportion is absorbed and converted into heat. This strong absorption prevents the light from penetrating much below the surface of the metal. If the surface is smooth, the light appears to be reflected. Most metals reflect all wavelengths equally, giving them a silver appearance. Copper and gold are notable exceptions, in that they do not reflect evenly across the visible spectrum. Powdered metals repeatedly reflect light but, as some is absorbed at each interaction, much less escapes from the mass, and most finely divided metals look black.

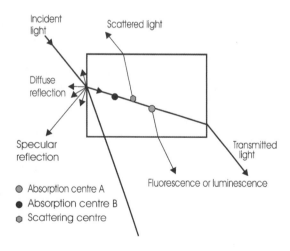

Figure 14.11 The interaction of light with a solid. The light can be transmitted, reflected, scattered or absorbed. Some absorption centres are able to re-emit light as fluorescence or luminescence. The appearance of the solid depends on which of these interactions occur

The appearance of a surface is bound up not only with the colour leaving it but also with the texture of the surface. A smooth surface reflects light and looks shiny, even if coloured, because a certain amount of specular reflection takes place. A rough surface exhibits diffuse reflection. Skin has a different texture and appearance from a plastic film. The rendition of objects showing different amounts of specular and diffuse reflection is a difficult task for artists and especially for those wishing to create computer graphics, and much effort has been devoted to this objective.

14.3.3 The Beer-Lambert law

The cause of light absorption in a solid is due to the presence of absorption centres. These are atoms or groups of atoms that have energy levels with a spacing equivalent to the energy of light. Ignoring processes such as fluorescence and phosphorescence, in which some light is re-radiated, the interactions of light with a solid can be expressed thus:

Incident intensity (I_0) = amount of light reflected (I_r)

$+$ amount scattered (I_s) + amount absorbed (I_a)

$+$ amount transmitted (I_t),

$$I_0 = I_r + I_s + I_a + I_t$$

or

$$1 = R + S + A + T$$

where R is the fraction of light reflected (I_r/I_0), S is the fraction of light scattered (I_s/I_0), A is the fraction of light absorbed (I_a/I_0), and T is the fraction of light transmitted (I_t/I_0). In good-quality optical materials, the amount of light scattered and absorbed is small and it is often adequate to write:

$$I_0 = I_r + I_t$$

or

$$1 = R + T$$

When absorption centres are distributed uniformly throughout the bulk of the sample, the amount of light absorbed in a transparent plate is given by Lambert's law:

$$I = I_0 \exp(-\alpha_a l) \qquad (14.1)$$

where I is the intensity leaving the plate, I_0 is the incident intensity, l is the thickness of the plate (in m), and α_a is the linear absorption coefficient (in m^{-1}). The amount of absorption will be a function of the concentration of the absorbing centres throughout the bulk of the material. This is taken into account in the Beer–Lambert law:

$$\log\left(\frac{I}{I_0}\right) = -\varepsilon[J]\, l$$

where I is the intensity after passage through a length of sample l, I_0 is the incident intensity, $[J]$ is the molar concentration of absorption centres or absorbing species, and ε is the molar absorption coefficient or extinction coefficient, which has units of $dm^3\,mol^{-1}\,m^{-1}$, or $l\,mol^{-1}\,m^{-1}$. The dimensionless product A, equal to $\varepsilon[J]l$, is called the absorbance or optical density. The ratio I/I_0 is the transmittance, T, hence:

$$\log T = -A$$

The Beer–Lambert law finds use in the measurement of concentrations. For example the clarity or otherwise of polluted air is often measured by comparing the intensity of light at a certain time (I) with the intensity on a fine day (I_0).

14.4 Refraction and dispersion

14.4.1 Refraction

When light enters a transparent and insulating material, it is refracted. Refraction is the cause of the apparent bending of a ray of light when it enters water or glass and is the physical effect used in lenses (Figure 14.12). The magnitude of the effect is given by the index of refraction, or refractive index, n, where:

$$n = \frac{\sin \theta_1}{\sin \theta_2}$$

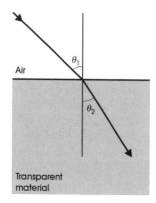

Figure 14.12   Refraction of a light beam on entering a transparent solid

θ_1 being called the angle of incidence, and θ_2 the angle of refraction. This equation is known as Snell's law (even though the originator was named Snel). The above equation is a special case of the more general relation that applies to light passing from a medium of refractive index n_1 to one of refractive index n_2:

$$\frac{\sin \theta_1}{\sin \theta_2} = \frac{n_2}{n_1}$$

Some refractive indices are listed in Table 14.3 and Table 11.1 (page 342). In many crystals, the index of refraction varies with direction. These are

called optically anisotropic materials. In amorphous materials such as glass, or in crystals with a cubic structure, the index of refraction is the same in all directions. These are called optically isotropic solids.

In effect, the refractive index is a manifestation of the fact that the light is slowed down on entering a transparent material. This is due to the interaction of the light with the electrons around the atoms that make up the solid. It is found that the refractive index, n, of a transparent substance is given by:

$$n = \frac{\text{velocity of light in a vacuum (c)}}{\text{velocity of light in the medium (v)}}$$

The frequency of the light does not alter when it enters a transparent medium and, because of the relationship between the velocity of a light wave and its frequency,

$$\nu \lambda = \text{velocity}$$

$$n = \frac{c}{v} = \frac{\lambda_{\text{vac}}}{\lambda_{\text{subs}}}$$

where λ_{vac} is the wavelength of the light wave in a vacuum, and λ_{subs} is the wavelength in the transparent substance. Light has a smaller wavelength in a transparent material than in vacuum (Figure 14.13).

Table 14.3   Some refractive indices

Substance	Refractive index[a]	Substance	Refractive index[a]
Vacuum	1.0[b]	Dry air, 1 atm, 15 °C	1.00027
Water	1.3324	Na_3AlF_6 (cryolite)	1.338[c]
MgF_2	1.382[c]	Fused silica (SiO_2)	1.4601
KCl (sylvite)	1.490	Crown glass	1.522
Extra-light flint glass[d]	1.543	NaCl (halite)	1.544
Flint glass[d]	1.607	MgO (periclase)	1.735
Dense flint glass[d]	1.746	Al_2O_3 (corundum)	1.765[c]
ZrO_2 (zirconia)	2.160[c]	C (diamond)	2.418
$CaTiO_3$ (perovskite)	2.740	TiO_2 (rutile)	2.755[c]

[a] A value appropriate to the yellow light emitted by sodium atoms, the sodium D-lines, with an average wavelength 589.3 nm, is given.
[b] Definition.
[c] The refractive index varies with direction; the average value is given.
[d] The flint glasses contain significant amounts of lead oxide, PbO, as follows: extra-light flint, 24 mass% PbO; flint, 44 mass% PbO; dense flint, 62 mass % PbO.

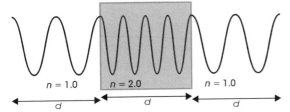

Figure 14.13 The effect of refractive index n on the wavelength of light. The wavelength is compressed in materials with a high refractive index

This can introduce confusion when a light ray traverses several different materials. To overcome this it is useful to define the optical path or optical thickness $[d]$, and distinguish it from the real or physical thickness of a material, d. The relationship is given by:

$$[d] = nd$$

where d is the physical thickness (in m) and n is the refractive index. The optical thickness of a material is frequently quoted as a number of wavelengths. Thus, a thin film with an optical thickness of λ has a real thickness given by the wavelength (in m) of the light involved divided by the refractive index. For several transparent materials traversed in sequence,

$$[d] = n_1 d_1 + n_2 d_2 + n_3 d_3 + \cdots$$

When light passes from a higher refractive index material such as glass to one of lower refractive index such as air, the refraction causes the emerging ray to bend towards the interface. As the angle, θ, at which the ray approaches the surface increases the angle of the emerging ray becomes closer to the surface, until, at the critical angle, θ_c the emerging ray actually travels exactly along the surface (Figure 14.14). If θ_c is exceeded then *no light escapes* and the light behaves as if it were reflected from the under surface. This effect is called total internal reflection. The critical angle follows from the general relation given above when θ_i is equal to 90°:

$$\sin \theta_c = \frac{n(\text{low})}{n(\text{high})}$$

where $n(\text{low}) < n(\text{high})$.

14.4.2 Refractive index and structure

The relationship between refractive index and the atomic or molecular structure of a material was considered in Sections 11.1.4 and 11.1.5. Recall that light can be treated as a varying electric field and this interacts with the internal charges on the solid. At the frequency of a light wave, only the electrons respond, and so contribute to the refractive

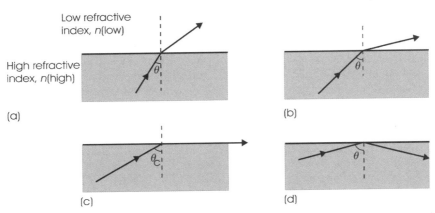

Figure 14.14 Total internal reflection of a beam of light in a solid with a higher refractive index than the surrounding medium. As the angle of the beam, θ, approaches the critical angle, θ_c, (parts a and b), the emerging beam approaches the external surface ever more closely. When the critical angle, θ_c, is reached the emerging beam lies along the surface (part c). If the critical angle is exceeded (part d) no light escapes, and it suffers total internal reflection

index. The refractive index of a material, n, is a reflection of the electronic polarisability of the solid and is related to the relative permittivity by the equation:

$$n \approx \sqrt{\varepsilon_r}$$

where ε_r is the relative permittivity of the material.

In general, strongly bound electrons, trapped at atomic nuclei or in strong chemical bonds, have a low polarisability, and this leads to a low refractive index. Loosely bound electrons, outer electrons on large atoms, or lone-pair electrons, are highly polarisable and so will yield materials with a larger refractive index.

The refractive index of a solid can be estimated via the Gladstone–Dale formula. It is especially useful for complex oxides, for which the Gladstone–Dale formula can be written:

$$n = 1 + \rho(p_1 k_1 + p_2 k_2 + p_3 k_3 \cdots)$$

or

$$n = 1 + \rho \sum p_i k_i$$

where ρ is the density of the complex oxide. The factors k_i, called the refractive coefficients, are empirically determined constants; some examples are given in Table 14.4. The amount of each oxide is taken into account by multiplying the refractive coefficient by its weight fraction in the compound, p.

The assumption underlying the formula is that the refractive index of a complex oxide is made up by adding together the contributions from a collection of simple oxides, oxide 1, oxide 2 and so on, for which optical data are known. The rule works well and usually gives answers within about 5 %. Note, however, that the value obtained is an average

refractive index. The Gladstone–Dale relationship ignores the fact that many oxides have refractive indices that vary according to crystallographic direction.

14.4.3 The refractive index of metals and semiconductors

The refractive index of transparent materials is mainly a function of electronic polarisation, arising from strongly bound electrons. In metals and many semiconductors there are considerable numbers of free electrons present. The refractive index of these materials is written as a complex number:

$$N = n + ik$$

where N is the complex refractive index of the solid, i is the square root of -1, n is the real part of the refractive index, and k is called the absorption index, absorption coefficient or attenuation coefficient. At optical frequencies, the complex refractive index of a metal manifests itself in the reflectivity (see Section 14.5.1).

14.4.4 Dispersion

The refractive index of a solid varies with wavelength (Figure 14.15). This is called dispersion. In general, the index of refraction of transparent materials increases as the wavelength decreases so that the refractive index of red light in a material is less than that of violet light. The dispersion can be formally defined as $dn/d\lambda$, which is the slope of the curve of refractive index, n, against wavelength, λ. Although the dispersion of many materials is

Table 14.4   Refractive coefficients, k, for some oxides

Oxide	k	Oxide	k	Oxide	k	Oxide	k	Oxide	k	Oxide	k	Oxide	k	Oxide	k
H_2O	0.34														
Li_2O	0.31	BeO	0.24												
Na_2O	0.18	MgO	0.20							B_2O_3	0.22	CO_2	0.22	N_2O_5	0.24
K_2O	0.19	CaO	0.23							Al_2O_3	0.20	SiO_2	0.21	P_2O_5	0.19
		SrO	0.14	Y_2O_3	0.14	TiO_2	0.40								
		BaO	0.13	La_2O_3	0.15	ZrO_2	0.20	Nb_2O_5	0.30			SnO_2	0.15		
												PbO	0.15	Bi_2O_3	0.16

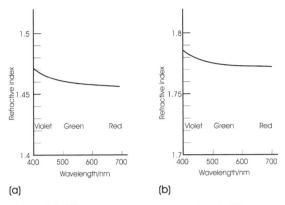

Figure 14.15 The dispersion of (a) fused silica glass and (b) corundum

rather small, it is important to include it when calculating the optical properties of optical components.

14.5 Reflection

14.5.1 Reflection from a surface

Light is reflected from smooth surfaces; following the law of reflection, the angle of incidence, θ_i, is equal to the angle of reflection, θ_r (Figure 14.16). The amount of light reflected from a surface at normal incidence (i.e. perpendicular to the surface) is given by the coefficient of reflection, r:

$$r = \frac{n_0 - n_1}{n_0 + n_1}$$

Figure 14.16 The reflection of a beam of light: (a) the angle of incidence, θ_i is equal to the angel of reflection, θ_r; (b) the coefficient of reflection at normal incidence, r, relates the amplitude reflected to the incident amplitude, a_0, whereas the reflectivity, R, relates the intensity reflected to the incident intensity, I_0

where n_0 is the refractive index of the entrance medium and n_1 is the refractive index of the material making up the reflecting surface (Figure 14.16). The coefficient of reflection is defined such that if a wave of amplitude a_0 falls on the surface, then the amplitude of the reflected wave is ra_0. For reflection at a surface between a substance of low refractive index and a substance of high refractive index, r is *negative*. This signifies a phase change of π radians on reflection, which means, in terms of a light wave, that a peak turns into a trough on reflection (Figure 14.17).

The eye detects intensity changes rather than amplitude changes, and so it is more convenient to work with the reflectivity or reflectance, R:

$$R = r^2 = \left(\frac{n_0 - n_1}{n_0 + n_1}\right)^2$$

This is because the intensity, I_0, is proportional to the square of the amplitude, $(a_0)^2$. The reflected intensity, $R(I_0)$, is proportional to $r^2(a_0)^2$. The reflectivity, R, for a plate of a transparent material of refractive index n in air is:

$$R = \left(\frac{1-n}{1+n}\right)^2$$
$$= \left(\frac{n-1}{n+1}\right)^2 \tag{14.2}$$

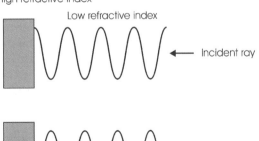

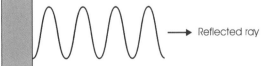

Figure 14.17 A light beam incident on a plate with a higher refractive index than the surrounding medium suffers a phase change on reflection so the incident peak is reflected as a trough

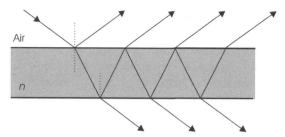

Figure 14.18 Multiple reflections at the upper and lower surfaces of a thin film of refractive index n

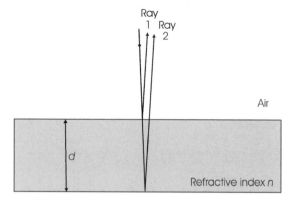

Figure 14.19 Reflection at a thin film in air

As n depends on wavelength, the reflectivity will vary across the spectrum. When the reflecting surface is a metal, it is necessary to use the complex refractive index, $N = N_1 + ik$. In this case, the reflectivity of a metallic surface at normal incidence is

$$R = \frac{(n_1 - n_0)^2 + k^2}{(n_1 + n_0)^2 + k^2}$$

For the case of a metallic film in air this becomes:

$$R = \frac{(n_1 - 1)^2 + k^2}{(n_1 + 1)^2 + k^2}$$

14.5.2 Reflection from a single thin film

Monochromatic light travelling through air, falling on a homogeneous thin film of refractive index n, will be reflected from the top surface to give a reflected ray. The light transmitted into the film will be repeatedly reflected from the bottom surface and the underside of the top surface (Figure 14.18). At each reflection, some of the light will escape to produce additional reflected and transmitted rays. As the reflectivity is rather small, the first reflected ray and the first transmitted ray are of most importance. Because of the difference in the paths taken by the repeatedly reflected rays, the waves will interfere with each other.

The appearance of the film will depend on the extent of this interference. In the case of an observer looking down on a film in a perpendicular direction, some of the light incident on the surface and seen by

the observer will have been reflected at the top surface (ray 1, Figure 14.19). In addition, some light travels through the film and is reflected from the bottom surface before reaching the observer (ray 2, Figure 14.19). In addition, because ray 1 is reflected at a surface of higher refractive index, a wave peak will turn into a trough. Interference between the two waves will occur, which will cause the film to look either dark or bright. The optical path difference between rays 1 and 2 will be [p]:

$$[p] = 2nd$$

where d is the physical thickness, and n is the refractive index of the film. If the path difference, [p], is equal to an integral number of wavelengths the film will appear dark as a result of destructive interference:

$$[p] = m\lambda, \quad m = 1, 2, 3, \ldots, \quad \text{minimum (dark)}$$

If the path difference, [p], is equal to a half-integral number of wavelengths the film will then appear bright, because constructive interference will occur:

$$[p] = \left(m + \tfrac{1}{2}\right)\lambda$$
$$m = 1, 2, 3, \ldots, \quad \text{maximum (bright)}$$

At other path differences, the film will appear to have an intermediate tone, depending on the exact phase difference between the rays.

Should the light beam fall on the surface at an angle of incidence θ, producing an angle of refraction θ', the path difference, $[p]$, between rays 1 and 2 is:

$$[p] = 2nd \cos \theta'$$

Thus

$$[p] = 2nd \cos \theta'$$
$$= \begin{cases} m\lambda \text{ gives a minimum (dark)}; \\ \left(m + \frac{1}{2}\right)\lambda \text{ gives a maximum (bright)}. \end{cases}$$

14.5.3   The reflectivity of a single thin film in air

The reflectivity of a thin film in air will be different from that for a thick plate, given in Section 14.5.1, as interference effects from the bottom surface also need to be considered.

For light at normal incidence on a transparent solid, the reflectivity is given by:

$$R = \frac{2r_1^2 - 2r_1^2 \cos 2\delta}{1 - 2r_1^2 \cos 2\delta + r_1^4}$$

where

$$r_1 = \frac{n_0 - n_f}{n_0 + n_f}$$

where n_0 is the refractive index of the surrounding medium, usually air ($n_0 = 1.0$) n_f is the refractive index of the film; and

$$\delta = \frac{2\pi[d]}{\lambda}$$

where $[d]$ is the optical thickness of the film, given by

$$[d] = n_f d$$

with d being the physical thickness of the film.

The reflectivity is found to vary in a cyclic fashion, with zero for values of $[d]$ equal to 0, $\lambda/2$, λ, etc. and a maximum for values of $[d]$ given by $\lambda/4$, $3\lambda/4$, etc. (Figure 14.20). Because

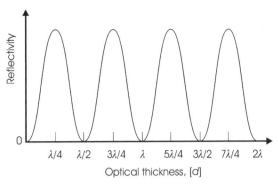

Figure 14.20   The reflectivity of a thin film varies sinusoidally with the film thickness, being zero when the optical thickness is a multiple of a half wavelength, and maximum when the thickness is an odd multiple of a quarter wavelength

the value of n depends on wavelength, the reflectivity will also vary across the spectrum.

14.5.4   The colour of a single thin film in air

When a thin transparent film is viewed in white light, the same reflection and interference discussed in Section 14.5.2 will occur, except that we have to take into account the effects of all of the different wavelengths present. In order to determine the reflected colour of a thin film when viewed in white light it is necessary to add the contributions of all of the wavelengths. The intensity pattern and perceived colour generated by adding the contributions as a function of the optical path difference of the film, $[p]$, is given in Section S4.11. The sequence of colours seen will repeat in a cyclical fashion as the film thickness increases or decreases, as certain colours are either reinforced or cancelled. Each sequence of spectral colours is called an order, which starts with the first order for the thinnest of films. A new order begins every 550 nm of retardation.

Since the fraction of incident white light that is reflected is coloured, it follows that the transmitted light will be depleted in this colour. The transmitted colour seen will therefore be the complementary colour to that strongly reflected (Section S4.11).

If the angle of viewing is not perpendicular to the film, the optical path difference is given by:

$$[p] = 2[d] \cos \theta'$$

where θ' is the angle of refraction. This formula indicates that as the viewing angle moves away from perpendicular to the film the colour observed will move towards that appropriate to lower optical path length. For example, second-order orange-red will change towards green and blue.

14.5.5 The colour of a single thin film on a substrate

The behaviour of a single thin film on a substrate is similar to that discussed for the case of a single thin film in air. Now, however, it is necessary to take into account any change of phase that might occur on reflection at the back surface of the film. If the substrate has a lower refractive index than the film on the surface then the treatment will be identical to that for a thin film in air. In this case, the reflected colours observed when the film is viewed at normal incidence in white light will be the same as those listed in the 'colour reflected' column of Table S4.1, Section S4.11.

If the refractive index of the substrate is greater than that of the film then a phase change will be introduced both at the air–film interface and at the film–substrate interface. In this case, the reflected colour seen at normal incidence when viewed in white light will be the complementary colour to that just described. These are listed in Section S4.11 in the column labelled 'colour transmitted'.

14.5.6 Low-reflectivity (antireflection) and high-reflectivity coatings

The reflectivity, R, of a homogeneous nonabsorbing thin film on a substrate, illuminated by light of one wavelength perpendicular to the surface, is given by:

$$R = \frac{r_1^2 + 2r_1r_2 \cos 2\delta + r_2^2}{1 + 2r_1r_2 \cos 2\delta + r_1^2 r_2^2}$$

where

$$r_1 = \frac{n_0 - n_f}{n_0 + n_f}$$
$$r_2 = \frac{n_f - n_s}{n_f + n_s}$$

and n_0 is the refractive index of the surrounding medium, n_f is the refractive index of the film and n_s is the refractive index of the substrate;

$$\delta = \frac{2\pi[d]}{\lambda}$$

where $[d]$ is the optical thickness of the film, given by

$$[d] = n_f d$$

where d is the physical thickness of the film.

For values of $[d]$ given by $[d] = \frac{\lambda}{2}, \lambda, \frac{3\lambda}{2}$, etc.,

$$R = \frac{(n_0 - n_s)^2}{(n_0 + n_s)^2}$$

This equation is interesting, because if n_0 is set as 1.0, for air, and n_s is set as the refractive index of the material, n, it is identical to the equation for as uncoated surface, Equation (14.2). Thus a layer of optical thickness $\lambda/2$, etc. can be considered to be optically absent and the surface has normal uncoated reflectivity.

When the optical thickness of the film is $[d] = \lambda/4, 3\lambda/4$, etc.,

$$R = \left(\frac{n_f^2 - n_0 n_s}{n_f^2 + n_0 n_s} \right)^2$$

The reflectance will be either a maximum or a minimum. When the film has a higher refractive index than the substrate the reflectivity will be a maximum, $n_0 < n_f > n_s$. When the film has a lower refractive index than the substrate the reflectivity will be a minimum, $n_0 < n_f < n_s$.

To make a nonreflective coating [antireflection (AR) coating] on a glass surface in air the value of n_f must lie between that of air and the glass. The reflectivity will be a minimum for a $\lambda/4$ film. Putting R equal to zero yields a value of the refractive index of a film that will give no reflection at all:

$$n_f = \sqrt{n_s}$$

For glass, n_S is about 1.5, so the antireflecting film must have a refractive index:

$$n_f = \sqrt{1.5} = 1.225$$

Very few solids have such a low index of refraction, and a compromise material often used is magnesium fluoride, MgF_2, for which n in the middle of the visible is 1.384.

A similar strategy can be used to optimise the reflectivity of a surface by coating it with a high-reflection coating. The object is to make the reflectivity, R, as close to 1 as possible. A film of thickness $\lambda/4$ will increase the reflectivity in the case when the refractive index of the film, n_f, is greater than the refractive index of the substrate, n_S. Two materials frequently used are SiO ($n = 2.0$) and TiO_2 ($n = 2.90$). A TiO_2 film of thickness $\lambda/4$ on glass will have a reflectivity of about 0.48 (48 %). As R for a single glass surface in air is about 0.04 (4 %), almost 48 % represents a great improvement. The effect is used in costume jewellery. Rhinestones are made of glass coated with an approximately $\lambda/4$ thickness film of TiO_2.

14.5.7 Multiple thin films and dielectric mirrors

Multiple thin films of transparent materials can be laid down one on top of the other in such a way as to form perfect mirrors. These are often called dielectric mirrors. The simplest formulae for the reflectance of such a mirror refers to the specific case in which all layers are $\lambda/4$ thick and of alternating high

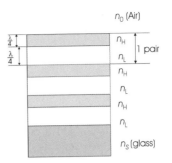

Figure 14.21 A quarter-wave stack of alternating-layers of high and low refractive index solids, n_H, and n_L, each of optical thickness one quarter of a wavelength, $\lambda/4$

(H) and low (L) refractive indices, n_H and n_L, illuminated by light falling *perpendicular* to the surface (Figure 14.21). The arrangement is called a quarter-wave stack. For a quarter-wave stack deposited on a substrate in the sequence:

substrate; L; H; L; H; L; H; ..., L; H; air

maximum reflectance is given by the formula:

$$R = \left(\frac{n_s f - n_0}{n_s f + n_0}\right)^2$$

where f is equal to $(n_H/n_L)^{2N}$, n_0 is the refractive index of the surrounding medium, usually air ($n_0 = 1.0$), n_s is the refractive index of the substrate, usually glass ($n_s \approx 1.5$), and N is the number of LH pairs of layers in the stack. For a stack in air:

$$R = \left[\frac{n_s - \left(\frac{n_L}{n_H}\right)^{2N}}{n_s + \left(\frac{n_L}{n_H}\right)^{2N}}\right]^2$$

The general approach used to make a dielectric mirror is to lay down a stack of thin films that have alternately higher and lower refractive indices. Manipulation of the thickness and the refractive index of each layer in the stack allows the optical properties to be modified at will to produce virtually perfect mirrors and virtually perfect antireflection coatings – both of which can be tuned to respond to very specific wavelengths – as well as a variety of optical filters. The fabrication of such devices falls into the area of photonic or thin-film engineering. Filters utilising multilayers, sometimes referred to as interference filters, fall into three different categories (Figure 14.22). Shortpass filters transmit visible wavelengths and cut out infrared radiation (Figure 14.22a). They are often used in surveillance cameras to eliminate heat radiation. Longpass filters block ultraviolet radiation and transmit the visible (Figure 14.22b). Bandpass filters pass only a limited section (or band) of the electromagnetic spectrum, (Figure 14.22c).

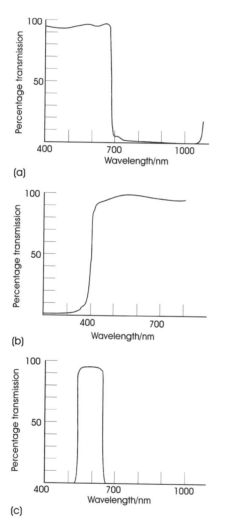

(a)

(b)

(c)

Figure 14.22   The transmission characteristics of interference filters: (a) a shortpass filter, (b) a longpass filter, and (c) a bandpass filter

The reflectivity of a stack of transparent thin films that are not carefully engineered in terms of thickness or refractive index will give a reflection that appears to be metallic silver. This can be seen in a less-than-perfect fashion with a stack of microscope slides or a roll of thin transparent plastic film, such as 'cling film'.

14.6   Scattering

14.6.1   Rayleigh scattering

If a transparent medium contains scattering centres, the intensity of light traversing the medium in the incident direction will gradually fall as the light is scattered into other directions. The reduction in the intensity of such a beam can be written as:

$$I = I_0 \exp\left(-\alpha_s l\right)$$

where I_0 is the incident beam intensity, I is the intensity after travelling a distance l in the turbid medium, and α_s is an experimentally determined linear scattering coefficient. The form of this equation is identical to that of Lambert's law for absorption.

Rayleigh scattering applies to spherical insulating particles with a diameter less than about a tenth of the wavelength of the incident light. When a beam of unpolarised light of intensity I_0 is scattered *once only* by the scattering centre, the intensity of the scattered light, I_s at a distance r from the scattering centre is:

$$I_s = I_0 \left(\frac{9\pi^2 V^2}{2r^2\lambda^4}\right) \left(\frac{m^2 - 1}{m^2 + 2}\right)^2 (1 + \cos^2\theta)$$

where V is the volume of the scattering particle, λ is the wavelength of the light, θ is the angle between the incident beam and the direction of the scattered beam, and m is the relative refractive index of the particle:

$$m = \frac{n_{particle}}{n_{medium}}$$

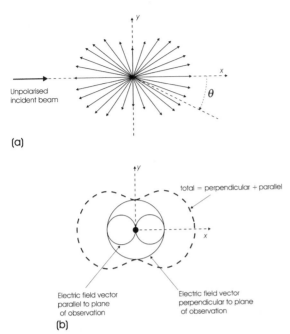

(a)

(b)

Figure 14.23   (a) The Rayleigh scattering pattern from a small spherical particle; (b) the scattering pattern is the sum of radiation scattered with its electric field vector parallel and perpendicular to the plane of observation

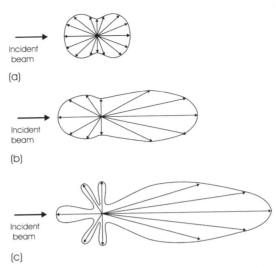

(a)

(b)

(c)

Figure 14.24   (a) The Rayleigh scattering pattern from a small particle. As the particle size increases, the scattering pattern becomes asymmetrical parts (b) and develops side lobes (part c). The scattering in parts (b) and (c), by particles larger than that in (part a), is called Mie scattering

In this case, $n_{particle}$ is the refractive index of the particle, and n_{medium} is the refractive index of the surrounding medium. For air, n_{medium} is 1.0.

As much light is scattered backwards as forwards, and only half as much intensity is scattered normal to the beam direction (Figure 14.23). All wavelengths scatter in this pattern, but the shorter wavelengths are more strongly scattered than are the longer wavelengths. Because the scattering of light is proportional to $1/\lambda^4$, violet light is scattered far more than is red light. The blue appearance of the sky on a sunny day is the result of the preferential scattering of violet light combined with the maximum sensitivity of the eye, which lies in the green-yellow region. Similarly, the red skies visible at dawn and dusk, and the rare blue moon, are caused by scattering of light from small particles in the upper atmosphere.

14.6.2   Mie scattering

The term Mie scattering is generally reserved for scattering by particles that are somewhat larger than those for which Rayleigh scattering is valid, about a third the wavelength of light or more. As the particle size increases from that appropriate to Rayleigh scattering, forward-scattering begins to dominate over backward scattering (Figure 14.24). As particle size passes the wavelength of light, the forward-scattering lobes increase further, and side bands develop, due to maxima and minima of scattering at definite angles. The position of these lobes depends on the wavelength of the scattered light and so they are strongly coloured. These coloured bands, referred to as higher-order Tyndall spectra, are dependent on the particle size. Mie scattering is responsible for the colours produced in ruby glass, which contains a dispersion of gold particles as the scattering centres.

With even larger particles, white light becomes reflected (rather than scattered, as we are discussing here) evenly in all directions. This situation holds in fogs and mists.

14.7   Diffraction

Diffraction effects occur when waves interact with objects having a size similar to the wavelength of the radiation. In general, two regimes have been explored in most detail: (1) diffraction quite close to the object which interacts with the light, called Fresnel diffraction; and (2) the effects of diffraction far from the object which interacts with the light, called Fraunhofer diffraction. The result of diffraction is a set of bright and dark fringes, due to constructive and destructive interference, called a diffraction pattern.

14.7.1   Diffraction by an aperture

If a long narrow slit is illuminated by monochromatic light the intensity pattern observed far from the slit (the Fraunhofer diffraction pattern) is given by the expression:

$$I_x = I_0 \left[\frac{\sin x}{x} \right]^2$$

where

$$x = \frac{\pi w \sin \theta}{\lambda}$$

and w is the width of the slit, θ is the angular deviation from the 'straight-through' position, and λ is the wavelength of the light. This produces a set of bright and dark fringes with minima given by:

$$\sin \theta_{min} = \frac{m\lambda}{w}$$

where m takes values 1, 2, 3, etc. For θ_{min} to be appreciable, w must be close to λ. In fact, the formula shows that the spacing between the minima will be proportional to the reciprocal of the slit width, so that the narrower the opening the wider

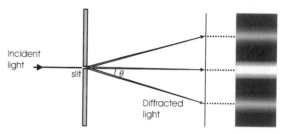

Figure 14.25 The diffraction pattern from a thin slit shows regions of higher and lower intensity

the fringe spacing (Figure 14.25). The positions of the maxima between these dark bands are not given by a simple formula, but are approximately midway between the minima.

The sine of the angle through which a ray is diffracted is related to its wavelength. This indicates that each wavelength in white light will be diffracted through a slightly different angle and that red light will be diffracted through a greater angle than will violet light. In this way, white light will produce a set of diffraction patterns, each belonging to a different wavelength. These patterns look like, and are called, spectra. They are referred to as first-order, second-order and so on as they are recorded further and further from the undeviated beam.

When the slit is shortened to form a rectangular aperture the diffraction maxima will take the form of small rectangular spots running in two perpendicular directions. White light will produce coloured spots via the same mechanism as described above.

The form of the diffraction pattern produced by a circular aperture consists of a series of bright and dark circles concentric with the original aperture. The spacing of the maxima and minima is given by:

$$\sin \theta = \frac{n\lambda}{d}$$

where θ is the angle between the directly transmitted ray and the diffraction ring, λ is the wavelength of the light, and d is the diameter of the aperture. The computation of n requires rather sophisticated mathematics, the results of which show that n takes the values 0 (central bright spot), 1.220 (first dark ring), 1.635 (first bright ring), 2.333

(second dark ring), 2.679 (second bright ring) and 3.238 (third dark ring).

Just as with the slit, the dependence of the diffraction angle on wavelength means that a circular aperture illuminated with white light will produce a set of coloured rings, rather like miniature circular rainbows. The formula indicates that each ring will have a violet inner edge and a red outer edge.

Diffraction by circular apertures plays an important part in the overall performance of many optical instruments such as telescopes and microscopes. The resolution of such instruments, which is, roughly speaking, equivalent to the separation of two points which can just be distinguished as separate objects, is controlled by diffraction. It is of the order of the wavelength of the observing radiation. Because of this limitation, optical microscopes are unable to image atoms. Electron microscopes, using radiation with a wavelength of the order of 0.002 nm, are able to do so.

14.7.2 Diffraction gratings

Planar diffraction gratings consist of an object inscribed with a set of parallel lines with spacing similar to that of the wavelength of light. A transmission grating has alternating clear and opaque lines, and diffraction effects are observed in light transmitted by the clear strips (Figure 14.26a). A reflection grating consists of a set of grooves or blazes, and diffraction effects are observed in the light reflected from the patterned surface (Figure 14.26b). The effectiveness of a grating is the same whether light is transmitted through it or reflected from it.

The positions of the diffraction maxima from a transmission grating illuminated by monochromatic light normal to the surface is given by:

$$\sin \theta = \frac{n\lambda}{d}$$

where d is the repeat spacing of the grating, λ is the wavelength of the radiation, and θ is the angle through which the beam has been diffracted (Figure 14.26). The positions of the maxima for light at

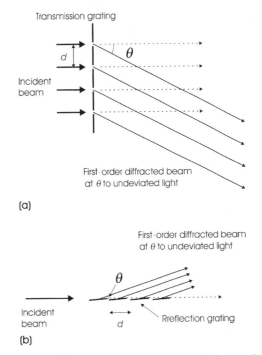

Figure 14.26   (a) Diffraction by a transmission grating and (b) diffraction by a reflection grating

grazing incidence to a reflection grating are given by:

$$1 - \cos \theta = \frac{n\lambda}{d}$$

where d is the repeat spacing of the grating, λ is the wavelength of the radiation, and θ is the angle through which the beam in question has been diffracted.

The term n in both formulae can take integer values of 0, ± 1, ± 2 and so on. Each of these corresponds to a different diffraction maximum, called an order. When illuminated by white light, each wavelength will be diffracted through a slightly different angle so that each order will consist of a spectrum, similar to those produced by a long narrow slit but, because each line on the grating acts as a contributing slit, they are of much greater intensity.

14.7.3 Diffraction from crystal-like structures

Crystal structures can be determined by X-ray diffraction (Section 5.2.2). The position of the strongly diffracted beams is given by Bragg's law:

$$n\lambda = 2d \sin\theta$$

where d is the separation of the planes of atoms that are responsible for the diffraction, λ is the X-ray wavelength, and θ is the angle between the X-ray beam and the atom planes (Section 5.2.2). The theory holds for any three-dimensional array no matter the size of the 'atoms'. Thus, any arrangement of particles, or even voids, which are spaced by distances similar to the wavelength of light, will diffract light according to Bragg's law. When white light is used, each wavelength will diffract at a slightly different angle, and colours will be produced.

The colour of precious opal is due to the diffraction of white light. The regions producing the colours are made up of an ordered packing of spheres of silica (SiO_2), which are embedded in amorphous silica or a matrix of disordered spheres (Figure 14.27). These small volumes of ordered spheres resemble small crystallites. They interact with light because the spacing of the ordered spheres is similar to that of the wavelength of light.

The conditions under which diffraction takes place are the same as those specified by the Bragg equation. However, because the diffraction takes place within a silica matrix, it is necessary to use the optical path instead of the vacuum path. The layer spacing, d, must be replaced by the optical thickness, $[d]$, equal to $n_s d$, where n_s is the refractive index of the silica in opal, about 1.45. The correct equation to use for opal is thus:

$$n\lambda = 2n_s d \sin\theta$$
$$\approx 2.9 d \sin\theta$$

with

$$\lambda_{max} = 2n_s d$$
$$\approx 2.9d$$

The relationship between the radius of the spheres, r, and the distance between the layers, d, will depend on the exact geometry of the packing. If each layer of spheres is arranged in hexagonal closest packing, the relationship between the sphere radius and the layer spacing is:

$$d = \frac{2\sqrt{2}r}{\sqrt{3}} = 1.633\,r$$

A useful general relationship is that the radius of the spheres is given, to a reasonable approximation, by one fifth of the wavelength of the colour observed at normal incidence.

14.7.4 Photonic crystals

Photonic crystals are artificial structures that diffract light in specified ways. The dimensions of the diffracting centres in the 'crystals' are approximately the same as the wavelength of light, and the diffraction can generally be understood in terms of the Bragg equation. For example, artificial opals are photonic crystals, and Section 14.7.3 has explained how these diffract light of certain wavelengths. However, the terminology employed to describe diffraction in photonic crystals is that of semiconductor physics. The transition from a diffraction description to a physical description can be illustrated with respect to a one-dimensional photonic crystal.

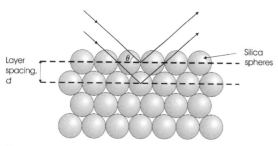

Figure 14.27 The diffraction of light by ordered arrays of silica spheres gives colour to precious opals. The geometry of the diffraction is identical to that of the diffraction of X-rays

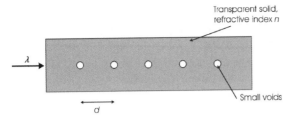

Figure 14.28 A one-dimensional photonic crystal can be thought of as a regularly spaced linear array of diffracting particles, such as small voids in a transparent solid

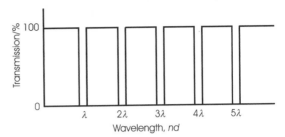

Figure 14.29 The idealised transmission profile of a one-dimensional photonic crystal. Wavelengths that are a multiple of the optical thickness between the scattering centres (voids) are completely diffracted back on themselves and not transmitted. Other wavelengths pass unhindered

A one-dimensional photonic crystal is simply a stack of transparent layers of differing refractive indices. They are also called Bragg stacks or, when built into an optical fibre, fibre Bragg gratings. The simplest model is that of a transparent material containing 'atoms' consisting of regularly spaced air voids (Figure 14.28). When a beam of light is incident on such a grating, a wavelength, λ, will be diffracted when the Bragg law is obeyed:

$$\lambda = 2\,[d]\,\sin\theta$$

where n is the refractive index of the material, and $[d]$ is the repeat spacing. (As for opal, it is necessary to use the optical thickness, $[d] = nd$, and not the physical thickness, d, for the repeat spacing.) For a beam normal to the 'atoms', $\sin\theta = 1$, hence

$$\lambda = 2\,n[d]$$

and the light will be diffracted back on itself and not be transmitted (Figure 14.29).

This same idea was used in the description of Brillouin zones in crystals, in Section 2.3.9. The result of which was the creation of an energy band gap for the electrons. In terms of semiconductor physics, the array of voids has opened a photonic band gap (PBG) in the material. A photonic band gap blocks transmission of the light wave with an energy equal to the band gap. In real materials, the 'atoms' have thickness, and a small range of wavelengths is blocked rather than just one. The result is similar to that described for the multilayer interference filters, and the range of wavelengths blocked increases as the difference between the refractive indices of the atoms and the surrounding medium increases.

Two-dimensional PBG crystals can be thought of as a two-dimensional array of 'atoms' in a transparent medium, and opal is an example of a three-dimensional PBG crystal. The light reflected by an opal gives a measure of the photonic band gap of the gemstone. Many insects also use 'photonic crystal' structures for the production of vivid colours, and a study of insect colours, especially of iridescent butterfly wings, has led to advances in understanding structures similar to those in PBG materials.

14.8 Fibre optics

14.8.1 Optical communications

The transmission of light along thin fibres of glass, plastic or other transparent materials is referred to as fibre optics. Data are carried by a series of pulses of light encoded so that information can be stored and retrieved. In this brief survey, the properties of the materials used in the fibre will be outlined. The engineering of an optical communications system can be explored by reference to some of the sources listed in the Further Reading section.

The transparent optical wave carrier used for communications is silica (SiO_2) glass. The light pulses launched into the fibre are constrained to stay within the fibre by total internal reflection.

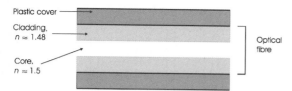

Figure 14.30 The structure of an optical fibre. The fibre is made of glass with a higher refractive index – the core – and glass of a slightly lower refractive index – the cladding. The fibre is covered in a plastic coating to protect the glass from damage

Thus, the core of the fibre, along which light travels, must possess a higher refractive index than the outer surface of the fibre. Moreover, a glass surface at which the total internal reflection is to occur is easily damaged, and needs protection. Both of these objectives are met by providing a surface cladding of lower refractive index glass compared with the core of the fibre. The core and the cladding make up a single glass fibre (Figure 14.30). The cladding should not be confused with a plastic protective covering, which has no optical role to play.

14.8.2 Attenuation in glass fibres

Attenuation describes the loss of light intensity as the signal is transmitted along the fibre. This is of major concern, as any degradation of the signal must be minimised.

The unit of loss is the decibel, dB – the base unit, the bel, is almost never used. The loss is defined as:

$$\text{loss (dB)} = -10 \log_{10}\left(\frac{\text{power in}}{\text{power out}}\right)$$

$$= -10 \log_{10}\left[\frac{P(x)}{P(0)}\right]$$

where $P(0)$ is the power input at $x = 0$, $P(x)$ is the power at a remote point, x.

The attenuation is defined as the loss per kilometre, thus

$$\text{attenuation} = \frac{-10}{x} \log_{10}\left[\frac{P(x)}{P(0)}\right]$$

where $P(x)$ is the power at a point x kilometres along the cable. The units of attenuation are

dB km^{-1}. For a material showing an attenuation of 1 dB km^{-1} an input power of 10 W would give an output power of 7.9 W after 1 km. Ordinary window glass has an attenuation of about 10 000 dB km^{-1}. Attenuation, like dispersion, varies with wavelength. The spectral response of a fibre defines the way in which the fibre attenuation changes with the frequency of the radiation being transmitted.

Attenuation is caused by a combination of absorption and scattering within the glass. Extrinsic attenuation is due to poor processing or fabrication techniques, and may be due to artefacts such as bubbles, particles, impurities and variable fibre dimensions. These problems have been eliminated in modern optical fibre manufacture. Intrinsic attenuation is a property of the pure material itself and cannot be removed by processing. It is the ultimate limit on the performance of the fibre and mainly arises from two factors, Rayleigh scattering and lattice vibrations.

Rayleigh scattering arises from small inhomogeneities in the glass, which cause changes in refractive index. This variation is an inevitable feature of the noncrystalline state and cannot be removed by processing. As Rayleigh scattering is proportional to λ^{-4}, where λ is the wavelength of the optical pulse, the effect is more important for short-wavelength radiation. For any particular glass, most of the factors affecting Rayleigh scattering are constant and cannot be easily changed. However, materials with a low refractive index and glass transition temperature tend to exhibit low Rayleigh scattering.

Absorption due to lattice vibrations, referred to as phonon absorption, occurs when the lattice vibrations of the solid match the energy of the radiation. This occurs for infrared wavelengths and converts the signal energy into heat. It is a function of the mass of the atoms in the glass and the strength of the chemical bonds between them and results in a decrease in the transparency of the glass at long wavelengths.

Absorption due to electronic transitions, mostly at high energies and associated with ultraviolet wavelengths, do not figure significantly in present-day applications but may become important if shorter signal wavelengths are to be used in the future. The dependence on wavelength of absorption due to electronic transitions can often be expressed by a formula of the type:

$$\text{Electronic absorption} = B_1 \exp\left(\frac{B_2}{\lambda}\right)$$

where B_1 and B_2 are constants relating to the glass used, and λ is the wavelength of the radiation.

Attenuation in early fibres was mainly due to metallic impurities. The gravest problem was iron, present as Fe^{2+} – the ion that also imparts a greenish tint to window glass. Even a concentration as low as 1 part per million (ppm) of iron can result in an attenuation of 15 dB km^{-1}. The presence of transition metal cations was avoided by the preparation of silica from very high-purity chemicals made available by the semiconductor industry.

The most important impurity in silica fibres today is hydroxyl (−OH). Hydroxyl arises from water or hydrogen incorporation into the glass during fabrication. An impurity level of 1 ppm can give an attenuation of 10^4 dB km^{-1} at a 1.4 μm signal wavelength. Despite careful processing, fibres currently in production still contain significant amounts of this impurity, and hydroxyl remains an important source of attenuation.

By 1979, the best silica fibres showed only intrinsic attenuation and had a loss of about 0.2 dB km^{-1} at 1.5 μm wavelength. This is currently the industry standard.

14.8.3 Dispersion and optical fibre design

A short pulse of light launched into a fibre will tend to spread out, as a result of dispersion. In optical fibres, the dispersion is defined as the delay between the arrival time of the start of a light pulse and its finish time relative to that of the initial pulse. It is measured at half peak amplitude. If the initial pulse has a spread of t_i seconds at 50 % amplitude and the final pulse a spread of t_f seconds at 50 % amplitude after having travelled d kilometres, the dispersion is given by:

$$\text{dispersion} = \frac{t_f - t_i}{d}$$

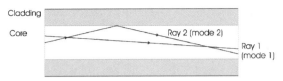

Figure 14.31 The allowed paths that light beams can take in the core region of an optical fibre are called modes; although drawn here as ray paths, in reality they are alternative light wave patterns in the core

The units of dispersion in optical fibres are ns km^{-1}.

Dispersion will result if the light source is not strictly monochromatic. An initially sharp pulse consisting of a group of wavelengths will spread out as it travels down the fibre, because the refractive index depends on wavelength. Thus, different wavelengths will travel at different speeds. This effect is known as wavelength dispersion.

Even with completely monochromatic light, pulse spreading can still occur, because the radiation can take various paths, or modes, through the fibre, as sketched in Figure 14.31. It is apparent that a ray that travels along the axis of a fibre will travel less than one that is continually reflected on its journey. [In fact, the dispersion that results cannot be properly understood in terms of the transmission of light rays, and the various modes are better described in terms of the allowed wave patterns that can travel down the fibre.] The resultant pulse broadening, due to the various modes present, is called modal (or intermodal) dispersion. In order to overcome modal dispersion a number of different fibre types have evolved.

The earliest fibres were called stepped index multimode fibres. These fibres have a large core region, allowing many modes to propagate (Figure 14.32a). The ray labelled H in Figure 14.32(a) is known as a high-order mode, whereas the ray L is a low-order mode. Stepped index multimode fibres are easy to make and join but have a lower performance compared with those described below.

The first advance on stepped index fibres was the graded index fibre. In this design, the refractive index of the fibre varies smoothly, from high at the centre to low at the periphery of the core region (Figure 14.32b). The refractive index gradient

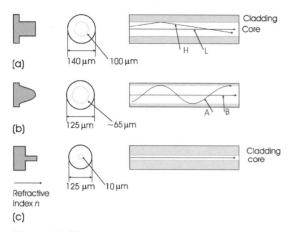

Figure 14.32 Types of optical fibre: (a) stepped index fibre, (b) graded index fibre and (c) monomode fibre

means that light travels faster as it approaches the edge regions of the fibre. The velocity of mode A will vary smoothly from lowest at the fibre centre to greatest near to the fibre edge. The velocity of mode B will be fairly constant and lower on average than mode A. The differences in path length between high-order and low-order modes is thus minimised by this velocity variation.

For best results, monomode fibres (Figure 14.32c) are now used. The number of possible modes is reduced by decreasing the diameter of the core. When the core diameter reaches 10 µm or less only one mode can propagate and, in principle, modal dispersion is zero for these fibres. Monomode fibres have a high performance but are harder to make and join.

The material used for optical communications fibre is highly purified silica (SiO_2) glass. The cladding and core regions are created by doping with carefully chosen impurities. A commonly used production method utilises a tube of pure silica glass. Layers of germanium dioxide, GeO_2, are laid down in the centre of the tube. Germinium dioxide is chemically and physically very similar to silica and readily forms a solid solution with the silica glass. As the germanium atoms are heavier than silicon atoms, they increase the refractive index of the doped inner region relative to the undoped outer region of the tube. When a sufficient amount

of GeO_2 has been laid down, the tube is heated until it collapses into a solid rod called a preform. The preform has a germanium-rich higher refractive index core and a lower refractive index periphery. The preform is drawn into a fibre and, because of the nature of the way in which glass flows at elevated temperatures, the refractive index profile of the preform is maintained in the fibre.

Although fibres in commercial use are made of silica glass, it is not perfect. The dispersion is lowest at 1.3 µm, but the minimum attenuation occurs at 1.5 µm, leading to some sacrifice of performance irrespective of the signal wavelength chosen. The search for new materials to resolve this conflict continues in many research laboratories.

14.8.4 Optical amplification

The amplification of signals in fibre optic transmission systems is of great importance. Amplification uses a section of optical fibre doped with erbium (Er^{3+}) as the activator. The amplifying section consists of about 30 m of monomode fibre core containing just a few hundred parts per million of Er^{3+} (Figure 14.33a). This section of the fibre is illuminated by a semiconductor diode laser at the frequency related to that of the carrier signal. The commonest wavelengths used are 980 nm and 1480 nm. The erbium ions have the remarkable ability to transfer energy from the laser to the signal pulses as they traverse this section of fibre.

The energy transfer comes about in the following way. With reference to the schematic energy-level diagram of the Er^{3+} ion (Figure 14.33b), illumination of the erbium-containing section of fibre with energy of wavelength 980 nm excites the ions from the ground state ($^4I_{15/2}$) to the upper state ($^4I_{11/2}$) from whence they rapidly decay to the $^4I_{13/2}$ level shown by the dashed diagonal line. The use of radiation of 1480 nm wavelength excites the Er^{3+} ions directly from the ground state to the $^4I_{13/2}$ level. This process is referred to as pumping, and the laser involved is described as the pump. The excited state has quite a long lifetime, and so a passing light pulse, with a wavelength close to 1480 nm, can empty it via stimulated emission. This achieves

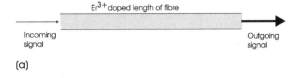

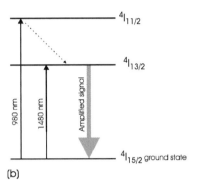

(a)

(b)

Figure 14.33 Signal amplification in an optical fibre using erbium doping: (a) an incoming weak signal is amplified on passage through a length of fibre in which the core has been doped with Er^{3+} ions. (b) The principle energy levels used by the Er^{3+} amplifier. Pump wavelengths of 980 nm and 1480 nm (narrow, upward-pointing arrows) excite the Er^{3+} ions. This energy is transferred to the signal as the Er^{3+} ions lose energy (wide, downward-pointing arrow)

signal amplification while retaining the coherence of the pulse constituting the signal.

14.9   Nonlinear optical materials

14.9.1   Nonlinear optics

The electric field of a light beam induces a polarisation in a solid. For light beams of ordinary intensity the polarisation, P, is a linear function of the electric field, E:

$$P = \varepsilon_0 \chi E$$

where ε_0 is the permittivity of free space, and χ is the dielectric susceptibility of the material, which is proportional to the relative permittivity and refractive index of the substance. This approximation is perfectly adequate for normal optics, but laser beams in particular can be associated with very high electric fields and, in this case, it is necessary to write the polarisation as a series:

$$P = \varepsilon_0 \chi^{(1)} E + \varepsilon_0 \chi^{(2)} E^2 + \varepsilon_0 \chi^{(3)} E^3 + \cdots \quad (14.3)$$

where $\chi^{(1)}$ is the linear dielectric susceptibility, $\chi^{(2)}$ is the second-order dielectric susceptibility, $\chi^{(3)}$ is the third-order dielectric susceptibility and so on. The polarisation is no longer a simple linear function of the electric field.

The extra 'nonlinear' terms are only high enough to be of importance in relatively few materials. In general, the magnitudes of the values of the dielectric susceptibilities decrease rapidly as the order increases, so that the second-order, $\chi^{(2)}$, term is the most important nonlinear coefficient. Moreover, all even-order terms, including the second order, $\chi^{(2)}$, term, are zero in centrosymmetric crystals. The second-order term, $\chi^{(2)}$, has a nonzero value in noncentrosymmetric crystals, and it is these that are generally known as nonlinear optical materials.

The nonlinear terms in the polarisation equation allow photons to be added and subtracted in certain specific ways. For example, if the crystal is irradiated with laser light characterised by angular frequencies ω_1 and ω_2 a collection of frequencies $2\omega_1$, $2\omega_2$, $\omega_1 + \omega_2$ and $\omega_1 - \omega_2$ can all be produced. The production of a frequency $2\omega_1$, the second harmonic, from a single input frequency ω_1, is known as frequency doubling or second harmonic generation (Figure 14.34a). The production of the other frequencies, which can also occur, is known as frequency mixing (Figure 14.34b).

Second harmonic generation comes about in the following way. The sinusoidally varying electric field associated with a light beam can be written:

$$E = E_0 \cos \omega t$$

where ω is the angular frequency of the light. If this is substituted into equation (14.3):

$$P = \varepsilon_0 \chi^{(1)} E_0 \cos \omega t + \varepsilon_0 \chi^{(2)} (E_0 \cos \omega t)^2 + \varepsilon_0 \chi^{(3)} (E_0 \cos \omega t)^3 + \cdots$$

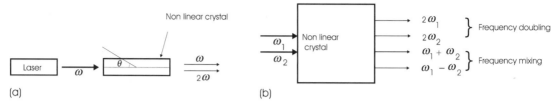

Figure 14.34   (a) Frequency doubling by a nonlinear crystal: an input signal of angular frequency ω, from a laser, is partly converted into a signal with angular frequency 2ω on passing through the crystal. (b) Frequency mixing by a nonlinear crystal: signals of angular frequencies ω_1 and ω_2 are partly mixed to produce signals $(\omega_1 + \omega_2)$ and $(\omega_1 - \omega_2)$ on passing through a nonlinear crystal as well as producing frequency-doubled signals

This series can be written more simply as:

$$P = A + B \cos \omega t + C \cos 2\omega t + D \cos 3\omega t + \cdots$$

where A, B, C and D are constants. At field strengths of the order of those found in laser light the wave that emerges from the crystal can have both the ω and the higher 2ω, 3ω and even greater angular frequencies present if the symmetry is suitable. [Because $\chi^{(2)}$ is zero in centrosymmetric crystals, a second harmonic is not produced, and the generation of a second harmonic from a crystal when illuminated by a laser is usually taken as a good test for the lack of a centre of symmetry.]

If two input waves are used, frequency mixing can occur. Suppose the crystal is irradiated with two beams simultaneously:

$$E_1 = E_{01} \cos \omega_1 t$$
$$E_2 = E_{02} \cos \omega_2 t$$

The electric field in the sample is then:

$$E = E_1 + E_2 = E_{01} \cos \omega_1 t + E_{02} \cos \omega_2 t$$

Substituting this into Equation (14.3) will yield a second-order polarisation, P:

$$P = \varepsilon_0 \chi^{(2)} [(E_{01} \cos \omega_1 t)(E_{02} \cos \omega_2 t)]$$
$$= \varepsilon_0 \chi^{(2)} (E_{01} E_{02} \tfrac{1}{2} [\cos(\omega_1 + \omega_2)t + \cos(\omega_1 - \omega_2)t]$$

The term

$$\tfrac{1}{2} [\cos(\omega_1 + \omega_2)t + \cos(\omega_1 - \omega_2)t]$$

corresponds to the production of two output waves, one of frequency $(\omega_1 + \omega_2)$, the sum wave, and one of frequency $(\omega_1 - \omega_2)$, the difference wave.

14.10   Energy conversion

14.10.1   Photoconductivity and photovoltaic solar cells

If radiation of a suitable wavelength falls on a semiconductor it will excite electrons across the band gap. The most obvious effect of this is that the conductivity of the material increases. The magnitude of the effect is roughly proportional to the light intensity. This effect, called the photoconductive effect, has been used in light meters, exposure meters and automatic shutters in cameras and many other devices. In practice, a dc voltage is applied to the ends of a semiconductor. On illumination the resistance of the semiconductor falls, which provides a means of measurement.

A p–n junction can act in a similar way to a single piece of semiconductor. However, the control afforded by the junction makes the device, called a photodiode, far more flexible and, as a result, photodiodes are used in a number of devices, including solar cells. A solar cell is a large-area p–n junction. It is fabricated so that the depletion region is approximately 500 nm thick, and the electric field across the junction is high (Figure 14.35a). The junction is not connected to any

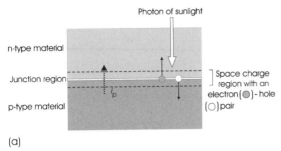

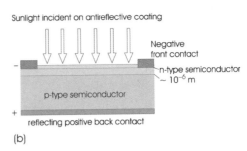

Figure 14.35 (a) Sunlight falling on a p–n junction creates an electron – hole pair. These are swept into the external circuit by the field in the junction region, generating a photocurrent, I_p. (b) A solar cell needs a thin antireflection coating on the front surface, a thin n-type layer and a junction region near to the front surface. A reflecting layer below the cell helps to increase efficiency by reflecting back photons that have not been absorbed in the semiconductor

external power source. Holes and electrons produced in the junction region by sunlight are swept across the depletion region by the high field present, the electrons going from p to n and the holes from n to p. This charges the p region more positive and the n region more negative, and produces a current, I_p, across the junction which generates a photovoltage. Materials that allow a voltage to be produced on illumination are called photovoltaic materials. The photovoltage corresponds to a forward bias, and so will cause a current I to flow. At equilibrium $I = I_p$. Should an external load, R, be connected, some current can flow through it, and so do useful work.

A number of practical considerations influence the design of solar cells (Figure 14.35b). Obviously, the band gap of the semiconductor must be such that as much as possible of the solar spectrum is absorbed. At sea level, the energy available in sunlight amounts to approximately $1000 \, W \, m^{-2}$, and has a wavelength spread of approximately 400 nm to 2500 nm, with a peak in the yellow-green at 550 nm. Indirect band gap materials have a lower efficiency than direct band gap materials, which, from this perspective, are preferred. Materials used include copper indium selenide (CuInSe), cadmium telluride (CdTe), gallium arsenide (GaAs) and amorphous silicon. Because impurities and defects trap mobile electron and holes, which greatly reduces the efficiency of the cell, high-purity materials are mandatory.

Solar cells must have a large area, to collect as much sunlight as possible. Some systems use reflec-

tors or lenses to 'concentrate' the sunlight on the cell surface. Moreover, all incoming photons must be utilised, and it is necessary to coat the front surface with an antireflection layer and to make the lower surface reflecting so that transmitted photons cross the junction again. Similarly, carriers must be produced in the junction region, so that they can separate and not recombine. The junction region must therefore be placed close to the upper surface, and is generally within 1000 nm of the upper surface.

Recently, much effort has been put into the construction of solar cells with use of polymers. These have the great advantages of low weight and flexibility. However, efficiencies are not yet adequate for commercial purposes.

14.10.2 Photoelectrochemical cells

In a conventional solar cell the conversion of the light to charge carriers is carried out by the solid semiconductor, which then has to move these away from the junction in order to obtain energy. The method of conversion of sunlight to energy of most importance on the Earth, photosynthesis, uses slightly different methods of achieving the same objective. The central reactions are oxidation and reduction. Photoelectrochemical cells aim to mimic this process. A number of cells have been devised to utilise redox reactions in a liquid electrolyte for energy conversion.

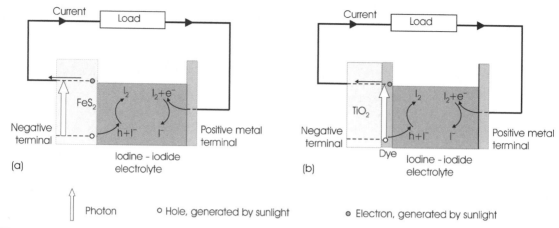

Figure 14.36 Schematic construction of (a) a pyrite–iodine (FeS$_2$–I$_2$) photoelectrochemical cell and (b) a dye sensitised titanium dioxide (TiO$_2$) photoelectrochemical cell. In both cells the generation of electron and hole pairs by the interaction of photons from the sun, with the semiconducting pyrite crystal in part (a) and with the dye coated onto the TiO$_2$ in part (b), leads to power output. The cell is completed by an internal oxidation–reduction cycle involving iodide ions. The thin arrows show the electron and hole paths in the cell

The iron pyrites cell uses FeS$_2$, iron pyrites, as the light absorber, and the interconversion of iodine – iodide as the redox reaction (Figure 14.36a). The pyrites crystal has a broad absorption spectrum and is able to absorb photons from across the visible spectrum to generate a supply of electrons and holes. The electrons are pulled through the crystal by the internal field, and the holes are pushed into the electrolyte. At the pyrite surface the holes (h) react with iodide ions in the following way:

$$2I^-(aq) + 2h \rightarrow I_2(aq)$$

Electrons (e^-) travel the external circuit, doing work, before entering the electrolyte to regenerate iodide ions:

$$I_2(aq) + 2e^- \rightarrow 2I^-(aq)$$

A cell that is rather closer in design to a natural photosynthesis system uses a dye to absorb the radiation. This type of cell uses titanium dioxide as the semiconductor that separates the charges. The dye is used as a more efficient photon absorber than pure titanium dioxide. The cell is shown schematically in Figure 14.36b. Because the charge separation takes place in the dye, the purity and defect structure of the solid is not crucial to satisfactory operation. The holes are pushed into the electrolyte, and the electrons travel the external circuit and power any external devices. The titanium dioxide is prepared in nanocrystalline form in order to achieve a large surface area in a compact cell. The circuit is completed with an electrolyte that can undergo oxidation and reduction. The iodine – iodide electrolyte is often used; the reactions are as above.

14.11   Nanostructures

Nanostructures are structures at a scale that gives properties to the solid that are noticeably different from those of bulk material. In optical terms, this implies dimensions of the order of the wavelength of light. Thus, these effects have been already been discussed with respect to diffraction by photonic crystals and reflection by multiple thin films. In this section the optical consequences of quantum wells, quantum wires and quantum dots, will be outlined.

14.11.1   The optical properties of quantum wells

The physical and electronic structure of quantum wells has been outlined in Sections 3.2.5 and 13.3. In a quantum well, the electrons and holes occupy energy levels that are approximately given by Equation (13.7) in Section 13.3.1, as shown in Figure 14.37. The electrons in the upper energy levels can drop to the lower hole levels and emit a photon. The energy separation of these levels is greater than that of the bulk conduction band – valence band energy gap, E_g, and hence the photons will be of higher energy, or shorter wavelength, than the bulk. The emission is said to be blue-shifted compared with the bulk, and the transitions are called interband transitions.

The photon energy derived from an interband transition is:

$$E(\text{photon}) = E_g + E_{\text{electron}} + E_{\text{hole}}$$

$$h\nu = E_g + \left(\frac{h^2}{8a^2}\right)\left(\frac{n^2}{m_e^*} + \frac{n^2}{m_h^*}\right)$$

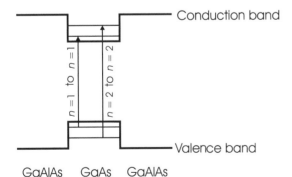

GaAlAs   GaAs   GaAlAs

Figure 14.37 The energy levels in a quantum well. Electron transitions between the sharp energy levels resemble atomic transitions rather than the broad transitions normally seen in solids

where h is the Planck constant, ν is the frequency of the radiation, E_g is the band gap of the bulk well material, a is the dimension of the quantum well, m_e^* is the electron effective mass, and m_h^* is the hole effective mass. In the approximation that the effective mass of the electron and the hole are identical:

$$h\nu = E_g + \frac{n^2 h^2}{4a^2 m^*}$$

The selection rule for the transition is $\Delta n = 0$, that is, transitions can take place only between levels with the same quantum number. (As with all selection rules, these are never perfectly obeyed, and transitions between levels with differing n values do occur infrequently, giving rise to weak lines in the emission spectrum.)

Electrons can also be excited from one electron level, say $n = 1$, to another electron level, say $n = 2$, both levels lying in the electron subband. Holes can make similar transitions between levels in the hole subband. These transitions, which give rise to extra peaks in the emission spectrum, are known as intersubband transitions.

Because the dimensions of the quantum well can be varied, the emission spectrum can be varied or tuned. This feature, in quantum wells and in quantum wires and dots, discussed below, is called band-gap engineering.

14.11.2   Quantum wires and quantum dots

Carbon nanotubes behave as quantum wires, and emission spectra from isolated carbon nanotubes have confirmed a blue shift in the radiation as expected.

The optical properties of quantum dots have been extensively investigated. They have discrete atomic-like energy levels that alter predictably as the size of the particle changes, so that the light emitted can be tuned. This finds application in many areas, including displays and lasers.

Quantum dots can be prepared in a number of ways. Conventional semiconductor techniques can be used to grow small islands of the dot on the surface of a semiconductor crystal. Isolated quantum dots are synonymous with nanoparticles, and small particles precipitated in glasses, colloids or solutions have been obtained. The most studied of these are the compounds cadmium sulphide (CdS), zinc sulphide (ZnS), cadmium selenide (CdSe) and zinc selenide (ZnSe), all of which find applications in display technology. Layers of nanocrystals, laid down by size, will give a rainbow-like emission, as each layer will emit light at a frequency appropriate to the size of the particles in the layer.

Answers to introductory questions

What are lasers?

The word laser is an acronym for the expression light amplification by stimulated emission of radiation. Lasers are devices that emit light by the process of stimulated emission. This produces light that is coherent and often polarised. The key to laser action is to obtain atoms or molecules in an excited state and to keep them there long enough for photons to pass and trigger stimulated emission. This situation is called a population inversion. A population inversion can be achieved in several ways. For solid-state lasers the high energy state that is populated is often protected by having a low transition probability to the ground state. This happens in ruby lasers. In semiconductor lasers, doping and current levels are chosen to ensure that electron numbers in the conduction band outnumber the population of holes in the valence band.

Why are thin films often brightly coloured?

Thin films are often coloured because of interference between light that is reflected from the upper surface of the film and the lower surface of the film. The extent of the interference is dependent on the film thickness. When a film of gradually increasing thickness is illuminated with monochromatic light, alternating strips of dark and light contrast are seen as a result of alternating regions of constructive and destructive interference. When white light is used, each wavelength undergoes constructive and destructive interference at slightly different film thickness. The addition of these effects results in bands of rainbow-like colours, called orders, running parallel to the thickness contours of the film.

What produces the colour in opal?

The colour of precious opal is due to the diffraction of white light by ordered crystallite-like regions in the gemstone. The regions producing the colours are made up of an ordered packing of spheres of silica (SiO_2), which are embedded in amorphous silica or a matrix of disordered spheres. They interact with light because the spacing of the ordered spheres is similar to that of the wavelength of light. The conditions under which diffraction takes place are the same as those specified by the Bragg equation. However, because the diffraction takes place within a silica matrix, it is necessary to use the optical path instead of the vacuum path. A useful general relationship is that the radius of the spheres is given, to a reasonable approximation, by one fifth of the wavelength of the colour observed at normal incidence.

Further reading

The properties of light in general are covered in:

M. Fox, 2001, *Optical Properties of Solids*, Oxford University Press, Oxford.
O.S. Heavens, R.W. Ditchburn, 1993, *Insight into Optics*, John Wiley & Sons, Chichester, Sussex.

The properties of light with respect to colour are found in:

K. Nassau, 2001, *The Physics and Chemistry of Colour*, 2nd edn, John Wiley & Sons, Chichester, Sussex.
R.J.D. Tilley, 2000, *Colour and the Optical Properties of Materials*, John Wiley & Sons, Chichester, Sussex.

A series of articles on photovoltaics is to be found in *Materials Research Society Bulletin* **XVIII** (October 1993).

A series of articles on photonic materials for optical communications is to be found in *Materials Research Society Bulletin* **28** (May 3002).

Problems and exercises

Quick quiz

1 The long wavelength part of the electromagnetic spectrum is associated with:
   (a) Radiowaves

   (b) X-rays

   (c) Infrared radiation

2 The short wavelength region of the visible spectrum is associated with the colour:
   (a) Red

   (b) Violet

   (c) Green

3 A beam of light is said to be coherent when:
   (a) All of the waves have the same wavelength

   (b) All of the waves are in step

   (c) All of the waves travel at the same speed

4 The emission of light by a solid at ordinary temperatures is called:
   (a) Incandescence

   (b) Phosphorescence

   (c) Luminescence

5 When light is emitted slowly from an excited solid it is called:
   (a) Fluorescence

   (b) Phosphorescence

   (c) Electroluminescence

6 Some sugar is placed in a bowl in a dark room and is ground. Light is clearly emitted during this process, which is called:
   (a) Triboluminescence

   (b) Chemoluminescence

   (c) Bioluminescence

7 The emission of light by a solid at high temperatures, for example the Sun, is called:
   (a) Luminescence

   (b) Incandescence

   (c) Phosphorescence

8 The word laser is an acronym derived from:
   (a) Light amplification by spontaneous emission of radiation

   (b) Light amplification by stimulated emission of radiation

   (c) Light amplification by simultaneous emission of radiation

9 The light produced by spontaneous emission is:
   (a) Coherent

   (b) Incoherent

   (c) Partly coherent

10 The centres that produce the laser action in a ruby crystal are:
   (a) Cr^{3+} ions

   (b) Al^{3+} ions

   (c) O^{2-} ions

11 Electron transitions involved in colour (optical transitions) are:
   (a) Spin-allowed transitions

   (b) Phonon-assisted transitions

   (c) Radiationless transitions

12 The four-level neodymium laser utilises:
   (a) p-electron transitions

   (b) d-electron transitions

   (c) f-electron transitions

13 The acronym LED stands for:
   (a) Light emitting device

   (b) Light emitting diode

   (c) Laser emitting diode

14 Mixing the three additive primary colours gives:
   (a) Black

   (b) White

   (c) No colour (colourless)

15 Mixing the three subtractive primary colours gives:
   (a) Black

   (b) White

   (c) No colour (colourless)

16 Which of the following subtractive primary colours absorbs *green*:
   (a) Cyan
   (b) Yellow
   (c) Magenta

17 A ray of light passes from a material A, with a refractive index 1.33, to material B, with refractive index 1.5. The wavelength of the light is:
   (a) Longer in A than in B
   (b) Longer in B than in A
   (c) The same in both materials

18 The optical thickness of a film of transparent material is:
   (a) Less than the physical thickness
   (b) Greater than the physical thickness
   (c) The same as the physical thickness

19 The refractive index of a transparent material is mainly attributable to:
   (a) Ions in the material
   (b) Dipoles in the material
   (c) Electrons in the material

20 The refractive index of a transparent material:
   (a) Increases as the wavelength of the light increases
   (b) Decreases as the wavelength of the light decreases
   (c) Does not change as the wavelength of the light changes

21 A simple glass lens is used to form an image of a white circular object. The edges of the image appear coloured. Which colour is outermost:
   (a) White?
   (b) Violet?
   (c) Red?

22 Rayleigh scattering applies to particles of:
   (a) Diameter less than half the wavelength of the radiation
   (b) Diameter less than a third of the wavelength of the radiation
   (c) Diameter less than a tenth of the wavelength of the radiation

23 Mie scattering obscures a distant object:
   (a) It will be clearer if imaged in ultraviolet light
   (b) It will be clearer if imaged in infrared light
   (c) There will be no difference between imaging in ultraviolet or infrared light

24 Diffraction is a form of scattering of light that occurs when the light has:
   (a) A wavelength comparable in size to the object
   (b) A wavelength much larger than the object
   (c) A wavelength much smaller than the object

25 A beam of white light is diffracted during passage through a small circular aperture. The colour of the diffracted rings will have:
   (a) Red on the inside and violet on the outside
   (b) Violet on the inside and red on the outside
   (c) White on the outside and red on the inside

26 An opal strongly diffracts green light of wavelength 550 nm. The approximate diameter of the silica spheres in the opal is:
   (a) 110 nm
   (b) 220 nm
   (c) 550 nm

27 The cladding of an optical fibre consists of:
   (a) Air
   (b) Plastic
   (c) Glass

28 Intrinsic attenuation in an optical fibre can be caused by:
   (a) Minute air bubbles in the glass
   (b) Impurities in the glass
   (c) Density fluctuations in the glass

29 The 'spreading out' of a pulse in an optical communications fibre is called:
   (a) Pulse dispersion
   (b) Modal dispersion
   (c) Wavelength dispersion

30 The addition of germanium dioxide to a silica fibre:

(a) Increases the refractive index of the glass

(b) Lowers the refractive index of the glass

(c) Lowers the dispersion of the glass

31 Amplification of signals in optical communications fibres uses:
(a) Yttrium ions doped into the glass

(b) Ytterbium ions doped into the glass

(c) Erbium ions doped into the glass

32 Nonlinear optical materials are mostly:
(a) Centrosymmetric crystals

(b) Noncentrosymmetric crystals

(c) Noncrystalline

33 Photovoltaic materials:
(a) Produce a voltage when illuminated

(b) Produce light when a voltage is applied

(c) Produce chemical (oxidation–reduction) energy when illuminated

Calculations and questions

14.1 Calculate the frequency and energy of photons associated with wavelengths 425 nm, 575 nm and 630 nm. What colour is attributed to these wavelengths?

14.2 Light of wavelength 400 nm is shone through a gas of absorbing molecules. Calculate the energy absorbed by one mole of gas if each molecule absorbs one photon.

14.3 Carbon dioxide, CO_2, is a greenhouse gas that absorbs infrared radiation escaping from the Earth. What is the energy per mole absorbed by the gas if each CO_2 molecule absorbs one photon of wavelength 15 μm?

14.4 The energy required to break the bond linking the two oxygen atoms in a molecule of O_2 is 495 kJ mol^{-1}. What is the longest wavelength light that could cause this decomposition to occur?

14.5 The energy required to dissociate ozone, O_3, into O_2 plus O is 142.7 kJ mol^{-1}. What is the longest wavelength light that will dissociate ozone in the upper atmosphere?

14.6 The light absorbed by the complex ion $[FeF_6]^{3-}$ peaks at 719 nm. This absorption is due to the promotion of an electron from the lower (t_{2g}) to the upper (e_g), state in the Fe^{3+} ion.
(a) Calculate the magnitude of the crystal field splitting of the Fe^{3+} d orbitals due to F^- (see Section S4.5 for information on crystal field splitting in magnetic ions).

(b) What is the relative population of the two levels at 300 K?

14.7 The light absorbed by the complex ion $[Fe(CN)_6]^{3-}$ peaks at 333 nm. This absorption is due to the promotion of an electron from the lower (t_{2g}) to the upper (e_g) state in the Fe^{3+} ion.
(a) Calculate the magnitude of the crystal field splitting of the Fe^{3+} d orbitals due to CN^- (see Section S4.5 for information on crystal field splitting in magnetic ions).

(b) What is the relative population of the two levels at 300 K?

14.8 Calculate the wavelength at which the rates of spontaneous and stimulated emission become equal at 300 K.

14.9 The optical transitions in ruby are:
● ground state $E_0 \rightarrow E_2$, 556 nm;

● ground state $E_0 \rightarrow E_3$, 407 nm.

What are the energies of these states above the ground state? What are the relative populations N_2/N_0 and N_3/N_0 at 300 K?

14.10 The laser light from a ruby laser is at 694.3 nm.
(a) What is the energy of the lasing state, E_1, above the ground state, E_0?

(b) Estimate the fraction of Cr^{3+} ions in this upper state due to thermal equilibrium alone at 300 K.

14.11 What is the separation of the energy levels in neodymium ions that give rise to laser lines at (a) 0.914 μm and (b) 1.06 μm?

14.12 Gallium arsenide, GaAs, has a band gap of 1.35 eV, and aluminium arsenide, AlAs, has a band gap of 2.16 eV. What is the wavelength and colour of photons emitted by these solids? In order to make an orange LED with an emission at a wavelength of 600 nm, it is proposed to make a solid solution $Ga_xAl_{1-x}As$. Taking the variation in band gap with composition as linear, what value of x is required?

14.13 Gallium nitride, GaN, has a band gap of 3.34 eV, and indium nitride, InN, has a band gap of 2.0 eV. What is the wavelength and colour of photons emitted by these solids? In order to make a green LED with an emission at a wavelength of 525 nm, it is proposed to make a solid solution $Ga_xIn_{1-x}N$. Taking the variation in band gap with composition as linear, what value of x is required?

14.14 A solid solution of indium phosphide, InP, and aluminium phosphide, AlP, $In_xAl_{1-x}P$ is made up. At what value of x will the light emitted by an LED made from this compound just be visible? What colour will it be? The band gap of InP is 1.27 eV and that of AlP is 2.45 eV.

14.15 A solution is quoted as having a 22 % transmittance. What is the absorbance?

14.16 The linear absorption coefficient of zinc metal for X-rays from a nickel target is $5.187\,m^{-1}$ and for cadmium metal is $18.418\,m^{-1}$. What thickness of plates of these metals is needed to reduce the intensity of the radiation passing through a plate to 0.1 of the incident radiation? What will be the transmittance and absorbance of the plates?

14.17 A plate of a cadmium – zinc alloy 21.7 cm thick is used to reduce the X-radiation from a nickel target to 0.05 of its incident value. What is the transmittance and absorbance of the plate? Assuming that the absorption coefficients, given in Question 14.16, can be added, what is the composition of the alloy, in atom%?

14.18 A lead glass fibre has a refractive index of 1.682. What is the critical angle for total internal reflection at the interface with an acrylic coating with refractive index 1.498?

14.19 A ray of light passing through water (refractive index, $n = 1.33$) in a glass tank (refractive index 1.58) hits the water/glass surface at an angle of 23°.
   (a) What angle does it make with the surface as it continues through the glass?
   (b) What is the critical angle for light passing through the water striking the water/glass interface?
   (c) What is the critical angle for light passing through the glass striking the glass/water interface?

14.20 Estimate the refractive indices of the ceramics barium titanate, $BaTiO_3$, and lead titanate, $PbTiO_3$, using the Gladstone – Dale formula. Densities are as follows: $BaTiO_3$, $6017\,kg\,m^{-3}$; $PbTiO_3$, $8230\,kg\,m^{-3}$.

14.21 Estimate the refractive index of the minerals spinel, $MgAl_2O_4$, and akermanite, $CaMgSl_2O_7$, using the Gladstone – Dale formula. Densities are as follows: spinel, $3600\,kg\,m^{-3}$; akermanite, $2940\,kg\,m^{-3}$.

14.22 Estimate the refractive index of the minerals beryl, $Be_3Al_2(SiO_3)_6$, and garnet, $Mg_3Al_2Si_3O_{12}$, using the Gladstone – Dale formula. Densities are as follows: beryl, $2640\,kg\,m^{-3}$; garnet, $3560\,kg\,m^{-3}$.

14.23 Estimate the refractive coefficient of Al_2O_3 using the information that the mineral andalusite, Al_2SiO_5, has a density of $3150\,kg\,m^{-3}$ and a refractive index of 1.639. The refractive coefficient of SiO_2 is 0.21.

14.24 Calculate the reflectivity of the surfaces of the following transparent materials in air: cryolite, Na_3AlF_6, $n = 1.35$; glass, $n = 1.537$; corundum, Al_2O_3, n (average) $= 1.63$; 1.63; cubic zirconia, ZrO_2, $n = 2.05$; tantala, Ta_2O_5, $n = 2.15$ (average).

14.25 Calculate the reflectivity of the surfaces of the following metals in air (the refractive indices are for a wavelength of 550 nm): aluminium, $n = 0.82$, $k = 5.99$; silver, $n = 0.255$, $k = 3.32$; gold, $n = 0.33$, $k = 2.32$; chromium, $n = 2.51$, $k = 2.66$; nickel, $n = 1.85$, $k = 3.27$.

14.26 A thin film on a substrate is viewed in air and has an optical thickness of $\lambda/4$. Will the film be reflecting or not if (a) the substrate has a lower refractive index than the film and (b) the substrate has a higher refractive index than the film?

14.27 A thin film on a substrate is viewed in air and has an optical thickness of $\lambda/2$. Will the film be reflecting or not if (a) the substrate has a lower refractive index than the film and (b) the substrate has a higher refractive index than the film?

14.28 Describe the differences between the reflectivity of a soap film, thickness $\lambda/4$, in air, compared with the same film on an oil surface. Assume that the refractive index of the soap film is very slightly more than that of pure water ($n = 1.33$), the refractive index of the oil is 1.44.

14.29 (a) What is the minimum real (physical) thickness of a film of titanium dioxide, TiO_2, in air for constructive interference of green light, $\lambda = 550$ nm? The refractive index of TiO_2 is 2.875 (average). Estimate the colour that the film would appear when viewed in white light by (b) reflection and (c) transmission. Use the information in Section S4.11.

14.30 Derive the relationship

$$n_f = \sqrt{n_s}$$

for an antireflection coating, refractive index n_f, on a substrate, refractive index n_s, in air. [Note: derivation is not provided in the answers at the end of this book.]

14.31 Determine the reflectivity of a $\lambda/4$ film of (a) silicon oxide, SiO, $n = 2.0$, and (b) titanium dioxide, TiO_2, $n = 2.504$, on glass, $n = 1.504$, in air.

14.32 Determine the reflectivity of a $\lambda/4$ and a $\lambda/2$ film of magnesium difluoride, MgF_2 ($n = $

1.384), on glass ($n = 1.504$), in air. What changes would occur if the substrate was titanium dioxide, TiO_2, $n = 2.875$ (average)?

14.33 Plot a graph of reflectivity versus the number of pairs of layers for a quarter wave stack on a glass substrate ($n = 1.495$), in air, using alternating layers of magnesium fluoride, MgF_2 ($n = 1.384$) and titanium dioxide ($n = 2.875$, average). How many pairs are needed to achieve a reflectivity of at least 99.9 %? [Note: graph is not shown in the answers at the end of this book.]

14.34 A quarter wave stack on a glass substrate ($n = 1.545$), in air, is required to reflect light of 650 nm from a laser. The materials chosen have refractive indices of 2.15 (tantalum pentoxide, Ta_2O_5) and 1.35 (cryolite, Na_3AlF_6). What are the real (physical) thicknesses of the layers?

14.35 The intensity of a light beam traversing a solution placed into a cell of 10 cm path length drops to 80.3 % of the incident intensity. Calculate the linear absorption coefficient of the solution.

14.36 The visibility of the atmosphere is reported as being 10 km. Assuming that at this distance an object can just be perceived, with 1 % of the initial light falling on the object reaching the observer, determine the linear absorption coefficient of the atmosphere.

14.37 Plot the relative scattering of light by the Rayleigh model across the visible spectrum, 400–700 nm. [Note: graph is not shown in the answers at the end of this book.]

14.38 Determine the amount of light scattered by dust particles with a refractive index of 1.45 relative to that of water droplets with a refractive index 1.33 if scattering takes place in accordance with the Rayleigh formula.

14.39 Will air visibility be improved if water droplets responsible for scattering are replaced by similar sized limestone dust particles? The refractive index of limestone, $CaCO_3$, is 1.53.

14.40 The linear scattering coefficient of limestone dust in the air is $0.0002194 \, \text{m}^{-1}$. Over what distance will the intensity of a light beam diminish to 10 % of its initial value?

14.41 What is the relative amount of light scattered (a) in the incident direction, (b) at 45° to the incident direction, (c) perpendicular to the incident direction and (d) in the reverse direction, for a monochromatic beam of light, according to the Rayleigh model of scattering?

14.42 (a) What slit width is needed in a transmission diffraction grating to cause the red light, $\lambda = 700 \, \text{nm}$, to be deviated by 7.5° to the normal? (b) What will be the deviation of the violet light, $\lambda = 400 \, \text{nm}$?

14.43 The first photonic crystal made was formed by drilling an array of holes in a block of material with a refractive index of 3.6 so as to form a face-centred cubic array of holes throughout the block, with a unit cell parameter of 2 mm. What wavelength radiation will not pass in the [1 0 0] direction?

14.44 A photonic crystal was made by laying down close-packed layers of polystyrene spheres of refractive index 1.595 to form a hexagonal closest packed array. The sphere diameter was 250 nm, and the layer separation normal to the layers can be taken as 0.8 of the sphere diameter.
   (a) What is the wavelength that will not be transmitted in light normal to the close-packed layers?
   (b) What colour does this correspond to?

(c) If the ordering is not perfect and the close-packed layers are sometimes in hexagonal (AB) packing and sometimes in cubic (ABC) packing, how will the result change?

14.45 Estimate the attenuation of an optical fibre in which 0.5 % of the initial power is lost in a distance of 1 km.

14.46 The attenuation of ordinary window glass is of the order of $10^6 \, \text{dB km}^{-1}$. What thickness of glass would cause the incident light intensity to fall to 50 %?

14.47 An instantaneous pulse of wavelength $1000 \pm 60 \, \mu\text{m}$ is introduced into a silica optical fibre. What will be the pulse spread in km after 1 second? The refractive indices of silica are as follows: $n(400 \, \text{nm}) = 1.47000$; $n(\text{average}) = 1.46265$; $n(700 \, \text{nm}) = 1.45530$.

14.48 A thin film of silver oxide, Ag_2O, forms on a silver surface. The band gap for silver oxide is 2.25 eV. What colour would be emitted by bulk silver oxide? Treating the thin film as a quantum well, what film thickness is needed to obtain an emission in the blue region of the spectrum, at a wavelength of 413 nm? Assume that the effective mass of electrons and holes in Ag_2O is equal to $0.3m_e$.

14.49 A film of zinc sulphide, ZnS, 4 nm in thickness, forms on metallic zinc. Treating the film as a quantum well, what is the wavelength of the transition $\Delta n = 2$? The band-gap of ZnS is 3.54 eV, and the effective mass of electrons and holes is $0.4m_e$.

15

Thermal properties

- What is zero-point energy?

- What solids are named high thermal conductivity materials?

- What physical property does thermoelectric refrigeration utilise?

Heat is a form of energy. Thermal properties describe how a solid responds to changes in this thermal energy. In a normal solid, the constituent atoms are in constant vibration, and the vibrations constitute the thermal energy of the material. The physical aspects of thermal properties – temperature, heat capacity, thermal expansion and thermal conductivity – are the external manifestation of these vibrations. Chemical aspects, such as phase changes and rates of reaction, are equally important. Indeed, it will be apparent that chemical aspects of the thermal properties of solids have been considered already throughout this book. For example, the phase equilibria of Chapter 4 are an expression of the thermal properties of solids. The way in which materials respond to changes in thermal energy is defined as the subject of thermodynamics, some aspects of which have been touched on earlier. In this chapter, aspects of the physical properties of solids not covered previously will be outlined and some new concepts introduced.

15.1 Temperature effects

15.1.1 Heat capacity

The heat capacity of a solid quantifies the relationship between the temperature of a body and the energy supplied to it. The ideas date from times when heat was thought to be a fluid that could be transferred from one object to another. If a large amount of heat (energy) supplied to a body produced only a small temperature increase, the solid was said to have a large heat capacity. The heat capacity of a material is defined as:

$$C = \frac{dQ}{dT}$$

where dQ is the amount of energy needed to produce a temperature change of dT. The units of heat capacity are $J\,K^{-1}$ per amount of sample, frequently quoted as the molar heat capacity, $J\,K^{-1}\,mol^{-1}$, or the specific heat capacity, $J\,K^{-1}\,g^{-1}$. The heat capacity of a solid generally increases with temperature.

The measured value of the heat capacity is found to depend on whether the measurement is made at

Understanding solids: the science of materials. Richard J. D. Tilley
© 2004 John Wiley & Sons, Ltd ISBNs: 0 470 85275 5 (Hbk) 0 470 85276 3 (Pbk)

Table 15.1 The Debye temperature, θ_D, and room temperature heat capacity, C_p, of some elements

Element	θ_D/K	$C_p/J\,K^{-1}\,mol^{-1}$
Li	345	24.9
Na	158	28.2
K	91	29.6
Rb	56	31.1
Cs	38	32.2
C[a]	2340	13.0
Si	645	19.8
Ge	375	23.2
Sn[b]	230	27.1
Pb	105	26.7

[a] Diamond.
[b] White tin (β-tin), tetragonal.

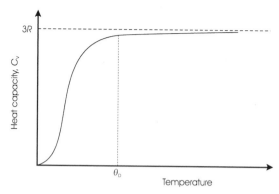

Figure 15.1 The variation of the heat capacity of a solid with temperature

constant volume, C_v, or at constant pressure, C_p. The relationship between them is:

$$C_v = \frac{dU}{dT}$$

$$C_p = \frac{dH}{dT}$$

$$C_p = C_v + \text{constant} \times T$$

where U is the internal energy of the material, and H is the enthalpy of the material (see Section S3.2). For solids, the difference between C_p and C_v is very small at room temperature and below. Some room-temperature values of C_p are given in Table 15.1.

15.1.2 Theory of heat capacity

Thermal energy is mainly taken up as the vibrations of the atoms in the solid. Classically, the calculation of the heat capacity of a solid was made by assuming that each atom vibrated quite independently of the others. The heat capacity was then the sum of all of the identical atomic contributions, and independent of temperature. The result was

$$C_v(\text{classical}) = 3R = 25\,J\,K^{-1}\,mol^{-1}$$

where R is the gas constant, 8.3145 J K^{-1} mol^{-1}. This value is reasonable for high temperatures but is completely incorrect at low temperatures (Figure 15.1). This discrepancy was resolved by realising that the energy of the vibrating atoms, E, was quantised in the following way:

$$E = \left(n + \frac{1}{2}\right)h\nu \qquad (15.1)$$

where n is an integer quantum number, h is Planck's constant, and ν is the vibration frequency. The quantum of vibrational energy is called a phonon. The resulting calculation, by Einstein, which assumed that each atom vibrated independently of the others, was a good overall fit to the data but was not in perfect accord with the experimental values at low temperatures.

Debye improved on the calculation by including the fact that the atomic vibrations throughout the crystal are coupled together, via chemical bonding. The lattice vibrations are then likened to waves throughout the whole of the solid body. The waves must fit into the dimensions of the solid, in a similar way that electron waves must, as illustrated in Figure 2.27. In this model, the lattice vibration waves are equivalent to sound waves of high frequency travelling through the solid. The wavelength, frequency and energy of the vibrations are quantised, and the term phonon is often used for the waves themselves. The heat capacity calculated by Debye is:

$$C_v = 9R \left(\frac{T}{\theta_D}\right)^3 \int_0^{\theta_D/T} \frac{x^4 e^x}{(e^x - 1)^2} dx$$

$$x = \frac{h\nu}{kT}$$

$$\theta_D = \frac{h\nu_D}{k}$$

where θ_D is called the Debye temperature, ν_D is the Debye frequency, R is the gas constant, and T is the temperature (in K). Values of the Debye temperature are given in Table 15.1. Note that the Debye temperature drops systematically on moving down a periodic table group.

For temperatures well above the Debye temperature, the value of C_v is $3R$, where R is the gas constant, $8.3145 \, \mathrm{J \, K^{-1} \, mol^{-1}}$. Thus, all solids would be expected to have a high-temperature heat capacity of about $25 \, \mathrm{J \, K^{-1} \, mol^{-1}}$, equal to the classical value. As the value of the Debye temperature is below room temperature for many solids, the room-temperature heat capacity can be approximated to $3R$ (Table 15.1).

At temperatures well below the Debye temperature, the value of C_v is given by:

$$C_v = \frac{12\pi^4}{5} R \left(\frac{T}{\theta_D}\right)^3$$

That is,

$$C_v = \text{constant} \times T^3$$

The phonon contribution to the heat capacity is the most important one, but others also occur. As noted in Section 2.3.7, the heat capacity due to free electrons in metals is small but significant heat capacity changes accompany phase changes, such as order–disorder changes of the type noted with respect to ferroelectrics (Sections 11.3.5 and 11.3.6), or when a ferromagnetic solid becomes paramagnetic (Sections 12.1.2 and 12.3.1).

15.1.3   Quantum and classical statistics

The energy of the system of vibrating atoms, given by Equation (15.1), shows that the lowest energy possible is not zero, as it is for a classical system, but is equal to $\frac{1}{2} h\nu$. This energy is called zero-point energy. Any quantum mechanical system will possess this amount of energy even when in the lowest energy state.

To calculate the total energy of a system it is necessary to allocate the phonons to the available energy levels (as with electrons, see Section 2.3.7). Phonons are distributed according to Bose–Einstein statistics. This accounting system places no constraints on the number of phonons that can occupy a single energy level (Section S4.12). At low temperatures, all phonons can be in the lowest energy state, giving rise to an appreciable zero-point energy.

Phonons interact with photons, electrons and neutrons, causing scattering. This causes a beam of radiation incident on a crystal to spread out, and a diffracted beam will be broadened by this extra contribution. The extent of the spreading is related to the phonon spectrum in the crystal, and its measurement gives information on the phonon distribution in the solid.

15.1.4   Thermal conductivity

When the two ends of a solid are held at different temperatures, heat flows across it from the hot to the cold side (Figure 15.2a). The amount of heat transferred, Q, per unit time, depends on the cross-sectional area over which the heat is conducted, A, the temperature difference between the hot and cold regions, $T_1 - T_2$, and the separation, $x_1 - x_2$:

$$Q = \frac{\kappa A (T_1 - T_2)}{x_1 - x_2}$$

or

$$\frac{dQ}{dt} = \kappa A \frac{dT}{dx}$$

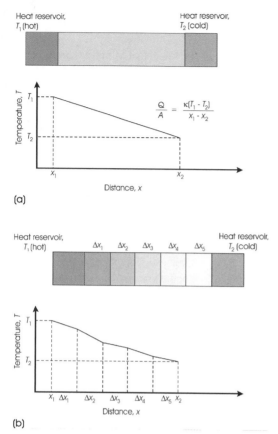

(a)

(b)

Figure 15.2 (a) The thermal conductivity of a material is the heat flow from a hot to a cold region. The temperature gradient between the hot and cold ends is a measure of the thermal conductivity of the material. (b) The temperature drop across a series of different materials will vary with the thermal conductivity of each, although the total temperature drop is the same. Note: Q, heat transferred; A, cross-sectional area

where dQ/dt is the rate of energy transfer, and dT/dx is the temperature gradient. The constant of proportionality, κ, is called the thermal conductivity, with units $W\,m^{-1}\,K^{-1}$. When the heat transfer is across a number of materials (Figure 15.2b):

$$\frac{dQ}{dt} = A\Delta T \left[\sum_{i=1}^{n} \frac{\Delta x_i}{\kappa_i} \right]^{-1}$$

where ΔT is the total temperature drop, $T_1 - T_2$, and Δx_i is the thickness of a slab of thermal conductivity κ.

The thermal conductivity of solids varies considerably (Table 15.2). Metals have a high thermal conductivity, with silver having the highest room-temperature thermal conductivity, at $430\,W\,m^{-1}\,K^{-1}$. Alloys have lower thermal conductivities than pure metals. Ceramics are even lower, especially porous porcelains or fired clay products (Figure 15.3). The lowest thermal conductivities are shown by plastic foams such as foamed polystyrene. As would be expected, the thermal conductivity of crystals varies with direction. For example, the thermal conductivity of the hexagonal metal cadmium Cd, (A3 structure), is $83\,W\,m^{-1}\,K^{-1}$ parallel to the c axis and $104\,W\,m^{-1}\,K^{-1}$ parallel to the a axis. At $25\,°C$, the oxide quartz, which has a hexagonal unit cell, has a thermal conductivity parallel to the c axis of $11\,W\,m^{-1}\,K^{-1}$, and $6.5\,W\,m^{-1}\,K^{-1}$ parallel to the a axis.

A number of nonmetallic materials are called high thermal conductivity materials. The most notable of these is diamond, with a thermal conductivity of $2000\,W\,m^{-1}\,K^{-1}$. All of the others have a diamond-like structure, and include boron nitride, BN, and aluminium nitride, AlN (Table 15.2).

Thermal conductivity is attributed mainly to the mobile electrons present and the vibration waves in the structure, phonons:

$$\kappa = \kappa(\text{electrons}) + \kappa(\text{phonons})$$

Mobile electrons make the greatest contribution, and so metals would be expected to show a much higher thermal conductivity than insulators. At the simplest level the electrons can be imagined as a free-electron gas, moving with a velocity that is higher at the hot end of the solid than at the cold end. (The same model was mentioned in Section 2.3.5, with respect to electrical conductivity.) The kinetic energy is gradually transferred to the cold end by collisions between the electrons themselves and with the atoms in the structure. The thermal conductivity increases as the number of free electrons increases. The model is successful in some ways. For example, it predicts that thermal

Table 15.2 Coefficients of thermal conductivity, κ, and thermal expansion

Material	$\kappa/\mathrm{W\ m^{-1}\ K^{-1}}$	$\alpha/10^{-6}\ \mathrm{K^{-1}}$
Metals:		
Silver	428	18.9
Copper	403	16.5
Gold	319	14.2
Iron	83.5	11.8
Nickel	94	13.4
Titanium	22	8.6[a]
Alloys:		
Brass	106	17.5
Bronze	53	17.3
Carbon steel	~50	~10.7
Monel[b]	21	~14
Lead–tin solder	~50	~24
Refractories:		
Alumina	38	5.5
Magnesia	40	9.5
Silica	1.6	0.49
Porcelain	~2	~4.5
Polymers:		
Nylon 6,6	0.25	80
Polyethylene	~0.4	~200
Polystyrene foam	~0.04	–
High thermal conductivity materials:		
Diamond	2000	1
Graphite[c]	2000	−0.6
Cubic BN	1300	–
SiC	490	3.3
BeO	370	–
BP	360	–
AlN	320	–
BeS	300	–
BAs	210	–
Si_3N_4	200	2.5
GaN	170	–
Si	160	2.6
AlP	130	–
GaP	100	4.7

– Data could not be located.

[a] Mean value.
[b] 67 wt% Ni, 29 wt% Cu, 4 wt% Fe.
[c] Perpendicular to the c axis.

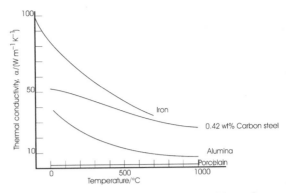

Figure 15.3 The approximate variation of thermal conductivity with temperature for some common materials

conductivity will be proportional to electrical conductivity. However, it does not explain the differences between one metal and another and, for this, the dynamics of the electrons at the Fermi surface must be employed.

Thermal conductivity in insulators, ceramics and polymers can be explained in terms of a phonon gas. The solid is imagined to contain phonons travelling with a range of kinetic energies, similar to the molecules in a gas or the electron gas just mentioned. At the hot end of a solid the kinetic energy of the phonons is greater than at the cold end. This energy is gradually transferred from hot to cold by phonon–phonon interactions and by interactions between the phonons and the solid structure. The differences between the thermal conductivities of different insulators reflect these interactions.

The thermal conductivity depends on the mean free path of the phonons, which is the distance between collisions of the phonons in the structure. A short mean free path correlates with a low thermal conductivity. Defects in a structure drastically shorten the mean free path and reduce thermal conductivity significantly. An inherent problem with ceramic materials is that they are usually formed by sintering. This process naturally leads to the formation of many internal defects such as grain boundaries, pores and voids. Because of this, the thermal conductivity of sintered bodies is usually much lower than the intrinsic thermal conductivity. Fired-clay ceramics have very high porosity and a very low thermal conductivity. At present, the best sintered ceramic solids have about $\frac{3}{4}$ of the intrinsic thermal conductivity of the parent phase.

Point defects can drastically lower the thermal conductivity of the important carbide and nitride high thermal conductivity ceramics. In this respect, oxygen, which is a common impurity, has been found to be very important. For example, silica (SiO_2), an impurity in silicon nitride (Si_3N_4), formed by oxidation at high temperatures in air, can react to produce substitutional defects and vacancies in the following way:

$$2SiO_2\,(s) \rightarrow 2Si_{Si} + 4O_N + V_{Si}$$

where Si_{Si} represents silicon atoms on normal silicon sites (not defects), O_N represents oxygen atoms on nitrogen atom sites (substitutional defects) and V_{Si} represents a vacancy on a silicon site (vacancy defects). Thus, two SiO_2 units produce five defects in the silicon nitride. Purification therefore forms an important step in the manufacture of high thermal conductivity ceramic materials.

The thermal conductivity of polymers is found to depend on the degree of crystallinity. The crystalline portions of the structure have a higher thermal conductivity than do the disordered regions of the solid. Materials with high porosity, such as plastic foams, have particularly low thermal conductivities. Foamed plastics such as polystyrene are widely used as insulating materials.

15.1.5   Heat transfer

When a solid is heated or cooled, heat is transferred through the structure. The equations of heat transfer were initially formulated by Fourier. They predate and are of identical mathematical form to Fick's laws of diffusion (Sections 7.1 and 7.3). In the case of steady-state heat transfer, the one-dimensional heat transfer equation is:

$$J_Q = -K\frac{dT}{dx}$$

where J_Q is the heat flux (units $J\,m^{-1}\,s^{-1}$), dT/dx is the temperature gradient along the direction of heat flow, taken as x, and K is the thermal diffusivity, defined by:

$$K = \frac{\kappa}{\rho C_p}$$

where κ is the thermal conductivity, ρ is the density, and C_p is the specific heat capacity (the heat capacity per unit weight) at constant pressure. The units of K are the same as those of the diffusion coefficient ($m^2\,s^{-1}$).

In the case of nonsteady-state heat transfer, the equation is analogous to the diffusion equation:

$$\frac{dT}{dt} = \frac{K d^2 T}{dx^2}$$

where dT/dt is the change of temperature with time at a point x in the solid. The solutions to this equation are identical in form to those given in Section 7.2 for the diffusion equation.

15.1.6   Thermal expansion

Most materials increase in volume as the temperature is increased, a feature called thermal expansion. There are, though, an increasing number of solids known that contract as the temperature increases.

Thermal expansion is a most important property in practice. It is used in most everyday thermometers. The shattering of ordinary glass on being cooled rapidly is due to the thermal contraction of the outer layers, and is prevented in special glasses such as Pyrex® glass or fused silica, which have low thermal expansion. The thermal expansion of components in electronic devices is important, and the difference in thermal expansion of materials in a construction can lead to grave difficulties. The coincidence of the thermal expansion of steel and concrete at normal temperatures allows the use of steel-reinforced concrete in buildings.

The mean coefficient of linear thermal expansion, α_{mean}, of a material is the increase in length per unit length over a temperature interval from T_i to T_f

$$\alpha_{\text{mean}} = \frac{l_f - l_i}{l_i}$$

where T_i is the initial temperature, T_f is the final temperature, l_f is the final length, and l_i is the initial length. The units are K^{-1}. In general, for a temperature interval ΔT:

$$\alpha_{\text{mean}} = \frac{\Delta l}{l_i}$$

where Δl is the length increment. A similar expression can be written for the mean coefficient of volume expansion, also called the cubical expansion coefficient, β_{mean}, over a temperature interval ΔT:

$$\beta_{\text{mean}} = \frac{\Delta V}{V_i}$$

where ΔV is the volume change, and V_i is the original volume at T_i. The units are K^{-1}.

A reasonable approximation is

$$3\alpha_{\text{mean}} \approx \beta_{\text{mean}}$$

In the case of a solid with different mean thermal expansion coefficients along x, y and and z axes (α_x, α_y and α_z, respectively):

$$\alpha_x + \alpha_y + \alpha_z \approx \beta_{\text{mean}}$$

The linear expansivity of a solid, α, is defined as the increase in length per unit length at a given temperature:

$$\alpha = \frac{1}{l}\frac{dl}{dT}$$

The expansivity is the slope of the dl/l versus temperature curve at a temperature T (Figure 15.4a) and is often different in value from α_{mean}. The value of α_{mean} tends to the expansivity as Δl and ΔT become small, and both have units of K^{-1}. The expansivity of many solids tends to increase as the temperature increases (Figure 15.4b).

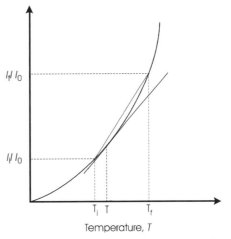

Mean expansion coefficient, $\alpha_{\text{mean}} = \dfrac{l_f - l_i}{l_0}$

Expansivity, α, at T is the slope of the curve at T

(a)

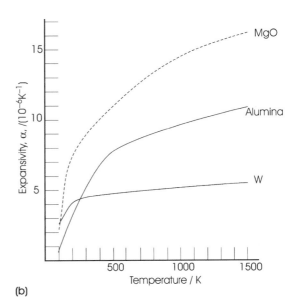

(b)

Figure 15.4 (a) The mean coefficient of thermal expansion, α_{mean}, is not identical to the expansivity, α, at all temperatures over the range within which α_{mean} is measured. (b) The approximate variation of thermal expansion with temperature for a metal (tungsten, W) and two ceramics (magnesium oxide, MgO, and alumina, Al_2O_3)

The thermal expansion of a multiphase solid depends on the expansivity of the individual components and the ratios present. Thus, the thermal expansion of alloys, glasses and glass ceramics can be tailored by changing the bulk composition of the material. For many applications, a very small coefficient of expansion is desirable, and in cooking ware, for example, glass ceramics with negligible thermal expansion over the temperature ranges encountered in cooking are widely available.

15.1.7   Thermal expansion and interatomic potentials

An idea of the origin of thermal expansion can be obtained from a consideration of the potential energy of a pair of atoms as a function of their spacing (Figure 15.5; see also Section S4.1.7). The extent of the vibrational energy at a low temperature, T_0, and energy, E_0, leads to a mean separation of the atoms, r_0. As the temperature increases, the energy of vibration increases to E_1, E_2 and so on, and the mean separation to r_1 and r_2, and so on. Because of the asymmetrical nature of the potential energy curve, the midpoint of the vibration has

increased. Further temperature increases magnify this off-centre displacement, and the net result is an expansion.

The shape of the interatomic energy curve is related to the chemical bond strength between the atoms. Strong bonds result in a steep potential energy curve that is reasonably symmetrical close to the minimum. Weak bonding results in a flatter curve that is very unsymmetrical (Figure 15.6). This suggests that strongly bonded solids, such as silica and other ceramics, would have low expansivity, whereas polymers, in which the chains are linked by weak chemical bonds, would have high expansivity. This is the case (Table 15.2).

In a real solid, account has to be taken of all the atoms in the unit cell, and the interatomic potentials between each pair of atoms has to be evaluated, so as to obtain the mean change of expansion of the unit cell as a whole. Crystals expand less along directions corresponding to strong bonds and more along directions corresponding to weak bonds. The chain silicates, for example, have a higher coefficient of expansion perpendicular to the chains than parallel to them. In general, crystals of lower than cubic symmetry have different expansivities along the different crystal axes.

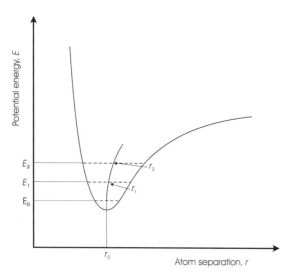

Figure 15.5   The variation of the potential energy between two atoms linked by a strong chemical bond, as a function of the interatomic spacing

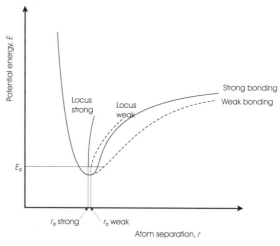

Figure 15.6   Potential energy curves for a pair of atoms linked by weak and strong bonds. The locus of the centre of the bond between the atoms increases more rapidly in the case of weakly bonded atoms

15.1.8 Thermal contraction

The thermal expansion of a material depends on the overall balance between all of the interatomic and intermolecular forces present. In some cases, this can produce materials that contract as the temperature increases.

The best-known material that behaves in this anomalous fashion is water, between $0\,°C$ and $3.98°C$ (Figure 15.7). This feature arises from the hydrogen bonding between the molecules and the molecular structure. At lower temperatures, the hydrogen bonds pull the water molecules closer together as the thermal vibrations of the fluid decrease. However, the angular structure prevents them packing closely together and they maintain an open structure that is similar to the structure of ice, which is also open and has a lower density than water. Above a temperature of $3.98°C$, the thermal vibrations begin to dominate, molecular rotation in the liquid increases and the molecules are effectively spherical. As the temperature increases, normal expansion is found, as depicted in Figure 15.6. The consequences of this feature of water are important for life on Earth.

Many anisotropic crystals are known that show a contraction along one or two axes as the temperature increases, although other axes may show normal thermal expansion (Figure 15.8). Some of the most important compounds that reveal this behaviour are cordierite ($Mg_2Al_4Si_5O_{12}$), β-eucryptite ($LiAlSiO_4$), β-spodumene ($LiAlSi_2O_6$) and $NaZr_2P_3O_{12}$ (NZP). One reason for their importance is that all of these materials exist over a wide composition range. For example, β-spodumene can take compositions $Li_2Al_2Si_nO_{4+2n}$, in which n can take values from 4 to 9. Similarly, NPZ can form solid solutions in which phosphorus is replaced by silicon, for example, $Na_{1+x}Zr_2P_{3-x}Si_xO_{12}$, and sodium by calcium and strontium, for example, $Ca_{1-x}Sr_xZr_4P_6O_{24}$. This ability allows the thermal expansion and contraction to be carefully tailored, and materials with almost zero thermal expansion produced.

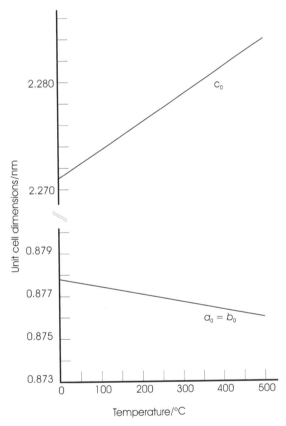

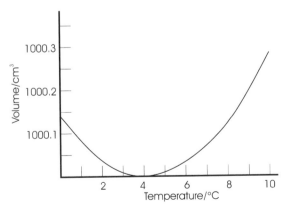

Figure 15.7 The thermal expansion of water close to $0\,°C$

Figure 15.8 The thermal expansion of the c axis and contraction of the a and b axes for $CaZr_4P_6O_{24}$, redrawn from data in D.K. Agrawal 1994, '[NPZ]: A New Family of Real Materials for Low Thermal Expansion Applications', *Journal of Materials Education* **16** 139–165

There are also materials known in which all axes contract as the temperature rises, including cubic ZrW_2O_8 and a silica polymorph with the faujasite structure.

Thermal contraction is not the result of a single mechanism. However, all of the changes can be related to the cation polyhedra that build up the structure. For example, in many ferroelectric perovskite structure compounds, a distortion of the metal–oxygen MO_6 octahedra is responsible for the ferroelectric effect (see Section 11.3.7). Frequently, the distortion, which is caused by an off-centring of the cation, causes one of the octahedron diagonals to lengthen, transforming the cubic unit cell into a tetragonal cell. As the temperature increases, the distortion tends to decrease, because of changes in vibrational energy and a decrease in anion–anion repulsion, so that the long diagonal shortens. At the same time, the undistorted diagonals expand normally. In some cases, such as for $PbTiO_3$, the contraction is greater than the expansion, and a polycrystalline sample of $PbTiO_3$ shows overall thermal contraction (Figure 15.9).

In the families of cordierite, β-eucryptite, β-cordierite and NZP, a mechanism similar to that giving rise to auxetic (negative Poisson's ratio) materials seems to occur (Section 10.3.2). The structure is built from inflexible layers, similar to those found in clay minerals (see Sub-section 6.2.1.2) linked by Si—O—Si and O—Si—O bonds (Figure 15.10). As the temperature rises the layers expand, mostly laterally. The Si—O bonds linking the layers are strong and do not break to relieve the stress generated. Instead, the bond angles change, and the groups act as hinges that pull the layers closer, giving rise to thermal contraction in a direction normal to the layers.

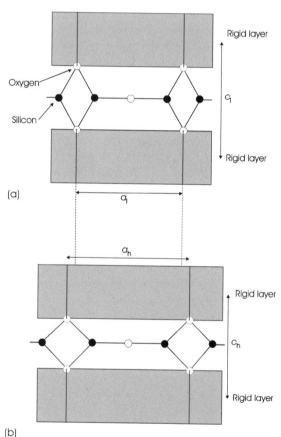

(a)

(b)

Figure 15.10 Thermal contraction brought about by silicon–oxygen 'hinges' connecting rigid sheets within a crystal structure: (a) the low-temperature structure and (b) the high-temperature structure. As the rigid sheets expand normally, due to an increase in temperature (from the configuration shown in part a to that shown in part b), the strong silicon–oxygen bonds act as hinges and pull the sheets together (part b). The material shows thermal expansion parallel to the sheets and contraction normal to the sheets

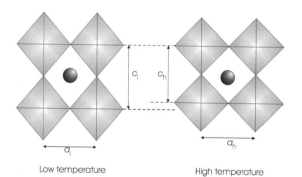

Low temperature

High temperature

Figure 15.9 Thermal contraction due to the relief of octahedral distortion in the perovskite structure type. The low-temperature distortions (left) are lost at high temperatures (right) leading to a smaller unit cell

A third group, which includes cubic Zr_2WO_8 and $NbOPO_4$, achieves the contraction via rocking of the polyhedra (Figure 15.11). In this group, the polyhedra remain the same shape, but simply tilt in a cooperative fashion as the temperature increases, so that one or more unit cell edges contract while the others expand. In cases where the contraction is sufficient, an overall thermal contraction is observed in polycrystalline samples.

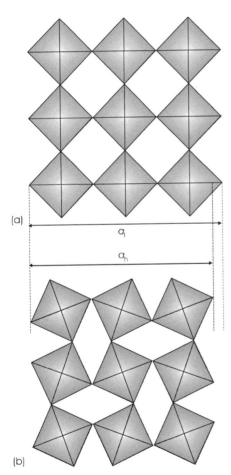

Figure 15.11 Thermal contraction due to the rotation of polyhedra in the *perovskite* structure type: (a) the low-temperature structure and (b) the high-temperature structure. As the temperature increases, (from part a) to (part b), increased vibration allows the polyhedra to rotate, thus producing a contraction of the unit cell

15.2 Thermoelectric effects

15.2.1 Thermoelectric coefficients

The first thermoelectric effect to be discovered was the Seebeck effect. In this phenomenon, a current flow is induced in a circuit made of two different conductors A and B when the junctions between the materials are held at different temperatures. The effect is generally observed by breaking the circuit and observing the voltage generated with a potentiometer (Figure 15.12a). This voltage is given by:

$$\Delta V_{AB} = \Sigma_{AB}\Delta T = \Sigma_{AB}(T_H - T_C)$$

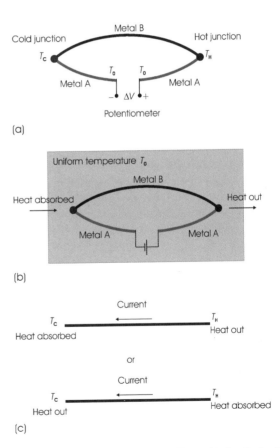

Figure 15.12 Thermoelectric effects: (a) the Seebeck effect, (b) the Peltier effect and (c) the Thompson effect

where Σ_{AB} is called the thermoelectric power or Seebeck coefficient, ΔV_{AB} is the voltage measured, and ΔT is the temperature difference between the hot junction (temperature T_H) and the cold junction (temperature T_C). The Seebeck coefficient, Σ, has units of $V\,K^{-1}$ and varies significantly with temperature. The voltage depends only on the two materials chosen and the temperature difference between the junctions. The Seebeck coefficient for metals is of the order of $10\,\mu V\,K^{-1}$, and for semiconductors it is about $200\,\mu V\,K^{-1}$.

The complementary effect, in which a current flow induces a temperature difference between the two junctions, is called the Peltier effect (Figure 15.12b). Heat is liberated at one junction and absorbed at the other. If the direction of the current is reversed, the heat output and input occur at the opposite junctions. The amount of heat produced or absorbed, ΔQ, is given by:

$$\Delta Q = \Pi_{AB} I t$$

where Π_{AB} is the Peltier coefficient, I is the current flowing in the circuit, and t is the time. The units of Π are $J\,A^{-1}\,s^{-1}$. Peltier heat production is quite different from Joule heating, which occurs along the length of the conductor instead of at a junction between two materials.

The Seebeck and Peltier effects were shown to be related by Thomson (later Lord Kelvin). The relationship is:

$$\Pi = \Sigma T$$

where the temperature, T, is in K. Thomson also predicted the existence of a third thermoelectric effect, now known as the Thomson effect, in which a reversible heating or cooling is observed when a current flows along a (single) conductor that has one end at a different temperature from the other (Figure 15.12c). The amount of heat energy absorbed or given out, ΔQ, is given by:

$$\Delta Q = \tau I t \Delta T$$

where τ is the Thompson coefficient, I is the current flowing for a time t, and ΔT is the temperature

difference between the points of measurement. The units of τ are $J\,A^{-1}\,s^{-1}\,K^{-1}$.

Although observed at junctions in circuits, the Seebeck and Peltier coefficients are not caused by the junctions themselves. All materials that contain mobile charge carriers show thermoelectric effects when heated. That is, temperature gradients produce electrical effects, and electrical effects produce thermal effects. The appearance of these thermoelectric effects are properties of pure materials, and a material is characterised by an absolute Seebeck coefficient, σ_S, an absolute Peltier coefficient, π, and the Thomson coefficient, τ, which only refers to a single material. The Seebeck coefficient, Σ, and the Peltier coefficient, Π, apparent in circuits made of two different electronically conducting materials, are relative coefficients, that is, the difference between the absolute coefficients of the two materials. For the arrangement drawn in Figure 15.12a, in which the positive terminal of metal A is connected to the hot junction, the Seebeck coefficient of two the materials Σ_{AB}, at a temperature T, is given by:

$$\Sigma_{AB} = \sigma_S(A) - \sigma_S(B)$$

15.2.2 Thermoelectric effects and charge carriers

Thermoelectric effects can be explained by considering the electron, hole and phonon distributions in a material. It is apparent that the charge carriers near to the Fermi surface in the hot region of a single material will have a higher kinetic energy, and hence a higher velocity, than those in the cold region. This means that the net velocity of the charge carriers at the hot end moving towards the cold end will be higher than the net velocity of the charge carriers at the cold end moving towards the hot end. In this situation, more carriers will flow from the hot end towards the cold end than vice versa. This will cause a voltage to build up between the hot and cold ends of the sample. Eventually, equilibrium will be established and a potential will be set up. The same is true for the phonons. As phonons interact strongly with electrons and holes, they will drag these particles along with them, to create an additional potential. The magnitude of the measured thermopower is a complex function of both of these features and varies considerably with temperature. For example, the value of σ_S for platinum is approximately $+5\ \mu\text{V K}^{-1}$ at 100 K and $-2\ \mu\text{V K}^{-1}$ at 200 K. The thermopower also varies with direction in noncubic crystals.

15.2.3 Thermocouples, power generation and refrigeration

Thermocouples are a widely used application of the Seebeck effect, and they are the main means of temperature monitoring and regulation for measurements of temperatures above about 150 °C (Figure 15.13). This is because the potential generated in the circuit is easily measured, and metal thermocouples capable of operating up to temperatures of almost 2000 °C are available. In practice, one junction is maintained at 0 °C. The voltage generated by a thermocouple is related to the temperature difference between the junctions by a polynomial function, such as

$$T_\text{H} = a_0 + a_1\Delta V + A_2(\Delta V)^2 + \cdots + a_n(\Delta V)^n$$

where T_H is the temperature of the hot junction, ΔV is the measured voltage of the thermocouple, and a_0, a_1 and so on are constants. These coefficients depend on the reference junction temperature and the materials used in the device. The relationship between temperature and voltage is usually found by reference to 'thermocouple tables' supplied by manufacturers or located in handbooks.

A series of thermocouples linked in series form a thermopile (Figure 15.14). This has increased sensitivity compared with a single thermocouple when used to measure temperatures. The same arrangement can be used as a power generator. For this purpose, the low-temperature junctions (temperature T_C) are at a fixed temperature by connecting them to a heat sink, and the high-temperature junctions (temperature T_H) are in contact with a heat source, such as a radioactive sample. The potentiometer in Figure 15.14 is replaced by whatever needs to be powered by the electricity generated by the Seebeck effect.

Thermoelectric materials are also used for the generation of electricity and for refrigeration

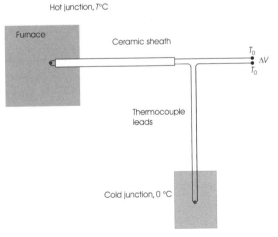

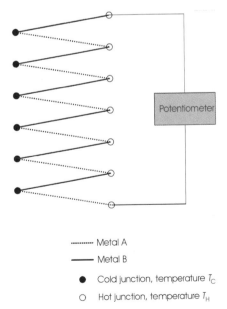

Figure 15.13 A thermocouple consists of a loop made of two different metals. One junction is kept at 0 °C and the other at the temperature to be measured, for example in a furnace. The voltage developed in the circuit, ΔV, is a measure of the temperature difference between the junctions

Figure 15.14 A thermopile, consisting of a number of thermocouples connected in series

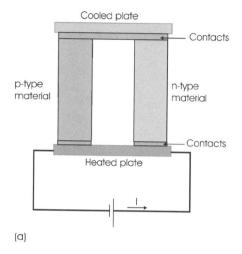

(a)

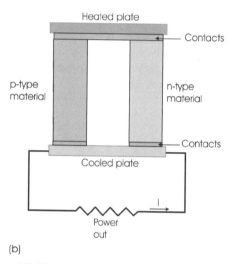

(b)

Figure 15.15 (a) Use of the Peltier effect for thermo-electric heating or cooling. A current, I, passed through a circuit containing n-type and p-type thermoelectric materials will cause one plate to become warmer and one to become cooler. Reversal of the current reverses the warm and cool plate. (b) Use of the Peltier effect for thermo-electric power generation. Current, I, will flow in a circuit containing a heated plate and a cooled plate connected by an n-type and a p-type thermoelectric material

utilising the Peltier effect. Two thermoelectric materials are coupled by metal plates, which act as the junctions (Figure 15.15a). A current passed through the circuit in one direction will heat one plate and cool the other. If the temperature of the hot junction

is constant, maintained by connection to a heat sink, continuous cooling will occur at the cold junction. Alternatively, if the temperature of the cold junction is fixed, continuous heating will occur at the hot junction. A reversal of the current will change the hot plate and cold plate. Such a device is called a heat pump. These are widely used in food and drinks coolers powered by car batteries.

The arrangement can also be used as a power source (Figure 15.15b). If one plate is continuously maintained hotter than the other, a current will flow in the circuit, and power is generated. In this format, these devices are used in space probes that operate too far from the sun for photoelectricity to be used for power supplies. In such cases, heat is generated by the slow decay of radioactive isotopes.

The effectiveness of devices using thermoelectric effects depends on the magnitude of the relative Peltier coefficient, Π, or its equivalent, the relative Seebeck coefficient, Σ. However, these are not the only material parameters of importance. As an example, consider the operation of a heat pump. The amount of heat produced or absorbed is

$$\Pi I = \Sigma T I$$

where I is the current flowing. The requirement for a large Σ acts to rule out metals as components, as metals have very low Seebeck coefficients. However, a low electrical resistivity is needed, to cut down on Joule heating, which points towards metals. Additionally, the thermal conductivity of the thermoelectric elements must be low to reduce the flow of heat from the hot to the cold region. This suggests an insulator, but these have high values of electrical resistivity. All of these conflicting factors are taken into account by using a figure of merit, ZT, for the material, given by:

$$ZT = \frac{T\Sigma^2}{\kappa\rho}$$

where Σ is the relative Seebeck coefficient of the thermoelectric elements, κ is the thermal conductivity, and ρ is the electrical resistivity. The best compromise is given by the material with the highest figure of merit. The figure of merit varies considerably with temperature and, although the best materials have a figure of merit of about 1.0, when combined in a device an overall energy conversion efficiency of only a few percent is realised at present. For small portable coolers, solid solutions of the semiconductors bismuth telluride (Bi_2Te_3) and antimony telluride (Sb_2Te_3), $Bi_xSb_{1-x}Te_3$, doped p-type and n-type, are used. Space vehicles use silicon–germanium alloys, Si_xGe_{1-x}.

Answers to introductory questions

What is zero-point energy?

The energy of a system of vibrating atoms, given by Equation (15.1),

$$E = \left(n + \frac{1}{2}\right) h\nu$$

shows that the lowest energy possible, when $n = 0$, is not zero, as it is for a classical system, but is equal to $\frac{1}{2}h\nu$ per atom. There is no restriction on the number of atoms that can occupy this lowest energy level and, at low temperatures, most will occupy this lowest energy state. At the lowest temperature, 0 K, all atoms will be in this lowest energy state and the solid will possess an equivalent amount of energy. This energy is the zero-point energy.

What solids are named high thermal conductivity materials?

Diamond has long been known to have a very high thermal conductivity. It has been realised that the high thermal conductivity is a reflection of the bonding and structure in diamond. Much effort is currently directed towards reproducing this high thermal conductivity in other materials by synthesis of nonmetallic compounds with the same structure and bonding as in diamond. Recently, a number of such structures have been synthesised. All have high thermal conductivity, and form the high thermal conductivity materials. Notable examples are boron nitride, BN, and aluminium nitride, AlN.

What physical property does thermoelectric refrigeration utilise?

Thermoelectric materials used for refrigeration utilise the Peltier effect. Two thermoelectric materials are coupled by metal plates, which act as the junctions. A current passed through the circuit in one direction will heat one plate and cool the other. For refrigeration to occur, the temperature of the hot junction is maintained constant by connection to a heat sink, which may simply be a fan. Continuous cooling will occur at the cold junction. These are widely used in food and drinks coolers powered by a car battery.

Further reading

D.K. Agrawal, 1994, '[NPZ]: A New Family of Real Materials for Low Thermal Expansion Applications', *Journal of Materials Education* **16** 139–165.

R.E. Hummel, 2001, *Electronic Properties of Materials* 3rd edn, Springer, New York.

The *Materials Research Society Bulletin* **26** (June 2001) contains a series of articles on thermal conductivity, including:

G.P. Srivastava, 'Theory of Thermal Conduction in Nonmetals', p. 445.

See also:

D.M. Rowe (Ed.), 1995, *CRC Handbook of Thermoelectrics*, CRC Press, Boca Raton, FL.

B.C. Sales, 1998, 'Electron Crystals and Phonon Glasses: A New Path to Improved Thermoelectric Materials'. *Materials Research Society Bulletin* **23** (January) 15–21.

A. W. Sleight, 1998, 'Compounds that Contract on Heating', *Inorganic Chemistry* **37** 2854–2860.

Problems and exercises

Quick quiz

1 The heat capacity at constant volume, C_v:
   (a) Is greater than the heat capacity at constant pressure, C_p

(b) Is less than the heat capacity at constant pressure, C_p

(c) Is equal to the heat capacity at constant pressure, C_p

2 The low-temperature heat capacity of a solid is proportional to:
(a) T^3
(b) T^{-3}
(c) T^{-2}

3 The high-temperature heat capacity of a solid is approximately:
(a) $2.5 \, \text{J K}^{-1} \, \text{mol}^{-1}$
(b) $25 \, \text{J K}^{-1} \, \text{mol}^{-1}$
(c) $250 \, \text{J K}^{-1} \, \text{mol}^{-1}$

4 The main contribution to the heat capacity of a solid is from:
(a) Phonons (lattice vibrations)
(b) Electrons
(c) Phonons plus electrons

5 The thermal conductivity of a solid is mainly due to:
(a) Phonons (lattice vibrations)
(b) Defects
(c) Free electrons

6 Alloys generally have:
(a) A higher thermal conductivity than the parent metals
(b) A lower thermal conductivity than the parent metals
(c) About the same thermal conductivity as the parent metals

7 Compared with a poorly crystalline polymer, a highly crystalline polymer has:
(a) A lower thermal conductivity
(b) A higher thermal conductivity
(c) About the same thermal conductivity

8 The mean volume expansivity of liquid mercury is $18.2 \times 10^{-5} \, \text{K}^{-1}$. The mean linear expansivity is:
(a) $54.6 \times 10^{-5} \, \text{K}^{-1}$

(b) $6.06 \times 10^{-5} \, \text{K}^{-1}$
(c) $2.63 \times 10^{-5} \, \text{K}^{-1}$

9 Solids linked with strong chemical bonds have:
(a) A lower thermal expansivity than weakly bonded solids
(b) A greater thermal expansivity than weakly bonded solids
(c) About the same thermal expansivity as weakly bonded solids

10 A current flow induced in a circuit made of two different conductors by holding the junctions between the materials at different temperatures is called:
(a) The Peltier effect
(b) The Thomson effect
(c) The Seebeck effect

11 A temperature difference between the two junctions in a circuit made of two different conductors induced by a current flow is called:
(a) The Seebeck effect
(b) The Peltier effect
(c) The Thomson effect

12 A thermocouple makes use of:
(a) The Seebeck effect
(b) The Peltier effect
(c) The Thomson effect

13 Thermoelectric refrigerators utilise:
(a) The Thomson effect
(b) The Seebeck effect
(c) The Peltier effect

Calculations and questions

15.1 How much energy is needed to raise the temperature of 2.5 moles of alumina from $0 \, °\text{C}$ to $120 \, °\text{C}$, taking the specific heat capacity of alumina, $0.907 \, \text{J K}^{-1} \, \text{g}^{-1}$, to be independent of temperature?

15.2   How much energy has to be extracted to lower the temperature of 15 g tungsten metal from 2500 K to 1500 K? The molar heat capacity at 2000 K, $32.26\,\mathrm{J\,K^{-1}\,mol^{-1}}$, can be considered to apply across the whole of this temperature range.

15.3   The specific heat of silicon at 50 K is $2.162\,\mathrm{J\,K^{-1}\,mol^{-1}}$. Estimate the value at the boiling point of neon, 27.07 K.

15.4   Calculate the specific heat of silicon at the boiling point of neon by using the fact that the Debye temperature of silicon is 645 K. Compare this with the result in Question 15.3.

15.5   A styrofoam box is used to transport 10 kg meat. The box is $50\,\mathrm{cm} \times 30\,\mathrm{cm} \times 20\,\mathrm{cm}$, and the foam thickness is 5 cm. How long will it take the contents to increase in temperature from the initial $0\,^\circ\mathrm{C}$ to $5\,^\circ\mathrm{C}$, assuming that the heat capacity of meat is $4.22\,\mathrm{J\,K^{-1}\,g^{-1}}$ and the thermal conductivity of styrofoam is $0.035\,\mathrm{W\,m^{-1}\,K^{-1}}$?

15.6   Show that the equation for the heat transfer across a number of slabs of material with the same surface area, A, is

$$\frac{dQ}{dt} = A\Delta T \left[\sum_{i=1}^{n} \frac{\Delta x_i}{\kappa_i} \right]^{-1},$$

where ΔT is the total temperature drop, $T_1 - T_2$, and Δx_i is the thickness of slab i, of thermal conductivity κ_i. [Note: derivation is not given in the answers at the end of this book.]

15.7   A cooking pot of 15 cm diameter and a base of 3 mm copper and 1 mm stainless steel, contains 2 l of water at $20\,^\circ\mathrm{C}$. (a) What is the initial rate of heat transfer if the hot plate is at $150\,^\circ\mathrm{C}$? How does this compare with a pan of the same dimensions with (b) a solid copper bottom 4 mm thick and (c) a solid stainless steel bottom, 4 mm thick. The thermal conductivity of copper is

$403\,\mathrm{W\,m^{-1}\,K^{-1}}$ and that of stainless steel is $18\,\mathrm{W\,m^{-1}\,K^{-1}}$.

15.8   A window of area $2\,\mathrm{m} \times 1.30\,\mathrm{m}$ is glazed with a single sheet of glass 5 mm thick. (a) What is the heat loss per hour from a room at $25\,^\circ\mathrm{C}$ when the outside temperature is $4.5\,^\circ\mathrm{C}$? (b) If the area is double glazed with two such sheets, separated by an air gap of 1 cm, what will the heat loss per hour be? The thermal conductivity of the glass is $0.96\,\mathrm{W\,m^{-1}\,K^{-1}}$ and that of air is $2.41 \times 10^{-2}\,\mathrm{W\,m^{-1}\,K^{-1}}$.

15.9   A gap is left between rails in a railway so that the rails can expand without causing track buckling at high temperatures. What gap needs to be left between 10 m rail lengths installed at $10\,^\circ\mathrm{C}$ if the ground temperature might reach $50\,^\circ\mathrm{C}$? The expansivity of steel is $10.7 \times 10^{-6}\,\mathrm{K^{-1}}$.

15.10   A volume of mercury of $10^{-6}\,\mathrm{m^3}$ at $20\,^\circ\mathrm{C}$ is contained in glass bulb, with expansion taken up by the mercury moving into a capillary 0.5 mm diameter, similar to a mercury thermometer. The aim is to allow the mercury to expand and complete an electrical circuit and activate a cooling device. If the circuit contact is 5 mm above the mercury level at $20\,^\circ\mathrm{C}$, what temperature will activate the device? The mean volume expansivity of liquid mercury is $18.2 \times 10^{-5}\,\mathrm{K^{-1}}$.

15.11   (a) Estimate the mean coefficient of linear expansion for the a and c axes of $CaZr_4P_6O_{24}$, using the data in Figure 15.8. (b) What is the mean coefficient of volume expansion of this material?

15.12   The voltage generated across a Pt–Au thermocouple when the cold junction is at $0\,^\circ\mathrm{C}$ and the hot junction is at $100\,^\circ\mathrm{C}$ is $+780\,\mu\mathrm{V}$. The experimental setup is as shown in Figure 15.12a, with Pt as metal B, and Au as metal A. Calculate the average value of Σ_{AuPt} over the temperature range. Assuming that the absolute value of the Seebeck coefficient for Pt, σ_S, is $-6.95 \, \mu\text{V K}^{-1}$ over the whole of this temperature range, estimate the average value of the absolute Seebeck coefficient for Au.

15.13 The voltage generated across a Pt–Al thermocouple when the cold junction is at $0\,^\circ\text{C}$ and the hot junction is at $100\,^\circ\text{C}$ is $+420 \, \mu\text{V}$. The experimental setup is as shown in Figure 15.12a, with Pt as metal B, and Al as metal A. Calculate the average value of Σ_{AlPt} over the temperature range. Assuming that the absolute value of the Seebeck coef-

ficient for Pt, σ_S, is $-6.95 \, \mu\text{V K}^{-1}$ over the whole of this temperature range, estimate the average value of the absolute Seebeck coefficient for Al.

15.14 The absolute Seebeck coefficients for lead, σ_{Pb}, at 300 K is $-1.047 \, \mu\text{K}^{-1}$, and for platinum, σ_{Pt}, at 300 K is $-5.05 \, \mu\text{V K}^{-1}$. A thermocouple is constructed as in Figure 15.12a, with Pb as metal A, and Pt as metal B. Estimate the voltage generated when the cold junction is at $0\,^\circ\text{C}$ and the hot junction at 300 K.

PART 5

Nuclear properties of solids

16

Radioactivity and nuclear reactions

- What is the difference between an isotope and a nuclide?

- What chemical or physical procedures can be used to accelerate radioactive decay?

- Why does nuclear fission release energy?

16.1 Radioactivity

The properties described earlier in this book are, in principle, a function of the outer electron configuration of each individual atom, and the nucleus of each atom can be regarded as simply providing mass. However, some of the heaviest atomic nuclei have been found to be unstable. The elements that show this phenomenon are said to be radioactive. It is now known that radioactivity is an external manifestation of the spontaneous disintegration of a nucleus.

At the turn of the 20th century only the two heaviest naturally occurring elements known at that epoch, thorium, Th, with atomic number 90, and uranium, U, with an atomic number of 92, were known to be radioactive, but research has revealed that all elements with an atomic number of more than 83, bismuth, are naturally radioactive. Since then many of the lighter, normally stable, atoms have also been made artificially radioactive in the laboratory.

16.1.1 Radioactive elements

The term radioactivity was coined by Marie Curie to describe the phenomenon of an atomic species constantly emitting 'penetrating radiation', that is, radiation capable of passing through matter and ionising gas. The radiation is emitted by unstable atomic nuclei in order to gain stability. The earliest studies indicated that the penetrating radiation was found to be composed of three components:

- α-particles, which were subsequently found to be helium nuclei: these have a very low penetrating power and can be stopped by thin card;

- β-particles, which were subsequently found to be electrons: these have a medium penetrating power;

- γ-rays, which were subsequently found to be high-energy photons: these have a high penetrating power and can pass through 1 m of concrete.

Although other components of 'penetrating radiation' have since been discovered, the three listed,

Understanding solids: the science of materials.   Richard J. D. Tilley
© 2004 John Wiley & Sons, Ltd   ISBNs: 0 470 85275 5 (Hbk) 0 470 85276 3 (Pbk)

together with neutrons (see below), make up the most important categories.

Very soon after the radioactivity of thorium and uranium had been discovered it was found that pure samples of both of these elements were only very weakly radioactive. However, such pure samples became more and more radioactive with time until they reached a steady level identical to that in the original samples before purification. This suggested that the uranium or thorium atoms were transforming or decaying into other radioactive 'daughter elements' and that hitherto undiscovered series of such elements might exist. The search for the radioactive products of uranium by Marie and Pierre Curie led to the characterisation of two new elements, which were named polonium, Po, and radium, Ra. Both elements are far more radioactive than uranium and decay so rapidly that no ore deposits are formed. They exist only because they are formed constantly from naturally occurring uranium.

Since that time, in the early years of the 20th century, many radioactive elements have been prepared, and radioactive forms of the majority of the elements in the periodic table are readily available for research purposes.

16.1.2 Isotopes and nuclides

For the present purposes the nucleus of an atom can be considered to be made up of protons and neutrons, collectively known as nucleons. A proton has a charge of $+1$ and the number of protons in a nucleus determines the proton number (also called the atomic number), Z, of the atom. The neutron bears no charge but has a similar mass to that of the proton. The number of electrons in a neutral atom is also equal to Z. Atoms with the same value of Z are chemically identical. The total number of nucleons in the nucleus is called the nucleon number (also called the mass number), A. The number of neutrons in a nucleus need not be the same as the number of protons. Atoms with the same value of Z but different numbers of neutrons are known as isotopes of the element in question. Isotopes are represented by the symbol $^A_Z X$. For example, the radioactive

carbon isotope used in radiocarbon dating has a symbol $^{14}_6 C$. Because the name of the element or its symbol already contains information about the atomic number of the atom, this is often omitted. Thus the radioactive carbon atom is often referred to as carbon-14 or ^{14}C. An isotope that is radioactive is called a radioisotope. A nuclide is any atomic species that has a specified nucleon number and proton number. Hence $^9_4 Be$ is a nuclide.

16.1.3 Nuclear equations

In order to describe radioactive transformations, nuclear equations are needed. These are very similar to chemical equations, except that the nucleon numbers and proton numbers of each reactant must also be specified, that is, the reactions are written with nuclides. A typical nuclear equation, representing the decay of an isotope of uranium, uranium-238, is:

$$^{238}_{92}U \rightarrow ^{234}_{90}Th + ^4_2 He$$

The reactant is uranium-238 ($^{238}_{92}U$), a naturally occurring radioisotope that emits α-particles, $^4_2 He$. The change in the proton number, from 92 to 90, specifies that a different chemical element has been produced; a fact confirmed by the chemical symbol Th. The other product, the α-particle, is specified in a similar way. As it is a helium nucleus the chemical symbol is He. The mass number, 4, and the atomic number, 2, are added to the symbol to complete matters. A reaction of this type, in which an α-particle is given out, is called an α-decay.

A similar reaction, but involving a β-particle, is:

$$^{234}_{90}Th \rightarrow ^{234}_{91}Pa + ^{\ 0}_{-1} e$$

Here, radioactive thorium-234 emits an electron, $^{\ 0}_{-1} e$, also called a β-particle, and written β^-. There is no change in the nucleon number, but the proton number has increased by one in the transformation. [Actually, this reaction is not completely correct as written. In addition to the particles listed, an anti-neutrino, a particle with (apparently) no mass or charge, is needed to conserve spin and is always emitted during a β-decay. These particles need not concern us further as they have no chemical effects under normal circumstances.] A reaction of this type, in which a β-particle is given out, is called a β-decay.

There are a number of rules that must be followed in writing nuclear equations.

- The sum of the proton numbers on the left-hand side of the equation must equal the sum of the proton numbers on the right-hand side.

- The sum of the nucleon numbers on the left-hand side must equal the sum of the nucleon numbers on the right-hand side.

Additionally, we can note that:

- When a radioactive element emits an α-particle the daughter element has a nucleon number 4 less than that of the parent, and a proton number of 2 less.

- When a radioactive element emits a β-particle the daughter element has a nucleon number the same as the parent and an proton number 1 greater than that of the parent.

- Electrons are not present in the nucleus and are produced (at least schematically) via 'decomposition' of neutrons, following the nuclear reaction:

$$^1_0 n \rightarrow {}^1_1 H + {}^{\,0}_{-1} e$$

In all β-decay transformations an antineutrino is also emitted but not written into the equation.

Often, nuclear equations are written in a shorthand form. An example of this nomenclature is given below for the reaction between uranium-238 and a deuteron, the nucleus deuterium or heavy hydrogen, which produces neptunium-238 and two neutrons:

$$^{238}_{92} U + {}^2_1 H \rightarrow {}^{238}_{93} Np + 2\,{}^1_0 n$$

we may also write this as

$$^{238}U(d, 2n)^{238}Np$$

In this form of shorthand, the particles involved are placed in parenthesis and given letter symbols, d for deuteron, which is a nucleus of deuterium, the isotope of hydrogen with nucleon number of 2, and n for neutron. Further examples will be given in this chapter.

16.1.4 Radioactive series

The series of transformations that take place as a radioactive element changes into successive daughter elements halts when a stable (nonradioactive) element forms. The sequence of transformations is called a radioactive series. Four different radioactive series have been described, three of which occur naturally.

16.1.4.1 The uranium series

The parent nuclide is the naturally occurring isotope uranium-238, ^{238}U, and the series ends with the stable nuclide lead-206, ^{206}Pb (Figure 16.1). The parent nuclide is shown at the top right-hand side of the figure. An α-decay is represented by a diagonal displacement of 2 proton number units to the left. A β-decay is shown as a horizontal displacement of 1 proton number unit to the right. The nucleon numbers, A, of the members of the series conform to the formula $4x + 2$, where x takes values between 51 for lead-206 and 59 for uranium-238.

Of the atomic species taking part in this cascade, only uranium-238 has a long lifetime. Apart from the stable, nonradioactive end-product lead-206, all other species decay rapidly. These include two other radioactive lead nuclides, Pb-210 and Pb-214, which have a transitory existence. Some nuclides have two alternative ways of decay and so the path occasionally branches. For example, Bi-214 can form Po-214 by a β-decay, or Tl-210 by an α-decay. [Not all of the possible branching reactions

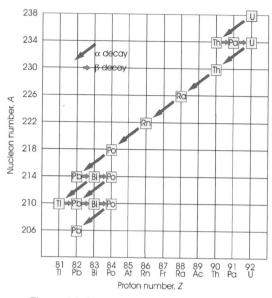

Figure 16.1 The uranium-238 decay series

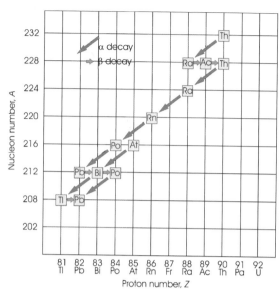

Figure 16.2 The thorium-232 decay series

that have been observed are included in Figure 16.1.] All of the daughter elements in the series are metals, with the exception of the noble gas radon. Radon is a naturally occurring radioactive gas, especially associated with granite in nature. The radon isotope formed as part of the uranium series is an α-emitter that decays according to the following reaction:

$$^{222}_{86}\text{Rn} \rightarrow {}^{218}_{84}\text{Po} + {}^{4}_{2}\text{He}$$

It is produced by the decay of radium-226 thus:

$$^{226}_{88}\text{Ra} \rightarrow {}^{222}_{86}\text{Ra} + {}^{4}_{2}\text{He}$$

Different isotopes of radon form in the radioactive series described below.

16.1.4.2 The thorium series

This series (Figure 16.2) starts with naturally occurring thorium-232 and ends with the stable isotope lead-208. The daughter elements all have nucleon numbers divisible by four, and so the series formula is $4x + 0$. The first reaction in the series is:

$$^{232}_{90}\text{Th} \rightarrow {}^{228}_{88}\text{Ra} + {}^{4}_{2}\text{He}$$

16.1.4.3 The actinium series

This series (Figure 16.3) was so called because it was originally thought that actinium-227 was the parent element. However, actinium-227 decays too quickly for the series to persist for any length of time, and eventually uranium-235 was proved to be the true parent element. Uranium-235 is the less common naturally occurring isotope of uranium. The series ends with yet another stable lead isotope, lead-207, and the series formula is $4x + 3$.

16.1.4.4 The neptunium series

The existence of the three naturally occurring series characterised by the formulae $4x$, $4x + 2$ and

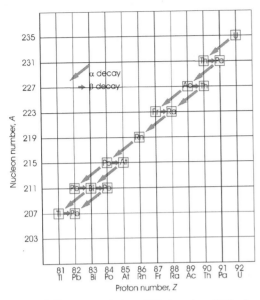

Figure 16.3 The actinium-227 (uranium-235) decay series

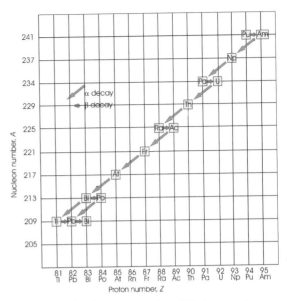

Figure 16.4 The plutonium-241 decay series

$4x + 3$, lead to the expectation of a fourth series, with isotopic weights given by $4x + 1$. No isotope with a sufficiently long lifetime is found in nature for this series to exist outside of the laboratory. Eventually, the discovery of the transuranic element neptunium, Np (see Section 16.1.5), enabled much of the series to be constructed (Figure 16.4). As can be seen, the parent nuclide is not neptunium-237, which is the longest-lived radioactive isotope in the series, but plutonium-241. This series differs from the three naturally occurring series by ending not with an isotope of lead but with bismuth-209.

16.1.5 Transuranic elements

Energetic particles, either from naturally radioactive materials or from particle accelerators, can be used to bring about atomic transmutations and so form new isotopes. In this way, the number of known elements has been extended above uranium, the heaviest naturally occurring element, with atomic number 92. All these artificial heavy elements are radioactive, and many have very short lifetimes.

They form part of a series in which the 5f electron orbitals become occupied in the ground state of the elements. In this sense they can be considered to be analogous to the lanthanides, in which the 4f orbitals are occupied in the ground state of the elements, and have been named the actinides. The actinides are listed in Table 16.1. They appear in the periodic table after element 88, radium, Ra, and end with element 103, lawrencium, Lr.

16.1.6 Artificial radioactivity

When ordinary stable nuclei are bombarded by sufficiently energetic radiation they can be transformed into artificially radioactive species. Many of the elements with atomic numbers lighter than 82, bismuth, Bi, which are not normally radioactive, have now been made in radioactive forms.

Nuclear reactions of this type were observed in the earliest years of the 20th century, when research into radioactive materials was shrouded in mystery. The first nuclear transformation described as such was interpreted by Rutherford in 1919. In this

Table 16.1 The actinide elements

Name	Symbol	Proton number, Z	Nucleon number, A^a	Half-life^b
Actinium	Ac	89	227	21.8 years
Thorium	Th	90	232	1.41×10^{10} years
Protactinium	Pa	91	231	2.38×10^4 years
Uranium	U	92	238	4.47×10^9 years
Neptunium	Np	93	237	2.14×10^6 years
Plutonium	Pu	94	244	8.1×10^7 years
Americium	Am	95	243	7.38×10^3 years
Curium	Cm	96	247	1.6×10^7 years
Berkelium	Bk	97	247	1.38×10^3 years
Californium	Cf	98	249	350 years
Einsteinium	Es	99	254	277 days
Fermium	Fm	100	257	100 days
Mendelevium	Md	101	258	55 days
Nobelium	No	102	259	1 hour
Lawrencium	Lr	103	260	3 minutes

^{a} Nucleon number of the most stable isotope.
^{b} Half-life of the most stable isotope.

reaction energetic α-particles collide with nitrogen atoms to produce an isotope of oxygen and a proton:

$$^{14}_{7}\text{N} + ^4_2\text{He} \rightarrow ^{17}_8\text{O} + ^1_1\text{H}$$

The reaction that Chadwick used to establish the existence of the neutron, in 1932, was similar, and involved the bombardment of beryllium, Be, with energetic α-particles:

$$^9_4\text{Be} + ^4_2\text{He} \rightarrow ^{12}_6\text{C} + ^1_0\text{n}$$

The first artificial radioisotope to be produced was made by Irene Curie and Joliot in 1934. The reaction again involved the use of energetic α-particles, colliding this time with boron, B. The product of the reaction, an isotope of nitrogen, decayed by emission of a positive electron or positron, thus:

$$^{10}_5\text{B} + ^4_2\text{He} \rightarrow ^{13}_7\text{N} + ^1_0\text{n}$$
$$^{13}_7\text{N} \rightarrow ^{13}_6\text{C} + ^0_{+1}\text{e}$$

Release of a positron is called β^+-decay and, as in the case of β-decay, a neutrino is also given out. Two further examples of this process are:

$$^{38}_{19}\text{K} \rightarrow ^{38}_{18}\text{Ar} + ^0_{+1}\text{e}$$
$$^{120}_{51}\text{Sb} \rightarrow ^{120}_{50}\text{Sn} + ^0_{+1}\text{e}$$

Positrons, like electrons, do not exist in the nucleus, and are believed to be generated by the transformation of a proton into a neutron thus:

$$^1_1\text{H} \rightarrow ^1_0\text{n} + ^0_{+1}\text{e}$$

Positrons have a lifetime of about 10^{-9} s, and are annihilated by combination with electrons to produce γ-radiation, which is high-energy electromagnetic radiation (see Section 14.1.1).

A wide variety of reactions are now utilised for the production of artificial radioisotopes. Some examples are:

$$^6_3\text{Li} + ^1_0\text{n} \rightarrow ^4_2\text{He} + ^3_1\text{H}$$
$$^{14}_7\text{N} + ^1_1\text{H} \rightarrow ^{11}_6\text{C} + ^4_2\text{He}$$
$$^{58}_{26}\text{Fe} + 2^1_0\text{n} \rightarrow ^{60}_{27}\text{Co} + ^0_{-1}\text{e}$$

Cobalt-60 is used in cancer therapy. It breaks down according to the reaction:

$$^{60}_{27}\text{Co} \rightarrow ^{60}_{28}\text{Ni} + ^0_{-1}\text{e}$$

The emission of γ-rays during a nuclear reaction does not change either mass number or atomic number, because they are high-energy photons. An example is:

$$^{16}_{7}N \rightarrow\ ^{16}_{8}O +\ ^{0}_{-1}e + \gamma$$

As in this reaction, the production of γ-rays often takes place after an α-emission or β-emission. These processes frequently leave the daughter nucleus in an excited state, which subsequently loses energy by way of γ-emission. Although γ-emission is of importance, it is usually not indicated when nuclear equations are written.

16.2   Rates of decay

16.2.1   Nuclear stability

Although many radioactive isotopes have been discovered or prepared, most known nuclei are stable. Although it is not possible to explain nuclear stability theoretically, empirical observations allow one to guess the likely stability of any particular nuclide. A useful guide is a graph of the number of protons, Z, versus the number of neutrons in the stable nuclei, or the nucleon number (Figure 16.5). A narrow strip, the band of stability, is found, within which the nuclei are stable. Nuclei above the band of stability tend to emit positrons or to incorporate an outer electron into the nucleus (called electron capture) so as to move the product nucleus nearer to the band of stability. Nuclei below the band of stability tend to emit electrons for the same reason. Thus carbon-14 has a proton-to-neutron ratio of 6/8 (i.e. 0.75). This is well below the band of stability, and carbon-14 would be expected to decay by β-emission, which is in accord with experimental evidence. All nuclei with more than 82 protons are radioactive. Only a small number of stable nuclei have odd numbers of protons or neutrons. Most stable nuclei contain an even number of protons or neutrons. There are only four stable nuclei that contain an odd number of both protons and neutrons: $^{2}_{1}H$, $^{6}_{3}Li$, $^{10}_{5}B$, $^{14}_{7}N$. Nuclei that contain 2, 8, 20, 28, 50, 82 and 126 protons or neutrons

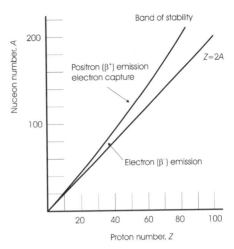

Figure 16.5 The band of stability of the elements: isotopes close to the band are stable; those above it tend to decay via positron (β^{+}) emission or electron capture, whereas those below it tend to decay via electron (β^{-}) emission. The isotopes with equal numbers of protons and neutrons, the $2Z = A$ line, are generally unstable except for the lightest elements

are particularly stable, and these numbers are called magic numbers.

16.2.2   The rate of nuclear decay

One of the most important properties of a radioactive nuclide is its lifetime. At present it is not possible to predict theoretically when any particular nucleus in a sample will decay. However, the number of nuclides in a sizeable sample that will decompose in a given time can be measured, and it is found that this rate of decay is characteristic of a given isotope. In fact, the rate of decay of an isotope is constant and unvarying. That is, if a fraction of a radioactive nuclide decays in a certain time interval t, then the same fraction of the remainder will decay in another increment of time t, irrespective of external conditions. Nuclear reactions are not affected by outside influences such as temperature and pressure and it is not possible to significantly alter the constant rate of radioactive decay. For example, radioactive strontium-90, an important product of nuclear fission (see Section 16.3.2), decays by β-decay to yield the daughter atom yttrium, Y:

$$^{90}_{38}Sr \rightarrow \, ^{90}_{39}Y + \, ^{0}_{-1}e$$

It is found that it will take 28.5 years for half of the sample to decay and another 28.5 years for half of the remaining strontium-90 to decay and so on:

Time (years)	0	28.5	57	85.5
Amount of Sr-90 (g)	1	$\frac{1}{2}$	$\frac{1}{4}$	$\frac{1}{8}$

The lifetime of a radioactive substance is usually quoted in terms of the time in which half of the sample decays, called its half-life, $t_{1/2}$. The half-lives of some radioisotopes are given in Table 16.2.

The number of atomic disintegrations that occur in a radioactive material per second is called its activity. A rate of decay of 1 disintegration per second is the SI unit of activity, 1 becquerel (1 Bq). [The equivalent non-SI unit is the curie (Ci):

Table 16.2   Half-lives of some radioisotopes

Nuclide	Half-life	Decay mode
$^{3}_{1}H$ (tritium)	12.33 years	β^-
$^{14}_{6}C$	5730 years	β^-
$^{22}_{11}Na$	2.602 years	β^+, γ
$^{47}_{20}Ca$	4.536 days	β^-, γ
$^{59}_{26}Fe$	44.496 days	β^-, γ
$^{60}_{27}Co$	5.271 years	β^-, γ
$^{90}_{38}Sr$	28.5 years	β^-
$^{131}_{53}I$	8.040 days	β^-, γ
$^{133}_{54}Xe$	5.245 days	β^-, γ
$^{137}_{55}Cs$	30.1 years	β^-, γ
$^{198}_{79}Au$	2.6395 days	β^-, γ
$^{222}_{86}Rn$	2.825 days	α, γ
$^{226}_{88}Ra$	1600 years	α, γ, X^a
$^{228}_{88}Ra$	5.75 years	β^-, γ
$^{235}_{92}U$	7.037×10^8 years	α, γ, X^a
$^{238}_{92}U$	4.468×10^9 years	α, γ

^{a} X-ray emission.

1 Ci = 3.7×10^{10} disintegrations per second, equal to 3.7×10^{10} Bq. A curie is the radioactivity of 1 g of radium-226.] The specific activity of a material is the activity of 1 gram of that material. Radium has a specific activity of 3.7×10^{10} Bq [1 Ci].

The disintegration of atomic nuclei can be expressed by the rate equation:

$$\text{Activity} = \text{rate of decay} = \frac{-dN}{dt} = kN \quad (16.1)$$

where dN/dt is the rate of disintegration in units of kilograms, moles, atoms, etc., per unit time (seconds, minutes, years) and N is the number of kilograms, moles, atoms, etc., present at any given instant. The proportionality constant, k, is different for each radioactive isotope and has units of 'per time' (i.e. s^{-1}, min^{-1}, y^{-1}, etc.). This type of rate equation is called a first-order rate law.

To relate the concentration N_0 that exists at time $t = 0$ to the amount present at any later time, N, Equation (16.1) must be integrated, to give:

$$N = N_0 \, e^{-kt} \quad (16.2)$$

This is shown graphically in Figure 16.6. Equation (16.2) is often written in the logarithmic form:

$$\ln N - \ln N_0 = -kt$$

thus

$$\ln \left(\frac{N_0}{N} \right) = kt \quad (16.3)$$

The value of the rate constant, k, can be determined from a plot of ln N against t, the slope of the straight-line graph being $-k$.

Substitution of $N = N_0/2$ into Equation (16.3) shows that the half-life, $t_{1/2}$, is given by:

$$t_{1/2} = \frac{0.693}{k} \quad (16.4)$$

The fraction of the radioisotope remaining after n half-lives have elapsed is:

$$\frac{N}{N_0} = \left(\frac{1}{2} \right)^n \quad (16.5)$$

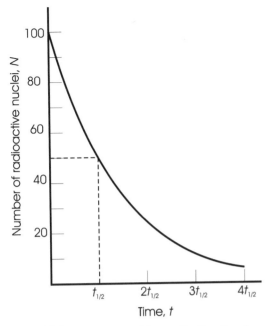

Figure 16.6 The exponential rate of radioactive decay. The time for the number of radioactive nuclei present in any sample to fall to one half of the original number is the half-life, $t_{1/2}$

The specific activity of a material can be related to its half-life in the following way:

$$\text{activity} = \text{rate of decay} = \frac{-\mathrm{d}N}{\mathrm{d}t} = kN$$

The specific activity is the activity divided by the mass in grams, m:

$$\text{specific activity} = \frac{\text{activity}}{m} = \frac{kN}{m}$$

Substituting for k from Equation (16.4) gives:

$$\text{specific activity} = \frac{0.693N}{t_{1/2}\,m}$$

The number of atoms, N, remaining at any time, and their mass, m, are related in the following way. The number of moles present is given by:

$$\text{moles of material} = \frac{N}{N_{\mathrm{A}}}$$

where N_{A} is the Avagadro constant. The mass of this quantity is:

$$m = M\left(\frac{N}{N_{\mathrm{A}}}\right)$$

where M is the isotopic mass. Hence,

$$\frac{N}{m} = \frac{N_{\mathrm{A}}}{M}$$

and

$$\text{specific activity} = \frac{0.693N_{\mathrm{A}}}{Mt_{1/2}}$$
$$= \frac{4.2 \times 10^{23}}{Mt_{1/2}} \text{ disintegrations s}^{-1}\,\text{g}^{-1}$$

16.2.3 Radioactive dating

One of the principle problems confronting geologists and archaeologists is the accurate dating of the materials under examination. Because of the fixed rate of decay of radioactive isotopes, naturally occurring radioactive minerals can be used to date rocks. An example is afforded by the uranium series described in Sub-section 16.1.4.1. The starting nuclide, uranium-238, decays eventually to lead-206. The half-life of the most stable nuclide, uranium-238, is 4.51×10^9 years. A sample of rock containing uranium is analysed for the amounts of uranium-238 and lead-206 present. It is then assumed that all of the lead-206 has been derived from the decay of uranium-238. This allows the initial amount of uranium-238 to be calculated, and the time over which the decay has occurred is then obtained from Equations (16.1)–(16.5). The utility of the method depends on the types of rock that contains U-238, and the time-scale for accurate dating depends on the longest half-life. The U-238 – Pb-206 method can be used with minerals such as zircon, and gives dates in the range of 10 million years ago to 40 000 million years ago.

In order to extend the range of radioactive dating, other nuclides, with different half-lives, occurring in different mineral types have been sought. Two of the most widely used are:

$$^{87}_{37}\text{Rb} \rightarrow {}^{87}_{38}\text{Sr} + {}^{0}_{-1}\text{e} \quad t_{1/2} = 4.9 \times 10^{10} \text{ years}$$

This series gives dates over the same time-range as the U-238 – Pb-208 analysis but can use mica, a mineral with a different distribution to uranium-containing minerals. This makes it applicable to artefacts not located adjacent to uranium minerals.

The reaction

$$^{40}_{19}\text{K} \rightarrow {}^{40}_{18}\text{Ar} + {}^{0}_{-1}\text{e} \quad t_{1/2} = 1.3 \times 10^{9} \text{ years}$$

gives dates over the time-scale 100 000 years to 30 000 million years, and uses minerals such as the mica biotite, $K(MgFe)_3(AlSi_3O_{10})(OH)_2$, that naturally contain potassium.

In order to date organic materials less than about 50 000 years old, the radioactive decay of carbon-14 is preferred. Carbon-14 is produced at a steady rate in the Earth's upper atmosphere as a result of the interaction of cosmic ray neutrons with nitrogen:

$$^{14}_{7}\text{N} + {}^{1}_{0}\text{n} \rightarrow {}^{14}_{6}\text{C} + {}^{1}_{1}\text{H}$$

Carbon-14 subsequently decays thus:

$$^{14}_{6}\text{C} \rightarrow {}^{14}_{7}\text{N} + {}^{0}_{-1}\text{e} \quad t_{1/2} = 5730 \text{ years}$$

The carbon-14 is distributed throughout the atmosphere in the form of carbon dioxide (CO_2) molecules and, because of atmospheric diffusion, a fairly constant proportion of all CO_2 is therefore radioactive owing to the presence of carbon-14. Living plants absorb CO_2 and so incorporate carbon-14 into their tissues. The relative quantity of carbon-14 in an organism remains constant until that organism dies. At this point the carbon-14 begins to decay at its normal rate. A measurement of the radioactivity of the once-living samples can then be used to determine the age of the sample itself.

16.3   Nuclear power

16.3.1   The binding energy of nuclides

When protons and neutrons are (conceptually) brought together to form an atomic nucleus, the mass of the resulting nucleus is less than the combined masses of the protons and neutrons. This mass difference, when expressed as energy, is called the binding energy of the nucleus. Writing the formation of a nucleus $^{A}_{Z}\text{X}$ as a pseudo-chemical reaction,

$$Z \text{ protons} + (A - Z) \text{ neutrons} \rightarrow {}^{A}_{Z}\text{X}$$

the mass difference, Δm, is given by

$$\Delta m = \text{mass of products} - \text{mass of reactants}$$
$$= \text{mass} [{}^{A}_{Z}\text{X}] - \text{mass} [Z \text{ protons}$$
$$+ (A - Z) \text{ neutrons}]$$

This difference mass is converted into energy by using the Einstein equation:

$$E = \Delta m \, c^2$$

Where c is the speed of light. In making these calculations it is simplest to use the masses of the particles in atomic mass units, u (see Chapter 1) and then to convert to kilograms later. Moreover, it is usual to use the mass of the hydrogen isotope $^{1}_{1}\text{H}$ rather than the mass of the proton in making these calculations. This is because the mass of isotopes is obtained from the relative molar mass of atoms, and so the mass of the electrons present on the $^{1}_{1}\text{H}$ isotopes cancels neatly with the mass of the electrons on the $^{A}_{Z}\text{X}$ isotope.

The isotopic binding energy is often quoted per nucleon (Figure 16.7). The curve is smooth over much of the range. For the 'earliest' elements there is a series of peaks that occurs for the isotopes ^4He, ^{12}C and ^{16}O. The curve rises to a maximum at ^{56}Fe and decreases slightly thereafter.

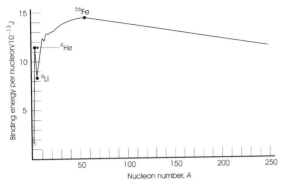

Figure 16.7 The isotopic binding energy per nucleon. The most stable isotope is ^{56}Fe. Isotopes of lower nucleon number than ^{56}Fe will release energy in fusion reactions. Isotopes of higher nucleon number than ^{56}Fe will release energy on fission

16.3.2 Nuclear fission

Nuclear fission is the breaking apart of atomic nuclei into two or more pieces. This can take place spontaneously in the case of the heaviest atoms. Neutron bombardment of atoms can also cause the nuclei to break apart. This process is called induced nuclear fission. Atoms that undergo this process are called fissionable. Some nuclides can undergo fission with slow (not very energetic, or thermal) neutrons. These atoms are called fissile.

The neutron, being uncharged, is not repelled by the positive charge on the nucleus, and makes an ideal nuclear probe. Soon after the discovery of this property many experiments were carried out to try to make new elements that were more massive than uranium by bombarding heavy atoms, notably U itself, with neutrons. Two such elements that can be made this way are neptunium, Np, and plutonium, Pu:

$$^{238}_{92}U + ^{1}_{0}n \rightarrow ^{239}_{93}Np + ^{0}_{-1}e$$

Neptunium has a half-life of about 2 days and decays to plutonium-239 by way of β-decay:

$$^{239}_{93}Np \rightarrow ^{239}_{94}Pu + ^{0}_{-1}e$$

Plutonium has a half-life of about 24 000 years.

However, the experiments most often resulted in fission of the heavy nuclei into two more-or-less equal parts during the bombardment. For example, fission of uranium-235 can produce krypton, Kr, and barium, Ba:

$$^{235}_{92}U + ^{1}_{0}n \rightarrow ^{92}_{36}Kr + ^{141}_{56}Ba + 3^{1}_{0}n \qquad [16.1]$$

In practice, a range of fission products with masses similar to krypton and barium are formed when uranium is irradiated with neutrons. This reaction is important, as it is used to produce nuclear power. There are two vital features that make this application possible; first, the amount of energy liberated and, second, the number of neutrons produced.

Consider the first of these points, the energy balance. The energy produced in all fission reactions is derived from the difference in masses of the reactants and the products. This mass difference is liberated as heat. The mass difference for Reaction [16.1] is -3.198×10^{-28} kg. The negative value arises because the reactants are heavier than the products, and this is the amount of mass that is converted into energy. It amounts to 2.878×10^{-11} J. The amount of energy per gram of U-235 is then calculated to be 7.5×10^{10} J. In contrast to this, one gram of coal burnt in air produces about 3×10^4 J. That is, uranium-235 fission produces about a million times more energy than the burning of fossil fuels.

The second important feature of Reaction [16.1] is the number of neutrons emitted. When more neutrons are emitted than are produced, an ever-accelerating reaction, called a chain reaction, can result (Figure 16.8). Suppose that the first disintegrating nucleus is surrounded by sufficient U-235 so that all of the neutrons are absorbed and none is lost through the surface of the material. These will then undergo fission to produce more neutrons, and so on, in a rapidly escalating fashion. Unless controlled, in a fraction of a second all of the U-235 will have transformed into fission products, with the liberation of huge amounts of energy. This is the principle on which atomic bombs operate. The smallest amount of material for which more neutrons are produced than are lost through the surface

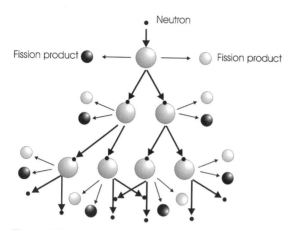

Figure 16.8 Schematic illustration of a fission chain reaction. The key to the continuation of the reaction is that each dividing nucleus must emit more than one neutron that reacts with another nucleus. In the chain shown, each nucleus emits two neutrons that cause further fission

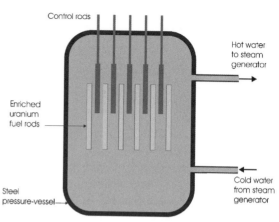

Figure 16.9 Layout of a pressurised water nuclear reactor for power generation. The moderator is pressurised water, which also acts as a coolant, transferring heat to steam generators

is called the critical mass. A quantity of U-235 smaller than the critical mass will not support a chain reaction because too many neutrons escape without hitting another U-235 nucleus.

For a chain reaction to occur it is not enough simply to have a quantity of material greater than the critical mass. It is also necessary for it to be pure, or at least not to contain appreciable quantities of substances that absorb neutrons. Moreover, the isotope U-235 is fissile and reacts best with thermal neutrons, and the arrangement of the uranium-235 must incorporate a mechanism for slowing down the energetic neutrons released by the fission. Use of a moderator slows the neutrons down.

16.3.3 Thermal reactors for power generation

The fission of U-235 is used in a nuclear reactor to produce nuclear power. Because U-235 interacts with low-energy thermal neutrons the reactors are usually called thermal reactors (Figure 16.9). The fuel used in almost all nuclear power reactors is uranium dioxide, UO_2. Natural uranium ores consist of approximately 99 % of the U-238 isotope, about 0.71 % U-235 and small amounts of U-234. For

power production, the amount of U-235 in the UO_2 is increased to 2–4 %, the resulting material being *enriched* uranium dioxide. This is achieved by converting the uranium present in the ores to a gas, uranium hexafluoride. The hexafluoride is then spun at high speeds in a gas centrifuge. This has the effect of separating the lighter, $^{235}UF_6$, and heavier, $^{238}UF_6$, molecules, on the basis of the slight difference in their masses. The hexafluoride mixture that is enriched in the U-235 isotope is then converted back into UO_2 and this is made into fuel pellets.

Uranium dioxide has a number of properties that make it suitable for a fuel. The crystal structure is the *fluorite* (CaF_2) type, similar to that of calcia-stabilised zirconia, and is stable to temperatures in excess of 2000 °C. Because it is a ceramic oxide, the material is refractory, chemically inert and resistant to corrosion Enrichment does not change these features. The oxide powder is pressed into pellets and sintered to a density of about 95 % maximum by traditional ceramic processing technology but is carried out in conditions that minimise risks from radiation effects. The pellets are contained in zirconium alloy (zircaloy) containers, which are then introduced into the reactor. The moderator, which surrounds the fuel rods, can be graphite, ordinary water or heavy water. [Heavy water is water in which the ordinary hydrogen isotope, ^1H, is replaced by the heavier deuterium isotope, ^2H, symbol D, which has a nucleus containing one proton and one neutron. Heavy water is often written as D_2O.] In addition, the reactor contains control rods, which are constructed from a material that absorbs neutrons strongly. Cadmium and boron are used for this purpose. In the event of an accident, the control rods can be inserted into the reactor. This has the effect of stopping the chain reaction and hence closes the reactor down.

16.3.4 Fuel for space exploration

Space exploration relies heavily on solar energy when the spacecraft is in the inner solar system. However, solar power is insufficient for spacecraft that have to journey to the outer planets. Chemical energy sources, typically batteries, tend to be relatively heavy and have rather short lifetimes for missions that are to last many years. The solution adopted to date is to combine a radioactive solid with a thermoelectric generator (see Section 15.2.3). The advantages of this solution are that there are no liquids to spill and no moving parts to wear, and a nuclear isotope with a long half-life will continue to provide power over the lifetime of the craft.

A suitable fuel, used in the *Galileo* Jupiter explorer, which was finally destroyed in 2003, is the isotope ^{238}Pu. This is an α-emitter, which provides about $0.5\,Wg^{-1}$. The half-life is 87.4 years. The fuel is the solid oxide plutonium dioxide, $^{238}PuO_2$. Chemically, it is similar to the uranium dioxide used in thermal reactors, and adopts the same *fluorite* (CaF_2) crystal structure, similar to that of calcia-stabilised zirconia and UO_2. This structure is inert chemically and stable up to the melting point of approximately 2500 °C. The oxide is pressed and sintered into pellets under conditions that lead to high density and low, but not zero, porosity. This is to ensure dimensional stability of the pellets over the lifetime of the spacecraft because, as ^{238}Pu is an α-emitter, the resulting helium gas must be allowed to escape.

16.3.5 Fast-breeder reactors

The amount of U-235 present in a nuclear fuel rod is gradually depleted, and ultimately there is insufficient present for the economic generation of power. A fast-breeder reactor uses the interaction of U-238 with energetic (fast) neutrons to generate the plutonium isotope Pu-239. As Pu-239 can be used as a nuclear fuel, a breeder reactor produces more fuel than it consumes. The sequence of steps is:

$$^{238}_{92}U + {}^{1}_{0}n \rightarrow {}^{239}_{92}U, \qquad t_{1/2} = 24 \text{ minutes}$$

$$^{239}_{92}U \rightarrow {}^{239}_{93}Np + {}^{0}_{-1}e \quad t_{1/2} = 2.35 \text{ days}$$

$$^{239}_{93}Np \rightarrow {}^{239}_{94}Pu + {}^{0}_{-1}e \quad t_{1/2} = 24\,000 \text{ years}$$

The Pu-239 has a half-life of 24 000 years and can be collected for use in fission reactions. The high neutron flux needed is obtained from a reactor using uranium-235 and no moderator. As each decay from the uranium produces more than 2 neutrons it is possible for the reactor to produce more plutonium-239 than it consumes uranium-238. However, the returns are not great, and it would take about 20 breeder reactors running for one year to produce enough plutonium to run a further reactor for one year.

The fuel in fast-breeder reactors is an oxide, as with a thermal reactor. The material chosen is a solid solution of uranium and plutonium dioxides, $U_xPu_{1-x}O_2$. This material shares the same *fluorite* (CaF_2) structure-type as uranium dioxide and plutonium dioxide.

16.3.6 Fusion

The energy released during fission is the difference between the mass of the heavy atom and the masses of the product atoms. The binding energy curve (Figure 16.7) shows that far more energy should be released when the lightest atoms are built up into heavier ones than when heavy atoms are broken down into lighter ones. These building-up reactions are called fusion reactions. Of those available, the production of helium from hydrogen would appear to be the reaction that would yield the most energy per atom.

As hydrogen is commonly available, and as fusion reactions do not produce problems with radioactive waste, there has been considerable research into the construction of controlled fusion reactors. In order for fusion reactions to occur, extremes of temperature must be achieved, some millions of degrees, while keeping the reactants confined and under control. Intensive study of this problem has not yet produced a working reactor, but the possibility of producing limitless clean power will continue to be a goal for research.

Uncontrolled fusion reactions form the basis of the hydrogen bomb. The temperatures needed to initiate fusion are brought about by a fission bomb. A number of competing and cooperative reactions take place during fusion. Typical reaction schemes for fusion reactions are:

$$^2_1H + {}^2_1H \rightarrow {}^3_2He + {}^1_0n$$
$$^2_1H + {}^3_1H \rightarrow {}^4_2He + {}^1_0n$$

These involve the hydrogen isotopes deuterium, ^2H, and tritium, ^3H, as reactants. Tritium has a low natural abundance and is generated in the reaction by use of the isotope lithium-6:

$$^6_3Li + {}^1_0n \rightarrow {}^3_1H + {}^4_2He$$

The reacting material is lithium-6 deuteride, $^6Li^2H$, which forms the core inside a fission bomb. The energy released by such reactions can be calculated by the methods described in Section 16.3.1.

16.3.7   Solar cycles

The process that powers the Sun and all stars is fusion. It seems that the early universe contained clouds of hydrogen atoms dispersed throughout space. Gravitational forces gradually caused these to collapse into dense regions. Ultimately, it is supposed that when the collapse produced a sufficiently high-pressure core, the temperature reached the order of $10^7\,°C$ and hydrogen fusion started. This process is taking place in the Sun today. Under

the intense conditions within the Sun's core the fusion of protons into helium nuclei takes place following the reaction scheme

$$4\,{}^1H \rightarrow {}^4He + 2\,{}^0_{+1}e$$

The energy production is enormous, about $10^{26}\,J\,s^{-1}$.

In older stars the hydrogen is eventually consumed and the rate of fusion slows. At this stage, gravitational collapse again occurs, resulting in an increase in pressure and temperature until He fusion starts. The products of this reaction are mostly carbon and oxygen.

When the He supply becomes depleted smaller stars may explode under gravitational collapse or follow other pathways. Stars that are larger than about 20 solar masses (i.e. about 20 times the mass of the Sun, where 1 solar mass is about $2 \times 10^{30}\,kg$) can collapse in a relatively controlled fashion until core temperatures reach the order of $10^9\,°C$. At this point carbon and oxygen fusion begins. The products are now the elements close to silicon in the periodic table, known as the silicon peak elements.

Further exhaustion and collapse raises the temperature to the order of $3 \times 10^9\,°C$, at which point the silicon elements fuse to give the iron peak elements, chromium manganese, iron, cobalt and nickel. The binding energy curve (Figure 16.7) indicates that at this point fusion no longer produces energy as ^{56}Fe is the element with the greatest binding energy and so further fusion will consume rather than realise energy. At this point in the life of a star, the nuclear reactions diminish, gravitational collapse raises the internal pressure and, ultimately, the star explodes as a supernova. At this point vast numbers of nuclear reactions take place and all of the other atoms are synthesised. The most abundant elements in the universe – which are, in order of abundance, H, He, O, Ne, N, C, Mg, Si, Fe and S – mirror this process.

16.4   Nuclear waste

The problems with unwanted nuclear material, irrespective of its origin, are the same. Although some nuclear materials are poisonous, at the heart of the matter are the facts that they behave chemically in an identical way to nonradioactive counterparts, and they emit damaging radiation that can disrupt cells and lead to genetic problems and cancers. Even these features would not pose problems, though, if radioisotopes were short-lived, or the decay process could be accelerated. Unfortunately, some of the more important by-products of fission are long-lived, and nuclear decay rates cannot be altered by any physical or chemical means at our disposal.

In a nuclear accident, these problems have to be dealt with as an emergency whereas in the case of the decommissioning of a nuclear plant or the disposal of laboratory chemicals the problems can be approached within a longer time-scale.

16.4.1 Nuclear accidents

There have been few serious accidents, the most recent being at Chernobyl, in the (then) USSR, when a nuclear reactor caught fire on 26 April 1986. The lid of the reactor was blown off and the explosion and fire sent radioactive material high into the atmosphere. The reactor cooling system was closed down and the chain reaction came close to an uncontrolled nuclear explosion. In the event, steam generated in the accident blew open the reactor. Subsequent reactions between steam and graphite and between steam and zirconium produced hydrogen, which then caught fire. The reactor core and the graphite burnt with a temperature of about $1600\,°C$. As the fuel was uranium dioxide, UO_2, with a melting point of $2000\,°C$, no melting occurred.

The major problem, apart from local radioactivity, in this and the other accidents, was that large amounts of volatile fission products were widely distributed high in the atmosphere. These subsequently polluted large land areas. The main volatile elements released at Chernobyl were the noble gases (especially krypton, Kr, and xenon, Xe), iodine, I, and caesium, Cs. The caesium isotopes, caesium-134 and caesium-137, posed a particular problem. The half-life of caesium-134 is 2.06 years,

decaying via β-emission. The half-life of caesium-137 is 30.17 years, which also decays via β-emission. Plants readily took up the caesium, which is chemically similar to its neighbouring alkali metal potassium, and which is a trace element vital to plant growth. The caesium was then transferred to meat, milk, eggs and so on by way of grazing animals, and so entered the human food-chain.

The only practical solution in cases of this nature is to isolate the area as far as possible and to prevent contaminated products from reaching market.

16.4.2 The storage of nuclear waste

The storage of nuclear waste is one of the more significant challenges facing materials engineers. A storage facility must last for thousands of years, and even small amounts of leakage are unacceptable. Moreover, radioactive waste is not a passive material. The heat generated by the radioisotopes is significant, damage from radiation greatly increases the rates of chemical changes, and the products of nuclear decay can cause significant pressure and volume changes that can lead to container damage.

Nuclear waste is divided into three categories. High-level waste, which is the most radioactive component, forms about 0.2 % of the whole. It is derived mainly from weapons applications and spent nuclear fuel rods. In addition there is about 20 % intermediate-level waste, which arises from similar sources and is increased by materials used in reprocessing. This component is not very radioactive and does not liberate large amounts of heat. The remainder, described as low-level waste, is material that is slightly radioactive. Apart from military and nuclear energy sources, this material comes from hospitals, research laboratories and industry, and includes contaminated paper towels, gloves and laboratory equipment.

Spent fuel rods from nuclear power stations are a major source of nuclear waste. Nuclear fuel is composed of uranium dioxide, UO_2. After some years of use, when 1–4 % of the uranium has undergone fission, the performance of the fuel rods falls, and these are then replaced. The spent fuel rods consist of uranium dioxide together with fission products, typically the gases krypton and xenon, volatile elements such as caesium and iodine, and metals such as barium, technetium, molybdenum and ruthenium and the lanthanides. In addition, the transuranium elements plutonium, Pu, and americium, Am, generated by neutron capture, are also present in significant quantities.

The spent fuel rods are far more radioactive than are the unused rods. On removal from the reactor, these 'hot' fuel rods are placed into ponds of water for 10 years or so, to cool down. During this period, many of the radioactive elements decay, as most have short half-lives; for example

$$^{142}_{56}Ba \rightarrow {}^{142}_{57}La + \beta^- \quad t_{1/2} = 11 \text{ minutes}$$

After 10 years, the major radioactive materials present are the long-lived isotopes ^{90}Sr ($t_{1/2} = 28.5$ years) and ^{137}Cs ($t_{1/2} = 30.1$ years), as well as 239 Pu ($t_{1/2} = 24\,000$ years). At this stage, the fuel can be reprocessed to regain uranium and plutonium and to reduce the amount of material that has to be safely stored to manageable amounts. The result of this is to produce a relatively small amount of high-level waste. In addition there is a considerable amount of intermediate-level waste, which arises from the zircaloy cases, graphite, stainless-steel containers and components and materials used in the reprocessing.

The major materials problems in radioactive waste disposal are associated with the storage of the high-level waste. Generally, waste disposal is broken down into three stages. The initial stage, immobilisation of the radioactive isotopes, involves trapping the radioactive isotopes in a stable solid, such as glass, cement or a ceramic. The second stage is to seal the solid into a metal container. Finally, the container must be buried in a geologically stable area and isolated with an impermeable barrier material.

There are materials problems associated with all of these steps. The most widely explored solid for waste immobilisation is borosilicate glass. Unfortunately, glass is damaged by radiation effects, which accelerate devitrification and cause volume changes leading to cracking or erosion. Radiation, combined with the heat produced by radioactive decay, can enhance chemical reactivity, even in such nonreactive materials as cement or ceramic oxides, again leading to physical disintegration.

The durability of the canister material also poses a problem. At present, the favoured material is stainless steel. The conditions that the canister must tolerate include: temperatures of up to 200 °C, water vapour, water, and corrosion products from the immobilising solid. The time-scale of thousands of years increases the durability problem enormously. Additionally, the presence of other metallic elements, both inside and outside of the container, can lead to electrochemical corrosion.

The ideal backfill material to isolate sealed canisters is a clay-like substance. This is because clays are able to absorb cations, which are then incorporated in the crystal structure. Thus, the clay acts as a further barrier to dispersal in the event that a storage canister is breached and a radioactive solution is formed. In addition, both cation and water absorption cause the clay particles to swell, thus increasing the pressure of the backfilling material, so further impeding the movement of solutions containing radioactive ions.

The many problems associated with the safe storage of high-level radioactive waste is an ongoing area of intensive materials research.

Answers to introductory questions

What is the difference between an isotope and a nuclide?

The nucleus of an atom can be considered to be made up of protons and neutrons, collectively known as nucleons. A proton has a charge of $+1$ and the number of protons in a nucleus determines the proton number (atomic number), Z, of the atom. The neutron bears no charge but has a similar mass to that of the proton. The number of electrons in a neutral atom is also equal to Z. Atoms with the same value of Z are chemically identical. The total number of nucleons in the nucleus is called the nucleon number (mass number), A. The number of neutrons in a nucleus need not be the same as the number of protons.

Atoms with the same value of Z but different numbers of neutrons are known as isotopes of the element in question. Isotopes are represented by the symbol $^A_Z X$; for example, the radioactive carbon isotope has a symbol $^{14}_6 C$.

A nuclide is any atomic species that has a specified nucleon number and proton number. Hence $^9_4 Be$ is a nuclide.

What chemical or physical procedures can be used to accelerate radioactive decay?

There are no known chemical or physical procedures that can change the rate of decay of a radioactive isotope.

Why does nuclear fission release energy?

The energy produced in all fission reactions is derived from the difference in masses of the reactants and the products. For example, fission of uranium-235 can produce krypton, Kr, and barium, Ba according to Reaction [16.1]:

$$^{235}_{92}U + ^1_0 n \rightarrow ^{92}_{36}Kr + ^{141}_{56}Ba + ^3_0 n$$

[In practice, a range of fission products with masses similar to krypton and barium are formed when uranium is irradiated with neutrons.] This mass difference is liberated as heat. The mass difference for Reaction [16.1] is -3.198×10^{-28} kg. The negative value arises because the reactants are heavier than the products, and this is the amount of mass that is converted into energy. It amounts to 2.878×10^{-11} J. The amount of energy per gram of U-235 is then calculated to be 7.5×10^{10} J. In contrast to this, one gram of coal burnt in air produces about 3×10^4 J. That is, uranium-235 fission produces about a million times more energy than the burning of fossil fuels.

Further reading

The history of the birth and development of atomic physics and radiochemistry, from the original dis-covery of radioactivity to the production of the hydrogen bomb, makes fascinating reading. A unique insight to this epoch can be gained by reading the lectures given by the key Nobel prizewinners, in book form and also available at www. nobel.se.

M. Chown, 1999, *The Magic Furnace*, Jonathan Cape, London.

R. Corfield, 2001, *Architects of Eternity*, Hodder Headline, London.

C. Lewis, 2000, *The Dating Game*, Cambridge University Press, Cambridge.

D.A. McQuarrie and P.A. Rock, 1991, *General Chemistry*, 3rd edn, Freeman, New York, Ch 24.

The materials science of radioactive waste forms several articles in the *Materials Research Society Bulletin* **XIX** (December 1994). See also:

A.F. Gardiner, R.S. Gillett, P.S. Phillips, 1992, 'The Menace Under the Floorboards', *Chemistry in Britain* **28** 344.

M. Segal, C. Morris, 1991, 'The legacy of Chernobyl', *Chemistry in Britain* **27** 904.

Problems and exercises

Quick quiz

1 Which of the following is *not* a nuclide:
   (a) Protons
   (b) Neutrons
   (c) Electrons

2 The atomic number of an element defines:
   (a) The number of protons in a nucleus
   (b) The number of neutrons in a nucleus
   (c) The number of protons plus neutrons in the nucleus

3 The number of electrons surrounding a neutral atom is the same as the:
   (a) Mass number
   (b) Proton number
   (c) Nucleon number

4 Isotopes of an element have the same numbers of:
   (a) Neutrons

(b) Nucleons

(c) Protons

5 An atomic species that has specified nucleon number and proton number is named:
   (a) An isotope
   (b) A nuclide
   (c) a radioisotope

6 When a radioactive element emits an α-particle the daughter element has a nucleon number:
   (a) 4 less that the parent atom
   (b) 2 less than the parent atom
   (c) The same as the parent atom

7 When a radioactive element emits a β-particle the daughter element has a nucleon number:
   (a) 4 less that the parent atom
   (b) 2 less than the parent atom
   (c) The same as the parent atom

8 The number of different radioactive series that have been described is:
   (a) Two
   (b) Three
   (c) Four

9 The total number of different gaseous elements produced in all the different radioactive series is:
   (a) One
   (b) Two
   (c) Four

10 The only nonnaturally occurring radioactive decay series, the neptunium series, has a formula:
   (a) $4x + 1$
   (b) $4x + 2$
   (c) $4x + 3$

11 The elements with a partly filled 5f electron shell are named the:
   (a) Lanthanides

(b) Actinides

(c) Uranides

12 During β^+ decay, the particle ejected from the nucleus is:
   (a) A neutron
   (b) An electron
   (c) A positron

13 Radioactive nuclei below the band of stability tend to emit:
   (a) Positrons
   (b) Electrons
   (c) Neutrons

14 The number of atomic disintegrations that occur in a radioactive material per second is called its:
   (a) Activity
   (b) Specific activity
   (c) Specific radioactivity

15 The half-life of strontium-90 is 28.5 years. A person ingests 0.01 g of this isotope, which is incorporated into the bones. How long will it take before the level reaches 0.00125 g:
   (a) 28.5 years?
   (b) 57.0 years?
   (c) 85.5 years?

16 The binding energy of a nucleus is:
   (a) The chemical bonding energy between the protons and neutrons
   (b) The energy of interaction between the protons in the nucleus
   (c) This mass of the nucleus minus the total masses of the neutrons and protons

17 Fissile materials undergo nuclear fission:
   (a) Spontaneously
   (b) Under bombardment by not very energetic neutrons
   (c) Under bombardment by very energetic neutrons

18 The purpose of a moderator in a nuclear reactor is to:
    (a) Slow the neutrons down
    (b) Initiate the chain reaction
    (c) Control the energy produced

19 Thermal reactors for the production of nuclear power use a fuel of:
    (a) Uranium
    (b) Uranium dioxide
    (c) Uranium hexafluoride

20 The purpose of the control rods in a nuclear reactor is to:
    (a) Slow down neutrons
    (b) Generate neutrons
    (c) Absorb neutrons

21 A fast-breeder reactor produces:
    (a) Uranium
    (b) Plutonium
    (c) Neptunium

22 Fusion reactions for the production of power envisage utilisation of the reaction of:
    (a) Helium to form hydrogen
    (b) Hydrogen to form helium
    (c) Lithium-6 to form hydrogen

23 The process that powers the Sun and all stars is:
    (a) Fusion
    (b) Fission
    (c) A mixture of fusion and fission

24 Intermediate-level nuclear waste constitutes:
    (a) Less than 1 % of the total
    (b) Approximately 20 % of the total
    (c) Approximately 80 % of the total

Calculations and questions

16.1 Write the nuclear symbol, and the number of protons, neutrons and nucleons in: (a) rhenium-166; (b) barium-140; (c) oxygen-18; and (d) boron-14.

16.2 Write the nuclear symbol, and the number of protons, neutrons and nucleons in: (a) mercury-181; (b) iridium-169; (c) iodine-117; and (d) zirconium-98.

16.3 Write nuclear equations for the decay of the following radioactive isotopes (the particles emitted in the decay are given in brackets): (a) $^{27}_{14}\text{Si}$ (positron); (b) $^{28}_{13}\text{Al}$ (electron); (c) $^{24}_{11}\text{Na}$ (electron); (d) $^{17}_{9}\text{F}$ (positron).

16.4 Write nuclear equations for the decay of the following radioactive isotopes (the particles emitted in the decay are given in brackets): (a) $^{24}_{8}\text{O}$ (electron); (b) $^{47}_{23}\text{V}$ (positron); (c) $^{32}_{15}\text{P}$ (electron); and (d) $^{39}_{20}\text{Ca}$ (positron).

16.5 Write nuclear equations for the decay of the following radioactive isotopes (the particles emitted in the decay are given in brackets): (a) $^{243}_{100}\text{Fm}$ (alpha); (b) $^{241}_{95}\text{Am}$ (alpha); (c) $^{241}_{94}\text{Pu}$ (electron); and (d) $^{237}_{92}\text{U}$ (electron).

16.6 Determine the half-life of a radioactive isotope from the variation of the number of radioactive disintegrations observed, in counts per minute, over a period of time, as given is Table 16.3.

16.7 A sample of radioactive sodium-24 with a half-life of 15.0 h is used to measure the diffusion coefficient of Na in NaCl. How long will it take for the activity to drop to 0.1 of its original activity?

Table 16.3 Data for Question 16.6

Time/min	0	2	4	6	8	10	12	14
Activity/counts per min	3160	2512	1778	1512	1147	834	603	579

16.8   What is the specific activity of radium-226, which has a half-life of 1600 years?

16.9   What is the specific activity of plutonium-241, which has a half-life of 14.35 years?

16.10   What is the specific activity of neptunium-233, which has a half-life of 36.2 minutes?

16.11   A purified sample of a radioactive compound is found to have an activity of 1365 counts per minute at 10 am but only 832 counts per minute at 1 pm. What is the half-life of the sample?

16.12   A 250 mg piece of carbon from an ancient hearth showed 1530 carbon-14 disintegrations in 36 h; 250 mg of fresh carbon from charcoal gave 8280 disintegrations in the same time. What is the date of the carbon sample? The half-life of carbon-14 is 5.73×10^3 years.

16.13   A cellar of dimensions $2m \times 3m \times 2.5$ m is found to show an activity of $0.37 \, \text{Bq dm}^{-3}$, owing to the presence of radon. (a) How many nuclei decay per minute in the cellar? (b) How long will it take for the activity to fall to 0.015 Bq, if no more radon leaks into it? The half-life of radon is 3.8 days.

16.14   The suggested upper limit for radon concentration in a building is equivalent to an activity of $200 \, \text{Bq m}^{-3}$. (a) How many radon atoms are needed per dm^3 of air to give this figure? (b) How long will it take for the activity to fall to one tenth of this value in a sealed room? The half-life of radon is 3.8 days.

16.15   In 1986 a nuclear reactor at Chernobyl exploded, depositing caesium-137 over large areas of northern Europe. An initial activity of $1000 \, \text{Bq m}^{-2}$ was found on vegetation over parts of Scotland. (a) How many grams of caesium-137 were deposited per square metre of vegetation? (b) What was the activity after 10 years? The half-life of caesium-137 is 30.1 years.

16.16   Calculate the binding energy per nucleon for ^4_2He. The masses are: ^1_1H, 1.0078 u; ^1_0n, 1.0087 u; ^4_2He, 4.0026 u.

16.17   Calculate the binding energy per nucleon for $^{23}_{12}\text{Mg}$. The masses are: $^{23}_{12}\text{Mg}$, 22.9941 u; ^1_1H, 1.0078 u; ^1_0n, 1.0087 u.

16.18   Calculate the binding energy per nucleon for $^{34}_{16}\text{S}$. The masses are: $^{34}_{16}\text{S}$, 33.967865 u; ^1_1H, 1.0078 u; ^1_0n, 1.0087 u.

16.19   On decay, an atom of radium-226 emits one α-particle and is converted into an atom of radon-222. A quantity of radium-226 produced $4.48 \times 10^{-6} \, \text{dm}^3$ of helium at 273 K and 1 atm pressure. Determine (a) the mass of radium-226 that decayed and (b) the mass of radon-222 produced, if no radon decays. One mole of helium gas occupies $22.4 \, \text{dm}^3$ at 273 K and 1 atm pressure.

16.20   Calculate the energy released per mole in the fission reaction:

$$^{235}_{92}\text{U} + ^1_0\text{n} \rightarrow ^{102}_{42}\text{Mo} + ^{128}_{50}\text{Sn} + 6 \, ^1_0\text{n}$$

The masses are: $^{235}_{92}\text{U}$, 235.0439 u; $^{102}_{42}\text{Mo}$, 101.91025 u; $^{131}_{50}\text{Sn}$, 127.91047 u; ^1_0n, 1.0087 u.

16.21   Energy generation in the Sun is by way of the reaction of hydrogen to form helium. One reaction is:

$$4 \, ^1\text{H} \rightarrow ^4\text{He} + 2 \, ^0_{+1}\text{e}$$

The masses are: ^1_1H, 1.0078 u; ^4_2He, 4.0026 u; $^0_{+1}\text{e}$, 5.486×10^{-4} u.

(a) Calculate the energy liberated per fusion reaction.

(b) It is estimated that the Sun produces approximately $10^{26} \, \text{J s}^{-1}$. How many fusion reactions per second are required for this?

Supplementary material

S1

Supplementary material to Part 1: structures and microstructures

S1.1 Chemical equations and units

Chemical reactions are summarised by writing a chemical equation. By convention the reactants are on the left-hand side and the products are written on the right-hand side in the following format:

$$\text{Reactants} \rightarrow \text{Products}$$

Some reactions can proceed in either direction and it is not then possible to specify the products or the reactants uniquely. These are written:

$$\text{Reactants} \rightleftharpoons \text{Products}$$

Chemical equations are devices for 'accounting'. In the course of chemical reactions, no atoms must be lost or gained. Chemical equations should therefore be 'balanced' to ensure that this rule holds. Typical balanced equations are:

$$CaCO_3 \rightarrow CaO + CO_2$$
$$Fe_2O_3 + 3CO \rightarrow 2Fe + 3CO_2$$

Sometimes it is desirable to specify the state of the reactants, whether solid, liquid or gas, or in solution. In such cases a symbol enclosed in brackets after the species gives this data; s = solid; l = liquid; g = gas; aq = water (aqueous) solution. These are only labels and are not a part of the reaction. For example:

$$CaCO_3 \, (s) \rightarrow CaO \, (s) + CO_2 \, (g)$$

Chemical calculations are easiest if one starts with the appropriate chemical equation. The most important aspect in specifying the reaction is not the weight of the reactants or products but the number of (real or hypothetical) molecules taking part. To make the conversion from equations to weights a unit called the mole is utilised.

One mole (mol) is the number of atoms in exactly 12 g of carbon-12. It is equal to the Avogadro constant, N_A, 6.022×10^{23} mol^{-1}. Carbon-12 is the isotope of the chemical element carbon that has 6 protons and 6 neutrons in its nucleus. It has an atomic number of 6.

The molar mass of an element is the mass of 1 mole of atoms (in grams). The molar mass of a molecule is the mass of 1 mole of molecules (in grams). The molar mass of a compound is the mass of 1 mole of one formula unit of the compound.

The lightest element is hydrogen, with a molar mass of 1.0079 g mol^{-1} and atomic number of 1. The heaviest naturally occurring element is uranium, with a molar mass of 238.03 g mol^{-1} and an atomic number of 92. Molar mass can be converted to real mass in grams by multiplying by the conversion

Understanding solids: the science of materials.   Richard J. D. Tilley
© 2004 John Wiley & Sons, Ltd   ISBNs: 0 470 85275 5 (Hbk) 0 470 85276 3 (Pbk)

factor 1.66054×10^{-24}. For example, the average mass of a uranium atom is

$$\frac{238.03 \, \text{g mol}^{-1}}{6.022 \times 10^{23} \, \text{atoms mol}^{-1}}$$

$$= 238.03 \times 1.66054 \times 10^{-24} \, \text{atom}^{-1}$$

$$= 3.953 \times 10^{-22} \, \text{g atom}^{-1}.$$

The mass of an atom in atomic mass units, u, is numerically equal to its molar mass or isotopic mass. Thus, the mass of hydrogen is 1.0079 u. An atomic mass unit is defined as follows:

$$1 \, \text{u} = \tfrac{1}{12} \, \text{mass of one atom of the isotope carbon-12.}$$

As the mass of one atom of carbon-12 is 1.9926×10^{-23} g, $1 \, \text{u} = 1.6605 \times 10^{-24}$ g.

The number of moles in a given mass of the compound is:

$$\frac{\text{mass (g)}}{\text{molar mass of compound (g mol}^{-1}).}$$

To find the molar mass of a compound, one adds the molar masses of its constituent elements.

The mass percent of an element A in a compound is:

$$\frac{\text{mass of element A in sample}}{\text{total mass of sample}} \times 100.$$

The weight percent of an element A in a compound is:

$$\frac{\text{weight of element A in sample}}{\text{total weight of sample}} \times 100.$$

The atom percent (at %) of an element A in a compound is:

$$\frac{\text{number of atoms of element A in sample}}{\text{total number of atoms in sample}} \times 100.$$

S1.2   Electron configurations

S1.2.1   The electron configurations of the lighter atoms

The electron configuration of an atom or an ion describes the arrangement of the electrons around

Table S1.1   The electronic configurations of the lighter atoms

Element	Symbol	Configuration	Alternative
Hydrogen	H	$1s^1$	
Helium	He	$1s^2$	
Lithium	Li	$1s^2 \, 2s^1$	[He] $2s^1$
Beryllium	Be	$1s^2 \, 2s^2$	[He] $2s^2$
Boron	B	$1s^2 \, 2s^2 \, 2p^1$	[He] $2s^2 \, 2p^1$
Carbon	C	$1s^2 \, 2s^2 \, 2p^2$	[He] $2s^2 \, 2p^2$
Nitrogen	N	$1s^2 \, 2s^2 \, 2p^3$	[He] $2s^2 \, 2p^3$
Oxygen	O	$1s^2 \, 2s^2 \, 2p^4$	[He] $2s^2 \, 2p^4$
Fluorine	F	$1s^2 \, 2s^2 \, 2p^5$	[He] $2s^2 \, 2p^5$
Neon	Ne	$1s^2 \, 2s^2 \, 2p^6$	[He] $2s^2 \, 2p^6$
Sodium	Na	$1s^2 \, 2s^2 \, 2p^6 \, 3s^1$	[Ne] $3s^1$
Magnesium	Mg		[Ne] $3s^2$
Aluminium	Al		[Ne] $3s^2 \, 3p^1$
Silicon	Si		[Ne] $3s^2 \, 3p^2$
Phosphorus	P		[Ne] $3s^2 \, 3p^3$
Sulphur	S		[Ne] $3s^2 \, 3p^4$
Chlorine	Cl		[Ne] $3s^2 \, 3p^5$
Argon	Ar		[Ne] $3s^2 \, 3p^6$
Potassium	K		[Ar] $3s^1$
Calcium	Ca		[Ar] $3s^2$
Gallium	Ga		[Ar] $3s^2 \, 3p^1$
Germanium	Ge		[Ar] $3s^2 \, 3p^2$
Arsenic	As		[Ar] $3s^2 \, 3p^3$
Selenium	Se		[Ar] $3s^2 \, 3p^4$
Bromine	Br		[Ar] $3s^2 \, 3p^5$
Krypton	Kr		[Ar] $3s^2 \, 3p^6$

Note: [He] $= 1s^2$; [Ne] $=$ [He] $2s^2 \, 2p^6$; [Ar] $=$ [Ne] $3s^2 \, 3p^6$.

the nucleus based upon the orbital model. Table S1.1 lists the configuration of selected elements; for the configurations of the 3d transition metals, see Table S1.2.

S1.2.2   The electron configurations of the 3d transition metals

The 10 3d transition metal atoms (shown in bold below) are found in period 4 of the periodic table:

K, Ca, **Sc**, **Ti**, **V**, **Cr**, **Mn**, **Fe**, **Co**, **Ni**, **Cu**, **Zn**, Ga, Ge, As, Se, Br, Kr.

They are characterised by having partly filled 3d atomic orbitals. Zinc does not behave as a typical

Table S1.2   The electronic configuration of the 3d transition metals and their ions

Atom	Symbol	Configuration	Ion	Configuration	d electron configuration
Scandium	Sc	[Ne] $3s^2 3p^6 3d^2 4s^1$	Sc^{3+}	[Ne] $3s^2 3p^6$	d^0
Titanium	Ti	[Ne] $3s^2 3p^6 3d^2 4s^2$	Ti^{4+}	[Ne] $3s^2 3p^6$	d^0
			Ti^{3+}	[Ne] $3s^2 3p^6 3d^1$	d^1
			Ti^{2+}	[Ne] $3s^2 3p^6 3d^2$	d^2
Vanadium	V	[Ne] $3s^2 3p^6 3d^3 4s^2$	V^{5+}	[Ne] $3s^2 3p^6$	d^0
			V^{4+}	[Ne] $3s^2 3p^6 3d^1$	d^1
			V^{3+}	[Ne] $3s^2 3p^6 3d^2$	d^2
Chromium	Cr	[Ne] $3s^2 3p^6 3d^5 4s^1$	Cr^{3+}	[Ne] $3s^2 3p^6 3d^3$	d^3
Manganese	Mn	[Ne] $3s^2 3p^6 3d^5 4s^2$	Mn^{4+}	[Ne] $3s^2 3p^6 3d^3$	d^3
			Mn^{3+}	[Ne] $3s^2 3p^6 3d^4$	d^4
			Mn^{2+}	[Ne] $3s^2 3p^6 3d^5$	d^5
Iron	Fe	[Ne] $3s^2 3p^6 3d^6 4s^2$	Fe^{3+}	[Ne] $3s^2 3p^6 3d^5$	d^5
			Fe^{2+}	[Ne] $3s^2 3p^6 3d^5$	d^6
Cobalt	Co	[Ne] $3s^2 3p^6 3d^7 4s^2$	Co^{4+}	[Ne] $3s^2 3p^6 3d^5$	d^5
			Co^{3+}	[Ne] $3s^2 3p^6 3d^6$	d^6
			Co^{2+}	[Ne] $3s^2 3p^6 3d^7$	d^7
Nickel	Ni	[Ne] $3s^2 3p^6 3d^8 4s^2$	Ni^{2+}	[Ne] $3s^2 3p^6 3d^8$	d^8
Copper	Cu	[Ne] $3s^2 3p^6 3d^{10} 4s^1$	Cu^{2+}	[Ne] $3s^2 3p^6 3d^9$	d^9
			Cu^+	[Ne] $3s^2 3p^6 3d^{10}$	d^{10}
Zinc	Zn	[Ne] $3s^2 3p^6 3d^{10} 4s^2$	Zn^{2+}	[Ne] $3s^2 3p^6 3d^{10}$	d^{10}

Note: [Ne] = $1s^2\ 2s^2\ 2p^6$.

transition metal as it has full d orbitals. The configurations of the 3d transition metals are listed in Table S1.2.

S1.2.3   The electron configurations of the lanthanides

The 14 lanthanides, or rare earth elements (shown in bold below), are found in period 6 of the periodic table:

Cs, Ba, **(La)**, **Ce**, **Pr**, **Nd**, **Pm**, **Sm**, **Eu**, **Gd**, **Tb**, **Dy**, **Ho**, **Er**, **Tm**, **Yb**, **(Lu)**, Hf, Ta, W, ...

The electron configuration of some of these atoms is uncertain and neither lanthanum nor lutetium behave as typical lanthanides, although both are frequently included in the group. The configurations of the lanthanides are listed in Table S1.3.

S1.3   Energy levels and term schemes

S1.3.1   Energy levels and term schemes of many-electron atoms

The Russell–Saunders terms of an atom are derived by adding the individual spin quantum numbers of the electrons to yield a total spin quantum number, S, and by adding the individual orbital angular momentum quantum numbers of the electrons to give a total orbital angular momentum quantum number, L. For example, the total spin angular momentum quantum number, $S(2)$, for two electrons, is given by adding the individual quantum numbers, thus:

$$S(2) = (s_1 + s_2),\ (s_1 + s_2 - 1),\ \ldots |s_1 - s_2|$$

As s_1 and s_2 are both equal to 1/2,

$$S(2) = 1 \text{ or } 0$$

Table S1.3 The electronic configurations of the lanthanides (rare earth elements) and their ions

Atom	Symbol	Configuration	Ion	Configuration	f electron configuration
Lanthanum	La	[Xe] $5d^1$ $6s^2$, or [Xe] $4f^1$ $6s^2$	La^{3+}	[Xe]	f^0
Cerium	Ce	[Xe] $4f^1$ $5d^1$ $6s^2$, or [Xe] $4f^2$ $6s^2$	Ce^{4+}	[Xe]	f^0
			Ce^{3+}	[Xe]$4f^1$	f^1
Praseodymium	Pr	[Xe] $4f^3$ $6s^2$	Pr^{4+}	[Xe]$4f^1$	f^1
			Pr^{3+}	[Xe]$4f^2$	f^2
Neodymium	Nd	[Xe] $4f^4$ $6s^2$	Nd^{3+}	[Xe]$4f^3$	f^3
Promethium	Pm	[Xe] $4f^5$ $6s^2$	Pm^{3+}	[Xe]$4f^4$	f^4
Samarium	Sm	[Xe] $4f^6$ $6s^2$	Sm^{3+}	[Xe]$4f^5$	f^5
			Sm^{2+}	[Xe]$4f^6$	f^6
Europium	Eu	[Xe] $4f^7$ $6s^2$	Eu^{3+}	[Xe]$4f^6$	f^6
			Eu^{2+}	[Xe]$4f^7$	f^7
Gadolinium	Gd	[Xe] $4f^7$ $5d^1$ $6s^2$	Gd^{3+}	[Xe]$4f^7$	f^7
Terbium	Tb	[Xe] $4f^9$ $6s^2$	Tb^{4+}	[Xe]$4f^7$	f^7
			Tb^{3+}	[Xe]$4f^8$	f^8
Dysprosium	Dy	[Xe] $4f^{10}$ $6s^2$	Dy^{3+}	[Xe]$4f^9$	f^9
Holmium	Ho	[Xe] $4f^{11}$ $6s^2$	Ho^{3+}	[Xe]$4f^{10}$	f^{10}
Erbium	Er	[Xe] $4f^{12}$ $6s^2$	Er^{3+}	[Xe]$4f^{11}$	f^{11}
Thulium	Tm	[Xe] $4f^{13}$ $6s^2$	Tm^{3+}	[Xe]$4f^{12}$	f^{12}
Ytterbium	Yb	[Xe] $4f^{14}$ $6s^2$	Yb^{3+}	[Xe]$4f^{13}$	f^{13}
			Yb^{2+}	[Xe]$4f^{14}$	f^{14}
Lutetium	Lu	[Xe] $4f^{14}$ $5d^1$ $6s^2$	Lu^{3+}	[Xe]$4f^{14}$	f^{14}

Note: [Xe] = $1s^2$ $2s^2$ $2p^6$ $3s^2$ $3p^6$ $3d^{10}$ $4s^2$ $4p^6$ $4d^{10}$ $5s^2$ $5p^6$.

In order to obtain the value of S for three electrons, $S(3)$, the value for two electrons, $S(2)$, is added to the spin quantum number of the third electron, s_3 thus:

$$S(3) = [S(2) + s_3], [S(2) + s_3 - 1], \ldots |S(2) - s_3|$$

Both of the values for $S(2)$ are permitted, so we obtain:

(a) $S(2) = 1$, $S(3) = 1 + \frac{1}{2}$, $1 + \frac{1}{2} - 1$

$$= \tfrac{3}{2}, \tfrac{1}{2}$$

(b) $S(2) = 0$, $S(3) = 0 + \frac{1}{2}$

$$= \tfrac{1}{2}$$

Thus

$$S(3) = \tfrac{3}{2}, \tfrac{1}{2}$$

This procedure is called the Clebsch–Gordon rule. It is used to obtain the S values for increasing numbers of spins. It is found that: for an even number of electrons, S values are integers; for an odd number of electrons, S values are half-integers.

The total angular momentum quantum number, L, is obtained in a similar fashion. For two electrons with individual angular momentum quantum numbers l_1 and l_2, the total angular momentum quantum number, $L(2)$ is:

$$L(2) = (l_1 + l_2), (l_1 + l_2 - 1), \ldots, |l_1 - l_2|$$

In the case of three electrons, the Clebsch–Gordon rule is applied thus:

$$L(3) = [L(2) + l_3], [L(2) + l_3 - 1], \ldots, |L(2) - l_3|$$

using every value of $L(2)$ obtained previously. Values of the quantum number L are given letter symbols as described in Section 1.4.3.

For example, the term schemes arising from the two p electrons on carbon, C, with $l_1 = l_2 = 1$, are obtained in the following way:

$$S = \tfrac{1}{2} + \tfrac{1}{2}, \tfrac{1}{2} - \tfrac{1}{2}$$

$$= 1, 0$$

$$2S + 1 = 3, 1;$$

$$L = (1 + 1), (1 + 1 - 1) 1, (1 - 1)$$

$$= 2, 1, 0 \ (D, P, S)$$

The total number of possible terms for the two p electrons is given by combining these values. The possible terms for two p electrons are therefore:

$$^3D, ^3P, ^3S, ^1D, ^1P, ^1S$$

Not all of these possibilities are allowed for any particular configuration, because the Pauli exclusion principle limits the number of electrons in each orbital to two with opposed spins. When this is taken into account, the allowed terms are:

$$^3P \, ^1D, \, ^1S$$

The energies of the terms are difficult to obtain and must be calculated by using quantum mechanical procedures, except for the ground-state term, as described below.

S1.3.2 The ground-state term of an atom

The lowest-energy term, the ground-state term, is easily found by using Hund's first and second rules:

1. The term with the lowest energy has the highest multiplicity, equivalent to the highest total spin quantum number, S.

2. For terms with the same value of multiplicity, the term with the highest value of L is lowest in energy.

There is a simple method of determining the ground state of any atom or ion. The procedure is as follows.

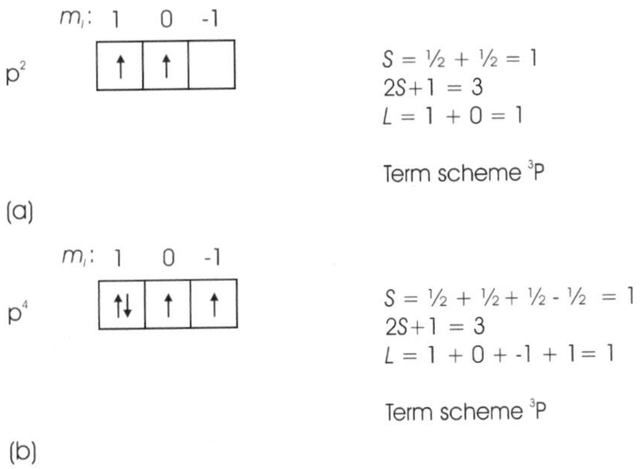

m_l: 1 0 -1

p^2 [↑ | ↑ |]

$S = \tfrac{1}{2} + \tfrac{1}{2} = 1$
$2S + 1 = 3$
$L = 1 + 0 = 1$

Term scheme ^3P

(a)

m_l: 1 0 -1

p^4 [↑↓ | ↑ | ↑]

$S = \tfrac{1}{2} + \tfrac{1}{2} + \tfrac{1}{2} - \tfrac{1}{2} = 1$
$2S + 1 = 3$
$L = 1 + 0 + -1 + 1 = 1$

Term scheme ^3P

(b)

Figure S1.1 The derivation of ground-state term symbols for (a) p^2 and (b) p^4 electron configurations

- Step 1: draw a set of boxes corresponding to the number of orbitals available. For a p electron, this is three (Figure S1.1).

- Step 2: label each box with the value of m_l, highest on the left and lowest on the right.

- Step 3: fill the boxes with unpaired electrons, from left to right. When each box contains one electron, start again at the left.

- Step 4: sum the m_s values of each electron, $+1/2$ or $-1/2$. This is equal to the maximum value of S.

- Step 5: sum the m_l values of each electron to give a maximum value of L.

- Step 6: write the ground term ^{2S+1}L.

Using this technique, set out in Figure S1.1, we find the ground term of $2p^2$ and $2p^4$ configurations is ^3P.

S1.4 Madelung constants

A number of Madelung constants for a compound M_mX_n appear in the literature. These usually differ from each other in the molecular unit chosen for evaluation (e.g. M_mX_n or $MX_{n/m}$) and whether the ionic charges are included in the constant. If the differing charges on the ions, Z_M and Z_X, are included in the evaluation (step 1 described in Section 2.1.3), the Madelung constant, A, calculated, is not then a purely geometric term but includes information on the charges present.

Alternatively, the charge contribution may be separated out, and a reduced Madelung constant, α, derived. This is a purely geometric term and is the option used here. To convert between the Madelung constant, A, and the reduced Madelung constant, α, for a compound of formula M_mX_n:

$$\tfrac{1}{2}\,\alpha Z_M Z_X (m + n) = A$$

S1.5   The phase rule

S1.5.1   The phase rule for one-component (unary) systems

The number of phases that can coexist at equilibrium, P, is specified by the (Gibbs) phase rule:

$$P + F = C + 2 \qquad (S1.1)$$

where C is the number of components (i.e. elements or compounds) needed to form the system, and F is the number of degrees of freedom or variance. The variance specifies the number of variables in the system, such as composition, temperature and pressure, that can be altered independently without changing the state of the system.

In a one-component, or unitary, system, only one substance is present. In this case, the phase rule becomes:

$$P + F = 3 \qquad (S1.2)$$

The phases that can exist in a one-component system are limited to vapour, liquid and solid (Figure 4.1). There are many one-component systems, including all of the pure elements and compounds. The one-component carbon system is of interest, as one of the phases present, diamond, has

considerable value commercially and technologically. When only one phase is present (e.g. diamond) Equation (S1.2) becomes:

$$F = 2 \qquad (S1.3)$$

That is, the region over which the single phase is stable has to be specified in terms of two variables. These are taken as temperature, normally specified in degrees centigrade, and pressure, specified in atmospheres (1 atm $= 1.01325 \times 10^5$ Pa).

A slightly simplified phase diagram for water is drawn in Figure 4.2. The ordinate (y axis) specifies pressure, and the abscissa (x axis) the temperature. The particular phase present at a specified temperature and pressure can be read from the diagram. The areas over which single phases occur are bounded by lines called phase boundaries. The conditions limiting the existence and coexistence of phases are limited by the phase rule. The following examples explain how this comes about.

One point in the phase diagram of water, found at 0.01 °C and 0.006 atm (611 Pa), indicates that three phases, liquid (water), solid (ice) and vapour, occur together. This point is called the triple point, and at any triple point, three phases coexist at equilibrium. Thus, $P = 3$, and, from Equation (S1.2), $F = 0$. That is, there are no degrees of freedom available. That is to say, it is not possible to change either the temperature or the pressure and still keep three phases present. At this point, the system is said to be invariant. The phase diagram shows that a change in either temperature or pressure is most likely to lead to the formation of a single phase, and the system will change to all solid, all liquid or all vapour.

Along the phase boundary separating ice and water – the liquid–solid freezing curve – the number of phases present, solid and liquid, is two ($P = 2$), so, from Equation (S1.2), $F = 1$. Thus, this phase boundary (and all phase boundaries in a one-component system) is characterised by one degree of freedom. This means that one variable, either pressure or temperature, can be changed and two phases in equilibrium can still be found. However, the two variables are closely connected. If the pressure is changed then the temperature must also change, by exactly the amount specified in the phase diagram, to maintain two phases in coexistence.

Over much of the phase diagram, only one phase is present for any combination of temperature and pressure values. With only one phase present ($P = 1$), $F = 2$. In this situation, there are two degrees of freedom, available. That is, both the pressure and the temperature can be changed independently of one another in the presence of a single phase. For example, liquid water exists over a range of temperatures and pressures, and either quantity can be varied (within the limits given on the phase diagram) without changing the situation.

The pattern of temperature change observed as a substance cools (Figure 4.3) follows directly from the phase rule. When two phases are present, there is only one degree of freedom available, and the temperature can vary only if the pressure also varies. As the pressure is constant at one atmosphere, the temperature cannot change. The temperature must remain constant until the phase change is complete. A mixture of ice and water will therefore have a constant temperature that will not change until either the ice has melted or the water has frozen. This form of cooling curve will be found in any one-component system as a sample is cooled through a phase boundary.

S1.5.2 Two phase rule for two-component (binary) systems

The phase rule for a binary system is given by Equation (S1.4):

$$P + F = 4 \qquad \text{(S1.4)}$$

When only one phase is present, $P = 1$, so $F = 3$. The number of degrees of freedom (the variance) is three. Three variables are therefore needed to construct a binary phase diagram. The variables are usually chosen as temperature, pressure and composition. A single phase will be represented by a volume in the diagram. However, as most experiments are carried out at atmospheric pressure, a planar diagram, using temperature and composition

as variables, is usually sufficient. A single phase then occurs over an area in the figure.

On a phase boundary, two phases are present ($P = 2$) and, from equation (S1.4), $F = 2$. The variance is seen to be equal to two. Thus, phase boundaries form two-dimensional surfaces in the -pressure–temperature–composition representation. A two-component system containing two phases is called a bivariant system. In a bivariant system, it is possible to change any two of the three variables temperature, pressure and composition independently but the third will be fixed by the values selected for the other two. On a phase diagram drawn for a system at atmospheric pressure, phase boundaries are drawn as lines.

The maximum number of phases that can coexist is three (e.g. solid, liquid and gas). In this case, $P = 3$, and Equation (S1.4) gives $F = 1$, and there is only one degree of freedom available to the system. Three phases will coexist along a line in the phase diagram. On a phase diagram drawn for one atmosphere pressure, three phases occur at a point.

The addition of a second component thus changes the appearance of the one-component system as set out in Table S1.4.

Of the many two-component systems that occur, those comprising a mixture of a metal and oxygen gas are of great technological and commercial importance. The phases present in these systems may be metal, metal oxide [if only one oxide forms, such as in the magnesium–oxygen ($Mg-O_2$) or aluminium–oxygen ($Al-O_2$) systems] and oxygen gas. When a metal is sealed into a tube containing oxygen at a high temperature it will react to form oxide. At this stage, three phases are present, metal, metal oxide and oxygen gas, and the regime has

Table S1.4 A comparison of phase diagrams of one-component and two-component systems

	Unary	Binary
Axes	2	3
Single-phase region	Area	Volume
Phase boundary	Line	Surface
Three-phase coexistence	Point	Line

only one degree of freedom. The reaction will continue until the amounts of metal and oxide and the pressure of the oxygen gas all conform to the equilibrium phase diagram. Because there is only one degree of freedom available, the conditions for this situation to hold correspond to a line in the pressure – temperature – composition diagram. For example, a change in temperature will cause more metal to oxidise, thus reducing the oxygen pressure and changing the composition in one direction, or will cause some oxide to decompose, thus increasing the oxygen pressure and changing the composition in the other direction, until equilibrium is again restored.

Similarly, a mixture of metal and metal oxide sealed into an evacuated tube and heated to a high temperature will react to generate oxygen gas. In effect, some of the oxide will decompose and three phases will again be present in the tube. This decomposition will continue until the equilibrium pressure of oxygen is reached. The single degree of freedom means that when this stage is reached the amounts of metal and metal oxide present – the composition of the system – have adjusted to generate the correct oxygen pressure. A mixture of a metal and its oxide is often used to generate a fixed oxygen partial pressure in a closed vessel. Such systems are called oxygen buffers, because they can absorb or release significant quantities of oxygen gas without a change in the oxygen pressure.

In the open atmosphere, identical reactions will take place, without the constraints imposed by the sealed tube. Should the oxygen pressure in the atmosphere be above the equilibrium pressure, as it is for almost all metals, oxidation will continue unabated, because the amount of oxygen over the metal is continually replenished and, ultimately, the metal will be completely oxidised (see Section S3.2).

S1.6   Miller indices

The Miller indices, $(h\,k\,l)$, of a plane in a lattice or a crystal define the positions where the plane cuts the unit cell edges, a_0, b_0 and c_0. A set of parallel planes has the same Miller indices. Negative intersections are represented by bar h, $(\bar{h})$, bar k, $(\bar{k})$, or bar $l(\bar{l})$.

In order to find the Miller indices of a plane, such as those shown in Figures S1.2(a), and S1.2(c), first draw other members of the set, if not already present. This is done by drawing parallel planes through each lattice point, as shown in Figure S1.2(b) and S1.2(d). Move along the unit cell vector a, and count the number of spaces between the planes crossed in moving a distance of one cell edge, a_0. If the spaces lie in the $+a$ direction, this gives the index h. If the spaces lie in the $-a$ direction this gives $\bar{h}$. Repeat this along b to get index k, and along c to get index l.

S1.7   Interplanar spacing and unit cell volume

The interplanar spacing and unit cell volumes for the seven crystal systems are given in Table S1.5.

The spacing of the layers of $\{h\,k\,l\}$ planes, d_{hkl}, in a cubic crystal is given by:

$$d_{hkl} = \frac{a}{(h^2 + k^2 + l^2)^{1/2}}$$

$$= a, \frac{a}{\sqrt{2}}, \frac{a}{\sqrt{3}}, \frac{a}{\sqrt{4}}, \ldots$$

The angle, θ, between two planes $(h_1\,k_1\,l_1)$ and $(h_2\,k_2\,l_2)$ in a cubic crystal is given by:

$$\cos\theta = \frac{h_1 h_2 + k_1 k_2 + l_1 l_2}{[(h_1^2 + k_1^2 + l_1^2)(h_2^2 + k_2^2 + l_2^2)]^{1/2}}$$

S1.8   Construction of a reciprocal lattice

The steps in the construction of a reciprocal lattice are given below and are illustrated for a section of a monoclinic unit cell.

1. Draw the unit cell (Figure S1.3a).

2. Draw lines perpendicular to the $(1\,0\,0)$, $(0\,1\,0)$ and $(0\,0\,1)$ planes. These lines are perpendicular to the end faces of the unit cell and form the axes of the reciprocal lattice (Figures S1.3b and S1.3c).

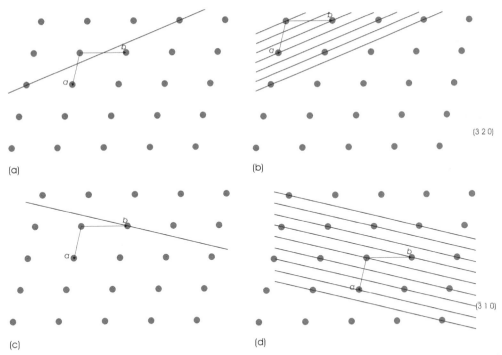

(a)   (b)

(3 2 0)

(c)   (d)

($\bar{3}$ 1 0)

Figure S1.2   Determination of Miller indices: for the plane shown (in parts a and c) draw parallel planes through all the lattice points (as in parts b and d) and count the spaces between the planes in a distance of $1a$ and $1b$. Spaces counted in a negative direction are given negative indices.

Table S1.5   Interplanar spacing, d_{hkl} and unit cell volume, V

System	$\dfrac{1}{d_{hkl}^2}$	Unit cell volume
Cubic	$\dfrac{h^2 + k^2 + l^2}{a^2}$	a^3
Tetragonal	$\dfrac{h^2 + k^2}{a^2} + \dfrac{l^2}{c^2}$	$a^2 c$
Orthorhombic	$\dfrac{h^2}{a^2} + \dfrac{k^2}{b^2} + \dfrac{l^2}{c^2}$	abc
Monoclinic	$\dfrac{h^2}{a^2 \sin^2 \beta} + \dfrac{k^2}{b^2} + \dfrac{l^2}{c^2 \sin^2 \beta} - \dfrac{2hl \cos \beta}{ac \sin^2 \beta}$	$abc \sin \beta$
Triclinic[a]	$\dfrac{1}{V^2}\,(S_{11}h^2 + S_{22}k^2 + S_{33}l^2 + 2S_{12}hk$ $+ 2S_{23}kl + 2S_{13}hl)$	$abc(1 - \cos^2 \alpha - \cos^2 \beta - \cos^2 \gamma$ $+ 2\cos \alpha \cos \beta \cos \gamma)^{1/2}$
Hexagonal	$\dfrac{4}{3}\left(\dfrac{h^2 + hk + k^2}{a^2}\right) + \dfrac{k^2}{b^2} + \dfrac{l^2}{c^2}$	$\dfrac{\sqrt{3}}{2} a^2 c = 0.866\, a^2 c$
Rhombohedral	$\dfrac{h^2 + k^2 + l^2 \sin^2 \alpha + 2(hk + kl + hl)\,(\cos^2 \alpha - \cos \alpha)}{a^2(1 - 3\cos^2 \alpha + 2\cos^3 \alpha)}$	$a^3 + (1 - 3\cos^2 \alpha + 2\cos^3 \alpha)^{1/2}$

[a] $S_{11} = b^2 c^2 \sin^2 \alpha$, $S_{22} = a^2 c^2 \sin^2 \beta$, $S_{33} = a^2 b^2 \sin^2 \gamma$, $S_{12} = abc^2(\cos \alpha \cos \beta - \cos \gamma)$, $S_{23} = a^2 bc(\cos \beta \cos \gamma - \cos \alpha)$, $S_{13} = ab^2 c(\cos \gamma \cos \alpha - \cos \beta)$.

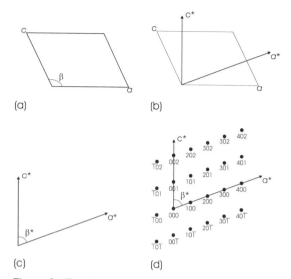

Figure S1.3 The construction of a reciprocal lattice: draw the unit cell (part a); draw lines perpendicular to the end faces of the unit cell (part b) to give the axes of the reciprocal lattice (part c); mark lattice points at $1/d_{100}$, $1/d_{010}$, and $1/d_{001}$ and fill in the lattice by extending these over the required region (part d)

3. Mark the lattice points at distances of $(1/d_{100}) = a^*$, $(1/d_{010}) = b^*$, and $(1/d_{001}) = c^*$, and fill in the lattice by extending these over the required region of reciprocal space (Figure S1.3d).

4. Index the planes hkl (Figure S1.3d).

The intensity of radiation diffracted from a set of hkl planes depends on the relative phases of the waves from adjacent planes, which in turn depends on the symmetry of the structure and the types of atoms that are in the unit cell. When these waves are completely in step the diffracted beam is intense. When they are completely out of step, the diffracted intensity is zero (see also Section 14.7). The relative intensity of the diffracted radiation from a set of planes is portrayed by the weighted reciprocal lattice.

The reflections not present because of the symmetry of the structure are termed systematic absences. These are:

- F Bravais lattice: points are present only for h, k and l all even or all odd and are absent for mixed even and odd combinations.

- I Bravais lattice: points are absent when $h + k + l$ is odd.

- C Bravais lattice: points are absent when $h + k$ is odd.

For example, the first five reflections from cubic crystals that are based on an F Bravais lattice, such as NaCl, KCl, $MgAl_2O_4$, are $\{111\}$, $\{220\}$, $\{311\}$, $\{400\}$ and $\{422\}$. The actual intensities of the reflections will depend on the atoms present in the crystal. Thus, the X-ray powder diffraction pattern of KCl (Figure 5.12) shows that the $\{111\}$ reflection is absent, although it would be expected to be present from symmetry grounds alone.

S2

Supplementary material to Part 2: classes of materials

S2.1 Summary of organic chemical nomenclature

Organic chemistry is the chemistry of carbon and its compounds, excluding a few small molecules such as carbon monoxide (CO), carbon dioxide (CO_2), graphite and a small number of other materials. Organic molecules generally contain large numbers of atoms, and conventional molecular formulae are often ambiguous. Even a simple molecular formula, such as C_2H_4O, represents two quite different compounds (acetaldehyde or ethylene oxide). This is because the way in which the atoms are linked in an organic molecule is as important as the constituents. To overcome this difficulty, organic formulae are often represented in a diagrammatic fashion, called structural formulae. In addition, an extensive formal naming procedure has been devised, which leads to an unambiguous structure for the molecule.

The atoms in organic compounds are linked by covalent bonds. Each atom has a characteristic and invariable valence, given in Table S2.1.

S2.1.1 Hydrocarbons

Hydrocarbons are compounds that contain only carbon and hydrogen. The hydrocarbons have been subdivided into alkanes, alkenes, alkynes – which are collectively called aliphatic hydrocarbons – and aromatic hydrocarbons (also called arenes, see below).

S2.1.1.1 Aliphatic hydrocarbons

Alkanes Alkanes contain only single bonds. They are all members of a homologous series with a series formula of C_nH_{2n+2}, where n is the number of carbon atoms in the molecule. Each member has an extra (CH_2) compared with its 'lower' neighbour. The simplest alkanes are listed in Table S2.2.

There are several ways of representing these compounds. The molecular formula is of least value. For example, the molecular formula of two compounds, butane and methyl propane, are the same. The structural formula is a two-dimensional representation of the three-dimensional molecular structure. This representation is ambiguous because the molecules are not planar, and the geometry of the bonds around the carbon atoms must be kept in mind when interpreting these formulae. Four single bonds are arranged tetrahedrally, with interbond angles of about 109°. Two single bonds and one double bond are coplanar and at angles of 120°. One single bond and a triple bond are linear. The structural formulae of the hydrocarbons are shown in Scheme S2.1.

Understanding solids: the science of materials.   Richard J. D. Tilley
© 2004 John Wiley & Sons, Ltd   ISBNs: 0 470 85275 5 (Hbk) 0 470 85276 3 (Pbk)

Table S2.1   Elements of importance in organic chemistry

Element	Symbol	Valence
Hydrogen	H	1
Carbon	C	4
Nitrogen	N	3
Phosphorus	P	3
Oxygen	O	2
Sulphur	S	2
Fluorine	F	1
Chlorine	Cl	1
Bromine	Br	1

Table S2.2   The simplest alkanes

Name	Molecular formula	Condensed structural formula	Alkyl group
Methane	CH_4	CH_4	Methyl
Ethane	C_2H_6	CH_3CH_3	Ethyl
Propane	C_3H_8	$CH_3CH_2CH_3$	Propyl
Butane	C_4H_{10}	$CH_3CH_2CH_2CH_3$	Butyl
Methyl propane	C_4H_{10}	$CH_3CH(CH_3)CH_3$	–
Pentane	C_5H_{12}	$CH_3(CH_2)_3CH_3$	Pentyl
Hexane	C_6H_{14}	$CH_3(CH_2)_4CH_3$	Hexyl
Heptane	C_7H_{16}	$CH_3(CH_2)_5CH_3$	Heptyl
Octane	C_8H_{18}	$CH_3(CH_2)_6CH_3$	Octyl

The structural formulae are often written in a condensed fashion, as listed in Table S2.2. Atom groups that are on side-chains are written in parentheses, for example, (CH_3). In order to distinguish between a group in the main chain of the molecule and a group on a side-chain it is necessary to recall that carbon has a valence of four. Thus, atoms in the main chain are linked to two other carbon atoms whereas those in a side-group are linked only to one.

The simplest representation of an organic molecule is a stick structure. The structure is drawn as a series of lines that represent the bonds between carbon atoms. The line is drawn in a zigzag fashion to represent the tetrahedral geometry of carbon–carbon single bonds; the carbon atoms are located at the end of each short line, and hydrogen atoms are omitted (Scheme S2.1). Structures in this book are

represented by a combination of stick structures and structural formulae. The aim is simply to present the structure in an unambiguous way.

The single bond linking the atoms is of a fixed length, but the atoms are free to rotate around the bond. A molecule can have many different *conformations* because of this bond rotation. The conformation of large molecules is important as it influences physical properties and chemical reactivity.

The formula C_4H_{10} corresponds to two different molecules – butane, which has a 'straight line' of carbon atoms, and methyl propane, which has three (CH_3) (methyl) groups joined to a single carbon atom. Compounds with the same formula, but different structures, are called structural isomers. Note that different isomers will have different names. Isomers correspond to different *configurations*. To change configuration it is necessary to break bonds. To change conformation it is necessary only to rotate atoms about the single bonds; no bond breaking is needed.

The nomenclature of organic chemistry derives from the names of straight-chain hydrocarbons, the most important of which are given in Table S2.2. To name a hydrocarbon:

1. Identify the longest unbranched carbon chain. This is given the name of the corresponding alkane.

2. Identify the branches by using the names for the side-chains given in Table S2.2. The alkane name ending -*ane* is changed to -*yl*. Alkyl groups are often given the general symbol —R.

3. Indicate the positions of the alkyl groups from the end of the chain, starting at the end that gives the lowest numbers.

4. Cyclic alkanes, those that have the carbon atoms joined in a nonplanar ring (not aromatic), have the prefix *cyclo*-attached.

Alkenes   Alkenes, hydrocarbons containing one or more double bonds, are also known as

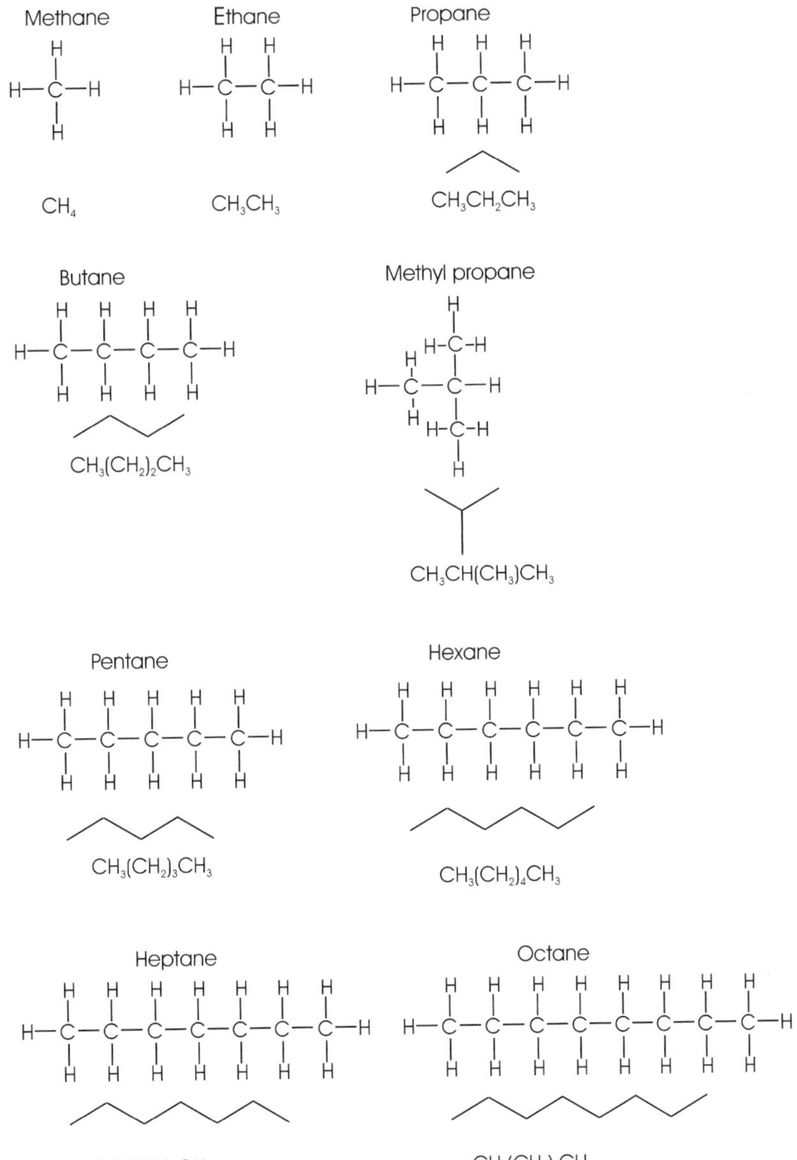

Methane

CH_4

Ethane

CH_3CH_3

Propane

$CH_3CH_2CH_3$

Butane

$CH_3(CH_2)_2CH_3$

Methyl propane

$CH_3CH(CH_3)CH_3$

Pentane

$CH_3(CH_2)_3CH_3$

Hexane

$CH_3(CH_2)_4CH_3$

Heptane

$CH_3(CH_2)_5CH_3$

Octane

$CH_3(CH_2)_6CH_3$

Scheme S2.1 The name (uppermost), structural formula (shown below the name), stick formula (shown below the structural formula) and condensed formula (lowermost) for some alkanes

unsaturated hydrocarbons. They form a homologous series with a formula C_nH_{2n-2}. The name ending -*ane* (alkane) changes to -*ene* (alkene); that is, $CH_2=CH_2$ is ethene; $CH_2=CHCH_3$ is propene; and $CH_2=CHCH_2CH_3$ is (1-) butene. To name an alkene, find the longest chain of consecutive carbon-atoms containing the C=C double bond. The parent name is derived from the alkane by changing -ane to -ene. The carbon atoms are numbered to give the double bond the lowest possible number.

Table S2.3 Names of organic compounds

Old name	Systematic name
Ethylene	Ethene
Propylene	Propene
Acetylene	Ethyne
Ortho-xylene (*o*-xylene)	1,2-Dimethylbenzene
Meta-xylene (*m*-xylene)	1,3-Dimethylbenzene
Para-xylene (*p*-xylene)	1,4-Dimethylbenzene
Toluene	Methylbenzene
Acetaldehyde	Ethanal
Acetone	Propanone
Formic acid	Methanoic acid
Acetic acid	Ethanoic acid
Malic acid	*Cis*-butenedioic acid
Fumaric acid	*Trans*-butenedioic acid
Vinyl chloride	Chloroethene
Vinylidene chloride	1,1-Dichloroethene

Many of these compounds have older, nonsystematic, names that are still widely used. Some of these are collected in Table S2.3.

The double bond consists of one σ bond and one π bond. This bond is rigid and, unlike a single bond, rotation is not possible. This leads to geometrical isomerism. Geometrical isomers have the atoms or groups at each end of a double bond arranged differently in space, which endows the molecules with different chemical and physical properties. The geometrical isomers are distinguished by the prefixes *cis*-, for those on which the important groups are on the same side of the double bond, and *trans*-, for those on opposite sides (Scheme S2.2).

Alkynes   The alkynes are a homologous series of hydrocarbons containing a triple bond. The triple bond is formed from one σ bond and two π bonds. The bond is rigid and allows no rotation. Alkynes with one triple bond have the formula C_nH_{2n}. The

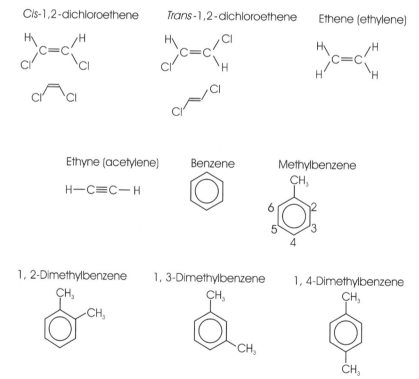

Scheme S2.2 The names and formulae of some organic compounds. When one substituent is attached to a benzene ring it sets position 1. Other substituents then take the positional numbers indicated on the diagram for methyl benzene

Table S2.4 Functional groups

Functional group	Name ending	Compound name	Example
—OH	-ol	Alcohol	CH_3CH_2OH, ethanol
—NH$_2$	-amine	Amine	CH_3NH_2, methylamine
—CHO	-al	Aldehyde	CH_3CHO, ethanal
=CO	-one	Ketone	$(CH_3)_2O$, propanone
—COOH	-oic acid	Carboxylic acid	CH_3COOH, ethanoic acid
=O	ether	Ether	$(CH_3CH_2)_2O$, diethyl ether
—X	(Prefix)[a]	Halide (F, Cl, Br, l)	$CH_3(CH_2)_2Cl$, 1-chloropropane
—CN	-nitrile	Nitrile (cyanide)	$CH_2{=}CH{-}CN$, propenenitrile
—COOR	-oate	Ester	CH_3COOCH_3, methyl ethanoate

[a]The presence of a halogen is indicated by a prefix, fluoro- (F), chloro- (Cl), bromo- (Br) or iodo- (I) and not by an ending.

simplest member is ethyne (Scheme S2.2). The nomenclature is similar to that for the alkenes. The longest chain of consecutive carbon atoms containing the triple bond gives the parent alkane name. Replace the termination *-ane* by *-yne*. The position of the triple bond from the end of the chain is indicated by a positional number.

S2.1.1.2 Aromatic hydrocarbons

Aromatic hydrocarbons (arenes) contain a planar ring of six carbon atoms called a benzene ring. The simplest aromatic compound is benzene, C_6H_6. The structure consists of a ring of six carbon atoms linked by sp^2 hybrid orbitals forming σ bonds. The remaining p orbitals on the carbon atoms form a delocalised π bond above and below the plane of the carbon atoms. The structure is usually drawn in 'stick' form (Scheme S2.2). Note that 'ordinary' hydrocarbons can have a cyclic form. These are not aromatic unless they also have delocalised electron orbitals.

To name substituents, the positions on the benzene ring are numbered 1 to 6 (see 1,2-dimethylbenzene, 1,3-dimethylbenzene and 1,4-dimethylbenzene in Scheme S2.2).

S2.1.2 Functional groups

Despite the bewildering complexity of organic compounds, the reactions and behaviour can be systematically studied because all compounds are made up of a relatively inert hydrocarbon skeleton together with one or more reactive 'functional groups'. These functional groups show characteristic chemical reactivity. Knowledge of the chemistry of a small number of functional groups allows one to understand the behaviour of vast numbers of organic compounds. The principle functional groups are listed in Table S2.4.

The nomenclature of compounds containing functional groups is derived from that of the longest carbon chain in the molecule, which is given the parent alkane name. The presence of the functional group is indicated by adding a termination that specifies the functional group, as in Table S2.4, except for halides. For example, an aldehyde has the ending *-al* substituted for the final *-e* in the parent name. Similarly, the general formula of the carboxylic acids is RCOOH, and the —COOH functional group, called the carboxyl group, is indicated by the addition of the termination *-oic acid*. Amines are named by adding the suffix *-amine* to the alkyl name.

S3

Supplementary material to Part 3: reactions and transformations

S3.1 Diffusion

S3.1.1 The relationship between D and diffusion distance

Suppose an atom is moving from one stable site to the next in the x direction by way of a random walk, as in self-diffusion. The net displacement of a diffusing atom after N jumps will be the algebraic sum of the individual jumps. If x_i is the distance moved along the x axis in the ith jump, the distance moved after a total of N jumps, x, will simply be the sum of all the individual steps; that is,

$$x = x_1 + x_2 + x_3 \cdots x_N = \Sigma x_i$$

In a linear 'crystal' in which the sites are separated by a distance a, each individual value of x_i can be $+a$ or $-a$.

As the jumps take place with an equal probability in both directions, after N jumps the total displacement may have any value between zero and $\pm Na$. In order to proceed, use a mathematical shortcut. If the jump distances are squared, the negative quantities are removed. Thus:

$$x^2 = (x_1 + x_2 + x_3 \cdots x_N)(x_1 + x_2 + x_3 \cdots x_N)$$
$$= (x_1 x_1 + x_1 x_2 + x_1 x_3 \cdots x_1 x_N)$$
$$+ x_2 x_1 + x_2 x_2 + x_2 x_3 \cdots x_2 x_N)$$
$$+ \cdots$$
$$+ x_N x_1 + x_N x_2 + x_N x_3 \cdots x_N x_N)$$

This can be written:

$$x^2 = \Sigma x_i^2 + 2\Sigma x_i x_{i+1} + 2\Sigma x_i x_{i+2} + \cdots$$
$$= \Sigma x_i^2 + 2\Sigma\Sigma x_i x_{i+j}$$

In the limit of a large number of jumps, knowing that each jump may be either positive or negative, the double sum terms average to zero. The equation therefore reduces to the manageable form:

$$\langle x^2 \rangle = \Sigma x_i^2$$

The result is called the mean square displacement, $\langle x^2 \rangle$. As each jump, x_i, can be equal to $+a$ or $-a$,

$$\langle x^2 \rangle = x_1^2 + x_2^2 + x_3^2 \cdots + x_N^2$$
$$= a^2 + a^2 + a^2 \cdots + a^2$$
$$= Na^2$$

The frequency with which an atom jumps from one site to another along the x direction is defined as Γ, so that the total number of jumps, N, will be given by Γ jumps per second multiplied by the time, t, over which the diffusion experiment has lasted, that is,

$$N = \Gamma t$$

Understanding solids: the science of materials.   Richard J. D. Tilley
© 2004 John Wiley & Sons, Ltd   ISBNs: 0 470 85275 5 (Hbk) 0 470 85276 3 (Pbk)

Hence:

$$\langle x^2 \rangle = \Gamma t a^2$$

However, the term Γa^2 is equal to $2D$, so that:

$$\langle x^2 \rangle = 2Dt$$

This is the Einstein diffusion equation.

The average distance that an atom will travel in time t is the square root of $\langle x^2 \rangle$, a quantity called the root mean square displacement of x, which is given by:

$$\sqrt{\langle x^2 \rangle} = \sqrt{2Dt}$$

In the case of diffusion in a three-dimensional crystal, with equal probability of a jump along x, y or z,

$$\sqrt{\langle x^2 \rangle} = \sqrt{6Dt}$$

S3.1.2 Atomic migration and the diffusion coefficient

Expressions for J and dc/dx can be derived by reference to Figure S3.1. Adjacent planes in a crystal, numbered 1 and 2, are separated by the atomic jump distance, a. Take n_1 and n_2 to be the numbers of diffusing atoms per unit area in planes 1 and 2, respectively. If Γ_{12} is the frequency with which an atom moves from plane 1 to plane 2, then the numbers of atoms moving from plane 1 to plane 2 per second is j_{12}, where

$$j_{12} = n_1 \Gamma_{12}$$

Similarly, the number moving from plane 2 to plane 1 is j_{21}, where

$$j_{21} = n_2 \Gamma_{21}$$

The net movement, often called the flux, between the planes, J, is given by:

$$J = j_{12} - j_{21} = n_1 \Gamma_{12} - n_2 \Gamma_{21}$$

If the process is random, the jump frequency is independent of direction and we can set Γ_{12} equal to Γ_{21}. Moreover, if the jump frequency is independent of direction, then half of the jumps, on average, will be in one direction and half will be in the opposite direction, so we can write

$$\Gamma_{12} = \Gamma_{21} = \tfrac{1}{2}\Gamma$$

where Γ represents the overall jump frequency of the diffusion atoms. Thus:

$$J = \tfrac{1}{2}(n_1 - n_2)\,\Gamma$$

The number of mobile atoms on plane 1 is n_1 per unit area, so that the concentration per unit volume at plane 1 is n_1/a, or c_1. Similarly, the number of mobile atoms per unit area on plane 2 is n_2, so that the concentration per unit volume at plane 2 is n_2/a, or c_2. Thus:

$$n_1 - n_2 = a(c_1 - c_2)$$

Hence:

$$J = \tfrac{1}{2}a(c_1 - c_2)\,\Gamma$$

The concentration gradient, dc/dx, is given by the change in concentration between planes 1 and 2

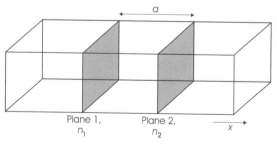

Figure S3.1 Two adjacent planes, 1 and 2, in a crystal, separated by the jump distance of the diffusing ion, a. The numbers of diffusing atoms on planes 1 and 2 are n_1 and n_2, respectively divided by the distance between planes 1 and 2; that is,

$$-\frac{dc}{dx} = \frac{c_1 - c_2}{a}$$

where a $-$ sign is introduced, as the concentration falls as we move from plane 1 to plane 2. Hence,

$$c_1 - c_2 = -a \frac{dc}{dx}$$

and

$$J = -\tfrac{1}{2}\Gamma a^2 \frac{dc}{dx}$$

A comparison with Fick's first law,

$$J = -D \frac{dc}{dx}$$

shows that:

$$D = \tfrac{1}{2}\, \Gamma a^2$$

Substituting the expression for the jump frequency, Γ, in terms of the barrier height to be negotiated, E, yields:

$$D = \tfrac{1}{2}a^2\, \nu \exp\left(\frac{-E}{kT}\right)$$

S3.1.3   Ionic conductivity

The ionic conductivity, σ, is defined in terms of the following equation:

$$\sigma = ne\mu$$

where n is the number of migrating monovalent ions per unit volume, each carrying a charge e, and μ is the mobility of the ion. The number of jumps that an ion will make in the direction of the field per second is given by

$$\Gamma_+ = \nu \exp\left[-\left(E - \tfrac{1}{2}\, eaV\right)/kT\right]$$

where the potential barrier is $(E - \tfrac{1}{2}eaV)$. In a direction against the field the number of successful jumps will be given by:

$$\Gamma_- = \nu \exp\left[-\left(E + \tfrac{1}{2}\, eaV\right)/kT\right]$$

where the potential barrier is $(E + \tfrac{1}{2}\, eaV)$.

The overall jump rate in the direction of the field is $(\Gamma_+ - \Gamma_-)$ and, as the net velocity of the ions in the direction of the field, v, is given by the net jump rate multiplied by the distance moved at each jump, we can write:

$$\begin{aligned}
\text{v} &= \nu a \exp\left[-\left(E - \tfrac{1}{2}\, eaV\right)/kT\right] \\
&\quad - \exp\left[-\left(E + \tfrac{1}{2}\, eaV\right)/kT\right] \\
&= \nu a \exp\left(\frac{-E}{kT}\right)\left[\exp\left(\frac{eaV}{2kT}\right) - \exp\left(\frac{-eaV}{2kT}\right)\right]
\end{aligned}$$

For low field strengths eaV is much less than kT, and

$$\exp\left(\frac{eaV}{2kT}\right) - \exp\left(\frac{-eaV}{2kT}\right)$$

may be replaced by eaV/kT (Table S3.1). Using this approximation:

$$\text{v} = \frac{\nu a^2 eV}{kT}\exp\left(\frac{-E}{kT}\right)$$

The mobility, μ, of the ion is defined as the velocity when the value of V is unity, so:

$$\mu = \frac{\nu a^2 e}{kT}\exp\left(\frac{-E}{kT}\right)$$

Substituting for μ in the equation

$$\sigma = ne\mu$$

we obtain

$$\sigma = \frac{n\nu a^2 e^2}{kT}\exp\left(\frac{-E}{kT}\right)$$

Table S3.1   The equivalence of eaV and $[\exp(eaV/2kT) - \exp(-eaV/2kT)]$

$eaV/2kT$	$\exp(eaV/2kT)$, A	$\exp(-eaV/2kT)$, B	Difference, $A - B$
0.001	1.00100	0.99900	0.00200
0.01	1.01005	0.99005	0.02000
0.1	1.10517	0.90484	0.20033
1.00	2.71828	0.36788	2.35040
10.00	22026.47	4.54×10^{-5}	22026.47

Note: to see just what sort of field strength this approximation corresponds to, take a value of $eaV/2kT$ equal to 1. Taking a temperature of 500 K and a value of a of about 0.3×10^{-9} m we obtain a value for V of 2.87×10^{4} V m^{-1}. Thus, the approximation is reasonable for field strength up to about 30 000 V m^{-1}.

This equation takes the form:

$$\sigma = \frac{\sigma_0}{T} \exp\left(\frac{-E}{kT}\right)$$

where n is the number of mobile species present in the crystal, ν is the vibration frequency of the solid, a is the jump distance between stable sites, e is the electron charge, k is the Boltzmann constant, and T is the temperature (in kelvin).

S3.2   Phase transformations and thermodynamics

S3.2.1   Phase stability

The direction (but not the rate) of a phase transformation can be determined from the equilibrium thermodynamic properties of the phases involved. For this purpose, the most important thermodynamic parameter is the Gibbs energy (or Gibbs free energy) of the phases. The Gibbs energy, G (units J mol^{-1}), of a pure material is defined by the equation:

$$G = H - TS$$

where H is the enthalpy of the phase (units J mol^{-1}), S is the entropy (units J K mol^{-1}), and T is the (absolute) temperature. The Gibbs energy of a pure substance depends on the temperature, pressure and the quantity of substance present. For many purposes, the pressure variation can be ignored, and

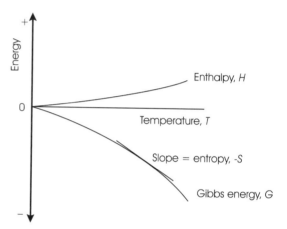

Figure S3.2   Variation of Gibbs energy, G, and enthalpy, H, with temperature, T. The slope of the Gibbs energy curve is equal to the negative value of the entropy, $-S$

the variation with temperature is the main criterion for phase stability (Figure S3.2).

The stable form of any pure phase at any defined temperature is the form with the lowest Gibbs energy. When the Gibbs energies of two polymorphs are plotted they will intersect at the transition temperature (Figure S3.3). Because the slope of the Gibbs energy curve is equal to $-S$, higher temperatures favour the phase with the higher entropy.

S3.2.2   Reactions

A reaction at constant pressure will take place spontaneously (should the kinetics allow) if the Gibbs energy of the system decreases because of

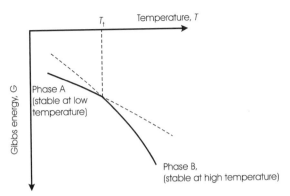

Figure S3.3 The variation of Gibbs energy, G, for two phases, one of which (A) is stable at low temperature, and one of which (B) is stable at high temperature. The transition temperature is T_t

the change. In order to determine if a reaction is spontaneous, it is simply necessary to add together the Gibbs energies of all of the reactants and all of the products and note the difference. Note that the Gibbs energy change depends on the quantities, in moles, of compounds present:

Gibbs energy change

= (total Gibbs energy of products)

− (total Gibbs energy of reactants)

The Gibbs energy of formation of a compound, ΔG_f, is used in these calculations. The Gibbs energy change for a reaction is therefore given by:

$$\Delta G_r = \Sigma n \, \Delta G_f(\text{products}) - \Sigma n \Delta G_f(\text{reactants})$$

Because the Gibbs energy varies with temperature, the free energy values have to be adjusted accordingly. This is especially true for reactions involving solids, as these are carried out at high temperatures.

For many purposes, the standard Gibbs energy of formation, ΔG_f^0, is adequate. The standard Gibbs energy of formation of a compound is the free energy of reaction when 1 mole of compound is formed from the constituent elements, in their most stable states, at 25 °C and 1 atm pressure. In this case, the standard Gibbs energy of reaction is given by:

$$\Delta G_r^0 = \Sigma n \Delta G_f^0(\text{products}) - \Sigma n \Delta G_f^0(\text{reactants})$$

S3.2.3 Oxidation

Two typical oxidation reactions are:

$$4Al + 3O_2 \rightarrow 2Al_2O_3, \quad \Delta G_f^0 = -1582.3 \, \text{kJ mol}^{-1}$$
$$Ca + \tfrac{1}{2}O_2 \rightarrow CaO, \quad \Delta G_f^0 = -604.0 \, \text{kJ mol}^{-1}$$

Note that the standard Gibbs energy of the reaction is equal to the formation energy of the oxide, ΔG_f^0. The values quoted are per mole of oxide. Because ΔG_f^0 is negative, the oxides are stable. This means that the reaction to form a metal oxide is spontaneous in air. This is true for most metals and is why native metals do not occur (with the exception of gold, silver and, very rarely, copper). In favourable cases, such as that of aluminium and chromium, the product of corrosion may provide a protective coating, preventing further corrosion from taking place. In less favourable circumstances, such as the corrosion of iron, the product may be loose or flake off, to expose new surfaces for corrosion.

The relationship between the Gibbs energy of formation and the equilibrium oxygen pressure surrounding a metal-plus-oxide mixture is also given by standard thermodynamic equations. For an oxidation reaction

$$\text{metal} + \text{oxygen} \rightarrow \text{metal oxide}$$

at equilibrium

$$\Delta G_f^0 = -RT \ln\left(\frac{1}{p_{O_2}}\right) = +RT \ln(p_{O_2})$$

where (p_{O_2}) is the partial pressure of oxygen reached at equilibrium. Calculation of the equilibrium partial pressures over metal oxides shows that values lie between approximately 10^{-7} atm to 10^{-40} atm.

If a mixture of a metal and its oxide is sealed into an evacuated tube and heated to a suitable temperature, oxygen gas will be evolved until the equilibrium pressure is reached. In air, the oxygen partial pressure is about a fifth of an atmosphere. It is clear, therefore, that metals will have a tendency to oxidise in the open. From the point of view of thermodynamics, there is always a considerable driving force for reaction.

S3.2.4 Temperature

The temperature changes in many chemical reactions are classified in terms of enthalpy. An endothermic reaction is one in which heat is consumed. This is signified by recording the enthalpy change of the reaction as positive:

$$CaCO_3(s) \rightleftharpoons CaO(s) + CO_2(g) \quad \Delta H_r^0 = +158 \, kJ$$

where ΔH_r^0 is the standard enthalpy change taking place during the reaction. This means that $158 \, kJ$ are consumed for each mole of CaO produced, when the reactants and products are in the thermodynamic standard state, signified by the superscript $\circ$ (the heat consumed will be different under other conditions). A reaction in which heat is liberated, denoted by a negative enthalpy change, is called an exothermic reaction. Combustion and the setting of cement are exothermic reactions.

S3.2.5 Activity

Equations such as those for the equilibrium constant, the reaction quotient and the Nernst equation are written mostly in terms of concentrations and partial pressures. This is correct only for very dilute solutions or ideal gases. In general, the amount of 'active material' present appears to be less than the nominal amount measured in the customary units (in moles per litre, or in atmospheres). In order to retain the form of the equations used here and yet increase precision, thermodynamics replaces the concentration terms by the activity, which is an effective concentration, and the pressure by the fugacity, which is an effective pressure. Thus, the equilibrium constant for the reaction

$$aA + bB \rightleftharpoons cC + dD$$

is then written

$$K_c = \frac{a_C^c a_D^d}{a_A^a b_B^b}$$

where the terms a_A, etc., represent the activities of the components. The reaction quotient, Q_c is written in the same way. The Nernst equation is written:

$$E = E^\circ - \frac{RT}{nF} \ln \left(\frac{a_C^c a_D^d}{a_A^a b_B^b} \right)$$

In order that the two systems of equations are equivalent, the activity of a solute is written as the product of concentration and an activity coefficient, γ:

$$a = c \, \gamma$$

where c is the molar concentration, in moles per litre of solution. The activity coefficient tends towards 1 at low concentrations, so that a becomes approximately equal to c.

The activity of pure liquids and solids is close to 1 over wide ranges of pressures, which is why pure solids and liquids are omitted from the above equations.

In equations using partial pressures, the pressure is replaced by the fugacity, f, which is an effective pressure. In this case,

$$f = \phi p$$

where ϕ is the fugacity coefficient. In gas mixtures, the partial fugacity, f_i, measured in atmospheres, is used. Thus, for the reaction

$$H_2 + Cl_2 \rightleftharpoons 2HCl$$

$$K_p \approx \frac{(p_{HCl})^2}{(p_{H_2})(p_{Cl_2})}$$

where (p_i) is the partial pressure of component i; a more accurate expression is

$$K_p = \frac{f_{HCl}^2}{f_{H_2} f_{Cl_2}}$$

where f_i is the partial fugacity of component i. At low pressures the fugacity coefficient approaches 1, and fugacity and pressure are approximately equal. As most reactions involve gases at relatively low pressures, fugacity is not often used.

S3.3   Oxidation numbers

Redox reactions involve oxidation and reduction, that is, electron transfer. Oxidation is equivalent to electron loss and reduction is equivalent to electron gain. An oxidising agent removes electrons, becoming reduced in the process, whereas a reducing agent supplies electrons and becomes oxidised. It is not always easy to detect whether oxidation or reduction have taken place during a reaction. The best way to do this is to look for a change in the oxidation number of the elements involved:

- an increase in oxidation number is equivalent to oxidation;

- a decrease in oxidation number is equivalent to reduction;

- oxidation must be accompanied by reduction, and vice versa.

There are simple rules for assigning oxidation numbers. For most purposes, just two suffice:

- elements have an oxidation number of zero;

- monatomic ions have an oxidation number equal to the charge on the ion.

Hence,

$$Fe \rightarrow Fe^{3+}$$

is an oxidation as the oxidation number changes from 0 to $+3$, and

$$Br \rightarrow Br^-$$

is a reduction as the oxidation number changes from 0 to -1.

In less straightforward cases, work through the rules below in order and stop when the oxidation number has been assigned.

1. The sum of all the oxidation numbers of all the atoms in the species is equal to the total charge on the species.

2. Atoms in elemental form have oxidation number 0.

3. Ions have an oxidation number equal to the charge on the ion.

4. H, in combination with nonmetals, has an oxidation number of $+1$; H, in combination with metals, has an oxidation number of -1.

5. F has an oxidation number of -1.

6. O, unless combined with F, has an oxidation number of -2; O in peroxides, O_2^{2-}, has an oxidation number of -1; O in ozonides, O_3^-, has an oxidation number of $-1/3$.

Table S3.2   Activity series

	Reduced form	Oxidised form
Strongly reducing	K	K^+
	Na	Na^+
	Mg	Mg^{2+}
	Cr	Cr^{2+}
	Zn	Zn^{2+}
	Cr	Cr^{3+}
	Fe	Fe^{2+}
	Ni	Ni^{2+}
	Sn	Sn^{2+}
	Fe	Fe^{3+}
	Pb	Pb^{2+}
	H	H^+
	Cu	Cu^{2+}
	Hg	Hg_2^{2+}
	Ag	Ag^+
	Pt	Pt^{2+}
Strongly oxidising	Au	Au^+

Metals can be arranged in a series such that those higher in the series reduce those below them. This is called the activity series (see Table S3.2).

Metals above H can reduce H^+ ions in acids to yield $H_2(g)$.

S3.4   Cell notation

In order to specify the makeup of an electrochemical cell a compact notation has been developed.

1. Each compartment of a cell contains the components of a redox half-reaction. This is often condensed into redox couple notation, as shown in the following example.
Cathode half-reaction:

$$Cu^2(aq) + 2e^- \rightarrow Cu(s) \quad Cu^{2+}/Cu$$

Anode half-reaction:

$$Zn(s) \rightarrow Zn^{2+}(aq) + 2e^- \quad Zn^{2+}/Zn$$

Half-reactions are always written:

$$oxidised\ state + ne^- \rightarrow reduced\ state$$

The redox couple is always written:

$$oxidised\ species\ /\ reduced\ species$$

2. The H^+/H_2 redox couple is used as a standard of potential; often called the 'hydrogen electrode'.
Cathode half reaction:

$$2H^+(aq) + 2e^- \rightarrow H_2(g); H^+/H_2$$

Anode half reaction:

$$H_2(g) \rightarrow 2H^+(aq) + 2e^-; H^+/H_2$$

3. Whether a species is actually oxidised or reduced depends on the other half-reaction in the cell.

4. Electrodes are written in the following form.
Anode:

$$Pt|H_2(g)|H^+(aq)$$

$$Zn|Zn^{2+}$$

Cathode:

$$H^+(aq)|H_2(g)|Pt$$

$$Cu^{2+}(aq)|Cu$$

The junctions between phases are represented by $||$

5. Cells are written in a diagrammatic form:

$$Zn(s)|Zn^{2+}(aq)||Cu^{2+}(aq)|Cu(s)$$

where $||$ represents the separation of the two electrodes or electrode compartments. The anode is always to the left and the cathode to the right.

S3.5   The stability field of water

The starting point in plotting the stability field of water is the Nernst equation, in the form:

$$E = E^\circ - \frac{0.05916}{n} \log Q$$

at 25 °C.

The oxidation of water is defined by the half-reaction:

$$O_2(g) + 4H^+(aq) + 4e^- \rightarrow 2H_2O(l), \quad E^\circ = +1.23\ V$$

In this case, $n = 4$, and

$$Q = \frac{1}{p_{O_2}[H^+]^4}$$

Substituting into the Nernst equation:

$$E = E^\circ - \frac{0.05916}{4} \log \left\{ \frac{1}{p_{O_2}[H^+]^4} \right\}$$

$$= E^\circ + 0.01479 \log \left\{ p_{O_2}[H^+]^4 \right\}$$

$$= 1.23 + 0.01479 \log \left\{ p_{O_2}[H^+]^4 \right\}$$

$$= 1.23 + 0.01479 \log p_{O_2} + 0.01479 \log[H^+]^4$$

For the normal oxygen partial pressure, p_{O_2}, of 0.21 atm, we find:

$$E = 1.23 - 0.01 + 0.05916 \log [H^+]$$

Remembering

$$pH = -\log_{10}[H^+]$$
$$E = 1.22 - 0.05916\,pH\ (volts)$$

For reasonable variation in oxygen partial pressure, this equation hardly changes. This line is plotted on the Pourbaix diagram. It signifies that any voltage/pH combination lying above the line will cause water to decompose according to the initial half-reaction given.

The reduction of water is defined by the half-reaction

$$2H_2O(l) + 2e^- \rightarrow H_2(g) + 2OH^-(aq)$$
$$E° = -0.83\,V$$

In this case $n = 2$ and $Q = p_{H_2}[OH^-]^2$

$$E = E° - \frac{0.05916}{2}\log Q$$
$$= -0.833 - 0.02958\log\{p_{H_2}[OH^-]^2\}$$

Taking the partial pressure of H_2 to be 1 atm,

$$E = -0.83 - 0.02958\log[OH^-]^2$$
$$= -0.83 - 0.05916\log[OH^-]$$

Now,

$$\log[OH^-] = pH - 14$$

thus

$$E = -0.83 + 0.05916 \times 14 - 0.05916\,pH$$
$$\approx -0.05916\,pH$$

This line is plotted on the diagram. It means that any combination of voltage and pH lying below it will be sufficient to decompose water according to the half-reaction given.

S3.6   Corrosion and the calculation of the Pourbaix diagram for iron

Some of the critical reactions of iron in air and water are, in order of reduction potential:

$$O_2(g) + 4H^+(aq) + 4e^- \rightarrow 2H_2O(l)$$
$$E° = +1.23\,V\,(+0.81\,V\,\text{at pH 7})$$
[S3.1]

(oxidation of water);

$$Fe^{3+}(aq) + e^- \rightarrow Fe^{2+}(aq)\quad E° = +0.77\,V\quad [S3.2]$$

$$2H_2O(l) + O_2(g) + 4e^- \rightarrow 4OH^-(aq)$$
$$E° = +0.40\,V\,(+0.81\,V\,\text{at pH 7})$$
[S3.3]

$$2H^+(aq) + 2e^- \rightarrow H_2(g)\quad E° = 0\,V\quad [S3.4]$$

(reduction of acid);

$$Fe^{3+}(aq) + 3e^- \rightarrow Fe(s)\quad E° = -0.04\,V\quad [S3.5]$$

$$H_2O(l) + O_2(g) + 2e^- \rightarrow HO_2^-(aq)$$
$$+ OH^-(aq)\quad E° = -0.08\,V$$
[S3.6]

$$Fe^{2+}(aq) + 2e^- \rightarrow Fe(s)\quad E° = -0.44\,V\quad [S3.7]$$

$$2H_2O(l) + 2e^- \rightarrow H_2(g) + 2OH^-(aq)$$
$$E° = -0.83\,V\,(-0.42\,V\,\text{at pH 7})$$
[S3.8]

(reduction of water).

S3.6.1   Corrosion of iron

The reaction of iron with acids can be treated in terms of these reactions. Any metal below $E° = 0\,V$ in the series is able to reduce acidic $H^+(aq)$ to H_2. Both of the half-reactions involving Fe metal come into this category and so reaction of Fe with aqueous acids will produce H_2. Because $E°$ for the formation of Fe^{2+} is lower than that for Fe^{3+}, the formation of Fe^{2+} will be favoured.

The reaction of Fe with water can be treated in a similar way. Consider Reaction [S3.8], involving water in the absence of oxygen. $E°$ is $-0.83\,V$ and so Fe metal will not be oxidised by oxygen-free water in the standard conditions implied by the table. (Reaction [S3.8] is below the Fe half-reactions). However, standard conditions refer to an OH^- concentration of $1\,mol\,L^{-1}$, with a pH of 14. At pH 7, which corresponds to normal neutral water, $E = -0.42\,V$. This is just above the Fe^{2+}/Fe couple, so that an oxidation of Fe to Fe^{2+} by pure water is just possible. That is, Fe will have only a very slight tendency to be oxidised by water in the absence of acid.

When we consider damp air, the requisite reaction is Reaction [S3.3]. At pH 7 we see that this reaction will cause Fe to oxidise to Fe^{2+} or Fe^{3+} and that Fe^{2+} will also oxidise to Fe^{3+}. Thus, as is well known, moist air will cause Fe metal to rust. The range of pH and oxidation conditions over which iron will react with acidic or alkaline water are summarised on a Pourbaix diagram.

S3.6.2   The simplified Pourbaix diagram for iron in water and air

The steps to be followed to produce a Pourbaix diagram of the system are as follows.

- Step 1: reactions that involve oxidation and reduction but no pH changes will not show a dependence on pH. In these reactions, electrons appear in the half-reaction equations, but no H^+ or OH^- ions are present. These plot as straight lines parallel to the abscissa (the pH axis) see Figure S3.4a).
  - Step 1(a):

$$Fe^{3+} + e^- \rightarrow Fe^{2+} \quad E^\circ = +0.77\,V \quad [S3.9]$$

The dependence of the cell voltage (which is a measure of the oxidising potential as well as the free energy of the reaction) as a function of the concentrations of the species present is expressed via the Nernst equation as:

$$E = E^\circ - \frac{RT}{nF}\ln Q$$

which in this case is best written as

$$E = 0.77 - 0.05916\,\log\left\{\frac{[Fe^{2+}]}{[Fe^{3+}]}\right\}$$

$$= 0.77 + 0.05916\,\log\left\{\frac{[Fe^{3+}]}{[Fe^{2+}]}\right\}$$

For standard conditions, $[Fe^{3+}] = [Fe^{2+}] = 1$ mol dm^{-3}, thus

$$E = E^\circ = 0.77\,V$$

Note that as long as the Fe^{3+} concentration is fairly close to the Fe^{2+} concentration this line will always lie close to 0.77 V.

- Step 1(b):

$$Fe^{2+} + 2e^- \rightarrow Fe(s) \quad E^\circ = -0.44\,V$$

$$Q = \frac{1}{[Fe^{2+}]} \quad n = 2$$

$$E = -0.44 - \frac{0.05916}{2}\log\left\{\frac{1}{[Fe^{2+}]}\right\}$$

$$= -0.44 + 0.02958\,\log[Fe^{2+}]$$

Under standard conditions $[Fe^{2+}] = 1$ mol dm^{-3}, thus

$$E = -0.44\,V$$

The concentration dependence of the position of this line can be calculated as follows. If $[Fe^{2+}] = 10^{-7}$ mol dm^{-3}, $E = -0.44 - 0.21 = 0.65$ V; if $[Fe^{2+}] = 10^{-4}$ mol dm^{-3}, $E = -0.44 - 0.12 = 0.56$ V; and so on.

- Step 2: reactions in which there is a change of pH but no oxidation or reduction will plot as vertical lines parallel to the ordinate (the voltage axis; see Figure S3.4b). In these equations, H^+ or OH^- appears in the equations but no electrons.

  - Step 2(a):

$$Fe^{2+}(aq) + 2H_2O(l) \rightarrow Fe(OH)_2(s) + 2H^+(aq)$$

To plot this line, use the equilibrium constant for the reaction, given by

$$K = \frac{[H^+]^2}{[Fe^{2+}]}$$

so

$$\log K = 2\log[H^+] - \log[Fe^{2+}]$$

$$= -2pH - \log[Fe^{2+}]$$

Obtain a value for $\log K$ from thermodynamic tables, or use the equation

$$\Delta G^\circ = -RT\ln K = -2.303\,RT\log K$$

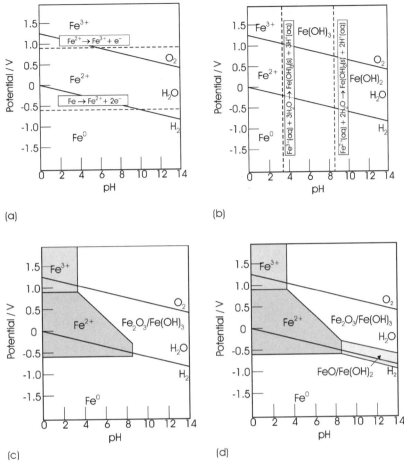

Figure S3.4 Simplified Pourbaix diagram showing the stable species in the iron–water–oxygen system: (a) oxidation–reduction boundaries that are not pH-dependent; (b) acid–base boundaries that are not oxidation-dependent; (c) as part (b), including a sloping boundary that is dependent on pH and oxidation potential; (d) complete diagram, showing further sloping boundaries that delineate regions that are dependent on pH and oxidation potential

so

$$\log K = \frac{-\Delta G^\circ}{2.303RT}$$

From the literature, $\log K = -11.75$, hence

$$-11.75 = -2\,\mathrm{pH} - \log[\mathrm{Fe}^{2+}]$$

$$\mathrm{pH} = \frac{11.75 - \log[\mathrm{Fe}^{2+}]}{2}$$

For $[\mathrm{Fe}^{2+}] = 10^{-4}\,\mathrm{mol\,dm}^{-3}$, $\mathrm{pH} = 5.9 - (-4)/2 = 7.9$; for $[\mathrm{Fe}^{2+}] = 10^{-7}\,\mathrm{mol\,dm}^{-3}$, $\mathrm{pH} = 5.9 - (-7)/2 = 9.4$.

○ Step 2(b): the same procedure is used for Fe^{3+}, using the equation:

$$\mathrm{Fe}^{3+}(\mathrm{aq}) + 3\mathrm{H}_2\mathrm{O}(\mathrm{l}) \rightarrow \mathrm{Fe}(\mathrm{OH})_3(\mathrm{s}) + 3\mathrm{H}^+(\mathrm{aq})$$

These lines are plotted on Figure S3.4b. In practice, the precipitate that forms is an oxy-hydroxide of uncertain composition, with a gel-like consistency, rather than a hydroxide of fixed formula. These precipitates may age and form hydroxides or the oxides FeO or $\mathrm{Fe}_2\mathrm{O}_3$ over time.

- When Figures A9.1a and b are combined, the area is divided into rectangular stability fields, with acidic reducing conditions towards the lower left-hand side and oxidising alkaline conditions towards the top right-hand side.

- Step 3: Reactions in which both oxidation/reduction and pH are important will plot as sloping lines.
  - Step 3(a):

$$2Fe(OH)_3(s) + 6H^+(aq) + 2e^- \rightarrow 2Fe^{2+}(aq)$$

$$+ 6H_2O(l) \quad E^\circ = 0.94\,V$$

The dependence of the cell voltage as a function of the concentrations of the species present is expressed via the Nernst equation, with $n = 2$, and Q given by

$$Q = \frac{[Fe^{2+}]^2}{[H^+]^6}$$

Thus

$$E = E^\circ - \frac{0.05916}{2} \log \left\{ \frac{[Fe^{2+}]^2}{[H^+]^6} \right\}$$

$$\log Q = 2\log [Fe^{2+}] - 6\log[H^+]$$

$$= 2\log [Fe^{2+}] + 6pH$$

hence

$$E = 0.94 - 0.1775\,pH - 0.05916 \log [Fe^{2+}]$$

For $[Fe^{2+}] = 10^{-6}$ molar,

$$E = 0.94 - 0.1775\,pH - 0.05916 \log [1 \times 10^{-6}]$$
$$= 1.295 + 0.1775\,pH.$$

This is a straight line with an intercept on the E-axis of 1.295 and a slope of 0.1775 (Figure S3.4c).

- Step 3(b): the diagram can be improved by taking into account more reactions. For example, Figure S3.4(d) is constructed with two further equations:

$$Fe(OH)_3(s) + H^+(aq) + e^- \rightarrow Fe(OH)_2(s) + H_2O(l)$$

which gives the boundary between the $Fe(OH)_3$ and $Fe(OH)_2$ stability fields, and

$$Fe(OH)_2(s) + 2H^+(aq) + 2e^- \rightarrow Fe(s) + 2H_2O(l)$$

for the boundary between the stability fields of $Fe(OH)_2$ and Fe. In essence, Figure S3.4(d) is appropriate for a single concentration. In order to display how the stability fields change with concentration, a third axis, concentration, must be added normal to the E and pH axes. The areas then become volumes in this representation.

S4

Supplementary material to Part 4: physical properties

S4.1 Elastic and bulk moduli

The response of an isotropic, homogeneous solid to a force is expressed in terms of the elastic constants or elastic moduli. [Unfortunately, a standard set of symbols for these constants is not in use.] Four elastic constants are frequently defined but, as they are interrelated, the elastic properties of a solid can be defined in terms of any two. They are most conveniently defined with respect to the stress, which is the force per unit area applied to the body, and the strain, which is the deformation of the body produced by the force.

S4.1.1 Young's modulus or the modulus of elasticity, Y or E

Young's modulus defines the response of a body, the strain, to a linear stress tending to stretch or compress it (Figure S4.1). The relationship between these quantities is:

$$\sigma = E\varepsilon \qquad (S4.1)$$

where σ is the stress on the body, and ε is the strain. Equation (S4.1) is known as Hooke's Law. The stress is given by:

$$\sigma = \frac{\text{force}}{\text{cross-sectional area}} = \frac{F}{A}$$

The force is measured in newtons, N, and the area in square metres, m^2, to give units of $N\,m^{-2}$, or Pa. The values of stress are usually expressed as $MPa = 10^6\,Pa$.

The linear strain is measured as:

$$\varepsilon = \frac{(l - l_0)}{l_0}$$

where l is the length after the stress has been imposed, and l_0 is the original unstressed length. Strain has no units and is often expressed as a percentage of l_0. In the case of compression, the linear strain is negative.

The modulus of elasticity is a measure of the stiffness of the solid. The reciprocal of the stiffness is called the compliance of the solid. Both the stiffness and the compliance are properly defined in terms of the elastic properties of anisotropic solids, in which the stress and strain are vector quantities.

S4.1.2 The shear modulus or modulus of rigidity, G

The shear stress and resulting shear strain can be most easily defined for a rectangular block of material (Figure S4.2). The shear stress, τ, is

Understanding solids: the science of materials. Richard J. D. Tilley
© 2004 John Wiley & Sons, Ltd ISBNs: 0 470 85275 5 (Hbk) 0 470 85276 3 (Pbk)

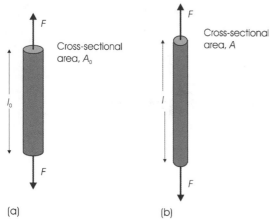

(a)   (b)

Figure S4.1 Young's modulus or the modulus of elasticity, Y or E: (a) sample before application of a tensile force, F and (b) sample after application of a tensile force, F

given by the ratio of the force applied to one face of the block to the area of these faces:

$$\tau = \frac{F}{A_0}$$

where F is the force (or load) applied, and A_0 is the area of each of the opposed faces. When a block is subject to a shear stress, it will become elongated in the direction of the forces. The (shear) strain, γ, is defined as the tangent of the angle of deformation, ϕ:

$$\gamma \text{ (shear strain)} = \frac{a}{b} = \tan \phi$$

The shear modulus defines the response of a body, the shear strain, to a shear stress tending to distort it. The relationship between these quantities is analogous to Hooke's law:

$$\tau = G\gamma \qquad \text{(S4.2)}$$

where τ is the shear stress, and γ is the shear strain.

S4.1.3 The bulk modulus, K or B

The isothermal bulk modulus, K (often also written B), relates the change in the volume of a solid, ΔV, to the hydrostatic strain, when subjected to a uniform pressure or hydrostatic stress, p (Figure S4.3):

$$\begin{aligned} p &= K\Delta V \\ &= \frac{\text{force applied}}{\text{area of face}} \end{aligned} \qquad \text{(S4.3)}$$

where

$$\Delta V = \frac{v_f - v_0}{v_0}$$

where v_f is the final volume, and v_0 is the initial volume. The hydrostatic strain can be positive or negative, depending on whether the pressure increases or decreases the volume of the solid.

The isothermal compressibility, written as K or κ, is the reciprocal of the bulk modulus. Beware of the

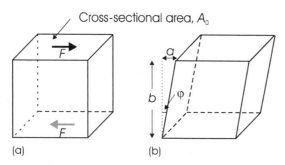

(a)   (b)

Figure S4.2 The shear modulus or modulus of rigidity, G: (a) shear forces, F, acting on a block of material, and (b) shear deformation

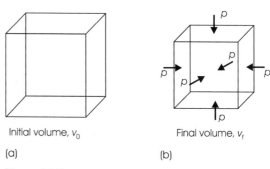

Initial volume, v_0 Final volume, v_f

(a)   (b)

Figure S4.3 The bulk modulus, K or B: (a) volume before application of pressure, p, and (b) volume after application of pressure, p

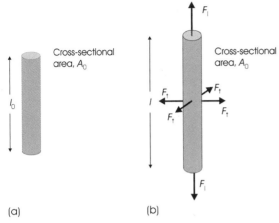

(a)   (b)

Figure S4.4 The longitudinal modulus, M: (a) before application of forces and (b) after application of longitudinal forces, F_l, and transverse forces, F_t

possible confusion between the use of K for the bulk modulus and for the compressibility.

S4.1.4   The longitudinal or axial modulus, M

The longitudinal modulus is the linear stress required to produce an elongation in a solid without any change in the lateral dimensions of the object (Figure S4.4). It is equivalent to a Young's modulus for zero transverse strain. That is, a linear 'Young's modulus' stress, F_l, must be accompanied by two perpendicular lateral stresses, F_t, to prevent any dimensional change. Thus:

$$\sigma_{\text{long}} = M\varepsilon \qquad (S4.4)$$

where the strain is still defined as:

$$\varepsilon = \frac{l - l_0}{l_0}$$

This modulus determines the velocity of ultrasonic stress pulses through a solid.

S4.1.5   Poisson's ratio, ν

Poison's ratio, ν, describes the lateral contractions $-\varepsilon_x$ and $-\varepsilon_y$, due to a tensile stress σ_z that also

produces an axial strain $+\varepsilon_z$ (in isotropic materials $-\varepsilon_x = -\varepsilon_y$:

$$\nu = \text{Poisson's ratio} = \frac{-\varepsilon \, (\text{lateral})}{\varepsilon \, (\text{longitudinal})} \qquad (S4.5)$$
$$= \frac{-\varepsilon_x}{\varepsilon_z}$$

The negative sign is to ensure that the numerical value of Poisson's ratio is positive for a 'normal' material that becomes 'thinner' as it is stressed. Auxetic materials have a negative value of ν, and become 'fatter' when stressed.

S4.1.6   Relations between the elastic moduli

The following relations between the elastic moduli hold:

$$E = 2G(1 + \nu)$$
$$G = \frac{E}{2(1 + \nu)}$$
$$K = \frac{E}{3(1 - 2\nu)}$$
$$K = \frac{EG}{3(3G - E)}$$
$$M = K + \frac{4G}{3}$$

where K is the bulk modulus in all cases.

S4.1.7   The calculation of elastic and bulk moduli

Calculation is one of the simplest ways to determine the bulk modulus, as experiments are difficult to perform. At the heart of the calculation is an analytical expression for the potential energy of a crystal in terms of interatomic distances and bond angles. The Born–Meyer function, Equation (2.9), is a simple example of such a function. The potential energy is calculated as a function of interatomic distance, bond angles and other parameters included in the potential energy equation. The interatomic distances and bond angles corresponding to the minimum energy yield the lattice constants of the unit cell and the unit cell volume. The force needed to compress the unit cell along any of the unit cell edges can be calculated by differentiating the potential energy function with respect to the interatomic distance. An estimation of the displacement caused by the application of a known force then gives the elastic moduli.

The principals can be explained by using a simple potential energy function. Take an equation such as Equation (2.6) as the starting point, but to make matters clearer, write it as:

$$U_L = \frac{-C_1}{r} + \frac{C_2}{r^n}$$

where U_L is the lattice potential energy, and C_1, C_2 and n are empirical constants. The first term in the equation represents the attractive energy between the atoms, and the second term the repulsive energy (Figure S4.5).

The force between the atoms is defined as F, where:

$$F = \frac{-dU_L}{dr} = \frac{-C_1}{r^2} + \frac{nC_2}{r^{n+1}}$$

This is also the sum of an attractive and repulsive term (Figure S4.6a). At the equilibrium separation, r_0, the force between the atoms will be zero, hence:

$$0 = \frac{-C_1}{r_0^2} + \frac{nC_2}{r_0^{n+1}}$$

thus

$$C_2 = \frac{C_1 r_0^{n-1}}{n}$$

For small values of r lying close to r_0, this curve can be approximated as a straight line (Figure S4.6b). When a crystal is subjected to an elastic strain, the force applied, ΔF, causes a displacement, Δr. Thus:

$$\sigma \, (\text{stress}) \approx \frac{\Delta F}{r_0^2}$$

$$\varepsilon \, (\text{strain}) \approx \frac{\Delta r}{r_0}$$

Hooke's law gives:

$$E = \frac{\sigma}{\varepsilon} = \frac{1}{r_0} \left(\frac{\partial F}{\partial r} \right)_{r=r_0}$$

thus

$$\left(\frac{\partial F}{\partial r} \right)_{r=r_0} = \frac{2C_1}{r_0^3} - \frac{n(n+1)C_2}{r_0^{n+2}} = \frac{C_1(1-n)}{r_0^3}$$

Hence:

$$E = \frac{C_1(1-n)}{r_0^4}$$

In terms of this simple model, Young's modulus depends sensitively on the interatomic distance in the solid.

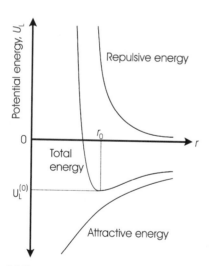

Figure S4.5 The potential energy, U_L, between atoms as a function of the atomic separation, r. The total energy is the sum of attractive and repulsive potential energy terms. The lattice energy, $U_L^{(0)}$, corresponds to the minimum in the total energy curve, reached at an interatomic separation of r_0

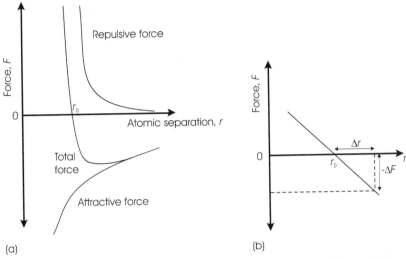

Figure S4.6   (a) Force–separation curve and (b) portion of graph shown in part (a). The equilibrium separation of the atoms, r_0, occurs when the total force is zero

The bulk modulus may be obtained in a similar way. In this case, the force ΔF needed to compress the solid by a distance Δr is computed for the three axial directions. The pressure is obtained from ΔF, and the corresponding volume decrease is obtained from Δr.

Computer calculations, which are far more sophisticated, give very accurate values of bulk modulus and elastic constants.

S4.2   Estimation of fracture strength

S4.2.1   *Estimation of the fracture strength of a brittle solid*

The stress, σ, on a solid as a function of the interatomic spacing, r, can be estimated from the interactions given in Section S4.1.7. A typical form for such a curve is sketched in Figure S4.7a. This can be understood as follows. The stress is zero at the equilibrium spacing, r_0. When the atoms are stretched, the stress will increase rapidly. In the case when the atoms are very far apart, there will be no interatomic force and thus the stress will become zero. This reasoning indicates that the curve must pass though a maximum, at some value σ_c. This

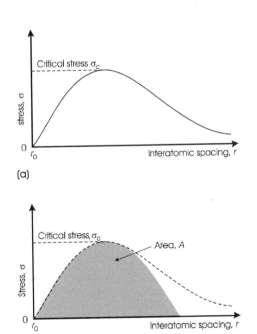

Figure S4.7   (a) Ideal curve of stress, σ, against interatomic spacing, r, and (b) approximation of the area under the curve to a half-sine-wave, of area A

value represents the cohesive stress or critical stress, because if σ_c is exceeded even slightly no more force is needed to separate the crystal into two parts. The energy to separate the two pieces is equal to the energy required to form the two new surfaces. This energy will be equal to the energy under the stress–distance curve (Figure S4.7b).

To simplify the computation, suppose that this energy can be approximately equated to the area under a half-sine-wave curve (shaded in Figure S4.7b). The equation for this part of the stress–distance curve is:

$$\sigma = \sigma_c \sin\left(\frac{2\pi r}{\lambda}\right) \qquad \text{(S4.6)}$$

The work necessary to separate two crystal planes in this approximation is:

$$\int_0^{\lambda/2} \sigma_c \sin\left(\frac{2\pi r}{\lambda}\right) \mathrm{d}x = \frac{\lambda \sigma_c}{\pi}$$

Equating this with the surface energy, 2γ,

$$\sigma_c = \frac{2\pi\gamma}{\lambda} \qquad \text{(S4.7)}$$

The slope of the initial part of the curve, near r_0, must conform to Hooke's law. The strain at r_0 is given by x/r_0, and

$$\sigma = \frac{Er}{r_0} \qquad \text{(S4.8)}$$

From Equation (S4.6),

$$\frac{\mathrm{d}\sigma}{\mathrm{d}r} = \frac{2\pi\,\sigma_c}{\lambda} \cos\left(\frac{2\pi r}{\lambda}\right)$$
$$= \left(\frac{2\pi\sigma_c}{\lambda}\right) \quad \text{at } r = 0 \qquad \text{(S4.9)}$$

From Equation (S4.8),

$$\frac{\mathrm{d}\sigma}{\mathrm{d}r} = \frac{E}{r_0} \qquad \text{(S4.10)}$$

Equating Equations (S4.9) and (S4.10), we obtain

$$\sigma_c = \frac{E\lambda}{2\pi r_0}$$

and, using Equation (S4.7), we find

$$\sigma_c = \left(\frac{E\gamma}{r_0}\right)^{1/2}$$

The value of the critical stress calculated from this equation is about $E/6$ for most solids.

S4.2.2 Estimation of the fracture strength of a brittle solid containing a crack

A crack (or other defect) in a solid will increase the stress in the region of the crack tip by a considerable amount. The stress concentration factor, K_t, is the ratio of the maximum stress due to a crack to the mean stress in the absence of the crack. The value of K_t at the end of the major axis of an elliptical hole or elliptical crack tip in a plane sheet of solid stressed as in Figure S4.8, was calculated early in the 20th century to be given by:

$$K_t = 1 + 2\left(\frac{a}{\rho}\right)^{1/2}$$

where the length of the crack is $2a$, defined in Figure S4.8, and ρ is the radius of curvature of the crack tip. For a sharp crack, a is much greater than ρ and hence

$$K_t \approx 2\left(\frac{a}{\rho}\right)^{1/2}$$

For a small crack in a large sheet, the mean stress is simply the stress, σ_0, far away from the crack. The stress at the crack tip is therefore:

$$\sigma_{\text{tip}} = K_t\sigma_0 = 2\sigma_0\left(\frac{a}{\rho}\right)^{1/2}$$

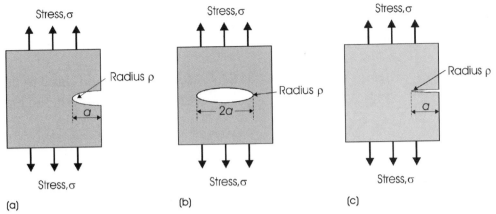

Figure S4.8   Stress at (a) an elliptical groove in a solid, (b) an elliptical pore in a solid and (c) a sharp crack in a solid

When the stress at the tip of the crack is equal to the cohesive stress calculated in Section S4.2.1, the bonds will break, so increasing the length of the crack and causing fracture to occur. Equating these values, we obtain

$$\left(\frac{E\gamma}{r_0}\right)^{1/2} = 2\sigma_c \left(\frac{a}{\rho}\right)^{1/2}$$

Thus, the material will fracture when the stress is given by:

$$\sigma_c = \frac{1}{2}\left(\frac{E\gamma\rho}{r_0 a}\right)^{1/2}$$

A growing crack does not have a smooth elliptical cross-section. However, a value of the radius can be estimated by considering that the forces between the atoms is very small at separations of about $4r_0$, giving ρ a value of about $2r_0$. Using this value, we obtain

$$\sigma_c = \left(\frac{2E\gamma}{a}\right)^{1/2}$$

Griffith originally made an estimate of this type (but in a slightly different way) and deduced

$$\sigma_c = \left(\frac{2E\sigma}{\pi a}\right)^{1/2}$$

This is the Griffith equation.

S4.3 Formulae and units used to describe the electrical properties of insulators

The electric dipole moment of a pair of charges $\pm q$ is given by p, where:

$$p = qr_1 - qr_2 = qr \qquad (S4.11)$$

As p is a quantity with both magnitude and direction, it is expressed as a vector. The direction of p is from the negative to the positive charge. The units are coulomb metre, C m. When direction is unimportant, the dipole moment is simply given by $p = qr$. (The dipole moment, p, of a gaseous water molecule is 6.1×10^{30} C m).

The bulk polarisation, P, is due to induced or permanent dipoles in the solid. P is a vector, pointing from the (induced) negative charge to the (induced) positive charge. P and p are parallel. The units of bulk polarisation, P, due to aligned internal dipoles, p, are C m/m^3 = C m^{-2}.

A dipole will give rise to an electric field, E, which is also a vector quantity. The direction of the electric field vector is from positive to negative (i.e. pointing away from a positive charge and towards a negative charge). The units of electric field are volts per metre, V m^{-1}. The unit of electric field can also be quoted in $N\,C^{-1}$:

$$\frac{N}{C} = \frac{N\,m}{C\,m}$$

As

$$J = N\,m$$

and

$$V = \frac{J}{C}$$

$$\frac{N}{C} = \frac{J}{C\,m} = \frac{V}{m}$$

Electric fields are often represented by electric field lines. These are drawn so that the tangent to the line represents the electric field vector.

- capacitance is in coulombs per volt, $C\,V^{-1}$, or farads ($1\,F = 1\,C\,V^{-1}$);

- potential is in volts, V, where $1\,V = 1\,J\,C^{-1}$;

- flux density is in $C\,m^{-2}$;

- relative permittivity, ε_r, has no units.

- The pyroelectric coefficient, π_i, has units of $C\,m^{-2}\,K^{-1}$.

The SI units of polarisability, α, can be derived from the equation

$$\alpha = \frac{p}{E}$$

Thus

$$\text{units of } \alpha = \frac{C\,m}{V\,m^{-1}} = \frac{C\,m^2}{V} = \frac{C^2\,m^2}{N\,m} = F\,m^2$$

The most commonly quoted units for polarisability in the literature are the older non-SI units of m^3, cm^3 or $Å^3$. To convert between these units note that:

$$1\,Å^3 = 10^{-24}\,cm^3 = 10^{-30}\,m^3;$$

$$\alpha\,(C\,m^2/V) = 4\pi\varepsilon_0 \times 10^{-6}\alpha\,(cm^{-3})$$

$$= 1.11265 \times 10^{-6}\alpha\,(cm^{-3});$$

$$\alpha\,(C\,m^2/V) = 4\pi\varepsilon_0\,\alpha\,(m^{-3}) = 1.11265\,\alpha\,(m^{-3});$$

ε_0, the permittivity of free space, is $\varepsilon_0 = 8.854 \times 10^{-12}\,J^{-1}\,C^2\,m^{-1} = C\,V^{-1}\,m^{-1} = F\,m^{-1}$.

S4.4   Formulae and units used to describe the magnetic properties of materials

Magnetic effects are generated by an electric current, and magnetic units are defined in terms of electric current. A small closed loop of current is called a magnetic dipole. It has a magnetic field (which can be plotted using iron filings) similar to that of a small bar magnet (i.e. a small rod of a ferromagnetic substance), and the two can be regarded as equal. The magnetic dipole moment of the current loop is defined by:

$$\mu = (\text{current in loop}) \times (\text{area of loop}) = I\pi r^2$$

where I is the current, and r is the radius of the loop. The units of magnetic dipole moment are $A\,m^2$.

A spinning electron has a magnetic dipole moment, the value of which is $9.2741 \times 10^{-24}\,A\,m^2$ (also quoted as $9.2741 \times 10^{-24}\,J\,T^{-1}$). This can be considered as the atomic unit of magnetism. It is called the Bohr magneton, μ_B, given by:

$$\mu_B = \frac{eh}{4\pi m}$$

where e is the charge on the electron, h is Planck's constant, and m is the mass of the electron.

The Earth also acts as magnetic dipole, with a magnetic dipole moment of approximately $8 \times 10^{22}\,A\,m^2$. The magnetic field of the Earth causes magnetic dipoles to align with one end pointing roughly north, and one south. A magnetic dipole is frequently represented by an arrow, with the arrowhead at the 'north-seeking', or north end of the dipole.

Two other important magnetic quantities, the magnetic induction, a vector quantity, B, and the magnetic field strength, a vector, H, are defined by reference to a solenoid, which is a cylindrical coil carrying an electric current. In a vacuum, the magnetic induction and magnetic field within the solenoid, due to the current in the windings, are given by:

$$B_0 = \mu_0 \, H$$

where μ_0 is a fundamental constant, the vacuum permeability, equal to $4\pi \times 10^{-7}$ henries per metre ($H\,m^{-1}$). The magnetic field within the solenoid is given by:

$$H = I \, n$$

where I is the current in the solenoid, and n is the number of turns of the coil per meter. The units of magnetic field, H, are ampere turns per metre ($A\,m^{-1}$). The units of magnetic induction, B, are tesla (T).

In the case when the solenoid is filled with a material, the magnetic induction and the magnetic field are given by:

$$B = \mu H$$

where μ is the permeability of the material. The units of μ are $H\,m^{-1}$. The induction is now composed of two parts, the vacuum value, B_0, and a part due to the magnetic behaviour of the sample. It is possible to write:

$$\mu = \mu_r \mu_0$$

where μ_r is the relative permeability of the material, and has no units.

The contribution of the material to the magnetic induction can also be defined by the following equation:

$$\begin{aligned} B &= B_0 + \mu_0 M \\ &= \mu_0 H + \mu_0 M \qquad \text{(S4.11)} \\ &= \mu_0 (H + M) \end{aligned}$$

where M is the magnetisation of the sample, which is the magnetic dipole moment per unit volume, with units $A\,m^{-1}$. If M opposes B_0, B is less than B_0 and the sample is diamagnetic. If M is in the same direction as B_0, B is increased and the substance is paramagnetic or ferromagnetic. For many para-

magnetic materials M is small, and it is possible to write:

$$B \approx \mu_0 H$$

For many ferromagnetic materials, B is much greater than H, and it is possible to write:

$$B \approx \mu_0 M$$

For isotropic materials, the magnetisation is related to the magnetic field, H, by the expression

$$M = \chi H \qquad \text{(S4.12)}$$

where the constant of proportionality, χ, is called the magnetic susceptibility. The magnetic susceptibility is a dimensionless quantity.

Combining Equations (S4.11) and (S4.12) we obtain

$$B = \mu_0 (1 + \chi) \, H$$

hence

$$\mu = \mu_0 (1 + \chi)$$

and

$$\mu_r = 1 + \chi$$

To summarise the units and notation:

- B, magnetic induction, T;

- H, magnetic field strength, $A\,m^{-1}$;

- M, magnetisation, $A\,m^{-1}$;

- μ, magnetic moment, $H\,m^{-1}$;

- μ_0, vacuum permeability, $H\,m^{-1} = T^2\,J^{-1}\,m^3 = J\,s^2\,C^{-2}\,m^{-1}$.

S4.4.1   Conversion factors for superconductivity

Magnetic fields quoted in studies of superconductivity are frequently in older units.

$$1\,Oe = 79.58\,A\,m^{-1}$$

$$1\,G\,(Gauss) = 10^{-4}\,T$$

$$\mu_0 = 4\pi \times 10^{-7}\,N\,A^{-2}$$

$$= 12.57 \times 10^{-7}\,N\,A^{-2}$$

A 1 T (Tesla) 'field' is equivalent to $\mu_0 H$. 1 T equates to a field, H, of $0.7955 \times 10^6 \, A \, m^{-1}$.

S4.5 Crystal field theory and ligand field theory

The optical spectra and magnetic properties of transition metal ions in solids, especially the 3d series, cannot be explained by using the information given in Chapter 1. To understand these phenomena it is necessary to take into account the effect that the surrounding atoms in the solid have on the energy of the d orbitals. There are two approaches used. The earlier is called crystal field theory. In this, the atoms in the surrounding structure are replaced by point electric charges. The interactions are only electrostatic in nature. Crystal field theory ignores all bonding between the atoms and the transition metal ion. For some purposes, crystal field theory is inadequate and has been replaced by ligand field theory, in which a molecular orbital approach to bonding is employed.

For an understanding of the magnetic properties of transition metal ions, crystal field theory is sufficient. In this theory, the surrounding point charges split the five d orbitals into groups with differing energies. This comes about in an interesting way and involves the shapes of the d orbitals. The d orbitals point along or between the x, y and z axes (Figure S4.9). In an isolated transition metal atom or ion the d orbitals all have the same energy. When the atom or ion is placed into a crystal the energy of the d orbitals increases as a result of repulsion between the surrounding electrons and the electrons in the d orbitals. If these surrounding electrons were distributed evenly over the surface of a sphere the five d orbitals would still have the same energy as each other, although much higher than in the isolated state. This is not true if the surrounding electrons are arranged differently.

The effect can be illustrated by considering a transition metal ion in an oxide. The commonest coordination polyhedron is an octahedron, made up of six negative O^{2-} ions. For the purposes of crystal field theory, these ions are replaced by six point charges, to generate an octahedral crystal field (Figure S4.10). The effect of this is that the d

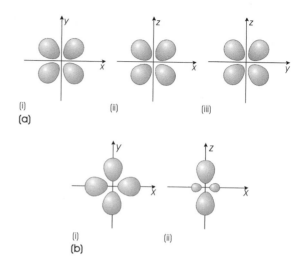

(a)

(b)

Figure S4.9 The five d orbitals: (a) one group [(i) d_{xy}, (ii) d_{xz} and (iii) d_{yz}] have lobes of electron density lying between the axes; (b) the other group [(i)$d_{x^2-y^2}$ and (ii) d_{z^2} have electron density lying along the axes

orbitals pointing directly towards the point charges, $d_{x^2-y^2}$ and d_{z^2}, will be raised in energy (Figure S4.11). The orbitals pointing between the oxygen ions, the d_{xy}, d_{xz} and d_{yz} orbitals, are favourably located and will have a lower energy. The octahedral arrangement has caused the d orbitals to split into two groups. The lower-energy group, called the

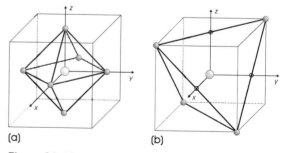

(a)

(b)

Figure S4.10 (a) A cation (large sphere) surrounded by six point charges (small spheres) arranged at the vertices of an octahedron; (b) a cation surrounded by four point charges at the vertices of a tetrahedron. The reference cation centred cubic outline indicates that the tetrahedral field is smaller than the octahedral crystal field because of the greater cation–charge distances. In (a) the anions are at the cube face centres whereas in (b) they are at four of the cube vertices

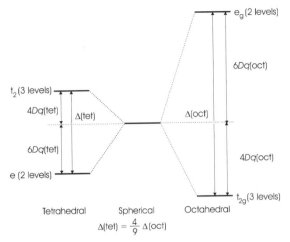

$$\Delta(\text{tet}) = \frac{4}{9}\,\Delta(\text{oct})$$

Figure S4.11 The crystal field splitting, Δ(tet) or 10Dq(tet), and Δ(oct) or 10Dq(oct), of the energy of the five d orbitals in tetrahedral (tet) and octahedral (oct) crystal fields with respect to a spherical distribution of surrounding charges. The tetrahedral configuration increases the energy of three d orbitals – d_{xy}, d_{xz} and d_{yz}, the t_2 group – relative to $d_{x^2-y^2}$ and d_{z^2}, the e group. The octahedral configuration increases the energy of two d orbitals – $d_{x^2-y^2}$ and d_{z^2}, the e_g group – relative to d_{xy}, d_{xz} and d_{yz}, the t_{2g} group

t_{2g} set, contains three orbitals of equal energy. The upper-energy group, called the e_g set, contains two orbitals of equal energy. [Note that lower-case letters are used as these energy levels are derived by assuming that only one d electron is involved.] The crystal field generates an energy gap, called the crystal field splitting, between the lower t_{2g} group of orbitals and the upper e_g group, which is written as Δ or 10Dq.

The magnitude of the crystal field splitting will depend on the geometry of the surrounding charges and how close they are to the cation. In a strong crystal field, produced when the surrounding charges are large and close to the cation, the crystal field splitting is large. In a weak crystal field, produced when the surrounding charges are small and further away from the cation, the splitting is smaller.

When a transition metal ion is surrounded by a tetrahedron of oxygen ions (Figure S4.10), the crystal field splitting is reversed. In this case, the d_{xy}, d_{xz} and d_{yz} orbitals, now called the t_2 group,

are raised in energy relative to $d_{x^2-y^2}$ and d_{z^2}, the e group. The magnitude of the splitting for ions in a tetrahedron will be less than that for ions in an octahedron because there are only four surrounding anions instead of six and because they are further away from the central ion. Calculations give the result that the tetrahedral crystal field splitting is 4/9 of the octahedral splitting.

In order to build up the electron configurations of the 3d transition metal ions, electrons are placed into the now-split d orbitals. As an example, consider the sequence for an ion in an octahedral crystal field. In the case of the ions $3d^1$ to $3d^3$, the electrons keep apart and retain parallel spins, in accordance with Hund's rules (Figure S4.12). Two options are possible for a $3d^4$ ion. The forth electron can keep apart from the others, and enter an empty e_g orbital, or it can spin-pair with an electron in the t_{2g} orbitals (Figure S4.12). Which alternative is chosen will depend on the strength of the crystal field splitting. When this is small, the weak-field case, separation is preferred. When this is large, the strong-field case, spin pairing is the energetically favoured option. These are also called the high-spin (HS) and low-spin (LS) configurations, respectively. Low-spin and high-spin alternatives exist for the ions from d^4 to d^7. In the cases of d^8 and d^9 ions, only one configuration is possible.

Only high-spin configurations are found for ions in a tetrahedral geometry, because the crystal field splitting is always much smaller than that in octahedral geometry.

Naturally, the magnetic properties of many transition metal ions will depend on whether they are in a high-spin or low-spin configuration.

S4.6   Electrical resistance and conductivity

Many materials (but not all) obey Ohm's law,

$$V = IR$$

where V is the voltage applied to either end of the material, I is the resultant current and R is the resistance. The units of voltage are volts (V), the units of current are amperes (A), and the units of

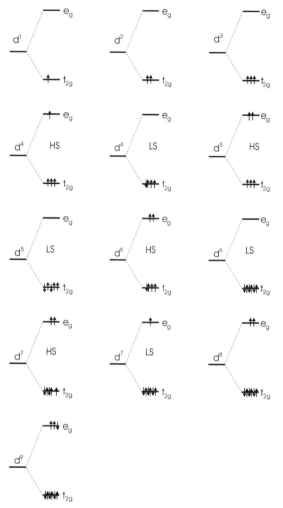

Figure S4.12 The electron configurations possible for d^n cations in an octahedral crystal field. For the ions d^4 to d^7, two configurations are possible. When the crystal field splitting is small, as a result of a weak crystal field, electrons avoid each other, and produce a high-spin (HS) configuration. When the crystal field splitting is large, as a result of a strong crystal field, electrons pair up and lead to a low-spin (LS) configuration

where ρ is the constant of proportionality, called the resistivity, with units of ohm metre (Ω m). The resistivity of a solid is an intrinsic property (resistance depends on the dimensions of the sample). Resistivity values span a range from about $10^{14}\,\Omega$ m for a polymer such as polythene, to about $10^{-8}\,\Omega$ m for a metal such as silver. Ceramics and polymers are generally regarded as insulators. Metals and alloys are regarded as conductors. Those materials with an intermediate value for the resistivity are semiconductors.

The conductivity of a solid is the inverse of the resistivity:

$$\sigma = \frac{1}{\rho}$$

The units of conductivity are $\text{ohm}^{-1}\,\text{metre}^{-1}$ ($\Omega^{-1}\,\text{m}^{-1}$), or siemens per metre (S m^{-1}), where the siemen is equivalent to ohm^{-1}.

The electric field in a material produced by the application of a voltage, V, across it is given by:

$$E = \frac{V}{L}$$

where the length of the conductor is L. The units of E are V m^{-1}.

S4.7   Current flow

Electrical conductivity is due to moving charges in a solid. For most materials, these are electrons. They move with a wide range of velocities but the average number of electrons moving in one direction is equal to the number moving in the opposite direction in the absence of an electric field. The imposition of an electric field causes the velocity distribution to shift so that overall the electrons drift in one direction. This drift constitutes an electric current.

The magnitude of the current is defined as the amount of charge that passes through unit cross-sectional area of the conductor per second. That is:

$$I = neA\mathrm{v}$$

resistance are ohms (Ω). [Solids that do not obey Ohm's law are called nonohmic materials.] The resistance is proportional to the length of the material, L, and the cross-sectional area, A:

$$R = \rho\left(\frac{L}{A}\right)$$

where I is the current, n is the number of mobile electrons per unit volume, A is the cross-sectional area of the conductor, and v is the drift velocity.

The force exerted on an electron by an electric field, E, is given by:

$$\text{force} = -eE$$

and the acceleration, a, imposed on an electron is then given by

$$a = \frac{eE}{m_e} \qquad (S4.13)$$

where m_e is the mass of the electron. The current in a conductor is steady, and not ever increasing, as would be expected if the electrons were accelerating. To account for this, it is assumed that the electrons constantly collide with the structure of the solid and that each collision resets the drift velocity to zero. If the time between collisions is τ, the drift velocity is found to be:

$$v = a\tau$$

Substituting for the acceleration from Equation (S4.13):

$$v = \frac{eE\tau}{m_e} \qquad (S4.14)$$

The total current flowing is therefore:

$$I = \frac{e^2 nAE\tau}{m_e}$$

If the length of the conductor is L, the electric field can be replaced by V/L, where V is the voltage applied to the conductor (see Section S4.6) to give:

$$I = \left(\frac{e^2 n\tau}{m_e}\right)\left(\frac{A}{L}\right)V$$

Now Ohm's law can be written

$$I = \left(\frac{A}{L\rho}\right)V$$

so that:

$$\frac{1}{\rho} = \sigma = \frac{e^2 n\tau}{m_e} \qquad (S4.15)$$

where σ is the conductivity, and ρ is the resistivity of the solid.

The mobility of the electrons, μ, is defined as the drift velocity gained per unit electric field; that is,

$$v = \mu E$$

Comparing this with Equation (S4.14) makes it apparent that:

$$\mu = \frac{e\tau}{m_e} \qquad (S4.16)$$

This is sometimes called the drift mobility, to distinguish it from mobility measured via the Hall effect. The conductivity and the mobility are then related by substituting τ of Equation (S4.16) into Equation (S4.15) to give:

$$\sigma = ne\mu$$

S4.8 The electron and hole concentrations in intrinsic semiconductors

Intrinsic semiconductors are those that show semiconducting behaviour when pure. The best-known examples are silicon and germanium. The simplest one-dimensional flat-band picture of such a semiconductor is drawn in Figure 13.7 (page 396). The number of electrons in the conduction band will be:

$$n = \int_{E_c}^{E_{ct}} N(E)F(E)dE$$

where E_c is the energy of the bottom of the conduction band, and E_{ct} is the energy at the top of the conduction band, $N(E)$ is the density of states in the conduction band, and $F(E)$ is the Fermi–Dirac distribution function. For ordinary semiconductors at normal temperature, this integral can be evaluated to give:

$$n = N_c \exp\left[-\frac{(E_c - E_F)}{kT}\right] \qquad (S4.17)$$

where E_F is the Fermi energy, and N_c is the effective density-of-states function in the conduction band, given by:

$$N_c = 2\left(\frac{2\pi m_e^* kT}{h^2}\right)^{3/2}$$

where m_e^* is the effective electron mass for electrons in the bottom of the conduction band.

By using a similar method, the number of holes at the top of the valence band is found to be:

$$p = N_v \exp\left[\frac{-(E_F - E_v)}{kT}\right] \qquad (S4.18)$$

where E_v is the energy of the top of the valence band, and N_v is the effective density-of-states function in the valence band, given by:

$$N_v = 2\left(\frac{2\pi m_h^* kT}{h^2}\right)^{3/2}$$

where m_h^* is the effective mass of the holes at the top of the valence band and is generally different from that of electrons in a semiconductor.

Sometimes the energy at the top of the valence band, E_v, is set at zero, and the energy difference between the top of the valence band and the bottom of the conduction band is written as E_g, the band-gap energy. In this case, Equations (S4.17) and (S4.18) are written as, respectively,

$$n = N_c \exp\left[\frac{-(E_g - E_F)}{kT}\right] \qquad (S4.19)$$

and

$$p = N_v \exp\left(\frac{-E_F}{kT}\right) \qquad (S4.20)$$

In an intrinsic semiconductor, the number of holes is equal to the number of electrons, so that:

$$n = p$$

equating Equations (S4.17) and (S4.18) we obtain:

$$N_c \exp\left[\frac{-(E_c - E_F)}{kT}\right] = N_v \exp\left[\frac{-(E_F - E_v)}{kT}\right]$$

Taking logarithms and rearranging, we can write the Fermi energy as:

$$E_F = \tfrac{1}{2}(E_v + E_c) + \tfrac{1}{2}\left[kT \ln\left(\frac{N_v}{N_c}\right)\right]$$

$$= \tfrac{1}{2}(E_v + E_c) + \tfrac{3}{4}kT \ln\left(\frac{m_h^*}{m_e^*}\right)$$

$$\approx \tfrac{1}{2}(E_v + E_c) = \tfrac{1}{2}E_g$$

In the case when the effective masses are identical, E_F is at the centre of the band gap, and, in general, the Fermi level is always close to the centre of the band gap in an intrinsic semiconductor.

Equations (S4.17) and (S4.18) are equal only when m_e^* is equal to m_h^*. When this is not so, the number of electrons and holes, the intrinsic carrier density, n_i, can be found in the following way. Substituting $\tfrac{1}{2}E_g$ for E_F into equations (S4.19) and (S4.20) we obtain

$$n = N_c \exp\left(\frac{-E_g}{2kT}\right) \qquad (S4.21)$$

$$p = N_v \exp\left(\frac{-E_g}{2kT}\right) \qquad (S4.22)$$

$$np = n_i^2 = N_c N_v \exp\left(\frac{-E_g}{kT}\right)$$

and

$$n = p = n_i = p_i = (N_c N_v)^{1/2} \exp\left(\frac{-E_g}{2kT}\right)$$

$$= 2\left(\frac{2\pi kT}{h^2}\right)^{3/2} (m_e^* m_h^*)^{3/4} \exp\left(\frac{-E_g}{2kT}\right)$$

Inserting values for the constants, and separating the effective mass of electrons and holes, and the temperature, we obtain:

$$n_i = p_i = 4.826 \times 10^{21} \left(\frac{m_e^* m_h^*}{m_e^2}\right)^{3/4} T^{3/2} \exp\left(\frac{-E_g}{2kT}\right)$$

in units of m^{-3}.

Writing the conductivity, $\sigma(\text{total})$, as

$$\sigma(\text{total}) = n e \mu_e + p e \mu_h$$

$$\sigma(\text{total}) = 4.826 \times 10^{21} \left(\frac{m_e^* m_h^*}{m_e^2}\right)^{3/4} T^{3/2} e\,(\mu_e + \mu_h)$$

$$\times \exp\left(\frac{-E_g}{2kT}\right)$$

$$= 773.1 \left(\frac{m_e^* m_h^*}{m_e^2}\right)^{3/4} T^{3/2}\,(\mu_e + \mu_h)$$

$$\times \exp\left(\frac{-E_g}{2kT}\right)$$

The mobility of holes and electrons falls with increasing temperature as a result of interactions with the crystal structure, but at all but the lowest temperatures the exponential term in the number of electrons and holes is dominant and the overall conductivity increases with temperature.

S4.9   Energy and wavelength conversions

A wide variety of energy units is used in the literature connected with light. A common nonstandard unit of energy is the electron volt (eV), but spectroscopists more often use reciprocal centimetres (cm^{-1}). In wavelength designations, a common nonstandard unit is the Ångström (Å).

- To convert a wavelength in nm to Å, multiply the values given by 10.

- To convert a wavelength in Å to nm divide the values given by 10.

- To convert energy in J to eV, divide the values given by 1.60219×10^{-19}.

- To convert energy in eV to J, multiply the values given by 1.60219×10^{-19}.

- To convert energy in cm^{-1} to eV, multiply the value given by 1.2399×10^{-4}.

- To convert energy in cm^{-1} to J, multiply the value given by 1.9865×10^{-23}.

- To convert an energy in eV to the equivalent wavelength, use:

$$\text{wavelength (nm)} = 1239.9/\text{energy (eV)}$$

- To convert wavelength in nm to μm, divide the values given by 1000.

- To convert wavelength in μm to nm, multiply the values given by 1000.

The relationship between angular frequency, ω (units radians per unit time, and frequency, ν (often written f; units Hz or s^{-1}), is:

$$\omega = 2\pi\nu$$

where

$$\nu = \frac{c}{\lambda}$$

where c is the velocity of light and λ is its wavelength.

S4.10   Rates of absorption and emission of energy

The rate of depopulation of an upper energy level, at energy E_1, $(-dN_1/dt)$, by spontaneous emission will be given by:

$$\frac{-dN_1}{dt} = A_{10}N_1$$

where the negative sign denotes that the number of atoms in the upper state per cubic metre, N_1, is decreasing with time. The rate is proportional to the number of atoms in the state, N_1. The rate constant, denoted here as A_{10}, is called the Einstein coefficient for spontaneous emission, and the subscript [1 0] means that a transition from the excited state, E_1 to the ground state E_0 is considered. The number of downward transitions due to spontaneous emission, per second, is given by:

$$A_{10}N_1$$

In similar fashion, two other rates can be defined for the cases of stimulated emission and for absorption. These can be expressed in terms of two rate constants (or Einstein coefficients), one for stimulated emission and one for absorption. The rates are proportional to the numbers of atoms in the relevant state and the number of photons present. Thus the rate at which atoms in state E_0 are excited to state E_1 is then given by:

$$\frac{-\mathrm{d}N_0}{\mathrm{d}t} = B_{01}\,\rho(\nu_{01})N_0$$

where N_0 is the number of atoms in state E_0, per cubic metre, $\rho(\nu_{01})$ is the radiation density responsible for absorption, which is the number of quanta per cubic metre incident per second at the correct excitation frequency ν_{01}, and B_{01} is the Einstein coefficient for absorption of radiation. Similarly, the rate of depopulation of state E_1 by stimulated emission is given by:

$$\frac{-\mathrm{d}N_1}{\mathrm{d}t} = B_{10}\,\rho(\nu_{10})N_1$$

where N_1 is the number of atoms in state E_1, per cubic metre, $\rho(\nu_{10})$ is the radiation density responsible for depopulation, which is the number of quanta per cubic metre incident per second at the correct frequency ν_{10}, and B_{10} is the Einstein coefficient for stimulated emission of radiation. The frequency for excitation will be the same as that for depopulation, so that

$$\nu_{10} = \nu_{01}$$

which can be writen ν, and the radiation density will be the same in each case, hence:

$$\rho(\nu_{10}) = \rho(\nu_{01}) = \rho(\nu)$$

The number of stimulated downward transitions per second will be given by:

$$N_1 B_{10}\,\rho(\nu)$$

and the total number of upward transitions in the same time will be given by:

$$N_0 B_{01}\,\rho(\nu)$$

At equilibrium, the total number of transitions in each direction must be equal, hence:

$$N_0 B_{01}\,\rho(\nu) = N_1 A_{10} + N_1 B_{10}\rho(\nu)$$

so

$$\rho(\nu) = \frac{N_1 A_{10}}{N_0 B_{01} - N_1 B_{10}}$$

At equilibrium, the Boltzmann distribution applies, thus:

$$\frac{N_1}{N_0} = \exp\left(\frac{-h\nu}{kT}\right)$$

and, by making this substitution, we obtain

$$\rho(\nu) = A_{10}\left[\exp\left(\frac{-h\nu}{kT}\right)B_{01} - B_{10}\right]^{-1}$$

This expression represents the net radiation emitted or absorbed by the material. This should be equal to the radiation density in a black body, derived by Planck:

$$\rho(\nu) = 8\pi h\nu^3\left\{c^3\left[\exp\left(\frac{-h\nu}{kT}\right)-1\right]\right\}^{-1}$$

leading to the conclusion that:

$$B_{01} = B_{10}$$

which will be replaced by the single symbol B, and

$$\frac{A_{10}}{B} = \frac{8\pi h\nu^3}{c^3}$$

The *ratio* of the rate of spontaneous emission to stimulated emission under conditions of thermal equilibrium is given by:

$$R = \frac{A_{10}}{\rho(\nu)B} = \exp\left(\frac{h\nu}{kT}\right)-1$$

At 300 K, at visible wavelengths, R is much greater than 1. This shows that, for light, stimulated emission will be negligible compared with spontaneous emission. However, if the wavelength increases beyond the infrared into the microwave and radiowave regions of the electromagnetic spectrum, R becomes much less than 1 and all emission will be stimulated. Hence, radiowaves and microwaves arise almost entirely from stimulated emission and are always coherent.

S4.11 The colour of a thin film in white light

Table S4.1 lists the reflected and transmitted colours from a thin film in white light.

Table S4.1 The reflected and transmitted colours from a thin film in white light

Optical path difference, [p]/nm	Colour reflected[a]	Colour transmitted[b]
Start of 1st order		
0	Black	Bright white
40	Iron grey	White
97	Lavender-grey	Yellowish white
158	Grey-blue	Brownish white
218	Grey	Brownish yellow
234	Green-white	Brown
259	White	Bright red
267	Yellow-white	Carmine red
281	Straw yellow	Deep violet
306	Bright yellow	Indigo
332	Yellow	Blue
430	Yellow-brown	Grey-blue
505	Orange-red	Blue-green
536	Red	Green
551	Deep red	Yellow green
555: End of 1st order; start of 2nd order		
565	Purple	Bright green
575	Violet	Green yellow
589	Indigo	Gold
664	Sky blue	Orange
680	Blue	Orange brown
728	Blue-green	Brown orange
747	Green	Carmine red
826	Bright green	Purple red
843	Yellow-green	Violet purple
866	Green-yellow	Violet
910	Yellow	Indigo

Table S4.1 *(Continued)*

Optical path difference, [p]/nm	Colour reflected[a]	Colour transmitted[b]
948	Orange	Dark blue
998	Orange-red	Green blue
1050	Violet-red	Yellow green
1100	Dark violet-red	Green
1120: End of 2nd order; start of 3rd order		
1128	Blue-violet	Yellow green
1151	Indigo	Off-yellow
1258	Blue-green	Pink
1334	Sea-green	Brown red
1350	Green	Purple violet
1376	Dull green	Violet
1400	Yellow green	Violet grey
1426	Green-yellow	Grey blue
1450	Yellow	Indigo
1495	Rose pink	Sea green
1534	Carmine red	Green
1621	Dull purple	Dull sea green
1650	Violet-grey	Yellow green
1665: End of 3rd order; start of 4th order		
1682	Blue-grey	Green yellow
1710	Dull sea green	Yellow grey
1750	Blue-green	Lilac
1800	Green-brown	Purple red
1811	Green	Carmine
1900	Pale green	Red
1927	Greenish-grey	Grey red
2000	Pale grey	Blue grey
2200	Very pale red-violet	Blue-green
2040	Red	Green
2240: End of 4th order; start of 5th order		
~2500	Green	–
~2700	Pink	–
2800: End of 5th order		

[a] This colour is the same as that shown in transmission by a thin transparent plate of an anisotropic crystal viewed in white light between crossed polars.

[b] This colour is the complementary colour to that reflected. It is the same as that shown in transmission by a thin transparent plate of an anisotropic crystal viewed in white light between parallel polars. In addition, these colours are seen in reflection when a thin transparent film on a substrate with a greater refractive index is viewed in white light.

Note: No entries are given in the 'colour transmitted' column for the 5th order as the colours are 'washed out' (this is also the case for much of the 4th order)

S4.12   Classical and quantum statistics

The energy of a thermodynamic system consisting of a collection of localised particles was determined by Bolzmann, and the resulting particle statistics using this framework is called classical, Boltzmann or Maxwell–Boltzmann statistics. For a system at equilibrium, containing N particles, n_1 with energy E_1, n_2 with energy E_2, n_i at energy E_i, and so on, at a temperature T, the mean energy, E(mean), of the system is given by:

$$E \text{ (mean)} = \left[\sum_i E_i \exp\left(\frac{-E_i}{kT}\right) \right] \left[\sum_i \exp\left(\frac{-E_i}{kT}\right) \right]^{-1}$$

The second expression in square brackets is called the partition function, Q:

$$Q = \sum_i \exp\left(\frac{-E_i}{kT}\right)$$

This function acts as a normalisation factor.

The probability of the occupation of an energy level E_i is given by:

$$p_i = \frac{1}{Q} \exp\left(\frac{-E_i}{kT}\right)$$

and the number of particles in energy level E_i is given by:

$$N_i = \frac{N}{Q} \exp\left(\frac{-E_i}{kT}\right)$$

From this equation it follows that the fraction of particles in two energy levels, a and b, separated by an energy ε, is given by:

$$\frac{N_a}{N_b} = \exp\left(\frac{-\varepsilon}{kT}\right)$$

There is no constraint on either the allowed energy values or the number of particles that can have any energy value.

In quantum statistics, in which the allowed energies are quantised, these conditions no longer apply,

and different statistics must be used. Two different systems are needed: Bose–Einstein statistics and Fermi–Dirac statistics.

Bose–Einstein statistics applies to particles such as photons, phonons and gases at very low temperatures. These particles are called bosons. Although the available energies are quantised, any number of bosons can occupy an energy level. The number of bosons with energy E_i is given by:

$$N_i = \left[\exp\left(\frac{E_i - \mu}{kT}\right) - 1 \right]^{-1}$$

where μ is the chemical potential of the bosons, which is always negative. N_i is a mean occupation number, and can be greater than 1. The chemical potential for phonons is zero and, in this case, the equation becomes:

$$N_i = \left[\exp\left(\frac{E_i}{kT}\right) - 1 \right]^{-1}$$

Fermi–Dirac statistics applies to particles with a half-integral spin, such as electrons, protons and neutrons. Such particles are called fermions. No two identical fermions can occupy a single (quantised) energy level. As spin can differentiate between two otherwise identical fermions, two fermions, with opposed spins, can occupy each energy level. Fermi–Dirac statistics specify the the probability, p_i, that an energy level, E_i, will be occupied is given by:

$$p_i = \left[\exp\left(\frac{E_i - \mu}{kT}\right) + 1 \right]^{-1}$$

where μ is the chemical potential of the fermions. The chemical potential of electrons in a metal is equal to the Fermi energy (strictly speaking, at absolute zero), E_F, hence:

$$p_i = \left[\exp\left(\frac{E_i - E_F}{kT}\right) + 1 \right]^{-1}$$

This is the Fermi function.

S4.13   Physical properties and vectors

Physical properties measure the response of a material to an imposed impulse and define the relationship between the two. For example, the imposition of an electric field on a metal gives rise to an electric current. The relationship between these, the physical property, is the electrical conductivity. For some properties, such as density, which defines the relationship between the mass of an object and its volume, direction is irrelevant, whereas for others, such as electrical conductivity itself, measurements made in different directions may produce different values. In these cases the physical property is sensitive to direction. Because of this, it is often necessary to describe imposed impulses and responses that are directional in nature by vectors, which are defined in terms of a magnitude and a direction. In cases where direction is unimportant and only magnitudes are important, such as in the case of density, the impulse and response are described as simple numbers, called scalars.

The precision with which this relationship between the imposed impulse and the physical property needs to be expressed depends on a number of factors. Of these, the nature of the sample is of considerable importance. Gases, liquids, amorphous solids and glasses, and polycrystalline arrays are isotropic. That is, the physical property is the same in all directions. In these cases there is little to be gained by specifying the imposed impulse and response as vectors, and scalars give all the information about the physical property that is needed. In many other cases the directional nature of the processes becomes all important, and the material is regarded as anisotropic. This happens when measuring molecular properties, properties of objects such as nanotubes, and most crystals. In this case it is necessary to specify the direction of the imposed impulse and the direction of the response. For example, the electrical conductivity of a carbon nanotube is unlikely to be the same along the tube axis as it is normal to the axis. However, the electrical conductivity normal to the tube axis is likely to be the same in all directions.

In Chapters 10–15, which deal with physical properties, the imposed impulse and response are sometimes described simply as numbers (scalars), which are written in normal type (i.e. F), and sometimes as vectors, which are written in bold type (i.e. $\boldsymbol{F}$). The magnitude of a vector $\boldsymbol{p}$ is written in normal type, p, as it is a number.

The decision on which description to use has been made on the basis of the common practice. For example, the mechanical properties of solids (Chapter 10) are most often determined for polycrystalline samples. In this case, an imposed force or load, which has direction as well as magnitude, is just as well specified by a scalar as a vector. [A discussion of the mechanical properties of single crystals would require greater precision, and the imposed force and resultant deformation of the crystal are best described as vectors.] The effects of an electric field on an insulator (Chapter 11), or of a magnetic field on a solid (Chapter 12) are conveniently described in terms of vectors. In describing electrical (Chapter 13) and optical (Chapter 14) properties it is convenient to use both approaches, largely depending on the nature of the solid discussed. Thus, the optical properties of a glass need only scalars, whereas the nonlinear optical properties of crystals require the greater precision of vector notation. The introductory discussion of thermal properties (Chapter 15) is in terms of scalars, and directional responses to the effects of temperature are described nonmathematically.

Answers to Problems and Exercises

Chapter 1

Quick quiz

1c, 2a, 3b, 4a, 5c, 6c,
7a, 8a, 9b, 10a, 11c, 12b, 13b,
14c, 15a, 16c, 17b, 18b, 19b, 20b,
21c, 22a, 23c, 24a, 25b, 26b, 27c

Calculations and questions

1.1 $5.79 \, \mu m$. **1.2** $462 \, ms^{-1}$.

1.3 $5.96 \times 10^7 \, m \, s^{-1}$. **1.4** $0.198 \, nm$.

1.5 $1.13 \, m \, s^{-1}$. **1.6** $2.52 \times 10^{-11} \, m$.

1.7 $1.20 \times 10^{-11} \, m$. **1.8** $3.53 \times 10^{-19} \, J$.

1.9 $4.84 \times 10^{-19} \, J$; **1.10** He^+, $8.72 \times 10^{-18} \, J$;

Li^{2+}, $1.90 \times 10^{-17} \, J$. **1.11** To $n = 1$,

$\lambda = 97.3 \, nm$, $\nu = 3.08 \times 10^{15} \, s^{-1}$; to $n = 2$,

$\lambda = 486 \, nm$, $\nu = 6.17 \times 10^{14} \, s^{-1}$; to $n = 3$,

$\lambda = 1.88 \, \mu m$, $\nu = 1.60 \times 10^{14} \, s^{-1}$.

1.12 To $n = 1$, $\lambda = 95.0 \, nm$, $\nu = 3.16 \times 10^{15} \, s^{-1}$;

to $n = 2$, $\lambda = 434 \, nm$, $\nu = 6.91 \times 10^{14} \, s^{-1}$;

to $n = 3$, $\lambda = 1.28 \, \mu m$, $\nu = 2.34 \times 10^{14} \, s^{-1}$.

1.13 To $n = 1$, $\lambda = 10.8 \, nm$, $\nu = 2.79 \times 10^{16} \, s^{-1}$;

to $n = 2$, $\lambda = 68.9 \, nm$, $\nu = 4.35 \times 10^{15} \, s^{-1}$.

1.14 To $n = 1$, $\lambda = 25.6 \, nm$, $\nu = 1.17 \times 10^{16} \, s^{-1}$;

to $n = 2$, $\lambda = 164 \, nm$, $\nu = 1.83 \times 10^{15} \, s^{-1}$.

1.15 $3.37 \times 10^{-19} \, J$. **1.16** $4.56 \times 10^{-19} \, J$.

1.17 $n = 2$; $l = 1$; $m_l = 1, 0, -1$;

$m_s = +1/2, -1/2$. **1.18** C, [He] $2s^2 \, 2p^2$;

P, [Ne] $3s^2 \, 3p^3$; Fe, [Ar] $3d^6 \, 4s^2$; Sr, [Kr] $5s^2$;

W, [Xe], $4f^{14} \, 5d^3 \, 6s^2$. **1.19** $J = 4, 3, 2$; $^3F_2$.

1.20 $J = 3/2$; $^4S_{3/2}$. **1.21** $J = 5/2, 3/2$; $^2D_{3/2}$.

1.22 $J = 3/2, 1/2$; $^2P_{1/2}$.

1.23 $7/12$. **1.24** $21/40$. **1.25–1.27** Extended

answers or illustrations required.

Chapter 2

Quick quiz

1c, 2b, 3b, 4a, 5b, 6a, 7c,
8b, 9b, 10b, 11a, 12b, 13a,
14a, 15c, 16c, 17b, 18b, 19a,
20c, 21a, 22b, 23c, 24b, 25c,
26c, 27c, 28a, 29a, 30b, 31c,
32a, 33a, 34a, 35b

Calculations and questions

2.1 Cl^-, [Ar]; Na^+, [Ne]; Mg^{2+}, [Ne]; S^{2-}, [Ar];

N^{3-}, [Ne]; Fe^{3+}, [Ar] $3d^5$. **2.2** F^-, [Ne];

Li^+, [He]; O^{2-}, [Ne]; P^{3-}, [Ar]; Co^{3+}, [Ar] $3d^6$.

2.3 O^{2-}; H^+; Na^+; Ca^{2+}; Zr^{4+}; W^{6+}, W^{4+}.

2.4 Fe^{3+}, Fe^{2+}; Cl^-; Al^{3+}; S^{2-}; La^{3+}; Ta^{5+}.

2.5 $3600 \, kJ \, mol^{-1}$. **2.6** NaCl, $769 \, kJ \, mol^{-1}$;

Understanding solids: the science of materials. Richard J. D. Tilley

© 2004 John Wiley & Sons, Ltd ISBNs: 0 470 85275 5 (Hbk) 0 470 85276 3 (Pbk)

KCl, 688 kJ mol^{-1}.   **2.7** NaBr, 725 kJ mol^{-1};
KBr, 657 kJ mol^{-1}.   **2.8** 5.89×10^{28} m^{-3}.
2.9 1.83×10^{29} m^{-3}.   **2.10** 8.46×10^{28} m^{-3}.
2.11 8.61×10^{28} m^{-3}.   **2.12** 1.70×10^{29} m^{-3}.
2.13 1.81×10^{-33} J; $kT = 4.14 \times 10^{-21}$ J
at 300 K.   **2.14** 8.81×10^{-19} J.
2.15 5.04×10^{-19} J.   **2.16** 7.04×10^{-19} J.
2.17 1.87×10^{-18} J.   **2.18–2.22** Extended
answers or illustration required.

Chapter 3

Quick quiz

1c,   2a,   3b,   4c,   5b,   6c,   7b,
8c,   9a,   10b,   11a,   12b,   13a,
14a,   15b,   16a,   17b,   18c,   19b,
20c,   21a,   22c,   23b,   24a,   25b,   26c.

Calculations and questions

3.1 1.64×10^{-21} J; 3.86×10^{-10} m.   **3.2** 118 K.
3.3 5.94 kJ mol^{-1}.   **3.4** 4.97×10^{-22} J;
3.13×10^{-10} m.   **3.5** $A = -6.72 \times 10^{-134}$ J m^{12};
$B = -2.48 \times 10^{-77}$ J m^6.   **3.6** Derivation; not
provided here.   **3.7** 1.99×10^{-4}.   **3.8** 4.52×10^{-4}.   **3.9** 5.05×10^{23} m^{-3}.   **3.10** 2.80×10^{24} m^{-3}.   **3.11** 500 K, 1.80×10^{-13}; 1000 K,
4.24×10^{-7}.   **3.12** 1.74×10^{20} m^{-3}.
3.13 3.26×10^{-10}.   **3.14** 1.30×10^{-9} m^{-3}.
3.15 3.27×10^{-19} J.   **3.16** 2.98×10^{16} m^{-3};
6.20×10^{21} m^{-3}; the number of vacancies is double
these figures.   **3.17** 2.03×10^{27} m^{-3}.
3.18 2.61×10^{14} m^{-3}.   **3.19** 5.62×10^{25} m^{-3}.
3.20 6.9×10^{24} m^{-3}.   **3.21** 225.2 kJ mol^{-1}.
3.23 $Y_{0.165}Zr_{0.835}O_{1.917}$: $x = 0.165$; $y = 0.835$;
$z = 1.917$; Y^{3+} substitutes for Zr^{4+}; for every two
Y^{3+} there is one oxygen vacancy.
3.24 $Ca_{0.105}Bi_{1.895}O_{2.947}$; Ca^{2+} substitutes for Bi^{3+};

there is one oxygen vacancy for every two Ca^{2+}
added.   **3.25** (a) Li^+ substitutes for Ca^{2+}, Br^-
vacancies; (b) Ca^{2+} substitutes for Li^+, Li^+ vacancies; (c) Mg^{2+} substitutes for Fe^{3+}, O^{2-} vacancies;
(d) Mg^{2+} substitutes for Ni^{2+}.   **3.26** (a) Cd^{2+}
substitutes for Na^+, Na^+ vacancies; (b) Na^+ substitutes for Cd^{2+}, Cl^- vacancies; (c) Sc^{3+} substitutes for Zr^{4+}, O^{2-} vacancies; (d) Zr^{4+} substitutes
for Hf^{4+}.   **3.27** (a) 2.24×10^{28} m^{-3};
(b) 2.23×10^{28} m^{-3}.

Chapter 4

Quick quiz

1c,   2c,   3b,   4c,   5b,   6a,   7b,
8a,   9b,   10c,   11b,   12c,   13a,
14b,   15a,   16b,   17c,   18c,
19a,   20b,   21b,   22c,   23c,   24c,
25a,   26b,   27a,   28c,   29a,   30a,
31c,   32b.

Calculations and questions

4.1 73.6 at%.   **4.2** 73.5 g.   **4.3** 64.3 at%.
4.4 63.6 at% Sn; 36.4 at% Pb.   **4.5** Ti$_2$Al.
4.6 (a) 33.2 at%; (b) $\sim$1322 °C; (c) $\sim$53 wt% Ni,
$\sim$55 at% Ni; (d) $\sim$1360 °C; (e) $\sim$74 wt% Ni,
$\sim$76 at% Ni.   **4.7** (a) solid + liquid;
(b) $\sim$64 wt%; (c) $\sim$36 wt%.   **4.8** (a) 53 wt% liquid,
47 wt% solid; (b) 55.6 vol% liquid, 44.4 vol%
solid.   **4.9** (a) 40.1 wt%;
(b) $\sim$54 mol% Cr_2O_3; (c) $\sim$18% liquid;
(d) $\sim$2215 °C; (e) $\sim$80 mol% Cr_2O_3.
4.10 (a) $\sim$2150 °C; (b) $\sim$59 mol% Cr_2O_3;
(c) $\sim$2090 °C; (d) $\sim$12.5 mol% Cr_2O_3.
4.11 (a) 30 mol% Cr_2O_3; (b) no liquid phase
present; (c) 100 % solid.   **4.12** (a) $\sim$36 mol%
Cr_2O_3; (b) $\sim$17 mol% Cr_2O_3; (c) 30 % liquid, 70 %
solid.   **4.13** (a) 3186 °C, Re, 2334 °C, Ru;

(b) 55.9 g Ru, 44.1 g Re; (c) solid α, $Ru_{70}Re_{30}$; (d) liquid, $Ru_{70}Re_{30}$.   **4.14** (a) ~42 at% Ru; (b) ~63 at% Ru; (c) ~86 % liquid, ~14 % solid.   **4.15** Extended answer required.

4.16 (a) 47.0 g BeO, 53.0 g Y_2O_3; (b) ~46 %; (c) BeO; (d) ~54 %; (e) ~37.5 mol% BeO.

4.17 (a) BeO + Y_2O_3; (b) BeO + Y_2O_3; (c) ratio: 20 mol% 1/2 (Y_2O_3) to 80 mol% BeO.

4.18 (a) 27.6 g Sn + 72.4 g Pb; (b) liquid; (c) ratio: 40 at% Sn to 60 at% Pb; (d) 100 % liquid.

4.19 (a) α + liquid; (b) α, ~ 2 at% Sn; liquid, ~46 at% Sn; (c) α, ~25 %, liquid, ~75 %.

4.20 (a) α + β (solid); (b) α, 9 at% Sn; β, ~99 at% Sn; (c) α, ~65.5 %; β, ~34.5 %.

4.21 (a) γ (austenite); (b) 100 % γ; (c) 1.5 wt% C.   **4.22** (a) γ (austenite) +cementite; (b) γ, ~91.4 %; cementite, ~8.6 %; (c) γ, ~1 wt% C; cementite, 6.70 wt% C.   **4.23** (a) ~8.6 at% C; (b) ~27 % liquid; (c) ~4 at% C; (d) ~73 %.

4.24 (a) WO_2 + ZrO_2 + $W_{18}O_{49}$; (b) ~33.3 % ZrO_2, ~20.5 % WO_2, ~46.3 % $W_{18}O_{49}$.

4.25 (a) $W_{18}O_{49}$ + ZrW_2O_8 + $W_{20}O_{58}$; (b) $W_{18}O_{49}$, ~19 %; ZrW_2O_8, ~30 %; $W_{20}O_{58}$, ~51 %.

4.26 Extended answer required.

Chapter 5

Quick quiz

1b,	2a,	3a,	4c,	5c,	6b,	7a,
8b,	9a,	10c,	11b,	12b,	13a,	
14c,	15b,	16b,	17c,	18a,	19a,	
20c,	21b,	22b,	23a,	24c,	25b,	
26a,	27b,	28a,	29a,	30b,		
31a,	32c,	33b				

Calculations and questions

5.1 Illustration required. **5.2** (a) (1 0 0); (b) (1 1 0); (c) (1 2 0); (d) (1 $\bar{1}$ 0).   **5.3** (a) (3 2 0); (b) (0 4 0);

(c) ($\bar{2}$ 2 0); (d) (1 3 0).   **5.4** (a) (3 $\bar{2}$ 0); (b) (2 $\bar{1}$ 0); (c) (1 4 0); (d) (1 0 0).   **5.5** (a) (2 0 0); (b) (3 $\bar{2}$ 0); (c) (3 1 0); (d) (0 2 0).   **5.6** Illustration required.   **5.7** Six planes, plus six equivalent planes: (1 1 0), (1 0 1), (0 1 1), (1 $\bar{1}$ 0), ($\bar{1}$ 0 1), (0 1 $\bar{1}$); (1 1 0) is equivalent to ($\bar{1}$ $\bar{1}$ 0), etc.   **5.8** Four, plus four equivalent planes: (1 1 1), (1 1 $\bar{1}$), (1 $\bar{1}$ 1), ($\bar{1}$ 1 1); ($\bar{1}$ $\bar{1}$ $\bar{1}$) is equivalent to (1 1 1), etc.   **5.9** Six, plus six equivalent planes: ($h h 0$), ($0 h h$), ($h 0 h$), ($h \bar{h} 0$), ($0 h \bar{h}$), ($\bar{h} 0 h$), etc.   **5.10** Twelve, plus twelve equivalent planes: ($h k 0$), ($0 h k$), ($h 0 k$), ($h \bar{k} 0$), ($h 0 \bar{k}$), ($k h 0$), ($0 k h$), ($k 0 h$), ($k \bar{h} 0$), ($0 k \bar{h}$), ($\bar{h} 0 k$), etc.   **5.11** (a) [$\bar{1}$ 1 0]; (b) [1 1 0]; (c) [$\bar{1}$ 3 0]; (d) [1 $\bar{1}$ 0]; (e) [$\bar{2}$ $\bar{1}$ 0].   **5.12** (a) [0 $\bar{1}$ 0]; (b) [$\bar{3}$ $\bar{1}$ 0]; (c) [$\bar{1}$ $\bar{1}$ 0]; (d) [$\bar{1}$ 4 0]; (e) [3 5 0].   **5.13** (a) [$\bar{1}$ $\bar{1}$ 0]; (b) [0 $\bar{1}$ 0]; (c) [$\bar{1}$ 3 0]; (d) [2 3 0]; (e) [$\bar{1}$ 2 0].   **5.14** (a) [0 1 0]; (b) [$\bar{1}$ 0 0]; (c) [$\bar{1}$ 2 0]; (d) [1 $\bar{2}$ 0]; (e) [$\bar{1}$ 3 0].   **5.15** 6: [$\bar{1}$ 0 0], [1 0 0], [0 1 0], [0 $\bar{1}$ 0]; [0 0 1]; [0 0 $\bar{1}$].

5.16 12: [1 1 0], [$\bar{1}$ $\bar{1}$ 0], [0 1 1], [0 $\bar{1}$ 1], [1 0 1], [$\bar{1}$ 0 $\bar{1}$], [1 $\bar{1}$ 0], [$\bar{1}$ 1 0], [0 1 $\bar{1}$], [0 $\bar{1}$ 1], [$\bar{1}$ 0 1], [1 0 $\bar{1}$].   **5.17** 90°.   **5.18** 90°.   **5.19** Illustration required.   **5.20** (1 1 1), 44.6°; (2 2 0), 76.6°; (4 0 0), 122.4°.   **5.21** (1 1 0), 38.6°; (2 1 1), 69.7°; (3 1 0), 95.1°.   **5.22** 0.294 nm.

5.23 0.8077 nm.   **5.24** Confirmation required.

5.25 $Ca_{0.45}Sr_{0.55}O$.   **5.26** $x = 0.45$, $ZnAl_{1.54}Ga_{0.45}O_4$.   **5.27** (0, 0, 0); (1/2, 1/2, 0); (0, 1/2, 1/2); (1/2, 0, 1/2); cell type A1 (face-centred cubic).   **5.28** Cl^-, (0, 0, 0); Cs^+, (1/2, 1/2, 1/2).

5.29 Ca, (1/2, 1/2, 1/2); Ti, (0, 0, 0), O (1/2, 0, 0), (0, 1/2, 0), (0, 0, 1/2).   **5.30** NiO, B1; sketch not shown.   **5.31** $AuCu_3$; sketch not shown.

5.32 2699 $kg\ m^{-3}$.   **5.33** 19260 $kg\ m^{-3}$.

5.34 1750 $kg\ m^{-3}$.   **5.35** 0.3619 nm.

5.36 Density: anion vacancies, 5522 $kg\ m^{-3}$; interstitials, 5972 $kg\ m^{-3}$; substitution plus anion vacancies.   **5.37** Density: zirconium vacancies, 5004 $kg\ m^{-3}$; sulphur interstitials, 6500 $kg\ m^{-3}$; zirconium vacancies predominate.

5.38 (a) no change; (b) no change; (c) down; (d) up. **5.39** FeTiO$_3$, Al$_2$O$_3$ (*corundum* type).

Chapter 6

Quick quiz

1b, 2a, 3b, 4c, 5b, 6a, 7b,
8a, 9b, 10c, 11a, 12c, 13a,
14b, 15a, 16c, 17a, 18c, 19b,
20c, 21a, 22c, 23b, 24a, 25b,
26c, 27b, 28a, 29a, 30c, 31b,
32a, 33c, 34a, 35c, 36b, 37a,
38b, 39b, 40c, 41b, 42a, 43b,
44b, 45c.

Calculations and questions

6.1 0.4073 nm. **6.2** 0.4950 nm.
6.3 0.3892 nm. **6.4** 0.3838 nm.
6.5 0.125 nm. **6.6** 0.134 nm.
6.7 0.518 nm. **6.8** 0.3395 nm.
6.9 0.131 nm. **6.10** 0.230 nm.
6.11 (a) 0.124 nm; (b) 0.3606 nm.
6.12 (a) 0.192 nm; (b) 0.4435 nm.
6.13 (a) 0.6084 nm; (b) 0.4825 nm.
6.14 a_0, 0.3200 nm; c_0, 0.5227 nm.
6.15 a_0, 0.2760 nm; c_0, 0.4508 nm.
6.16 r, 0.148 nm; c_0, 0.4820 nm.
6.17 r, 0.114 nm; c_0, 0.3734 nm.
6.18 (a) 0.160 nm; (b) 0.3588 nm.
6.19 (a) 0.182 nm; (b) 0.4095 nm.
6.20 1984 kg m^{-3}. **6.21** 0.3652 nm.
6.22 Platinum. **6.23** Vanadium.
6.24–6.27 Extended answers required.
6.28 $\approx$ 775 °C. **6.29** (a) 9.85×10^{11} Pa s;
(b) 2.96×10^7 Pa s. **6.30** $\approx$ 447 kJ mol^{-1}.
6.31 $\approx$ 554 kJ mol^{-1}. **6.32** $\approx$ 434 kJ mol^{-1}.
6.33 Low temperature, $\approx$ 310 kJ mol^{-1}; high temperature, $\approx$ 276 kJ mol^{-1}; knee $\approx$ 840 °C; The plot

is nonlinear; it is possible to approximate the low-temperature and the high-temperature portions to straight lines and thus to estimate the corresponding activation energies. **6.34** Low temperature, $\approx$ 455 kJ mol^{-1}; high temperature, $\approx$ 258 kJ mol^{-1}; knee, $\approx$ 900 °C; The plot is nonlinear; it is possible to approximate the low-temperature and the high-temperature portions to straight lines and thus to estimate the corresponding activation energies.
6.35 Reaction equation not given here; (a) 100 kg; (b) 5.78×10^{26}; (c) 2401.
6.36 Reaction equation not given here; (a) 100 kg; (b) 6.02×10^{26}; (c) 1998. **6.37** Reaction equation not given here; (a) 50.48 kg butadiene, 49.52 kg acrylonitrile; (b) 5.62×10^{26} of each.
6.38. Reaction equation not given here; (a) 72.8 kg terephthalic acid, 27.2 kg ethylene glycol; (b) 2.64×10^{26} of each.
6.39 See Scheme 6.14. **6.40** See Scheme 6.15; two different pairs of monomers are drawn, both of which are possible. **6.41** (a) 26600 g mol^{-1}; (b) 632. **6.42** (a) 26500 g mol^{-1}; (b) 630.
6.43 (a) 4; (b) 2. **6.44** Low-density, 58.5 %; medium-density, 65.4 %; high-density, 74.5 %.
6.45 (a) $H(3\,\mathrm{d}) = 261$ J g^{-1}, $H(1\,\mathrm{yr}) = 471$ J g^{-1}; (b) $H(3\,\mathrm{d}) = 219$ J g^{-1}, $H(1\mathrm{yr}) = 414$ J g^{-1}; (c) $H(3\,\mathrm{d}) = 263$ J g^{-1}, $H(1\,\mathrm{yr}) = 474$ J g^{-1}.

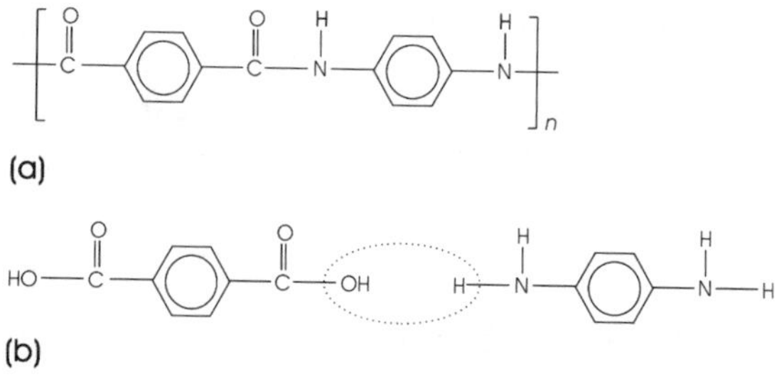

(a)

(b)

Scheme 6.14 (a) Kevlar®; (b) monomers of Kevlar®; answer to Question 6.39

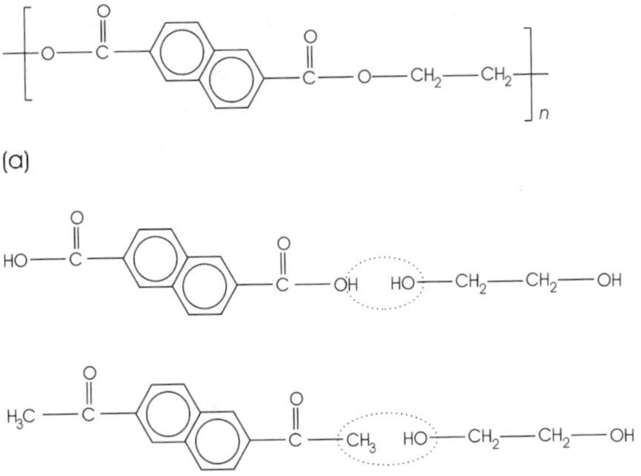

(a)

(b)

Scheme 6.15 (a) Poly(ethylene naphthalate); (b) monomers of poly(ethylene naphthalate); answer to Question 6.40

Chapter 7

Quick quiz

1b,   2a,   3c,   4a,   5b,   6b,   7c,
8 b,   9a,   10b,   11c,   12c,
13a,   14b,   15b,   16a,   17b,   18c,
19c,   20a,   21a,   22c

Calculations and questions

7.1   and   **7.2** Derivations required.
7.3 1.56×10^{-14} m^2 s^{-1}.   **7.4** 9.14×10^{-14} m^2 s^{-1}.   **7.5** $\sim 2.0 \times 10^{-10}$ m^2 s^{-1}.
7.6 0.30 wt%.   **7.7** 3.77 hr.   **7.8** 0.135 mm.
7.9 4.19×10^{-6} kg.   **7.10** 0.53 kg per hour.
7.11 460 kJ mol^{-1}.   **7.12** 1350 °C.
7.13 423 kJ mol^{-1}.   **7.14** 1230 °C.
7.15 162 kJ mol^{-1}.   **7.16** 289 kJ mol^{-1}.
7.17 158 kJ mol^{-1}.   **7.18** 3.29×10^{-14} m^2 s^{-1}.
7.19 2.10×10^{-9} m^2 s^{-1}.   **7.20** 4.64×10^{-16} m^2 s^{-1}.   **7.21** 1.95×10^{-18} m^2 s^{-1}.

7.22 1.12×10^{-12} m^2 s^{-1}.   **7.23** 1.9 µm.
7.24 250 s.   **7.25** 0.138 at 500 °C; 0.301 at 1000 °C.   **7.26** 1.68×10^4 C^2 J^{-1} m^{-3}.
7.27 1.39×10^{-14} m^2 s^{-1}.   **7.28** 2.71×10^{-9} m^2 s^{-1}.   **7.29** and **7.30** Extended answers required.

Chapter 8

Quick quiz

1c,   2b,   3b,   4a,   5b,   6c,   7a,
8b,   9b,   10a,   11c,   12c,   13b,
14a,   15c,   16a,   17b,   18a,   19b,
20b,   21c,   22c,   23a,   24b,   25c,
26a,   27b,   28c,   29b.

Calculations and questions

8.1 (a) [CO$_2$], p_{CO_2}; (b) [CH$_4$]/[H$_2$]^2, $p_{CH_4}/p_{H_2}^2$; (c) [SO$_3$]^2/{[O$_2$][SO$_2$]^2}, $p_{SO_3}^2/(p_{O_2}p_{SO_2}^2)$ (d) [CO]/[CO$_2$]; $(p_{CO})/(p_{CO_2})$.   **8.2** (a) Increase; (b) 0.039 atm; (c) 0.039 atm; (d) More CaCO$_3$ will form; (e) evenly distributed between CaCO$_3$ and CO$_2$; (f) in a kiln, equilibrium is not established, and CO$_2$ continually escapes.
8.3 $K_c = K_p/RT$.
8.4 (a)

$$CuSO_4 \cdot 5H_2O \rightleftharpoons CuSO_4 \cdot 3H_2O + 2H_2O;$$

$$CuSO_4 \cdot 3H_2O \rightleftharpoons CuSO_4 \cdot H_2O + 2H_2O;$$

$$CuSO_4 \cdot H_2O \rightleftharpoons CuSO_4 + H_2O;$$

(b) Sketch required. (c) (i) 3.93×10^7 Pa^2; (ii) 1.60×10^7 Pa^2; (iii) 600 Pa.   **8.5** p, 5.75×10^{16} m^{-3}; n, 1.00×10^{22} m^{-3}.   **8.6** p, 1.04×10^{19} m^{-3}; n, 1.62×10^{16} m^{-3}.   **8.7** (a) 5.75×10^{21} m^{-3}; (b) 1.16×10^{-5} %.   **8.8** (a) Gold; (b) and (c) Sketches required.   **8.9** (a) Rhenium.

(b) and (c) Sketches required.   **8.10** (a) Chromia, Cr_2O_3; (b) and (c) Sketches required.
8.11 Sketches required.   **8.12** (a) Sketch required; (b) Sketch required, 92 % cementite (Fe_3C), 8% austenite; (c) the austenite transforms to pearlite.
8.13 Sketches required.   **8.14** 1.027.
8.15 1.0037.   **8.16** Derivation required.
8.17 (a) 6.79×10^{-8} m diameter; (b) 2.15×10^{-7} m diameter; (c) 6.80×10^{-5} m diameter.
8.18 1.386×10^{-5} m diameter.
8.19 5.95×10^{-5} m diameter.   **8.20** Cu, Fe, Ti, Al.   **8.21** $\sim 6.23 \times 10^{-17}$ m^2 s^{-1}.
8.22 (a) 0.0044 m; (b) 0.0264 g cm^{-2}; (c) 0.00295 g.   **8.23** 7.15×10^{-18} m^2 s^{-1}.

Chapter 9

Quick quiz

1b,   2b,   3a,   4c,   5b,   6a,
7c,   8b,   9c,   10a,   11a,   12c,
13b,   14b,   15b,   16a,   17a,   18c,
19b,   20a,   21a,   22b,   23c,   24c,   25b.

Calculations and questions

9.1 (a) ox.; (b) red.; (c) redox; (d) red.; (e) redox; (f) red.   **9.2** (a) ox.; (b) red.; (c) redox; (d) red.; (e) redox; (f) red.   **9.3** (a) anode, zinc; cathode, silver; (b) anode reaction, $Zn(s) \rightarrow Zn^{2+}(aq) + 2e^-$; cathode reaction, $2Ag^+(aq) + 2e^- \rightarrow 2Ag(s)$; cell reaction, $Zn(s) + 2Ag^+(aq) \rightarrow Zn^{2+}(aq) + 2\,Ag(s)$; (c) 9.48 V.   **9.4** -212 kJ mol^{-1}.   **9.5** Monovalent, 25.85 mV; divalent, 12.93 mV; trivalent, 8.62 mV.   **9.6** (a) Anode reaction, $Zn(s) \rightarrow Zn^{2+}(aq) + 2e^-$; cathode reaction, $Ni^{2+}(aq) + 2e^- \rightarrow Ni(s)$; cell reaction, $Zn(s) + Ni^{2+}(aq) \rightarrow Zn^{2+}(aq) + Ni(s)$; (b) 0.53 V; (c) 0.55 V.
9.7 0.577 V.   **9.8** (a) Anode reaction, $H_2(g) \rightarrow 2H^+(aq) + 2e^-$; cathode reaction, $Cu^{2+}(aq) + 2e^- \rightarrow Cu$; cell reaction, $Cu^{2+}(aq) + H_2(g) \rightarrow$

$Cu(s) + 2H^+(aq)$; (b) $E = 0.34 + 0.02958\{2pH + \log p_{H_2} + \log[Cu^{2+}]\}$; (c) 8.7.   **9.9** (a) Anode reaction, $Zn(s) \rightarrow Zn^{2+}(aq) + 2e^-$; cathode reaction, $Fe^{2+}(aq) + 2e^- \rightarrow Fe(s)$; (b) 0.32 V; (c) -61.8 kJ mol^{-1}.   **9.10** (a) 0.306 V; (b) 0.293 V.   **9.11** $E = E^0 - 0.0592 \log [Cl^-] + 0.0592$ pH, or $E = E' + 0.0592$ pH.   **9.12** (a) 8.3; (b) 13.1.   **9.13** 3.2.   **9.14** -0.81 V.
9.15 (a) Cold working makes the head and point anodic with respect to the body of the nail. The tap water serves as the electrolyte. Corrosion is a result of galvanic action. (b) Pure deoxygenated water is a poor electrolyte, so very little corrosion occurs. (c) Differential aeration at the water surface causes corrosion. The cathode is at the water surface, and the anode, where pitting occurs, is just below the water surface.   **9.16** Metals with standard reduction potentials in Table 9.1 above the given voltage, -1.63 V (i.e. Zn, Fe and Sn), will be cathodic with respect to the Ti, and will enhance Ti corrosion. Metals below the given voltage (i.e. Mg, Al, Be, La and Sc) will be anodic with respect to the Ti and so inhibit corrosion of the container. [Note: plating or making bolts with these metals may not be practical.]   **9.17** (a) Diagram required. (b) Anode reaction, $Fe^{2+}(aq) + 2e^- \rightarrow Fe(s)$; cathode reaction, $Mg(s) \rightarrow Mg^{2+}(aq) + 2e^-$; cell reaction, $Fe^{2+}(aq) + Mg(s) \rightarrow Fe(s) + Mg^{2+}(aq)$; (c) 1.92 V.
9.18 Extended answer required.   **9.19** (a) Reduction of $Cu^{2+}(aq)$ to $Cu(s)$, with the formation of $Fe^{2+}(aq)$ (most favoured), and $Fe^{3+}(aq)$ (least favoured): $Fe(s) \rightarrow Fe^{2+}(aq) + 2e^-$, $Fe(s) \rightarrow Fe^{3+}(aq) + 3e^-$, $Cu^{2+}(aq) + 2e^- \rightarrow Cu(s)$; overall: $Cu^{2+}(aq) + Fe(s) \rightarrow Cu(s) + Fe^{2+}(aq)$ or $3Cu^{2+}(aq) + 2Fe(s) \rightarrow 3Cu(s) + 2Fe^{3+}(aq)$. (b) No significant reaction. (c) Reduction of $Pb^{2+}(aq)$ to $Pb(s)$, with the formation of $Fe^{2+}(aq)$; no formation of $Fe^{3+}(aq)$; $Fe(s) \rightarrow Fe^{2+}(aq) + 2e^-$, $Pb^{2+}(aq) + 2e^- \rightarrow Pb(s)$; overall : $Pb^{2+}(aq) + Fe(s) \rightarrow Pb(s) + Fe^{2+}(aq)$. (d) As for part (c), with the formation of Sn(s), but not

such a strong reaction: $Fe(s) \rightarrow Fe^{2+}(aq)+$
$2e^-$, $Sn^{2+}(aq) + 2e^- \rightarrow Sn(s)$; overall: $Sn^{2+}(aq)+$
$Fe(s) \rightarrow Sn(s) + Fe^{2+}(aq)$.   **9.20** (a) Anode
reaction, $2Cl^-(l) \rightarrow Cl_2(g) + 2e^-$; cathode reaction,
$Mg^{2+}(l) + 2e^- \rightarrow Mg(s)$; electrolysis reaction,
$MgCl_2(l) \rightarrow Mg(s) + Cl_2(g)$; (b) 40.8 g Mg;
119.2 g Cl_2.   **9.21** (a) 5.79×10^5 C;
(b) 2.15×10^7 C; (c) for 2 moles, 6.4 hours; for 2 kg
239.9 hours.   **9.22** 58.7; nickel.   **9.23** Extended
answer required.   **9.24** (a) 3+ (III); (b) no change
of oxidation state occurs; (c) yes; (d) $Al(OH)_3$;
(e) $\sim4.3 - 9.3$.   **9.25** (a) 2+ (II), 1+ (I); (b)
between oxidation potentials of ~0.2 and 1.0; (c)
corrosion in the Cu^{2+} and CuO_2^{2-} stability fields,
passivation in the CuO and Cu_2O stability fields,
immunity in the Cu stability field; (d) $Cu^{2+}(aq) +$
$H_2O(l) \rightleftharpoons CuO(s) + 2H^+(aq)$;
$CuO(s) + H_2O(l) \rightleftharpoons CuO_2^{2-}(aq) + 2H^+(aq)$;
(e) $Cu^{2+} + 2e^- \rightleftharpoons Cu(s)$; (f) $Cu(s) +$
$H_2O(l) \rightleftharpoons Cu_2O(s) + 2H^+(aq) + 2e^-$;
(g) $Cu_2O(s) + H_2O(l) \rightleftharpoons 2CuO(s) + 2H^+(aq) + 2e^-$.
9.26 (a) $CuO(s)$; (b) $Cu(s)$; (c) $Cu(s)$; (d) $Cu(s)$;
(e) $Cu^{2+}(aq)$; (f) $CuO(s)$; g. $Cu(s)$.

Chapter 10

Quick quiz

1a,   2b,   3a,   4c,   5a,   6b,   7b,
8c,   9c,   10b,   11a,   12c,   13a,
14b,   15a,   16c,   17b,   18a,   19b,
20a,   21b,   22c,   23a,   24c,   25b,
26a,   27b,   28c,   29a,   30c,   31b,
32b,   33c,   34a,   35a,   36c,   37c,   38b.

Calculations and questions

10.1 15.6 MPa.   **10.2** 194.3 MPa.
10.3 85.4 cm.   **10.4** 156.0 cm.
10.5 60.14 cm.   **10.6** 100.36 cm.
10.7 152.2 GPa.   **10.8** 104.8 GPa.   **10.9** 0.5.
10.10 102 mm × 9.93 mm × 9.93 mm.

10.11 12.46 mm.   **10.12** 21.2 kN.
10.13 (a) 0.29 mm; (b) −0.055 mm.
10.14 (a) 111.1 kPa; (b) 0.32 mm;
(c) 0.0011 (0.11 %); (d) 6.5×10^{-6} m.
10.15 (a) 733.4 kPa; (b) 0.63 mm; (c) 0.0018
(0.18 %); (d) 0.0062 mm.   **10.16** (a) 0.126;
(b) 12.6 %.   **10.17** (a) 0.210; (b) 21 %.
10.18 (a) Sketch required; (b) ~70 GPa;
(c) ~1.2 GPa; (d) ~6500 N; (e) 6.4 %.
10.19 (a) Sketch required; (b) ~138 GPa;
(c) ~1.3 GPa; (d) ~973 MPa; (e) 4.1 %.
10.20 (a) 1–2 GPa; (b) ~350 MPa; (c) ~55.8 kN;
37.5 mm.   **10.21** (a) ~1.9 GPa; (b) ~1.1 GPa;
(c) ~0.7 GPa; (d) ~0.9 GPa; (e) ~0.6 GPa;
(e) 1.4 %.   **10.22** (a) 45.5 kN; 100.27 mm.
10.23 116 MPa.   **10.24** 1.63 μm.
10.25 0.45 N m^{-2}.   **10.26** (a) 94.4 GPa;
(b) 58.5 GPa.   **10.27** (a) 202.7 GPa;
(b) 125.4 GPa.   **10.28** (a) 39.5 GPa;
(b) 58.5 GPa.   **10.29** (a) 155.2 GPa;
(b) 6.72 GPa.   **10.30** (a) 287.7 GPa;
(b) 167.1 GPa.   **10.31** (a) 114.2 GPa;
(b) 90.4 GPa.   **10.32** 167.2 GPa.   **10.33** 18.6 %.

Chapter 11

Quick quiz

1a,   2c,   3c,   4a,   5b,   6b,
7a,   8b,   9b,   10c,   11b,   12a,   13c,
14c,   15b,   16a,   17a,   18b,   19a,
20c,   21b,   22a,   23c,   24b,   25c,
26a,   27c.

Calculations and questions

11.1 (a) 8.854×10^{-12} F; (b) 2.04×10^{-11} F.
11.2 3.75.   **11.3** 9035.   **11.4** (a) 4.35×10^{-12} C
$(0.03e)$; (b) nitrogen.   **11.5** oxygen,
1.06×10^{-19} C, $0.66e$; hydrogen, 0.53×10^{-19} C,
$0.33e$.   **11.6** (a) 250 pm; (b) 0.074 C m^{-2}.
11.7 CO, $0.014e$; N_2O, $0.012e$; NH_3, $0.79e$;

SO_2, 0.37e.   **11.8** Derivation required.

11.9. (a) (i) 1.99×10^{-30} C m^2 V^{-1},

(ii) 1.79×10^{-30} m^3; (b) (i) 1.69×10^{-30} C m^2 V^{-1},

(ii) 1.52×10^{-30} m^3.   **11.10** (a) (i) $3.22 \times$

10^{-30} C m^2 V^{-1}, (ii) 2.89×10^{-30} m^3;

(b) (i) 2.16×10^{-30} C m^2 V^{-1},

(ii) 1.94×10^{-30} m^3.

11.11 (a) 5.54×10^{-30} C m^2 V^{-1};

(b) 7.20×10^{-30} C m^2 V^{-1};

(c) 12.74×10^{-30} C m^2 V^{-1}.

11.12 7.07×10^{-30} C m^2 V^{-1}.

11.13 (a) 7.20×10^{-30} C m^2 V^{-1}; (b) 9.83.

11.14 (a) 15.4×10^{-30} C m^2 V^{-1};

(b) 13.8×10^{-30} m^3.

11.15 (a) 14.0×10^{-30} C m^2 V^{-1};

(b) 12.5×10^{-30} m^3; (c) not a good fit, maybe the

value for Al_2O_3 is inaccurate.   **11.16** (a) $20.3 \times$

10^{-30} C m^2 V^{-1}; (b) 18.2×10^{-30} m^3;

(c) a good fit, giving confidence in the data.

11.17 1.39×10^{-30} C m^2 V^{-1}.

11.18 Equivalence of units to be shown

11.19 2.26×10^{-9} C m^{-2}.

11.20 85 nm.   **11.21** 4×10^{-4} C m^{-2}.

11.22 (a) $\sim$239 °C; (b) $\sim$1.3 $\times 10^5$ K.

11.23 (a) $\sim$190 K (-80 °C); (b) $\sim$3.14 $\times 10^5$ K.

11.24 (a) $\sim$47 °C; (b) $\sim$4.24 $\times 10^3$ K.

11.25 (a) Sketch required; (b) 2.68 C m^{-2}.

11.26 0.0097 C m^{-2}.

11.27 (a) 7.69×10^{-30} C m; (b) 0.12 C m^{-2}.

11.28 (a) 1.36×10^{-29} C m; (b) 0.21 C m^{-2}.

11.29 (a) 1.92×10^{-29} C m; (b) 0.30 C m^{-2}.

11.30 and **11.31** Extended answers required.

Chapter 12

Quick quiz

1c,	2b,	3a,	4a,	5b,	6a,
7c,	8c,	9b,	10c,	11c,	12b,	13a,
14a,	15a,	16b,	17b,	18a,	19c,
20b,	21c,	22a,	23c,	24a,	25a,
26b,	27c,	28b,	29c,	30a,	31c,
32c,	33a.				

Calculations and questions

12.1 $S = 3/2$, $L = 6$, $J = 9/2$, $m = 3.62\mu_B$.

12.2 $S = 7/2$, $L = 0$, $J = 7/2$, $m = 7.94\mu_B$.

12.3 (a) $S = 2$, $L = 2$, $J = 0$, $m = 4.90\mu_B$;

(b) $S = 1$, $L = 3$, $J = 2$, $m = 2.82\mu_B$.

12.4 (a) $S = 3/2$, $L = 3$, $J = 9/2$, $m = 3.87\mu_B$;

(b) $S = 1/2$, $L = 2$, $J = 5/2$, $m = 1.41\mu_B$.

12.5 Extended answer required.   **12.6** Derivation

required.   **12.7** (a) 2.4×10^6 A m^{-1}; (b) $4.92 \times$

10^5 A m^{-1}.   **12.8** (a) 3.24×10^6 A m^{-1};

(b) 1.44×10^5 A m^{-1}.   **12.9** $2.19\mu_B$.

12.10 $0.58\mu_B$.   **12.11** high-spin, $t_{2g}^3 e_g^2$.

12.12 low-spin, t_{2g}^5; the orbital component is not

completely quenched.   **12.13** (a) $[Co(NH_3)_6]^{3+}$:

low-spin, t_{2g}^6; $[CoF_6]^{3-}$: high-spin, $t_{2g}^4\,e_g^2$;

(b) $4.9\mu_B$.   **12.14** (a) $[Fe(CN)_6]^{4-}$: low-spin, t_{2g}^6;

$[Fe(NH_3)_6]^{2+}$: high-spin, $t_{2g}^4\,e_g^2$; (b) $4.9\mu_B$.

12.15 435 nm.   **12.16** 303 nm.

12.17 (a) 3.71×10^{-24} J; (b) 5.35 cm.

12.18 (a) 8.69×10^{-24} J; (b) 2.28 cm.

12.19 (a) 4.63×10^{-24} J; (b) 4.30 cm.

12.20 (a) 1.11×10^{-23} J; (b) 1.78 cm.

12.21 1.53×10^{-7} kg^{-1}.

12.22 6.44×10^{-8} kg^{-1}.

12.23 1.67×10^{-3} m^{-3}.   **12.24** (a) 1.6792/T;

(b) 168 K.   **12.25** (a) (i) 0.3023/T; (ii) 60.5 K;

(b) (i) 0.3627/T; (ii) 72.5 K.

12.26 $2.22\mu_B$.   **12.27** $0.60\mu_B$.

12.28 (a) 9.07×10^{28} m^{-3}; (b) 0.222 nm.

12.29 (a) 4.84×10^5 Am^{-1}; (b) 0.608 T.

12.30 (a) 3.62×10^5 Am^{-1}; (b) 0.455 T.

12.31 (a) 2.55×10^5 Am^{-1}; (b) 0.32 T.

12.32 (a) 2.63×10^5 Am^{-1}; (b) 0.33 T.

Chapter 13

Quick quiz

1c,   2b,   3c,   4c,   5b,   6a,   7b,
8a,   9b,   10c,   11b,   12a,   13c,
14a,   15b,   16c,   17a,   18b,   19c,
20a,   21a,   22b,   23c,   24c,   25b,   26c,
27a,   28c,   29a,   30b,   31a,   32c.

Calculations and questions

13.1 $\tau, 2.93 \times 10^{-14}$ s; $\Lambda, 0.411$ nm;
≈ 1.4 Au–Au.   **13.2** $\tau, 4.12 \times 10^{-14}$
$\Lambda, 0.573$ nm; ≈ 2Ag–Ag.   **13.3** $\tau, 2.86 \times 10^{-14}$ s;
$\Lambda \approx 2320$ nm; much greater than Rb–Rb;
13.4 $\tau, 1.02 \times 10^{-14}$ s; $\Lambda, 0161$ nm; ≈ 0.5 Mg–Mg.
13.5 $\Lambda(0°C), 0.7095$ nm; $\Lambda(100°C), 0.645$ nm;
$\Lambda(300°C), 0.522$ nm; $\Lambda(700°C), 0.312$ mm;
$\Lambda(1200°C), 0.106$ nm.   **13.6** Parallel to a,
0.00103 m^2 V^{-1} s^{-1}; parallel to c,
0.000865 m^2 V^{-1} s^{-1}.   **13.7** Parallel to a,
0.000814 m^2 V^{-1} s^{-1}; parallel to c,
0.000785 m^2 V^{-1} s^{-1}.   **13.8** 4.46×10^{-8} Ω m.
13.9 1.21×10^{-7} Ω m.   **13.10** [Graph required.]
A new phase seems to form between 77.3 wt% and
67.1 wt% Ni.   **13.11** [Graph required.] A new
phase seems to form between 89 wt% and 87 wt%
Al.   **13.12** [Graph required.] A new phase seems
to form between 84 wt% and 70 wt% Cu.
13.13 $n = p = 9.73 \times 10^{15}$ m^{-3};
$np = 9.47 \times 10^{31}$ m^{-6}.   **13.14** $n = p = 2.93 \times$
10^{13} m^{-3}, $np = 8.58 \times 10^{26}$ m^{-6}.   **13.15** $n =$
$p = 5.91 \times 10^{11}$ m^{-3}, $np = 3.49 \times 10^{23}$ m^{-6}.
13.16 $n = p = 1.55 \times 10^{10}$ m^{-3} $np =$
2.40×10^{20} m^{-6}.   **13.17** $E_g = 2.26 \times$
10^{-19} J $= 1.41$ eV; $\nu = 3.40 \times 10^{14}$ s^{-1}.
13.18 $E_g \approx 1.19 \times 10^{-19}$ J $= 0.74$ eV; $\nu = 1.79 \times$
10^{14} s^{-1}.   **13.19** 5.26×10^{-21} J $= 0.033$ eV.
13.20 1.36×10^{-21} J $= 0.0085$ eV.
13.21 9.50×10^{-22} J $= 0.0059$ eV.

13.22 1.03×10^{-21} J $= 0.0064$ eV.
13.23 $+6.24 \times 10^{-22}$ J $= +0.0039$ eV.
13.24 -9.70×10^{-22} J $= -0.0061$ eV.
13.25 μ_h, 0.026 m^2 V^{-1} s^{-1}; μ_e,
0.26 m^2 V^{-1} s^{-1}.   **13.26** 0.136 Ω^{-1} m^{-1}.
13.27 6.09×10^{-3} m^3 C^{-1}.
13.28 (a) R_H, 402.6 m^3 C^{-1};
(b) $V = 0.8$ mV.   **13.29** R_H, 0.011 m^3 C^{-1};
$p, 5.62 \times 10^{20}$ m^{-3}.   **13.30** (a) $1.57 \times$
10^{-21} J $= 0.0088$ eV; (b) 3.25×10^{-21} J
$= 0.020$ eV; (c) 1.04×10^{-20} J
$= 0.065$ eV.   **13.31** TiO$_{2-x}$, n-type;
Ti$_2$O$_{3+x}$, p-type.   **13.32** La^{3+}Co^{3+}O^{2-};
La$_{1-x}^{3+}$Sr$_x^{2+}$Co$_{1-x}^{3+}$Co$_x^{4+}$O^{2-}; p-type.
13.33 (Mg^{2+})[Ti^{4+} Mg^{2+}]O^{2-}; (Mg^{2+})[Ti$_2^{3+}$]O^{2-};
(Mg^{2+})[Ti^{4+}Mg$_{1-x}^{2+}$Ti$_x^{3+}$]O^{2-}; n-type.
13.34 (a) 0.417 nm; (b) 0.377 nm; (c) about 1
layer.   **13.35** (a) 0.119 nm; (b) 0.139 nm;
(c) about 1/2 a layer.   **13.36** 10 nm, $6.025 \times$
10^{-21} J $= 0.038$ eV; 1 cm, 6.025×10^{-33} J
$= 3.76 \times 10^{-14}$ eV.   **13.37** $n =$
1, 9.34×10^{-21} J $= 0.058$ eV; $n = 2, 3.74 \times$
10^{-20} J $= 0.232$ eV; $n = 3, 8.41 \times 10^{-20}$ J
$= 0.522$ eV.   **13.38** $7.95 \times$
10^{-22} J $= 0.0050$ eV.   **13.39** $4.77 \times$
10^7 A m^{-1}.   **13.40** 22.9 T.   **13.41** 10.4 T.
13.42 Cu$_{1-x}^{2+}$, Cu$_x^+$; electron superconductor.
13.43 La$_2^{3+}$Sr^{2+}Cu$_2^{2+}$O$_6^{2-}$, not a superconductor;
La$_2^{3+}$Sr^{2+}Cu$_{1.6}^{2+}$Cu$_{0.4}^{3+}$O$_{6.2}^{2-}$, hole superconductor.

Chapter 14

Quick quiz

1a,   2b,   3b,   4c,   5b,   6a,   7b,
8b,   9c,   10a,   11a,   12c,   13b,
14b,   15a,   16c,   17a,   18b,   19c,
20a,   21c,   22c,   23b,   24a,   25b,
26a,   27c,   28c,   29b,   30a,   31c,
32b,   33a.

Calculations and questions

14.1 $\lambda = 425$ nm, $\nu = 7.05 \times 10^{14}$ Hz, $E = 4.67 \times 10^{-19}$ J, violet; $\lambda = 575$ nm, $\nu = 5.21 \times 10^{14}$ Hz, $E = 3.45 \times 10^{-19}$ J, yellow-green; $\lambda = 630$ nm, $\nu = 4.76 \times 10^{14}$ Hz, $E = 3.15 \times 10^{-19}$ J, orange-red.
14.2 299 kJ mol^{-1}.   **14.3** 7.98 kJ mol^{-1}.
14.4 242 nm.   **14.5** 1140 nm.
14.6 (a) 2.76×10^{-19} J; (b) 1.11×10^{-29}.
14.7 (a) 5.96×10^{-19} J; (b) 3.01×10^{-63}.
14.8 6.9×10^{-5} m.   **14.9** $E_0 \rightarrow E_2$, 3.57×10^{-19} J, $N_2/N_0 = 3.60 \times 10^{-38}$; $E_0 \rightarrow E_3$, 4.88×10^{-19} J, $N_3/N_0 = 7.04 \times 10^{-52}$.
14.10 (a) $E_0 \rightarrow E_1$, 2.68×10^{-19} J; (b) $N_1/N_0 = 1.03 \times 10^{-30}$.   **14.11** (a) 2.17×10^{-19} J; (b) 1.87×10^{-19} J.   **14.12** GaAs, 918.4 nm, infrared; AlAs, 574 nm, yellow; $x = 0.925$.   **14.13** GaN, 371.2 nm, ultraviolet; InN, 620 nm, orange-red; $x = 0.62$.   **14.14** InP, 976.3 nm, infrared; AlP, 506.1 nm, blue-green; $x = 0.59$.   **14.15** 0.66.   **14.16** Zn, 44 cm; Cd, 12.5 cm; transmittance, 0.1; absorbance, 1.
14.17 Transmittance, 0.05; absorbance, 1.3010; 35 at% Zn:65 at% Cd.   **14.18** 62.9°
14.19 (a) 19.2°; (b) no critical angle; (c) 57.3°.
14.20 BaTiO$_3$, 2.15; PbTiO$_3$, 2.77.   **14.21** Spinel, 1.72; akermanite, 1.64.   **14.22** Beryl, 1.56; garnet, 1.71.   **14.23** 0.20.   **14.24** Cryolite, 0.022; glass, 0.045; corundum, 0.06; zirconia, 0.12; tantala, 0.13.   **14.25** Al, 0.916; Ag, 0.919; Au, 0.817; Cr, 0.48, Ni, 0.76.   **14.26** (a) Bright, reflecting; (b) dark, nonreflecting.   **14.27** (a) Dark, non-reflecting; (b) bright, reflecting.   **14.28** Reflecting in air, nonreflecting on oil.   **14.29** (a) 47.4 nm; (b) reflected colour yellow-white to straw yellow; (c) transmitted colour, carmine red to deep violet.
14.30 Derivation required.   **14.31** (a) SiO, 0.21; (b) TiO$_2$, 0.49.   **14.32** MgF$_2$: $\lambda/4$, 0.015; $\lambda/2$, 0.026. TiO2: $\lambda/4$, 0.04; $\lambda/2$, 0.23.   **14.33** Graph

required. 6 layers.   **14.34** Ta$_2$O$_5$, 75.6 nm; cryolite, 120.4 nm.   **14.35** 2.20 m^{-1}.
14.36 4.6×10^{-4} m^{-1}.   **14.37** Graph required.
14.38 1.7 times as much light is scattered by the dust.   **14.39** Limestone will cause about twice as much scattering as the water.   **14.40** 10.5 km.
14.41 (a) 2; (b) 1.5; (c) 1; (d) 2.   **14.42** (a) 5.36×10^{-6} m; (b) 4.28°.   **14.43** 14.4 mm.
14.44 (a) 638 nm; (b) orange-red; (c) no change.   **14.45** 0.0218 dB km^{-1}.
14.46 3.01 mm.   **14.47** 370 km.   **14.48** Yellow-green; 1.83 nm.   **14.49** 335.4 nm.

Chapter 15

Quick quiz

1b,   2a,   3b,   4a,   5c,   6a,   7b,
8b,   9a,   10c,   11b,   12a,   13c.

Calculations and questions

15.1 11.1 kJ.   **15.2** 2.36 kJ.   **15.3** 0.343 J K^{-1} mol^{-1}.   **15.4** 0.144 J K^{-1} mol^{-1}.   **15.5** 5.40 h.
15.6 Derivation required.   **15.7** (a) 1.16×10^4 W; (b) 2.314×10^5 W (Cu); (c) 1.03×10^4 W (steel).
15.8 (a) 6.14×10^5 J (glass); (b) 7.52×10^3 J (double glazing).
15.9 4.28 mm;   **15.10** 5.39 °C.   **15.11** a_0, -4.34×10^{-6} K^{-1}, c_0, $+1.15 \times 10^{-5}$ K^{-1}; (b) 2.82×10^{-6} K^{-1}.   **15.12** 7.80 µV K^{-1}; $+0.85$ µV K^{-1}.   **15.13** 4.20 µV K^{-1}; -2.75 µV K^{-1}.   **15.14** 105.4 µV.

Chapter 16

Quick quiz

1c,   2a,   3b,   4c,   5b,   6a,   7c,
8c,   9a,   10a,   11b,   12c,   13b,
14a,   15c,   16c,   17b,   18a,   19b,
20c,   21b,   22b,   23a,   24b.

Calculations and questions

16.1 (a) $^{166}_{75}$Re, 75 protons, 91 neutrons, 166 nucleons; (b) $^{140}_{56}$Ba, 56 protons, 84 neutrons, 140 nucleons; (c) $^{18}_{8}$O, 8 protons, 10 neutrons, 18 nucleons; (d) $^{14}_{5}$B, 5 protons, 9 neutrons, 14 nucleons.   **16.2** (a) $^{181}_{80}$Hg, 80 protons, 101 neutrons, 181 nucleons; (b) $^{169}_{77}$Ir, 77 protons, 92 neutrons, 169 nucleons; (c) $^{117}_{53}$I, 53 protons, 64 neutrons, 117 nucleons; (d) $^{98}_{40}$Zr, 40 protons, 58 neutrons, 98 nucleons.   **16.3**(a) $^{27}_{14}$Si $\rightarrow$ $^{27}_{13}$Al $+ ^{0}_{1}$e; (b) $^{28}_{13}$Al $\rightarrow ^{28}_{14}$Si $+ ^{0}_{-1}$e; (c) $^{24}_{11}$Na $\rightarrow$ $^{24}_{12}$Mg $+ ^{0}_{-1}$e; (d) $^{17}_{9}$F $\rightarrow ^{17}_{8}$O $+ ^{0}_{1}$e.
16.4 (a) $^{24}_{8}$O $\rightarrow ^{24}_{9}$F $+ ^{0}_{-1}$e; (b) $^{47}_{23}$V $\rightarrow ^{47}_{22}$Ti $+ ^{0}_{1}$e; (c) $^{32}_{15}$P $\rightarrow ^{32}_{16}$S $+ ^{0}_{-1}$e; (d) $^{39}_{20}$Ca $\rightarrow ^{39}_{19}$K $+ ^{0}_{1}$e.
16.5 (a) $^{243}_{100}$Fm $\rightarrow ^{239}_{98}$Cf $+ ^{4}_{2}$He;

(b) $^{241}_{95}$Am $\rightarrow ^{237}_{93}$Np $+ ^{4}_{2}$He; (c) $^{241}_{94}$Pu $\rightarrow$ $^{241}_{94}$Am $+ ^{0}_{-1}$e; (d) $^{237}_{92}$U $\rightarrow ^{237}_{93}$Np $+ ^{0}_{-1}$ e.
16.6 5.21 min.   **16.7** 49.8 h.   **16.8** 3.7 $\times$ 10^{10} disintegrations $s^{-1} g^{-1}$.   **16.9** 3.85 $\times$ 10^{12} disintegrations $s^{-1} g^{-1}$.   **16.10** 8.30 $\times$ 10^{17} disintegrations $s^{-1} g^{-1}$.   **16.11** 4.20 h.
16.12 13.9 $\times 10^3$ years.   **16.13** (a) 22.2 disintegrations $min^{-1} dm^{-3}$; (b) 17.6 days.
16.14 (a) 9.5 $\times 10^5$ atoms dm^{-3}; (b) 12.6 days.
16.15 (a) 9.88 $\times 10^{-18}$ g m^{-2}; (b) 794.3 Bq.   **16.16** 1.38 $\times 10^{-12}$ J.
16.17 1.27$\times 10^{-12}$ J.   **16.18** 1.38 $\times 10^{-12}$ J.
16.19 (a) 5.32 $\times 10^{-8}$ kg; (b) 4.44 $\times 10^{-8}$ kg.   **16.20** 1.61 $\times$ 10^{10} kJ mol^{-1}.   **16.21** (a) 4.103 $\times 10^{-12}$ J; (b) 2.44 $\times 10^{37}$ fusions s^{-1} required.

Chemical Index

Page numbers in italic, e.g. *497*, refer to figures. Page numbers in bold, e.g. **174**, signify entries in tables.

Subject Index

Page numbers in italic, e.g. *4*, refer to figures. Page numbers in bold, e.g. **344**, signify entries in tables.

Understanding solids: the science of materials.   Richard J. D. Tilley
© 2004 John Wiley & Sons, Ltd   ISBNs: 0 470 85275 5 (Hbk) 0 470 85276 3 (Pbk)

Conversion factors and other relationships

atmosphere (atm): $1(atm) = 101.325$ kPa

electron volt (eV): $1(eV) = 96.485$ kJ mol^{-1}

$$(eV) \times 1.60218 \times 10^{-19} \longrightarrow (J)$$

$$(J) \times 6.24150 \times 10^{18} \longrightarrow (eV)$$

atomic mass unit (u): $1(u) = 9.31494 \times 10^8$ eV

electron mass $= 5.48580 \times 10^{-4}$ u

neutron mass $= 1.00866$ u

proton mass $= 1.00728$ u

calorie (cal): $1(cal) = 4.184$ J

$RT = 2.4790$ kJ mol^{-1}

$RT/F = 25.693$ mV

$hc = 1.98645 \times 10^{-25}$ J m

Constants

Quantity	Symbol	Value	Units
atomic mass unit	$u = m[^{12}C]/12$	1.66054×10^{-27}	kg
Avogadro constant	N_A	6.02214×10^{23}	mol^{-1}
Bohr magneton	$\mu_B = eh/4\pi m_e$	9.27402×10^{-24}	J T^{-1}
Bohr radius	$a_0 = \varepsilon_0 h^2/\pi m_e e^2$	5.29177×10^{-11}	m
Boltzmann constant	k	1.38066×10^{-23}	J K^{-1}
elementary charge	e	1.60218×10^{-19}	C
electron mass	m_e	9.10939×10^{-31}	kg
Faraday constant	$F = N_A e$	9.6485×10^4	C mol^{-1}
gas constant	$R = N_A k$	8.31451	J K^{-1} mol^{-1}
neutron mass	m_n	1.67493×10^{-27}	kg
Planck constant	h	6.62608×10^{-34}	J s
	$\hbar = h/2\pi$	1.05457×10^{-34}	J s
proton mass	m_p	1.67262×10^{-27}	kg
Standard acceleration due to gravity	g	9.80665	m s^{-2}
vacuum permeability	μ_0	$4\pi \times 10^{-7}$	H m^{-1}
vacuum permittivity	ε_0	8.85419×10^{-12}	F m^{-1}
velocity of light	c	2.99792×10^8	m s^{-1}